Excluded Volume Effects in Polymer Solutions

Springer
Berlin
Heidelberg
New York
Barcelona
Hong Kong
London
Milan
Paris
Singapore
Tokyo

Lothar Schäfer

Excluded Volume Effects in Polymer Solutions

as Explained by the Renormalization Group

With 117 Figures

Springer

Professor Dr. Lothar Schäfer
Universität GH Essen
FB7 Physik
Universitätsstraße 5
D - 45141 Essen
Germany

ISBN-13: 978-3-642-64254-8 Springer-Verlag Berlin Heidelberg New York

CIP-data applied for
Die Deutsche Bibliothek - CIP-Einheitsaufnahme
Schäfer, Lothar: Excluded volume effects in polymer solutions as explained by the renormalization group : with 7 tables / Lothar Schäfer. -
Berlin ; Heidelberg ; New York ; Barcelona ; Hong Kong ; London ; Milan ; Paris ; Singapore ; Tokyo : Springer, 1999
ISBN-13: 978-3-642-64254-8 e-ISBN-13: 978-3-642-60093-7
DOI: 10.1007/978-3-642-60093-7

Coverdesign: E. Kirchner, Heidelberg
Typesetting: camera-ready by author
SPIN:10076621 02/3020-5 4 3 2 1 0

Preface

Among the many interesting and complicated topics of polymer physics this book concentrates on a comparatively simple and restricted task. It deals with physical properties of fairly dilute solutions of large flexible macromolecules. The solutions are taken to be in thermodynamic equilibrium, and only time independent properties like the osmotic pressure or the spatial correlations among polymer segments will be considered. Furthermore we concentrate on systems where different polymer segments repel each other, leading to a swelling of the average configuration of the polymer molecule. This is the domain of the so-called 'two parameter theories', and this book can be read as a critical reexamination and extension of that traditional approach.

Despite its restriction this topic comprises a wealth of interesting and surprising phenomena. During the last two decades it has developed considerably, due to the application of both new theoretical concepts and new experimental tools. In particular, the use of laser light scattering allows for a very precise determination of the segment density correlations in the solution. Also neutron scattering has been employed, yielding information on the configuration of a few isotopically marked macromolecules immersed in a solution of a finite concentration of other polymers. Another noteworthy 'experimental' development is the use of large scale simulations providing data on properties like the spatial correlations among internal chain segments or the density distribution relative to the center of mass. Such properties are not easily accessible to normal experiments. With present day computers we can simulate simplified models of large macromolecules in solution and extract detailed information on the internal structure or the overall shape of a typical configuration.

On the theoretical side it has been discovered that sufficiently dilute solutions of sufficiently large macromolecules make for a beautiful example of a 'critical system'. For such systems it is characteristic that they look the same on a large range of spatial scales. Looking at a polymer molecule with a resolving power of, say, 30 Å, we cannot distinguish it from a molecule ten times larger in spatial extend, looked at with a resolving power of 300 Å. It is only if we use a resolving power high enough to identify the microstructure, i.e. the individual monomers, that we can fix the spatial scale from the picture we see. This property of 'scale invariance' has significant consequences for the physics of polymer solutions. Its importance may well be compared to the importance of rotational invariance in atomic physics. On the qualitative level it implies that macroscopic observables obey 'scaling', which means that after choosing appropriate units of concentrations or scattering vectors, i.e., after 'rescaling', the data for a given quantity fall on a common curve for a lot of chemically different solutions: The resulting 'scaling functions' are universal. On the quantitative level scale invariance in its implementation via the 'Renormalization Group' and 'renormalized perturbation theory' allows for approximate calculations of the scaling functions. These quantitative predictions of the theory explain many scattering or thermodynamic experiments within the scatter of the data. They thus provide us with a firm basis for analyzing the experiments.

In principle the property of scale invariance has been recognized even in the first days of polymer physics. As is well known, we in the simplest approach may ignore any repulsion among the segments and model the configuration of a polymer chain in solution as a random walk of n steps of length ℓ. Then macroscopic observables like the average end-to-end distance of long chains depend only on the combination $\ell n^{1/2}$. A change of ℓ therefore can be compensated by a 'rescaling' of n – which is the simplest example of scale invariance. In the early days this freedom has been used to adjust other properties of the model, like the length of the totally stretched chain. The modern theory, however, noting that such additional conventions have nothing to do with the physics of polymer solutions, takes scale invariance seriously and evaluates its consequences.

Compared to the classical approach, the new theory implies a major change in our way of looking at the problem. It is one purpose of this book to help in popularizing this new view by explaining the basic features of the theory. As a second, but no less important goal I want to present a complete and consistent set of quantitative results suitable for an analysis of the experiments. Such a unified set of results does not seem to have been presented before. Published work mostly concentrates on a small subset of experimental quantities, often stressing the asymptotically valid power laws. However, only a consistent evaluation of all scaling functions also outside the power law regime makes full use of all the strength of the theory and supplies us with a firm basis for data analysis. Indeed, the interpretation of data in terms of asymptotic power laws – though being very popular by now – in the range covered by the experiments is only a first approximation and might be quite misleading.

This book is divided into five major parts. The first part centers around the standard cluster expansion of statistical mechanics, as used for the evaluation of interaction effects in a macromolecular solution. It thus recalls results derived prior to the discovery of the central role of scale invariance and sets the basis for further development. I formulate the cluster expansion for a type of spring and bead model of the polymer chains, which allows for a conceptually simple mathematical treatment. This part essentially recalls material covered by the well known book of Yamakawa.

The second part is devoted to the renormalization group and scaling. I explain the basic ideas, discuss the general structure of the results on a qualitative level, and indicate how quantitative calculations in principle could proceed.

These first two parts are hoped to give a concise picture of the basic features of the theory, exhibiting an appealing, even fascinating, structure of long macromolecules in solutions.

The methods of Part II are not optimized for actual quantitative calculations, and in Part III I present a more powerful approach using 'renormalized perturbation theory'. It is based on results of quantum field theory, which are cited without proof. However the method is explained to the level where actual calculations can be carried out. Several implementations of the general scheme are available, and I discuss one formulation in detail – 'minimal subtraction' – which is generally thought to be the most elegant and powerful.

In Part IV I work out the quantitative results on the so-called 'one loop' level. The results presented cover all the important experimental quantities. Both asymptotic power laws and crossover scaling functions interpolating among different asymptotic limits are calculated. The influence of polydispersity, i.e. of a distribution of molecular weights of the polymer sample, is considered throughout. I devote particular attention to a comparison with experimental results, critically examining heuristic extrapolation prescriptions used in data analysis. We will find that in general the experimental results support the theory; however at present the set of internally consistent data is not large enough for a really stringent test of all features of the theory.

The first four parts are concerned with the two parameter theory of binary solutions. It is based on the model of a random walk chain, where the segments pairwise repel each other. In Part V I consider two extensions of the theory. We will discuss the θ-point, where the effective two-body repulsion vanishes. We then have to take into account a repulsion among triplets of segments, which leads to the tricritical theory of the θ-point. The theory can be extended also to solutions containing several macromolecular species, which differ in the chemical structure of the monomer units. Also copolymers or macromolecules of more complicated structure (stars etc.) can be considered. Of this wide range of applications, which has not been explored to the same extent as binary solutions, I here discuss the theory of multicomponent solutions.

Throughout the book I tried to keep the mathematics as simple as possible. Besides some familiarity with the formalism of statistical mechanics mainly Taylor expansion and integration over Gaussian functions will be needed. This does not exclude some lengthy but straightforward calculations. Longer derivations or more mathematical topics are shifted to appendices. Some chapters of more complicated nature are indicated by an asterix in the table of contents. The book is written such that these chapters are no prerequisite for the remainder, except that I advise reading of the introductory sections summarizing the main ideas.

Though this book is written in the language of polymer physics and concentrates on quantities of interest in that context, it also can be read as a presentation of the modern (field) theory of critical phenomena in general. A reader somewhat familiar with that theory will easily translate most arguments and derivations into that other language, with only small modifications. At some places hopefully he might be amused to see the simple physical interpretation of general results that arises in the context of polymer theory.

This book, in particular in its latter parts, summarizes work to which numerous coworkers have contributed over the years. I especially should mention Dr. B. Krüger and Dr. Ch. Kappeler, who carried out the hard work for some of the topics covered. Thanks also are due to U. Lehr, Dr. U. Ebert, J. Hager, Dr. F. Rother, Dr. P. Sutter, and S. Müller for a critical reading of earlier versions of the manuscript, resulting in many useful suggestions. Mrs. Kowalski I want to thank for the careful typing, including lengthy formulae. Finally I want to acknowledge the support of the Deutsche Forschungsgemeinschaft via the Sonderforschungsbereich 'Unordnung und große Fluktuationen'.

November 1998 *Lothar Schäfer*

Table of Contents

Part III. Field Theoretic Formulation of the Theory

Part V. Special Topics

1. Power Laws, Scaling, Universal Crossover – a Survey of Phenomenology

We consider long flexible macromolecules, which, if immersed in a solvent, take a random coil-like configuration (see Fig. 1.1). We assume the solvent to be nonpolar and the macromolecules to be uncharged, so that any long range electric forces are absent and the intermolecular forces are of short range comparable to the size ℓ_0 of a monomer. For macroscopic properties like the osmotic pressure or the average size of a polymer coil, only some spatial average of these short range forces matters, and this average can be either attractive or repulsive. Indeed, it depends on the temperature T. The most direct measure of the average interaction among macromolecules is provided by the second virial coefficient A_2^{Π} of the osmotic pressure $\hat{\Pi}$, which, deviating from the standard definition, we define in terms of the expansion of $\hat{\Pi}/k_BT$ in powers of the number concentration c_p of polymer molecules in the solution. (k_B is Boltzmann's constant.)

$$\frac{\hat{\Pi}}{k_BT} = c_p + \frac{1}{2}A_2^{\Pi}c_p^2 + \frac{1}{6}A_3^{\Pi}c_p^3 + \dots \quad . \tag{1.1}$$

In this expansion the first term gives the behavior of an ideal solution (van't Hoff's law). The interaction influences only the higher order terms. A_2^{Π}, in

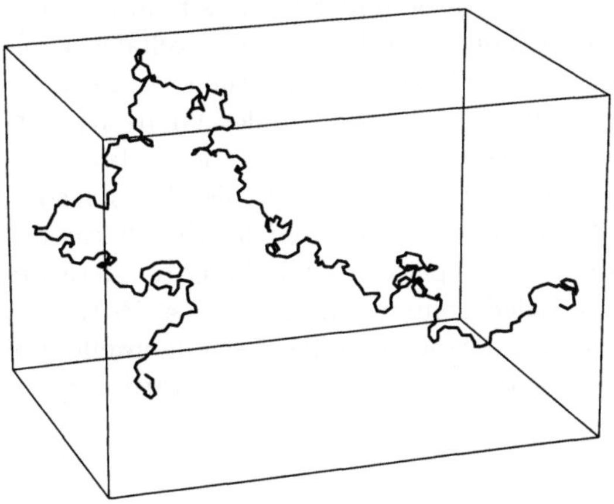

Fig. 1.1. A typical configuration of a self-avoiding chain

particular, involves a pair of interacting macromolecules. For repulsive interactions A_2^{Π} is positive: repulsive interactions increase the pressure. Attractive interactions yield a negative contribution $\frac{1}{2}A_2^{\Pi}c_p^2$, and eventually lead to a separation of the solution into a dense phase in equilibrium with a very dilute solution. No such separation can occur in the region of repulsive interactions.

We often can find a temperature where A_2^{Π} changes sign. This temperature clearly depends on the chemical composition of the solution, and in addition it shows a weak dependence on the molecular weight of the macromolecules. For high molecular weight it tends to a well defined limit, the 'Θ-temperature', which is characteristic for the given chemical system. Sometimes A_2^{Π} changes sign twice, leading to the general phase diagram of Fig. 1.2. Note that this diagram is schematic, since neither Θ_ℓ nor Θ_u have to be accessible for a given system. The solution might boil or freeze first.

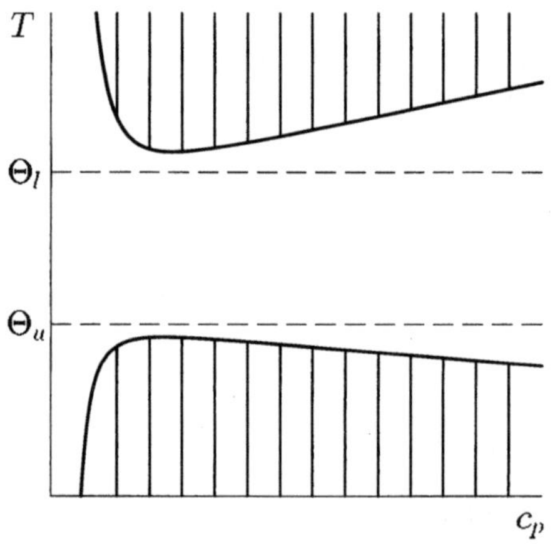

Fig. 1.2. Schematic phase diagram $T - c_p$ of a polymer solution, showing two two-phase regions (*shaded parts*) with their asociated Θ-points

The temperature dependence of A_2^{Π} can be understood as follows. Since two monomers cannot sit on top of each other, each monomer excludes a volume of size $\sim \ell_o^3$ to all other monomers. This strong repulsion on distances $r \lesssim \ell_o$ for larger distance $r \gtrsim \ell_o$ is followed by some van der Waals type attraction. Roughly speaking, each monomer creates an energetically favorable surrounding of solvent molecules, which then is favorable also for a second monomer. Strength and range of the attraction depend on the thermodynamic state of the solvent and thus on temperature. At the Θ-temperature the attraction just balances the excluded volume. For $\Theta_u < T < \Theta_l$ the excluded volume dominates, giving rise to phenomena like a strong swelling of the polymer coil in solution. All interaction effects in this temperature region are commonly known as *'excluded volume effects'*.

A typical excluded volume effect is the occurence of nontrivial *power laws.* As a first example let us consider the average size of an isolated polymer coil consisting of n segments. We assume each segment to consist of several monomers, so that any bond angle constraints among the directions of subsequent segments are absent. The size of an isolated coil can be measured in scattering experiments on extremely dilute solutions, $c_p \to 0$, which yield the 'radius of gyration' $R_g(n)$, to be defined precisely in Chap. 3. (Rigorously speaking the experiment yields some average over the radii of the chains in the solution, a feature that is ignored in this qualitative discussion but fully treated in later chapters.) Neglecting any excluded volume interaction we can model the typical configuration of the chain as a random walk of n steps. Such a walk on the average spans a distance of order $\ell\, n^{1/2}$, where ℓ is the length of the elementary step. Taking this distance as a measure of the size of the polymer coil we predict

$$R_g(n) \sim \ell n^{1/2} \quad , \tag{1.2}$$

a result which indeed is found experimentally for the radius of a chain at the Θ-point. Clearly ℓ depends on the chemical microstructure, but the power law is the same in all Θ-solvents. Now repeating the experiment in the region of strong excluded volume forces, we find a similar law, but with a nontrivial exponent:

$$R_g(n) = Bn^{\nu}, \quad \nu \approx 0.59 \quad . \tag{1.3}$$

This is illustrated in Fig. 1.3. *The remarkable fact is that the exponent is 'universal', it is a number independent of temperature or chemistry.* Indeed, the same law is also found for self-repelling chains generated in a computer experiment. All effects of chemical microstructure or temperature are contained in the nonuniversal prefactor B. More precisely this law holds in the *'excluded volume limit'*, which is reached for long chains as long as the effective interaction is repulsive. As we approach the Θ-temperature, the repulsion decreases, and the chains have to be longer to reach the excluded volume law (1.3). This will be discussed in more detail below.

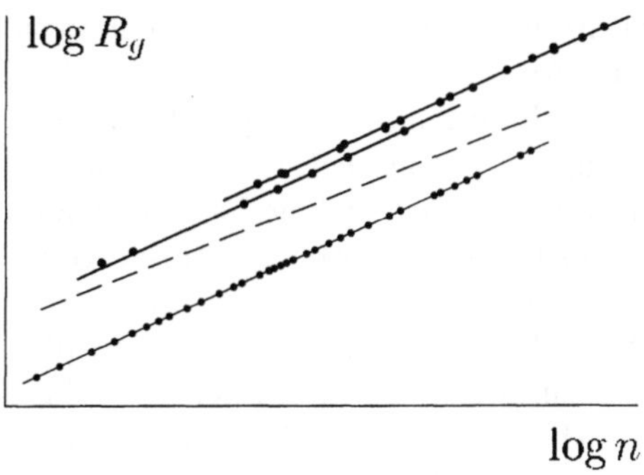

Fig. 1.3. Data for $\log_{10} R_g$ as function of $\log_{10} n$. From above: polystyrene-toluene; P-methylmethacrylate-acetone; self avoiding walks on a simple cubic lattice. The data collect results from different groups. *Full lines* are fits to Eq. (1.3). The *broken line* represents the behavior at the Θ-point (Eq. (1.2))

Equation (1.3) is by no means the only power law found in the excluded volume limit. For the second virial coefficient A_2^{Π}, for instance, we find

$$A_2^{\Pi} \sim const\ n^{3\nu} \quad . \tag{1.4}$$

Another celebrated example is des Cloizeaux's power law holding for the osmotic pressure $\hat{\Pi}$ in the so-called 'semidilute limit'. That limit is attained by keeping the segment concentration

$$c = c_p\, n \tag{1.5}$$

fixed at some small value, but making the chains longer and longer. Since with increasing n the coil volume rapidly increases, this limit implies that different coils strongly interpenetrate, the degree of interpenetration being measured by the 'overlap'

$$s = c_p R_g^3 \quad . \tag{1.6}$$

In the excluded volume limit for $s \gg 1$ we find the power law

$$\frac{\hat{\Pi}}{k_B T} \sim const\ c^{\frac{3\nu}{3\nu-1}} \quad , \tag{1.7}$$

governing the behavior for small c, but large overlap. In all these laws the exponent is universal and can be expressed in terms of ν, which thus plays a central role.

The occurence of power laws is not the only universal feature. Rather these laws just are limiting forms of more general *scaling laws*, which state that physical quantities depend only on certain 'scaling combinations' of their variables. To give an example, the osmotic pressure $\hat{\Pi}$ a priori depends on chain concentration c_p, chain length n, temperature T and chemistry of polymer and solvent. In the excluded volume limit it is found that $\hat{\Pi}$ obeys the scaling law

$$\frac{\hat{\Pi}}{k_B T} = c_p[1 + \mathcal{P}(s)] \quad , \tag{1.8}$$

where the *'scaling variable' s* is the overlap introduced above. $\mathcal{P}(s)$ is known as *'scaling function'*. This notion originates from the observation that the factor R_g^3 in Eq. (1.6) can be interpreted as a chain length-, temperature-, and chemistry-dependent scale of c_p. Measuring c_p on this scale, all data for different n should fall on the same curve. To demonstrate scaling, let us consider data on the osmotic pressure of the system P(methylmethacrylate)-acetone. Figure 1.4a shows the data plotted as function of the segment concentration c, as given in the literature. In Fig. 1.4b we use the same data to plot $\hat{\Pi}/(k_B T c_p)$ as function of s. Within the accuracy of the experiment, the data for the different chain lengths now fall on the same curve. This demonstrates the validity of the scaling law. Another remarkable feature is the fact that

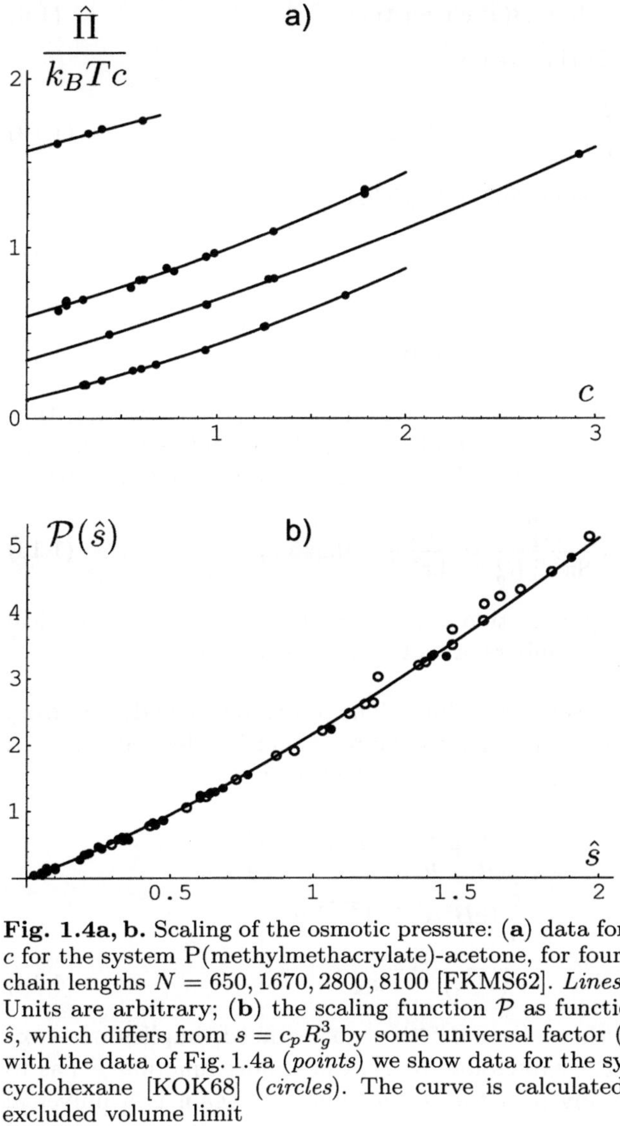

Fig. 1.4a, b. Scaling of the osmotic pressure: (**a**) data for $\hat{\Pi}/(k_B T c)$ as function of c for the system P(methylmethacrylate)-acetone, for four different number-average chain lengths $N = 650, 1670, 2800, 8100$ [FKMS62]. *Lines* serve for guiding the eye. Units are arbitrary; (**b**) the scaling function $\mathcal{P}$ as function of the overlap variable $\hat{s}$, which differs from $s = c_p R_g^3$ by some universal factor (cf. Sect. 13.3.2). Together with the data of Fig. 1.4a (*points*) we show data for the system P-dimethylsiloxane-cyclohexane [KOK68] (*circles*). The curve is calculated from our theory in the excluded volume limit

the scaling function again is universal, independent of chemistry and temperature. This is demonstrated by also including in Fig. 1.4b data for another chemical system.

The scaling laws imply the power laws as limiting forms, even establishing relations among the nonuniversal prefactors. Let us illustrate this for R_g and $\hat{\Pi}$. We recall the power law (1.3) $R_g = B n^{\nu}$, which involves the nonuniversal constant

$$B = B(T, \text{chemistry}) \quad . \tag{1.9}$$

The scaling law (1.8), (1.6) then reads

$$\frac{\hat{\Pi}}{k_B T} = c_p \left[1 + \mathcal{P}\left(c_p B^3 n^{3\nu}\right)\right] \quad . \tag{1.10}$$

Comparison to the virial expansion (1.1) yields

$$\frac{1}{2} A_2^{\Pi} = a_2 B^3 n^{3\nu} \quad , \tag{1.11}$$

where

$$a_2 = \left.\frac{d}{ds}\right|_o \mathcal{P}(s) \tag{1.12}$$

is a universal number. The nonuniversal constant in the power law (1.4) thus can be expressed in terms of B. The result can also be formulated as showing that the ratio A_2^{Π}/R_g^3 in the excluded volume limit is a universal number:

$$\psi = \frac{A_2^{\Pi}}{8\pi^{3/2} R_g^3} = \frac{a_2}{4\pi^{3/2}} \approx 0.245 \quad . \tag{1.13}$$

ψ is known as the 'interpenetration ratio', and in the context of critical phenomena such universal numbers are addressed as *'critical ratios'*.

Let us now consider the semidilute limit. The power law (1.7) derives from the assumption that $\hat{\Pi}$ for a strongly overlapping solution depends on c_p, n only via the segment concentration $c = c_p n$. We thus write

$$\begin{aligned} \frac{\hat{\Pi}}{k_B T} &= c\left[\frac{1}{n} + \frac{1}{n}\mathcal{P}\left(cB^3 n^{3\nu-1}\right)\right] \\ &\underset{s\to\infty}{\longrightarrow} \frac{c}{n}\left(cB^3 n^{3\nu-1}\right)^{\frac{1}{3\nu-1}} a_\infty \\ &= c\left(B^3 c\right)^{\frac{1}{3\nu-1}} a_\infty \quad , \end{aligned} \tag{1.14}$$

where we used the fact that $s \sim c\, n^{3\nu-1} \to \infty$, c fixed, implies $n \to \infty$. For n to drop out from the expression for $\hat{\Pi}$, the scaling function $\mathcal{P}$ has to reduce to the power law given in the second line. a_∞ again is a universal number derived from the large overlap limit of $\mathcal{P}(s)$. Again the nonuniversal coefficient of the power law (1.7) is expressed in terms of B, giving rise to a corresponding critical ratio.

In the excluded volume limit similar scaling behavior is found for other observables like the scattering functions, and it turns out that all nonuniversal temperature and chemistry dependence can be absorbed into the single parameter B. Since the scaling functions typically interpolate among asymptotic limits ($s \to 0$ or $s \to \infty$, for instance), they also are known as *'crossover functions'*.

So far we have considered phenomena in the excluded volume limit of long chains and temperatures not too close to $T = \Theta$. Approaching the Θ-temperature keeping n fixed or decreasing the chain length for fixed T we leave the excluded volume limit and simple power laws are no longer valid. However, some type of scaling still persists. The temperature- and chain length-dependence often combine into a single scaling variable

$$z = const \frac{T - \Theta}{\Theta} n^{1/2} \quad . \tag{1.15}$$

(We assume to be close to the lower bound of the excluded volume regime, so that $T \geq \Theta$.) To give an example, we consider the swelling factor

$$\alpha_g(T, n) = \frac{R_g(T, n)}{R_g(\Theta, n)} \quad , \tag{1.16}$$

which measures the excluded volume swelling of the chain relative to its size at the Θ-point. As illustrated in Fig. 1.5, for a large class of chemically different systems α_g is found to be a function of z only:

$$\alpha_g(T, n) = \alpha_g(z) \quad . \tag{1.17}$$

This function furthermore is independent of chemistry and thus universal. It describes the crossover from Θ-conditions $z = 0, \alpha_g(0) \equiv 1$ to the excluded volume limit, which is reached for $z \gg 1$. Indeed

$$\alpha_g(z) \underset{z \to \infty}{\longrightarrow} const\ z^{2\nu - 1} \quad ,$$

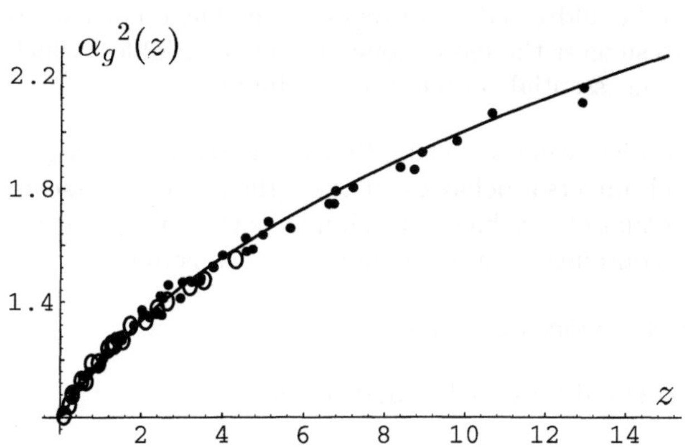

Fig. 1.5. The swelling factor α_g^2 (Eq. (1.17)) as function of z (Eq. (1.16)). *Points*: polystyrene-cyclohexane [MF81] [YFTY71]. *Circles*: P-isobutylene-iso-amylisovalerate [MNF72]. Each set of data comprises several chain lengths. The theoretical crossover curve is also given. The precise definition of the z-variable used here, which later will be called $\tilde{z}$, is given in Sect. 13.3

which via Eqs. (1.15): $z \sim n^{1/2}$ and (1.2): $R_g(\Theta, n) \sim n^{1/2}$ reproduces the excluded volume behavior (1.3). (We should immediately note that the scaling law (1.17) is not exact, the rigorous result being given in Chap. 15.)

For sake of clarity it may be helpful to collect the definitions of the different parameter regions. Throughout this book we are concerned with solutions of long chains

$$n \gg 1 \tag{1.18 i}$$

and small segment concentrations

$$\ell^3 c \ll 1 \quad , \tag{1.18 ii}$$

in the temperature range

$$\Theta_u \leq T \leq \Theta_\ell \quad , \tag{1.18 iii}$$

where the monomer interaction is repulsive. The following limiting situations are of special interest:

Θ-limit: s, n fixed, $T \to \Theta$, i.e. $z \to 0$.

Excluded volume limit: $s, T \neq \Theta$ fixed, $n \to \infty$, i.e. $z \to \infty$.

The crossover among these limits will be addressed as *'temperature crossover'*.

Dilute limit: z, n fixed, $c \to 0$, i.e. $s \to 0$.

Semidilute limit: z, c fixed, $n \to \infty$, i.e. $s \to \infty$.

The associated crossover will be called *'concentration crossover'*. The neighborhood of the Θ-limit will be addressed as 'Θ-region', etc. The renormalization group will be found to suggest the use of some modified variables, which however does not change the essential contents of the limits.

We may summarize our discussion by noting that in the parameter region given by Eqs. (1.18) we find universal behavior. It takes the form of crossover scaling laws, in limiting situations reducing to simple power laws. Specific microstructure effects are contained in a few nonuniversal constants.

There arise a number of obvious questions:

(i) What is the reason behind universality and scaling?

(ii) What is the precise form of the scaling laws comprising chain length, temperature, concentration and scattering angles?

(iii) In what parameter regions do the universal laws apply?

(iv) Can we calculate exponents and crossover scaling functions explicitly?

All these questions will be taken up in the following chapters. We here briefly note the answers which will emerge.

(i) Universal scaling reflects invariance of the system under spatial dilatations. (Chaps. 8, 10.)

(ii) The form of the scaling laws results from the general structure of the dilatation group, in this context called 'renormalization group'. (Chaps. 8, 10, 11.)

(iii) The results are correct up to corrections roughly of order $1/n$ or $c = c_p n$. (Chaps. 8, 11.)

(iv) The exponents can be calculated with high precision (Chap. 13). Scaling functions are known only in a low order approximation of perturbation theory (Part IV).

We should note that in this introductory overview we generously ignored a number of important complications. For instance, a typical solution will be polydisperse, comprising chains of different lengths. Also the approach to the excluded volume limit can follow two different paths, leading to two different, but universal, branches for functions like $\alpha_g(z)$. All such complications will be discussed in later chapters, where a more detailed data analysis will also be presented.

I want to finish this introduction by a short sketch of the history of the subject. The physics of dilute polymer solutions by now has been an active field for about 70 years. Much of the early work is connected to the name of Flory and summarized in his classic books [Flo53, Flo69]. Up to about 1970 much theoretical or experimental work concentrated on the behavior in the dilute limit, where via virial expansions the problem can be reduced to considering only a few interacting chains. The development led to the so-called 'two parameter theories', which essentially expand quantities like R_g or A_2 in powers of z. In 1971 these developments were most carefully reviewed in a book by Yamakawa [Yam71].

Being restricted to $z \lesssim 1$, this approach a priori could not deal with the excluded volume limit $z \to \infty$. One approach towards an understanding of the excluded volume region $z \gg 1$ was based on a mean field picture, which views the polymer coil as a cloud of uncorrelated monomers. A celebrated example is Flory's argument predicting $R_g(n) \sim n^{3/5}$. Another approach, mainly due to Edwards, concentrated on selfconsistent decoupling schemes for the equations governing the correlation functions. In all this older work the scaling or power law behavior in particular in semidilute systems went unnoticed.

In 1972 the development was turned to a new direction by de Gennes' discovery that correlations of a single polymer chain are closely related to critical correlations occurring near the critical point of a phase transition in ferromagnets. Des Cloizeaux soon generalized this observation to solutions of finite concentration, deriving the scaling law (1.8) for the osmotic pressure. On a more qualitative level the importance of scaling was realized, and many predictions were worked out and tested experimentally. The scaling theory is summarized in de Gennes' book [dG79]. At the more quantitative level the work first concentrated on translating to the polymer system the results established for ferromagnets, but since about 1980 we have learned how to formulate the relevant theoretical concepts, renormalization group and renormalized perturbation theory, directly for the polymer system. Meanwhile the theory has been worked out in great detail. The theoretical development has been reviewed in books by Freed [Fre87] or Des Cloizeaux and Jannink [CJ90]. Besides in advocating a somewhat different form of the renormalization group, the present treatise differs from these previous publications by the stress it puts on detailed results and extensive discussions for observable scaling functions. Finally we should note that from the point of view of an experimentalist the state of the art recently has been reviewed in a book by Fujita [Fuj90]. Furthermore Eisenriegler [Eis93] presented a detailed account of polymer adsorption, a topic which is not covered here.

Part I

Basic Theoretical Techniques

We here define our model and present a self-contained introduction to perturbation theory, deriving the Feynman graph representation of the cluster expansion. To deal with solutions of finite concentration we introduce the grand-canonical ensemble and resum the cluster expansion to construct the loop expansion. We then show that without further insight the expansions can be applied only in the Θ-region or for concentrated solutions since they diverge term by term in the excluded volume limit.

All this analysis will be based on the mathematically most simple spring and bead model. Standard two parameter theory is based on the continuous chain model, which can be derived from our model by a not quite trivial limiting process. The derivation stresses that standard two parameter theory is expected to hold only close to the Θ-temperature. In this context we also exhibit the relation of polymer theory to a special quantum field theory.

2. From a Microscopic Description to Simple Models: Some Folk-Lore

This chapter shall give a feeling for the status of the simple model we are analyzing in the sequel. This seems necessary since a priori this model has little to do with a realistic microscopic description of macromolecules in a solvent. Indeed we use a kind of of 'spring and bead' model, where the macromolecule is replaced by a sequence of elastic springs connecting small (point-like) beads, which repel each other. Of all the microscopic structure present in a real polymer molecule we thus keep only the features of chain connectedness and repulsion among chain units. (The reader interested in short range properties of other model chains might consult the books of Flory, Yamakawa, or Freed.)

To justify such a simplistic model, first of all we may invoke the experimental observation of universality, which tells us that details of the microstructure cannot be important for quantities like the size of the coil, quantities that characterize the large scale features of the macromolecules. However, we also can find simple heuristic arguments suggesting that our model indeed captures the essential features. In this chapter we first will define our model and characterize the quantities which can be described within our approach (Sect. 2.1). We then will present some general ideas on the relation to a more realistic description (Sect. 2.2). These ideas will be made somewhat more precise later (Chap. 10).

2.1 The model

We replace the macromolecule by a sequence of linear segments. The configuration of a chain is fixed by the positions of all the segment endpoints. (See Fig. 2.1.) For a system of M polymer chains, the m-th chain consisting of n_m segments, the set of these segment coordinates is written as

$$\{\mathbf{r}_j^{(m)}\} = \{\mathbf{r}_0^{(1)}, \mathbf{r}_1^{(1)}, \dots, \mathbf{r}_{n_1}^{(1)}; \mathbf{r}_0^{(2)}, \dots, \mathbf{r}_{n_2}^{(2)}; \dots; \mathbf{r}_0^{(M)}, \dots, \mathbf{r}_{n_M}^{(M)}\} . \tag{2.1}$$

The probability of finding the system in a specific configuration is given by the configurational weight $\mathcal{F}$, which we write as

$$\mathcal{F} = \frac{1}{\mathcal{Z}(M)} \exp[-V\{\mathbf{r}_j^{(m)}\}] \ , \tag{2.2}$$

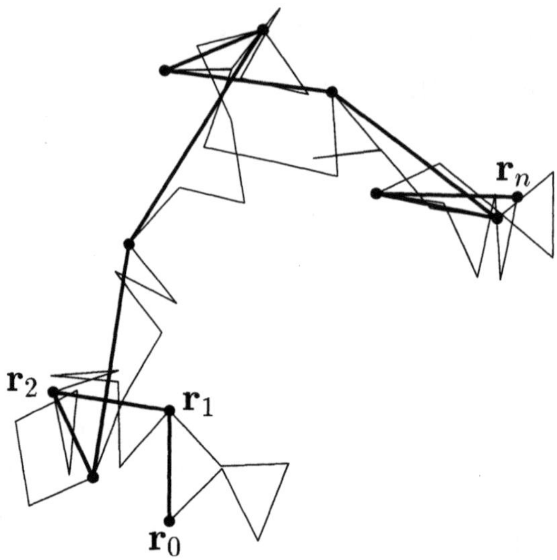

Fig. 2.1. A random walk chain of 50 'monomers' of fixed length (*thin polygon*). Taking the position of every fifth monomer as segment coordinates, we find our model chain (*fat polygon*), which is approximated by a Gaussian coil

$$\exp\{-V\} = \exp\left\{-\frac{1}{4\ell^2}\sum_{m=1}^{M}\sum_{j=1}^{n_m}\left(\mathbf{r}_j^{(m)} - \mathbf{r}_{j-1}^{(m)}\right)^2\right\}$$
$$\cdot \prod_{(mj,m'j')}{}' \left[1 - (4\pi\ell^2)^{\frac{3}{2}}\beta_e\delta^3(\mathbf{r}_j^{(m)} - \mathbf{r}_{j'}^{(m')})\right] \quad . \tag{2.3}$$

In Eq. (2.2) the normalizing factor $\mathcal{Z}^{(M)}$ is the 'reduced partition function' defined as

$$\mathcal{Z}^{(M)} = \mathcal{Z}^{(M)}(n_1, \ldots, n_M) = \int_\Omega \prod_{m=1}^{M}\prod_{j=0}^{n_m} \frac{d^3r_j^{(m)}}{(4\pi\ell^2)^{\frac{3}{2}}} \exp[-V\{\mathbf{r}_j^{(m)}\}] \quad . \tag{2.4}$$

Here all segment coordinates are integrated independently over the volume Ω of the system.

This model is known as 'spring and bead' model or as model of 'self-repelling Gaussian chains'. It incorporates chain connectedness via the exponential part in Eq. (2.3), which couples the endpoints of each segment by a harmonic potential. Thus each segment can be viewed as an elastic spring. The related model parameter is the microscopic length ℓ, which determines the average size of a segment, as will be shown in Sect. 3.3 (Eq. 3.43). At the endpoints of the segments there are point-like beads, which repel each other. The spatial shape of this repulsion is given by the δ-function $\delta^3(\mathbf{r})$, which is defined to be zero for $\mathbf{r} \neq 0$ and to tend to infinity for $\mathbf{r} = 0$ in such a way that $\int d^3r\delta^3(\mathbf{r}) = 1$. (See Appendix A 3.1 for a more precise definition, also illustrating how these 'point-like' beads can be understood as a limit of repelling spheres (Fig. 3.1).) The effective strength of the repulsion is measured by the excluded volume constant β_e. Factors $4\pi\ell^2$ have been introduced for later convenience. One virtue is to make $\mathcal{Z}^{(M)}$ or β_e dimensionless. The primed product in Eq. (2.3) extends over all pairs of segments in the system, with the restriction that in multiplying out the product to yield a series in powers of β_e we keep only terms in which any segment coordinate occurs at most once. This restriction, which is necessary to make the model well defined, will be explained in Sect. 2.2.

What physics can we describe by this model? We can define a free energy

$$F = -k_B T \ \ln \mathcal{Z}^{(M)} \ , \tag{2.5}$$

but this is not the true free energy $\tilde{F}$ of the solution. In a realistic description of a polymer chain the segments would show a rich internal structure, which is due to the chemical composition of the monomers. Eliminating all this structure we get an additional contribution $F_1 : \tilde{F} = F + F_1$. This additive 'background' will depend on temperature, chain lengths n_m, and all the other parameters of a realistic description, but is assumed to be a smooth function, not contributing to excluded volume effects. The idea is that the eliminated degrees of freedom describe physical behavior taking place on a local scale. The interesting excluded volume behavior, however, should be associated with the overall configuration of the chain, which is fixed by the segment coordinates. Indeed, universal scaling behavior is found for properties related to the chains as a whole, like the osmotic pressure or the coil radius. Thus excluded volume behavior should be embodied in the 'singular part' (2.5) of the free energy. Note however, that in comparing to experiments we might have to take care of the existence of the smoothly varying background F_1.

Other observables described by our model take the form of expectation values of quantities $\mathcal{O}\{\mathbf{r}_j^{(m)}\}$ depending only on the segment coordinates, but not on other degrees of freedom. Such expectation values $\langle \mathcal{O} \rangle$ can be calculated as

$$\langle \mathcal{O} \rangle = \int \prod_{k=1}^{M} \prod_{i=0}^{n_k} \frac{d^3 r_i^{(k)}}{(4\pi\ell^2)^{\frac{3}{2}}} \mathcal{O}\{\mathbf{r}_j^{(m)}\} \mathcal{F}\{\mathbf{r}_j^{(m)}\} \ . \tag{2.6}$$

The simplest example is provided by the end-to-end distribution of a given chain ($m = 1$, for definiteness)

$$\hat{P}(\mathbf{r}; n_1) = \left\langle \delta^3(\mathbf{r}_0^{(1)} - \mathbf{r}_{n_1}^{(1)} - \mathbf{r}) \right\rangle \ . \tag{2.7}$$

The δ-function guarantees that only those configurations contribute to the integral (2.6), for which $\mathbf{r}_0^{(1)} - \mathbf{r}_{n_1}^{(1)} = \mathbf{r}$ holds. Thus $\hat{P}(\mathbf{r}, n)$ gives the probability density of finding a configuration spanning distance $\mathbf{r}$. It is measurable in computer experiments. We expect that all observables showing interesting universal behavior in the excluded volume limit can be expressed in terms of the overall configuration of the chains and thus in the form (2.6).

2.2 Relation to a realistic microscopic description

The guiding principle in writing down the self-repelling Gaussian chain model is mathematical simplicity, not microscopic faithfulness. Do we have any idea why such a primitive model properly can explain the experiments? To investigate this question we consider how a realistic microscopic description could be reduced to our model.

Ignoring quantum effects in a microscopic description we fix the configuration of a many particle system by prescribing spatial coordinates, momenta, and internal degrees of freedom of all the particles. In statistical mechanics we then define the probability of finding a specific configuration as

$$\tilde{\mathcal{F}}(\mathit{configuration}) = \tilde{\mathcal{Z}}^{-1} \exp\left[-\frac{1}{k_B T}\mathcal{H}(\mathit{configuration})\right] . \qquad (2.8)$$

Here $\mathcal{H}$ is the Hamiltonian of the system, which gives the energy of the configuration considered, and the normalization factor $\tilde{\mathcal{Z}}$ is the (canonical) partition function. By construction, $\tilde{\mathcal{F}}$ is a probability density, normalized to 1 with respect to integration over all the degrees of freedom.

For our polymer system we can now, among the set of spatial coordinates of the polymer chains, single out a set of 'segment coordinates', as described in the previous section. Going along the chain we could for instance as segment coordinates choose the center of mass positions of every fifth monomer. (See Fig. 2.1.) Keeping these positions fixed and integrating Eq. (2.8) over all the other degrees of freedom including those of the solvent, we find a result of the form (2.2). V, however, will be a most complicated function of the segment coordinates. Still, this quantity is well defined and should in principle be calculated from the microscopic theory (2.8) by carrying through the elimination process. In practice this task is prohibitively difficult, and we replace it by speculating on the result.

We first note that after fixing the segment coordinates each of the other internal coordinates of the chains can range only over a very restricted volume of typical linear size ℓ. Furthermore in general also the correlations among the solvent molecules will be of microscopic scale. The integration over such short range fluctuations should not give rise to long range contributions in $V\{\mathbf{r}_i^{(m)}\}$. We thus write

$$V = V_0 + \sum_{\mu \geq 2} V_\mu \ , \qquad (2.9)$$

where V_0 is the 'chain part' tying together subsequent segments of the chains. V_μ is an effective μ-segment interaction, and we may assume all V_μ to be of microscopic range of the order of the segment size ℓ. We then omit all V_μ for $\mu \geq 3$. To defend this, we invoke a simple estimate. Due to its short range, the effect of V_μ should be proportional to the number $n_c^{(\mu)}$ of μ-segment contacts in a typical configuration of the polymer coils. In terms of the average density $\ell^3 n/R_g^3$ of polymer segments in the coil volume, this number is estimated as

$$n_c^{(\mu)} \sim n \left(\frac{\ell^3 n}{R_g^3}\right)^{\mu-1} \sim n^{\mu-(\mu-1)3\nu} \quad . \tag{2.10 i}$$

In the last expression we used the power law $R_g \sim n^\nu$. For $\nu \sim 0.59$ this estimate suggests that $n_c^{(\mu)}, \mu \geq 3$, vanishes for $n \to \infty$: All many-body forces should be irrelevant in the excluded volume limit, just because the number of μ-body contacts, $\mu \geq 3$, is negligibly small for a long chain. In the Θ-region, $\nu = \frac{1}{2}$, the same result is found for $\mu \geq 4$, but $n_c^{(3)}$ becomes independent of n. Indeed, three-body potentials play some role near the Θ-temperature, as will be discussed in Chap. 21, but they can be ignored in the excluded volume region.

Note that this estimate strictly speaking refers to the contacts among segments of a single chain. For a many-chain system of concentration c_p it can be generalized to yield the number of many-segment contacts of a given chain with all the others. Keeping the overlap $s = c_p R_g^3$ fixed, for long chains we again find the behavior (2.10). Indeed, in terms of the total segment concentration $c = c_p n$ we may estimate the number of μ-body contacts of a given chain as

$$\begin{aligned} n_c^{(\mu)} \sim n\left(\ell^3 c\right)^{\mu-1} &= n\left(c_p R_g^3 \frac{\ell^3 n}{R_g^3}\right)^{\mu-1} \\ &\sim s^{\mu-1} n^{\mu-(\mu-1)3\nu} \quad . \end{aligned} \tag{2.10 ii}$$

Keeping s fixed, we obviously can apply the above discussion.

We now turn to our choice of the chain part V_0, which clearly reduces to a sum over all chains in the solution.

$$V_0 = \sum_{m=1}^{M} V_0^{(m)}\{\mathbf{r}_0^{(m)}, \ldots, \mathbf{r}_{n_m}^{(m)}\} \quad . \tag{2.11}$$

In our model (2.3) we use the simple ansatz ('ansatz' is a common word in such texts)

$$\begin{aligned} V_0^{(m)}\{\mathbf{r}_0, \ldots, \mathbf{r}_n\} &= \sum_{j=1}^{n} v_0(\mathbf{r}_j - \mathbf{r}_{j-1}) \\ &= \frac{1}{4\ell^2} \sum_{j=1}^{n} (\mathbf{r}_j - \mathbf{r}_{j-1})^2 \quad . \end{aligned} \tag{2.12}$$

To defend this form, we may invoke the *central limit theorem*. Assume that we are given a more realistic form of v_0, which depends not only on the endpoints $\mathbf{r}_j, \mathbf{r}_{j-1}$ of the segment considered, but also on the coordinates of a finite number of neighboring segments: $v_0(\mathbf{r}_j, \mathbf{r}_{j-1}) \to v_0(\mathbf{r}_{j-1-k}, \mathbf{r}_{j-k}, \ldots, \mathbf{r}_{j+k})$. The central limit theorem states that for a noninteracting chain ($V_\mu \equiv 0$ for all $\mu \geq 2$) the endpoint distribution $\hat{P}_0(\mathbf{r}; n)$ in the limit of large n takes the form

$$\hat{P}_0(\mathbf{r}; n) \sim \exp\left[-\frac{3\mathbf{r}^2}{2n\ell_{eff}^2}\right] , \qquad (2.13)$$

irrespective of the precise form of v_0 , which fixes only the effective segment size ℓ_{eff}. We may thus use the mathematically simplest form, which is the 'Gaussian chain model' (2.12). We should note, however, that the result (2.13) is valid only for distances $|\mathbf{r}|$ small compared to the extension $\sim n\ell_{eff}$ of the fully stretched chain. This poses no problem, since such stretched configurations are extremely rare and contribute only negligibly to thermodynamic expectation values. Thus the Gaussian model yields a valid description for large scale properties of noninteracting chains.

To argue in favor of the Gaussian model for *interacting* chains we again invoke the estimate (2.10) which suggests that the *relative* number of μ-body contacts $n_c^{(\mu)}/n$ vanishes for long chains even for $\mu = 2$. We thus expect that two successive interacting segments are connected by long noninteracting parts of the chain. According to the central limit theorem these parts should adequately be described by a Gaussian chain. (This is a very naive argument; see the end of this section.)

We finally have to defend our choice of the two-body potential

$$V_2 = \sum_{(mj,m'j')} v_2(\mathbf{r}_j^{(m)} - \mathbf{r}_{j'}^{(m')}) . \qquad (2.14)$$

The summation extends over all pairs of segments in the solution. We may specify $v_2(\mathbf{r})$ in terms of the cluster function $\beta(\mathbf{r})$ defined as

$$\exp(-v_2(\mathbf{r})) = 1 - \beta(\mathbf{r}) . \qquad (2.15)$$

For the typical form of $v_2(\mathbf{r})$ consisting of a strong short range repulsion followed by some weak attraction, $\beta(\mathbf{r})$ has a shape as shown in Fig. 2.2. In our model (2.3) we replace this by a simple local repulsion

$$\beta(\mathbf{r}) = (4\pi\ell^2)^{\frac{3}{2}}\beta_e\delta^3(\mathbf{r}) . \qquad (2.16)$$

This is based on the belief that for an analysis of universal excluded volume effects the detailed shape of $\beta(\mathbf{r})$ is not relevant as long as it simulates a short range repulsive interaction. Indeed, the integral over all configurations averages over the factors of $\beta(\mathbf{r}_i - \mathbf{r}_j)$, and for a Gaussian chain the probability

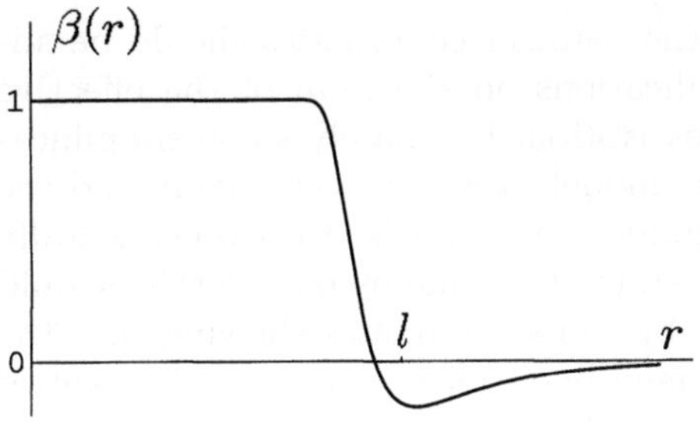

Fig. 2.2. Typical form of the cluster function $\beta(r)$

that two segments i, j, $(i-j) \gg 1$ approach each other on distance $|(\mathbf{r}_i - \mathbf{r}_j)| \sim \ell \ll \ell\,(i-j)^{1/2}$ essentially is independent of $r = |\mathbf{r}_i - \mathbf{r}_j|$. Thus only the spatial average of $\beta(\mathbf{r})$ matters.

There however remains a delicate point. Keeping only two-body forces we a priori should write the interaction as

$$\exp\Big\{-\sum_{(i,j)} v_2(\mathbf{r}_i - \mathbf{r}_j)\Big\} = \prod_{(i,j)} (1 - \beta(\mathbf{r}_i - \mathbf{r}_j)) \quad , \tag{2.17}$$

where the chain indices are suppressed for simplicity. Expanding the product we also find terms of type

$$\beta(\mathbf{r}_{i_1} - \mathbf{r}_{i_2})\beta(\mathbf{r}_{i_2} - \mathbf{r}_{i_3}) \ ; \beta(\mathbf{r}_{i_1} - \mathbf{r}_{i_2})\beta(\mathbf{r}_{i_2} - \mathbf{r}_{i_3})\beta(\mathbf{r}_{i_3} - \mathbf{r}_{i_1}) \quad ; \ \text{etc.} \quad , \tag{2.18}$$

where three (or more) segments close together in space interact simultaneously. In such terms the substitution (2.16) may be invalid. For instance for the second term it yields

$$\left(4\pi\ell^2\right)^{9/2} \beta_e^3 \delta^3(\mathbf{r}_{i_1} - \mathbf{r}_{i_2})\delta^3(\mathbf{r}_{i_2} - \mathbf{r}_{i_3})\delta^3(\mathbf{r}_{i_3} - \mathbf{r}_{i_1}) \quad .$$

Since the first two δ-functions imply

$$\mathbf{r}_{i_1} = \mathbf{r}_{i_2}, \, \mathbf{r}_{i_3} = \mathbf{r}_{i_2} \ ,$$

the argument of the third δ-function in fact is zero, leading to a divergent expression: $\delta^3(0) = \infty$. Now terms of type (2.18) contribute only for configurations $\mathbf{r}_{i_1} \approx \mathbf{r}_{i_2} \approx \mathbf{r}_{i_3}$, where also three-body interactions do not vanish. Neglecting the latter, we should suppress the former, thus eliminating the problem. We thus impose the rule that in expanding the product (2.17) all those terms have to be omitted, in which any segment coordinate occurs in two or more factors of $\beta(\mathbf{r})$. This rule is consistent with the neglect of many-body forces.

Let us summarize this discussion. We first note that the excluded volume behavior should be a property of the global chain configurations. Thus a reduced description involving only the segment coordinates should be adequate. We then make drastic simplifications on the form of the effective potential occurring in this reduced description. Ultimately these simplifications are justified by the success of the model. If we want to understand the reason for this success we, as a first argument, may invoke the experimentally verified concept of universality, which states that microscopic details should be unimportant. We can support this by simple estimates showing that the two-body repulsion dominates over many-body effects, and by the central limit theorem suggesting that a Gaussian chain model is adequate. To some extent this is a step in the direction of proving universality.

Our argument is heavily based on the estimate (2.10). This is a 'Flory-type' estimate, which views the chain as a cloud of uncorrelated monomers, extending over a volume $\sim R_g^3$. It thus completely neglects the chain structure. This can and must be criticised. Indeed, due to chain structure, close to any segment the local segment density is enhanced and taken naively the estimate (2.10) is wrong. Still the argument basically is correct, but it needs a much more careful analysis, which in fact introduces the concept of 'additive renormalization'. This will change the interpretation, not the form of the terms included in our model, showing that β_e parameterizes the interaction of whole strands of the chain rather than an interaction of individual segments. As a consequence, the relation of β_e to a realistic microscopic model is much less direct than our simplistic arguments might suggest. In Chap. 10 we will discuss this topic, which is important for a more precise understanding of the model, but can be ignored in the actual evaluation of physical results, where in any case β_e or ℓ are treated as effective fit parameters.

3. The Noninteracting Gaussian Chain

This chapter deals only with chain connectedness and ignores any interaction among segments which are well separated along their chain or which are part of different chains. We thus ignore all excluded volume effects, essentially treating a model chain at its Θ-point. We first consider the correlations among individual segments of the chain (Sect. 3.1), which are all found to be given by simple Gaussian functions. We then discuss the segment density correlation function (Sect. 3.2), which can be measured in scattering experiments. We show how this function for scattering vectors $\mathbf{q}$, $q\ell \ll 1$, reduces to the well known Debye function. For $q\ell \sim 1$ the result is more complicated, but also meaningless: scattering vectors in that range see the unphysical microstructure of the spring and bead model. We then turn to a more detailed evaluation of the endpoint Greensfunction, which is closely related to the endpoint correlation function (Sect. 3.3). This serves to derive results used in Sects. 3.1 and 3.2, and it is a necessary prerequisite of perturbation theory.

Mathematical appendices introduce δ-functions, Fourier transform, or d-dimensional integration over Gaussian functions, respectively.

3.1 Endpoint correlations

Ignoring the excluded volume interaction we deal with independent chains. It thus is sufficient to consider a single chain system. The spatial dimension is found to play an important role in the theory, and we therefore work with general dimension d. We write the configurational probability for a noninteracting Gaussian chain as

$$\mathcal{F}_0(\mathbf{r}_0, \ldots, \mathbf{r}_n) = \frac{1}{\mathcal{Z}_0(n)} \exp\{-\frac{1}{4\ell^2}\sum_{j=1}^{n}(\mathbf{r}_j - \mathbf{r}_{j-1})^2\} \tag{3.1}$$

$$\mathcal{Z}_0(n) = \int_\Omega D[\mathbf{r}] \exp\{-\frac{1}{4\ell^2}\sum_{j=1}^{n}(\mathbf{r}_j - \mathbf{r}_{j-1})^2\} \ . \tag{3.2}$$

Here

$$\mathcal{D}[\mathbf{r}] = \prod_{j=0}^{n} \frac{d^d r_j}{(4\pi\ell^2)^{\frac{d}{2}}} \quad , \tag{3.3}$$

and $\mathcal{F}_0$ is normalized with respect to the integration measure $\mathcal{D}[\mathbf{r}]$.

For this model only the spatial correlations among segments are of interest. Let us first consider the distribution function of the end-to-end distance, which is defined as

$$\hat{P}_0(\mathbf{r}, n) = \left\langle \delta^d(\mathbf{r}_n - \mathbf{r}_0 - \mathbf{r}) \right\rangle^0 \quad . \tag{3.4}$$

$d^d r \hat{P}_0(\mathbf{r}, n)$ gives the probability to find the end-to-end vector in volume element $d^d r$ centered at $\mathbf{r}$. The symbol $\langle \ldots \rangle^0$ denotes an expectation value with respect to $\mathcal{F}_0$:

$$\left\langle \delta^d(\mathbf{r}_n - \mathbf{r}_0 - \mathbf{r}) \right\rangle^0 = \int_{\Omega} D[\mathbf{r}]\, \delta^d(\mathbf{r}_n - \mathbf{r}_0 - \mathbf{r})\, \mathcal{F}_0(\mathbf{r}_0, \ldots, \mathbf{r}_n) \, . \tag{3.5}$$

By construction $\hat{P}_0(\mathbf{r}, n)$ is normalized

$$\int d^d r \hat{P}_0(\mathbf{r}, n) = 1 \tag{3.6}$$

and obeys the relation (see Appendix A 3.1)

$$\int d^d r \mathbf{r}^2 \hat{P}_0(\mathbf{r}, n) = \langle (\mathbf{r}_n - \mathbf{r}_0)^2 \rangle^0 = R_e^2 \quad , \tag{3.7}$$

which defines the mean-squared end-to-end distance R_e^2.

With our model the evaluation of $\hat{P}_0(\mathbf{r}; n)$ is an exercise in doing Gaussian integrals. (See Appendix A 3.3.) We only note here the results, leaving the explicit calculation to Sect. 3.3. We find

$$\hat{P}_0(\mathbf{r}; n) = (4\pi\ell^2 n)^{-\frac{d}{2}} \exp(-\frac{\mathbf{r}^2}{4\ell^2 n}) \quad , \tag{3.8}$$

$$R_e^2 = 2d\ell^2 n \quad . \tag{3.9}$$

We often will be concerned with the Fourier transform (see Appendix A 3.2)

$$P_0(\mathbf{p}; n) = \int d^d r e^{i\mathbf{pr}} \hat{P}_0(\mathbf{r}; n) \quad . \tag{3.10}$$

This again amounts to a Gaussian integral which yields

$$P_0(\mathbf{p}; n) = e^{-\mathbf{p}^2 \ell^2 n} \quad . \tag{3.11}$$

Note that the factor $1/(4\ell^2)$ in the definition (2.3) of our model was chosen with this simple result in mind. The Fourier variable $\mathbf{p}$, conjugate to $\mathbf{r}$, often will be called 'momentum', a notation taken from field theory. In a scattering experiment it equals the scattering vector.

These results immediately yield all the internal correlations among chain segments. The spatial distribution function for the pair of segments $k_1 < k_2$ is defined as

$$\hat{P}_0(\mathbf{r}; k_1, k_2; n) = \langle \delta^d(\mathbf{r}_{k_2} - \mathbf{r}_{k_1} - \mathbf{r}) \rangle^0 \quad . \tag{3.12}$$

Since the internal configurations of disjoint parts of a noninteracting Gaussian chain are uncorrelated, this distribution is identical to the endpoint distribution of a chain of length $k_2 - k_1$:

$$\hat{P}_0(\mathbf{r}; k_1, k_2; n) = \hat{P}_0(\mathbf{r}; k_2 - k_1) \quad . \tag{3.13}$$

Thus all segment correlations of our noninteracting model take the form of Gaussian functions. We stress that for our Gaussian chain model all these results are valid rigorously for all $n \geq 1$ or $k_2 - k_1 \geq 1$, respectively.

If we allow for correlations among segments $\mathbf{r}_{k_1} - \mathbf{r}_{k_2}, \mathbf{r}_{k_3} - \mathbf{r}_{k_4}$ closely spaced along the chain the results (3.8)–(3.13) remain valid for long chains, provided we replace $\sqrt{2d}\ell$ by some effective segment length ℓ_{eff}. This is a consequence of the central limit theorem. However, there are corrections of order $1/n$. Thus

$$R_e^2 = \ell_{eff}^2 n \left(1 + O\left(\frac{1}{n}\right)\right)$$
$$\hat{P}_0(\mathbf{r}, n) = \left(\frac{2\pi}{d} R_e^2\right)^{-d/2} \exp\left(-\frac{d\mathbf{r}^2}{2R_e^2}\right) \left[1 + O\left(\frac{1}{n}\right)\right] \quad .$$

It is the virtue of the Gaussian model to suppress these $1/n$ corrections.

3.2 Density correlations

Correlations among specific segments can be measured only in computer experiments. Normal scattering experiments measure the segment density correlation function, to which all segments contribute equally. To define this quantity we introduce the local segment density

$$\hat{\rho}(\mathbf{r}) = \sum_{j=1}^{n} \delta^d(\mathbf{r}_j - \mathbf{r}) \ , \tag{3.14}$$

which is nonzero only if an endpoint of some segment is found at point $\mathbf{r}$. (Recall that we are interested only in $n \gg 1$. Thus it is not important whether we define $\hat{\rho}(\mathbf{r})$ in terms of segment endpoints $\mathbf{r}_j, j \geq 1$ or in terms of the centers of the segments $(\mathbf{r}_j + \mathbf{r}_{j+1})/2$, $j \geq 0$. The present definition is more practical.) Integrating $\hat{\rho}(\mathbf{r})$ over the volume Ω we find the number n of segments:

$$\int_{\Omega} d^d r \hat{\rho}(\mathbf{r}) = \sum_{j=1}^{n} 1 = n \quad . \tag{3.15}$$

This allows for an easy evaluation of the expectation value $\langle \hat{\rho}(\mathbf{r}) \rangle^0$. Translational invariance guarantees that the expectation value is independent of $\mathbf{r}$, except for surface effects. We thus can write

$$\langle \hat{\rho}(\mathbf{r}) \rangle^0 = \frac{1}{\Omega} \int_{\Omega} d^d r \langle \hat{\rho}(\mathbf{r}) \rangle^0 = \frac{n}{\Omega} \quad . \tag{3.16}$$

The segment density correlation function is defined as

$$I_d(\mathbf{q}) = \int_{\Omega} d^d r d^d r' e^{i\mathbf{q}(\mathbf{r}-\mathbf{r}')} \langle \hat{\rho}(\mathbf{r}) \hat{\rho}(\mathbf{r}') \rangle^0 \quad , \tag{3.17}$$

where $\langle \hat{\rho}(\mathbf{r}) \hat{\rho}(\mathbf{r}') \rangle^0$ gives the probability to find a pair of segments at points $\mathbf{r}$ and $\mathbf{r}'$. Rewriting the product of segment densities as

$$\begin{aligned} \hat{\rho}(\mathbf{r}) \hat{\rho}(\mathbf{r}') = & \sum_{k_1 \neq k_2} \delta^d(\mathbf{r}_{k_1} - \mathbf{r}) \, \delta^d(\mathbf{r}_{k_2} - \mathbf{r}') \\ & + \sum_{k} \delta^d(\mathbf{r}_k - \mathbf{r}) \, \delta^d(\mathbf{r}_k - \mathbf{r}') \quad , \end{aligned} \tag{3.18}$$

we can evaluate the contribution of the second term to Eq. (3.17) to find

$$\left\langle \sum_k e^{i\mathbf{q}(\mathbf{r}_k - \mathbf{r}_k)} \right\rangle^0 = \left\langle \sum_k 1 \right\rangle^0 = n \quad . \tag{3.19}$$

This is the autocorrelation scattering due to n point-like scattering centers. The first term of Eq. (3.18) yields

$$\begin{aligned} \sum_{k_1 \neq k_2} \langle e^{i\mathbf{q}(\mathbf{r}_{k_1} - \mathbf{r}_{k_2})} \rangle^0 &= \sum_{k_1 \neq k_2} \int d^d r e^{i\mathbf{q}\mathbf{r}} \langle \delta^d(\mathbf{r}_{k_1} - \mathbf{r}_{k_2} - \mathbf{r}) \rangle^0 \\ &= 2 \sum_{0 < k_1 < k_2 \leq n} P_0(\mathbf{q}; k_2 - k_1) \quad . \end{aligned} \tag{3.20}$$

In the last line we used Eqs. (3.10)–(3.13), and we exploited the reflection symmetry of the chain: the chain is invariant under a relabeling $j \to n - j$. To evaluate the result (3.20), for small values of the momentum $\mathbf{q}$, such that $\mathbf{q}^2 R_e^2 / 2d = \mathbf{q}^2 \ell^2 n = O(1)$, we replace the summations by integrals. (This step will be discussed in more detail below.)

$$\frac{1}{n} \sum_{k=1}^{n} f(k) \to \int_0^1 dx f(nx)(1 + O(\frac{1}{n})), \qquad x = \frac{k}{n} \quad . \tag{3.21}$$

Using the explicit form (3.11) of $P_0(\mathbf{q}, k_2 - k_1)$ we find

$$\sum_{0<k_1<k_2\leq n} P_0(\mathbf{q};k_2-k_1) \to n^2 \int_0^1 dx_2 \int_0^{x_2} dx_1 e^{-\mathbf{q}^2\ell^2 n(x_2-x_1)}(1+O(\frac{1}{n}))$$
$$= \frac{n^2}{2} D(n\mathbf{q}^2\ell^2)(1+O(\frac{1}{n})) \quad , \tag{3.22}$$

where

$$D(x) = \frac{2}{x^2}(e^{-x} - 1 + x) \tag{3.23}$$

is called 'Debye function' in honour of P. Debye, who first calculated the scattering intensity due to a polymer chain. Combining these results we find that the autocorrelation scattering can be neglected, since it is proportional to n, whereas the contribution (3.22) is proportional to n^2. The result

$$I_d(\mathbf{q}) = n^2 D(n\mathbf{q}^2\ell^2) \tag{3.24}$$

follows. It is interesting to note that also for a self-repelling chain $I_d(\mathbf{q})$ for momenta not too large is found to be well approximated by the Debye function (cf. Chap. 15), a feature which by no means is obvious a priori.

Let us now consider more closely the replacement of the summations by integrals. The summations (3.20) are simple enough to be evaluated rigorously. In fact they can be reduced to a geometric series. Some algebra yields

$$I_d(\mathbf{q}) = \frac{2e^{-x}}{(1-e^{-x})^2}\left[e^{-nx} - 1 + n(1-e^{-x})\right]_{x=\mathbf{q}^2\ell^2} + n \quad , \tag{3.25}$$

which is the full result for our model. To derive Eq. (3.24) we write $x = (\mathbf{q}^2\ell^2 n)/n$, and we expand with respect to $1/n$ keeping $\mathbf{q}^2\ell^2 n$ fixed. Equation (3.24) gives the leading contribution in that expansion and therefore is valid for $\mathbf{q}^2\ell^2 = (\mathbf{q}^2\ell^2 n)/n \ll 1$. In real space such values of $\mathbf{q}$ correspond to distances $r \gg \ell$, i.e. we concentrate on correlations of large parts of the total chain. In fact it are only these correlations that we are interested in. Scattering in the q-range $q \sim 1/\ell$ tests the internal structure of the segments and thus the microstructure of the polymer molecule. This is badly mistreated by our model, and the experimental scattering intensity in that q-range will strongly deviate from the Debye function. *The analysis leading from Eq. (3.25) to Eq. (3.24) therefore is not just a simplification but a necessary step to get rid of all unphysical features. Only the result (3.24) really is meaningful.* Note further that this discussion implies that the result (3.24) up to corrections of order $1/n$ is valid also for more complicated microstructure of the chain.

A quantity of special interest, which can be derived from $I_d(\mathbf{q})$, is the radius of gyration.

$$R_g^2 = \frac{1}{n}\langle\sum_{k=1}^{n}(\mathbf{r}_k - \mathbf{R}_{cm})^2\rangle^0 \quad . \tag{3.26}$$

Here $\mathbf{R}_{cm}$ gives the position of the center of mass of the chain.

$$\mathbf{R}_{cm} = \frac{1}{n}\sum_{j=1}^{n}\mathbf{r}_j \quad . \tag{3.27}$$

A little algebra shows that R_g^2 also can be expressed as

$$R_g^2 = \frac{1}{2n^2}\langle\sum_{k_1 k_2}(\mathbf{r}_{k_1} - \mathbf{r}_{k_2})^2\rangle^0 \quad , \tag{3.28}$$

which in turn is related to $I_d(\mathbf{q})$. Rewriting Eq. (3.17) as

$$I_d(\mathbf{q}) = \langle\sum_{k_1 k_2} e^{i\mathbf{q}(\mathbf{r}_{k_1} - \mathbf{r}_{k_2})}\rangle^0 \quad , \tag{3.29}$$

we find

$$R_g^2 = -\frac{1}{2n^2}\Delta_q I_d(\mathbf{q})|_{\mathbf{q}=0} \quad , \tag{3.30}$$

where Δ_q is the d-dimensional Laplacian:

$$\Delta_q = \frac{\partial^2}{\partial q_1^2} + \ldots + \frac{\partial^2}{\partial q_d^2} \quad . \tag{3.31}$$

Equations (3.24), (3.30) yield

$$R_g^2 = \frac{d}{3}n\ell^2 = \frac{1}{6}R_e^2 \quad . \tag{3.32}$$

The result $R_e^2/R_g^2 = 6$ is characteristic of all noninteracting chains. It holds in the limit of large n irrespective of the microstructure as embodied in the chain part V_0. It is the first example of a universal critical ratio. Adding the excluded volume interaction, we will find that this ratio in the excluded volume limit of long chains again takes a universal value, close to but different from 6.

3.3 Calculation of the Greensfunction

To derive the results quoted in Sect. 3.1 for our Gaussian chain model, we here calculate the 'unperturbed' (since there is no interaction) Greensfunction defined as

$$G_0(\mathbf{p}; n) = \frac{(4\pi\ell^2)^{\frac{d}{2}}}{\Omega}\int_\Omega \mathcal{D}[\mathbf{r}]e^{-\frac{1}{4\ell^2}\sum_{j=1}^{n}(\mathbf{r}_j - \mathbf{r}_{j-1})^2 + i\mathbf{p}(\mathbf{r}_n - \mathbf{r}_0)} \quad . \tag{3.33}$$

In terms of G_0 the partition function takes the form

$$\mathcal{Z}_0(n) = \frac{\Omega}{(4\pi\ell^2)^{\frac{d}{2}}} G_0(\mathbf{0}; n) \quad , \tag{3.34}$$

and the endpoint distribution is given as

$$P_0(\mathbf{p}; n) = \frac{G_0(\mathbf{p}; n)}{G_0(\mathbf{0}; n)} \quad . \tag{3.35}$$

The integral (3.33) is evaluated by introducing the segment vectors $\mathbf{s}_j = \mathbf{r}_j - \mathbf{r}_{j-1}, j \geq 1; \mathbf{s}_0 = \mathbf{r}_0$. This yields

$$G_0(\mathbf{p}; n) = \frac{(4\pi\ell^2)^{\frac{d}{2}}}{\Omega} \int \mathcal{D}[\mathbf{s}] \prod_{j=1}^{n} \exp(-\frac{\mathbf{s}_j^2}{4\ell^2} + i\mathbf{p} \cdot \mathbf{s}_j) \quad . \tag{3.36}$$

A priori the $\mathbf{s}_j$ range over a complicated domain, reflecting the fact that the $\mathbf{r}_j$ are constrained to volume Ω. To overcome this problem we note that ultimately we are interested in the thermodynamic limit $\Omega \to \infty$, so that the linear size of the container $L \sim \Omega^{\frac{1}{d}}$ can be taken to be large compared to the average size $\sim R_e$ of the polymer coil. Taking $\mathbf{s}_0 = \mathbf{r}_0$ to be well inside the container we then find that for configurations leaving Ω, some $|\mathbf{s}_j|, j \geq 1$, must be of order L. Extending all $\mathbf{s}_j$-integrals over the infinite volume we therefore make an error of order $\exp(-constL^2)$, which rapidly vanishes in the thermodynamic limit. Furthermore after this extension the integrand becomes independent of $\mathbf{s}_0$ which is to be integrated over Ω. The leading correction to this result is due to configurations where the distance of $\mathbf{s}_0$ to the surface of Ω is of order R_e. Then the extension of the $\mathbf{s}_j$-integrals is unjustified. This is a surface effect resulting in a correction of order R_e/L. These considerations yield

$$\begin{aligned} G_0(\mathbf{p}; n) &= \frac{(4\pi\ell^2)^{\frac{d}{2}}}{\Omega} \int_\Omega \frac{d^d s_0}{(4\pi\ell^2)^{\frac{d}{2}}} \int_\infty \prod_{j=1}^{n} \left[\frac{d^d s_j}{(4\pi\ell^2)^{\frac{d}{2}}} \exp(-\frac{\mathbf{s}_j^2}{4\ell^2} + i\mathbf{p} \cdot \mathbf{s}_j) \right] \\ &\quad \cdot \left(1 + \mathcal{O}\left(\frac{R_e}{L}\right)\right) \\ &= \left[\int_\infty \frac{d^d s}{(4\pi\ell^2)^{\frac{d}{2}}} \exp(-\frac{\mathbf{s}^2}{4\ell^2} + i\mathbf{p} \cdot \mathbf{s}) \right]^n (1 + \mathcal{O}(\frac{R_e}{L})) \quad . \end{aligned} \tag{3.37}$$

Using Eq. (A 3.12) to evaluate the Gaussian integral we find

$$G_0(\mathbf{p}; n) = e^{-\mathbf{p}^2 \ell^2 n} \quad , \tag{3.38}$$

valid for all n. Equation (3.35) yields the result (3.11) for $P_0(\mathbf{p}; n)$ and shows that in fact P_0 and G_0 are identical: $P_0(\mathbf{p}, n) \equiv G_0(\mathbf{p}, n)$. The inverse Fourier transform yields Eq. (3.8) for $\hat{P}_0(\mathbf{r}; n)$, and from Eq. (3.34) we find

$$\mathcal{Z}_0(n) = \frac{\Omega}{(4\pi\ell^2)^{\frac{d}{2}}} \quad . \tag{3.39}$$

For later use we note another representation of $\hat{P}_0(\mathbf{r};n)$. From the definition (3.5) together with Eq. (3.39) we find

$$\hat{P}_0(\mathbf{r};n) = \frac{(4\pi\ell^2)^{\frac{d}{2}}}{\Omega} \int_\Omega D[\mathbf{r}] \exp\left[-\frac{1}{4\ell^2}\sum_{j=1}^{n}(\mathbf{r}_j - \mathbf{r}_{j-1})^2\right] \delta^d(\mathbf{r}_n - \mathbf{r}_0 - \mathbf{r}) \quad . \tag{3.40}$$

As above we ignore surface effects, and we use the shift $\mathbf{r}_j \to \mathbf{r}_j + \mathbf{r}_0, j = 1, \ldots, n$, to eliminate $\mathbf{r}_0$ from the integrand. As a result, the $\mathbf{r}_0$-integral cancels the prefactor, the other integrals being extended to infinity. The δ-function takes the form $\delta^d(\mathbf{r}_n - \mathbf{r})$ and is used to integrate $\mathbf{r}_n$. The form

$$\hat{P}_0(\mathbf{r};n) = (4\pi\ell^2)^{-\frac{d}{2}} \int \prod_{j=1}^{n-1} \frac{d^d r_j}{(4\pi\ell^2)^{\frac{d}{2}}} \exp[-\frac{1}{4\ell^2}\sum_{j=1}^{n}(\mathbf{r}_j - \mathbf{r}_{j-1})^2]|_{\mathbf{r}_0=0,\mathbf{r}_n=\mathbf{r}} \tag{3.41}$$

results, which will play a central role in constructing perturbation theory.

To close this section we calculate the effective segment size

$$\begin{aligned} \ell_{eff}^2 &= \left\langle (\mathbf{r}_k - \mathbf{r}_{k-1})^2 \right\rangle^0 \\ &= \frac{(4\pi\ell^2)^{d/2}}{\Omega} \int \mathcal{D}[\mathbf{s}] s_k^2 \prod_{j=1}^{n} \exp\left(\frac{s_j^2}{4\ell^2}\right) \quad . \end{aligned} \tag{3.42}$$

Proceeding as above we find

$$\begin{aligned} \ell_{eff}^2 = &\int_\infty \frac{d^d s_k}{(4\pi\ell^2)^{d/2}}\, s_k^2 \exp\left(-\frac{s_k^2}{4\ell^2}\right) \\ &\cdot \left[\int_\infty \frac{d^d s}{(4\pi\ell^2)^{d/2}} \exp\left(-\frac{s^2}{4\ell^2}\right)\right]^{n-1} \quad . \end{aligned}$$

With the results of Appendix A 3.3 this reduces to

$$\begin{aligned} \ell_{eff}^2 &= \int_\infty \frac{d^d s_k}{(4\pi\ell^2)^{d/2}}\, s_k^2 \exp\left(-\frac{s_k^2}{4\ell^2}\right) \\ &= 2d\ell^2 \quad , \end{aligned} \tag{3.43}$$

an expression independent of k.

A 3. Appendix

A 3.1 The δ-Function

For our purpose it is sufficient to define the (one-dimensional) δ-function as a formal object by the relation

$$\int dx f(x)\delta(x - x_0) = f(x_0) \quad , \tag{A 3.1}$$

which shall hold for any function $f(x)$ allowing for a Taylor expansion about x_0. It is understood that x_0 is contained in the interior of the range of integration. The normalization condition

$$\int dx\delta(x - x_0) = 1 \tag{A 3.2}$$

is a consequence of Eq. (A 3.1). $\delta(x - x_0)$ is not a function in the proper sense, but can be expressed in various ways as limit of a sequence of proper functions. To give an example we use the sequence (see Fig. 3.1)

$$\delta_a(x - x_0) = \begin{cases} a & \text{if } |x - x_0| < \frac{1}{2a} \\ 0 & \text{otherwise} \end{cases} \quad . \tag{A 3.3}$$

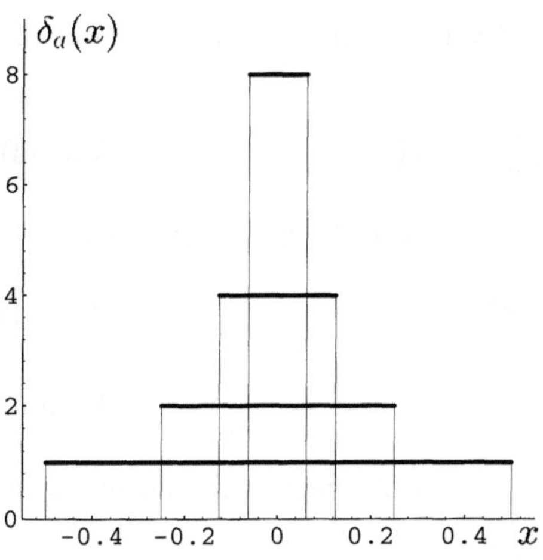

Fig. 3.1. The functions $\delta_a(x)$ (Eq. (A 3.3)), $a = 1, 2, 4, 8$. Note that appropriately generalizing this sequence to three dimensions:

$$\delta_a^{(3)}(\mathbf{r}) = \begin{cases} \frac{3}{4\pi} R_a^{-3} & r < R_a \\ 0 & r > R_a \end{cases} \quad , \quad R_a = a\, R_1 \quad ,$$

we can understand the local two-body repulsion in Eq. (2.3) as a limit of mutually repelling spheres

It is easily checked that for smooth functions $f(x)$ the relation $\lim_{a\to\infty}\int dx f(x)\delta_a(x-x_0) = f(x_0)$ holds. However, $\lim_{a\to\infty}\delta_a(x-x_0) = \delta(x-x_0)$ is no well-defined function: it vanishes for all $x \neq x_0$ but is infinite for $x = x_0$. Nevertheless we can manipulate this object very much like normal functions. We will mainly use the d-dimensional δ-function defined as

$$\delta^d(\mathbf{x}-\mathbf{x}_0) = \prod_{\alpha=1}^{d} \delta(x_\alpha - x_{0\alpha}) \quad . \tag{A 3.4}$$

As an example for the manipulation of δ-functions we note Eq. (3.7):

$$\int d^d r \mathbf{r}^2 \hat{P}_0(\mathbf{r};n) = \int d^d r \mathbf{r}^2 \langle \delta^d(\mathbf{r}_n - \mathbf{r}_0 - \mathbf{r})\rangle^0 = \langle \int d^d r \mathbf{r}^2 \delta^d(\mathbf{r}_n - \mathbf{r}_0 - \mathbf{r})\rangle^0$$
$$= \langle (\mathbf{r}_n - \mathbf{r}_0)^2 \rangle^0 \quad .$$

A 3.2 The Fourier transform

Consider a function $\hat{f}(x)$, defined for all x: $-\infty < x < \infty$. The Fourier transform $f(p)$ of $\hat{f}(x)$ is defined by the integral

$$f(p) = \int_{-\infty}^{+\infty} dx e^{ipx} \hat{f}(x), \tag{A 3.5 i}$$

provided this integral exists. The transformation may be inverted, and the basic theorem states that

$$\hat{f}(x) = \int_{-\infty}^{+\infty} \frac{dp}{2\pi} e^{-ipx} f(p) \quad . \tag{A 3.5 ii}$$

These relations immediately can be generalized to d-dimensional space:

$$f(\mathbf{p}) = \int d^d r e^{i\mathbf{pr}} \hat{f}(\mathbf{r}) \tag{A 3.6 i}$$

$$\hat{f}(\mathbf{r}) = \int \frac{d^d p}{(2\pi)^d} e^{-i\mathbf{pr}} f(\mathbf{p}) \quad . \tag{A 3.6 ii}$$

Combining these equations we find an important representation of the δ-function

$$\delta^d(\mathbf{r}-\mathbf{r}') = \int \frac{d^d p}{(2\pi)^d} e^{i\mathbf{p}(\mathbf{r}-\mathbf{r}')} \quad . \tag{A 3.7}$$

This representation again is formal since the integral does not exist in the proper sense. It has to be used under an integral with respect to $\mathbf{r}$ or $\mathbf{r}'$, which is carried out first.

We note that we have defined the Fourier transform in terms of an integral over infinite space. In all applications therefore the thermodynamic limit has to be taken, in the sense explained in Sect. 3.3. We finally note that we often will use the short-hand notation

$$\int \frac{d^d p}{(2\pi)^d} f(\mathbf{p}) = \int_p f(\mathbf{p}) \quad . \tag{A 3.8}$$

A 3.3 The Gaussian integral and integration in *d*-dimensional space

The standard Gaussian integral takes the form

$$\int_{-\infty}^{+\infty} dx e^{-ax^2} = \sqrt{\frac{\pi}{a}} \quad . \tag{A 3.9}$$

The result for d-dimensional space immediately follows:

$$\int d^d x e^{-a\mathbf{x}^2} = \prod_{\alpha=1}^{d} \left[\int_{-\infty}^{+\infty} dx_\alpha e^{-ax_\alpha^2} \right] = \left(\frac{\pi}{a} \right)^{\frac{d}{2}} \quad . \tag{A 3.10}$$

We will often need the more general integral

$$I = \int d^d x e^{-a\mathbf{x}^2 + 2\mathbf{x}\cdot\mathbf{y}} , \tag{A 3.11}$$

which is evaluated by completing the square in the exponent: $-a\mathbf{x}^2 + 2\mathbf{x}\cdot\mathbf{y} = -a(\mathbf{x} - \mathbf{y}/a)^2 + \mathbf{y}^2/a$, and substituting $\mathbf{x}' = \mathbf{x} - \mathbf{y}/a$:

$$\begin{aligned} I &= \int d^d x' \exp\{-a\mathbf{x}'^2 + \frac{\mathbf{y}^2}{a}\} \\ &= \left(\frac{\pi}{a} \right)^{\frac{d}{2}} e^{\frac{\mathbf{y}^2}{a}} \quad . \end{aligned} \tag{A 3.12}$$

Integrals of the form $\int d^d x (\mathbf{x}^2)^m \exp(-a\mathbf{x}^2)$ are easily evaluated by taking derivatives with respect to a. We for instance find

$$\int d^d x \mathbf{x}^2 e^{-a\mathbf{x}^2} = -\frac{\partial}{\partial a} \left(\frac{\pi}{a} \right)^{\frac{d}{2}} = \frac{d}{2a} \left(\frac{\pi}{a} \right)^{\frac{d}{2}} , \tag{A 3.13}$$

a result used to derive Eq. (3.43).

The r.h.s. of Eqs. (A3.10), (A3.12) are well defined also for noninteger d and can be taken to give a meaning to the Gaussian integral in noninteger dimensions. It will turn out to be important to define the theory for arbitrary

$d > 0$, which essentially amounts to assign a meaning to integrals of the general form

$$\int d^d x f(\mathbf{x}^2), \quad \int d^d x\, g(\mathbf{x}^2, \mathbf{x}\cdot\mathbf{y}), \text{ etc.} \quad .$$

We first consider the integral over $f(\mathbf{x}^2)$. Exploiting rotational invariance and introducing $t = \mathbf{x}^2$, we clearly can write this as

$$\int d^d x f(\mathbf{x}^2) = \frac{1}{2} O_d \int_0^\infty dt\, t^{d/2-1} f(t) \quad , \tag{A 3.14}$$

where O_d denotes the surface of the unit sphere in d dimensions. To determine this quantity we combine Eqs. (A3.14), (A3.10) to find

$$\pi^{d/2} = \int d^d x\, e^{-\mathbf{x}^2} = \frac{1}{2} O_d \int_0^\infty dt\, t^{d/2-1} e^{-t} \quad .$$

The remaining integral defines the Gamma-function

$$\Gamma(\alpha) = \int_0^\infty dt\, t^{\alpha-1} e^{-t} \quad , \tag{A 3.15}$$

which exists for all $\alpha > 0$ and generalizes the factorial to real (or even complex) numbers:

$$\Gamma(\alpha + 1) = \alpha!, \quad \alpha = 0, 1, 2, \ldots \quad .$$

We thus find

$$O_d = \frac{2\pi^{d/2}}{\Gamma(d/2)} \quad . \tag{A 3.16}$$

Equations (A3.14), (A3.16) define the d-dimensional integral over $f(x^2)$.

To define integrals over $g(\mathbf{x}^2, \mathbf{x}\cdot\mathbf{y})$ we write

$$\int d^d x\, g(\mathbf{x}^2, \mathbf{x}\cdot\mathbf{y}) = \int d^{d-1} x_\perp \int dx_\parallel g\left(\mathbf{x}_\perp^2 + x_\parallel^2, x_\parallel |\mathbf{y}|\right) \quad , \tag{A 3.17}$$

where $x_\parallel$ or $\mathbf{x}_\perp$ denote the component of $\mathbf{x}$ parallel or perpendicular to $\mathbf{y}$, respectively. We then can apply the above rule to the integral over $\mathbf{x}_\perp$ to find

$$\int d^d x\, g(\mathbf{x}^2, \mathbf{x}\cdot\mathbf{y}) = \frac{1}{2} O_{d-1} \int_o^\infty dt\, t^{\frac{d}{2}-\frac{3}{2}} \int_{-\infty}^{+\infty} dx_\parallel g(t + x_\parallel^2, x_\parallel |\mathbf{y}|) \; . \tag{A 3.18}$$

More complicated integrals can be defined in the corresponding way.

4. The Cluster Expansion

The central topic of this chapter is the derivation of diagrammatic perturbation theory, which lays the basis for all the subsequent analysis.

The only systematic method available for calculating properties of an interacting many-body system consists in some kind of expansion in powers of the interaction. Consider, for instance, the partition function of a single chain, which in our model takes the form

$$\mathcal{Z}^{(1)}(n) = \int_{\Omega} \prod_{j=0}^{n} \frac{d^3 r_j}{(4\pi\ell^2)^{3/2}} \exp\Big\{ -\frac{1}{4\ell^2} \sum_{j=1}^{n} (\mathbf{r}_j - \mathbf{r}_{j-1})^2 \Big\}$$
$$\prod_{(j,j')}' \Big[1 - \left(4\pi\ell^2\right)^{3/2} \beta_e \delta^3 \left(\mathbf{r}_j - \mathbf{r}_{j'}\right)\Big] \quad . \tag{4.1}$$

This is a complicated $3(n+1)$-fold integral, and there is no hope of an exact evaluation. Our only chance consists in expanding the product in powers of β_e, a series known as the 'cluster expansion':

$$\mathcal{Z}^{(1)}(n) = \int_{\Omega} \prod_{j=0}^{n} \frac{d^3 r_j}{(4\pi\ell^2)^{3/2}} \exp\Big\{ -\frac{1}{4\ell^2} \sum_{j=1}^{n} (\mathbf{r}_j - \mathbf{r}_{j-1})^2 \Big\}$$
$$\Big\{1 - \left(4\pi\ell^2\right)^{3/2} \beta_e \sum_{j_1<j_2} \delta^3 \left(\mathbf{r}_{j_1} - \mathbf{r}_{j_2}\right) \tag{4.2}$$
$$+ \left(4\pi\ell^2\right)^3 \beta_e^2 \sum_{\substack{j_1<j_2,\, j_1<j_3 \\ j_3<j_4}} \delta^3 \left(\mathbf{r}_{j_1} - \mathbf{r}_{j_2}\right) \delta^3 \left(\mathbf{r}_{j_3} - \mathbf{r}_{j_4}\right) + O(\beta_e^3)\Big\}.$$

In principle this expansion by brute force can be pushed to any desired order, but higher orders become increasingly cumbersome, in particular if many chains are involved. What is needed is a lucid way to organize the expansion. This leads us to introduce Feynman diagrams (Feynman graphs).

Basically in the present context a Feynman graph is a schematic picture of the polymer chain, indicating which segments interact in the perturbative term considered. We draw the chain as a straight line and connect each

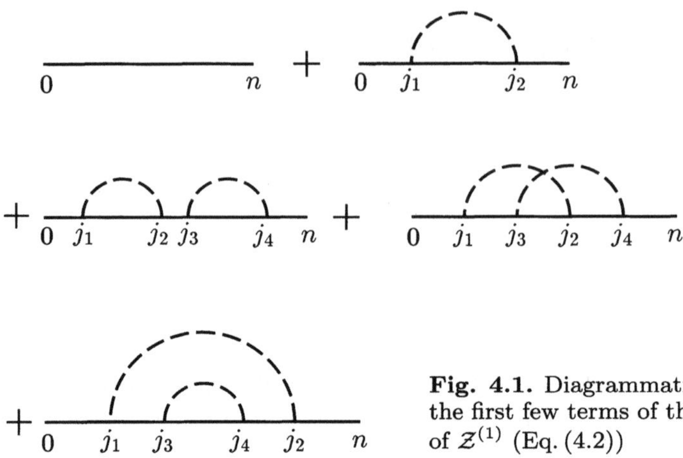

Fig. 4.1. Diagrammatic representation of the first few terms of the cluster expansion of $\mathcal{Z}^{(1)}$ (Eq. (4.2))

pair of interacting segments by a broken line. Equation (4.2), for instance, is represented by the graphs of Fig. 4.1, where the last three graphs represent the three different possibilities to order the segment indices in the term of order β_e^2. A priori, this is nothing but a bookkeeping device. However, we will find that it greatly facilitates the evaluation of the theory, provided we can answer the following questions:

(i) What are the rules for constructing a valid Feynman graph for a given physical observable?

(ii) How to evaluate the contribution of a given Feynman graph?

We will find that for the contribution of a given graph we can easily carry out the multiple integral over the segment coordinates. This is a major achievement, reducing the evaluation of low order terms of the cluster expansion to a few integrations and segment summations. Furthermore the diagrammatic representation is particularly well suited for a general analysis and partial resummation of the cluster expansion. In the next chapter, dealing with solutions of finite concentration, this aspect of the diagrammatic method will become essential. One example, namely the reduction of correlation functions to so-called 'cumulants' will be discussed in this chapter.

The organization of this chapter is as follows: Sect. 4.1 illustrates the basic ideas with a first order calculation of the average squared end-to-end distance. The general rules of constructing and evaluating Feynman graphs are derived in Sect. 4.2. Section 4.3 introduces the cumulants, and in Sect. 4.4 we apply the general rules to an evaluation of the second virial coefficient. In two appendices of more mathematical nature (A 4.1, A 4.2) we derive a formal expression for the second virial coefficient and analyze some finite sums occurring in low order perturbation theory.

4.1 First order corrections to endpoint correlations

The interacting Gaussian chain model is defined by Eqs. (2.2)–(2.4). As before we consider the generalization to d-dimensional space, and we evaluate expectation values by expanding in powers of β_e. We demonstrate the method with the example of the endpoint Greensfunction $G(\mathbf{p};n)$ of a single chain system, which generalizes $G_0(\mathbf{p};n)$ (Eq. 3.33).

$$G(\mathbf{p};n) = \frac{(4\pi\ell^2)^{\frac{d}{2}}}{\Omega} \int_\Omega D[\mathbf{r}] \exp\{-\frac{1}{4\ell^2}\sum_{j=1}^{n}(\mathbf{r}_j - \mathbf{r}_{j-1})^2 + i\mathbf{p}(\mathbf{r}_n - \mathbf{r}_0)\}$$
$$\prod_{(j_1,j_2)}{}' [1 - (4\pi\ell^2)^{\frac{d}{2}} \beta_e \delta^d(\mathbf{r}_{j_2} - \mathbf{r}_{j_1})] \quad . \tag{4.3}$$

The partition function and the endpoint distribution function can be expressed in terms of $G(\mathbf{p};n)$, the results being identical in form to Eqs. (3.34), (3.35).

$$\mathcal{Z}(n) = \frac{\Omega}{(4\pi\ell^2)^{\frac{d}{2}}} G(\mathbf{0};n) \tag{4.4}$$

$$P(\mathbf{p};n) = \frac{G(\mathbf{p};n)}{G(\mathbf{0};n)} \tag{4.5}$$

Evaluating the product in Eq. (4.3) we get back the Greensfunction $G_0(\mathbf{p};n)$ of the noninteracting model to zero order in β_e. Writing

$$G(\mathbf{p};n) = G_0(\mathbf{p};n) - \beta_e G_1(\mathbf{p};n) + O(\beta_e^2) \quad , \tag{4.6}$$

we find as first order correction

$$-\beta_e \frac{\Omega}{(4\pi\ell^2)^{\frac{d}{2}}} G_1(\mathbf{p};n)$$
$$= -\beta_e (4\pi\ell^2)^{\frac{d}{2}} \sum_{0<j_1<j_2<n} \int D[\mathbf{r}] \delta^d(\mathbf{r}_{j_2} - \mathbf{r}_{j_1})$$
$$\exp\{-\frac{1}{4\ell^2}\sum_{j=1}^{n}(\mathbf{r}_j - \mathbf{r}_{j-1})^2 + i\mathbf{p}(\mathbf{r}_n - \mathbf{r}_0)\} \quad . \tag{4.7}$$

(To simplify the general rules given in the next section we do not allow the endpoints to take part in an interaction, which is a negligible $1/n$ effect.) Equation (4.7) can be read in the following way. (See Fig. 4.2a.) The chain consists of three random walks:

(1)	going from	$\mathbf{r}_0$ to	$\mathbf{r}_{j_1}$	in	j_1	steps
(2)	going from	$\mathbf{r}_{j_1}$ to	$\mathbf{r}_{j_2} = \mathbf{r}_{j_1}$	in	$j_2 - j_1$	steps
(3)	going from	$\mathbf{r}_{j_2}$ to	$\mathbf{r}_n$	in	$n - j_2$	steps .

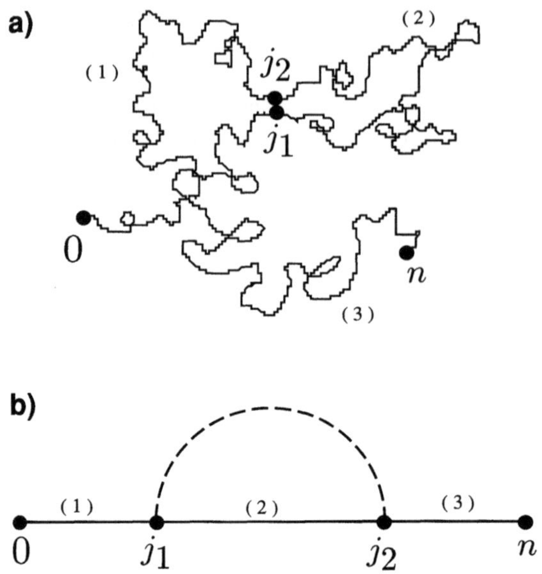

Fig. 4.2. (**a**) A chain configuration, where the interaction of segments j_1, j_2 is taken into account. (**b**) The corresponding Feynman-diagram. Here for clarity we indicated the distinguished segments by *heavy dots*

Note that these random walks can have any number of intersections, but only the interaction of the pair $\mathbf{r}_{j_1} = \mathbf{r}_{j_2}$ is taken into account. This situation is represented schematically by the 'Feynman diagram' of Fig. 4.2b. The chain is drawn as a straight line, and the two interacting segments are connected by a broken line. This graphic representation generalizes to all orders of the cluster expansion.

Now the part of the chain between two successive heavy dots representing segments j, j' represents a random walk going from $\mathbf{r}_j$ to $\mathbf{r}_{j'}$ in $j' - j$ steps. Integrating over the internal segment coordinates of this random walk we, according to Eq. (3.41), find the end-to-end distribution $(4\pi\ell^2)^{\frac{d}{2}} \hat{P}_0(\mathbf{r}_j - \mathbf{r}_{j'}; j' - j)$ of this part of the chain. Thus the first order contribution (4.7) takes the form

$$
\begin{aligned}
&-\beta_e \frac{\Omega}{(4\pi\ell^2)^{\frac{d}{2}}} G_1(\mathbf{p}; n) \\
&= -\beta_e \sum_{0<j_1<j_2<n} \int d^d r_0 d^d r_{j_1} d^d r_{j_2} d^d r_n \delta^d(\mathbf{r}_{j_2} - \mathbf{r}_{j_1}) e^{i\mathbf{p}(\mathbf{r}_n - \mathbf{r}_o)} \\
&\cdot \hat{P}_0(\mathbf{r}_{j_1} - \mathbf{r}_0; j_1) \hat{P}_0(\mathbf{r}_{j_2} - \mathbf{r}_{j_1}; j_2 - j_1) \hat{P}_0(\mathbf{r}_n - \mathbf{r}_{j_2}; n - j_2) \,. \qquad (4.8)
\end{aligned}
$$

(The factors $(4\pi\ell^2)^{d/2}$ cancel with the factors from the integration measure.) We next introduce the Fourier transform

$$
\hat{P}_0(\mathbf{r}_{j'} - \mathbf{r}_j; j' - j) = \int_k e^{-i\mathbf{k}(\mathbf{r}_{j'} - \mathbf{r}_j)} G_0(\mathbf{k}; j' - j) \;, \qquad (4.9)
$$

where we use the shorthand notation (A 3.8):

$$\int_k = \int \frac{d^d k}{(2\pi)^d} \quad .$$

Using the results explained in Appendices A 3.1, A 3.2 we then carry through the remaining **r**-integrals to find

$$-\beta_e G_1(\mathbf{p}; n) = -\beta_e (4\pi\ell^2)^{\frac{d}{2}} \sum_{0<j_1<j_2<n} G_0(\mathbf{p}; j_1) \int_k G_0(\mathbf{k}; j_2 - j_1) G_0(\mathbf{p}; n - j_2). \tag{4.10}$$

The **k**-integral is easily evaluated to yield

$$-\beta_e G_1(\mathbf{p}; n) = -\beta_e e^{-p^2\ell^2 n} \sum_{0<j_1<j_2<n} (j_2 - j_1)^{-\frac{d}{2}} e^{p^2\ell^2(j_2-j_1)} \quad . \tag{4.11}$$

It is the purpose of the 'Feynman rules', to be derived in Sect. 4.2, to shorten this calculation. These rules immediately yield expressions of the type (4.10) for any contribution represented in the form of a Feynman diagram. The example considered here exhibits the basic idea of the diagrammatic method: parts of the chain between two successive interacting segments are just random walks represented by the appropriate factor of $\hat{P}_0$. The interactions via their δ-functions tie together end points of these walks, and finally the interacting segments have to be summed over, respecting their ordering along the chains.

We now proceed with the evaluation of $G(\mathbf{p}, n)$, (Eq. (4.6)), which we write as

$$G(\mathbf{p}, n) = e^{-p^2\ell^2 n}[1 - \beta_e \Sigma_1(p^2\ell^2) + O(\beta_e^2)] \tag{4.12}$$

$$\Sigma_1(y) = \sum_{0<j_1<j_2<n} (j_2 - j_1)^{-\frac{d}{2}} e^{y(j_2-j_1)} \quad . \tag{4.13}$$

We are in particular interested in the averaged squared end-to-end distance, which is determined as (compare the discussion of R_g^2 given in Sect. 3.2)

$$R_e^2 = -\Delta_p \Big|_0 \frac{G(\mathbf{p}; n)}{G(\mathbf{0}; n)} = 2d\ell^2[n + \beta_e \Sigma_1' + O(\beta_e^2)] \tag{4.14}$$

$$\Sigma_1' = \frac{d}{dy}\Big|_0 \Sigma_1(y) = \sum_{0<j_1<j_2<n} (j_2 - j_1)^{1-\frac{d}{2}} \quad . \tag{4.15 i}$$

For $d < 4$, $n \gg 1$, the leading n-dependence of Σ_1' is found by replacing the summation by integration.

$$\frac{1}{n^2}\Sigma_1' \to n^{1-\frac{d}{2}} \int_0^1 dx_1 \int_{x_1}^1 dx_2 (x_2 - x_1)^{1-\frac{d}{2}}$$
$$= \frac{4n^{1-\frac{d}{2}}}{(4-d)(6-d)} \tag{4.15 ii}$$

This yields

$$\frac{R_e^2}{2d\ell^2 n} = 1 + \beta_e \frac{4n^{2-\frac{d}{2}}}{(4-d)(6-d)} + O(\beta_e^2) \quad ,(d < 4) \quad , \tag{4.16 i}$$

showing the expected increase of the radius resulting from segment repulsion. This result, keeping only the leading power of n in Σ_1', is not quite sufficient for our further discussion. A more careful evaluation, presented in Appendix A 4.2, also takes into account the next-to-leading term. It yields for arbitrary d

$$\frac{R_e^2}{2d\ell^2 n} = 1 + \beta_e \left[\frac{2}{\varepsilon}(n^{\frac{\varepsilon}{2}} - 1) - \frac{2}{2+\varepsilon} n^{\frac{\varepsilon}{2}} + \alpha_1(\varepsilon) + O(n^{-1+\frac{\varepsilon}{2}})\right] + O(\beta_e^2) \, . \tag{4.16 ii}$$

We introduced the standard notation

$$\varepsilon = 4 - d \quad , \tag{4.17}$$

which henceforth will be used throughout. $\alpha_1(\varepsilon)$ is some finite function of ε, explicitly given in Eq. (A 4.11).

The methods illustrated here can be used to evaluate all quantities of interest in the excluded volume problem. Another example (A_2^{Π}) will be considered after we derive the general Feynman rules.

4.2 The Feynman rules

We consider the M-chain Greensfunction defined as

$$G^{(M)}(\mathbf{p}_1, \ldots, \mathbf{p}_{2M}; n_1, \ldots, n_M) = \int D[\mathbf{r}^{(1)}, \ldots, \mathbf{r}^{(M)}] \exp\{-\sum_{m=1}^{M} V_0^{(m)}\}$$
$$\prod_{(j_1 m_1; j_2 m_2)}' \left[1 - \left(4\pi\ell^2\right)^{d/2} \beta_e \delta^d(\mathbf{r}_{j_2}^{(m_2)} - \mathbf{r}_{j_1}^{(m_1)})\right]$$
$$\prod_{m=1}^{M} \exp(-i\mathbf{p}_{2m-1}\mathbf{r}_0^{(m)} - i\mathbf{p}_{2m}\mathbf{r}_{n_m}^{(m)}) \quad . \tag{4.18}$$

Many physical quantities can be expressed in terms of the $G^{(M)}$. The M-chain partition function, for instance, reads

$$\mathcal{Z}^{(M)}(n_1, \dots, n_M) = G^{(M)}(\mathbf{0}, \dots, \mathbf{0}; n_1, \dots, n_M) \tag{4.19}$$

and a comparison to Eq. (4.3) yields

$$\frac{\Omega}{(4\pi\ell^2)^{\frac{d}{2}}} G(\mathbf{p}, n) = G^{(1)}(\mathbf{p}, -\mathbf{p}; n) \ . \tag{4.20}$$

We now analyze the expansion of the product $\prod'$ in powers of β_e. The individual terms of this expansion can be represented by Feynman diagrams: for each polymer chain we draw a straight line, and we represent the interaction $-\beta_e \left(4\pi\ell^2\right)^{\frac{d}{2}} \delta^d(\mathbf{r}_{j_2}^{(m_2)} - \mathbf{r}_{j_1}^{(m_1)})$ by a broken line connecting segment j_1 on line m_1 to segment j_2 on line m_2. Figure 4.3 shows some examples contributing in second order for a three-chain system. Figure 4.3a, for example, stands for the term

$$\left(-\left(4\pi\ell^2\right)^{\frac{d}{2}} \beta_e\right)^2 \delta^d(\mathbf{r}_6^{(1)} - \mathbf{r}_{130}^{(1)}) \delta^d(\mathbf{r}_{10}^{(1)} - \mathbf{r}_9^{(2)}).$$

In these 'labeled' diagrams the individual pairs of interacting segments are fixed. We now sum up all terms of the expansion having the same topological structure, which means that they can be identified by shifting interaction lines along their chains without interchanging the ordering of the interacting segments. The sum of all these terms is represented by an unlabeled diagram. As an example, diagrams 4.3a,b contribute to the unlabeled diagram of Fig. 4.4, which stands for the expression

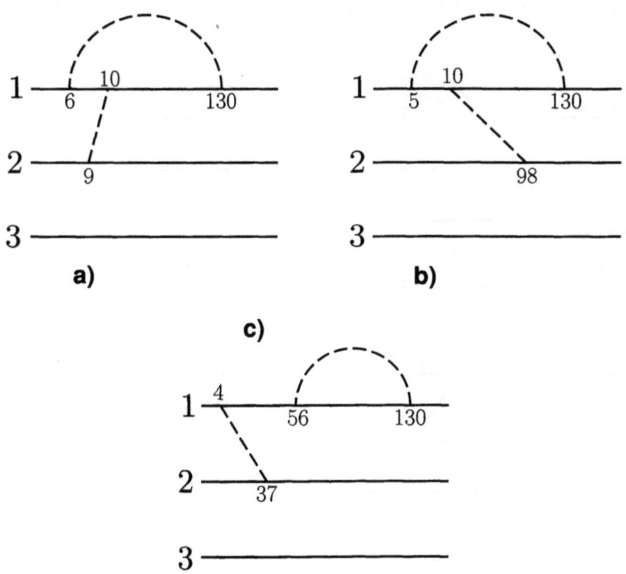

Fig. 4.3a–c. Some labeled diagrams for a three-chain system

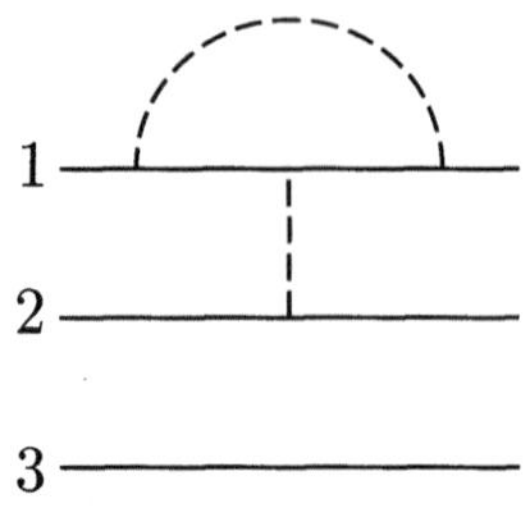

Fig. 4.4. The unlabeled diagram to which diagrams Fig. 4.3a, b contribute

$$\left(-\left(4\pi\ell^2\right)^{\frac{d}{2}}\beta_e\right)^2 \sum_{0<j_1<j_2<j_3<n_1} \sum_{0<j_4<n_2} \delta^d(\mathbf{r}_{j_1}^{(1)}-\mathbf{r}_{j_3}^{(1)})\delta^d(\mathbf{r}_{j_2}^{(1)}-\mathbf{r}_{j_4}^{(2)}) \quad .$$

Central to the diagrammatic method is the following statement: *The μ-th order term of the expansion of $\prod'$ is represented by the set of all topologically different unlabeled diagrams with M full lines and μ interaction lines.* For a proof it is sufficient to note that each diagram corresponds to a unique ordering of the segment indices occurring in the expansion, and vice versa. In Fig. 4.5 we present the 15 diagrams occurring for $M = 2, \mu = 2$. (The large number of diagrams contributing to such a low order of the expansion may be somewhat shocking, but we will find below that the number of diagrams to be explicitly considered can be reduced tremendously.)

We introduce some simplifying notation. We call 'polymer line' the full line representing a polymer. Interaction lines generally are known as 'vertices'. The endpoints of the polymer lines and the points where vertices are attached

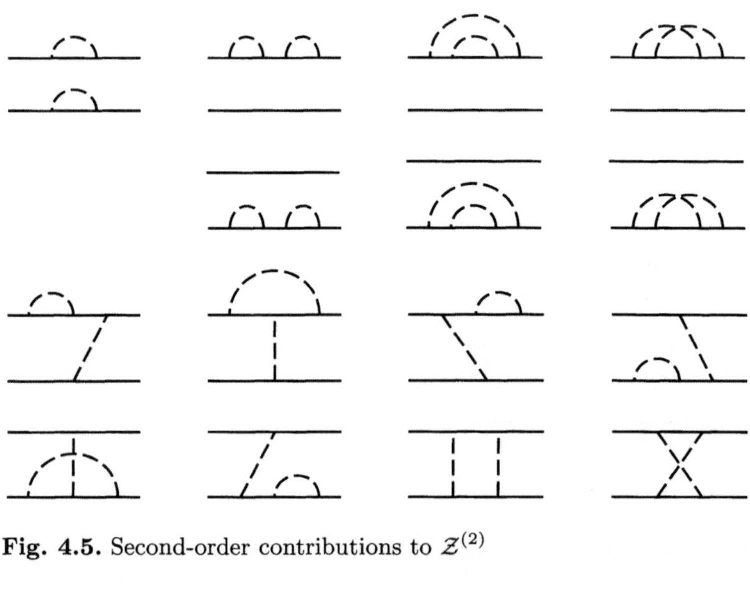

Fig. 4.5. Second-order contributions to $\mathcal{Z}^{(2)}$

to the polymer lines will be called 'distinguished points'. A piece of a polymer line connecting two subsequent distinguished points is called a 'propagator line'.

So far we have introduced diagrams to represent the expansion of the product $\prod'$. If we interpret a given diagram as contribution to the multiple integral (4.18), a propagator line connecting distinguished points $j < j'$ of the m-th line represents a Gaussian subchain. According to Eqs. (3.41), (4.9) the integration over the internal coordinates of this subchain yields

$$\left(4\pi\ell^2\right)^{\frac{d}{2}} \int_k e^{-i\mathbf{k}(\mathbf{r}_{j'}^{(m)}-\mathbf{r}_j^{(m)})} G_0(\mathbf{k}, j' - j) \quad . \tag{4.21}$$

The factors $\left(4\pi\ell^2\right)^{\frac{d}{2}}$ cancel factors $\left(4\pi\ell^2\right)^{-\frac{d}{2}}$ associated with the $\mathbf{r}_j$- integrals of the distinguished points, leaving a single factor $\left(4\pi\ell^2\right)^{-\frac{d}{2}}$ for each polymer line. To keep track of the sign in the exponential in expression (4.21), we, if necessary, attach to the propagator line an arrow pointing from j to j', and we count the momentum $\mathbf{k}$ as 'outgoing' from segment j and 'incoming' to segment j'.

We next carry through the $\mathbf{r}$-integrations associated with two interacting segments. The relevant labeled substructure of a diagram is shown in Fig. 4.6. It yields

$$\begin{aligned} &-\left(4\pi\ell^2\right)^{\frac{d}{2}} \beta_e \int d^d r_j^{(m)} d^d r_{j'}^{(m')} \delta^d\left(\mathbf{r}_j^{(m)} - \mathbf{r}_{j'}^{(m')}\right) e^{-i\mathbf{r}_j^{(m)}(\mathbf{k}-\mathbf{k}')} e^{-i\mathbf{r}_{j'}^{(m')}(\mathbf{k}''-\mathbf{k}''')} \\ &= -\left(4\pi\ell^2\right)^{\frac{d}{2}} \beta_e (2\pi)^d \delta^d(\mathbf{k} + \mathbf{k}'' - \mathbf{k}' - \mathbf{k}'''), \end{aligned} \tag{4.22}$$

the exponential factors being contributed by the propagators (4.21). The δ-function guarantees that the sum of the momenta incoming at the vertex equals the sum of the outgoing momenta. This is known as 'momentum conservation at the vertex'. The δ-function together with its accompanying factor $(2\pi)^d$ can be used to eliminate one k-integration together with its factor $(2\pi)^{-d}$.

Fig. 4.6. Diagrammatic structure underlying Eq. (4.22)

We finally integrate over the endpoint coordinates. Integrating $r_0^{(m)}$ we find

$$\int d^d r_0^{(m)} \exp[i\mathbf{r}_0^{(m)}(\mathbf{k}-\mathbf{p}_{2m-1})] = (2\pi)^d \delta^d(\mathbf{k}-\mathbf{p}_{2m-1}) \quad , \tag{4.23 i}$$

where $\mathbf{k}$ is the momentum of the propagator starting at segment $(m, 0)$. With our conventions the 'external momentum' $\mathbf{p}_{2m-1}$ flows into this endpoint. Graphically we represent this by an arrow as shown in Fig. 4.7a. Integrating $\mathbf{r}_{n_m}^{(m)}$ we in general find

$$\int d^d r_{n_m}^{(m)} \exp[-i\mathbf{r}_{n_m}^{(m)}(\mathbf{k}+\mathbf{p}_{2m})] = (2\pi)^d \delta^d(\mathbf{k}+\mathbf{p}_{2m}) \quad , \tag{4.23 ii}$$

where $\mathbf{k}$ is the momentum of the propagator ending at segment (m, n_m). (See Fig. 4.7b). Since incoming momentum $\mathbf{p}$ is equivalent to outgoing momentum $-\mathbf{p}$ we can also draw this situation in the form 4.7c.

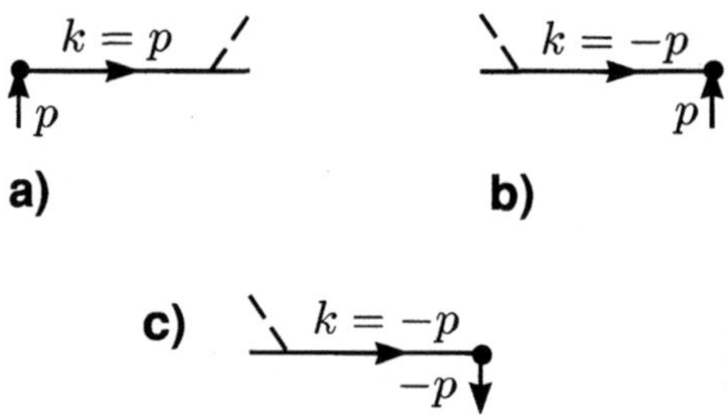

Fig. 4.7a–c. Diagrammatic representations of endpoint insertions (*arrows*)

In this discussion we missed some important point. Since momentum is conserved, a diagram yields a nonvanishing contribution only if the total external momentum flowing into any of its connected parts sums up to zero. The diagram of Fig. 4.8a, for instance, contributes only if $\mathbf{p}_1 + \mathbf{p}_2 + \mathbf{p}_3 + \mathbf{p}_4 = 0$, $\mathbf{p}_5 + \mathbf{p}_6 = 0$. For each connected component of a diagram the last integral over an endpoint takes the form

$$\int_\Omega d^d r_{n_m}^{(m)} = \Omega \tag{4.23 iii}$$

replacing Eq. (4.23 ii), since for this last endpoint $\mathbf{k} + \mathbf{p}_{2m} = 0$ is guaranteed by momentum conservation in the remainder of the connected component considered.

This discussion is summarized by the following rules for constructing or evaluating Feynman graphs for $G^{(M)}(\mathbf{p}_1, \ldots, \mathbf{p}_{2M}; n_1, \ldots, n_M)$.

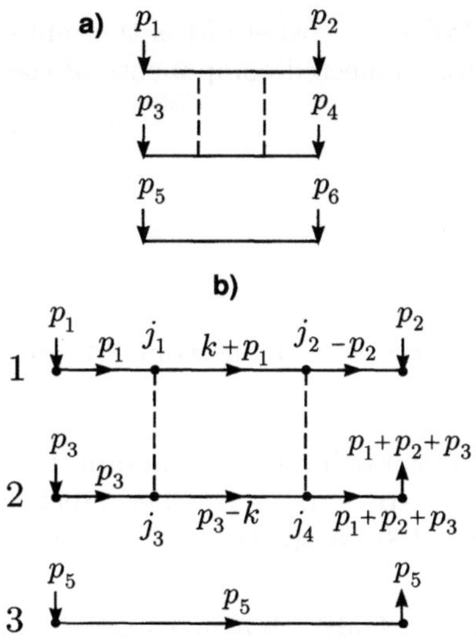

Fig. 4.8. (**a**) A typical diagram. (**b**) Its labeled form

Construction rules

(i) Draw M parallel straight lines.

(ii) Add any number of broken lines connecting pairs of points on these lines, so that no broken line is attached to an endpoint, nor are there two broken lines connected to the same point.

(iii) If nonvanishing momentum is associated with an end of the m-th chain, attach an arrow to the corresponding end point. Arrows for zero momenta need not be drawn.

Evaluation rules

(i) Label the distinguished points by some summation indices, and attach to each propagator line some momentum variable, respecting momentum conservation throughout. An arrow, attached to some endpoint, contributes the corresponding external momentum.

(ii) A propagator line of momentum $\mathbf{k}$, connecting distinguished points $j < j'$, yields a factor $G_0(\mathbf{k}; j' - j) = \exp[-\mathbf{k}^2\ell^2(j' - j)]$. A broken line yields a factor $\left(-\left(4\pi\ell^2\right)^{\frac{d}{2}} \beta_e\right)$.

(iii) Integrate over all internal momenta $\left(\text{recall} \int_k \equiv \int \frac{d^d k}{(2\pi)^d}\right)$ and sum over the segment labels of the distinguished points, respecting their ordering along their chain.

(iv) Multiply by an overall factor of $\Omega^c(4\pi\ell^2)^{-M\frac{d}{2}}$, where M is the number of polymer lines and c is the number of connected components of the diagram.

We then have the following

Diagrammatic representation of $G^{(M)}$

$G^{(M)}(\mathbf{p}_1,\ldots,\mathbf{p}_{2M};n_1,\ldots,n_M)$ is given by the sum of the contributions of all topologically distinct diagrams.

To exemplify the rules we write down the contribution of diagram 4.8a which contributes only for $\mathbf{p}_6 = -\mathbf{p}_5$, $\mathbf{p}_4 = -\mathbf{p}_1 - \mathbf{p}_2 - \mathbf{p}_3$. The labeled version resulting from step (i) is shown in Fig. 4.8b. Steps (ii) to (iv) yield

$$\left(4\pi\ell^2\right)^d \beta_e^2 \frac{\Omega^2}{(4\pi\ell^2)^{\frac{3}{2}d}} \sum_{0<j_1<j_2<n_1} \sum_{0<j_3<j_4<n_2} G_0(\mathbf{p}_1;j_1)G_0(\mathbf{p}_3;j_3)$$
$$\int_k G_0(\mathbf{k}+\mathbf{p}_1;j_2-j_1)G_0(\mathbf{p}_3-\mathbf{k};j_4-j_3)G_0(-\mathbf{p}_2;n_1-j_2)$$
$$G_0(\mathbf{p}_1+\mathbf{p}_2+\mathbf{p}_3;n_2-j_4)G_0(\mathbf{p}_5;n_3) \quad .$$

The diagrammatic method is easily generalized to deal with segment density insertions. We define a Greensfunction $G^{(L,M)}_{m_1\ldots m_L}$ containing L such insertions in polymer lines $m_1,\ldots,m_L$.

$$G^{(L,M)}_{m_1\ldots m_L}(\mathbf{q}_1,\ldots,\mathbf{q}_L;\mathbf{p}_1,\ldots,\mathbf{p}_{2M};n_1,\ldots,n_M) =$$
$$\int D[\mathbf{r}^{(1)},\ldots,\mathbf{r}^{(M)}]\exp\{-\sum_{m=1}^{M} V_0^{(m)}\}\prod{}' \left[1-\left(4\pi\ell^2\right)^{\frac{d}{2}}\beta_e\delta^d(\mathbf{r}^{(m_2)}_{j_2}-\mathbf{r}^{(m_1)}_{j_1})\right]$$
$$\prod_{m=1}^{M}\exp\left(-i\mathbf{p}_{2m-1}\mathbf{r}^{(m)}_0 - i\mathbf{p}_{2m}\mathbf{r}^{(m)}_{n_m}\right)\prod_{\nu=1}^{L}\rho_{m_\nu}(\mathbf{q}_\nu). \tag{4.24}$$

Here we introduced the Fourier transform of the local segment density (3.14) for segments of polymer line m:

$$\rho_m(\mathbf{q}) = \int d^dr e^{i\mathbf{q}\mathbf{r}}\hat{\rho}_m(\mathbf{r}) = \sum_{0<j<n_m} e^{i\mathbf{q}\mathbf{r}^{(m)}_j} \quad . \tag{4.25}$$

(Note that we again omit the endpoint contribution.) Graphically a factor of $\rho_m(\mathbf{q})$ is represented by an arrow of incoming momentum $-\mathbf{q}$ pointing to the new distinguished point j of the m-th polymer line (see Fig. 4.9). The Feynman rules for evaluating a diagram are not changed, except that we have to take into account the density insertions in an obvious way.

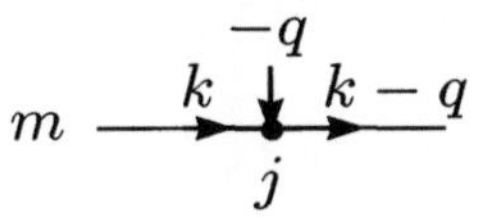

Fig. 4.9. Density insertion

Most physical quantities of interest can be expressed in terms of the $G^{(L,M)}$ and of the $G^{(M)} \equiv G^{(0,M)}$. To give an example we note the density-density correlation of a single chain system, which takes the form

$$I_d(\mathbf{q}) = \frac{G_{11}^{(2,1)}(\mathbf{q}, -\mathbf{q}; \mathbf{0}, \mathbf{0}; n)}{G^{(1)}(\mathbf{0}, \mathbf{0}; n)} \quad . \tag{4.26}$$

To lowest order the numerator is given by the two diagrams of Fig. 4.10, which, together with the result $G^{(1)}(\mathbf{0}, \mathbf{0}; n) = \Omega(4\pi\ell^2)^{-\frac{d}{2}}(1 + O(\beta_e))$ yield Eq. (3.20) found for the noninteracting chain and thus reproduce the Debye function.

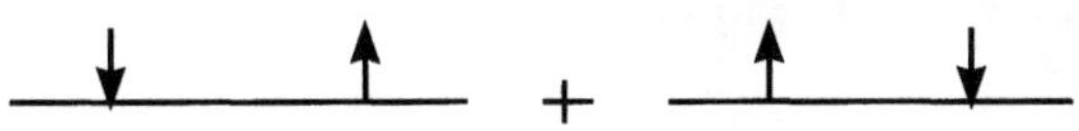

Fig. 4.10. Zero order representation of the density-density correlation function (Eq. (4.26))

We should finally note that the rules given here for evaluating diagrams are not of the most efficient form possible. There exists a formulation in which all internal **k**-integrations are carried through, leaving us with the summations over segments only. This amounts to going from expressions of type (4.10) to those of type (4.11). For low order calculations, however, the simplification gained this way does not outweigh the labor needed for deriving this most elegant representation. The interested reader therefore is referred to the literature [Nak71]. Furthermore we also may introduce symmetry operations to reduce the number of diagrams explicitly considered. For example, reflection symmetry of the chain can be invoked to combine the two diagrams of Fig. 4.10 into a single diagram, multiplied by a weight factor of 2. Such considerations will be carried through in due course.

4.3 Cumulants

An important simplification which we should introduce here, consists in the reduction of the Greensfunctions to their connected parts or 'cumulants'. By definition the cumulant $G_{c,m_1\ldots m_L}^{(L,M)}$ equals the sum of all totally connected diagrams contributing to $G_{m_1\ldots m_L}^{(L,M)}$. In Fig. 4.5, for example, only the last eight diagrams contribute to $G_c^{(0,2)}$. In general a diagram contributing to the Greensfunction $G^{(L,M)}$ may consist of several disconnected pieces. According to the Feynman rules the contribution of such a diagram is given by the product of the contributions of these pieces. We may now sum up all diagrams in which the individual polymer lines are distributed in a prescribed way to

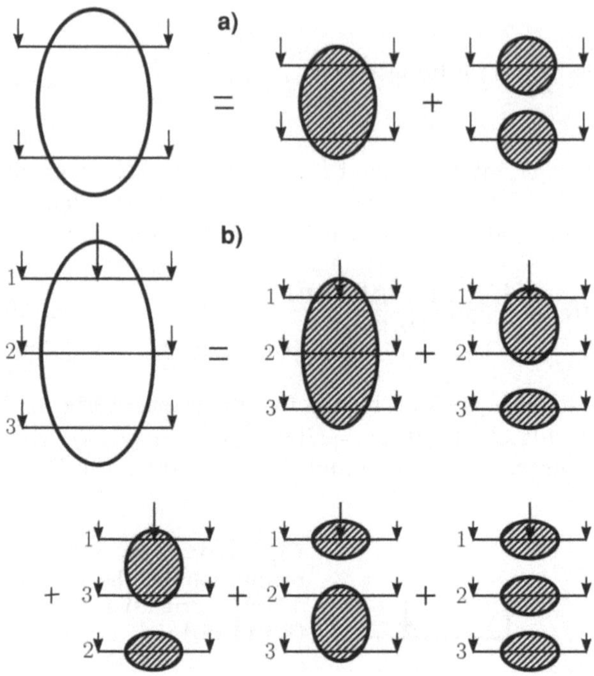

Fig. 4.11a, b. Examples of the reduction of Greensfunctions (*empty circles*) to cumulants (*shaded circles*); (**a**) $G^{(0,2)}(\mathbf{p}_1, \mathbf{p}_2, \mathbf{p}_3, \mathbf{p}_4)$; (**b**) $G_1^{(1,3)}(\mathbf{q}_1; \mathbf{p}_1, \mathbf{p}_2, \mathbf{p}_3, \mathbf{p}_4, \mathbf{p}_5, \mathbf{p}_6)$. Note the permutation of the chains in the first diagram of the last line

a fixed number of internally connected pieces. Clearly the result equals the product of the corresponding cumulants. Figure 4.11 gives two examples for the representation of $G^{(L,M)}$ in terms of cumulants. Analytically these results read

Fig. 4.11a:

$$\begin{aligned} &G^{(0,2)}(\mathbf{p}_1, \mathbf{p}_2, \mathbf{p}_3, \mathbf{p}_4; n_1, n_2) \\ &= G_c^{(0,2)}(\mathbf{p}_1, \mathbf{p}_2, \mathbf{p}_3, \mathbf{p}_4, n_1, n_2) + G_c^{(0,1)}(\mathbf{p}_1, \mathbf{p}_2, n_1) G_c^{(0,1)}(\mathbf{p}_3, \mathbf{p}_4; n_2) \end{aligned}$$

Fig. 4.11b:

$$\begin{aligned} &G_1^{(1,3)}(\mathbf{q}_1; \mathbf{p}_1, \ldots, \mathbf{p}_6; n_1, n_2, n_3) = G_{c,1}^{(1,3)}(\mathbf{q}_1; \mathbf{p}_1, \ldots, \mathbf{p}_6; n_1, n_2, n_3) \\ &+ G_{c,1}^{(1,2)}(\mathbf{q}_1; \mathbf{p}_1, \ldots, \mathbf{p}_4; n_1, n_2) G_c^{(0,1)}(\mathbf{p}_5, \mathbf{p}_6; n_3) \\ &+ G_{c,1}^{(1,2)}(\mathbf{q}_1, \mathbf{p}_1, \mathbf{p}_2, \mathbf{p}_5, \mathbf{p}_6; n_1, n_3) G_c^{(0,1)}(\mathbf{p}_3, \mathbf{p}_4, n_2) \\ &+ G_{c,1}^{(1,1)}(\mathbf{q}_1; \mathbf{p}_1, \mathbf{p}_2; n_1) G_c^{(0,2)}(\mathbf{p}_3, \mathbf{p}_4, \mathbf{p}_5, \mathbf{p}_6; n_2, n_3) \\ &+ G_{c,1}^{(1,1)}(\mathbf{q}_1; \mathbf{p}_1, \mathbf{p}_2; n_1) G_c^{(0,1)}(\mathbf{p}_3, \mathbf{p}_4, ; n_2) G_c^{(0,1)}(\mathbf{p}_5, \mathbf{p}_6; n_3) \end{aligned}$$

Note, that $G_c^{(L,1)} \equiv G^{(L,1)}$.

The reduction to cumulants tremendously reduces the number of diagrams to be explicitly considered. Furthermore, each cumulant yields a single factor of the volume Ω of the system. The cumulants therefore will play a prominent role in our construction of the thermodynamic limit $\Omega \to \infty$ for a many-chain system.

4.4 An example: the second virial coefficient

As a first application we calculate the second virial coefficient A_2^{Π} of the osmotic pressure for a monodisperse solution: $n_m \equiv n$. According to Eq. (1.1) it is defined as

$$A_2^{\Pi} = \frac{d^2}{dc_p^2}\bigg|_0 \frac{\hat{\Pi}}{k_B T} \; . \tag{4.27}$$

In Appendix A 4.1 we derive the representation

$$A_2^{\Pi} = -\Omega \frac{G^{(2)}(\mathbf{0},\mathbf{0},\mathbf{0},\mathbf{0};n,n) - (G^{(1)}(\mathbf{0},\mathbf{0};n))^2}{(G^{(1)}(\mathbf{0},\mathbf{0};n))^2} \; . \tag{4.28}$$

The evaluation of this expression makes for an instructive example of the manipulation of diagrams.

We first note that $G^{(1)}$ from Eqs. (4.20), (4.12) is found as

$$G^{(1)}(\mathbf{0},\mathbf{0};n) = \frac{\Omega}{(4\pi\ell^2)^{\frac{d}{2}}}[1 - \beta_e \Sigma_1(0) + O(\beta_e^2)] \; . \tag{4.29}$$

Furthermore according to Fig. 4.11a the numerator in Eq. (4.28) is just the cumulant $G_c^{(2)}$. Figure 4.12 gives the expansion of $G_c^{(2)}(\mathbf{0},\mathbf{0},\mathbf{0},\mathbf{0};n,n)$ up to order β_e^2, where we have taken into account the fact that interchanging the polymer lines for the present values $\mathbf{q}_i \equiv \mathbf{0}, n_m \equiv n$ does not change the contribution of the diagram.

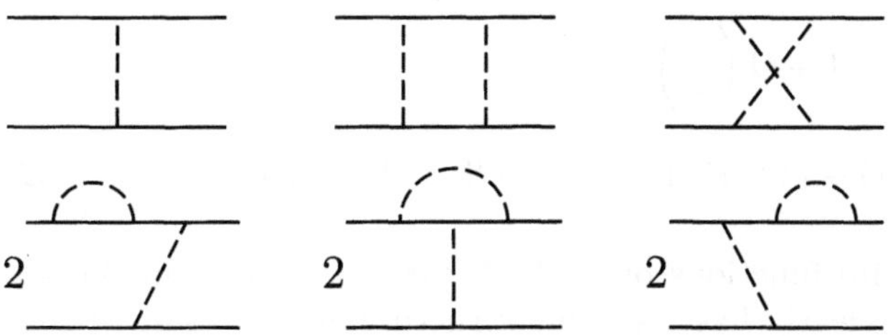

Fig. 4.12. Diagrams contributing to $G_c^{(2)}(\mathbf{0},\mathbf{0},\mathbf{0},\mathbf{0})$ to second order. Permutation of the polymer lines yields the weight factor of 2 for the last three contributions

In the diagrams of the second line of Fig. 4.12 the vertex connecting the polymer lines does not carry momentum. Therefore the two propagators attached to the upper end of this vertex (distinguished point j') combine to yield a single propagator

$$\exp(-k^2\ell^2(j'-j+j''-j')) = \exp(-k^2\ell^2(j''-j)).$$

The index j' survives only as part of the ordered summation. Adding the three diagrams we sum j' over all those segments of its chain not occupied by the other interaction. This yields a factor of $n-3$. The remaining contribution of the upper polymer line is identical to the first order contribution to $G^{(1)}(\mathbf{0},\mathbf{0};n)$. The sum of these three diagrams therefore is found as

$$-2\beta_e \frac{\Omega}{(4\pi\ell^2)^{d/2}}(n-1)(n-3)(-\beta_e\Sigma_1(0)) \quad .$$

By reflection symmetry of the chains the last two diagrams in the first line of Fig. 4.12 yield identical contributions which we write as

$$\begin{aligned} &-\beta_e \frac{\Omega}{(4\pi\ell^2)^{d/2}}(-\beta_e\Sigma_2(0)) \\ &\Sigma_2(0) = \sum_{0<j_1<j_2<n} \ \sum_{0<j_3<j_4<n} (j_2-j_1+j_4-j_3)^{-\frac{d}{2}} \quad . \end{aligned} \tag{4.30}$$

Combining all these results with the value of the trivial first order diagram we find

$$\begin{aligned} A_2^{\Pi} &= (4\pi\ell^2)^{d/2}\,\beta_e \frac{(n-1)^2 - 2\beta_e(n-1)(n-3)\Sigma_1(0) - 2\beta_e\Sigma_2(0) + O(\beta_e^2)}{1-2\beta_e\Sigma_1(0)+O(\beta_e^2)} \\ &= (4\pi\ell^2)^{d/2}\,\beta_e(n-1)^2\left[1+\frac{4\beta_e}{n-1}\Sigma_1(0) - \frac{2\beta_e}{(n-1)^2}\Sigma_2(0) + O(\beta_e^2)\right] \quad . \end{aligned} \tag{4.31}$$

The expressions for $\Sigma_1(0), \Sigma_2(0)$ are evaluated in Appendix A 4.2. The final result reads

$$\begin{aligned} (4\pi)^{-\frac{d}{2}} \frac{A_2^{\Pi}}{\beta_e \ell^d n^2} &= 1 + O\left(\frac{1}{n}\right) \\ &-\beta_e\left[\frac{4}{\varepsilon}(n^{\frac{\varepsilon}{2}}-1) - a(\varepsilon)n^{\frac{\varepsilon}{2}} + \alpha_2(\varepsilon) + O(n^{-1+\frac{\varepsilon}{2}})\right] + O(\beta_e^2) \,. \end{aligned} \tag{4.32}$$

$a(\varepsilon), \alpha_2(\varepsilon)$ again are finite functions of ε which can be taken from the appendix. This expression clearly has the same structure as the result (4.16ii) for R_e^2. It is again easily checked that the contributions proportional to $n^{\varepsilon/2}$ in Eq. (4.32) for $d<4$ can be derived by evaluating in Eq. (4.30) the segment summations as integrals.

A 4. Appendix

A 4.1 Formal expression for the second virial coefficient

Introducing the chemical potential μ_p of the polymer chains we can define the grand partition function $\mathcal{Z}_G$ of a monodisperse system of chains of length n as

$$\mathcal{Z}_G(\mu_p) = 1 + \sum_{M=1}^{\infty} \frac{1}{M!} e^{M\mu_p} \mathcal{Z}^{(M)}(n, \dots, n) \ . \tag{A 4.1}$$

Osmotic pressure $\hat{\Pi}$ and chain concentration c_p can be expressed in terms of $\mathcal{Z}_G$.

$$\frac{\hat{\Pi}}{k_B T} = \frac{1}{\Omega} \ln \mathcal{Z}_G(\mu_p) \tag{A 4.2}$$

$$c_p = \frac{1}{\Omega} \frac{\partial}{\partial \mu_p} \ln \mathcal{Z}_G(\mu_p) \ . \tag{A 4.3}$$

Equation (A 4.3) can be solved formally for e^{μ_p} as function of Ωc_p.

$$e^{\mu_p} = \frac{\Omega c_p}{\mathcal{Z}^{(1)}} \left[1 - \Omega c_p \frac{\mathcal{Z}^{(2)} - (\mathcal{Z}^{(1)})^2}{(\mathcal{Z}^{(1)})^2} + O(\Omega c_p)^2 \right] \ . \tag{A 4.4}$$

Substituting this result into Eq. (A 4.2) we find the virial expansion of the osmotic pressure

$$\frac{\hat{\Pi}}{k_B T} = c_p \left[1 - \frac{\Omega}{2} \frac{\mathcal{Z}^{(2)} - (\mathcal{Z}^{(1)})^2}{(\mathcal{Z}^{(1)})^2} c_p + O(c_p^2) \right] \ , \tag{A 4.5}$$

from which we can read off the second virial coefficient

$$A_2^{\Pi} = -\Omega \frac{\mathcal{Z}^{(2)}(n,n) - (\mathcal{Z}^{(1)}(n))^2}{(\mathcal{Z}^{(1)}(n))^2} \ . \tag{A 4.6}$$

The form (4.28) follows from Eq. (4.19).

A 4.2 Evaluation of some segment summations

The central tool is the asymptotic expansion ([Olv74] p 292)

$$\sum_{s=1}^{n-1} s^{-\alpha} = \zeta(\alpha) + \frac{n^{1-\alpha}}{1-\alpha} \left[\sum_{t=0}^{m} \binom{1-\alpha}{t} \frac{B_t}{n^t} + O\left(n^{-m-1}\right) \right] \ . \tag{A 4.7}$$

Here $\zeta(\alpha)$ denotes Riemann's ζ-function, and the B_t are Bernoulli's numbers: $B_0 = 1, B_1 = -\frac{1}{2}$, etc. Equation (A 4.7) is valid for $\alpha \neq 1$. For $\alpha = 1+\eta, \eta \to 0$, we have to use the expression

$$\zeta(1+\eta) = -2^{1+\eta}\pi^{\eta} \sin\left(\frac{\pi}{2}(1+\eta)\right)\zeta(-\eta)\frac{\Gamma(1-\eta)}{\eta} \underset{\eta\to 0}{\longrightarrow} \frac{1}{\eta} \quad , \qquad \text{(A 4.8)}$$

where $\Gamma(x)$ is the Γ-function. Substituting Eq. (A 4.8) into Eq. (A 4.7) it is easily checked that the limit $\alpha \to 1$ is finite.

Specifically we will need the expressions (recall $d = 4 - \varepsilon$)

$$\sum_{s=1}^{n-1} s^{-d/2} = \zeta\left(\frac{d}{2}\right) - \frac{n^{-1+\varepsilon/2}}{1-\varepsilon/2} + O\left(n^{-2+\varepsilon/2}\right) \quad , \qquad \text{(A 4.9)}$$

$$\begin{aligned}\sum_{s=1}^{n-1} s^{1-d/2} &= \zeta\left(1-\frac{\varepsilon}{2}\right) + \frac{n^{\varepsilon/2}}{\varepsilon/2} + O\left(n^{-1+\varepsilon/2}\right) \\ &= \frac{2}{\varepsilon}\left(n^{\varepsilon/2}-1\right) + \alpha_1(\varepsilon) + O\left(n^{-1+\varepsilon/2}\right) \quad , \qquad \text{(A 4.10)}\end{aligned}$$

$$\alpha_1(\varepsilon) = \frac{2}{\varepsilon} + \zeta\left(1-\frac{\varepsilon}{2}\right) \quad , \qquad \text{(A 4.11)}$$

$$\sum_{s=1}^{n-1} s^{2-d/2} = \frac{n^{1+\varepsilon/2}}{1+\varepsilon/2} + O\left(n^{\varepsilon/2}\right) \quad . \qquad \text{(A 4.12)}$$

Note that by virtue of Eq. (A 4.8) $\alpha_1(\varepsilon)$ is finite for $\varepsilon \to 0$.

We now turn to the evaluation of the sums occurring in low order Feynman diagrams.

$$\Sigma_1(0) = \sum_{0<j_1<j_2<n} (j_2 - j_1)^{-\frac{d}{2}}$$

Summing j_1, keeping $j_2 - j_1 = s$ fixed, we find

$$\Sigma_1(0) = \sum_{s=1}^{n-1}(n-1-s)s^{-d/2} = (n-1)\sum_{s=1}^{n-1} s^{-d/2} - \sum_{s=1}^{n-1} s^{1-d/2} \quad .$$

Equations (A 4.9), (A 4.10) yield

$$\Sigma_1(0) = (n-1)\zeta(\tfrac{d}{2}) - \frac{n^{\varepsilon/2}}{1-\varepsilon/2} - \frac{2}{\varepsilon}(n^{\varepsilon/2}-1) - \alpha_1(\varepsilon) + O(n^{-1+\varepsilon/2}). \qquad \text{(A 4.13)}$$

All the other summations can be evaluated in a similar fashion. We find

$$\begin{aligned}\Sigma_1' =& \sum_{0<j_1<j_2<n} (j_2-j_1)^{1-\frac{d}{2}} \\ =& \frac{2}{\varepsilon} n\left(n^{\varepsilon/2}-1\right) - \frac{n^{1+\varepsilon/2}}{1+\varepsilon/2} + n\alpha_1(\varepsilon) + O\left(n^{\varepsilon/2}\right) \quad ,\end{aligned} \tag{A 4.14}$$

$$\begin{aligned}\Sigma_2(0) =& \sum_{\substack{0<j_1<j_2<n \\ 0<j_3<j_4<n}} (j_2-j_1+j_4-j_3)^{-\frac{d}{2}} \\ =& \frac{2}{\varepsilon} n^2\left(n^{\varepsilon/2}-1\right) - \frac{1}{2}a(\varepsilon)n^{2+\varepsilon/2} + n^2\left(\alpha_1(\varepsilon)-\zeta(\tfrac{d}{2})\right) + O\left(n^{1+\varepsilon/2}\right),\end{aligned} \tag{A 4.15}$$

where

$$a(\varepsilon) = \frac{4}{2+\varepsilon} - \frac{2}{3(4+\varepsilon)} - 2\int_1^2 dx x^{-2+\varepsilon/2}\left(\frac{4}{3} - 2x + x^2 - \frac{x^3}{6}\right) \quad . \tag{A 4.16}$$

In Eq. (4.32) we furthermore introduced

$$\alpha_2(\varepsilon) = 2\alpha_1(\varepsilon) - 6\zeta(\tfrac{d}{2}) \quad . \tag{A 4.17}$$

5. Grand Canonical Description of Solutions at Finite Concentration

The statistical mechanics of solutions at nonvanishing concentration is best formulated within the grand canonical ensemble, where the number of solute molecules contained in volume Ω is allowed to fluctuate. We only control the average concentration c_p via the chemical potential μ_p. This has great technical advantages, allowing for a very simple analysis of the thermodynamic limit. In contrast, in the canonical ensemble, where the particle number $M = \Omega c_p$ is taken as basic variable, an analysis of the thermodynamic limit is more tricky. A short discussion is given in the appendix.

In this chapter we work out the general aspects of the grand canonical formalism as applied to polymer solutions, to the point where we can use it to evaluate specific observables. We first define the ensemble, and we exploit the so-called *'Linked Cluster Theorem'* to find expressions for the observables valid in the thermodynamic limit. The quantities to be considered can be grouped into two classes: thermodynamic quantities can be derived from the grand potential, which essentially is the logarithm of the partition function. The other class comprises correlation functions, which are defined as spatially localized expectation values. The density correlations constitute the most important example. Explicit expressions for all observables can be given in diagrammatic form, which needs some minor modification of the Feynman rules.

However, these formal expansions are not adapted to an actual evaluation valid outside the dilute region. A simple consideration reveals the problem. Consider some chain of length n in a solution of segment concentration c. Following Chap. 2, Eq. (2.10ii) we estimate the number of two-body contacts of that chain with all the other chains as

$$n_c^{(2)} \sim n\,\ell^d c \quad ,$$

which yields an interaction energy of order

$$W(n) = (4\pi)^{d/2}\ \beta_e \ell^d c\, n \quad . \tag{5.1}$$

This characteristic energy naturally shows up in the cluster expansion. Note that W can become large even for $\beta_e \ll 1$, $\ell^d c \ll 1$, just because the chains become very long. Since formally $W = O(\beta_e)$, straightforward cluster expansion automatically implies an expansion in powers of W. It is thus valid only for $W \lesssim 1$: either the chains must be isolated: $\ell^d c\, n \lesssim 1$, or the interaction constant β_e must be very small. A closely related statement is that the cluster expansion automatically implies that we treat the concentration dependence in terms of the virial expansion.

In order to apply the theory outside the dilute limit we have to carry through partial resummations, which sum up whole infinite series in W. Now partial resummation of perturbation theory for many-body systems is a delicate task. Perturbation theory in general is not absolutely convergent, and thus some arbitrary resummation may yield arbitrary unphysical results. Any valid resummation must be guided by some physical insight, identifying the dominant effects to be taken into account. In the present case this is quite simple: outside the dilute limit a given chain interacts with many others, which to zero order just produce a constant background concentration. This is to be corrected by the fact that the chain acts back on its surrounding and creates fluctuations of the local segment density, which then in turn affect the chain considered. We should thus expand in the effect of density fluctuations about the average segment density c. Technically this is achieved by resumming all terms of the cluster expansion, which have the same number of 'loops', i.e. of independent internal momentum integrations. This can be consistently done, and it is found that the leading zero loop or 'tree' approximation is identical to standard mean field theory. It reproduces such well known approximations as Flory-Huggins theory or the Random Phase Approximation (RPA).

This consideration alone, though physically most reasonable, does not guarantee that the loop expansion, i.e. the expansion in fluctuations about the average density, really eliminates all problems connected to $W \gg 1$. We have to add one further essential observation: in some sense the density fluctuations are small, due to the 'screening effect', which in the context of dense polymer solutions has been discovered by Edwards long ago [Edw65]. The physical mechanism is this. Consider a subchain of length n', being part of a larger chain. This subchain will extend over a volume of linear size $R(n') \sim \ell n'^{\nu}$, yielding a local segment density $\rho(n') \sim n'/R^d(n') \sim n'^{1-d\nu}$. For $n' < n^*(c) = (\ell^d c)^{-1/(\nu d-1)}$, $\rho(n')$ will exceed the average segment density c in the solution, whereas for $n' \gg n^*$ we find $\rho(n') \ll c$, so that on the corresponding scale $R(n')$ many chains interpenetrate to build up the average segment density c. As a consequence, for subchains $n' \ll n^*$ the intra-chain interaction will dominate, whereas the inter-chain interaction dominates for subchains $n' \gg n^*$. Now a chain swells in order to decrease the interaction energy. This, however, can work only for subchains $n' < n^*$, where swelling indeed decreases the locally measured segment density and thus reduces the number of (intra-chain) contacts. On scales $R(n') \gg R(n^*)$ the segment density and thus the total interaction energy cannot be reduced by swelling. As a consequence, on scale $r \gg R(n^*)$ the excluded volume interaction is ineffective or 'screened'. To first approximation it yields a constant background energy proportional to W, but cannot lead to nontrivial spatial correlations. Rather the chain configuration at large scales is dominated by the entropy. There emerges the picture of a chain swollen over short distances, but essentially random-walk-like over large distances.

How does this insight help us for the loop expansion? We have pointed out above that the basic objects of the loop expansion are the local density fluctuations. We now learn that these fluctuations are of finite range, independent of $n \to \infty$, i.e. of $W \to \infty$, u_0, c fixed. We should thus expect that the limit $W \to \infty$ can be taken without problems. On the technical level we will find that the resummation resulting in the loop expansion introduces a W- and q-dependent effective interaction. For $W > 0$ this interaction decreases with decreasing q, corresponding to increasing distances, and it vanishes in the limit $q \to 0, W \to \infty$. This behavior effectively suppresses the divergence found for $W \to \infty$.

We should immediately note that the picture described here in the context of scaling or renormalization theory leads to the important concept of 'concentration blobs'. (See Sect. 9.1.)

This discussion also allows for a more precise understanding of the notions 'dilute' or 'semidilute', introduced in Chap. 1. By definition, in the dilute region the interaction of a given chain with its surrounding is small, in the sense of $W \ll 1$. This does not imply that the chains are isolated geometrically! On the contrary, the geometrical overlap $s = c_p R_g^d(n)$ may be large, provided the excluded volume coupling β_e is small enough. Also $W \gg 1$ does not imply that we are concerned with a concentrated solution. It only defines a parameter region where a given chain interacts strongly with many others. This comprises both dense systems: $\ell^d c \approx 1$, or semidilute systems: $\ell^d c \ll 1$, $s \gg 1, \beta_e$ not too small. This adds an important aspect to the discussion given in Chap. 1: the strength of the two-body interaction strongly influences the size of the dilute or semidilute regions.

The organization of this chapter is as follows. In Sect. 5.1 we present the basic formalism and work out the Feynman rules for the grand canonical ensemble. Diagrammatic representations valid in the thermodynamic limit are derived for both thermodynamic quantities and correlation functions. The proof of the Linked Cluster Theorem is given in Appendix A 5.1. Section 5.2 is devoted to the resummation leading to the loop expansion. In the main text we use a diagrammatic approach. A more elegant but formal argument based on functional integration is sketched in Appendix A 5.3. This argument also clearly exhibits the physical interpretation of the loop expansion discussed above. As soon as we deal with a many-chain system there arises the problem of *'polydispersity'*, i.e. the solutions contain some distribution of chain lengths. Relevant concepts and notations are introduced in Sect. 5.3. The loop expansion is illustrated in Sect. 5.4 with a discussion of the important zero-loop ('tree') approximation for a variety of quantities. We also present an example of a one loop calculation explicitly exhibiting screening. Specifically, we calculate the one loop correction for the relation between chemical potential and concentration. Finally Appendix A 5.2 deals with some technical peculiarity occurring with the loop expansion of the grand potential, and in Appendix A 5.4 we briefly discuss the canonical formalism.

5.1 Basic features of the grand canonical formalism

5.1.1 Definitions

In Sect. 4.2, Eq. (4.19), we have introduced the canonical partition function $\mathcal{Z}^{(M)}(n_1, \ldots, n_M)$ of M chains of lengths $n_1, \ldots, n_M$. The grand partition function $\mathcal{Z}[\mu_p]$ is constructed as a superposition of the $\mathcal{Z}^{(M)}$.

$$\mathcal{Z}[\mu_p] = 1 + \sum_{M=1}^{\infty} \frac{1}{M!} \sum_{n_1 \ldots n_M = 1}^{\infty} \prod_{m=1}^{M} e^{\mu_p(n_m)} \mathcal{Z}^{(M)}(n_1, \ldots, n_M) \quad . \tag{5.2}$$

Here $\mu_p(n)$ is the chemical potential of chains of length n, where for brevity I suppress the factor $1/k_B T$. Thus $\mu_p(n) = \hat{\mu}_p(n)/k_B T$, where $\hat{\mu}_p(n)$ is the conventionally defined chemical potential. The corresponding notation will be used throughout.

Note that in the standard definition of the grand potential the summation independently extends over the particle numbers of the different molecular species. For the present problem this implies summation over the numbers $M(n)$ of 'macromolecules' of length $n = 1, 2, 3, \ldots$.

$$\mathcal{Z}[\mu_p] = \sum_{M(1), M(2), \ldots = 0}^{\infty} \left(\prod_{n=1}^{\infty} \frac{1}{M(n)!} \, e^{M(n)\mu_p(n)} \right) \mathcal{Z}^{[\Sigma_n M(n)]} \quad .$$

It is easily checked that this can be rewritten in the simpler form (5.2).

Thermodynamics is derived from the grand potential $(-\Omega \Pi)$ defined by

$$\Pi[\mu_p] = \frac{1}{\Omega} \ln \mathcal{Z}[\mu_p] \quad . \tag{5.3}$$

Physically $\hat{\Pi} = k_B T \Pi$ is nothing but the osmotic pressure of the solution. The concentration $c_p(n)$ of chains of length n is calculated as

$$c_p(n) = \frac{\partial}{\partial \mu_p(n)} \Pi[\mu_p] \quad . \tag{5.4}$$

This relation can be inverted to yield $\mu_p(n)$ as function of $c_p(n')$ (cf. Appendix A 4.1). Substituting the result into $\Pi[\mu_p]$, we find the osmotic pressure as function of the concentration, which is the standard form of the osmotic equation of state. Also all the other thermodynamic quantities can be calculated from $\Pi[\mu_p]$. The excess free energy due to the solute, for instance, takes the form

$$\frac{\hat{F}[c_p(n)]}{k_B T} = F[c_p(n)] = \sum_n c_p(n)\mu_p(n) - \Pi[\mu_p(n)] \ . \tag{5.5}$$

5.1.2 Linked cluster theorem and thermodynamic limit

The grand canonical formalism is useful only provided $\Pi[\mu_p]$ exists in the thermodynamic limit $\Omega \to \infty$. This property is not completely obvious, as becomes clear from an analysis of the canonical partition function $\mathcal{Z}^{(M)}$. Since the positions of all M chains are integrated independently over all the volume, we find

$$\mathcal{Z}^{(M)}(n_1, \ldots, n_M) = \prod_{m=1}^{M} \mathcal{Z}^{(1)}(n_m)\left(1 + O\left(\frac{1}{\Omega}\right)\right) \ , \tag{5.6}$$

where the leading term describes M independent chains, and the correction terms are due to the interactions among different chains. Since $\mathcal{Z}^{(1)} \sim \Omega$, this implies that the expansion (5.2), defining $\mathcal{Z}[\mu_p]$, does not exist in the thermodynamic limit. The M-th term diverges like Ω^M. This is a familiar problem, showing up in all many-body theories. To prove the existence of $\Pi[\mu_p]$ we must show that in $\ln \mathcal{Z}[\mu_p]$ all terms of order $\Omega^M, M > 1$, do cancel. This is not unconceivable, since substituting Eq. (5.6) into Eq. (5.2) and ignoring the corrections of relative order $1/\Omega$ we find

$$\mathcal{Z}[\mu_p] \approx \exp\left\{\sum_{n=1}^{\infty} e^{\mu_p(n)} \mathcal{Z}^{(1)}(n)\right\} \ ,$$

so that $\Pi[\mu_p] = \ln \mathcal{Z}[\mu_p]/\Omega$ exists for $\Omega \to \infty$.

Ignoring the correction terms in Eq. (5.6), this argument is valid only if we switch off the interaction among different chains, i.e. it holds for an 'ideal solution'. The generalization to mutually interacting chains follows by virtue of the Linked Cluster Theorem. In the present context this important theorem can be formulated as follows.

Linked Cluster Theorem

$$\ln \mathcal{Z}[\mu_p] = \sum_{M=1}^{\infty} \frac{1}{M!} \sum_{n_1 \ldots n_M = 1}^{\infty} \left(\prod_{m=1}^{M} e^{\mu_p(n_m)}\right) \mathcal{Z}_c^{(M)}(n_1, \ldots, n_M) \tag{5.7}$$

where

$$\mathcal{Z}_c^{(M)}(n_1,\ldots,n_M) = G_c^{(0,M)}(\mathbf{0},\ldots,\mathbf{0};n_1,\ldots,n_M) \tag{5.8}$$

is the M-chain cumulant, evaluated at zero momenta.

The proof is given in Appendix A 5.1, where it will become clear that basically the theorem relates the Taylor series expansion of any given function $f(x)$ to the expansion of $\ln f(x)$. As such it is a very general statement, of fundamental importance for the thermodynamic limit in any many-body theory. In the present context we note that $G_c^{(0,M)}$ by definition (Sect. 4.3) is given by the sum of all connected diagrams containing M polymer lines. From the Feynman rules we find $G_c^{(0,M)} \sim \Omega$, resulting in the statement

Corollary to the Linked Cluster Theorem

$$\Pi[\mu_p] = \lim_{\Omega\to\infty} \frac{1}{\Omega} \ln \mathcal{Z}[\mu_p]$$

exists and can be calculated perturbatively in terms of connected diagrams with any number of polymer lines.

Note that 'existence' here only means that all the individual terms of the expansion are finite. It does not imply a statement on the convergence of the expansion. This is a much harder problem, in general unsolved for the expansions we deal with. If not stated otherwise, we will always use the notion of 'existence' in that restricted sense.

5.1.3 Diagrammatics

It is useful to modify the Feynman rules to account for the chemical potential. For grand canonical diagrams we close each polymer line by crosses (Fig. 5.1), and we add the following rule.

Each closed line is to be weighted by a factor $2e^{\mu_p(n)}$, and is to be summed over n.

The factor 2 needs some explanation, which leads us to discuss symmetry classes of Feynman diagrams – a beautiful but nontrivial task. In the present formulation a diagram with M polymer lines has to be multiplied by a factor of $2^{-M}/M!$, the factor $1/M!$ being explicit in the definition of the grand canonical ensemble, and the factor 2^{-M} compensating for the factors of 2 introduced above. All the chains as well as the ends of any given chain are distinguishable, the diagrams in principle are labeled as shown for the first diagram in Fig. 5.1. As a consequence, all the eight graphs of Fig. 5.1 correspond to different contributions to perturbation theory and must be considered explicitly. On the other hand, in the grand canonical ensemble all chains are equivalent and are invariant under reflection (segment index

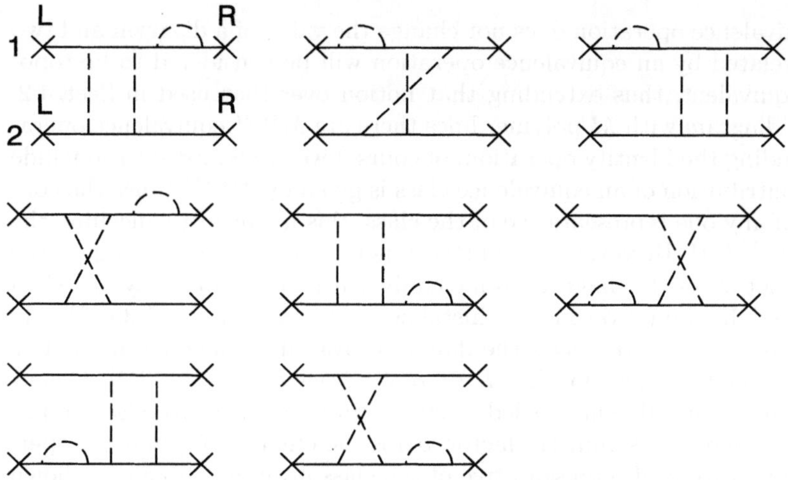

Fig. 5.1. A typical graph together with its equivalence class

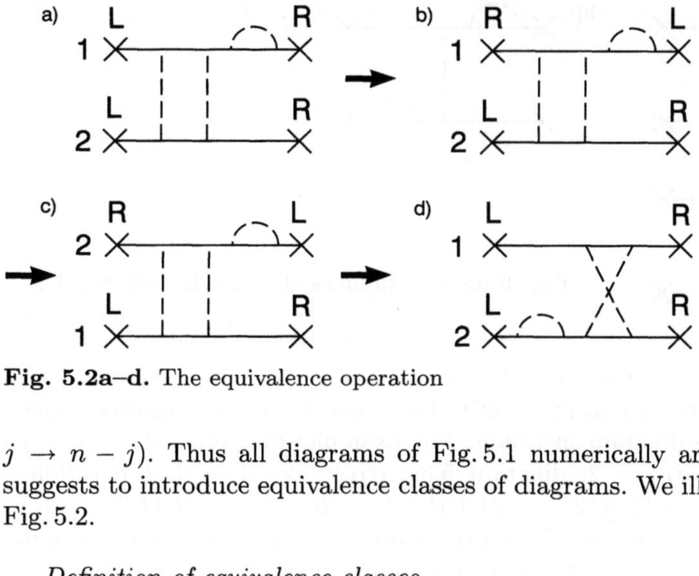

Fig. 5.2a–d. The equivalence operation

$j \to n - j$). Thus all diagrams of Fig. 5.1 numerically are identical. This suggests to introduce equivalence classes of diagrams. We illustrate this with Fig. 5.2.

Definition of equivalence classes

(i) Label the diagram in some canonical ordering (Fig. 5.2a).

(ii) Interchange labels L, R for any number of chains (Fig. 5.2b, $[1 : L \leftrightarrow R]$), and permute the chain labels in some arbitrary way (Fig. 5.2c, $[(12) \to (21)]$).

(iii) Redraw the diagram, so that the canonical ordering is restored (Fig. 5.2d).

An equivalence operation does not change the value of a diagram and two diagrams related by an equivalence operation will be considered to be topologically equivalent, thus extending that notion over that used in Sect. 4.2. Since for a diagram with M polymer lines there are $M!2^M$ equivalence operations (including the identity operation, of course) we are tempted to conclude that the contribution of an equivalence class is given by $M!2^M$ times the contribution of any one representative of the class, this factor just cancelling the prefactor $2^{-M}/M!$. However, the matter is not so simple. Depending on the specific structure of the diagram, some equivalence operations may not yield a new labeled diagram. Consider for instance the example in Fig. 5.3a. Clearly the operation $[2 : L \leftrightarrow R]$ leaves the diagram invariant. The diagram 5.3b is invariant under the operation $[1 : L \leftrightarrow R, 2 : L \leftrightarrow R, (12) \rightarrow (21)]$. Generally the number of different labeled diagrams in an equivalence class equals $M!2^M/\sigma_D$, where the 'symmetry factor' σ_D is the number of symmetry operations leaving a labeled representative of the class invariant. It can be shown that σ_D is independent of the representative graph chosen, as it must.

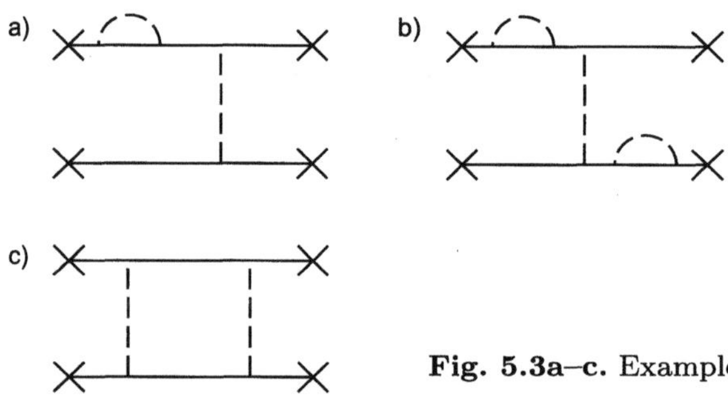

Fig. 5.3a–c. Examples of diagrams with $\sigma_D \neq 1$

For low order diagrams the determination of σ_D is quite simple. With a little experience one can easily identify by inspection those symmetry operations leaving the diagram invariant. For example: diagram 5.2a: $\sigma_D = 1$; diagrams 5.3a, b: $\sigma_D = 2$; diagram 5.3c: $\sigma_D = 2 \cdot 2 = 4$. (operations: $[1 : L \leftrightarrow R, 2 : L \leftrightarrow R], [(12) \rightarrow (21)], [1 : L \leftrightarrow R, 2 : L \leftrightarrow R, (12) \rightarrow (21)]$ and identity). For more complicated diagrams the determination of σ_D may be a formidable task, which in fact can be implemented on a computer.

This discussion can be summarized as follows:

Diagrammatic representation of the grand potential

$\Pi[\mu_p]$ is given by the sum of all topologically different connected diagrams obeying the following rules.

Construction

(i) Draw any number of polymer lines closed by crosses.

(ii) Add any number of (broken) interaction lines, so that no broken line is attached to an endpoint, nor are there two broken lines connected to the same point.

Evaluation

(i) Label the distinguished points by some summation indices and assign momentum variables to the propagators, respecting momentum conservation. Label the m-th polymer line by a chain length variable n_m.

(ii) A propagator line of momentum $\mathbf{k}$, connecting distinguished points $j < j'$, yields a factor $G_0(\mathbf{k}, j' - j)$. A broken line yields a factor $-(4\pi\ell^2)^{d/2}\beta_e$. The two crosses closing the m-th line yield a factor $(4\pi\ell^2)^{-d/2} 2 e^{\mu_p(n_m)}$.

(iii) Integrate over all internal momenta and sum over the segment labels, respecting their ordering along the chains. Sum over all chain length variables.

(iv) Divide by the appropriate symmetry factor σ_D.

It is now an easy task to derive also the diagrammatic representation of $c_p(n)$ (Eq. (5.4)). The derivative $\partial/\partial\mu_p(n)$ singles out one polymer line, fixing its chain length to n. We draw this as an open line, noting that it is no longer equivalent to the remaining closed lines. (Its two ends are still equivalent, of course.) The following representation results.

Diagrammatic representation of $c_p(n)$

$$I^{(1)}(n) = \frac{1}{2} e^{-\mu_p(n)} c_p(n)$$

is given by the sum of all topologically different connected diagrams with exactly one open line, all other lines being closed by crosses. The open line has fixed chain length n and carries only its weight factor $(4\pi\ell^2)^{-d/2}$. Its ends are equivalent.

Having constructed representations of the grand potential and of $c_p(n)$ we are in principle able to evaluate all thermodynamic quantities.

5.1.4 Correlation functions

Experimentally the density correlations are most important, and we therefore exemplify the construction of grand canonical correlation functions with the density autocorrelation function or the density cumulant. Recall the definition (3.14) of the local segment density of the m-th chain:

$$\hat{\rho}_m(\mathbf{r}) = \sum_{0<j<n_m} \delta^d(\mathbf{r} - \mathbf{r}_j^{(m)}) \quad . \tag{5.9}$$

(In the summation we omit the endpoint n_m as usual.) The density autocorrelation function is defined as

$$\begin{aligned}
&\hat{I}_{d,a}(\mathbf{r} - \mathbf{r}') = \\
&= \left\langle \sum_m \hat{\rho}_m(\mathbf{r})\hat{\rho}_m(\mathbf{r}') \right\rangle \\
&= \mathcal{Z}^{-1}[\mu_p] \sum_{M=1}^{\infty} \frac{1}{M!} \sum_{n_1 \dots n_M = 1}^{\infty} \prod_{m=1}^{M} e^{\mu_p(n_m)} \int e^{-\mathcal{H}} \sum_{m=1}^{M} \hat{\rho}_m(\mathbf{r})\hat{\rho}_m(\mathbf{r}') \ ,
\end{aligned} \tag{5.10}$$

where we use the symbolic notation

$$\begin{aligned}
\int e^{-\mathcal{H}} X\{r_j^{(m)}\} = \int D(\mathbf{r}^{(1)}, \dots, \mathbf{r}^{(M)}) \exp\Big\{ - \sum_{m=1}^{M} \sum_{j=1}^{n_M} \frac{(\mathbf{r}_j^m - \mathbf{r}_{j-1}^{(m)})^2}{4\ell^2} \Big\} \\
\cdot \prod_{(j_1 m_1, j_2 m_2)}{}' \left[1 - (4\pi\ell^2)^{d/2} \beta_e \delta^d(\mathbf{r}_{j_1}^{(m_1)} - \mathbf{r}_{j_2}^{(m_2)}) \right] X\{r_j^{(m)}\} \ .
\end{aligned}$$

Equation (5.10) sums over the autocorrelations of all the individual chains in the ensemble. It is a special case of a grand canonical expectation value, which for some quantity $X = X\{\mathbf{r}_j^{(m)}\}$ is generally defined as

$$\langle X \rangle = \mathcal{Z}^{-1}[\mu_p] \sum_{M=1}^{\infty} \frac{1}{M!} \sum_{n_1 \dots n_M = 1}^{\infty} \prod_{m=1}^{M} e^{\mu_p(n_m)} \int e^{-\mathcal{H}} X\{\mathbf{r}_j^{(m)}\} \ . \tag{5.11}$$

The density cumulant is defined as

$$\hat{I}_{d,c}(\mathbf{r} - \mathbf{r}') = \langle \hat{\rho}(\mathbf{r})\hat{\rho}(\mathbf{r}') \rangle - c^2 \quad , \tag{5.12}$$

where

$$\hat{\rho}(\mathbf{r}) = \sum_m \hat{\rho}_m(\mathbf{r}) \tag{5.13}$$

is the local segment density of all chains, and

$$c = \sum_n n\, c_p(n) \tag{5.14}$$

is the average segment concentration. To understand the definition (5.12) we note that

$$c = \langle \hat{\rho}(\mathbf{r}) \rangle \tag{5.15}$$

independent of $\mathbf{r}$, as will be shown below. Thus

$$\begin{aligned} \hat{I}_{d,c}(\mathbf{r}-\mathbf{r}') &= \langle \hat{\rho}(\mathbf{r})\hat{\rho}(\mathbf{r}') \rangle - \langle \hat{\rho}(\mathbf{r}) \rangle \langle \hat{\rho}(\mathbf{r}') \rangle \\ &= \Big\langle \Big(\hat{\rho}(\mathbf{r}) - \langle \hat{\rho}(\mathbf{r}) \rangle \Big) \Big(\hat{\rho}(\mathbf{r}') - \langle \hat{\rho}(\mathbf{r}') \rangle \Big) \Big\rangle \ . \end{aligned} \tag{5.16}$$

The density cumulant therefore measures the correlation among local density fluctuations $\hat{\rho}(\mathbf{r}) - \langle \hat{\rho}(\mathbf{r}) \rangle$. It is these fluctuations which lead to scattering.

We will mainly be concerned with the Fourier transform of the correlation functions. We use the definition

$$I_{d,a(c)}(\mathbf{q}) = \frac{1}{\Omega} \int d^d r d^d r' e^{i\mathbf{q}(\mathbf{r}-\mathbf{r}')} \hat{I}_{d,a(c)}(\mathbf{r}-\mathbf{r}') \ , \tag{5.17}$$

where the factor $1/\Omega$ deviates from our previous choice (3.17), adequate for a single chain system. This factor guarantees that $I_{d,a(c)}(\mathbf{q})$ is an intensive quantity, independent of Ω in the thermodynamic limit.

All the other correlation functions are to be defined according to Eq. (5.11). For instance, being interested in endpoint correlations, we introduce the local endpoint density of the m-th chain

$$\hat{\rho}_{e,m}(\mathbf{r}) = \delta^d(\mathbf{r}_0^{(m)} - \mathbf{r}) + \delta^d(\mathbf{r}_{n_m}^{(m)} - \mathbf{r}) \ , \tag{5.18}$$

and we define the endpoint autocorrelation function as

$$\hat{I}_{e,a}(\mathbf{r}-\mathbf{r}') = \left\langle \sum_m \hat{\rho}_{e,m}(\mathbf{r}) \hat{\rho}_{e,m}(\mathbf{r}') \right\rangle \ . \tag{5.19}$$

The endpoint cumulant is defined by replacing in Eq. (5.16) the total segment density $\hat{\rho}(\mathbf{r})$ by the total endpoint density

$$\hat{\rho}_e(\mathbf{r}) = \sum_m \hat{\rho}_{e,m}(\mathbf{r}) \ . \tag{5.20}$$

We now turn to the *'sum rules'*, which relate locally defined expectation values to thermodynamic quantities. Equation (5.15), identifying the average of the local segment density with the average segment concentration, presents the simplest example. From a physical point of view this relation is obvious. Yet it is instructive to construct the formal proof. We first recall that the ensemble is translational invariant, which implies

$$\langle\hat{\rho}(\mathbf{r})\rangle = \frac{1}{\Omega}\int_{\Omega} d^d r\,\langle\hat{\rho}(\mathbf{r})\rangle \quad , \tag{5.21}$$

up to surface effects. Furthermore $\int d^d r \hat{\rho}_m(\mathbf{r}) = n_m$. Thus Eqs. (5.11), (5.13) yield

$$\langle\hat{\rho}(\mathbf{r})\rangle = \frac{1}{\Omega}\mathcal{Z}^{-1}[\mu_p]\sum_{M=1}^{\infty}\frac{1}{M!}\sum_{n_1\ldots n_M=1}^{\infty}\prod_{m=1}^{M}e^{\mu_p(n_m)}\int e^{-\mathcal{H}}\cdot\sum_{m'=1}^{M}n_{m'}\,. \tag{5.22}$$

Now we have the identity

$$\begin{aligned}\sum_n n\frac{\partial}{\partial\mu_p(n)}\prod_{m=1}^{M}e^{\mu_p(n_m)} &= \sum_n n\left(\sum_{m'=1}^{M}\delta_{nn_{m'}}\right)\prod_{m=1}^{M}e^{\mu_p(n_m)} \\ &= \sum_{m'=1}^{M}n_{m'}\prod_{m=1}^{M}e^{\mu_p(n_m)}\,,\end{aligned}$$

where $\delta_{nn'}$ is Kronecker's symbol:

$$\delta_{nn'} = \begin{cases}1 \text{ if } & n = n' \\ 0 & \text{otherwise}\,.\end{cases}$$

We can thus write

$$\begin{aligned}\langle\hat{\rho}(\mathbf{r})\rangle &= \frac{1}{\Omega}\mathcal{Z}^{-1}[\mu_p]\sum_n n\frac{\partial}{\partial\mu_p(n)}\mathcal{Z}[\mu_p] \\ &= \frac{1}{\Omega}\sum_n n\frac{\partial}{\partial\mu_p(n)}\ln\mathcal{Z}[\mu_p] \quad ,\end{aligned} \tag{5.23}$$

where in the first line we used the definition (5.2) of $\mathcal{Z}[\mu_p]$. Equations (5.3), (5.4) yield

$$\langle\hat{\rho}(\mathbf{r})\rangle = \sum_n n c_p(n) = c \quad ,$$

which is the desired result.

Another important example is the sum rule for $I_{d,a}(\mathbf{q}=0)$:

$$I_{d,a}(\mathbf{0}) = \left\langle\sum_m n_m^2\right\rangle = \sum_n n^2 c_p(n) \quad . \tag{5.24}$$

The proof closely follows the derivation of the relation $c = \langle\hat{\rho}(\mathbf{r})\rangle$.

We finally need the diagrammatic representation of the correlation functions. As shown in Appendix A 5.1, as a simple consequence of the general Linked Cluster Theorem all correlation functions or cumulants defined here are given by the set of connected grand canonical diagrams with the appropriate density- or endpoint density-insertions. Due to the definition (5.17) of the Fourier transform, no explicit factor of Ω shows up. Specifically, $I_{d,a}(\mathbf{q})$ is given by the set of all connected diagrams with two density insertions in the same polymer line, whereas $I_{d,c}(\mathbf{q})$ is given by the set of all connected diagrams with two density insertions in the same or different lines. The discussion of symmetry operations is not changed; we must be aware, however, that the symmetry operation by definition does not exchange the arrows representing density insertions ('external vertices'). We summarize these considerations as

Diagrammatic representation of correlation functions.

Any correlation function is represented by the set of connected grand canonical diagrams with the appropriate segment density- or endpoint density-insertions.

5.2 Loop expansion, vertex irreducible graphs and screened interaction

5.2.1 The notion of 'loops'

We have argued in the introductory section that the cluster expansion in powers of β_e breaks down as soon as a chain strongly interacts with a lot of others. To overcome this problem we must resum the cluster expansion to arrive at the so-called 'loop expansion'. We first define the number of loops.

Definition

The number of loops n_L of a Feynman diagram is the number of internal momentum integrations which survive after all momentum conserving δ-functions have been exploited.

For the diagrams of Fig. 5.3 we, for instance, find $n_L(5.3a) = 1$, $n_L(5.3b) = 2$, $n_L(5.3c) = 1$. We may show that with each independent internal momentum we may associate a closed curve on the graph, through which that momentum flows. This explains the notion of 'loop'. A connected diagram without loops is called a 'tree'.

We now order the Feynman graph expansion according to the number of loops. Then in the grand canonical ensemble even the lowest order, i.e. tree, approximation amounts to resuming an infinity of diagrams, taking into account any number of chains. We will see in Sect. 5.4 that it essentially accounts for the average segment density in the solution. Higher loop orders then reflect the effect of local density fluctuations induced by the chain considered.

5.2.2 Vertex reducibility

To carry through the partial summation of perturbation theory we introduce the notion of vertex reducibility.

Definition

A graph is said to be vertex reducible, if it can be cut into two disconnected pieces by cutting one vertex.

For example, the graphs 5.3a,b are vertex reducible. The graphs 5.3c, 5.1 are vertex irreducible.

We now consider the density autocorrelation function $I_{d,a}(\mathbf{q})$, which is represented by the set of all connected diagrams with two density insertions of momenta $\mathbf{q}$ or $-\mathbf{q}$ in the same polymer line. Let $K_a^{(2,0)}(\mathbf{q})$ denote the subset of vertex irreducible graphs contributing to $I_{d,a}(\mathbf{q})$, an example being given by the first graph of Fig. 5.4. Any vertex reducible graph contributing to $I_{d,a}$ uniquely can be associated with a graph of $K_a^{(2,0)}$ by cutting off

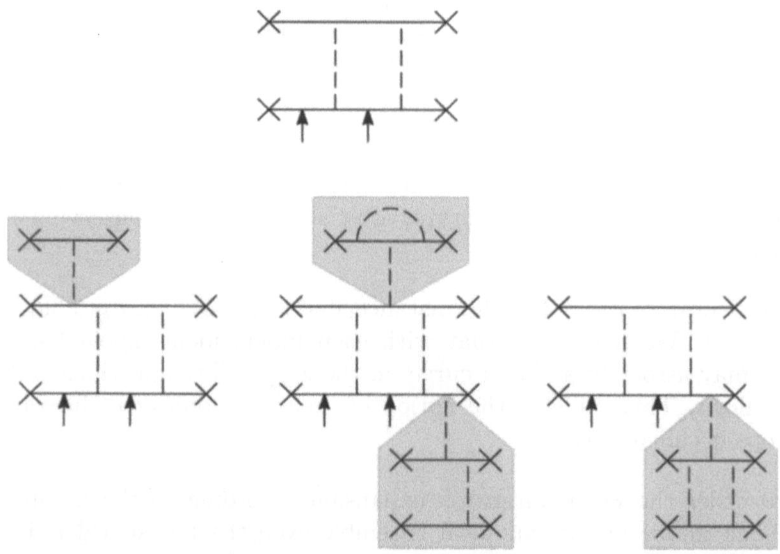

Fig. 5.4. An irreducible graph (*first line*) and some reducible graphs (*second line*) yielding that graph after cutting off all reducible parts. The reducible parts are indicated by *shaded boxes*

all reducible vertices. Figure 5.4 gives some examples. Now it is clear by momentum conservation in the diagram that the reducible vertices do not carry momentum. The subdiagram connected to $K_a^{(2,0)}$ by a reducible vertex therefore has the structure of a grand canonical diagram with a single density insertion of momentum zero. The sum of all such subdiagrams just yields $\langle\hat{\rho}(\mathbf{r})\rangle = c$. We therefore conclude that the subset of all diagrams contributing to $I_{d,a}(\mathbf{q})$, which are associated with the same graph of $K_a^{(2,0)}(\mathbf{q})$, just yields that graph with any number of insertions $C = -u_0 c$ in its polymer lines, where we introduced the shorthand notation

$$u_0 = (4\pi\ell^2)^{d/2}\beta_e \quad . \tag{5.25}$$

Diagrammatically we represent such an insertion as shown in Fig. 5.5.

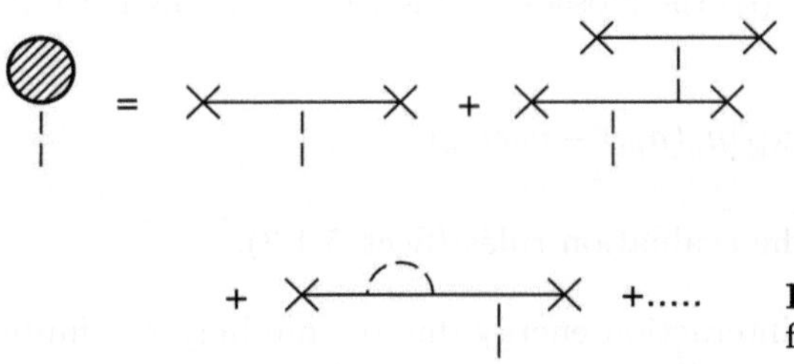

Fig. 5.5. Diagrammatic notation for the sum of all reducible parts

Now consider a propagator line of momentum $\mathbf{k}$, connecting segments j_1 and j_2, and decorate it with any number of such C-insertions (Fig. 5.6). Summing all such terms we find

$$\begin{aligned} &G_0(\mathbf{k}, j_2 - j_1)\left[1 + C \sum_{j_1<i_1<j_2} 1 + C^2 \sum_{j_1<i_1<i_2<j_2} 1 + \ldots\right] \\ &= G_0(\mathbf{k}, j_2 - j_1) \sum_{i=0}^{j_2-j_1-1} \binom{j_2 - j_1 - 1}{i} C^i \\ &= G_0(\mathbf{k}, j_2 - j_1)\,(1 + C)^{j_2-j_1-1} \quad . \end{aligned}$$

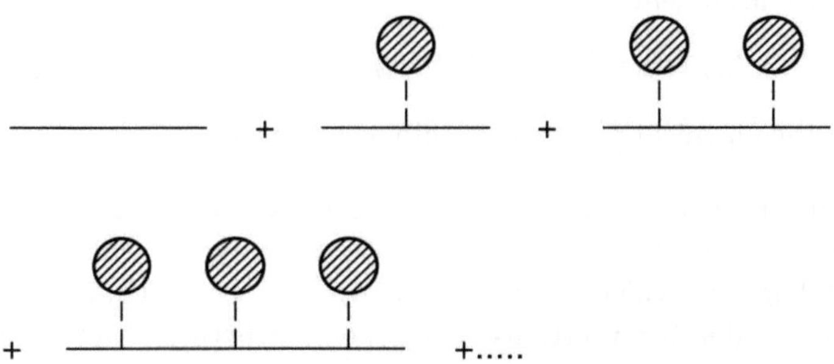

Fig. 5.6. Resummation leading to the shift of the chemical potential

Combining these factors for all propagators of a given polymer line we find an overall factor

$$(1+C)^{n_m(1+O(1/n_m))} = \exp\left[-n_m u_0 c\left(1+O\left(\frac{1}{n_m}, u_0 c\right)\right)\right] \quad ,$$

where the correction terms are small in the region of interest: $n_m \gg 1, u_0 c \ll 1$. Decorating a graph of $K_a^{(2,0)}$ with any number of C-insertions thus amounts to shifting the chemical potential $\mu_p(n)$ by $-n u_0 c$, this contribution being identical to the average interaction energy $W(n)$ introduced at the beginning of this chapter. We thus have the following result:

$$I_{d,a}(\mathbf{q}) = K_a^{(2,0)}(\mathbf{q}) \quad , \tag{5.26}$$

where in the evaluation of $K_a^{(2,0)}(\mathbf{q})$ the crosses closing the m-th polymer line stand for

$$\frac{2}{(4\pi\ell^2)^{d/2}} \exp\left[\mu_p(n_m) - u_0 c n_m\right] \quad .$$

This rule modifies part (ii) of the evaluation rules (Sect. 5.1.3).

Note that here the average interaction energy due to coupling to a finite concentration of other chains comes in quite naturally. Clearly its first effect must be a shift in the chemical potential.

In all this argument we tacitly assumed that the symmetry factors σ_D do not mess up the derivation. Now it is fairly obvious that the symmetry factor of a diagram contributing to $I_{d,a}$ factorizes into the symmetry factor of the corresponding graph of $K_a^{(2,0)}$ and independent factors for all C-type insertions. Clearly this is sufficient for the validity of the argument.

The argument exploits the fact that in each diagram contributing to $I_{d,a}$ we can uniquely identify some distinguished vertex irreducible part. It is easily transferred to any quantity showing that property. Consider for instance $c_p(n)$, expressed in terms of the set $I^{(1)}(n)$ of diagrams with one open line, according to Sect. 5.1.3. Denoting by $K^{(0,1)}(n)$ the subset of irreducible diagrams from the above argument we find

$$c_p(n) = 2e^{\mu_p(n) - u_0 c n} K^{(0,1)}(n) \quad , \tag{5.27}$$

where in $K^{(0,1)}(n)$ the chemical potential is shifted as above.

As a further important quantity we consider the density cumulant. Here we have to proceed in two steps. As intermediate quantity we define the subset $\tilde{K}_d^{(2,0)}(\mathbf{q})$ of diagrams which are irreducible with respect to cuts which do not separate the two density insertions. The general structure of $\tilde{K}^{(2,0)}$ is indicated in Fig. 5.7. Clearly

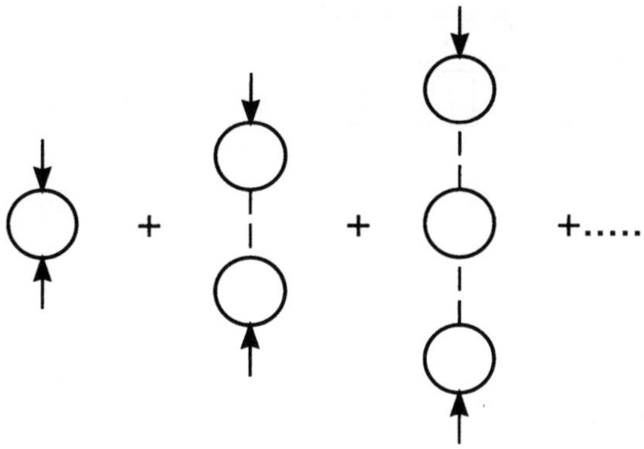

Fig. 5.7. Analysis of $\tilde{K}^{(2,0)}(\mathbf{q})$, representing Eq. (5.29). *Filled circles* stand for vertex-irreducible parts

$$I_{d,c}(\mathbf{q}) = \tilde{K}^{(2,0)}(\mathbf{q}) \quad , \tag{5.28}$$

where it is understood that in evaluating $\tilde{K}^{(2,0)}$ we shift the chemical potential. $\tilde{K}^{(2,0)}(\mathbf{q})$ can be reduced further. We denote by $K^{(2,0)}(\mathbf{q})$ the set of all vertex irreducible diagrams with two density insertions. Figure 5.7 yields

$$\begin{aligned}\tilde{K}^{(2,0)}(\mathbf{q}) &= K^{(2,0)}(\mathbf{q}) - K^{(2,0)}(\mathbf{q})u_0K^{(2,0)}(\mathbf{q}) \\ &\quad + K^{(2,0)}(\mathbf{q})u_0K^{(2,0)}(\mathbf{q})u_0K^{(2,0)}(\mathbf{q}) - \dots \quad .\end{aligned} \tag{5.29}$$

Upon resummation we find

$$I_{d,c}(\mathbf{q}) = \frac{K^{(2,0)}(\mathbf{q})}{1 + u_0K^{(2,0)}(\mathbf{q})} \quad . \tag{5.30}$$

This is a rigorous result generalizing the random phase approximation, as will be shown in Sect. 5.4.

A similar analysis can be carried through for all correlation functions. Only for the grand potential $\Pi[\mu_p]$ itself is the argument not so simple, since an insertion distinguishing some irreducible part is missing. The analysis, presented in Appendix A 5.2, yields

$$\Pi[\mu_p] = K^{(0,0)} + \frac{u_0}{2}c^2 \quad , \tag{5.31}$$

where $K^{(0,0)}$ denotes the set of all vertex irreducible graphs contributing to $\Pi[\mu_p]$ and the shift of $\mu_p(n)$ is understood.

5.2.3 The screened interaction

Having reduced all quantities to their vertex irreducible parts we can introduce the screened interaction. In each order of the loop expansion we have to resum a series of diagrams. For instance, the series of one loop graphs

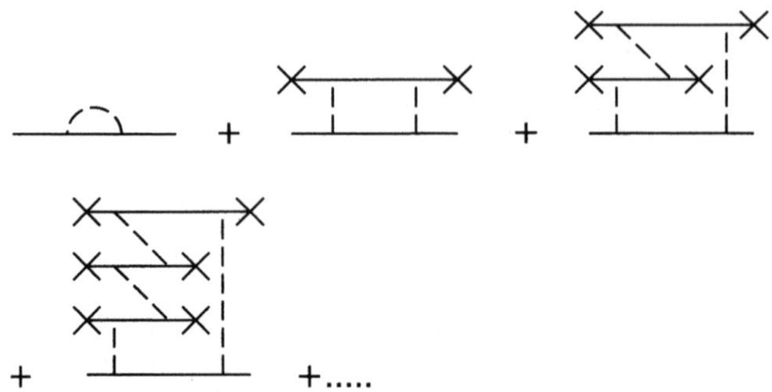

Fig. 5.8. One-loop graphs contributing to $K^{(0,1)}$

contributing to $K^{(0,1)}(n)$ is shown in Fig. 5.8. Clearly this series can be resummed in terms of the 'screened vertex' $\tilde{u}_0(\mathbf{q})$, which diagrammatically is defined in Fig. 5.9. Analytically we find

$$\begin{aligned} -\tilde{u}_0(\mathbf{q}) &= -u_0 + u_0 X(\mathbf{q}^2) u_0 - u_0 X(\mathbf{q}^2) u_0 X(\mathbf{q}^2) u_0 \\ &\quad + u_0 X(\mathbf{q}^2) u_0 X(\mathbf{q}^2) u_0 X(\mathbf{q}^2) u_0 \ldots \\ &= \frac{-u_0}{1 + u_0 X(\mathbf{q}^2)} \quad , \end{aligned} \tag{5.32}$$

where

$$\begin{aligned} X(\mathbf{q}^2) &= \sum_n \frac{2}{(4\pi\ell^2)^{d/2}} \, e^{\mu_p(n) - u_0 c\, n} \sum_{0<j_1<j_2<n} e^{-q^2\ell^2(j_2-j_1)} \\ &= \sum_n \frac{2}{(4\pi\ell^2)^{d/2}} \, e^{\mu_p(n) - u_0 c\, n} \frac{n^2}{2} D(\mathbf{q}^2\ell^2 n) \quad . \end{aligned} \tag{5.33}$$

In the last line we used Eq. (3.22) to introduce the Debye function. We will find in Sect. 5.4 that $\tilde{u}_0(\mathbf{q})$ shows all the aspects of screening. Here we want to show that $\tilde{u}_0(\mathbf{q})$ can be introduced to replace consistently all vertices

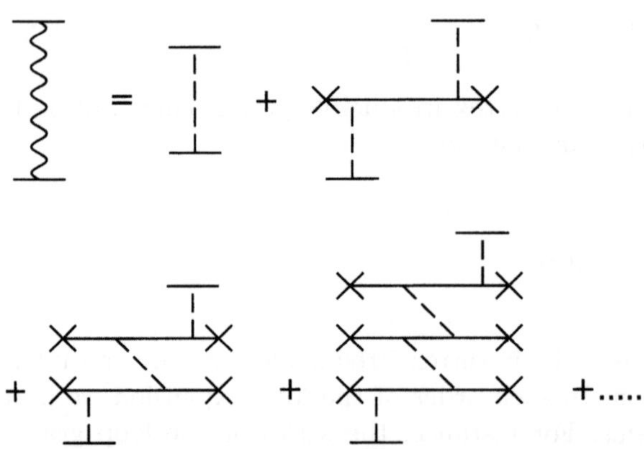

Fig. 5.9. Diagrammatic definition of $\tilde{u}_0(\mathbf{q})$

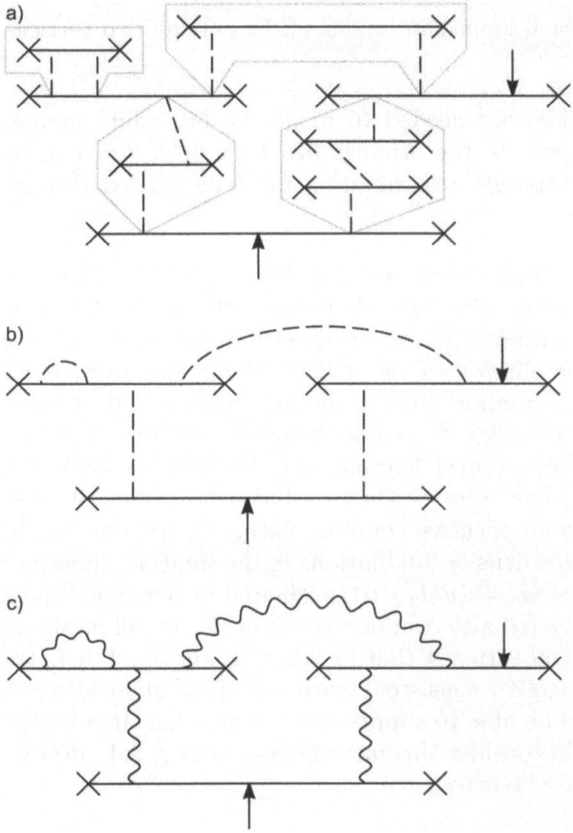

Fig. 5.10. (**a**) A graph containing $\tilde{u}_0(\mathbf{q})$-parts, indicated by the grey boxes. (**b**) The corresponding reduced diagram. (**c**) The corresponding graph containing screened vertices $\tilde{u}_0(\mathbf{q})$

u_0. We concentrate on quantities possessing external vertices. In the corresponding vertex irreducible diagrams we can identify chain-like structures not containing external vertices or internal loops, which can be separated from the diagram by cutting two vertices. Figure 5.10a shows a typical example, where the chain-like structures are included in boxes. Replacing these boxes by simple vertices, we uniquely define a reduced diagram (Fig. 5.10b). The loop expansion resums all diagrams in which each box independently can be replaced by any contribution to the screened vertex (Fig. 5.9). For our example this yields the diagram of Fig. 5.10c. We thus find the following result.

The loop expansion

The loop expansion of a vertex irreducible quantity containing external vertices is represented by all vertex irreducible diagrams containing only screened interactions. Diagrams, in which a chain-like structure not containing external vertices or internal loops can be cut off by cutting two vertices, are forbidden.

Again the external vertices are needed to make the procedure unique. Thus the result does not apply to the osmotic pressure, or $K^{(0,0)}$, equivalently. For this quantity a separate resummation must be carried through (See Appendix A 17.1).

What do we gain from these resummations? Recall that we aim at a reformulation of the cluster expansion such that it can also be used in a region where a given chain simultaneously interacts with many others, so that $W(n) = u_0 c n \gtrsim 1$. We have eliminated two effects giving rise to concern. Firstly we resummed the interaction with a random background of other chains, which yields a shift of order W in the chemical potential. This is a mean field effect, independent of local fluctuations. Secondly we took into account that two segments might interact via an arbitrary sequence of intervening chains, giving rise to an effective coupling $\tilde{u}_0(\mathbf{q})$. $\tilde{u}_0(\mathbf{q}) - u_0$ can be interpreted as being due to the density fluctuations in the solution. Indeed, it is easily checked that $\tilde{u}_0(\mathbf{q}) - u_0 = -u_0^2 I_{d,c}(\mathbf{q})$, evaluated in tree approximation (cf. Eqs. (5.56), (5.60)). $\tilde{u}_0(\mathbf{q})$ also resums a series in W, as will be shown in Sect. 5.4. Furthermore it will turn out that for large W and small $\mathbf{q}R_g$ the coupling $\tilde{u}_0(\mathbf{q})$ is of order u_0/W, consistent with the qualitative ideas on screening. $\tilde{u}_0(\mathbf{q})$ thus should be able to suppress any remaining singularities in the region $W \gg 1$. We will consider this more closely in Sect. 5.4, after we have discussed the tree approximation.

We finally note that an elegant, but more implicit, derivation of the loop expansion can be given in closed form with the help of the so-called auxiliary field method. This will be sketched in Appendix A 5.3.

5.3 Polydispersity: Notation and basic features

A physical solution of long chains always contains some distribution of chain lengths, and to control the resulting 'polydispersity effects' is an important, but experimentally quite difficult issue. In the theory, as we have formulated it up to now, polydispersity is contained in the function $c_p(n)$, giving the concentration of chains of length n. Introducing the total chain concentration

$$c_p = \sum_n c_p(n) \quad , \tag{5.34}$$

we may define the chain length distribution as

$$P(n) = \frac{c_p(n)}{c_p} \quad . \tag{5.35}$$

$P(n)$ gives the probability that a chain, randomly chosen from our solution, has length n. We will often use a 'reduced' distribution $p(y)$ defined as

$$P(n) = \frac{1}{N} p(n/N) \quad . \tag{5.36}$$

Here

$$N = \sum_n n P(n) \tag{5.37}$$

is the number averaged chain length. The reduced distribution obeys the normalization conditions

$$1 = \sum_n P(n) = \int_0^\infty dy\, p(y) \left(1 + O\left(\frac{1}{N}\right)\right) \tag{5.38 i}$$

$$1 = \int_0^\infty dy\, y\, p(y) \left(1 + O\left(\frac{1}{N}\right)\right) \tag{5.38 ii}$$

where in the sequel we will neglect the corrections of order $1/N$. It will turn out that $p(y)$ is the appropriate form of the chain length distribution in a renormalized formalism, since it is invariant under the renormalization group. Note further that the segment concentration c (Eq. (5.14)) can be written as

$$c = N c_p \quad . \tag{5.39}$$

In explicit perturbative expressions, $p(y)$ often occurs in terms of its Laplace transform

$$\tilde{p}(x) = \int_0^\infty dy e^{-xy} p(y) \quad . \tag{5.40}$$

Since any physically reasonable form of $p(y)$ for small y is bounded, $\tilde{p}(x)$ for $x \to \infty$ has to vanish at least as $const/x$. For small x the normalization conditions (5.38) imply

$$\tilde{p}(x) = 1 - x + \frac{\tilde{p}_2}{2} x^2 - \frac{\tilde{p}_3}{6} x^3 + \ldots \quad , \tag{5.41}$$

where $\tilde{p}_2$ etc. are the moments of the reduced chain length distribution. The low order moments are related to standard chain length averages introduced in the literature. In particular the weight average N_w is given by

$$N_w = \frac{1}{N} \sum_n P(n) n^2 = N \tilde{p}_2 \quad , \tag{5.42 i}$$

whereas the so-called z-average N_z takes the form

$$N_z = \frac{1}{N\,N_w}\sum_n P(n)n^3 = \frac{\tilde{p}_3}{\tilde{p}_2}N \ . \tag{5.42 ii}$$

All our results will hold for general $p(y)$, subject only to the condition that the Laplace transform exists. Experimentally $p(y)$ in general is not very well known, but is characterized by some low order moments. An often used ansatz is the so-called Schultz distribution

$$p_\sigma(y) = \frac{\sigma^\sigma}{\Gamma(\sigma)} y^{\sigma-1} e^{-\sigma y} \quad , \tag{5.43}$$

where $\Gamma(\sigma)$ denotes the Gamma function. $p_\sigma(y)$ is uniquely characterized by the second moment

$$\tilde{p}_{2\sigma} = 1 + \frac{1}{\sigma} \quad . \tag{5.44 i}$$

The third moment takes the form

$$\tilde{p}_{3\sigma} = \tilde{p}_{2\sigma}(2\tilde{p}_{2\sigma} - 1) \ . \tag{5.44 ii}$$

The Laplace transform reads

$$\tilde{p}_\sigma(x) = \left(1 + \frac{x}{\sigma}\right)^{-\sigma} \quad . \tag{5.45}$$

Figure 5.11 shows $p_\sigma(y)$ for a set of σ. For $\sigma = 1$, $p_\sigma(y)$ reduces to the exponential distribution, for which all calculations become particularly simple. For $\sigma \to \infty$ it reduces to a monodisperse system. We will often evaluate the analytical results for these limiting cases, which span the range of polydispersity which is relevant experimentally.

In view of the experimental difficulties in determining $p(y)$ it may be helpful to note some rigorous inequalities resulting from the fact that $p(y) \geq 0$, by definition. Using the Schwartz inequality it is easily checked that

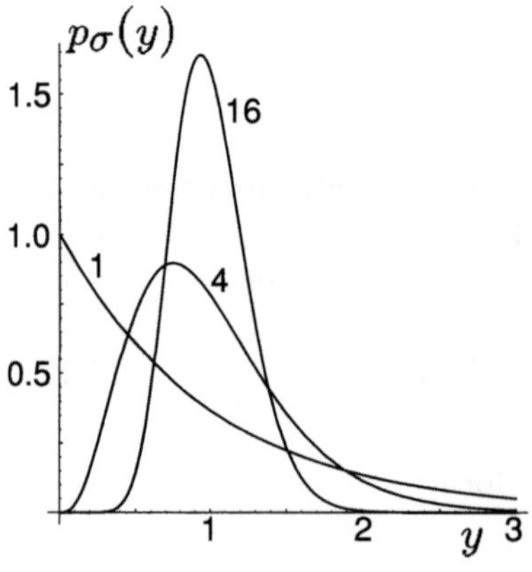

Fig. 5.11. Reduced Schultz-distribution $p_\sigma(y)$ (Eq. (5.43)) for values of $\sigma = 1, 4, 16$, as indicated

$$1 \le \tilde{p}_2, \quad \tilde{p}_2^2 \le \tilde{p}_3 \quad , \tag{5.46}$$

resulting in

$$N \le N_w \le N_z \quad . \tag{5.47}$$

Furthermore, if $p(y)$ is strictly symmetric about $y = 1$ we have

$$\tilde{p}_3 = 3\tilde{p}_2 - 2 \tag{5.48 i}$$

or

$$N_z = N\left(3 - \frac{2N}{N_w}\right) \quad , \tag{5.48 ii}$$

equivalently. Schultz distributions obey this relation up to corrections of order $(\tilde{p}_2 - 1)^2$. Equation (5.48) thus may be a good assumption for fairly sharp distributions.

In many perturbative results there naturally occurs the Debye function averaged over polydispersity. We thus introduce

$$D_p(Nq^2\ell^2) = \sum_n P(n)\left(\frac{n}{N}\right)^2 D(nq^2\ell^2) \quad . \tag{5.49}$$

Up to corrections of order $1/N$ it takes the form

$$D_p(t) = \int_0^\infty dx\, x^2 p(x) D(xt) = \frac{2}{t^2}(\tilde{p}(t) - 1 + t) \quad . \tag{5.50}$$

For small values of the argument we find the expansion

$$D_p(t) = \tilde{p}_2 - \frac{\tilde{p}_3}{3}\, t + O(t^2) \quad . \tag{5.51 i}$$

Due to the asymptotic behavior of $\tilde{p}(t)$, $D_p(t)$ for $t \to \infty$ behaves as

$$D_p(t) = \underset{t \to \infty}{\longrightarrow} \frac{2}{t}\left(1 - \frac{1}{t} + O\left(\frac{1}{t^2}\right)\right) \quad . \tag{5.51 ii}$$

This indicates that correlations deep inside a chain are independent of polydispersity, a feature obvious for noninteracting chains. We will find that this feature persists also if interactions are taken into account.

5.4 Tree approximation, screening, and a one loop example

5.4.1 Tree approximation

A number of well known expressions result from evaluating the theory to lowest order approximation. Figure 5.12 collects the trivial zero-loop diagrams occurring in the evaluation of the vertex irreducible parts in tree approximation. Let us start with the relation among chemical potential and concentrations (Eq. (5.27)). Figure 5.12a gives the tree approximation to $K^{(0,1)}(n)$. Momentum being zero, this diagram just yields the symmetry factor 1/2 and the weight $(4\pi\ell^2)^{-d/2}$. Thus

$$c_p(n) = (4\pi\ell^2)^{-d/2} e^{\mu_p(n) - u_0 cn}(1 + O(1Loop)) \quad , \tag{5.52}$$

or

$$\mu_p(n) - \frac{d}{2}\ln(4\pi) = \ln\left(\ell^d c_p(n)\right) + u_0 nc \quad . \tag{5.53}$$

This result has the form of the well known Flory-Huggins approximation, evaluated in the limit of small concentrations ($\ell^d c \ll 1$). (Recall that $\mu_p(n) = \hat{\mu}_p(n)/k_B T$, where $\hat{\mu}_p$ is the conventionally defined chemical potential.) The term $\ln\left(\ell^d c_p(n)\right)$ can be interpreted as an entropic contribution of noninteracting chains and $u_0 nc$ is the interaction energy with a homogeneous background of other chains. The constant shift $\ln(4\pi)d/2$ is due to our normalization of the spatial integrals in the partition function.

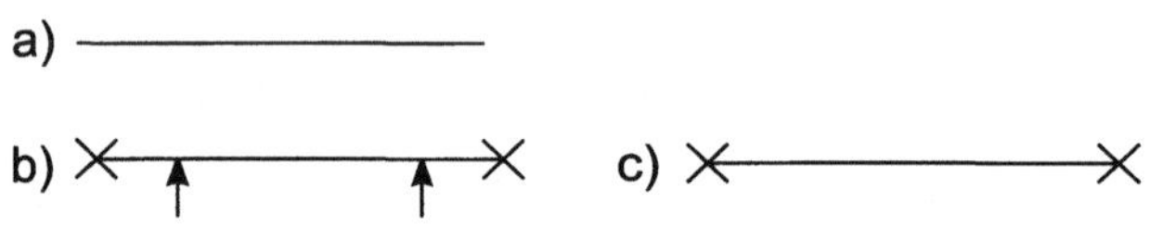

Fig. 5.12a–c. Tree graphs contributing to: **(a)** $K^{(0,1)}(n)$; **(b)** $K_a^{(2,0)}(\mathbf{q})$; **(c)** $K^{(0,0)}$

We now turn to the density autocorrelations. The relevant diagram (Fig. 5.12b) yields

$$\begin{aligned} K_a^{(2,0)}(\mathbf{q}) &= \frac{2}{(4\pi\ell^2)^{d/2}} \sum_n e^{\mu_p(n) - u_0 cn} \sum_{0<j_1<j_2<n} e^{-q^2\ell^2(j_2 - j_1)} \\ &= \sum_n c_p(n) n^2 D(nq^2\ell^2) \quad , \end{aligned}$$

where we used Eq. (5.52). In terms of the generalized Debye function (5.49) this result can be written as

$$I_{d,a}(\mathbf{q}) = K_a^{(2,0)}(\mathbf{q}) = cND_p(N\mathbf{q}^2\ell^2) \quad , \tag{5.54}$$

showing that to lowest order the form factor of a single chain is not modified by the presence of a background of other chains. In tree approximations the other chains do not act back on the chain considered.

Since in tree approximation

$$K_a^{(2,0)}(\mathbf{q}) = K^{(2,0)}(\mathbf{q}) \quad , \tag{5.55}$$

the results (5.54), (5.30) immediately yield the density cumulant:

$$I_{d,c}(\mathbf{q}) = \frac{cND_p(N\mathbf{q}^2\ell^2)}{1 + u_0 cND_p(N\mathbf{q}^2\ell^2)} \quad . \tag{5.56}$$

This is the well known random phase approximation, which plays an important role in the theory of more concentrated solutions.

We finally consider the osmotic pressure $\hat{\Pi}$. The tree approximation for $K^{(0,0)}$ (Fig. 5.12c) yields

$$K^{(0,0)} = \sum_n c_p(n) = c_p \quad . \tag{5.57}$$

Thus

$$\Pi = \frac{\hat{\Pi}}{k_B T} = c_p + \frac{u_0}{2} c^2 \quad . \tag{5.58}$$

In tree approximation the virial expansion of the osmotic pressure stops after the second term.

All these results of the tree approximation coincide with standard expressions, used widely to describe polymer solutions at moderate or higher concentrations. Typically these results can also be derived by other methods. The traditional derivation of the random phase approximation, for instance, uses a kind of self-consistency argument in coupling one chain to its surrounding (see, for instance, [dG79]). We will find later that renormalized perturbation theory to lowest order yields results of the same structure, but with renormalized parameters, depending on chain length, concentration, or temperature via the renormalization group equations. This 'renormalized tree approximation' turns out to be in good qualitative agreement with experiments on dilute or semidilute solutions.

To summarize, the lowest order approximation of the loop expansion yields a consistent derivation of some standard expressions, which are widely used in the description of polymer solutions.

5.4.2 Analysis of the screened interaction

To lowest order we may use Eq. (5.52) to eliminate the chemical potential from Eqs. (5.32), (5.33), which define the screened vertex. We find

$$\tilde{u}_0(\mathbf{q}) = u_0(\mathbf{q})\,(1 + O(1\text{Loop})) \tag{5.59}$$

where

$$u_0(\mathbf{q}) = \frac{u_0}{1 + u_0 c N D_p(\mathbf{q}^2 \ell^2 N)} \quad . \tag{5.60}$$

This expression strongly depends on the value of $W(N) \equiv W = u_0 c N$. In the dilute regime $W \lesssim 1$, we clearly can expand out the denominator to recover the normal expansion in powers of $u_0 \sim \beta_e : u_0(\mathbf{q}) = u_0 + O(u_0^2)$. The regime $W \gg 1$ is more interesting. We find

$$u_0(\mathbf{0}) \quad = \quad \frac{u_0}{1 + W\tilde{p}_2} \sim \frac{u_0}{W\tilde{p}_2} \ll u_0 \tag{5.61 i}$$

$$u_0(\mathbf{q}) \underset{\mathbf{q}\to\infty}{\longrightarrow} u_0 \ . \tag{5.61 ii}$$

Thus $u_0(\mathbf{q})$ varies from $u_0(\mathbf{0}) \ll u_0$ to $u_0(\mathbf{q} \to \infty) \sim u_0$. To extract the typical scale of that variation we determine that momentum q^* where $W D_p(q^{*2}\ell^2 N) \sim 1$. For $W \gg 1$ this momentum is found in the region where D_p can be replaced by its asymptotic behavior (5.51 ii). We thus find

$$q^{*2}\ell^2 N = 2W\left(1 + O\left(\frac{1}{W}\right)\right)$$

or

$$q^{*2}\ell^2 = 2u_0 c \quad , \tag{5.62}$$

corresponding to a typical length scale

$$\xi_E \approx \frac{1}{q^*} = \sqrt{\frac{\ell^2}{2u_0 c}} \quad , \tag{5.63}$$

known as Edward's 'screening length'. Note that ξ_E depends only on $u_0 c$, but is independent of N.

Figure 5.13 shows $u_0(\mathbf{q})/u_0$ for a monodisperse ensemble, using $u_0 c = .001$ and a series of N values. Note, that in the limit $u_0 c > 0, N \to \infty$, i.e. $W \to \infty, u_0(\mathbf{q})$ tends to a well defined limiting function

$$u_\Theta(\mathbf{q}) = \frac{u_0 \mathbf{q}^2 \ell^2}{\mathbf{q}^2 \ell^2 + 2u_0 c} = u_0 \frac{q^2}{q^2 + q^{*2}} \quad , \tag{5.64}$$

independent of polydispersity. In that limit a given coil simultaneously interacts with an infinite number of others.

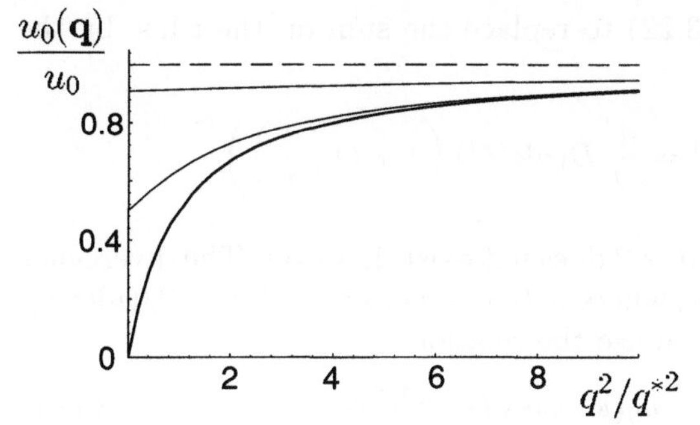

Fig. 5.13. The screened interaction $u_o(\mathbf{q})/u_o$ as function of q^2/q^{*2}. $u_0 c = 0.001; N = 100, 1000, \infty$ (from above). The *broken line* gives $u_0(\mathbf{q}) \equiv u_0$ (dilute limit)

We may summarize this discussion by noting that $u_0(\mathbf{q})$ shows all the aspects of screening. For dilute solutions, $W \lesssim 1$, it essentially equals the full interaction u_0. For $W \gg 1$ and $q \ll q^*$, corresponding to length scales $r \gg \xi_E$, $u_0(\mathbf{q})$ is small compared to u_0, strictly vanishing for $q = 0$ in the limit $W \to \infty$. For $q \gg q^*$, i.e. $r \ll \xi^*$, the full interaction u_0 is recovered. The essential feature is the suppression of $u_0(\mathbf{q})$ for $W \gg 1, |\mathbf{q}| < q^*$. This effect enables the loop expansion to deal with solutions of higher concentration, as will be illustrated now.

5.4.3 One loop correction to the Flory-Huggins equation

To show the loop expansion at work, we evaluate here the relation (5.27) among $\mu_p(n)$ and $c_p(n)$ to one loop order. The diagrams for $K^{(0,1)}(n)$ are shown in Fig. 5.14. The analytic expression reads

$$c_p(n) = \frac{2}{(4\pi\ell^2)^{d/2}} e^{\mu_p(n) - u_0\, cn} \frac{1}{2} \left[1 - \int_k \tilde{u}_0(\mathbf{k}) \sum_{0<j_1<j_2<n} e^{-\mathbf{k}^2\ell^2(j_2-j_1)} + 2\text{Loop} \right] . \tag{5.65}$$

To this order, we may replace $\tilde{u}_0(\mathbf{k})$ by $u_0(\mathbf{k})$ and solve for $\mu_p(n)$:

$$\begin{aligned} \mu_p(n) - \frac{d}{2}\ln(4\pi) &= \ln\left(\ell^d c_p(n)\right) + u_0 cn \\ &\quad + \int_k u_0(\mathbf{k}) \sum_{0<j_1<j_2<n} e^{-\mathbf{k}^2\ell^2(j_2-j_1)} + 2\text{Loop} . \end{aligned} \tag{5.66}$$

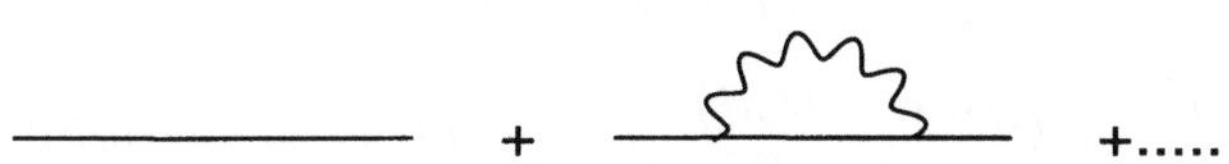

Fig. 5.14. $K^{(0,1)}(n)$ to order of one loop

It now is tempting to use Eq. (3.22) to replace the sum on the r.h.s. by the Debye function:

$$\sum_{0<j_1<j_2<n} e^{-\mathbf{k}^2\ell^2(j_2-j_1)} = \frac{n^2}{2} D(n\mathbf{k}^2\ell^2)\left(1+O\left(\frac{1}{n}\right)\right) .$$

The resulting integral over $\mathbf{k}$ for $d \geq 2$ does not exist, however. The divergence is due to integration over large $\mathbf{k}$, where $u_0(\mathbf{k}) \to u_0$, $D(n\mathbf{k}^2\ell^2) \to 2/(n\mathbf{k}^2\ell^2)$. To extract the dangerous part, we use the relation

$$u_0(\mathbf{k}) = u_0 - u_0(\mathbf{k}) u_0 c N D_p(\mathbf{k}^2\ell^2 N) \tag{5.67}$$

to write

$$\int_k u_0(\mathbf{k}) \sum_{0<j_1<j_2<n} e^{-\mathbf{k}^2\ell^2(j_2-j_1)} = u_0 \sum_{0<j_1<j_2<n} \int_k e^{-\mathbf{k}^2\ell^2(j_2-j_1)}$$
$$-u_0 c N \int_k u_0(\mathbf{k}) D_p(\mathbf{k}^2\ell^2 N) \sum_{0<j_1<j_2<n} e^{-\mathbf{k}^2\ell^2(j_2-j_1)} . \tag{5.68}$$

The first part on the r.h.s. takes the form

$$u_0 \sum_{0<j_1<j_2<n} \int_k e^{-\mathbf{k}^2\ell^2(j_2-j_1)} = u_0 \ell^{-d} \frac{\pi^{d/2}}{(2\pi)^d} \sum_{0<j_1<j_2<n} (j_2-j_1)^{-d/2}$$
$$= \beta_e \Sigma_1(0) ,$$

where we used the definitions (4.13), (5.25) of $\Sigma_1(y)$ or u_0, respectively. In the second term on the r.h.s. of Eq. (5.68) to leading order in n we may replace the sum by the Debye function, provided $d < 4$. We here restrict the analysis to such values of d. Using Eq. (A 4.13) for $\Sigma_1(0)$ and omitting terms of order n^0 we find

$$\mu_p(n) - n\beta_e\zeta\left(\frac{d}{2}\right) = \ln\left(\ell^d c_p(n)\right) - \beta_e n^{\varepsilon/2}\left(\frac{2}{\varepsilon} + \frac{1}{1-\varepsilon/2}\right)$$
$$+ u_0 c n \left[1 - \frac{Nn}{2} u_0 \int_k \frac{D_p(\mathbf{k}^2\ell^2 N) D(\mathbf{k}^2\ell^2 n)}{1 + W D_p(k^2\ell^2 N)}\right] ,$$

where we again use the notation $W = u_0 c N$. Introducing the integration variable $t = \mathbf{k}^2\ell^2 N$ and using the results of Appendix A 3.3 to evaluate the integral over the unit sphere we find the final result

$$\mu_p(n) - n\beta_e\zeta\left(\frac{d}{2}\right) = \ln\left(\ell^d c_p(n)\right) - \beta_e n^{\varepsilon/2}\left(\frac{2}{\varepsilon} + \frac{1}{1-\varepsilon/2}\right)$$
$$+ u_0 c n \left[1 - \beta_e \frac{N^{\varepsilon/2}}{2\Gamma(d/2)} J\left(\frac{n}{N}, W\right)\right] \tag{5.69}$$

$$J\left(\frac{n}{N}, W\right) = \frac{n}{N} \int_0^\infty dt\; t^{d/2-1} \frac{D_p(t)}{1 + W D_p(t)} D\left(\frac{n}{N} t\right) . \tag{5.70}$$

This result deserves some detailed discussion. It shows two different types of one loop corrections. Corrections of the first type are independent of the segment concentration c, i.e. they are due to the intrachain interactions. The leading term is a shift proportional to n, resulting from interactions among segments closely spaced along the chain. It is corrected by terms of the order $n^{\varepsilon/2}$. The other one loop corrections are due to interchain interactions, changing the coefficient of u_0cn. Recall that in Flory-Huggins theory this term is a mean field estimate representing the energy of interaction with a homogeneous background of monomers. The correction is thus due to fluctuations of the monomer density. Let us consider the integral J (Eq. (5.70)) more closely. In the dilute region $W \ll 1$ it can be expanded to yield a series in powers of W, which in view of $W \sim c$ is nothing but a virial expansion. Clearly this expansion could have been constructed directly from the results of Sect. 5.1, not ordering the diagrams in terms of the number of loops. The expansion breaks down in the limit $c = const$, $\frac{n}{N} = const$, $N \to \infty$, i.e. $W \to \infty$. Indeed, we might even suspect that the loop expansion breaks down in that limit, the correction J in Eq. (5.69) being multiplied by a divergent prefactor $N^{\varepsilon/2}$. However it turns out that the denominator in the integrand of J suppresses the integral so as to compensate this divergence. This is a consequence of screening, since the denominator results from $u_0(\mathbf{k})$. Basically we observe that $D_p(t)$ in the denominator of Eq. (5.70) must become small. Otherwise the contribution is suppressed by a factor $const/W$. For $D_p(t)$ to become small, t must become large, so that we can use the asymptotic expansion (5.51ii) of the Debye function. For $W \to \infty$ to leading order the integrand therefore can be approximated as

$$t^{d/2-1}\frac{D_p(t)}{1+WD_p(t)}D\left(\frac{n}{N}t\right) \to 4\frac{N}{n}\frac{t^{d/2-2}}{t+2W} \quad .$$

Thus

$$\begin{aligned} J\left(\frac{n}{N}, W\right) \underset{W \to \infty}{\longrightarrow} & \; 4\int_0^\infty dt\, \frac{t^{d/2-2}}{t+2W} \\ = & \; 4(2W)^{-\varepsilon/2}\int_0^\infty d\tau \frac{\tau^{-\varepsilon/2}}{\tau+1} \\ = & \; 4\Gamma(1-\varepsilon/2)\Gamma(\varepsilon/2)(2W)^{-\varepsilon/2} \quad . \end{aligned} \tag{5.71}$$

For $W \gg 1$ the factor correcting the interaction energy u_0cn in Eq. (5.69) therefore reduces to

$$1 - 2\beta_e\frac{\Gamma(\varepsilon/2)}{1-\varepsilon/2}(2u_0c)^{-\varepsilon/2} \quad , \tag{5.72}$$

exhibiting a finite and small correction for finite concentration c and small interaction constant β_e. We should note that in terms of the original integral over momentum $\mathbf{k}$ the asymptotic correction is dominated by the range

$\mathbf{k}^2\ell^2 N = t \sim 2W$. In view of Eqs. (5.62), (5.63) this range can also be characterized as $\mathbf{k}^2 \sim q^{*2}$ or $r \sim \xi_E \ll R_g$, equivalently, which is consistent with the qualitative discussion at the beginning of this chapter.

This calculation illustrates how the loop expansion cures the divergences which for a system of many strongly interacting chains plague the straightforward cluster expansion. It should be clear that the analysis is completely consistent with the qualitative ideas on screening. It must be noted, however, that here we have given no rigorous proof of the validity of the method, showing that the limit $W \to \infty$ exists to *all orders* of the loop expansion. A most detailed, but still incomplete, analysis using the field theoretic approach to polymer solutions, may be found in [Sch87]. A formal proof of the existence of the limit $W \to \infty$ to all orders of the loop expansion has been given in [Sch90] in a somewhat different context.

A 5. Appendix

A 5.1 Linked cluster theorem

Consider some function $G(x)$ having a formal Taylor series expansion

$$G(x) = 1 + \sum_{M=1}^{\infty} \frac{x^M}{M!} G^{(M)} \quad . \tag{A 5.1}$$

To define the 'cumulant' $G_c^{(M)}$, we represent $G^{(M)}$ by M points contained in a box, and we draw all diagrams in which these points are grouped together in an arbitrary way (Fig. 5.15). A group of m connected points stands for the cumulant $G_c^{(m)}$. Thus Fig. 5.15, for instance, stands for

$$G^{(4)} = (G_c^{(1)})^4 + 6(G_c^{(1)})^2 G_c^{(2)} + 3(G_c^{(2)})^2 + 4G_c^{(1)}G_c^{(3)} + G_c^{(4)} \quad . \tag{A 5.2}$$

With the initial condition

$$G_c^{(1)} = G^{(1)} \tag{A 5.3}$$

this defines a recursive scheme to calculate the $G_c^{(m)}$ in terms of the $G^{(M)}$, $m \leq M$.

Now, consider all contributions to $G^{(M)}$ consisting of $n_m, 1 \leq m \leq M$ m-point cumulants. Clearly

$$\sum_m n_m m = M \quad . \tag{A 5.4}$$

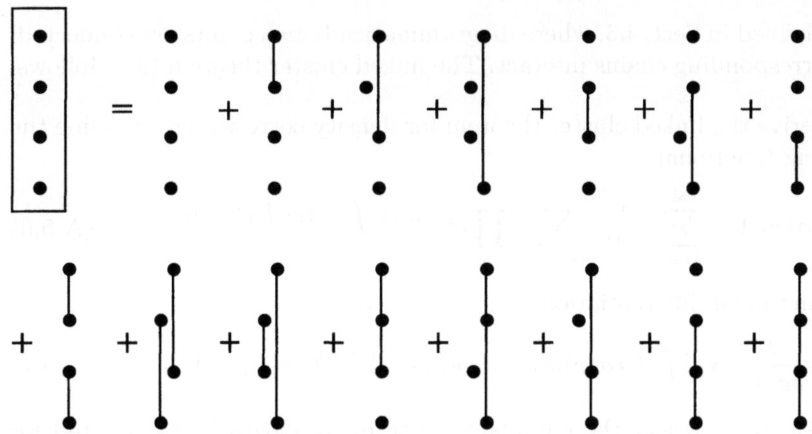

Fig. 5.15. Decomposition of $G^{(4)}$ into cumulants

Each such contribution yields a factor

$$\prod_m \left(G_c^{(m)}\right)^{n_m} ,$$

their number being

$$M! \prod_m \left((m!)^{-n_m} \frac{1}{n_m!}\right) .$$

Thus

$$G(x) = 1 + \sum_{M=1}^{\infty} x^M \sum_{\substack{n_1 \dots n_M \\ \sum_m n_m m = M}} \prod_m \left[\left(\frac{G_c^{(m)}}{m!}\right)^{n_m} \frac{1}{n_m!}\right] .$$

The summation over M eliminates the constraint $\Sigma_m n_m m = M$, and $G(x)$ takes the form

$$\begin{aligned} G(x) &= \sum_{n_1, n_2, \dots = 0}^{\infty} \prod_m \left[\left(\frac{x^m}{m!} G_c^{(m)}\right)^{n_m} \frac{1}{n_m!}\right] \\ &= \exp\left\{\sum_{M=1}^{\infty} \frac{x^M}{M!} G_c^{(M)}\right\} . \end{aligned} \tag{A 5.5}$$

Consequently the cumulants are the coefficients of the Taylor series of $\ln G(x)$.

The application to the grand partition function is obvious. We in Eq. (5.2) formally multiply $Z^{(M)}$ by x^M, thus defining a function $\mathcal{Z}(x, [\mu])$ reducing to $\mathcal{Z}[\mu]$ for $x = 1$. The cumulants as defined here are identical to the cumulants

$G_c^{(0,M)}$ defined in Sect. 4.3, where diagrammatically two points are connected, if the corresponding chains interact. The linked cluster theorem (5.7) follows.

To derive the linked cluster theorem for density correlations we define the generating functional

$$\mathcal{Z}[\mu_p, \sigma] = 1 + \sum_{M=1}^{\infty} \frac{1}{M!} \sum_{n_1 \ldots n_m} \prod_m e^{\mu_p(n_m)} \int e^{-\mathcal{H} + \int d^d \hat{r} \sigma(\hat{r}) \hat{\rho}(\hat{r})} \quad . \tag{A 5.6}$$

Using functional differentiation

$$\frac{\delta}{\delta \sigma(\mathbf{r})} \exp[\int d^d \hat{r} \sigma(\hat{r}) \rho(\hat{r})] = \rho(\mathbf{r}) \exp[\int d^d \hat{r} \sigma(\hat{r}) \hat{\rho}(\hat{r})] \tag{A 5.7}$$

we can easily represent the cumulants in terms of derivatives evaluated for $\sigma(\hat{r}) \equiv 0$. We find for instance

$$\begin{aligned} &\frac{\delta}{\delta \sigma(\mathbf{r})} \frac{\delta}{\delta \sigma(\mathbf{r}')} \ln \mathcal{Z}[\mu_p, \sigma] \Big|_{\sigma \equiv 0} \\ &= \left(\frac{\frac{\delta}{\delta \sigma(\mathbf{r})} \frac{\delta}{\delta \sigma(\mathbf{r}')} \mathcal{Z}[\mu_p, \sigma]}{\mathcal{Z}[\mu_p]} - \frac{\frac{\delta}{\delta \sigma(\mathbf{r})} \mathcal{Z}[\mu_p, \sigma]}{\mathcal{Z}[\mu_p]} \cdot \frac{\frac{\delta}{\delta \sigma(\mathbf{r}')} \mathcal{Z}[\mu_p, \sigma]}{\mathcal{Z}[\mu_p]} \right)_{\sigma \equiv 0} \\ &= \left\langle \hat{\rho}(\mathbf{r}) \hat{\rho}(\mathbf{r}') \right\rangle - \left\langle \hat{\rho}(\mathbf{r}) \right\rangle \left\langle \hat{\rho}(\mathbf{r}') \right\rangle = \hat{I}_{d,c}(\mathbf{r} - \mathbf{r}') \quad . \end{aligned} \tag{A 5.8}$$

The generating term $\int \sigma(\mathbf{r}) \hat{\rho}(\mathbf{r})$ is diagrammatically represented by a wiggly arrow (cf. Fig. 5.18) inserted into polymer lines. We then without any change can take over the discussion given above for $\mathcal{Z}[\mu_p]$ to find that $\ln \mathcal{Z}[\mu_p, \sigma]$ is given by the set of all connected diagrams with any number of $(\sigma \cdot \hat{\rho})$ - insertions. Functional differentiation reduces a $(\sigma \cdot \hat{\rho})$ - insertion to a density insertion. The result for $\hat{I}_{d,c}$ then is found from Eq. (A 5.8).

Defining an appropriate generating functional we can modify the argument to deal with any correlation function. For density autocorrelation, for instance, we may introduce a generating term $\int d^d \hat{r} d^d \hat{r}' \sigma(\hat{\mathbf{r}}, \hat{\mathbf{r}}') \sum_m \hat{\rho}_m(\hat{\mathbf{r}}) \hat{\rho}_m(\hat{\mathbf{r}}')$.

A 5.2 Vertex reducibility for the grand potential

The argument of Sect. 5.2.2 cannot straightforwardly be applied to $\Pi[\mu_p]$, since the diagrams are missing a distinguished part. Indeed, the general structure of a diagram contributing to $\Pi[\mu_p]$ is indicated in Fig. 5.16. It consists of vertex irreducible parts connected by reducible vertices to yield a tree-like structure. No irreducible part is distinguished.

To solve this problem we denote by n_I the number of irreducible parts and by n_R the number of reducible vertices, and we note that

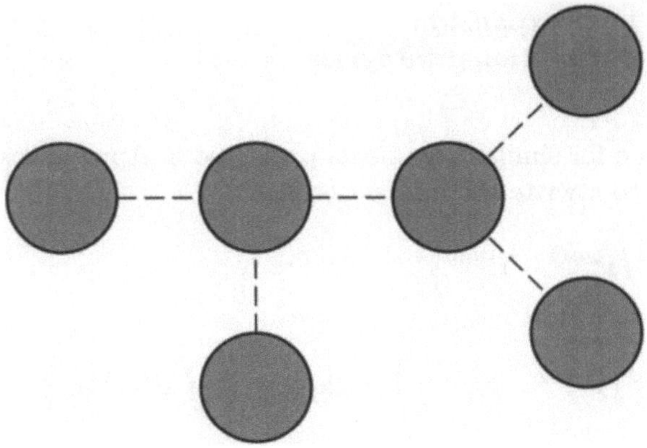

Fig. 5.16. Structure of a general diagram contribution to $\Pi[\mu_p]$. The *filled parts* are vertex-irreducible

$$n_I - n_R = 1 \quad , \tag{A 5.9}$$

by virtue of the tree structure. We now successively mark all irreducible parts or all reducible vertices. The sum of all diagrams with marked irreducible parts minus the sum of all vertex-marked diagrams by virtue of Eq. (A 5.9) just yields the original contribution (see Fig. 5.17). The marking distinguishes exactly one part. The sum of all diagrams where an irreducible part has been marked just yields $K^{(0,0)}$, the shift $\mu_p(n) \to \mu_p(n) - u_0 c n$ being understood. The sum of all diagrams with a marked vertex yields $-\frac{1}{2}u_0\langle\hat{\rho}(\mathbf{r})\rangle^2 = -\frac{1}{2}u_0 c^2$, where the factor 1/2 takes into account the symmetry of the vertex. Equation (5.31) results:

$$\Pi[\mu_p] = K^{(0,0)} + \frac{u_0}{2}\, c^2 \quad .$$

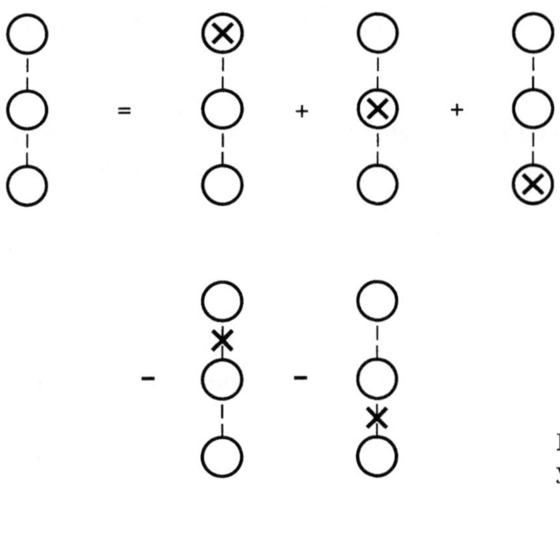

Fig. 5.17. Example of the analysis leading to Eq. (5.31)

A 5.3 Derivation of the loop expansion by formal manipulation of functional integrals

For the present derivation and for similar arguments presented in Appendices A 5.4, A 7.1 it is preferable to rewrite the interaction as

$$\prod_{(j_1 m_1, j_2 m_2)}{}' \left[1 - (4\pi\ell^2)^{d/2} \beta_e \delta^d \left(\mathbf{r}_{j_1}^{(m_1)} - \mathbf{r}_{j_2}^{(m_2)}\right)\right]$$

$$\to \exp\left(-\frac{u_0}{2} \sum_{(j_1,m_1,j_2,m_2)} \delta^d \left(\mathbf{r}_{j_2}^{(m_2)} - \mathbf{r}_{j_1}^{(m_1)}\right)\right) = \exp\left(-\frac{u_0}{2} \int d^d r \hat{\rho}^2(\mathbf{r})\right). \tag{A 5.10}$$

Summing over all pairs of segments in the solution we strictly speaking run into problems with the δ-function interaction, which could be avoided by using a potential of finite range (cf. Sect. 2.2). Furthermore the constant u_0 in principle differs from the definition (5.25). We here ignore all such complications, which do not affect the essential features of the derivation. We rather argue on a purely formal level.

We will derive here the loop expansion for the generating functional of density correlations, which for M chains is defined as

$$\mathcal{Z}^{(M)}[\sigma] = \int D\left[\mathbf{r}^{(1)}, \ldots, \mathbf{r}^{(M)}\right] \exp\left\{-\frac{1}{4\ell^2} \sum_{m=1}^{M} \sum_{j=1}^{n_m} \left(\mathbf{r}_j^{(m)} - \mathbf{r}_{j-1}^{(m)}\right)^2 \right.$$
$$\left. - \frac{u_0}{2} \int d^d r \hat{\rho}^2(\mathbf{r}) + \int d^d r \sigma(\mathbf{r}) \hat{\rho}(\mathbf{r})\right\} . \tag{A 5.11}$$

The density correlations can be derived by functional differentiation, as indicated in Sect. A 5.1.

A 5.3.1 The method of auxiliary fields. Our first aim is to decouple the chains by linearizing the interaction term. We introduce an auxiliary field $\varphi(\mathbf{r})$ obeying

$$\int D[\varphi] e^{-\frac{1}{2u_0} \int d^d r \varphi^2(\mathbf{r})} = 1 \ , \tag{A 5.12}$$

where the functional integral $\int D[\varphi]$ needs some explanation. We assume the system to be confined to a d-dimensional cube $\Omega = L^d$ of linear size L, and we impose periodic boundary conditions. Introducing the discrete set of basis functions

$$\frac{1}{\Omega^{1/2}} e^{-i\mathbf{k}\mathbf{r}}, \ \mathbf{k} = \frac{2\pi}{L}(\kappa_1, \ldots, \kappa_d), \ \kappa_\alpha \varepsilon \mathbb{Z} \ , \tag{A 5.13}$$

we can represent $\varphi(\mathbf{r})$ by a Fourier series

$$\varphi(\mathbf{r}) = \frac{1}{\Omega^{1/2}} \sum_{\mathbf{k}} \varphi_{\mathbf{k}} e^{-i\mathbf{k}\mathbf{r}} \quad . \tag{A 5.14}$$

The integral (A 5.12) is defined as the properly normalized integration over the Fourier components

$$D[\varphi] \equiv \frac{1}{\sqrt{2}} \prod_{\mathbf{k}} \frac{d\varphi_{\mathbf{k}}}{\sqrt{u_0 \pi}} \quad . \tag{A 5.15}$$

To complete the definition we note that $\varphi(\mathbf{r})$ is real, implying

$$\varphi^*_{\mathbf{k}} = \varphi_{-\mathbf{k}} \quad . \tag{A 5.16}$$

For $k \neq 0$ the definition

$$d\varphi_{\mathbf{k}} d\varphi_{-\mathbf{k}} \equiv dRe\varphi_{\mathbf{k}} dIm\varphi_{\mathbf{k}}$$

is understood.

In the functional integral (A 5.12) we may shift $\varphi(\mathbf{r})$ as in normal Gaussian integrals. Substituting

$$\varphi(\mathbf{r}) \to \varphi(\mathbf{r}) + iu_0 \left(\hat{\rho}(\mathbf{r}) - \frac{1}{u_0} \sigma(\mathbf{r}) \right)$$

we find

$$1 = \int D[\varphi] \exp\Big\{ -\frac{1}{2u_0} \int d^d r \Big[\varphi^2(\mathbf{r}) + 2iu_0 \varphi(\mathbf{r}) \left(\hat{\rho}(\mathbf{r}) - \frac{1}{u_0}\sigma(\mathbf{r}) \right) - u_0^2 \left(\hat{\rho}^2(\mathbf{r}) - \frac{2}{u_0} \hat{\rho}(\mathbf{r}) \sigma(\mathbf{r}) + \frac{1}{u_0^2} \sigma^2(\mathbf{r}) \right) \Big] \Big\} \quad .$$

Inserting this result into Eq. (A 5.11) we find

$$\mathcal{Z}^{(M)}[\sigma] = e^{\frac{1}{2u_0} \int d^d r \sigma^2(\mathbf{r})} \int D[\varphi] e^{-\frac{1}{2u_0} \int d^d r \varphi^2(\mathbf{r}) + \frac{i}{u_0} \int d^d r \varphi(\mathbf{r}) \sigma(\mathbf{r})} \cdot \prod_{m=1}^{M} \mathcal{Z}_0(n_m, [-i\varphi]) \quad , \tag{A 5.17}$$

where

$$\mathcal{Z}_0(n, [\sigma]) = \int D[\mathbf{r}] \exp\Big\{ -\frac{1}{4\ell^2} \sum_{j=1}^{n} (\mathbf{r}_j - \mathbf{r}_{j-1})^2 + \int d^d r \sigma(\mathbf{r}) \hat{\rho}(\mathbf{r}) \Big\} \tag{A 5.18}$$

is nothing but the generating functional of density correlations for a single Gaussian chain of length n.

Equation (A 5.17) factorizes in the M chains. We can therefore easily resum the grand canonical generating functional $\mathcal{Z}[\mu_p, \sigma]$ (Eq. (A 5.6)) to find

$$\mathcal{Z}[\mu_p, \sigma] = e^{\frac{1}{2u_0} \int d^d r \sigma^2(\mathbf{r})} \int D[\varphi] e^{-S[\varphi] + \frac{i}{u_0} \int d^d r \varphi(\mathbf{r}) \sigma(\mathbf{r})} \qquad \text{(A 5.19)}$$

$$S[\varphi] = \frac{1}{2u_0} \int d^d r \varphi^2(\mathbf{r}) - \sum_n e^{\mu_p(n)} \mathcal{Z}_0(n, [-i\varphi]) \ . \qquad \text{(A 5.20)}$$

To understand that result we note that $\sum_n e^{\mu_p(n)} \mathcal{Z}_0(\ldots)$ can be represented by all diagrams with a single polymer line closed by crosses and any number of density-type insertions coupled to $-i\varphi(\mathbf{r})$ (cf. Fig. 5.18). After expanding $\exp(-S[\varphi])$ in powers of $\sum e^{\mu_p} \mathcal{Z}_0$, the remaining integral over the Gaussian fluctuating field $\varphi(\mathbf{r})$ pairwise ties together these insertions, weighting each pair by a factor of u_0. The φ-integral thus creates the vertices. The field φ itself can be given a physical meaning by taking in $\mathcal{Z}[\mu_p, \sigma]$ the derivative with respect to $\sigma(\mathbf{r})$. Equations (A 5.6), (A 5.19) yield

$$\langle \hat{\rho}(\mathbf{r}) \rangle = \frac{\delta}{\delta \sigma(\mathbf{r})} \mathcal{Z}[\mu_p, \sigma] \Big|_{\sigma \equiv 0} = \frac{i}{u_0} \langle \varphi(\mathbf{r}) \rangle^S \quad , \qquad \text{(A 5.21)}$$

where the last expectation value is taken with respect to 'action' $S[\varphi]$. Thus the expectation value of φ up to a prefactor coincides with the segment density. A more direct interpretation is found from noting that in the derivation we could have shifted $\varphi(\mathbf{r})$ as $\varphi(\mathbf{r}) \to \varphi(\mathbf{r}) + i u_0 \hat{\rho}(\mathbf{r})$, omitting the contribution $\sim \sigma(\mathbf{r})$. As a result, $\sigma(\mathbf{r})$ is suppressed in Eq. (A 5.19), showing up only in Eq. (A 5.20) where $\mathcal{Z}_0(n, [-i\varphi])$ is to be replaced by $\mathcal{Z}_0(n, [-i\varphi + \sigma])$. Since $\sigma(\mathbf{r})$ in Eq. (A 5.11) plays the role of a (space-dependent) chemical potential this shows that $(-i\varphi(\mathbf{r}))$ can be viewed as a locally fluctuating contribution to the chemical potential of polymer segments.

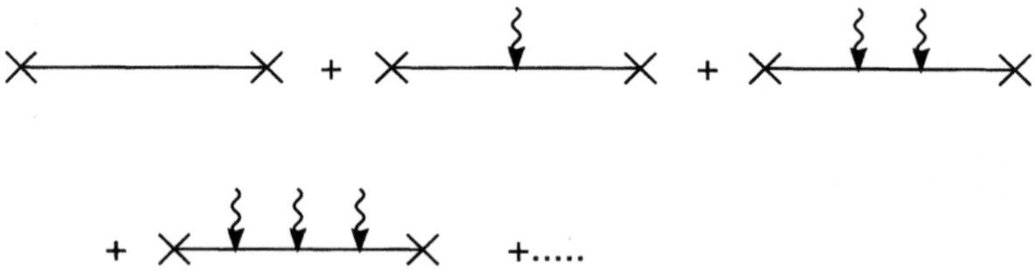

Fig. 5.18. Expansion of $\Sigma_n e^{\mu_p(n)} \mathcal{Z}_0(n, [-i\varphi])$ in powers of φ

A useful relation should be noted. Splitting $\varphi(\mathbf{r})$ into a constant part $\varphi^{(0)}$ and a remainder $\hat{\varphi}(\mathbf{r})$ we find from Eq. (3.18)

$$\int d^d r \varphi(\mathbf{r}) \hat{\rho}(\mathbf{r}) = \varphi^{(0)} n + \int d^d r \hat{\varphi}(\mathbf{r}) \hat{\rho}(\mathbf{r}) \ .$$

Thus

$$\mathcal{Z}_0(n, [-i\varphi^{(0)} - i\hat{\varphi}(\mathbf{r})]) = e^{-in\varphi^{(0)}} \mathcal{Z}_0(n, [-i\hat{\varphi}(\mathbf{r})]) \quad , \qquad \text{(A 5.22)}$$

a result to be used below.

The representations (A 5.17)–(A 5.20) are most useful in deriving the relation of polymer theory to field theory of critical phenomena and allow for a fairly simple analysis of polymer solutions in a canonical ensemble. They are also adapted to a derivation of the loop expansion, as will be discussed now.

A 5.3.2 Formal derivation of the loop expansion. To construct the loop expansion we look for the extrema of $S[\varphi]$, being guided by the idea that the functional integral is dominated by the neighborhood of the minimum of $S[\varphi]$, i.e. the maximum of $\exp(-S[\varphi])$. To shortcut the discussion from the outset we assume that the minimizing function is a constant

$$\varphi(\mathbf{r}) \equiv \varphi^{(0)} \quad ,$$

so that

$$S[\varphi^{(0)}] = \frac{\Omega}{2u_0}\varphi^{(0)2} - \frac{\Omega}{(4\pi\ell^2)^{d/2}} \sum_n e^{\mu_p(n) - in\varphi^{(0)}} \quad . \qquad \text{(A 5.23)}$$

The form of the last term results from Eq. (A 5.22) together with

$$\mathcal{Z}_0(n, [0]) \equiv \mathcal{Z}_0(n) = \frac{\Omega}{(4\pi\ell^2)^{d/2}}$$

(Eq. (3.39)). The condition for an extremum reads

$$\frac{1}{\Omega}\frac{\partial S}{\partial \varphi^{(0)}} = 0 = \frac{\varphi^{(0)}}{u_0} + i\left(4\pi\ell^2\right)^{-d/2} \sum_n n e^{\mu_p(n) - in\varphi^{(0)}} \quad . \qquad \text{(A 5.24)}$$

This should be compared to Eq. (5.52), multiplied by n and summed over n:

$$c = \left(4\pi\ell^2\right)^{-d/2} \sum_n n e^{\mu_p(n) - u_0 n c} \quad .$$

We conclude that

$$\varphi^{(0)} = -iu_0 c \qquad \text{(A 5.25)}$$

holds in tree approximation, this result being consistent with Eq. (A 5.21).

We now introduce the fluctuations

$$\chi(\mathbf{r}) = \varphi(\mathbf{r}) - \varphi^{(0)} \quad , \qquad \text{(A 5.26)}$$

and we rewrite the functional integral in (A 5.19) terms of χ. Using Eq. (A 5.22) we find

$$\mathcal{Z}[\mu_p, \sigma] = e^{\frac{1}{2u_0}\int d^d r \sigma^2(r) + i\frac{\varphi^{(0)}}{u_0}\int d^d r \sigma(\mathbf{r}) - S(\varphi^{(0)})} \cdot \int D[\chi] e^{-\hat{S}[\chi] + \frac{i}{u_0}\int d^d r \chi(\mathbf{r})\sigma(\mathbf{r})} , \quad \text{(A 5.27)}$$

where

$$\hat{S}[\chi] = \frac{1}{2u_0}\int d^d r \chi^2(r) - \sum_n e^{\mu_p(n) - in\varphi^{(0)}} \hat{\mathcal{Z}}_0(n, [-i\chi]) \ . \quad \text{(A 5.28)}$$

$\hat{\mathcal{Z}}_0$ derives from $\mathcal{Z}_0$ by subtracting the terms of order χ^0 or χ^1. The first one is contained in $S(\varphi_0)$, the second one cancels due to condition (A 5.24).

Now consider the expansion of $\hat{\mathcal{Z}}_0(n, [-i\chi])$. The generating term can be written as

$$-i \int d^d r \chi(\mathbf{r})\hat{\rho}(\mathbf{r}) = -i \sum_{j=1}^{n} \chi(r_j) \ .$$

Expanding in that term we find

$$\begin{aligned}
&\hat{\mathcal{Z}}_0(n, [-i\chi]) \\
&= \frac{1}{(4\pi\ell^2)^{d/2}} \Big\{ - \int d^d r_0 d^d r_1 d^d r_2 d^d r_n \sum_{j_1<j_2} G_0(\mathbf{r}_1 - \mathbf{r}_0, j_1)\chi(\mathbf{r}_1) \\
&\cdot G_0(\mathbf{r}_2 - \mathbf{r}_1, j_2 - j_1)\chi(\mathbf{r}_2) G_0(\mathbf{r}_n - \mathbf{r}_2, n - j_2) \\
&+ i \int d^d r_0 d^d r_1 d^d r_2 d^d r_3 d^d r_n \sum_{j_1<j_2<j_3} G_0(\mathbf{r}_1 - \mathbf{r}_0, j_1)\chi(\mathbf{r}_1) G_0(\mathbf{r}_2 - \mathbf{r}_1, j_2 - j_1) \\
&\chi(\mathbf{r}_2) G_0(\mathbf{r}_3 - \mathbf{r}_2, j_3 - j_2)\chi(\mathbf{r}_3) G_0(\mathbf{r}_n - \mathbf{r}_3, n - j_3) + \ldots \Big\} \ .
\end{aligned} \quad \text{(A 5.29)}$$

Using the Fourier series for χ analogous to Eq. (A 5.14) we can write the quadratic terms as

$$\frac{-1}{(4\pi\ell^2)^{-d/2}} \sum_{\mathbf{k}} \chi_{\mathbf{k}} \chi_{-\mathbf{k}} \frac{n^2}{2} D(\mathbf{k}^2 \ell^2 n) \ .$$

The contribution to $\hat{S}[\chi]$ quadratic in χ therefore takes the form

$$\frac{1}{2} \sum_{\mathbf{k}} \chi_{\mathbf{k}} \frac{1}{u_0} \left(1 + u_0 \sum_n \frac{e^{\mu_p(n) - in\varphi^{(0)}}}{(4\pi\ell^2)^{d/2}} n^2 D(\mathbf{k}^2 \ell^2 n) \right) \chi_{-\mathbf{k}} \quad \text{(A 5.30)}$$

Comparison to Eqs. (5.32), (5.33) reveals that the coefficient of $\chi_{\mathbf{k}}\chi_{-\mathbf{k}}$ is just the inverse of the screened vertex $\tilde{u}_0(\mathbf{k})$, with $u_0 c$ replaced by $i\varphi^{(0)}$ (cf. Eq. (A 5.25)).

The calculation of the functional integral now proceeds by expanding out all the higher contributions to $\hat{\mathcal{Z}}_0(n, [-i\chi])$. This yields a set of closed polymer lines with any number $t \geq 3$ of $(\chi\rho)$-insertions. The Gaussian integral over χ again ties these insertions together in pairs, the connection representing a screened vertex. We thus recover the loop expansion with the only modification that $i\varphi^{(0)}/u_0$ replaces c. This difference is compensated by some additional (so-called 'tadpole') graphs, the present expansion not being vertex irreducible. It, however, does not mess up the loop expansion, since the difference $c - i\varphi^{(0)}/u_0$ is at least of one loop order. The occurrence of that difference could be avoided by expanding $\varphi(\mathbf{r})$ about $-iu_0c$ instead of $\varphi^{(0)}$. In this formulation the basic fluctuating field $\chi(\vec{r})$ describes fluctuations of the local chemical potential about its average value $\varphi^{(0)}$, which is fixed by the concentration. This shows that the loop expansion indeed expands in local density fluctuations.

A 5.4 A short discussion of the canonical formalism

A 5.4.1 Manipulation of closed form expressions. We consider the canonical partition function $\mathcal{Z}^{(M)}$ of a system of M chains in volume Ω. The number of chains of length n is denoted by M_n, and clearly the relations

$$\begin{aligned}\sum_n M_n &= M \quad ; \\ \frac{1}{\Omega} M_n &= c_p(n) \\ \frac{1}{\Omega}\sum_n n\, M_n &= c \end{aligned} \qquad \text{(A 5.31)}$$

hold. Introducing factors $1/M_n!$ which take into account the equivalence of chains of the same length, and using the auxiliary field representation we can write

$$\begin{aligned}\mathcal{Z}^{(M)} = \prod_n \frac{1}{M_n!} \int D[\varphi] \exp\Big\{ &- \frac{1}{2u_0} \int d^d r \varphi^2(\mathbf{r}) \\ &+ \sum_n M_n \ln \mathcal{Z}_0(n, [-i\varphi])\Big\} \quad .\end{aligned} \qquad \text{(A 5.32)}$$

This expression corresponds to Eq. (A 5.17). Splitting $\varphi(\mathbf{r})$ (Eq. (A 5.14)) as

$$\varphi(\mathbf{r}) = \frac{\varphi_0}{\Omega^{1/2}} + \hat{\varphi}(\mathbf{r}) \qquad \text{(A 5.33)}$$

$$\hat{\varphi}(\mathbf{r}) = \frac{1}{\Omega^{1/2}} \sum_{\mathbf{k}\neq 0} \varphi_{\mathbf{k}} e^{-i\mathbf{k}\mathbf{r}} \qquad \text{(A 5.34)}$$

and using Eq. (A 5.22) we can separate off the φ_0 integral. Note that φ_0 is the amplitude of the $k = 0$ mode, not the saddle point $\varphi^{(0)}$.

$$\mathcal{Z}^{(M)} = \prod_n \frac{1}{M_n!} \exp\Big\{ \sum_n M_n \ln \frac{\Omega}{(4\pi\ell^2)^{d/2}} \Big\}$$
$$\frac{1}{\sqrt{2u_0\pi}} \int d\varphi_0 \exp\Big\{ -\frac{1}{2u_0} \Big(\varphi_0^2 + 2i\varphi_0 \Omega^{1/2} u_0\, c\Big) \Big\} \cdot \hat{\mathcal{Z}} \tag{A 5.35}$$

$$\hat{\mathcal{Z}} = \int D[\hat{\varphi}] \exp\Big\{ -\frac{1}{2u_0} \sum_{\mathbf{k}\neq 0} \varphi_{\mathbf{k}} \varphi_{-\mathbf{k}}$$
$$+ \sum_n M_n \ln \Big[\frac{(4\pi\ell^2)^{d/2}}{\Omega} \mathcal{Z}_0(n, [-i\hat{\varphi}]) \Big] \Big\} \,. \tag{A 5.36}$$

The Gaussian integral over φ_0 is easily carried out. Using Stirling's formula to evaluate $M_n!$ after a little calculation we find an expression for the density of the free energy in the large volume limit:

$$\frac{1}{\Omega} F = -\frac{1}{\Omega} \ln \mathcal{Z}^{(M)}$$
$$= f_0 - \frac{1}{\Omega} \ln \hat{\mathcal{Z}}^{(M)} \tag{A 5.37}$$

$$f_0 = c_p \left(\frac{d}{2} \ln(4\pi) - 1 \right) + \sum_n c_p(n) \ln(\ell^d c_p(n)) + \frac{u_0}{2} c^2 \,. \tag{A 5.38}$$

Clearly f_0 is the mean field result, corresponding to Flory-Huggins theory.

We now turn to the evaluation of $\hat{\mathcal{Z}}^{(M)}$. The argument of the logarithm in Eq. (A 5.36) can be expanded in $\hat{\varphi}$, closely following Eq. (A 5.29). No linear term occurs, since $\hat{\varphi}$ contains no $\mathbf{k} = 0$ mode. In Fourier representation the result can be written as

$$\frac{(4\pi\ell^2)^{d/2}}{\Omega} \mathcal{Z}_0(n, [-i\hat{\varphi}]) = 1 - \frac{1}{\Omega} \sum_{\mathbf{k}}{}' \varphi_{\mathbf{k}} \frac{n^2}{2} D(n\mathbf{k}^2\ell^2) \varphi_{-\mathbf{k}}$$
$$+ \frac{i}{\Omega^{3/2}} \sum_{\mathbf{k}_1\mathbf{k}_2\mathbf{k}_3}{}' \delta_{\mathbf{k}_1+\mathbf{k}_2+\mathbf{k}_3,0} \sum_{0<j_1<j_2<j_3<n} \varphi_{\mathbf{k}_1} G_0(\mathbf{k}_1, j_2 - j_1)$$
$$\varphi_{\mathbf{k}_2} G_0(\mathbf{k}_1 + \mathbf{k}_2, j_3 - j_2) \varphi_{\mathbf{k}_3} + \ldots \,. \tag{A 5.39}$$

The term of order s takes the form

$$\left(\frac{-i}{\Omega^{1/2}} \right)^s \sum_{\mathbf{k}_1 \ldots \mathbf{k}_s}{}' \delta_{\mathbf{k}_1+\ldots+\mathbf{k}_s,0} \sum_{0<j_1<\ldots<j_s<n} \varphi_{\mathbf{k}_1} G_0(\mathbf{k}_1, j_2 - j_1)$$
$$\varphi_{\mathbf{k}_2} G_0(\mathbf{k}_1+\mathbf{k}_2, j_3-j_2) \varphi_{\mathbf{k}_3} \cdots G_0(\mathbf{k}_1+\mathbf{k}_2+\ldots+\mathbf{k}_{s-1}, j_s-j_{s-1}) \varphi_{\mathbf{k}_s} . \tag{A 5.40}$$

Here the δ's are Kronecker symbols and the primed summations extend only over nonvanishing **k**-vectors. In deriving this result we used the orthonormality of the plane waves $\Omega^{-1/2}e^{-i\mathbf{kr}}$ and we freely shifted **r**-integrations exploiting the periodic boundary conditions.

The expansion of the logarithm in $\hat{\mathcal{Z}}$ to leading order yields a contribution quadratic in φ, which reads

$$-\frac{1}{\Omega}\sum_n M_n {\sum_{\mathbf{k}}}' \varphi_{\mathbf{k}}\, D(n\mathbf{k}^2\ell^2)\varphi_{-\mathbf{k}} = -\frac{1}{2}{\sum_{\mathbf{k}}}' \varphi_{\mathbf{k}} c_p N^2 D_p(N\mathbf{k}^2\ell^2)\varphi_{-\mathbf{k}} \ .$$

(See Eqs. (A 5.31), (A 5.49).) This term can be added to the explicit quadratic term in the exponent of Eq. (A 5.36) to yield the screened vertex in the form of $u_0(\mathbf{k})$ (Eq. (5.60)). Thus

$$\hat{\mathcal{Z}} = \int D[\hat{\varphi}] \exp\Big\{ -\frac{1}{2}{\sum_{\mathbf{k}}}' \frac{\varphi_{\mathbf{k}}\varphi_{-\mathbf{k}}}{u_0(\mathbf{k})} + \mathcal{W}[\hat{\varphi}]\Big\} \tag{A 5.41}$$

$$\begin{aligned}\mathcal{W}[\hat{\varphi}] = \Omega \sum_n c_p(n)\Big\{ &\ln\left[\frac{(4\pi\ell^2)^{d/2}}{\Omega}\mathcal{Z}_0(n,[-i\hat{\varphi}])\right] + \\ &+ \frac{1}{\Omega}{\sum_{\mathbf{k}}}' \varphi_{\mathbf{k}}\frac{n^2}{2}D(n\mathbf{k}^2\ell^2)\varphi_{-\mathbf{k}}\Big\} \ .\end{aligned} \tag{A 5.42}$$

Ignoring $\mathcal{W}[\hat{\varphi}]$ in Eq. (A 5.41) results in the one loop correction to the free energy:

$$\begin{aligned} f_1 &= -\frac{1}{\Omega}\ln\int D[\hat{\varphi}]\exp\Big\{-\frac{1}{2}{\sum_{k}}' \frac{\varphi_{\mathbf{k}}\varphi_{-\mathbf{k}}}{u_0(\mathbf{k})}\Big\} \\ &= -\frac{1}{\Omega}\ln\prod_{\mathbf{k}\neq 0}\left(\frac{u_o(\mathbf{k})}{u_o}\right)^{1/2} = \\ &= \frac{1}{2\Omega}{\sum_{\mathbf{k}}}' \ln\left(1 + WD_p(\mathbf{k}^2\ell^2 N)\right) \ . \end{aligned} \tag{A 5.43}$$

The higher loop corrections are contained in

$$\tilde{f} = -\frac{1}{\Omega}\ln\tilde{\mathcal{Z}}(x=1) \tag{A 5.44}$$

where

$$\tilde{\mathcal{Z}}(x) = e^{\Omega f_1}\int D[\hat{\varphi}]\exp\Big\{-\frac{1}{2}{\sum_{\mathbf{k}}}' \frac{\varphi_k\varphi_{-k}}{u_0(\mathbf{k})} + x\mathcal{W}[\hat{\varphi}]\Big\} \ . \tag{A 5.45}$$

We introduced the variable x for bookkeeping in the application of the linked cluster theorem.

A 5.4.2 Diagrammatic perturbation theory. In this formulation of the theory the screened vertex $u_0(\mathbf{k})$ is the 'propagator' of the $\hat{\varphi}$-field, the '$\hat{\varphi}$-vertices' being contributed by $\mathcal{W}[\hat{\varphi}]$. Inserting the expansion (A 5.39) and expanding the logarithm we can represent $\mathcal{W}[\hat{\varphi}]$ by a series of diagrams as shown in Fig. 5.19. The polymer lines with their '$\hat{\varphi}$- legs' stand for the terms of form (A 5.40), resulting from expanding $\left(4\pi\ell^2\right)^{d/2/\Omega}\, \mathcal{Z}_0(n, [-i\hat{\varphi}])$. The expansion of the logarithm creates products of terms of the form (A 5.40), tied together by the common summation over n. Diagrammatically we indicate this by including the corresponding polymer lines in a box, which stands for $\Omega \sum_n c_p(n)$. The second or third lines of Fig. 5.19 show examples arising from the second or third term in the Taylor expansions of the logarithm. We will call such $\hat{\varphi}$-vertices 'internally disconnected'.

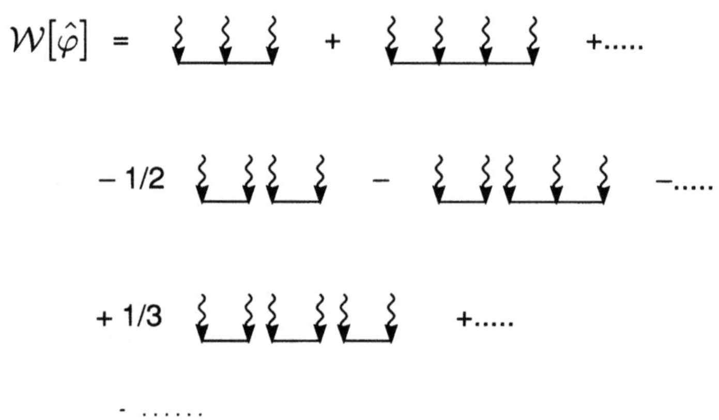

Fig. 5.19. Expansion of $\mathcal{W}[\hat{\varphi}]$

Expanding $\tilde{\mathcal{Z}}(x)$ in powers of x and carrying through the $\hat{\varphi}$-integration we pairwise tie together all $\hat{\varphi}$-legs by $\hat{\varphi}$-propagators, which are represented by a wiggly line. Figure 5.20 shows some of the resulting diagrams. $\tilde{\mathcal{Z}}(x)$ is given by the set of all such diagrams, where in the evaluation we have to take care of the weight factors due to the expansion of the logarithm, a complication which we will not consider further.

To the function $\tilde{\mathcal{Z}}(x)$ we may now apply the linked cluster theorem as formulated in Appendix A 5.1. The resulting expansion of $\ln \tilde{\mathcal{Z}}(x)$ is given by the set of all 'externally connected' diagrams, where we define a diagram to be externally disconnected if it separates into groups of disconnected boxes. Thus Fig. 5.20e gives an externally disconnected diagram not contributing to $\ln \tilde{\mathcal{Z}}$. All the other examples are externally connected, though diagrams 5.20b, c are internally disconnected.

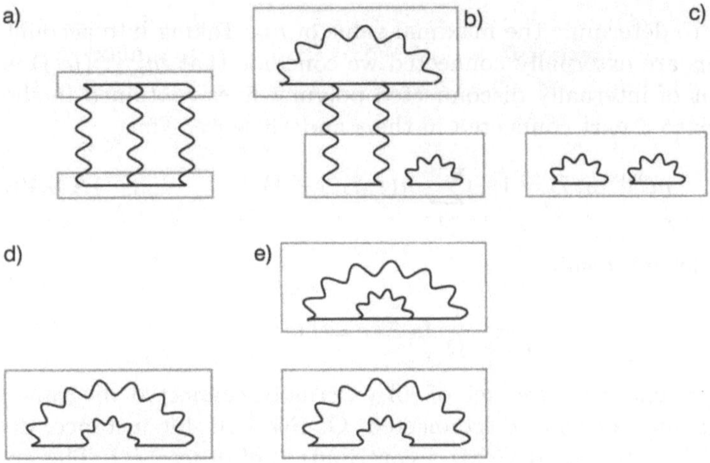

Fig. 5.20a–e. Typical canonical diagrams

We now have to analyze the behavior of the externally connected diagrams in the thermodynamic limit. In view of Eq. (A 5.37) we have to single out the contribution of order Ω, and we have to show that no diagram increases stronger than Ω. This is an exercise in power counting. Let us address as (s,j)-box a box containing s $\hat{\varphi}$-legs distributed on j polymer lines. From Eqs. (A 5.40), (A 5.42) such a box carries an explicit factor $\Omega^{1-s/2}$. Thus a diagram consisting of $m(s,j)$ (s,j)-boxes comes with the explicit power

$$\Omega^{\sum_{sj} m(s,j)(1-s/2)} \quad . \tag{A 5.46}$$

Implicitly powers of Ω are hidden in the summations over $\mathbf{k}$ according to

$$\frac{1}{\Omega}\sum_{\mathbf{k}}{}' \underset{\Omega\to\infty}{\longrightarrow} \int_k \quad .$$

We thus need to count the number of independent momentum summations, each contributing a factor of Ω. There are $\sum_{sj} m(s,j)s/2$ $\hat{\varphi}$-propagators, each carrying a momentum variable. For each polymer line one variable $\mathbf{k}$ is fixed by the Kronecker symbol (cf. Eq. (A 5.40)). However, for a diagram decomposing into m_c pieces mutually disconnected in the standard sense (i.e. ignoring the boxes), m_c of these constraints are ineffective, just expressing total momentum conservation for each disconnected piece. Thus the number of independent $\mathbf{k}$-summations is found to be

$$\sum_{sj} m(s,j)\left(\frac{s}{2}-j\right)+m_c \quad . \tag{A 5.47}$$

Combining results (A 5.46), (A 5.47) we find the total power of Ω:

$$\Omega^{m_c-\sum_{sj} m(s,j)(j-1)} \quad . \tag{A 5.48}$$

We still have to determine the maximal value of m_c. Taking into account that the diagrams are externally connected we conclude that $m_c^{\max}(\mu, j)$ is reached if no pair of internally disconnected polymer lines contained in the same box belongs to a part connected in the standard sense. Thus

$$m_c^{\max}(s,j) = 1 + \sum_{sj} m(s,j)(j-1) \quad , \tag{A 5.49}$$

leading to the following result:

$$\tilde{f} = -\lim_{\Omega \to \infty} \frac{1}{\Omega} \ln \tilde{\mathcal{Z}}(x=1)$$

exists and is represented by the set of all externally connected diagrams, which are maximally internally disconnected. Of Fig. 5.20, for instance, diagrams a, b, c contribute, but d yields a contribution of order $1/\Omega$. The argument is easily extended to an evaluation of the generating functional of density correlations or to endpoint correlation functions, the results being of the same form.

A 5.4.3 Comparison to the grand canonical formalism. In the thermodynamic limit the grand canonical and the canonical approach are known to be strictly equivalent, though the diagrammatic formulation of the grand canonical approach is much simpler. Indeed the grand canonical diagrams one by one can be mapped on that subset of canonical diagrams built from internally connected φ-vertices only. The prize paid is the occurrence of factors $(4\pi\ell^2)^{-d/2} \exp(\mu_p(n) - u_0 c\, n)$ weighting each polymer line occurring explicitly, or implicitly in the screened vertex. In the canonical approach these factors are replaced by $c_p(n)$. Being interested in physical quantities as function of $c_p(n)$ we in the grand canonical approach have to insert the expansion for the equation of state relating $\mu_p(n)$ to $c_p(n)$. The resulting contributions in the canonical formalism are just generated by the internally disconnected vertices.

For the free energy, disconnected vertices first contribute in two loop order (Fig. 5.20c). For correlation functions, however, disconnected vertices may contribute even to one loop order, and a calculation ignoring such contributions by keeping only naively connected diagrams would yield wrong results.

Due to its more complicated structure the canonical expansion is not well adapted for a discussion of general features like the properties of the theory under renormalization. Thus in the sequel we exclusively work with the grand canonical approach.

Notes

For a special choice of the chemical potential the grand canonical description as used here was suggested by Des Cloizeaux [Clo75]. The generalization to arbitrary $\mu_p(n)$ is due to Schäfer and Witten [SW80].

The importance of vertex irreducibility has been stressed by Des Cloizeaux [Clo80a]. Starting from a field theoretic formulation of the theory Knoll et al. [KSW81] independently came to the same conclusion. Our discussion follows that work which in turn follows a solution of the so-called 'Goldstone mode' problem in field theory of critical phenomena [SH78].

The auxiliary field method in the context of polymer physics has been introduced by Edwards [Edw66, Edw75] and has been widely used by Freed and coworkers. (See [Fre87]).

To order of one loop the canonical formalism has been worked out by Ohta and Nakanishi [ON83].

6. Breakdown of the Expansions in the Excluded Volume Region

In the previous chapters we have established a perturbative scheme, which in principle allows for a calculation of interaction effects to any order desired. We want to examine here whether this theory is indeed valid in all the parameter space of interest. We first consider quantities involving only a few chains, to be treated by the cluster expansion as derived in Chap. 4.

Inspecting the typical low order results (4.16), (4.32) we find that in the limit of long chains and for $d < 4$, i.e. $\varepsilon = 4 - d > 0$, the leading first order corrections are proportional to

$$z = \beta_e n^{\frac{\varepsilon}{2}} \quad , \tag{6.1}$$

this dominant contribution being derived by replacing the discrete segment summations by integrations. This result generalizes to higher orders of the expansion. A detailed analysis will be presented in the next chapter, where we are concerned with two parameter theory in its traditional form. Here we only give a rough argument, which exhibits the essential point. Without further analysis we assume that the substitution of the sums by integrals for $d < 4$ quite generally is a valid procedure to extract the leading n-dependence. Introducing variables $x = j/n$ to replace the segment indices, we thus write

$$\sum_{j=1}^{n} \rightarrow n \int_0^1 dx$$

The propagator takes the form

$$e^{-\mathbf{k}^2 \ell^2 (j-j')} \rightarrow e^{-\hat{\mathbf{k}}^2 \ell^2 (x-x')} \quad ,$$

where we introduced the variable $\hat{\mathbf{k}} = n^{1/2}\mathbf{k}$. With this substitution the measure of the momentum integration yields a factor of $n^{-d/2}$:

$$d^d k \rightarrow n^{-d/2} d^d \hat{k}$$

Since each additional interaction vertex in the cluster expansion contributes two segment summations and one momentum integral, with these substitutions a total factor of $n^{2-d/2} = n^{\varepsilon/2}$ accompanies the additional excluded volume coupling β_e. This yields a leading term $\sim z^\mu$ in μ-th order perturbation theory. *Concerning the leading n-dependence in each order of perturbation theory, the cluster expansion reduces to an expansion in z.* It is therefore valid only in the Θ-region $z \lesssim 1$, but diverges term by term in the excluded volume limit $z \to \infty$.

This result could have been forseen from the simple estimate of the number of μ-segment contacts in a Gaussian chain given in Sect. 2.2. Generalized to d dimensions it reads (note $\nu = 1/2$):

$$n_c^{(\mu)} \sim n^{\mu - \frac{d}{2}(\mu-1)} \tag{6.2}$$

Including only two-body contacts we estimate the interaction energy for a typical Gaussian configuration as

$$E_I \sim \beta_e n_c^{(2)} \sim \beta_e n^{2-\frac{d}{2}} = z \ . \tag{6.3}$$

As a result, in the partition function typical Gaussian configurations are suppressed by a factor of

$$\exp(-E_I) \sim \exp(-const.z) \quad ,$$

this factor decreasing rapidly with increasing z. For large z the partition function is dominated by those atypical random walk configurations showing an anomalously small number of intersections. This on the one hand explains the swelling. On the other hand it is quite clear that an expansion about the Gaussian chain model, which by construction puts most weight on the typical random walk configurations, is bound to fail, the quantities of interest being dominated by configurations in the extreme tail of the Gaussian configurational distribution.

So far we have considered the normal cluster expansion. What about the loop expansion? Clearly it cannot cure the problem in the dilute region, since there it coincides with the simple cluster expansion. However, in Sect. 5.4.3 we have found that for $W = u_0 c N \gg 1$ the corrections become independent of chain length and thus of z (cf. Eq. 5.72), so that the limit $z \to \infty$ can be taken. The correction term, however, is found to be proportional to $(\ell^d u_0 c)^{-\varepsilon/2}$ and becomes arbitrarily large in the limit of small segment concentrations. In other words, the loop expansion works for dense systems $\ell^d u_0 c \sim 1$, but breaks down in the semidilute region $\ell^d u_0\, c \ll 1$.

To summarize, we have found that the expansions yield a valid approximation scheme only in the Θ-region of not too long chains and small interaction: $z \lesssim 1$, or for a fairly dense solution: $\ell^d u_0 c \approx 1$. In both these regions the expansions have been applied with some success, but in the region of low density $\ell^d u_0 c \ll 1$ and strong excluded volume, $z \gg 1$, both the simple cluster expansion and the loop expansion break down. This is exactly the region, where the nontrivial scaling- or power laws are found. In its naive 'unrenormalized' form the theory cannot explain these laws. To solve this 'excluded volume problem' it needs additional insight and a new tool, which will be the renormalization group.

To some extent the excluded volume problem can be overcome if we are able to calculate enough orders of the expansions. The power series for the end-to end distance of an isolated chain, $R_e(z)$, has been calculated to sixth order, [MN84] and fourth order results are available for the radius of gyration (Shanes and Nickel, referred to in [Nic91]). The calculations exploit the field theoretic formulation of the continuous chain model (see Sect. 7.2), which strictly speaking is valid only close to the Θ-temperature. It is found that the coefficients of the series alternate in sign and increase strongly. In such a situation the direct evaluation of a finite number of terms does not make sense, but we may use most refined summation methods to extract the behavior for large z [MN87]. The outcome is consistent with renormalization group results and in fact it is the most accurate calculation of the swelling of a single chain available. This 'brute force' method, however, is restricted to those few quantities, where higher orders of the expansions can be calculated within reasonable effort. A calculation of for instance the concentration crossover of the osmotic pressure is beyond all reach. Furthermore, the resummation methods imply some prejudice concerning the outcome. Typically the existence of asymptotic power laws is assumed from the outset, or renormalization group results are used in the analysis.

Is there any situation in which the cluster expansion or the loop expansion solve the excluded volume problem, without recourse to higher order calculations or renormalization? Indeed, if we are willing to go to higher dimensional space from Eq. (6.1) we find that z vanishes as $n \to \infty, d > 4$. To understand this, we recall the estimate $n_c^{(2)} \sim n^{2-\frac{d}{2}}$, which suggests that for long chains in $d > 4$ dimensions the probability of close contacts among segments vanishes. The two-body interaction should be ineffective, and the chain should behave like a random walk. This basically is correct and can be proven rigorously.

To illustrate this, we consider the result (4.16ii). For $d > 4, n \to \infty$ it reduces to

$$R_e^2 = 2d\ell^2 n \left[1 + \beta_e(\alpha_1(\varepsilon) - \frac{2}{\varepsilon}) + O(\beta_e^2)\right] = \ell_{eff}^2 n \quad . \tag{6.4}$$

As expected, this expression is random walk like, which means that R_e^2 is proportional to n. The effective segment size depends on the interaction, and this effect can be calculated perturbatively. It is due to segments that are spaced closely along the chain. Similar results can be established for all other quantities, defined in the dilute region. In the semidilute region we again have to use the loop expansion. We however have to take into account the discrete nature of the chains more carefully than was done in Sect. 5.4.3. From the integral in Eq. (5.66), for instance, we have to separate both the single chain- and the two chain-contribution to $u_0(\mathbf{k})$, before we can replace the sum by the Debye function. The resulting correction to the Flory term $u_0 cn$ is finite also for $N \to \infty$, $\ell^d c \to 0$, in complete accord with Eq. (6.4).

To summarize, $d = 4$ makes a border line, the *'upper critical dimension'* for the excluded volume problem. For $d \leq 4$ both the cluster expansion and the loop expansion break down term by term in the excluded volume limit. For $d > 4$ the expansions are valid, the leading n- or c-dependence of the results being trivial, however. We may state that for $d > 4$ the random walk model or Flory-Huggins type mean field theories catch the essential physics of the problem. As will be explained more accurately in Chap. 10 the mechanism behind this is the fact that for $d > 4$ two uncorrelated random walks in general do not cross.

7*. Continuous Chain Model and Naive Two Parameter Theory

This chapter is concerned with the relation of the discrete Gaussian chain model to standard two parameter theory of polymer solutions [Yam71], which most naturally is formulated in terms of Edwards' continuous chain model [Edw65, Edw66]. In the context of our general presentation of renormalization group theory this is not an essential topic, but in explicit calculations to be presented later we make use of the simplifications brought about by the continuous chain limit. Furthermore the two parameter hypothesis is central to a lot of theoretical work, and it anticipates some of the scaling structure exhibited by renormalization. Indeed, the renormalization group can be interpreted as establishing a nontrivial version of two parameter theory. As a further topic we in the present chapter establish the connection to standard field theory of critical phenomena.

The two parameter hypothesis assumes that physical properties of dilute polymer solutions depend only on the two parameters

$$R_0^2 = \ell^2 \, n \tag{7.1}$$

and

$$z = \beta_e \, n^{\varepsilon/2} \quad , \tag{7.2}$$

where it is understood that the concentration c_p is treated by a virial expansion. (A naive interpretation of z as interaction energy of a typical Gaussian chain configuration was given in the previous chapter, Eq. (6.3).) Since β_e close to the Θ-temperature obeys $\beta_e \sim T - \Theta$, this hypothesis incorporates scaling as function of temperature and chain length, as discussed in Chap. 1. Close to the Θ-temperature this aspect is well supported by experiments. The hypothesis is easily extended to systems of larger overlap, where it amounts to the assumption that the concentration comes only in the form

$$s_0 = c_p \, R_0^d \quad , \tag{7.3}$$

a variable which we will call 'Θ-overlap', since it essentially involves the coil radius at the Θ-temperature. With this hypothesis two parameter theory results from simple dimensional analysis. Since R_0 is the only length scale of the theory we for instance find for the end-to-end distance

$$R_e(n, c_p, \beta_e, \ell) = R_0 \bar{R}_e(z, s_0) \quad . \tag{7.4}$$

In writing down such expressions we ignore polydispersity, assuming the system to be monodisperse so that all chains are of length $n = N$. In general $\bar{R}_e$ depends on the reduced chain length distribution $p(n/N)$ (Eq. (5.36)) and R_0, z, s_0 are to be defined in terms of the number averaged chain length N.

In two parameter theory we ignore the microscopic length scale ℓ. It is thus intimately related to the continuous chain limit $\ell \to 0$, where the discrete sequence of segments is replaced by a continuous space curve. On the level of the configurational probability $\mathcal{F} \sim \exp[-V]$ this leads to the Edwards model. It is found that the continuous chain limit from each order of the cluster expansion keeps only the leading n-dependence. Corrections vanish like $\left(\ell^2/R_0^2\right)^\varepsilon \sim n^{-\varepsilon}$. Provided this limit exists, this proves the basic assumption of the previous chapter: keeping only the leading n-dependence we find an expansion in powers of z. On the other hand the analysis of the previous chapter shows that the continuous chain limit cannot exist for $d \geq 4$, since here the discreteness of the chains cannot be ignored. Indeed, Eq. (6.4) for R_e^2 does not conform to the two parameter form (7.4). We thus immediately note a first important restriction: *two parameter theory and the continuous chain model are valid concepts only for* $d < 4$.

Even for $d < 4$ the question of existence of the continuous chain limit is not completely trivial. The problem is most easily analyzed by taking a Laplace transform with respect to the chain length, which results in the 'field theoretic representation' of polymer theory. In field theory it is not hard to show that the limit $\ell \to 0$ can be taken only after a so-called 'additive renormalization': we first have to extract some contributions which for $\ell \to 0$ would diverge. The extracted terms can be absorbed into a 'renormalization', i.e. a redefinition of the parameters of the model. Transforming back to polymer theory we find that this renormalization just shifts the chemical potential per segment. We thus can prove the following statement: *after an appropriate shift of the chemical potential the continuous chain limit for* $d < 4$ *can be taken order by order in perturbation theory.* In this sense the continuous chain model or two parameter theory are a well defined limit of our model of discrete Gaussian chains.

It turns out that this shift of $\mu_p(n)/n$ has a good physical reason. It identifies the chemical potential per segment of an infinitely long chain and thus is essential for reaching the critical limit $R_e \to \infty$.

We indeed encountered additive renormalization before. In our calculation of $\mu_p(n)$ (Sect. 5.4.3) we had to subtract the single chain part from the one loop correction, before we could replace the sum over discrete segments by the Debye function, which is calculated for a continuous chain. (See Eq. (3.22).) In the result (5.69) this introduced the term $-n\beta_e\zeta\left(\frac{d}{2}\right)$ to shift the chemical potential. All the other one loop corrections can be rewritten in terms of z or s_0. The variable $W = u_0 c N$, in particular, is identical to $W = (4\pi)^{d/2}\, z\, s_0$.

We must note also a second important restriction of the continuous chain model. As we will see, by construction it deals with infinitely long chains: $n \to \infty$, infinitesimally close to the Θ-point: $\beta_e \to 0$. Thus naive two parameter theory is valid only very close to the Θ -temperature. In later chapters we will see how further renormalization leads to a theory of excluded volume effects valid for all $\beta_e > 0$.

The organization of this chapter is as follows. In Sect. 7.1 we carefully define the continuous chain limit and we introduce the appropriate modification of the Feynman rules. We then establish the two parameter scheme by dimensional analysis. Section 7.2 is devoted to the question whether the continuous chain limit exists. The analysis is presented on the diagrammatic level. It exploits the field theoretic representation, which also is derived on the level of diagrams. All the analysis is based on the cluster expansion. Extension to the loop expansion is not difficult, but will not be considered, since it is not needed in the sequel.

Finally, for completeness in Appendix A 7.1 we consider the formal relation of the continuous chain model to a field theoretic Hamiltonian, used to describe critical phenomena in ferromagnets. It was this relation discovered by de Gennes [dG72] and extended by Des Cloizeaux [Clo75], which initiated the application of the renormalization group to polymer solutions and led to the embedding into the larger realm of critical phenomena.

7.1 Continuous chain limit and the Edwards model

In the sequel, for simplicity of notation, we take all chains to be of the same length n.

Definition

The continuous chain limit is defined by taking

$$\ell \to 0$$

but keeping

$$R_0^2 = \ell^2 n\,;\quad c_p$$

and

$$\tilde{u}_0 = (4\pi)^{d/2}\, \beta_e \ell^{-\varepsilon} \equiv \ell^{-4} u_0 \tag{7.5}$$

fixed.

We thus take the segment size smaller and smaller, at the same time increasing the number of segments so as to keep the unperturbed coil size $\sim R_0^2$ constant. Thus

$$n = \frac{R_0^2}{\ell^2} \to \infty \quad .$$

(We should note that R_0^2 is sometimes called the 'Gaussian surface' of the chain, a notion introduced by Des Cloizeaux.) Provided $d < 4$ the constraint $\tilde{u}_0 = const$ furthermore implies that we decrease β_e:

$$\beta_e = \frac{\tilde{u}_0}{(4\pi)^{d/2}} \, \ell^{\varepsilon} \to 0 \quad ,$$

such that the naively estimated total interaction energy stays constant:

$$E_I \sim \beta_e n_c^{(2)} \sim \beta_e \ell^d \frac{n^2}{R_0^d} = \frac{\tilde{u}_0}{(4\pi)^{d/2}} \, R_0^{\varepsilon} \quad .$$

Clearly it is not obvious that the physical quantities of interest exist in that limit. This problem will be discussed in the next section. At present we evaluate the consequences of the limit for $d < 4$, assuming its existence.

To carry through the limit on the diagrammatic level we introduce variables $s = \ell^2 j$ to replace the segment indices. (These segment variables should not be confused with the overlap.) The propagator takes the form $\exp\left(-\mathbf{k}^2(s - s')\right)$. Rewriting the summations coming with a vertex as

$$-\left(4\pi\ell^2\right)^{d/2} \beta_e \sum_{jj'=1}^{n} \Rightarrow -\tilde{u}_0 \sum_{j=1}^{R_0^2/\ell^2} \ell^2 \sum_{j'=1}^{R_0^2/\ell^2} \ell^2 \quad ,$$

we find that the only explicit ℓ-dependence is associated with the summations. The continuous chain limit formally transforms the sums into integrals

$$\sum_{j=1}^{R_0^2/\ell^2} \ell^2 \underset{\ell \to 0}{\longrightarrow} \int_0^{R_0^2} ds \quad .$$

We thus find the following modification of the Feynman rules as given in Sect. 4.2.

Evaluation rules in the continuous chain limit

(ii') A propagator of momentum $\mathbf{k}$, connecting distinguished points $s < s'$, yields a factor $G_{0E}(\mathbf{k}, s' - s) = \exp\left[-\mathbf{k}^2(s' - s)\right]$. A vertex yields a factor $-\tilde{u}_0$.

(iii') Integrate over all internal momenta. Integrate over the labels of the distinguished points, respecting their ordering along the chain. The upper bound of the ordered integration for the m-th chain is $R_{0,m}^2 (\hat{=}\ell^2 n_m)$.

(iv') Multiply by a factor Ω^c, where c is the number of connected components of the graph.

These rules define Greensfunctions $\tilde{G}_E^{(L,M)}$, where the subscript E stands for Edwards' continuous chain model. Besides transforming summations into integrals we suppressed some factors of ℓ. The proper correspondence between the discrete and the continuous model reads

$$\begin{aligned}&\left(4\pi\ell^2\right)^{M\frac{d}{2}} \ell^{-2L} G_{m_1\dots m_2}^{(L,M)}(\mathbf{q}_1,\dots,\mathbf{q}_2;\mathbf{p}_1,\dots,\mathbf{p}_{2M};n_1,\dots,n_M;\beta_e,\ell)\\ &\Rightarrow \tilde{G}_{E,m_1\dots m_2}^{(L,M)}(\mathbf{q}_1,\dots,\mathbf{q}_2;\mathbf{p}_1,\dots,\mathbf{p}_{2M};R_{0,1}^2,\dots,R_{0,m}^2;\tilde{u}_o)\ . \qquad (7.6)\end{aligned}$$

The ℓ-dependent prefactor either will be eliminated by some appropriate normalization of the physical quantity considered, or it will be absorbed into the chemical potential.

Formally the continuous chain limit may also be carried through on the level of the configurational weight $\mathcal{F} = \exp(-V)$. Taking into account that $\ell^d \beta_e$ becomes small, we rewrite $\exp(-V)$ (Eq. (2.3)) as

$$\begin{aligned}\exp(-V) \approx \exp\Big\{ &-\frac{1}{4}\sum_{m=1}^{M}\sum_{j=1}^{n} \ell^2 \left(\frac{\mathbf{r}_j^{(m)} - \mathbf{r}_{j-1}^{(m)}}{\ell^2}\right)^2 \\ &- \frac{\tilde{u}_0}{2} \sum_{m,m'=1}^{M} \sum_{j=1}^{n} \ell^2 \sum_{j'=1}^{n} \ell^2 \delta^d\left(\mathbf{r}_j^{(m)} - \mathbf{r}_{j'}^{(m')}\right)\Big\}\ ,\end{aligned}$$

where in this formal argument we again ignore problems related to self interactions of segments or to segments interacting more than once. Introducing functions $\mathbf{r}^{(m)}(s)$ interpolating among the segment coordinates of the m-th chain we can take the continuous chain limit to find

$$\exp(-V) \to \exp(-V_E)\ ,$$

where

$$\begin{aligned}V_E = &\frac{1}{4}\sum_{m=1}^{M}\int_0^{R_0^2} ds \left(\frac{d\mathbf{r}^{(m)}}{ds}\right)^2 \\ &+ \frac{\tilde{u}_0}{2}\sum_{mm'=1}^{M}\int_0^{R_0^2} ds \int_0^{R_0^2} ds'\, \delta^d\left(\mathbf{r}^{(m)}(s) - \mathbf{r}^{(m')}(s')\right) \qquad (7.7)\end{aligned}$$

is the Edwards Hamiltonian. Setting up perturbation theory with this Hamiltonian, we find the $\tilde{u}_0$-expansion as sketched above.

We should note that Eq. (7.7) differs from the standard form of the Edwards model by a trivial factor of 1/2 in the first term, which can be absorbed in a redefinition of s. The present form is most directly related to a special field theory (see Appendix A 7.1).

The Edwards Hamiltonian is an appealing but most formal object. To mention a simple fact, shrinking to zero the segment size of the discrete chain model as done in the continuous chain limit, we in general get a continuous but not differentiable space curve. Strictly speaking the first part of V_E does not exist. Further serious mathematical problems are connected to the δ-function interaction. The question in which sense V_E is a mathematically well defined object beyond its formal perturbation expansion is an interesting problem of mathematical physics.

Consider now some physical observable Q of length dimension d_Q, which for $\ell > 0$ can be written in the form

$$Q(\{\mathbf{p}_i\}; n, c_p, \beta_e, \ell) = R_0^{d_Q} \tilde{Q}(\{\mathbf{p}_i\}, R_0, c_p, \tilde{u}_0, \ell) \quad . \tag{7.8}$$

Here $\tilde{Q}$ is dimensionless. Assuming the existence of the continuous chain limit we find

$$\begin{aligned} \lim_{\ell \to 0} Q(\{\mathbf{p}_i\}, n, c_p, \beta_e, \ell) &= R_0^{d_Q} \tilde{Q}(\{\mathbf{p}_i\}, R_0, c_p, \tilde{u}_0, 0) \\ &= R_0^{d_Q} Q_E(\{\mathbf{p}_i R_0\}, z, s_0) \quad , \end{aligned} \tag{7.9}$$

where

$$z = \frac{\tilde{u}_0}{(4\pi)^{d/2}} R_0^{\varepsilon} \hat{=} \beta_e n^{\varepsilon/2} \tag{7.10}$$

and

$$s_0 = c_p R_0^d \quad .$$

The last line of Eq. (7.9) follows since dimensionless quantities can depend only on dimensionless combinations of their variables. As a result of this simple dimensional analysis in the continuous chain model we have found the two parameter theory: instead of the three independent parameters ℓ, β_e, n, physical observables involve only the two combinations R_0, z. Perturbation theory now proceeds in powers of z. Thus the continuous chain limit gives a precise meaning to the simple argument presented in Chap. 6.

Let us now reflect upon the physical meaning of these results. Within the two parameter scheme $R_e(c_p = 0)$ or A_2^{Π} are written as (cf. Eqs. (4.16), (4.32))

$$\frac{R_e^2}{2d} = R_0^2 R_{e,E}^2(z) = R_0^2 \left[1 + \frac{4z}{\varepsilon(2+\varepsilon)} + O(z^2)\right] \tag{7.11}$$

$$(4\pi)^{-d/2} A_2^{\Pi} = R_0^d A_{2,E}^{\Pi}(z) = R_0^d z \left[1 - \left(\frac{4}{\varepsilon} - a(\varepsilon)\right) z + O(z^2)\right] \quad . \tag{7.12}$$

The end-to-end distribution for a chain in a solution of arbitrary concentration takes the form

$$\hat{P}(\mathbf{r}; n, c_p, \beta_e, \ell) = \hat{P}_E(\frac{\mathbf{r}}{R_0}; z, s_0) \quad . \tag{7.13}$$

Such results look like scaling laws, measuring all spatial lengths on the chain length dependent scale R_0. In fact they can be viewed as a preliminary stage of scaling, being based on naive dimensional analysis, as shown above. They seriously differ in content however from the nontrivial scaling laws derived via the renormalization group. Indeed, resulting from the continuous chain limit they presume a very special limiting form of the microstructure, which can be described as an 'infinitely flexible' chain, flexible on arbitrarily small intervals of contour length. Furthermore, these infinitesimally small segments interact via an infinitesimally small excluded volume constant, as shown above. Now $\beta_e \to 0$ implies that we approach the Θ-point. *The continuous chain limit describes a very long chain close to the Θ-point.* This is clearly brought out by comparing Eqs. (7.11), (7.12) to the corresponding results for discrete chains (cf. Eq. (4.16 ii), for instance). The results only coincide for very long chains: $\beta_e n^{\varepsilon/2} \gg \beta_e$ and extremely small excluded volume: $\beta_e \ll 1$. *Universality of the naive two-parameter form holds only in the region $(T - \Theta)/\Theta \ll 1$.* Outside that region the continuous chain model is no valid representation of a polymer chain. (To avoid confusion we should recall that $(T - \Theta)/\Theta \ll 1$ does *not* imply $z \ll 1$.)

Universality and two-parameter scaling in the general case of finite excluded volume β_e comes about by the much more sophisticated mechanism of renormalization. As will be discussed in later chapters (see Chap. 11, in particular) both the discrete chain model and the continuous chain model can be mapped on the same 'renormalized' theory. The renormalized results superficially look similar to expressions like Eq. (7.13), but the definition of the scaling variables R_0, z is more complicated. Indeed, it is in the definition of R_0 and z in terms of the parameters of the original unrenormalized theory, that the difference in microstructure of the continuous or discrete chain models is absorbed.

7.2 Existence of the continuous chain limit

We still have to prove that the limit $\ell \to 0$ exists, a task which is not completely trivial. Indeed we will find that the limit naively exists only for quantities like A_2^{Π} (Eq. (4.28)) or $P(\mathbf{p}, n)$ (Eq. (4.5)), defined as *ratios* of partition functions. For the partition function itself or for the Greensfunctions $G^{(L,M)}(\ldots)$ (Eq. (4.24)) the continuous chain limit exists only after a microstructure dependent factor has been extracted.

To illustrate the problem we consider the expansion of the Greensfunction $G^{(0,1)}$ (Eq. (4.20)), which to first order is given by Eq. (4.6):

$$\frac{\left(4\pi\ell^2\right)^{d/2}}{\Omega} G^{(0,1)}(\mathbf{p}, -\mathbf{p}; n) = G_0(\mathbf{p}, n) - \beta_e G_1(\mathbf{p}, n) + O(\beta_e^2) \ .$$

The first order diagram is reproduced in Fig. 7.1a. Taking the continuous chain limit we from Eq. (4.11) find

$$\beta_e G_1(\mathbf{p}, n) \underset{\ell \to 0}{\longrightarrow}$$

$$(4\pi)^{-d/2}\, \tilde{u}_0\, e^{-\mathbf{p}^2 R_0^2} \int_0^{R_0^2} ds_2 \int_0^{s_2} ds_1\, (s_2 - s_1)^{-d/2}\, e^{-\mathbf{p}^2(s_1 - s_2)} \ . \qquad (7.14)$$

It is easily checked that this expression does not exist, the divergence being due to integration over small distances $s_2 - s_1$:

$$\int_\sigma ds\, s^{-d/2} \sim \sigma^{1-d/2} \underset{\sigma \to 0}{\longrightarrow} \infty,\ d > 2 \ .$$

Clearly this problem is a microstructure effect resulting from the continuous chain limit, which allows for self-interactions of arbitrarily small pieces of the chain. To avoid it, we introduce a *cut off*. We impose the rule that segment variables s_j are not allowed to approach each other along the chain on distances smaller than $\sigma > 0$. In some sense this again introduces a finite segment size $\ell \sim \sigma^{1/2}$. Imposing the same cut off at the chain ends we thus write

$$\beta_e G_1(\mathbf{p}, n) \to \tilde{u}_0 \tilde{G}_E^{(1)}(\mathbf{p}, R_0^2)$$

$$\tilde{G}_E^{(1)}(\mathbf{p}, R_0^2) = \int_\sigma^{R_0^2 - \sigma} ds_2 \int_\sigma^{s_2 - \sigma} ds_1 \int_k e^{-\mathbf{k}^2(s_2 - s_1) - \mathbf{p}^2(s_1 + R_0^2 - s_2)}$$

$$= \int_\sigma^{R_0^2 - \sigma} ds_2 \int_\sigma^{s_2 - \sigma} ds_1 \int_k G_{0E}(\mathbf{p}, s_1) G_{0E}(\mathbf{k}, s_2 - s_1) G_{0E}(\mathbf{p}, R_0^2 - s_2) \,. \qquad (7.15)$$

Recall the definition $G_{0E}(\mathbf{k}, s) = \exp(-\mathbf{k}^2 s)$. We took a step back here, again introducing the integral over the internal momentum.

The cut off renders the theory finite but ruins the two parameter structure. Indeed, it introduces a new dimensionless variable $\sigma/R_0^2 \sim 1/n$, resulting in perturbative expressions which closely resemble those of the discrete chain

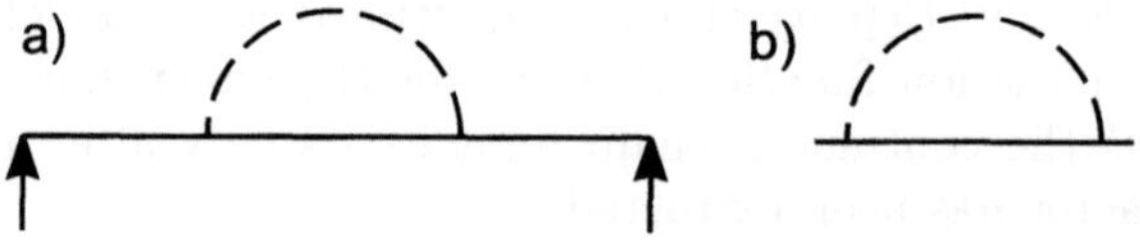

Fig. 7.1. (**a**) First order diagram contributing to $\tilde{G}_F^{(0,1)}(\mathbf{p}, -\mathbf{p}, \tau)$. (**b**) The divergent subgraph

model. Still a priori the cut off is necessary for the theory to be well defined. How then can we eliminate it again?

To answer this question an excursion to field theory is appropriate. We carry through a Laplace transform, defining

$$\tilde{G}_F^{(0,1)}(\mathbf{p}, -\mathbf{p}; \tau) = \int_\sigma^\infty dR_0^2 e^{-\tau R_0^2} \tilde{G}_E^{(0,1)}(\mathbf{p}, -\mathbf{p}; R_0^2) \quad , \tag{7.16}$$

$$\begin{aligned} G_{0F}(\mathbf{k}, \tau) &= \int_\sigma^\infty ds e^{-\tau s} G_{0E}(\mathbf{k}, s) \\ &= \frac{e^{-\sigma(k^2+\tau)}}{k^2+\tau} \quad , \end{aligned} \tag{7.17}$$

where the subscript (F) stands for field theory. This serves to factorize the convolutional integral (7.15):

$$\begin{aligned} \frac{1}{\Omega}\tilde{G}_F^{(0,1)}(\mathbf{p}, -\mathbf{p}; \tau) &= G_{0F}(\mathbf{p}, \tau) \\ &- \tilde{u}_0 G_{0F}(\mathbf{p}, \tau) \int_k G_{0F}(\mathbf{k}, \tau) G_{0F}(\mathbf{p}, \tau) + O(\tilde{u}_0^2) \ . \end{aligned} \tag{7.18}$$

Comparing to the diagram of Fig. 7.1a we find that now a propagator line of momentum $\mathbf{k}$ just yields a factor of $G_{0F}(\mathbf{k}, \tau)$, with all integrations over segment variables eliminated. It is easily checked that this result is completely general. After a Laplace transform over chain length each propagator line just yields a factor of G_{0F}. Furthermore the result immediately generalizes to correlation functions of M chains, each chain possesing its own variable $R_{0m}^2 = \ell^2 n_m, m = 1, \ldots, M$. In this case we carry through M independent Laplace transforms, introducing the conjugate variables $\tau_m, m = 1, \ldots, M$. (In field theory a variable of the type τ_m would be addressed as a 'mass'.) After the Laplace transform a propagator of momentum $\mathbf{k}$ in the m-th polymer line yields a factor $G_{0F}(\mathbf{k}, \tau_m)$. All segment integrations are eliminated.

We can summarize these considerations in the following Feynman rules.

Rules for evaluating $\tilde{G}_{F,m_1\ldots m_2}^{(L,M)}(\mathbf{q}_1, \ldots \mathbf{p}_{2M}; \tau_1, \ldots, \tau_M)$

(i) Label the propagator lines by some momentum variables, respecting momentum conservation at the vertices

(ii) A propagator line of momentum $\mathbf{k}$, being part of the m-th polymer line, yields $G_{0F}(\mathbf{k}, \tau_m)$. A broken line yields a factor $(-\tilde{u}_0)$

(iii) Integrate over all internal momenta. Factors of Ω are to be added as before.

These rules are identical to the Feynman rules of a special Euclidean field theory, extensively used in the theory of critical phenomena. (See, for instance, [Ami84].) This formulation therefore is known as the *field theoretic representation* of polymer theory. We will elaborate on the relation to field theory in Appendix A 7.1.

We now consider the limit $\sigma \to 0$, where $G_{0F}(\mathbf{k}, \tau)$ takes the form

$$G_{0F}(\mathbf{k}, \tau) \underset{\sigma \to 0}{\longrightarrow} \frac{1}{k^2 + \tau} \quad . \tag{7.19}$$

As a result the **k**-integral in Eq. (7.18) for $d \geq 2$ does not exist, the divergence being due to integration over large momenta. This is called an ultraviolet (u.v.) divergence. Since large momenta correspond to small spatial distances and thus to small sections of the chain, this is the same microstructure divergence as found in the polymer representation (7.14). In the field theoretic formulation its source clearly is identified as the diagrammatic structure of Fig. 7.1b, yielding the **k**-integral in Eq. (7.18). Below we will show quite generally that in the field theoretic representation the divergences can be attributed to specific 'subdiagrams', where Fig. 7.1b gives one example. It is this feature, which renders this formulation preferable for an analysis of the limit $\sigma \to 0$.

Turning again to our one loop example, we note another problem, seemingly unrelated to the limit $\sigma \to 0$. In standard field theories of interacting particles it can be shown that the Greensfunctions analogous to $\tilde{G}_F^{(0,1)}(\mathbf{p}, -\mathbf{p}; \tau)$ at $\mathbf{p} = 0$ have only simple poles and cuts as function of τ. In perturbation theory this property holds true for the zero-order term, a simple pole being found for $\tau = 0$. The first order correction, however, at $\tau = 0$ has a pole of second order, due to the two factors of $G_{0F}(\mathbf{p}, \tau)$. This signals that the interaction wants to shift the pole according to the mechanism

$$\frac{1}{\tau + \mathbf{p}^2} - \frac{1}{\tau + \mathbf{p}^2} \tilde{u}_0 \Delta\tau \frac{1}{\tau + \mathbf{p}^2} + O(\tilde{u}_0^2) = \frac{1}{\tau + \tilde{u}_0 \Delta\tau + \mathbf{p}^2} \quad . \tag{7.20}$$

This effect is most simply handled by considering the inverse of the Greensfunction. We introduce the so-called 'vertex function' $\tilde{\Gamma}_F^{(0,1)}$ defined as

$$\tilde{\Gamma}_F^{(0,1)}(\mathbf{p}, -\mathbf{p}; \tau) = \Omega \left(\tilde{G}_F^{(0,1)}(\mathbf{p}, -\mathbf{p}; \tau) \right)^{-1} \quad . \tag{7.21}$$

Equations (7.17), (7.18) yield

$$\tilde{\Gamma}_F^{(0,1)}(\mathbf{p}, -\mathbf{p}; \tau) = (\tau + \mathbf{p}^2) e^{(\tau + \mathbf{p}^2)\sigma} + \tilde{u}_0 \int_k \frac{e^{-(\tau + \mathbf{k}^2)\sigma}}{\tau + \mathbf{k}^2} + O(\tilde{u}_0^2) \quad . \tag{7.22}$$

We now can identify the pole of $\tilde{G}_F^{(0,1)}(0, 0; \tau)$ as zero of $\tilde{\Gamma}_F^{(0,1)}(0, 0; \tau)$.

$$0 = \tau_c + \tilde{u}_0 e^{-\tau_c \sigma} \int_k \frac{e^{-(\tau_c + \mathbf{k}^2)\sigma}}{\tau_c + \mathbf{k}^2} + O(\tilde{u}_0^2) \quad . \tag{7.23}$$

In field theory τ_c is known as 'critical mass'. Note that τ_c depends on σ, diverging for $\sigma \to 0$ (u.v. divergence).

How does this consideration help in an analysis of the limit $\sigma \to 0$? To understand its significance we calculate the correlation length

$$\begin{aligned} R_F^2 &= -\Delta_p|_0 \ln \tilde{G}_F^{(0,1)}(\mathbf{p}, -\mathbf{p}; \tau) \\ &= \frac{\Delta_p|_0 \tilde{\Gamma}_F^{(0,1)}(\mathbf{p}, -\mathbf{p}; \tau)}{\tilde{\Gamma}_F^{(0,1)}(\mathbf{0}, \mathbf{0}; \tau)} \\ &= \frac{2d + O(\tilde{u}_0^2)}{\tilde{\Gamma}_F^{(0,1)}(\mathbf{0}, \mathbf{0}; \tau)} \quad . \end{aligned} \tag{7.24}$$

By construction R_F corresponds to the end-to-end distance R_e of a single polymer chain. We are interested in the limit of long chains: $R_F \sim R_e \to \infty$, which according to Eq. (7.24) corresponds to $\tilde{\Gamma}_F^{(0,1)} \to 0$, i.e. $\tau \to \tau_c$. *It is the neighborhood of τ_c that corresponds to the region of long chains.* This suggests to introduce a shifted variable

$$\hat{\tau} = \tau - \tau_c \quad , \tag{7.25}$$

and to rewrite the theory in terms of $\hat{\tau}$. Subtracting Eq. (7.23) from Eq. (7.22) and neglecting some terms of order σ we find

$$\begin{aligned} &\Gamma_F^{(0,1)}(\mathbf{p}, -\mathbf{p}; \hat{\tau}) \equiv \tilde{\Gamma}_F^{(0,1)}(\mathbf{p}, -\mathbf{p}; \tau_c + \hat{\tau}) \\ &= \hat{\tau} + \mathbf{p}^2 + \tilde{u}_0 \int_k \left[\frac{e^{-(\hat{\tau} + \mathbf{k}^2)\sigma}}{\hat{\tau} + \mathbf{k}^2} - \frac{e^{-\mathbf{k}^2 \sigma}}{\mathbf{k}^2} \right] + O(\tilde{u}_0^2) \,, \end{aligned} \tag{7.26}$$

where we used $\tau_c = O(\tilde{u}_0)$ to justify the neglect of τ_c in the integral. In the result (7.26) the limit $\sigma \to 0$ can be taken for $2 < d < 4$. Introducing the distance $\hat{\tau}$ to the critical mass we thus eliminated the u.v. divergence of the integral. We should stress, however, that the shift of τ is not just a technical trick but is a necessary step if we want to analyse the critical region of long chains.

This is the most trivial example of an additive renormalization It can be shown that the mechanism is not restricted to the first order graph 7.1a but works generally. To give the essence of the argument we have to identify the u.v. divergent subgraphs: this is done by 'power counting' which determines the behavior of a subgraph when all its internal momenta go to infinity at the same rate. It is sufficient to consider only one-line-irreducible (1-l-i) subgraphs, i.e. subgraphs which cannot be cut into two independent parts by cutting a single propagator line. (See Fig. 7.2.) The Feynman rules of field

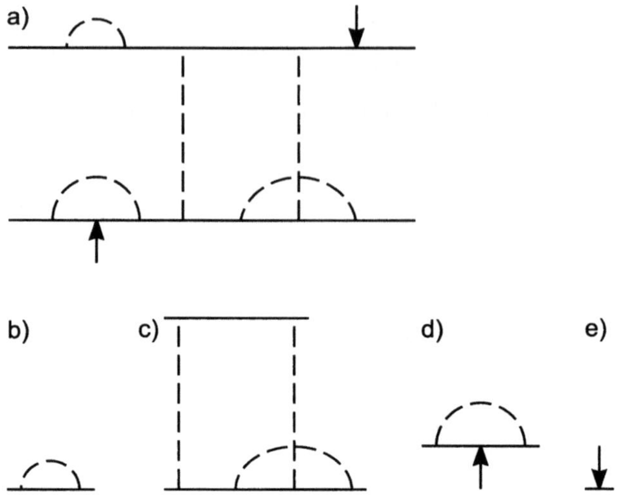

Fig. 7.2. (**a**) A typical diagram. (**b–e**) Its one-line-irreducible parts

theory guarantee that each graph factorizes into the contributions of its 1-l-i subgraphs, multiplied by the connecting propagators. Now consider a 1-l-i subgraph with μ vertices and $2M$ 'legs', where propagators have been cut off. In addition there may be $L \geq 0$ density insertions. It is easily checked that the number of internal propagator lines n_ℓ is given by

$$n_\ell = 2\mu - M + L \quad . \tag{7.27}$$

To each internal line there corresponds an integration over d-dimensional momentum, but momentum conservation at the vertices and density insertions fixes $L + \mu - 1$ of these integrations. (One combination of δ-functions is ineffective, fixing to zero the sum of all momenta flowing into the subdiagram.) Thus the number of independent internal integrations or 'loops' is found as

$$n_L = n_\ell - L - \mu + 1 = \mu - M + 1 \quad . \tag{7.28}$$

We can now calculate the 'superficial degree of divergence' δ of the subdiagram as

$$\begin{aligned} \delta &= d n_L - 2 n_\ell \quad , \\ &= (d-4)\mu - (d-2)M - 2L + d \quad . \end{aligned} \tag{7.29}$$

δ determines the behavior of the subdiagram for $\sigma \to 0$ in the u.v. limit (all $\mathbf{k}_i \to \infty$). Indeed, rescaling all internal momenta according to $\mathbf{k}_i \sigma^{\frac{1}{2}} \to \mathbf{k}_i$ and neglecting all τ_m, which is allowed for $\tau_m \ll \mathbf{k}_i^2$, from the subdiagram we extract a factor $\sigma^{-\delta/2}$. Thus the limit $\sigma \to 0$ exists for $\delta < 0$ and is singular for $\delta > 0$. For $\delta = 0$ a more detailed analysis exhibits a logarithmic singularity $\sim (\ln \sigma)^f$, where f depends on the structure of the subdiagram.

We now consider the theory for $d < 4$. In this case Eq. (7.29) shows that only subdiagrams with $M = 1, L = 0$ are u.v. divergent. Such structures are subdiagrams of $\tilde{G}_F^{(0,1)}$. Furthermore, analyzing the inverse $\tilde{\Gamma}_F^{(0,1)}$ (Eq. (7.21)) we easily find that it takes the form

$$\tilde{\Gamma}_F^{(0,1)}(\mathbf{p}, -\mathbf{p}; \tau_m) = (\tau_m + \mathbf{p}^2)e^{(\tau_m+\mathbf{p}^2)\sigma} - \Sigma(\mathbf{p}, \tau_m) \ , \tag{7.30}$$

where $\Sigma(\mathbf{p}, \tau)$ is given by the sum of all 1-l-i subdiagrams of $\tilde{G}_F^{(0,1)}$. Diagrammatically the argument is sketched in Fig. 7.3. Carrying through the shift $\tau_m \to \hat{\tau}_m + \tau_c$ and using the definition of τ_c:

$$\tilde{\Gamma}_F^{(0,1)}(\mathbf{0}, \mathbf{0}; \tau_c) = 0 = \tau_c e^{\tau_c \sigma} - \Sigma(\mathbf{0}, \tau_c) \ , \tag{7.31}$$

from each contribution to $\Sigma(\mathbf{p}, \tau_m)$ we subtract the corresponding diagram evaluated at $\mathbf{p} = 0, \tau_m = \tau_c$. Since the u.v. behavior of the integrand represented by the diagram is independent of $\mathbf{p}$ or τ_m this improves the convergence, effectively reducing δ by 2. (cf. Eq. (7.26)) We thus find that the 'mass subtracted' theory is finite for $\sigma \to 0, d < 4$.

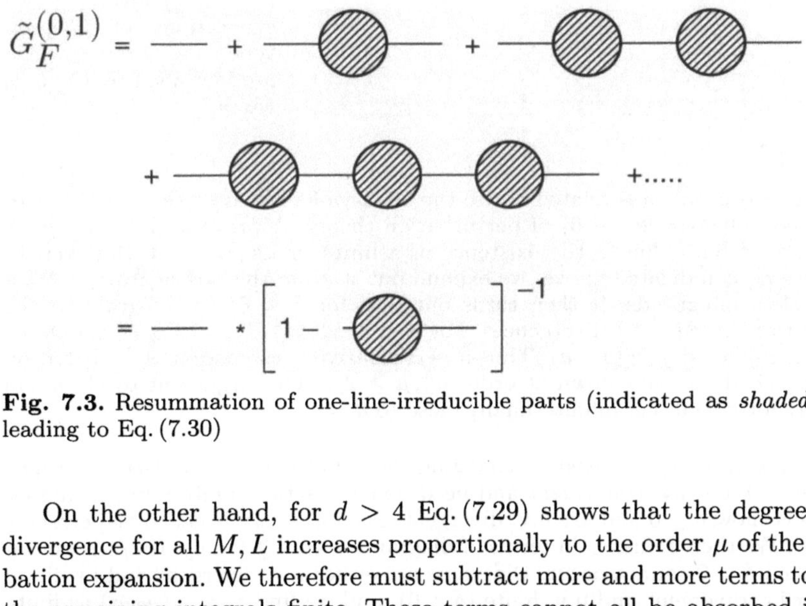

Fig. 7.3. Resummation of one-line-irreducible parts (indicated as *shaded circles*) leading to Eq. (7.30)

On the other hand, for $d > 4$ Eq. (7.29) shows that the degree of u.v. divergence for all M, L increases proportionally to the order μ of the perturbation expansion. We therefore must subtract more and more terms to render the remaining integrals finite. These terms cannot all be absorbed into the few constants of our original theory, and the continuous chain limit cannot be defined.

At the upper critical dimension $d = 4$ Eq. (7.29) shows that not only subdiagrams $(M = 1, L = 0)$ but also subdiagrams $(M = 1, L = 1)$, $(M = 2, L = 0)$ are divergent. Redefining all the parameters of the theory, we then can define a nontrivial form of the continuous chain limit, differing from the naive limit discussed in this chapter. This sophisticated 'multiplicative renormalization' underlies renormalized perturbation theory, which will be an important topic of later chapters. At present we may be content with the statement that a simple shift $\tau_m \to \tau_m + \tau_c$ is sufficient to give the continuous chain model and two parameter theory a valid meaning for $d < 4$.

Our discussion shows the basic features of mass subtraction, but a careful analysis exhibits some complications. As a first problem we note that a general contribution to $\Sigma(\mathbf{q}, \tau)$ involves internal divergent subdiagrams again of structure $\Sigma(\mathbf{q}', \tau)$. Figure 7.4 shows an example. Now this problem can be solved fairly easily by noting that the divergent substructures show a strict hierarchical ordering. The first generation of divergences may be defined as the set of divergent subdiagrams which are not part of a larger divergent subdiagram. These subdiagrams may contain divergent subdiagrams of the second generation, which in turn may contain a third generation, and so on. Expanding out τ_c from all propagators it is not hard to prove the existence of the u.v. limit $\sigma \to 0$ by mathematical induction over the hierarchy of generations.

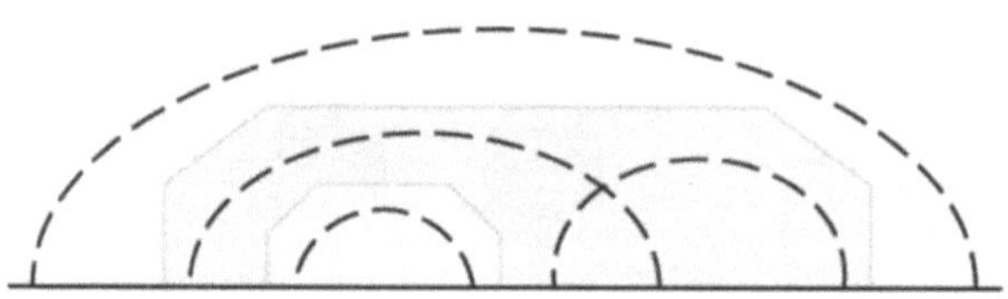

Fig. 7.4. A contribution to $\Sigma(\mathbf{q}, \tau)$ involving two generations of $\Sigma(\mathbf{q}, \tau)$-type subdiagrams, which are identified by the *grey boxes*

The second problem is related not to the ultra-violet behavior ($\mathbf{k}_i \to \infty$) but to the infrared behavior ($\mathbf{k}_i \to 0$) of perturbation theory. A priori all integrals exist in the infrared limit due to the existence of a finite 'mass' $\tau_m > 0$. However, in calculating τ_c as indicated above, we expand out τ_c from the propagators, dealing with massless integrands. It then turns out that for $\delta \le 0$ the integrals do not exist due to infrared (i.r.) divergences. For $d < 4, M = 1, L = 0$ Eq. (7.30) shows that $\delta \le 0$ implies $\mu \ge 2/(4-d)$. Thus a perturbative construction of τ_c, based on solving Eq. (7.31), breaks down at order $\tilde{u}_0^\mu, \mu \ge 2/\varepsilon$. Our argument to all orders works only for $\varepsilon \to 0$, i.e. infinitesimally close to $d = 4$.

The problem may be solved by changing the definition of τ_c. Indeed, diagrams showing $\delta < 0$ are u.v. convergent and need no subtraction. Only a finite number of diagrams show $\delta \ge 0$ and are u.v. divergent. Such a theory is called 'superrenormalizable'. We then define τ_c in terms of diagrams with $\delta \ge 0$ only. If $2/\varepsilon$ is not an integer, then $\delta = 0$ cannot occur. Diagrams are either u.v. divergent and i.r. finite ($\delta > 0$) or i.r. divergent and u.v. finite ($\delta < 0$), and we find $\tau_c = \tau_c(\tilde{u}_0, \sigma)$ as finite polynomial in $\tilde{u}_0$. If $2/\varepsilon$ is an integer in addition we find a logarithmic anomaly, i.e. a contribution to τ_c of the form $\tilde{u}_0^{2/\varepsilon} \ln \tilde{u}_0$. Since these problems are not essential in the sequel we do not go into further details.

Turning now to the polymer Greensfunction $\tilde{G}_E^{(L,M)}$ we note that it is related to the (not subtracted) field-theoretic Greensfunction $\tilde{G}_F^{(L,M)}$ by an inverse Laplace transform

$$\tilde{G}^{(L,M)}_{E,m_1\dots m_L}(\mathbf{q}_1,\dots \mathbf{q}_L;\mathbf{p}_1,\dots,\mathbf{p}_{2M};R^2_{01},\dots R^2_{0M}) =$$

$$\int_{a-i\infty}^{a+i\infty}\prod_{m=1}^{M}\left(\frac{d\tau_m}{2\pi i}e^{R^2_{0m}\tau_m}\right)\tilde{G}^{(L,M)}_{F,m_1\dots m_L}(\mathbf{q}_1,\dots \mathbf{q}_L;\mathbf{p}_1,\dots,\mathbf{p}_{2M};\tau_1,\dots,\tau_M)\,. \tag{7.32}$$

a is chosen such that the singularities of $\tilde{G}^{(L,M)}_F$ are found to the left of the integration contour. Substituting $\tau_m = \hat{\tau}_m + \tau_c$ and carrying through the mass subtraction we find

$$\begin{aligned} &G^{(L,M)}_E(\dots;R^2_{0,1},\dots,R^2_{0,M}) \\ &= \exp\left(-\tau_c\ell^2\sum_{m=1}^{M}n_m\right)\tilde{G}^{(L,M)}_E(\dots;R^2_{0,1},\dots,R^2_{0,M}) \\ &= \int_{a-i\infty}^{a+i\infty}\prod_{m=1}^{M}\left(\frac{d\hat{\tau}_m}{2\pi i}e^{R^2_{0,m}\hat{\tau}_m}\right)G^{(L,M)}_F(\dots;\hat{\tau}_1,\dots,\hat{\tau}_M)\ , \end{aligned} \tag{7.33}$$

where $G^{(L,M)}_F$ and thus $G^{(L,M)}_E$ exists for $\sigma = 0, d < 4$. We used relation (7.1): $R^2_{0m} = \ell^2 n_m$. Thus in M-chain Greensfunctions or cumulants we can take the continuous chain limit, after we extracted some well defined prefactor. Since this prefactor cancels in normalized expectation values, the continuous chain limit exists in the naive sense for quantities like $R^2_e(c_p = 0)$, $R^2_g(c_p = 0)$, A^{Π}_2, etc.

The argument is easily extended to the grand canonical formalism. We simply absorb the prefactor into the chemical potential, defining

$$e^{\mu_E(R^2_o)} = (4\pi)^{-d/2}\,\ell^{-d-2}e^{\mu_p(n)+\tau_c\ell^2 n}\ , \tag{7.34}$$

where the additional factor of ℓ^{-2} is due to transforming the summations over chain lengths n_m into integrals over R^2_{0m}. With Eqs. (7.33), (7.34), (7.6), Eq. (5.2) for the grand partition function can be rewritten as

$$\begin{aligned} \mathcal{Z}_E[\mu_E] = 1 + \sum_{m=1}^{\infty}\frac{1}{m!}\int_0^{\infty} dR^2_{01}\dots dR^2_{0M}\prod_{m=1}^{M}e^{\mu_E(R^2_{0m})} \\ G^{(0,1)}_E(0,\dots,0;R^2_{01},\dots,R^2_{0M})\ , \end{aligned} \tag{7.35}$$

showing that the continuous chain limit exists. Corresponding results are easily formulated for all grand canonical expectation values. As announced in the introductory section, our derivation shows that the shift of $\mu_p(n)$ by $\tau_c\ell^2 n$ is needed to keep the system in the critical region of long chains.

Also the construction of the loop expansion is not changed in any essential way. We only have to keep in mind that the divergent subdiagrams are properly subtracted. Since we will not need the explicit results in the sequel, we omit the detailed discussion.

A 7. Appendix

A 7.1 Formal relation to field theory

Much of our understanding of critical phenomena is based on the Landau-Ginzburg model of a ferromagnet. This model concentrates on the local magnetization, represented by an m-component vector field $S_\alpha(\mathbf{r}), \alpha = 1, \ldots, m$, often called a 'classical spin field'. The interaction of the spin field is described by the Landau-Ginzburg Hamiltonian

$$\mathcal{H}[\vec{S}] = \int d^d r \Big[\frac{\tau}{2}\vec{S}^2(\mathbf{r}) + \frac{1}{2}\left(\nabla\vec{S}(\mathbf{r})\right)^2 + \frac{u_0}{8}\left(\vec{S}^2(\mathbf{r})\right)^2 - \vec{h}\cdot\vec{S}(\mathbf{r})\Big] \quad , \qquad \text{(A 7.1)}$$

where

$$\vec{S}^2(\mathbf{r}) = \sum_{\alpha=1}^{m} S_\alpha^2(\mathbf{r}) \quad , \qquad \text{(A 7.2)}$$

and ∇ denotes the gradient in d dimensions. $\vec{h}$ is the constant external magnetic field. Temperature essentially affects the variable τ. The partition function is defined as

$$\mathcal{Z}_{LG}(\tau, \vec{h}) = \int D[\vec{S}(\mathbf{r})] e^{-\mathcal{H}[\vec{S}]} \quad , \qquad \text{(A 7.3)}$$

and correlation functions take the form

$$\langle S_\alpha(\mathbf{r}_1) S_\beta(\mathbf{r}_2)\rangle = \frac{1}{\mathcal{Z}_{LG}} \int D[\vec{S}(\mathbf{r})] S_\alpha(\mathbf{r}_1) S_\beta(\mathbf{r}_2) e^{-\mathcal{H}[\vec{S}]} \quad , \qquad \text{(A 7.4)}$$

where we used the spin-spin correlations as an example. Equations (A 7.3), (A 7.4) introduce functional integrals, summing over all configurations of the field $\vec{S}(\mathbf{r})$. These integrals for instance may be defined in terms of the Fourier components as in Appendix A 5.3, Eqs. (A 5.14), (A 5.15). As an alternative we may first formulate the theory on a lattice, which yields a discrete set of integration variables, and then we may take the limit of vanishing lattice constant. This gives rise to problems most similar to those discussed in Sect. 7.2 in the context of the continuous chain limit. In this appendix however we work on a purely formal level, ignoring all cut off effects or existence problems. The reader interested in a deeper understanding of field theory is refered to [Ami84, ZJ89].

Evaluating the model (A 7.1) one finds that it describes a second order phase transition occurring for $\vec{h} = 0$ and a critical value $\tau = \tau_c$. The transition is signaled by long range correlations of the spin field or power-type singularities occurring for $\vec{h} \to 0, \tau \to \tau_c$. Furthermore macroscopic observables like the free energy or the spin correlation functions show scaling in $\hat{\tau} = \tau - \tau_c$ and $\vec{h}$, most similar to the scaling laws in polymer solutions.

There exists a formal relation among polymer theory and the Landau-Ginzburg model. Specifically for the Greensfunction of a single continuous chain one finds

$$\begin{aligned} \hat{G}_F(\mathbf{r}, \mathbf{r}'; \tau) &= \int_0^\infty dR_0^2 e^{-\tau R_0^2} \hat{G}_E(\mathbf{r}, \mathbf{r}'; R_0^2) \\ &= \langle S_1(\mathbf{r}) S_1(\mathbf{r}') \rangle_{m=0}^{\vec{h}\equiv 0} \quad . \end{aligned} \tag{A 7.5}$$

Here $\hat{G}_E(\mathbf{r}, \mathbf{r}'; R_0^2)$ is the Greensfunction of the Edwards model in real space

$$\hat{G}_E(\mathbf{r}, \mathbf{r}'; R_0^2) = \int_{\mathbf{r}(0)=\mathbf{r}'}^{\mathbf{r}(R_0^2)=\mathbf{r}} D[\mathbf{r}(s)] e^{-V_E[\mathbf{r}(s)]} \quad . \tag{A 7.6}$$

The relation (A 7.5) first was derived by de Gennes, initiating the renormalization group or scaling approach to polymer solutions. Again these expressions need some explanation. Equation (A 7.6) defines $\hat{G}_E(\mathbf{r}, \mathbf{r}'; R_0^2)$ as 'path integral', summing over all continuous paths $\mathbf{r}(s), 0 \le s \le R_0^2$; $\mathbf{r}(0) = \mathbf{r}'$; $\mathbf{r}(R_0^2) = \mathbf{r}$. It is properly defined as the continuous chain limit of the discrete chain model. Equation (A 7.5) identifies the Laplace transform $\hat{G}_F(\mathbf{r}, \mathbf{r}'; \tau)$ as the spin correlation function of the Landau Ginzburg model in zero external field. We, however, have to take the formal limit of an $m = 0$ component spin field. Clearly such a field does not make any sense as a physical object, but the limit is well defined in perturbation theory in powers of $\tilde{u}_0$. Indeed, evaluating $\langle S_1(\mathbf{r}) S_1(\mathbf{r}') \rangle$ for arbitrary m, we find that each order of perturbation theory just is a polynomial of finite order in m, to be evaluated without harm for $m = 0$.

In view of the preceding remark the natural way to prove Eq. (A 7.5) is by comparing the perturbation expansions. Constructing diagrammatic perturbation theory for $m = 0$ we find Feynman rules identical to those established for the Laplace transformed continuous chain model. The derivation of perturbation theory for the Landau-Ginzburg model may be found, for instance, in [Ami84]. We present here a more formal, but also more elegant argument [Eme75].

Introducing the segment density generalized to a continuous chain

$$\rho_E(\mathbf{r}) = \int_0^{R_0^2} ds\, \delta^d(\mathbf{r} - \mathbf{r}(s)) \quad , \tag{A 7.7}$$

we write $\hat{G}_E(\mathbf{r}, \mathbf{r}'; R_0^2)$ as

$$\hat{G}_E(\mathbf{r},\mathbf{r}';R_0^2) = \int_{\mathbf{r}(0)=\mathbf{r}'}^{\mathbf{r}(R_0^2)=\mathbf{r}} D[\mathbf{r}(s)] \exp\Big\{ - \int_0^{R_0^2} ds \left(\frac{d\mathbf{r}(s)}{2ds}\right)^2 - \frac{u_0}{2}\int d^d r \rho_E^2(\mathbf{r})\Big\} \ . \tag{A 7.8}$$

We then again introduce the auxiliary field $\varphi(\mathbf{r})$ (Sect. A 5.3.1) to write

$$\hat{G}_E(\mathbf{r},\mathbf{r}';R_0^2) = \int D[\phi(\mathbf{r})] e^{-\frac{1}{2u_0}\int d^d r \phi^2(\mathbf{r})} G(\mathbf{r},\mathbf{r}';R_0^2,[\phi]) \ , \tag{A 7.9}$$

where

$$G(\mathbf{r},\mathbf{r}';R_0^2,[\phi]) = \int_{\mathbf{r}(0)=\mathbf{r}'}^{\mathbf{r}(R_0^2)=\mathbf{r}} D[\mathbf{r}(s)] \exp\Big\{ - \int_0^{R_0^2} ds \left[\left(\frac{d\mathbf{r}(s)}{2ds}\right)^2 + i\phi(\mathbf{r}(s))\right]\Big\}. \tag{A 7.10}$$

This is nothing but the continuous chain version of Eq. (A 5.18), except that the endpoints $\mathbf{r},\mathbf{r}'$ are held fixed. Equation (A 7.10) is the standard path integral representation of the Greensfunction of a diffusion equation, where R_0^2 plays the role of time:

$$\frac{\partial}{\partial R_0^2} G(\mathbf{r},\mathbf{r}';R_0^2,[\phi]) = [\Delta_r - i\phi(\mathbf{r})]\, G(\mathbf{r},\mathbf{r}',R_0^2,[\phi]) \tag{A 7.11}$$

$$G(\mathbf{r},\mathbf{r}';0,[\phi]) = \delta^d(\mathbf{r}-\mathbf{r}') \ . \tag{A 7.12}$$

To show this we recall that the path integral is defined as continuous chain limit of the discrete chain model:

$$\begin{aligned} &G_D(\mathbf{r},\mathbf{r}';n,[\ell^2\phi]) \\ &= (4\pi\ell^2)^{-d/2} \int_\Omega \prod_{j=1}^{n-1} \frac{d^d r_j}{(4\pi\ell^2)^{d/2}} \exp\Big\{ - \sum_{j=1}^{n} \Big[\frac{(\mathbf{r}_j - \mathbf{r}_{j-1})^2}{4\ell^2} + i\ell^2\phi(\mathbf{r}_j)\Big]\Big\}; \\ &\mathbf{r}_0 = \mathbf{r}'; \quad \mathbf{r}_n = \mathbf{r} \ . \end{aligned} \tag{A 7.13}$$

Singling out the integration over $\mathbf{r}_{n-1}$ and introducing $\mathbf{y} = (\mathbf{r}_n - \mathbf{r}_{n-1})/\ell$ we may write this as

$$e^{i\ell^2\phi(\mathbf{r})} G_D(\mathbf{r},\mathbf{r}';n,[\ell^2\phi]) = \int \frac{d^d y}{(4\pi)^{d/2}} e^{-y^2/4} G_D(\mathbf{r}-\ell\mathbf{y},\mathbf{r}';n-1,[\ell^2\phi]) \ . \tag{A 7.14}$$

Being interested in the continuous chain limit we expand the Greensfunction in the integral up to terms of order ℓ^2 and carry out the $\mathbf{y}$-integration to find

$$(1+i\ell^2\phi(\mathbf{r}))G_D(\mathbf{r},\mathbf{r}';n,[\ell^2\phi]) = (1+\ell^2\Delta_r)G_D(\mathbf{r},\mathbf{r}';n-1,[\ell^2\phi]) + O(\ell^4) \ ,$$

or

$$\frac{1}{\ell^2}\left(G_D(\mathbf{r},\mathbf{r}';n,[\ell^2\phi]) - G_D(\mathbf{r},\mathbf{r}';n-1,[\ell^2\phi])\right)$$
$$= \Delta_r G_D(\mathbf{r},\mathbf{r}';n-1,[\ell^2\phi]) - i\phi(\mathbf{r})G_D(\mathbf{r},\mathbf{r}';n,[\ell^2\phi]) + O(\ell^2) \quad . \quad \text{(A 7.15)}$$

Reintroducing $R_0^2 = n\ell^2, \partial R_0^2 = \ell^2$ and taking the continuous chain limit we find Eq. (A 7.11).

The Laplace transform now acts directly on $G(\mathbf{r},\mathbf{r}';R_0^2,[\phi])$:

$$\tilde{G}(\mathbf{r},\mathbf{r}';\tau,[\phi]) = \int_0^\infty dR_0^2 e^{-\tau R_0^2} G(\mathbf{r},\mathbf{r}';R_0^2,[\phi]) \quad , \quad \text{(A 7.16)}$$

and taking the transform of Eq. (A 7.11) we find

$$[\tau + i\phi(\mathbf{r}) - \Delta_r]\,\hat{G}(\mathbf{r},\mathbf{r}';\tau,[\phi]) = \delta^d(\mathbf{r}-\mathbf{r}') \quad . \quad \text{(A 7.17)}$$

This is a Schrödinger-type equation for the Greensfunction of a particle in the imaginary potential $i\phi(\mathbf{r})$, τ playing the role of the energy. It shows that $\hat{G}$ is given by the resolvent

$$\hat{G}(\mathbf{r},\mathbf{r}';\tau,[\phi]) = [\tau + i\phi - \Delta]^{-1}_{\mathbf{r},\mathbf{r}'} \quad , \quad \text{(A 7.18)}$$

where the $\mathbf{r},\mathbf{r}'$ matrix element of the operator inverse is understood. Summarizing the argument we have found the representation

$$\hat{G}_F(\mathbf{r},\mathbf{r}';\tau) = \int D[\phi(\mathbf{r})] e^{-\frac{1}{2u_0}\int d^d r \phi^2(\mathbf{r})} [\tau + i\phi - \Delta]^{-1}_{\mathbf{r},\mathbf{r}'} \quad . \quad \text{(A 7.19)}$$

We now turn to the field theoretic correlation function (A 7.4), where we first consider the numerator

$$G_{LG}(\mathbf{r}_1,\mathbf{r}_2) = \int D[\vec{S}(\mathbf{r})] S_1(\mathbf{r}_1) S_1(\mathbf{r}_2) e^{-\mathcal{H}[\vec{S}]} \quad . \quad \text{(A 7.20)}$$

Introducing the auxiliary field $\phi(\mathbf{r})$ to decouple the $(\vec{S}^2)^2$ term we find for $\vec{h} = 0$

$$G_{LG}(\mathbf{r}_1,\mathbf{r}_2) = \int D[\phi(\mathbf{r})] e^{-\frac{1}{2u_0}\int d^d r \phi^2(\mathbf{r})}$$
$$\cdot \int D[\vec{S}(\mathbf{r})] S_1(\mathbf{r}_1) S_1(\mathbf{r}_2) \exp\Big\{ -\frac{1}{2}\sum_{\alpha=1}^{m} \int d^d r S_\alpha(\mathbf{r}) \left[\tau + i\phi(\mathbf{r}) - \Delta\right] S_\alpha(\mathbf{r}) \Big\} \, . \quad \text{(A 7.21)}$$

The $(\nabla\vec{S})^2$ term was transformed by partial integration, assuming $\vec{S}(\mathbf{r})$ to vanish at the boundaries of the integration volume. Equation (A 7.21) has the form of a functional Gaussian integral yielding

$$G_{LG}(\mathbf{r}_1,\mathbf{r}_2) = \int D[\phi(\mathbf{r})] e^{-\frac{1}{2u_0}\int d^d r \phi^2(\mathbf{r})} \left(\mathrm{Det}\left[\tau + i\phi(\mathbf{r}) - \Delta\right]\right)^{-\frac{m}{2}}$$
$$\left[\tau + i\phi(\mathbf{r}) - \Delta\right]^{-1}_{\mathbf{r}_1,\mathbf{r}_2} \quad . \qquad \text{(A 7.22)}$$

This is easily shown by using a lattice in **r**-space to define the functional integral. Some necessary normalization factors are absorbed into the definition of the integration measure. A similar calculation yields

$$\mathcal{Z}_{LG} = \int D[\phi(\mathbf{r})] e^{-\frac{1}{2u_0}\int d^d r \phi^2(\mathbf{r})} \left(\mathrm{Det}\left[\tau + i\phi(\mathbf{r}) - \Delta\right]\right)^{-\frac{m}{2}} \quad . \qquad \text{(A 7.23)}$$

Taking $m = 0$ we now find

$$\mathcal{Z}_{LG} \equiv 1 \quad . \qquad \text{(A 7.24)}$$

by virtue of the normalization (A 5.12) of the ϕ-integral. Thus

$$\langle S_1(\mathbf{r}_1) S_1(\mathbf{r}_2)\rangle^{h=0}_{m=0} = G_{LG}(\mathbf{r}_1,\mathbf{r}_2)\Big|_{m=0} \quad . \qquad \text{(A 7.25)}$$

The basic result (A 7.5) follows by comparing Eqs. (A 7.19), (A 7.22).

The argument can easily be generalized to yield Des Cloizeaux's result: for $\vec{h} \neq 0, m = 0$ the Landau-Ginzburg model is equivalent to a grand canonical ensemble of polymer chains specified by a specific choice of the chemical potential $\mu_p(n)$ for chains of length n

$$\mu_p(n) = \ln\left(\frac{\vec{h}^2}{2}\left(4\pi\ell^2\right)^{d/2}\right) - \tau\ell^2 n \quad . \qquad \text{(A 7.26)}$$

This choice describes a polydisperse solution, in equilibrium with respect to a fictitious reaction of breaking and recombining chains:

$$[A]_n + [A]_{n'} \rightleftarrows [A]_{n+n'} \quad . \qquad \text{(A 7.27)}$$

$\vec{h}^2$ controls the chain concentration c_p, whereas τ determines the segment concentration c. We should mention that this ensemble plays some role experimentally. It describes 'living' solutions, i.e. solutions where the polymerisation reaction is not stopped, but the chains can prolong or shorten according to the surroundings.

Notes

The equivalence of the end-to-end correlations of a single chain to the spin correlations of an LGW-model has been observed in [dG72]. For the special choice (A 7.26) of the chemical potential this result was generalized to solutions of finite overlap in [Clo75], and was further generalized to density correlations in [SW77]. For arbitrary polydispersity the equivalence to a Landau-Ginzburg field theory was worked out in [SW80].

The Edwards model is discussed in detail in [Fre87].

A discussion of mass-subtraction in field theory can be found in standard textbooks [Ami84, ZJ89]. The equivalent argument directly in the polymer representation has been presented in [Clo80a]. The problems associated with mass-subtraction for $d < 4$ from a field theoretic point of view are discussed in [SD89]. For a polymer system in $d = 3$ a discussion has been presented in [Clo82]. (See also [CJ90], Sect. 12.3.2.)

The relation of the two-parameter limit to lattice models of polymer configurations has been discussed in [BD79].

Part II

Renormalization Group and Scaling

We discuss here the basic ideas of the renormalization group, using the discrete chain model. This is not the most elegant or powerful approach, and in Part III of this book we will present a much more efficient scheme. However, the present approach is conceptually the simplest, and it allows us to explain all the relevant features: dilatation symmetry and scaling, fixed points and universality, crossover. Furthermore, technical aspects like the ε-expansion also come up. We are then prepared to discuss the qualitative concept of scaling in its general form and to work out some consequences.

Another topic can also be taken up again. Can we justify our simple model somewhat better, not relying only on the very naive estimates given in Chap. 2? A rigorous argument is still beyond all reach, but the renormalization group at least provides us with a good and general framework to tackle such questions. Partial results presenting small steps towards a full understanding are available. They correct the naive arguments of Chap. 2, and they indicate that at least the effect of small deviations from our model rapidly dies out for long chains.

8. The Renormalization Group: Fundamental Aspects

This chapter is of central importance. We introduce here the renormalization group, based on the observation that our choice of the elementary segment involves a large amount of arbitrariness. Thus there should exist a mapping from segments of size $\ell = \ell_0$ to segments of size ℓ_1, which does not change macroscopic observables. It also should not affect the basic structure of the model, viz. Gaussian segments interacting via some local two-body repulsion. Our previous arguments on the structure of the model never refered to any distinguished value of ℓ. In Sect. 8.1 we formulate these considerations more precisely in terms of the *'Renormalization Group Hypothesis'*. This is to be amended by the *'Fixed Point Hypothesis'*, assuming that with increasing size of the elementary segment the excluded volume constant tends to some finite limiting value. Universal scaling and power laws follow naturally.

To corroborate these qualitative considerations in Sect. 8.2 we use our perturbative results for R_e^2, A_2^{Π} to construct a simple realization of the renormalization group, first concentrating on the excluded volume limit. In Sect. 8.3 we show how the crossover among Θ-point and excluded volume limit can be treated in principle.

Thus all the essential aspects of the renormalization group are explained in this chapter. We will see how the universality of the excluded volume limit results from the existence of a fixed point of the excluded volume coupling. Scaling and power laws in the excluded volume limit will be identified as fixed point properties. We furthermore will learn how the renormalization group flow equations, giving the change of chain length and excluded volume coupling under an infinitesimal change of ℓ, allow for a calculation of crossover effects outside the excluded volume limit. The somewhat strange concept of expanding in powers of $\varepsilon = 4 - d$ (recall d = spatial dimensionality) will emerge as a natural tool for organizing perturbation theory.

8.1 Basic ideas

In Chap. 6 we learned that in the excluded volume limit $\beta_e > 0, n \to \infty$, the cluster expansion breaks down, simply because it orders according to powers of $z = \beta_e n^{\varepsilon/2} \to \infty$. To proceed, we need a new idea, going beyond perturbation theory. The new concept is known as the 'Renormalization Group' (RG), which postulates, proves, and exploits the fascinating scale invariance property of the theory.

We recall that we are interested in universal features, i.e. properties that are independent of the microstructure. Our model – simple as it is – still has a definite microstructure: $n = n_0$ Gaussian segments of length $\ell = \ell_0$, interacting via an excluded volume of strength $\beta_e = \beta_{e0}$. Taking universality seriously we should be able to describe the same macroscopic physics by a chain of effective segments of length ℓ_1. Of course in the new description we have to adjust n and β_e. We thus may formulate the basic idea of the renormalization group.

RG hypothesis

Irrelevance of the microstructure implies the existence of a mapping

$$\begin{aligned} \ell_0 &\to \ell_1 = \ell_0/\lambda \\ n_0 &\to n_1 = A(\lambda, \beta_{e0}) n_0 \\ \beta_{e0} &\to \beta_{e1} = B(\lambda, \beta_{e0}) \beta_{e0} \ , \end{aligned} \tag{8.1}$$

which leaves macroscopic observables invariant.

λ will be called the scaling parameter, obeying $0 < \lambda < 1$. Note that the dimensionless functions $A(\ldots), B(\ldots)$ can depend on the length scale ℓ_0 only via the ratio $\lambda = \ell_0/\ell_1$. Note further that the functions A and B are independent of n_0. The mapping represents a redefinition of the microstructure, and as such should be independent of the total chain length.

It should be clear that for a noninteracting chain such a mapping exists and is trivial:

$$\ell_1 = \ell_0/\lambda, \ n_1 = n_0 \lambda^2, \ \beta_{e1} = \beta_{e0} = 0 \ . \tag{8.2}$$

The only relevant macroscopic parameter of the noninteracting model is the end-to-end distance, which is invariant under this transformation:

$$\frac{1}{2d} R_e^2 \equiv R_0^2 = \ell_0^2 n_0 = \ell_1^2 n_1 \ . \tag{8.3}$$

In this case the mapping just expresses the relation among segment size and segment number which holds for the Gaussian chain.

A trivial realization also exists in the two-parameter scheme:

$$\ell_1 = \ell_0/\lambda, \quad n_1 = n_0\lambda^2, \quad \beta_{e1} = \beta_{e0}\lambda^{-\varepsilon} \tag{8.4}$$

This transformation leaves both R_0^2 and $z = \beta_e n^{\varepsilon/2}$ (Eq. (7.10)) invariant. It just expresses naive dimensional analysis in the continuous chain model. The power of the RG-approach lies in the fact that we can construct nontrivial realizations. These take into account more than just the leading n-dependence of each order of perturbation theory and therefore obey the condition of invariance of the macroscopic observables up to much smaller corrections.

We now evaluate the consequences of the renormalization hypothesis. The mapping (8.1) has a *multiplicative structure.* First rescaling by a factor λ and in a second step rescaling by λ' we must find the same result as if we rescale by $\lambda \cdot \lambda'$ in a single step:

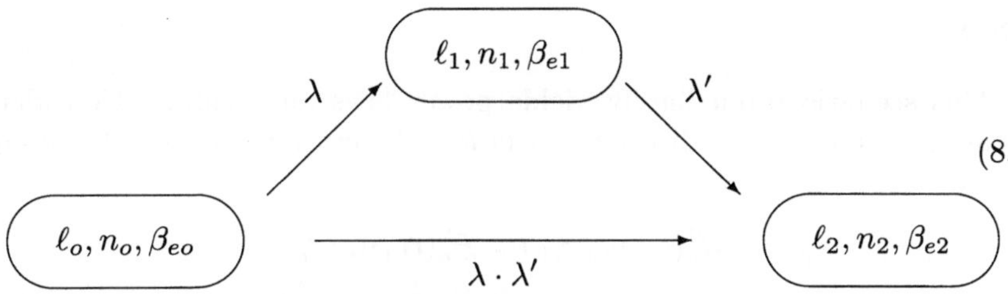

(8.5)

Mathematically this structure is expressed as

$$\beta_{e2} = B(\lambda\lambda', \beta_{e0})\beta_{e0} = B(\lambda', \beta_{e1})\beta_{e1} = B(\lambda', B(\lambda, \beta_{e0})\beta_{e0})B(\lambda, \beta_{e0})\beta_{e0} \ , \tag{8.6}$$

$$n_2 = A(\lambda\lambda', \beta_{e0})n_0 = A(\lambda', \beta_{e1})n_1 = A(\lambda', B(\lambda, \beta_{e0})\beta_{e0})A(\lambda, \beta_{e0})n_0 \ . \tag{8.7}$$

These equations have remarkable consequences if we add the so-called fixed point hypothesis. For a very long chain we can apply the mapping repeatedly, always rescaling by a factor of λ. In the limit of an infinitely long chain we can thus construct an infinite sequence of couplings $\beta_{e0}, \beta_{e1}, \beta_{e2}, \ldots$. We now make the crucial assumption that this sequence converges.

Fixed point hypothesis

There exists a fixed point β_e^* defined as

$$\lim_{k\to\infty} \beta_{ek} = \beta_e^*. \tag{8.8}$$

β_e^* is finite and independent of λ.

β_e^* necessarily obeys

$$\beta_e^* = B(\lambda, \beta_e^*)\beta_e^* \ , \tag{8.9}$$

or

$$B(\lambda, \beta_e^*) = 1, \ \text{for} \ \beta_e^* > 0 \ \ , \tag{8.10}$$

equivalently. Note that the trivial realization (8.4) does not obey the fixed point hypothesis for $\beta_{e0} \neq 0$.

As an immediate consequence we can determine the scale factor $A(\lambda, \beta_e)$ at the fixed point. Equation (8.7) yields

$$A(\lambda\lambda', \beta_e^*) = A(\lambda, \beta_e^*)A(\lambda', \beta_e^*) \ , \tag{8.11}$$

which is solved by

$$A(\lambda, \beta_e^*) = \lambda^{1/\nu} \tag{8.12}$$

for any ν. The exponent $1/\nu$ can depend on β_e^* only, but otherwise is unknown a priori.

This scenario immediately yields power laws and scaling. Consider for instance the end-to-end distance. After $k \gg 1$ iterations of the R.G. mapping we have

$$\begin{aligned} R_e^2(\ell_0, n_0, \beta_{e0}) &= R_e^2(\ell_k, n_k, \beta_{ek}) \\ &\approx R_e^2(\ell_k, n_k, \beta_e^*) \ , \end{aligned} \tag{8.13}$$

where we assume that k is so large that we can replace β_{ek} by the limiting value β_e^*. Dimensional analysis yields

$$R_e^2(\ell_k, n_k, \beta_e^*) = \ell_k^2 \bar{R}_e^2(n_k, \beta_e^*) \ . \tag{8.14}$$

Applying the mapping once more, this time using some arbitrary factor λ', we find from Eqs. (8.1), (8.9), (8.12), (8.13)

$$R_e^2(\ell_0, n_0, \beta_{e0}) \approx \frac{\ell_k^2}{\lambda'^2} \bar{R}_e^2(n_k \lambda'^{1/\nu}, \beta_e^*) \ . \tag{8.15}$$

Since the left-hand side is independent of λ', this variable must also drop out on the right-hand side. This yields the desired *power law*

$$R_e^2(\ell_0, n_0, \beta_{e0}) \approx \ell_k^2 n_k^{2\nu} a_e^*(\beta_e^*) = B^2 n_0^{2\nu} a_e^*(\beta_e^*) \ , \tag{8.16}$$

where

$$B = \ell_k \left(\frac{n_k}{n_0}\right)^\nu = \ell_0 \prod_{j=0}^{k-1} \left[\lambda^{-1}(A(\lambda, \beta_{ej}))^\nu\right] \ . \tag{8.17}$$

The exponent ν and the 'amplitude' a_e^* depend only on the fixed point coupling β_e^*, but are independent of ℓ_0 and β_{e0}. These microscopic parameters are absorbed into the nonuniversal scale factor B, which in turn is independent of n_0. B can be interpreted as an effective segment size, renormalized by interaction effects.

We expect the law (8.16) to be valid only for long chains. Which step of our argument makes use of this condition? A little thought reveals that it is hidden in Eq. (8.13). On the one hand we assume that k is so large that $\beta_{ek} \approx \beta_e^*$. On the other hand we assume that on this level we still deal with a proper chain of segments of size ℓ_k. This implies $R_e > \ell_k = \ell_0/\lambda^k \gg \ell_0$. The power law (8.16) is a valid approximation for large coils, becoming rigorously correct only in the limit $k \to \infty$, i.e. $R_e \to \infty$ or $n_0 \to \infty$, equivalently.

Other results easily follow. We, for instance, can carry through the very same argument for the radius of gyration to find the power law

$$R_g^2 \approx B^2 n_0^{2\nu} a_g^*(\beta_e^*) \,, \tag{8.18}$$

with the same nonuniversal scale factor B. This shows that

$$\lim_{n_0 \to \infty} \frac{R_g^2(\ell_0, n_0, \beta_{e0})}{R_e^2(\ell_0, n_0, \beta_{e0})} = \frac{a_g^*(\beta_e^*)}{a_e^*(\beta_e^*)} \tag{8.19}$$

is a *universal ratio*, independent of the microstructure, a result which for the noninteracting model was pointed out before (Eq. (3.32)).

As another example we consider the osmotic pressure $\hat{\Pi}$. The ratio

$$\frac{\hat{\Pi}}{k_B T c_p} = \bar{\Pi}(c_p \ell_0^d, n_0, \beta_{e0}) \tag{8.20}$$

is a dimensionless function. Choosing k as above, we find

$$\begin{aligned} \bar{\Pi}(c_p \ell_0^d, n_0, \beta_{e0}) &\approx \bar{\Pi}(c_p \ell_k^d, n_k, \beta_e^*) \\ &= \bar{\Pi}\left(c_p (\ell_k/\lambda')^d, n_k \lambda'^{1/\nu}, \beta_e^*\right) . \end{aligned} \tag{8.21}$$

Again λ' has to drop out, which yields the *scaling law*

$$\frac{\hat{\Pi}}{k_B T c_p} = 1 + \mathcal{P}(c_p B^d n_0^{\nu d}, \beta_e^*) \,, \tag{8.22}$$

where we split off a trivial part giving van't Hoff's law. We note that the single nonuniversal scale B (Eq. (8.17)) occurs in all these results.

Neglecting many-body forces, our model is restricted to small segment concentrations.

$$\ell^d c_p n \ll 1 \ . \tag{8.23}$$

This criterion must also hold for the renormalized theory: $\ell = \ell_k, n = n_k$. How does it behave under renormalization? From Eqs. (8.1), (8.17) we find

$$1 \gg \ell_k^d c_p n_k = \ell_k^d c_p n_0 \left(\frac{B}{\ell_k}\right)^{1/\nu} = \ell_0^d c_p n_0 \left(\frac{B}{\ell_0}\right)^{1/\nu} \lambda^{k(1/\nu - d)} \quad ,$$

which yields

$$\left(\frac{B}{\ell_0}\right)^{1/\nu} \ell_0^d c_p n_0 \ll \lambda^{k(d-1/\nu)} \underset{k\to\infty}{\longrightarrow} 0 \ . \tag{8.24}$$

Our theory is restricted to the limit of small segment concentrations and long chains. The overlap $s \sim c_p n_0^{\nu d}$ may take any value $s \geq 0$, however.

We can now state clearly what in the context of the renormalization group we mean by the *excluded volume limit.*

By definition the excluded volume limit is reached if under renormalization the coupling constant approaches the fixed point so closely that we can replace it by β_e^*.

By hypothesis, irrespective of the starting value $\beta_{e0} > 0$ this limit is attained for sufficiently long chains and sufficiently small segment concentration. Thus in principle this limit can also be reached for a system close to its Θ-temperature, and clearly this definition is completely equivalent to that given in Chap. 1. In the excluded volume limit we find universal scaling functions and power laws, all nonuniversality being absorbed into a single parameter B.

We should pause here a little to put the argument in a somewhat larger perspective. As basic step we may consider the redefinition of the segment size: $\ell_0 \to \ell_1$, which can be viewed as a *dilatation* of the microscopic length scale of our model. If macroscopic observables are to stay invariant, this dilatation must be compensated by a change of n or β_e. Dilatations form a group with a simple multiplicative structure: with $\ell \to \ell/\lambda, \ell \to \ell/\lambda'$ also $\ell \to \ell/\lambda\lambda'$ is a dilatation. The change of n or β_e must respect this structure, which imposes the strong conditions (8.6), (8.7). Mathematically speaking our postulated mapping is a 'representation' of the dilatation group in the space of parameters ℓ, n, β_e. (We will elaborate on this aspect in Chap. 10.) Groups and their representations play a prominent role in physics, governing qualitative features of a wealth of phenomena. Atomic spectra, for instance, in many features reflect the group of rotations. As in such other cases, the dilatation group here determines the qualitative features of the physical laws. If combined with the fixed point hypothesis it yields power laws and simple scaling laws. In Sect. 8.3 we will see that it also determines the crossover interpolating between the Θ-limit $\beta_{e0} = 0$ and the excluded volume limit $\beta_e \to \beta_e^*$.

It should be noted that the basic feature of dilatation invariance can hold only if the system shows no distinguished length scale. This is the defining property of a *'critical'* system. Other examples of such systems are fluids or solids near a second order phase transition. The polymer system strictly speaking is critical only in the limit $n \to \infty$, since in all our discussion we ignored terms of order $1/n$. For $n \gg 1$, finite, dilatation invariance holds up to small corrections. Dilatation invariance induces a specific 'selfsimilar' or 'fractal' structure of the segment density distribution inside the chain, which is not compatible with smoothed density models of the Flory type. This is at the root of the failure of earlier approaches to the excluded volume problem. We will discuss this aspect further in Chap. 16.

8.2 Explicit construction of a RG mapping

To substantiate our qualitative ideas we only have the cluster expansion at our disposal. Though this expansion is useless for a direct calculation of observables in the limit of large n, we can use it to construct the RG mapping, provided we make the additional assumption that the functions $A(\lambda, \beta_e)$, $B(\lambda, \beta_e)$ (Eq. (8.1)) allow for a Taylor expansion in powers of β_e:

$$A(\lambda, \beta_e) = A_0(\lambda)\left[1 + \beta_e A_1(\lambda) + O(\beta_e^2)\right] \tag{8.25 i}$$

$$B(\lambda, \beta_e) = B_0(\lambda)\left[1 + \beta_e B_1(\lambda) + O(\beta_e^2)\right] \quad . \tag{8.25 ii}$$

Recall that by virtue of the basic structure (8.1) of the RG mapping the functions $A(\lambda, \beta_e), B(\lambda, \beta_e)$ are *independent of n*. The expansions therefore order in powers of β_e and not in powers of $z \sim \beta_e n^{\varepsilon/2}$. They thus do not suffer from the breakdown of the cluster expansion as $z \to \infty$, discussed in Chap. 6.

The mapping is defined by the postulate that macroscopic observables are invariant, up to subleading corrections.

$$R_e^2(\ell, n, \beta_e) = R_e^2\left(\ell/\lambda, A(\lambda, \beta_e)n, B(\lambda, \beta_e)\beta_e\right)\left(1 + O(n^{-1})\right) \tag{8.26 i}$$

$$A_2(\ell, n, \beta_e) = A_2\left(\ell/\lambda, A(\lambda, \beta_e)n, B(\lambda, \beta_e)\beta_e\right)\left(1 + O(n^{-1})\right) \quad . \tag{8.26 ii}$$

We evaluate these equations, using the first order results (4.16 ii), (4.32), together with the ansatz (8.25). This yields

$$\begin{aligned}
\frac{R_e^2}{2d} &= \ell^2 n\Big\{1 + \beta_e\left[\frac{2}{\varepsilon}(n^{\varepsilon/2} - 1) - \frac{2}{2+\varepsilon}n^{\varepsilon/2} + \alpha_1(\varepsilon) + O(n^{-1+\varepsilon/2})\right] + O(\beta_e^2)\Big\} \\
&= \frac{\ell^2}{\lambda^2}A_0(\lambda)n\Big\{1 + \beta_e A_1(\lambda) + \beta_e B_0(\lambda)\Big[\frac{2}{\varepsilon}\left((A_0(\lambda)n)^{\varepsilon/2} - 1\right) \\
&\quad - \frac{2}{2+\varepsilon}\left(A_0(\lambda)n\right)^{\varepsilon/2} + \alpha_1(\varepsilon) + O(n^{-1+\varepsilon/2})\Big] + O(\beta_e^2)\Big\} \quad ,
\end{aligned} \tag{8.27}$$

$$\frac{A_2}{(4\pi)^{d/2}}$$
$$= \beta_e \ell^d n^2 \Big\{ 1 - \beta_e \Big[\frac{4}{\varepsilon}(n^{\varepsilon/2} - 1) - a(\varepsilon) n^{\varepsilon/2} + \alpha_2(\varepsilon) + O(n^{-1+\varepsilon/2}) \Big] + O(\beta_e^2) \Big\}$$
$$= \beta_e B_0(\lambda) \frac{\ell^d}{\lambda^d} A_0^2(\lambda) n^2 \Big\{ 1 + \beta_e B_1(\lambda) + 2\beta_e A_1(\lambda) -$$
$$-\beta_e B_0(\lambda) \Big[\frac{4}{\varepsilon} \Big((A_0(\lambda) n)^{\varepsilon/2} - 1 \Big) - a(\varepsilon)(A_0(\lambda) n)^{\varepsilon/2} + \alpha_2(\varepsilon)$$
$$+ O(n^{-1+\varepsilon/2}) \Big] + O(\beta_e^2) \Big\} \ . \tag{8.28}$$

Comparing leading powers of β_e we find

$$A_0(\lambda) = \lambda^2 \tag{8.29 i}$$

$$B_0(\lambda) = \lambda^{d-4} = \lambda^{-\varepsilon} \ . \tag{8.29 ii}$$

These results are familiar from the trivial invariance of the two-parameter model (cf. Eq. (8.4)). As a consequence, the terms $\sim n^{\varepsilon/2}$ taken along in the two-parameter model drop out from Eqs. (8.27), (8.28), *leaving the n-independent subleading terms to determine* $A_1(\lambda)$ *and* $B_1(\lambda)$. A little calculation yields

$$A_1(\lambda) = \left(\frac{2}{\varepsilon} - \alpha_1 \right) (\lambda^{-\varepsilon} - 1) \tag{8.30 i}$$

$$B_1(\lambda) = -\left(\frac{8}{\varepsilon} - 2\alpha_1 - \alpha_2 \right) (\lambda^{-\varepsilon} - 1) \ . \tag{8.30 ii}$$

Once we neglect corrections of order $n^{-1+\varepsilon/2}$, the mapping indeed is independent of n. This is a crucial feature, as stressed above. The mapping orders according to powers of β_e, which may be small, and not in powers of $z = \beta_e n^{\varepsilon/2}$, which becomes large! Thus it also makes sense in the excluded volume limit $\beta_e > 0, n \to \infty$.

We now look for a fixed point

$$\beta_e^* = B(\lambda, \beta_e^*) \beta_e^* = \beta_e^* \lambda^{-\varepsilon} \left[1 + \beta_e^* B_1(\lambda) + O(\beta_e^{*2}) \right] \ . \tag{8.31}$$

We always have the trivial solution $\beta_e^* = 0$ representing the chain at the Θ-point. Ignoring the contribution $O(\beta_e^{*2})$, we find as nontrivial solution

$$\beta_e^* = \frac{\lambda^\varepsilon - 1}{B_1(\lambda)} = \frac{\varepsilon \lambda^\varepsilon}{8 + \varepsilon \hat{\alpha}} \ , \tag{8.32}$$

where $\hat{\alpha}$ collects contributions to $B_1(\lambda)$ regular at four dimensions ($\varepsilon = 0$). As long as $\lambda^{-\varepsilon} < 3, B_1(\lambda) < 0$, the fixed point $\beta_e^* > 0$ is attractive. Under iteration of the mapping *any* starting value $\beta_{e,0} > 0$ converges to β_e^*. This is illustrated in Fig. 8.1, where we follow the path of $\beta_{e,0}$ under iterations of the

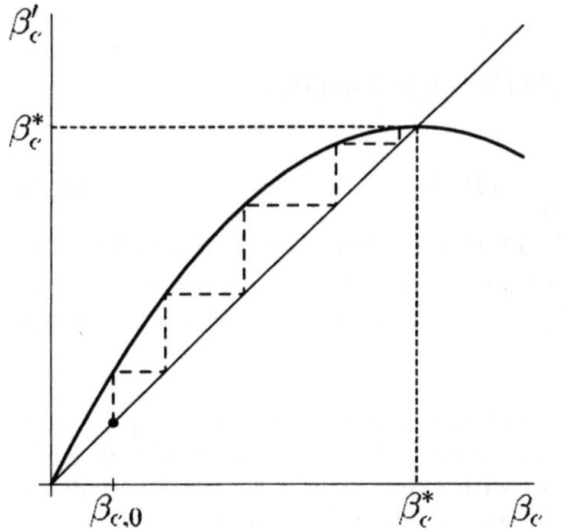

Fig. 8.1. Change of the coupling constant β_e under the RG-mapping. The *curved line* gives $\beta'_e = B(\lambda, \beta_e)\beta_e$. The *straight line* represents $\beta'_e \equiv \beta_e$. The *dotted flight of stairs* follows the change of some $\beta_{e,0}$

mapping. The fixed point $\beta_e = 0$ in contrast is always repulsive (provided $\varepsilon > 0$), as is clear from the same figure.

Thus all seems perfect. We have constructed an RG mapping, which indeed shows a fixed point. However, the expression (8.32) for β^*_e is not satisfactory. It must be independent of λ, otherwise dilatation by λ^2 does not lead to the same result as repeated dilatation by λ. Now Eq. (8.32) is only approximate since in Eq. (8.31) we omitted terms $O(\beta_e^{*2})$. This is justified only if β^*_e is small. We thus *need a parameter which allows us to make β^*_e arbitrarily small*, irrespective of λ. Only $\varepsilon = 4-d$ can take this role. In all our results the dimension of the system occurs only in the form of explicit factors of d or ε. It thus can be used formally as a continuous parameter. *To make our expansion a consistent theory, we have to introduce the formal trick of expanding in powers of $\varepsilon = 4-d$.* β^*_e vanishes for $\varepsilon = 0$, consistent with the observation (see Chap. 6) that the excluded volume is negligible above $d = 4$, not changing the Gaussian chain behavior qualitatively. For $\varepsilon > 0$ Eq. (8.32) to first order in ε yields

$$\beta^*_e = \frac{\varepsilon}{8} + O(\varepsilon^2) . \tag{8.33}$$

Since $\beta^*_e = O(\varepsilon)$, neglecting $O(\beta_e^{*2})$ in Eq. (8.31) is justified. Obviously the result (8.33) agrees with our general expectations since it neither depends on λ nor on microstructure effects, which in Eq. (8.32) are contained in $\hat{\alpha}$. (Recall that this parameter collects contributions $\alpha_1(\varepsilon), \alpha_2(\varepsilon)$, which in Sect. A 4.2 have been calculated by explicit evaluation of discrete segment summations.)

We can now determine the exponent ν to first order in ε. Equations (8.12), (8.25) yield

$$\frac{1}{\nu} \ln \lambda = \ln A(\lambda, \beta^*_e) = \ln A_0(\lambda) + \beta^*_e A_1(\lambda) + O(\beta_e^{*2}) \quad .$$

From Eqs. (8.29), (8.30) we find

$$\ln A_0(\lambda) = 2\ln\lambda, \quad A_1(\lambda) = -2\ln\lambda + O(\varepsilon) \quad ,$$

resulting in

$$\frac{1}{\nu} = 2 - \frac{\varepsilon}{4} + O(\varepsilon^2) \quad . \tag{8.34}$$

For $d = 3$ ($\varepsilon = 1$) we find $\nu \approx 0.57$. Thus even this first order calculation improves the exponent considerably. Higher orders have been calculated, and the result $\nu = 0.588$ [BNM78, GZJ85] has been established which is thought to be very precise.

Just a calculation of ν may not be considered as sufficient motivation to study a new theory. After all there exists the well known Flory argument [Flo49, FF51] giving $\nu = 3/5$ ($d = 3$), in good agreement with experimental data. (We recall and discuss this argument in Chap. 16.) However, we should be aware of two points. Firstly, for a theoretician the Flory argument is quite unsatisfactory. In fact it is demonstrably wrong. (See, for instance, [Clo76].) As a second and more important point we note that the present theory yields much more than just a value of ν. Having demonstrated the existence of an RG mapping leading to a fixed point, we can take over all the qualitative results of the previous section, like scaling and the existence of critical ratios. Furthermore the ε-expansion provides us with a consistent and unified scheme for quantitative calculations even outside the Θ-region. This is a great achievement compared to the traditional approach.

8.3 Flow equations and crossover

We can use the RG to interpolate between the excluded volume - and Θ-limits. Such interpolating behavior between asymptotic limits governed by fixed points ($\beta_e = 0$ or $\beta_e = \beta_e^*$, respectively) is known as 'crossover'. The basic idea is as follows. The RG effectively decreases the chain length. Iterating the mapping we thus can map a long chain $n_0 \gg 1$, for which perturbation theory breaks down, on a short chain $n' \sim 1$, for which perturbation theory holds. Under this mapping the coupling constant for chains of intermediate lengths n_0 may not yet reach the fixed point. We are then in the crossover regime, where asymptotic scaling laws are not yet valid.

To carry out this program with the form (8.6), (8.7) of the mapping is quite unwieldy, since the iteration yields expressions which can hardly be handled. It is preferable first to transform the mapping into a set of differential equations, which give the change of n or β_e under an infinitesimal change of ℓ. To derive these *'Renormalization Group Flow Equations'* we start from Eqs. (8.6), (8.7) written as

$$B(\lambda\lambda', \beta_{e0}) = B(\lambda', B(\lambda, \beta_{e0})\beta_{e0})B(\lambda, \beta_{e0})$$

$$A(\lambda\lambda', \beta_{e0}) = A(\lambda', B(\lambda, \beta_{e0})\beta_{e0})A(\lambda, \beta_{e0}) \quad .$$

Taking the derivative with respect to λ' evaluated at $\lambda' = 1$ we find from the first equation

$$\left.\frac{d}{d\lambda'}\right|_1 B(\lambda'\lambda, \beta_{e0}) \equiv \lambda \frac{d}{d\lambda} B(\lambda, \beta_{e0}) = \left.\frac{d}{d\lambda'}\right|_1 B\left(\lambda', B(\lambda, \beta_{e0})\beta_{e0}\right) B(\lambda, \beta_{e0}) \; . \tag{8.35}$$

With the notation

$$\beta_e(\lambda) = B(\lambda, \beta_{e0})\beta_{e0} \tag{8.36}$$

$$W(\beta_e) = \left.\frac{d}{d\lambda'}\right|_1 B(\lambda', \beta_e)\beta_e \tag{8.37}$$

this result takes the form

$$\lambda \frac{d}{d\lambda} \beta_e(\lambda) = W\left(\beta_e(\lambda)\right) \; . \tag{8.38}$$

This RG flow equation gives the change of the coupling under an infinitesimal dilatation of the elementary length scale ℓ. It is to be integrated with the initial condition

$$\beta_e(1) = \beta_{e0} \; . \tag{8.39}$$

Note that the function $W(\ldots)$ depends on λ via $\beta_e(\lambda)$ only, in contrast to $B(\ldots)$ which represents a finite RG step.

The flow equation for A is derived in the same way:

$$\left.\frac{d}{d\lambda'}\right|_1 \ln A(\lambda'\lambda, \beta_{e0}) = \lambda \frac{d}{d\lambda} \ln A(\lambda, \beta_{e0}) = \left.\frac{d}{d\lambda'}\right|_1 \ln A\left(\lambda', \beta_e(\lambda)\right) \; . \tag{8.40}$$

With the notation

$$\frac{1}{\nu(\beta_e)} = \left.\frac{d}{d\lambda'}\right|_1 \ln A(\lambda', \beta_e) \tag{8.41}$$

it takes the form

$$\lambda \frac{d}{d\lambda} \ln A(\lambda) = \frac{1}{\nu(\beta_e(\lambda))} \; , \tag{8.42}$$

the initial condition being

$$A(1) = 1 \; . \tag{8.43}$$

Combining Eqs. (8.38), (8.42) we find that $A(\lambda)$ depends on λ only via $\beta_e(\lambda)$: $A(\lambda) \to A(\beta_e(\lambda))$, the function $A(\beta_e)$ obeying

$$\frac{d}{d\beta_e} \ln A(\beta_e) = \frac{1}{W(\beta_e)\nu(\beta_e)} \; . \tag{8.44}$$

In this representation of the mapping the fixed points are found as zeros of $W(\beta_e)$:

$$W(\beta_e^*) = 0 \; . \tag{8.45}$$

At the fixed point, integration of Eq. (8.42) yields

$$A(\lambda) = \lambda^{1/\nu(\beta_e^*)} \quad , \tag{8.46}$$

identical to Eq. (8.12).

We now turn to the explicit construction of the flow equations to leading order in ε. Equations (8.25), (8.29), (8.30) yield

$$A(\lambda, \beta_e) = \lambda^2 \left[1 + \beta_e(\lambda^{-\varepsilon} - 1)\left(\frac{2}{\varepsilon} - \alpha_1\right) + O(\beta_e^2)\right] \tag{8.47}$$

$$B(\lambda, \beta_e) = \lambda^{-\varepsilon} \left[1 - \beta_e(\lambda^{-\varepsilon} - 1)\left(\frac{8}{\varepsilon} - 2\alpha_1 - \alpha_2\right) + O(\beta_e^2)\right] \quad . \tag{8.48}$$

Taking the derivative (8.37) we find

$$\begin{aligned} W(\beta_e) &= -\varepsilon\beta_e \left[1 - \beta_e\left(\frac{8}{\varepsilon} - 2\alpha_1 - \alpha_2\right) + O(\beta_e^2)\right] \\ &= -\beta_e \left[\varepsilon - 8\beta_e + O(\beta_e^2, \varepsilon\beta_e)\right] \quad . \end{aligned} \tag{8.49}$$

Equation (8.41) yields

$$\begin{aligned} \frac{1}{\nu(\beta_e)} &= 2 - \varepsilon\beta_e\left(\frac{2}{\varepsilon} - \alpha_1\right) + O(\beta_e^2) \\ &= 2 - 2\beta_e + O(\beta_e^2, \varepsilon\beta_e) \; . \end{aligned} \tag{8.50}$$

Using these approximations to determine β_e^* (Eq. (8.45)) and $1/\nu$ we recover our previous results: $\beta_e^* = \varepsilon/8, 1/\nu = 2 - \varepsilon/4$. Globally the RG flow results from integrating the flow equations. Equations (8.38), (8.49) yield

$$\lambda = \left(\frac{\beta_e^* - \beta_e(\lambda)}{\beta_e^* - \beta_{e0}}\right)^{1/\varepsilon} \left(\frac{\beta_{e0}}{\beta_e(\lambda)}\right)^{1/\varepsilon} \quad . \tag{8.51}$$

It is easily checked that this result obeys the initial condition (8.39). $A(\lambda, \beta_{e0}) = A(\beta_e(\lambda))$ is found from Eqs. (8.44), (8.50):

$$\frac{A(\beta_e(\lambda))}{A(\beta_{e0})} = \left(\frac{\beta_e^* - \beta_e(\lambda)}{\beta_e^* - \beta_{e0}}\right)^{1/\varepsilon\nu} \left(\frac{\beta_{e0}}{\beta_e(\lambda)}\right)^{2/\varepsilon} \quad . \tag{8.52}$$

Condition (8.43) yields $A(\beta_{e0}) = 1$. By virtue of Eq. (8.51) our result (8.52) can also be written in the equivalent forms

$$A(\beta_e(\lambda)) = \lambda^2 \left(\frac{\beta_e^* - \beta_e(\lambda)}{\beta_e^* - \beta_{e0}}\right)^{\frac{1}{\varepsilon}\left(\frac{1}{\nu} - 2\right)} \tag{8.53 i}$$

$$= \lambda^{1/\nu} \left(\frac{\beta_{e0}}{\beta_e(\lambda)}\right)^{\frac{1}{\varepsilon}\left(2 - \frac{1}{\nu}\right)} \quad . \tag{8.53 ii}$$

The excluded volume limit $\beta_e(\lambda) \to \beta_e^*$ is attained for $\lambda \to 0$, as is obvious from Eq. (8.51).

We should stress that once we derived the approximate flow equations we integrate them without further approximations. A general discussion of the integration of the flow equations will be given in Sect. 11.4. We will show there that higher order terms in the flow equations mainly correct the exponents in Eqs. (8.51), (8.52). The general structure of these equations is altered by weakly varying correction factors only.

Having constructed a global approximation to the RG mapping we are in a position to evaluate results like Eq. (8.26 i), which now takes the form

$$R_e^2(\ell_0, n_0, \beta_{e0}) = R_e^2(\ell_0/\lambda, A(\beta_e(\lambda))n_0, \beta_e(\lambda)) \,. \tag{8.54}$$

We use here an oversimplified argument (see below), fixing the parameter λ by the condition

$$n' = A(\beta_e(\lambda))n_0 = 1 \,. \tag{8.55}$$

The renormalized chain thus consists of a single segment, and its end-to-end distance may be assumed to take the simple form

$$R_e^2 = 2d\ell'^2 = 2d\frac{\ell_0^2}{\lambda^2} \,. \tag{8.56}$$

In the excluded volume limit $\beta_e(\lambda) \to \beta_e^*$ Eqs. (8.53 ii) (8.55) yield

$$R_e^2 = 2d\ell_0^2 \left(\frac{\beta_{e0}}{\beta_e^*}\right)^{\frac{2}{\varepsilon}(2\nu-1)} n_0^{2\nu} = 2dB^2 n_0^{2\nu} \,, \tag{8.57}$$

where the nonuniversal parameter B is now defined as

$$B = \ell_0 \left(\frac{\beta_{e0}}{\beta_e^*}\right)^{\frac{1}{\varepsilon}(2\nu-1)} \,. \tag{8.58}$$

The Θ-limit $\beta_{e0} \to 0, \beta_e(\lambda) \to \beta_{e0}\lambda^{-\varepsilon} \to 0$ is found as expected:

$$R_e^2 = 2d\ell_0^2 n_0 = 2dR_0^2 \,. \tag{8.59}$$

The crossover is traditionally described in terms of the swelling factor

$$\alpha_e^2 = \frac{R_e^2}{2dR_0^2} \,. \tag{8.60}$$

Equations (8.55), (8.56), (8.53 i) yield

$$\alpha_e^2 = \frac{1}{n_0\lambda^2} = \left(\frac{\beta_e^* - \beta_e(\lambda)}{\beta_e^* - \beta_{e0}}\right)^{\frac{1}{\varepsilon}\left(\frac{1}{\nu}-2\right)} \,. \tag{8.61}$$

The coupling constant $\beta_e(\lambda)$ is determined implicitly by Eqs. (8.52), (8.55):

$$\beta_e(\lambda)(\beta_e^* - \beta_e(\lambda))^{-\frac{1}{2\nu}} = (\beta_e^* - \beta_{e0})^{-\frac{1}{2\nu}}\beta_{e0}n_0^{\varepsilon/2}$$
$$\equiv \hat{z} \quad . \tag{8.62}$$

Our final result therefore reads

$$\alpha_e^2 = \left(1 - \frac{\beta_{e0}}{\beta_e^*}\right)^{\frac{1}{\varepsilon}(2-\frac{1}{\nu})} \bar{\alpha}_e^2(\hat{z}) \quad , \tag{8.63}$$

$$\bar{\alpha}_e^2(\hat{z}) = \left(1 - \frac{\beta_e(\lambda(\hat{z}))}{\beta_e^*}\right)^{\frac{1}{\varepsilon}(\frac{1}{\nu}-2)} \quad , \tag{8.64}$$

where $\beta_e(\lambda(\hat{z}))$ is determined by solving Eq. (8.62) numerically. We plot the result in Fig. 8.2, using for ν the value $\nu = 0.588$ valid for $\varepsilon = 1$. Clearly it closely resembles Fig. 1.5. Note, however, that our result does not conform to the two-parameter hypothesis, which states that α_e^2 is a function of the variable $z = \beta_{e0}n_0^{\varepsilon/2}$ only. Such a form is recovered only if we take the two-parameter limit $\beta_{e0} \to 0, \hat{z}$ fixed, hence $\hat{z} \to \beta_e^{*-\frac{1}{2\nu}} z, \alpha_e^2 \to \bar{\alpha}_e^2$. In general α_e^2 is not a universal function of $\hat{z}$, but in addition depends on temperature via the prefactor of $\bar{\alpha}_e^2$ in Eq. (8.63). A full discussion of the two-parameter scheme in the light of the R.G. will be presented in Sect. 11.5.

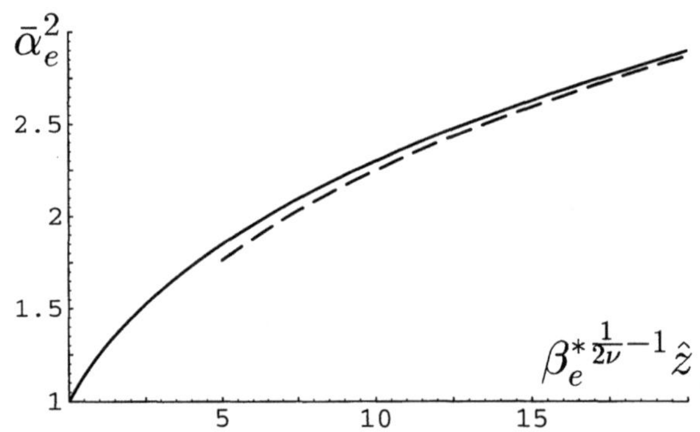

Fig. 8.2. The swelling $\bar{\alpha}_e^2$ (Eq. (8.64)) as function of $\beta_e^{*\frac{1}{2\nu}-1}\hat{z}$. The *broken line* represents the asymptotic power law

It is worthwhile to reconsider this calculation, since also on the technical level it exhibits the essential structure of all the results which will be presented later. We first derive an approximate form of the RG flow equations. Integrating this result without any further approximations, we find global RG equations, the one (Eq. (8.51)) implicitly relating the coupling $\beta_e(\lambda)$ to the scaling factor λ, the other (Eq. (8.52)) giving the rescaling factor A of the chain length as function of $\beta_e(\lambda)$. These two equations are to be supplemented by a condition of the type (8.55), which essentially fixes the renormalized length scale $\ell_R = \ell/\lambda$ via fixing λ. It must be chosen such that on the renormalized level physical observables can be calculated by simple perturbation theory (Eq. (8.56), for our example). The results of this calculation in general take the form of a power series in $\beta_e(\lambda)$, possibly multiplied by some explicit power of λ. The global RG map together with the choice of the renormalized length scale present a closed system of equations from which λ and $\beta_e(\lambda)$ are to be determined numerically as function of the unrenormalized or 'bare' parameters β_{e0}, n_0.

Our present discussion of the crossover is oversimplified since we mapped the long chain on a single segment, assuming Eq. (8.56) to hold on this level. Working consistently within our perturbative scheme we rather should map the chain of length n_0 on a chain of length n_1, where n_1 is a fixed chain length obeying $n_1 \gg 1, \beta_e^* n_1^{\varepsilon/2} \ll 1$. We can then use perturbation theory (Eq. (4.16 ii)) to evaluate R_e^2 for the renormalized chain, the condition $n_1 \gg 1$ guaranteeing that we still can neglect corrections of order $n_1^{-1+\varepsilon/2}$. A result of the type (8.63), (8.64) is found for a redefined swelling factor, giving the ratio of $R_e^2(n_0)$ to $R_e^2(n_1)$. Employing naive perturbation theory for $R_e^2(n_1)$, this method is unnecessarily complicated, however. Furthermore the form of $R_e^2(n_1)$ depends on the microscopic features of our model, so that by this 'matching' to $R_e^2(n_1)$ our final results mix universal and nonuniversal features in a complicated way. In Chaps. 11 and 12 we will construct a 'renormalized' version of perturbation theory, which allows for a clear separation of universal and nonuniversal aspects and yields a much more elegant and efficient calculational scheme, *with no matching to bare perturbation theory involved.* It should be noted, however, that the matching method has widely been used in early RG calculations.

Notes

The qualitative discussion of Sect. 8.1 follows an argument of de Gennes [dG79], which in turn is an adaption to polymer physics of Kadanoff's 'block spin' argument [Kad66], which played a great role in the early development of renormalization group theory of critical phenomena. It is discussed in detail in [Ma76]. The implementation of the RG presented here is technically known as a 'cut-off renormalization group', combined with a matching prescription. Such approaches were quite common in the early 1970s. The expansion in powers of $\varepsilon = 4 - d$ has been introduced in [WF72].

9. Scaling Theory

The scaling theory concentrates on a qualitative understanding of universal features of polymer solutions. This goal is approached in two steps. At first we identify the macroscopically relevant length scales, and we use dimensional analysis to write the observables as functions of dimensionless combinations of these variables. Here the essential assumption is that no microscopic scales like ℓ come in and that we identified all macroscopic scales. As a result we find the scaling laws. In a second step we add plausible assumptions on the behavior in certain limits to predict asymptotic power laws and to guess the shape of the scaling functions. A prominent role is played by the power law $R_g \sim n^{\nu}$, valid for isolated chains in the excluded volume limit. Within scaling theory this law is taken as input introducing the exponent ν.

Scaling theory is a simple way to arrive at testable predictions which otherwise can be derived only via the renormalization group. Since it does not properly consider crossover phenomena, it develops its full power if we are close to a fixed point, and it is particularly useful if there are several relevant length scales. Examples are the behavior of polymer solutions in restricted geometries like pores, where the pore geometry defines additional length scales, or the physics of solutions of two kinds of macromolecules of very different molecular weight. Here the different coil sizes define two length scales. In such or even more complicated situations a full renormalization group analysis may become very involved, but scaling considerations are still quite simple. The resulting scaling laws in general can be proven and sometimes are corrected by renormalization group theory. The assumptions on the limiting behavior of the scaling functions underlying simple power law behavior may be much harder to verify. They are not related to the general renormalization group structure, but can be examined only by an analysis of explicit perturbative results. In the later parts of this book, which are devoted to a detailed presentation of the results of renormalization theory, we will often refer to or compare to the results of the simple scaling approach. Indeed, it is the virtue of that approach that it focuses on the prominent role characteristic length scales play in the theory and provides us with intuitive pictures of the underlying physics. This will help us in constructing a successful perturbative approximation scheme.

We do not intend to give an overview over all results of scaling theory here. Rather we concentrate on topics relevant for the bulk behavior of normal polymer solutions. We discuss in particular the concentration dependence, introducing the 'blob'-picture (Sect. 9.1). Temperature dependence is discussed in Sect. 9.2. The results are summarized in the Daoud-Jannink diagram [DJ76] which separates parameter space into several regions, where different characteristic behavior is expected.

Not all the results of scaling theory are equally well supported by the more fundamental renormalization approach. In Sect. 9.3 we discuss which features should not be taken too literally. We furthermore should note that scaling theory in general ignores such complications as polydispersity effects. In the sequel we therefore explicitly consider a monodisperse system.

9.1 Concentration blobs and screening

We consider here the concentration dependence in the excluded volume limit. The scaling laws are easily derived by noting that the only macroscopic length scale characterizing a single polymer molecule is the average size of its coil. We may thus take the radius of gyration of an isolated coil: $R_g(n) \sim n^\nu$ as the relevant scale. It replaces the chain length n, which depends on our definition of the segments, i.e. on microscopic conventions. Chain concentration c_p can enter dimensionless observables only via the dimensionless combination

$$s = c_p R_g^d(n) \sim c\, n^{\nu d - 1} \quad , \tag{9.1}$$

which introduces the overlap. For the osmotic pressure we thus find the scaling law

$$\frac{\hat{\Pi}}{k_B T c_p} = 1 + \mathcal{P}^*(s) \quad . \tag{9.2}$$

To give another example we consider the density autocorrelation function $I_{d,a}(\mathbf{q})$ defined as Fourier transform of the segment density correlations within a chain (cf. Eq. (5.17)). According to Eq. (5.24) this function obeys the sum rule

$$I_{d,a}(\mathbf{0}) = n^2 c_p \quad . \tag{9.3}$$

We introduce the dimensionless normalized function

$$J_a = \frac{1}{c_p n^2} I_{d,a}(\mathbf{q}) \quad , \tag{9.4}$$

which can depend only on dimensionless combinations of $\mathbf{q}, c_p, R_g$. We thus predict the scaling law

$$J_a = J_a^*(\mathbf{q}R_g, s) \quad . \tag{9.5}$$

Other examples can be discussed in the same way. We note that all these results follow from simple dimensional analysis in terms of coil radius R_g, chain concentration c_p, and (if present) momentum variables. The method is based on the assumptions that

(i) microscopic length scales (ℓ) can be neglected

and that

(ii) we are in a fixed point region, so that the interaction strength β_e is taken as constant and does not appear as explicit variable.

The latter assumption distinguishes scaling theory from straightforward dimensional analysis in the continous chain model.

We further note that we necessarily have to take c_p as measure of the concentration. The segment concentration $c = c_p n$ depends on our definition of the segment and thus is no uniquely defined observable.

Scaling theory also derives such results in another, more intuitive way, based on some heuristic picture of the internal structure of the polymer solution. Consider some piece of length n_B within a chain of length n. It is natural to assume that this piece forms a subcoil, a 'blob', of typical extension R_B, which scales like the coil radius for a polymer molecule of n_B segments: $R_B \sim n_B^\nu$. Thus the local density of segments due to the blob is estimated as

$$\rho_B \sim \frac{n_B}{R_B^d} \sim n_B^{1-\nu d} \quad . \tag{9.6}$$

We may now define a distinguished blob size $n_B^* = n_B^*(c)$ by the condition

$$\rho_B(n_B^*) = c = c_p\, n \quad , \tag{9.7}$$

where c is the total segment density in the solution. Thus

$$n_B^* \sim c^{-\frac{1}{\nu d - 1}} \left(\sim c^{-1.309}\right); \quad R_B^* \sim c^{-\frac{\nu}{\nu d - 1}} \left(\sim c^{-.770}\right) \quad , \tag{9.8}$$

where the expressions for $d = 3, \nu = 0.588$ are given in brackets. For blobs $n_B \gg n_B^*$, the segment density due to a single blob is much smaller than c. Therefore on scales $R_B \gg R_B^*$ many different blobs must interpenetrate to build up the total segment density. In contrast, blobs $n_B < n_B^*$ show an enhanced local segment density. Clearly this is a consequence of the chain structure: it implies that two segments found at distance $r < R_B^*$ most likely are part of the same blob $n_B < n_B^*$. The blob size R_B^* defines a natural length scale connected to the segment concentration. In the spirit of scaling theory we may now take R_B^* as a measure of the concentration, for instance writing

$$\frac{\hat{\Pi}}{k_B T c_p} = 1 + \mathcal{P}_B(R_g/R_B^*) \quad . \tag{9.9}$$

Since

$$\frac{R_g}{R_B^*} \sim n^\nu c^{\frac{\nu}{\nu d-1}} \sim \left(c_p n\, n^{\nu d-1}\right)^{\frac{\nu}{\nu d-1}} \sim s^{\frac{\nu}{\nu d-1}} \quad , \tag{9.10}$$

this is equivalent to the form given above.

For the blob concept to make sense physically, a chain clearly must be larger than an n_B^*-blob: $n \gtrsim n_B^*$. We may then look at these blobs also from a different point of view. The concentration of n_B^*-blobs in the solution is given by

$$c_B^* = \frac{c}{n_B^*} \sim c^{\frac{\nu d}{\nu d-1}} \quad . \tag{9.11}$$

Thus

$$c_B^* R_B^{*d} \sim 1 \quad , \tag{9.12}$$

i.e. the n_B^*-blobs, seen naively, just fill the volume. The physical condition $n \gtrsim n_B^*$ then implies

$$s = c_p R_g^d(n) \sim c n^{\nu d-1} \gtrsim c n_B^{*\nu d-1} \sim c_B^* R_B^{*d} \sim 1 \quad ,$$

showing that the blob concept refers to semidilute solutions $s \gtrsim 1$.

Having established the scaling laws we now turn to the second step of the argument, viz. the evaluation of limiting cases. We indeed encountered n_B^* before in the context of the screening mechanism (see start of Chap. 5). There we argued that correlations on length scales $r \gg R_B^*$ are not affected by the interaction, since on such scales a swelling of the chain cannot reduce its interaction energy. Combining this idea with the blob concept we can view a chain in semidilute solution as a Gaussian chain of n_B^*-blobs, which internally are swollen by excluded volume effects. Indeed, R_B^* is the renormalized form of the screening length ξ_E (Sect. 5.4.2).

What does this picture predict for the scaling functions? Let us first consider the osmotic pressure. Clearly for small overlap the virial expansion holds

$$\frac{\hat{\Pi}}{k_B T c_p} = 1 + a_2 s + a_3 s^2 \ldots \tag{9.13}$$

Comparing to the virial expansion in the form of Eq. (1.1)

$$\frac{\hat{\Pi}}{k_B T} = c_p + \frac{A_2^\Pi}{2}\, c_p^2 + \frac{A_3^\Pi}{6}\, c_p^3 + \ldots \quad ,$$

we find

$$A_2^{\Pi} = 2a_2 R_g^d \sim n^{\nu d}$$
$$A_3^{\Pi} = 6a_3 R_g^{2d} \sim n^{2\nu d}$$

etc., where $a_2, a_3, \ldots$ are universal numbers. The virial expansion should be a good approximation in the dilute region $s \ll 1$.

For $s > 1$ the n_B^*-blob is smaller than the whole chain and the blob-concept starts to make sense. For large overlap in view of screening the number of concentration blobs per chain should not be important. Thus $\hat{\Pi}$ should reduce to a function of the blob concentration c_B^*, only. In view of Eq. (9.11) we therefore expect $\hat{\Pi}$ to become a function of c independent of n. With this assumption the scaling law (9.2) yields

$$\frac{\hat{\Pi}}{k_B T} \underset{s \gg 1}{\sim} c_p s^{\frac{1}{\nu d - 1}} \sim c^{\frac{\nu d}{\nu d - 1}} \left(\sim c^{2.309}\right) \quad . \tag{9.14}$$

In terms of c_B^* this law indeed becomes very simple. A comparison to Eq. (9.11) shows

$$\frac{\hat{\Pi}}{k_B T} \underset{s \gg 1}{\sim} c_B^* \quad .$$

With respect to the osmotic pressure a semidilute solution behaves like an ideal gas of n_B^*-blobs.

As a second example we consider the radius $R_g(n, c_p)$ of a coil in the solution. Dimensional analysis yields the scaling law

$$\frac{R_g^2(n, c_p)}{R_g^2(n, 0)} = \mathcal{R}_g^*(s) \quad . \tag{9.15}$$

For small overlap a virial expansion holds. For large overlap screening suggests that $R_g^2(n, c_p)$ is proportional to n (Gaussian chain behavior). Thus

$$R_g^2(n, c_p) \underset{s \gg 1}{\sim} n^{2\nu} s^{\frac{1-2\nu}{\nu d - 1}} \sim c^{\frac{1-2\nu}{\nu d - 1}} n$$
$$\left(\sim c^{-.230} n\right) \quad . \tag{9.16}$$

Again we can rewrite this in terms of quantities introduced in the blob picture:

$$R_g^2(n, c_p) \sim R_B^{*2}(c) \frac{n}{n_B^*} \quad . \tag{9.17}$$

Thus the chain can be viewed as a Gaussian chain of n/n_B^* n_B^*-blobs, consistent with the above discussion.

As a final example we consider the density autocorrelation function, where we first discuss the dilute limit

$$I_{d,a}(\mathbf{q}) = c_p n^2 J_a^*(\mathbf{q}R_g) \quad . \tag{9.18}$$

For $(\mathbf{q}R_g)^2 \ll 1$ we have

$$J_a^*(\mathbf{q}R_g) = 1 - \frac{1}{d}(\mathbf{q}R_g)^2 + \dots \quad , \tag{9.19}$$

by definition of R_g. For $(\mathbf{q}R_g)^2 \gg 1$ scattering is due to segment pairs of spatial distance $r \lesssim 1/|\mathbf{q}| \ll R_g$. Such pairs predominantly are part of blobs of lengths $n_q \sim |\mathbf{q}|^{-\frac{1}{\nu}}$. As a result $I_{d,a}$ should become proportional to the number of these blobs and thus proportional to n. Combining this information with the scaling law we predict

$$\frac{1}{c_p} I_{d,a}(\mathbf{q}) \underset{|\mathbf{q}|R_g \gg 1}{\sim} n^2 |\mathbf{q}n^\nu|^{-\frac{1}{\nu}} \sim n|\mathbf{q}|^{-\frac{1}{\nu}} \quad \left(\sim n|q|^{-1.701}\right) . \tag{9.20}$$

Turning now to semidilute solutions we note that $R_B^*(c)$ defines a new scale. For $|\mathbf{q}R_B^*(c)| \ll 1$, corresponding to distances $r \gg R_B^*(c)$ we expect to see the behavior of a Gaussian chain, as given by the Debye function. We thus predict

$$I_{d,a}(\mathbf{q}) \underset{|\mathbf{q}R_B^*| \lesssim 1}{\sim} c_p n^2 D\left(\frac{3}{d}\mathbf{q}^2 R_g^2(n, c_p)\right) \quad . \tag{9.21}$$

For $|\mathbf{q}R_B^*(c)| \gg 1$ scattering is due to segments inside an n_B^*-blob, which predominantly are part of the same chain. We thus expect

$$\frac{1}{c_p} I_{d,a}(\mathbf{q}, c_p) \underset{|\mathbf{q}R_B^*| \gg 1}{\longrightarrow} 1/c_p I_{d,a}(\mathbf{q}, c = 0) \sim n|\mathbf{q}|^{-\frac{1}{\nu}} \quad . \tag{9.22}$$

Similar scaling laws can be derived for density correlations among arbitrary chains.

These results of the scaling approach are consistent with the results of renormalization theory and are supported by experiment. We will discuss the experimental situation in later chapters.

9.2 The temperature-concentration diagram

So far we have considered scaling as function of $n, c_p, \mathbf{q}$ in the excluded volume limit *for fixed temperature.* We may extend the approach to include temperature variations close to $T = \Theta$, assuming

$$0 < t = \frac{T - \Theta}{\Theta} \ll 1 \quad . \tag{9.23}$$

As has been discussed in Chap. 7, in the region $t \ll 1$, implying $\beta_e \ll 1$, the two-parameter hypothesis is valid. Thus t should occur only in the combination

$$z \sim t\, n^{\varepsilon/2} \quad , \tag{9.24}$$

which also contains all n-dependence, except for trivial factors describing Θ-point behavior. Approaching the excluded volume limit within the two-parameter regime, we thus find for the radius of a single chain

$$\begin{aligned} R_g(n,t) \sim R_\Theta(n)\alpha_g(z) &\sim n^{1/2} z^{(2\nu-1)/\varepsilon} \sim t^{(2\nu-1)/\varepsilon} n^{\nu} \\ &\left(\sim t^{.176} n^{.588} \right) \quad . \end{aligned} \tag{9.25}$$

Note that this is completely consistent with our RG result (8.57). The crossover from the dilute to the semidilute region takes place at a concentration $c^* = c^*(n,t)$, where the chains just fill the volume:

$$1 = s = \frac{c^*(n,t)}{n} R_g^d(n,t) \quad . \tag{9.26}$$

Equation (9.25) yields

$$c^*(n,t) \sim t^{(1-2\nu)\frac{d}{\varepsilon}} n^{1-\nu d} \left(\sim t^{-.528} n^{-.764} \right) \quad . \tag{9.27}$$

This equation also can be read as defining a critical chain length $n = n_c^*$ of concentration crossover:

$$n_c^*(c,t) \sim t^{\frac{d}{\varepsilon}\frac{2\nu-1}{1-\nu d}} c^{\frac{1}{1-\nu d}} \sim \left(t^{-.691} c^{-1.309} \right) \quad . \tag{9.28}$$

Comparison of eqs. (9.26), (9.12) shows that n_c^* is nothing but the blob size n_B^*, so that we find the temperature dependence of R_B^*:

$$\begin{aligned} R_B^*(c,t) \sim R_g(n_c^*) &\sim t^{-\frac{2\nu-1}{\varepsilon(\nu d-1)}} c^{-\frac{\nu}{\nu d-1}} \\ &\left(\sim t^{-.230} c^{-.770} \right) \quad . \end{aligned} \tag{9.29}$$

We may now reconsider the arguments of the previous section. Equation (9.17), for instance, yields for the radius of a chain in semidilute solution

$$R_g^2(n, c_p, t) \sim R_B^{*2}(c, t)\frac{n}{n_c^*} \sim n t^{\frac{d-2}{\varepsilon}\frac{2\nu-1}{\nu d-1}} c^{\frac{1-2\nu}{\nu d-1}}$$
$$\left(\sim n\, t^{.230} c^{-.230}\right) \quad . \tag{9.30}$$

Other results follow in the same way.

So far we have considered only the excluded volume limit of the two parameter scheme: $z \gg 1$. We have found that in a $t - c$ diagram ($n \gg 1$, fixed), which generally is addressed as 'Daoud-Jannink diagram', there is a line $c^* \sim t^{-.528}$ separating a dilute regime I from a semidilute regime II. (See Fig. 9.1. We now restrict ourselves to $d = 3$.) Deep in these regimes, the blob concept predicts characteristic power laws. Now the line c^* cannot extend down to $t = 0$ since for fixed n we leave the region $z \gg 1$. We should thus draw another line $t \sim n^{-1/2}$ separating a Θ-region I' from the excluded volume region. In the Θ-region for instance we expect to find the law

$$R_g(n, c_p, t) \sim R_\Theta(n) \sim n^{1/2} \quad . \tag{9.31}$$

But close to $T = \Theta$ for $d = 3$ we have to take into account three-body forces, entering the so-called tricritical region. The most prominent effect of three-body forces is to replace β_e by a sum

$$\beta_e + \beta^{(3)} \ell^d c \quad . \tag{9.32}$$

This is indicated in Fig. 9.2. Using $\beta_e \sim t$ we thus find another line $\beta_e \approx \beta^{(3)} \ell^d c^{**}$, or

$$c^{**} \sim t \quad , \tag{9.33}$$

defining a region III where three-body forces are important. Concentrating on this region we can again define an overlap concentration c_3^* by the condition

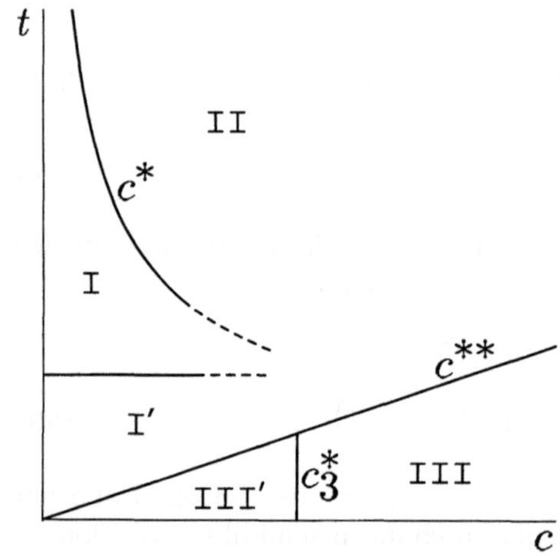

Fig. 9.1. Daoud-Jannink diagram. I: dilute excluded volume region; I′: dilute Θ-region; II: semidilute excluded volume region; III: semidilute tricritical region; III′: dilute tricritical region

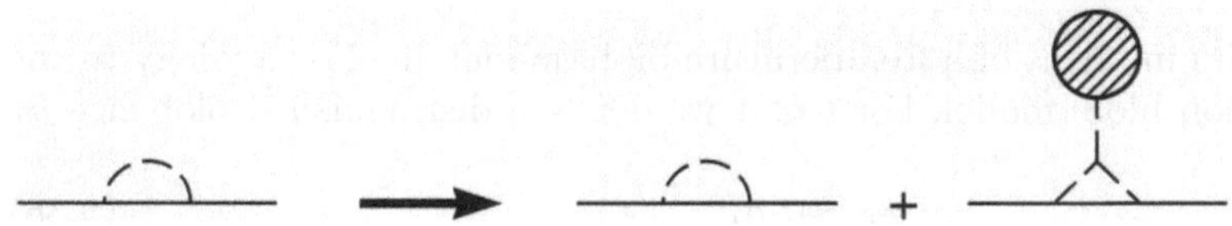

Fig. 9.2. Diagrammatic structure underlying Eq. (9.32). The *shaded blob* represents a factor of the average segment density coupling to a three-body interaction

$$1 \approx \frac{c_3^*}{n} R_\Theta^3 \quad , \tag{9.34}$$

which yields

$$c_3^* \sim n^{-1/2} \quad . \tag{9.35}$$

For $c \ll c_3^*, t \lesssim n^{-1/2}$ the chains are isolated, for $c \gg c_3^*$ they strongly overlap and feel the three-body forces. Writing in the Θ-region the scaling law for the osmotic pressure as

$$\frac{\hat{\Pi}}{k_B T} = c_p \tilde{\Pi} \left(c_p R_\Theta^3 \right) \tag{9.36}$$

and again assuming that $\tilde{\Pi}$ for large overlap depends on c only, we find

$$\frac{\hat{\Pi}}{k_B T} \sim c^3 \left(1 \gg \ell^3 c \gg \ell^3 c_3^*, \text{region III} \right) \quad , \tag{9.37}$$

a result perfectly reasonable for the osmotic pressure dominated by three-body forces.

It is clear that Fig. 9.1 gives only a rough overview of the structure of the parameter space. Simple power laws only hold deep inside the different regions, and the curves indicate broad crossover regimes. Furthermore we should note that a more accurate version of estimates like (9.8), (9.34) involves a factor of the couplings, which takes into account that noninteracting chains behave as isolated, irrespective of the geometrical overlap s. In Chap. 13 we reconsider the $t - c$ diagram in the absence of three-body forces, where we explicitly take into account the renormalized two-body coupling. The (scarce) experimental evidence is discussed in later chapters. The effects of a three-body force are discussed in Chap. 21.

9.3 Some words of caution

The blob concept is very successful in describing qualitative features of the 'concentration crossover' from dilute to semidilute solutions. This has led to a discussion of the 'temperature crossover' among the Θ- or excluded volume-limits in terms of a 'temperature blob' model, in close analogy to the 'concentration blob' model. For $t \ll 1$ we define a distinguished blob size n_t^* via

$$z_B^* \sim t\, n_t^{*\varepsilon/2} \sim 1 \quad . \tag{9.38}$$

Now consider an isolated chain of length $n \gg n_t^*$. Parts $n_t \lesssim n_t^*$ intrinsically should show only weak interaction effects, the effective z-parameter $z_B \sim t\, n_t^{\varepsilon/2}$ being small. (We ignore three-body interactions.) Thus n_t^*-blobs should be random walk-like intrinsically. A long chain should show a swollen configuration, built from Gaussian temperature blobs. This yields

$$R_g(n) \sim \left(\frac{n}{n_t^*}\right)^{\nu} n_t^{*1/2} \sim t^{(2\nu-1)/\varepsilon}\, n^{\nu} \quad ,$$

in agreement with Eq. (9.25). Clearly we can construct a similar argument, assuming that in a semidilute system concentration blobs are built as swollen configuration of Gaussian temperature blobs $n_t^* \ll n_c^*$. These ideas have been used to model crossover scaling functions for quantities like the swelling factor α_g^2. A critical discussion of this approach will be presented later in the context of an analysis of the internal correlations of a chain (Sect. 16.2), but we should note immediately that such attempts stretch the blob model too far. Indeed, by its very idea blob theory of scaling functions is concerned with behavior in certain limits. It should not be used in the crossover region.

To illustrate this point we consider a specific example. The scaling results for the radius of gyration in the excluded volume limit properly read

$$\lim_{s\to 0,n \text{ fixed}} R_g^2(n, c_p) = R_g^2(n) \sim n^{2\nu} \tag{9.39 i}$$

$$\lim_{s\to\infty,c \text{ fixed}} \frac{1}{n} R_g^2(n, c_p) = const\ c^{\frac{1-2\nu}{\nu d-1}} \quad . \tag{9.39 ii}$$

They should not be read as

$$R_g^2(n, c_p) \approx R_g^2(n, 0); \quad s \lesssim 1 \tag{9.40 i}$$

$$R_g^2(n, c_p) \approx const\, n\, c^{\frac{1-2\nu}{\nu d-1}}; \quad s \gtrsim 1 \quad , \tag{9.40 ii}$$

a form suggested by simplistic interpretation of the blob concept. Even though it may be plausible to assume that the radius does not change much as long as the chains do not overlap, it is false. This is illustrated in Fig. 9.3, comparing the form expected according to Eq. (9.40) to the renormalization group result. The result is typical in that renormalization group calculations predict extremely broad regions of gradual crossover, starting right at $s \to 0$. Thus blobs make no valid model for even only semiquantitative calculations of crossover functions. There is no s^* or c^* in the sense of an overlap where the scaling functions change qualitatively, nor is there a z^* or n_t^* where the chains suddenly begin to swell. *Scaling functions typically have no pronounced*

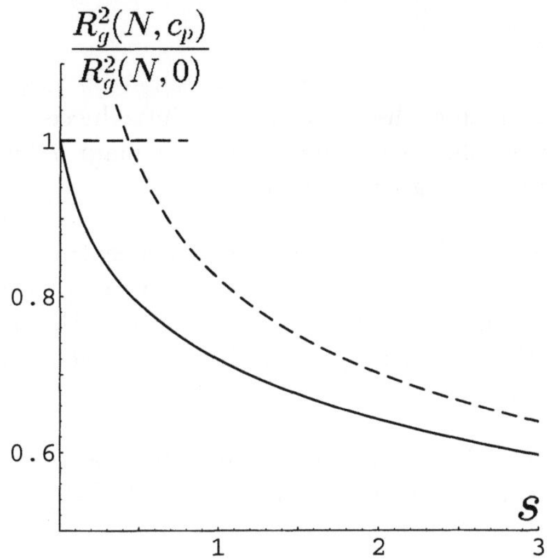

Fig. 9.3. The scaling function $R_g^2(N, c_p)/R_g^2(N, c_p = 0)$ (excluded volume limit, monodisperse) as function of $s = c_p R_g^3(N, c_p = 0)$. The *full line* is the renormalization group result. (See Chap. 18.) The *broken lines* give approximations as motivated by the blob model

structure. The blob model only defines scales relevant for the discussion of limiting situations. The actual region, where the limiting power laws hold to good approximation, for each quantity separately have to be determined from a full crossover calculation. These obvious facts are often masked by plotting the results in doubly logarithmic form, thus artificially blowing up regions like $s \ll 1$. A closer look typically reveals that the crossover region spans several decades in the parameter considered, a crossover value like c^*, for instance, being assigned to some point in the crossover regime according to fairly arbitrary criteria (cf. Eq. (9.26), for instance).

Furthermore we should stress that the length scales identified by the blob model concern characteristic features of the segment-segment correlations. Thus the appealing picture of a chain build up from well defined individual blobs should not be taken literally. Rather each segment can be taken as the center of its correlated blob, and translational invariance along the chain is not destroyed.

Notes

The concentration blob model of scaling theory has been introduced in [DCF+75, Far76], where it has been shown to qualitatively explain a lot of experiments. A comprehensive review is given in [dG79].

The work of Daoud and Jannink [DJ76] on the temperature-concentration diagram proceeds by translating results established in the theory of phase transitions to the polymer problem. They also consider the region $T \lesssim \Theta$. For $T \gtrsim \Theta$ their work is equivalent to the argument as given here and discusses an extensive set of physical observables.

Temperature blobs have been introduced in [DJ78]. They have been exploited to model the internal structure of a chain for instance in [WC79].

10*. More Complicated Interactions and the General Structure of the Renormalization Group

Having at hand the techniques of perturbation theory and the renormalization group we are prepared to consider once more a question of fundamental importance: can we justify our simple model? Are we really allowed to ignore many-body interactions or other features of a microscopically realistic description of the polymer solution? On a very superficial level we discussed this problem in Sect. 2.2. Let us critically reconsider that argument.

In Chap. 2 we have argued that many-body interactions can be neglected since the number $n_c^{(\mu)}$ of μ-body contacts ($\mu \geq 3$) in a typical configuration of a large coil becomes arbitrarily small. The argument was based on the estimate (2.10), generalized to arbitrary dimensions in Eq. (6.2):

$$n_c^{(\mu)}(n) \sim n^{\mu-(\mu-1)d/2} \quad . \tag{10.1}$$

This estimate views the chain as a cloud of segments, randomly distributed in a volume of order $R_0^d \sim n^{d/2}$. It completely ignores chain connectedness. A little thought reveals that this estimate, if interpreted naively, is just wrong. Splitting the chain into n/n' subchains of length n' we note that $n_c^{(\mu)}(n)$ cannot be less than the sum of the μ-segment contacts within the subchains:

$$n_c^{(\mu)}(n) \geq \frac{n}{n'} n_c^{(\mu)}(n') \quad . \tag{10.2}$$

Thus for all $\mu, n_c^{(\mu)}(n)$ increases at least proportionally to n, this fact being a consequence of the chain structure. The estimate (10.1) a priori is correct only for the number of simultaneous contacts among μ noninteracting and uncorrelated chains, occupying the same volume of extension R_0.

Is all of our preceding analysis based on an artificial model lacking physical relevance? This fortunately is not the case, but to show the validity of our model it needs a more careful analysis of the effects of additional interaction terms. This is the topic of the present chapter. A full analysis is prohibitively complicated and has not been given, neither within the context of polymer theory, nor within field theory. We therefore restrict our discussion to illustrating the basic ideas. As a first step we consider the discrete chain model, pointing out two features:

(i) in the limit of long chains the estimate (10.1) is correct for contacts among segments spaced a *finite fraction of the total chain length;*

(ii) the μ-body interaction among more closely spaced segments, which in fact dominates that considered in (i), can be absorbed into a redefinition of all μ'-body interactions with $\mu' < \mu$.

This redefinition is another example of additive renormalization. In principle it can be implemented perturbatively to all orders. As a result not the mathematics of our model but the interpretation of β_e changes. β_e sums up contributions from all higher interactions and thus should be viewed as a parameterization of the interaction among *whole strands of the chains* rather than as a true two-segment interaction. This also makes clear that there is no hope to calculate β_e from a realistic model.

The argument sketched above is based on the simple cluster expansion. It suggests that our model contains all terms of relevance for a description of large scale properties in the Θ-region $\beta_e N^{\varepsilon/2} \lesssim 1$ (with the exception of three-body forces for $d = 3$, which will be discussed separately in Chap. 21). We have thus established universality of the naive two-parameter theory close to the Θ-point. Being interested also in excluded volume behavior for $\beta_e N^{\varepsilon/2} \gg 1$, as a second step of the argument we have to discuss how (additively renormalized) higher interactions behave under the renormalization group. This leads us to introduce the general renormalization group framework as postulated by Wilson: the RG is a representation of the dilatation group in a space of infinitely many parameters specifying the microscopic Hamiltonian. Changing the microscopic length scale $\ell \to \ell/\lambda$ induces a flow in parameter space so that large scale properties are invariant along the flow lines. With decreasing λ the flow contracts on a manifold which is parameterized by only a few parameters corresponding to $N, \ell^d c, \beta_e$ in the present case. (The reduced chain length distribution $p(y)$ is invariant under the RG and need not be considered here.) Thus our model emerges as an asymptotic model, valid for λ small enough, i.e. for long chains and small concentrations, irrespective of the precise form of the microscopic Hamiltonian. Furthermore in the limit $\lambda \to 0$ the flow reaches a fixed point, a feature that can be viewed as a generalization of the central limit theorem.

This general scenario is easy to formulate. It is beautiful and most helpful for a general understanding, but it is extremely difficult to verify that a given mapping, constructed as a representation of the dilatation group according to some (complicated) prescription, indeed conforms to the picture globally, i.e. in all the infinite parameter space. In the theory of phase transitions there exist examples of approximate RG mappings, in which this question can be numerically analyzed in some detail. Furthermore, giving up the postulate of strict invariance under the RG flow we from the outset may restrict the parameter space to 'renormalized' counterparts of $\beta_e, N, \ell^d c$. Then the RG perturbatively can be constructed to all orders, in close analogy to our discussion of Sect. 8.2. This is the 'field theoretic' form of the RG, which will be discussed in Part III of this book. However in that field theoretic formulation we must be content with analyzing small corrections to our model and cannot follow the rigorous dilatation flow globally. Still it has been shown that if evaluated in first order ε-expansion the effect of all (additively renormalized) higher interactions vanishes under the renormalization group. Thus our picture is expected to be correct at least locally.

As mentioned above, the matter is quite involved and we will only sketch the arguments. Discussing some low order terms we will demonstrate the additive renormalizations due to higher order interactions (Sect. 10.1). Then we will discuss the general ideas on the structure of the renormalization group, defining important concepts like 'relevance' or 'irrelevance' of interactions or the critical manifold (Sect. 10.2). Concerning the field theoretic realization of the RG, we will summarize some results (Sect. 10.3).

We should stress finally that the arguments presented in this chapter are important for a deeper understanding of our model, but are no prerequisite for following the main thread of this treatise.

10.1 Higher interactions and additive renormalization

The inequality (10.2) shows that the estimate (10.1) does not count the number of μ-segment contacts correctly. To save the estimate we can interpret it as number of contacts among segments which are spaced along the chain by some finite fraction of the total chain length. Indeed, evaluating $n_c^{(\mu)}$, which is properly defined as

$$n_c^{(\mu)}(n) = \sum_{0<j_1<j_2<\ldots<j_\mu<n} \left\langle \prod_{m=2}^{\mu} \ell^d \delta^d \left(\mathbf{r}_{j_m} - \mathbf{r}_{j_{m-1}}\right) \right\rangle^0 , \tag{10.3}$$

we easily find

$$\begin{aligned} n_c^{(\mu)}(n) &= \sum_{0<j_1<\ldots<j_\mu<n} \prod_{m=2}^{\mu} [4\pi(j_m - j_{m-1})]^{-\frac{d}{2}} \\ &= (4\pi)^{-(\mu-1)\frac{d}{2}} \sum_{\substack{s_1+s_2+\ldots+s_{\mu-1}\le n-2 \\ s_m\ge 1, m=1,\ldots,\mu-1}} \Big(n-1-\sum_{m'} s_{m'}\Big) \prod_{m=1}^{\mu-1} s_m^{-\frac{d}{2}} . \end{aligned} \tag{10.4}$$

(Recall that the superscript 0 in Eq. (10.3) indicates an expectation value for noninteracting chains.) In the last line of Eq. (10.4) we introduced difference variables $s_{m-1} = j_m - j_{m-1}$ and we carried through the summation over j_1. Restricting the range of s_m to $s_m \geq \alpha n, \alpha > 0$ fixed, for large n we may replace the summations by integrals over $x_m = s_m/n$ to find

$$n_c^{(\mu)}(n,\alpha) = (4\pi)^{-(\mu-1)\frac{d}{2}} n^{\mu-(\mu-1)d/2}$$
$$\int\limits_{\substack{x_1+\ldots+x_{\mu-1}\leq 1 \\ x_m \geq \alpha, m=1,\ldots,\mu-1}} (1-\sum_{m'} x_{m'}) \prod_m x_m^{-\frac{d}{2}} \cdot \left[1+O\left(\frac{1}{n}\right)\right] \quad . \tag{10.5}$$

This shows that the estimate (10.1) is valid asymptotically, *provided the segments are spaced by at least a fraction α of the total chain length.* α may be arbitrarily small but must be greater than zero, since the integrals do not exist for $\alpha = 0$. Basically the constraint $s_m \geq \alpha n$ guarantees that in the limit $n \to \infty$ the distinguished segments, being spaced infinitely far along the chain, become uncorrelated, so that each just yields a factor of the density $\sim n/R^d$.

What is the effect of interactions within small fractions of the chain? Implicitly we have answered this question before, in the calculation of $\mu_p(n)$ (Sect. 5.4.3), for instance. We just have to recollect some results. We consider the free energy of a single chain, defined as

$$F(n) = -\ln \frac{\mathcal{Z}(n)}{\Omega} \quad . \tag{10.6}$$

Equations (4.4), (4.12) yield

$$F(n) = F_0(n) + \beta_e \Sigma_1(0) + O(\beta_e^2) \quad , \tag{10.7}$$

where $F_0(n)$ is the free energy of the noninteracting chain. $\Sigma_1(0)$ is given by Eq. (A 4.13):

$$\Sigma_1(0) = \sum_{0<j_1<j_2<n} (j_2 - j_1)^{-\frac{d}{2}} = n\,\zeta(\tfrac{d}{2}) - n^{\frac{\varepsilon}{2}} \left(\frac{2}{\varepsilon} + \frac{1}{1-\varepsilon/2}\right) + O(n^0) \quad . \tag{10.8}$$

The leading contribution $\sim n$ is dominated by the summation over small fractions $j_2 - j_1 \lesssim \alpha n$ of the chain. It can be interpreted as a correction to the segment chemical potential μ_s:

$$\mu_s = \frac{\partial}{\partial n} F(n) = \mu_{0s} + \beta_e \zeta(\tfrac{d}{2}) + O\left(n^{-1+\frac{\varepsilon}{2}}, \beta_e^2\right) \quad . \tag{10.9}$$

Now such terms also result from one-body potentials. Recall the general decomposition (2.9) of the effective segment potential:

$$V = V_0 + \sum_{\mu \geq 1} V_\mu \quad , \tag{10.10}$$

where we now allow for a one-body part

$$V_1 = \beta_1 \int d^3 r \hat{\rho}(\mathbf{r}) \equiv n\, \beta_1 \quad . \tag{10.11}$$

To the partition function V_1 contributes a trivial factor $\exp(-V_1) = \exp(-n\beta_1)$, resulting in an additive contribution $n\beta_1$ to F. We thus write

$$V_2 = V_{2,S} + \hat{V}_1 \quad , \tag{10.12}$$

adjusting the one-body part $\hat{V}_1$ so that it generates the term $\beta_e \xi \left(\frac{d}{2}\right) n$ in F:

$$\hat{\beta}_1 = \beta_e \xi(\frac{d}{2}) \quad . \tag{10.13}$$

$V_{2,S} = V_2 - \hat{V}_1$ is the additively renormalized (or 'subtracted') two-body interaction, which by construction generates only the term $\sim n^{\varepsilon/2} + O(n^0)$ and thus conforms to the estimate (10.1). The one-body potential $\hat{V}_1$ resulting from (10.12) can be absorbed into the chemical potential.

This argument is nothing but the additive renormalization discussed in Sect. 7.2 in the context of the continuous chain limit, now interpreted on the level of the interaction. From the field theoretic formulation we easily see that it holds to all orders. We thus may split the two-body interaction into a one-body part which effectively takes care of interactions within small fractions of the chains, and a remainder conforming to the estimate (10.1).

This analysis can be extended to many-body interactions. As an example we consider the first order contribution of the three-body cluster function

$$\beta_3(\mathbf{r}_{j_1}, \mathbf{r}_{j_2}, \mathbf{r}_{j_3}) = (4\pi)^d \beta_3 \ell^{2d} \delta^d(\mathbf{r}_{j_2} - \mathbf{r}_{j_1}) \delta^d(\mathbf{r}_{j_3} - \mathbf{r}_{j_2}) \tag{10.14}$$

to the two-chain cumulant $G_c^{(0,2)}(\mathbf{0}, \mathbf{0}, \mathbf{0}, \mathbf{0}; n, n)$. The three-body diagrams relevant to our discussion are shown in Fig. 10.1a,b. Straightforward application of the Feynman rules of Sect. 4.2, generalized to three-body interactions in an obvious way, yields

$$\begin{aligned} G_c^{(0,2)}(\mathbf{0}, \ldots, \mathbf{0}; n, n) &= -\Omega \left(4\pi\ell^2\right)^{-d} \Big[(4\pi)^{d/2} \ell^d \beta_2 (n-1)^2 \\ &+ 2(n-1)\,(4\pi)^d \beta_3 \ell^{2d} \sum_{0<j_1<j_2<n} \int_k e^{-k^2\ell^2(j_2-j_1)} + O(\beta_2^2, \beta_2\beta_3, \beta_3^2) \Big] \\ &= -\frac{\Omega}{\left(4\pi\ell^2\right)^{d/2}} \left[\beta_2(n-1)^2 + 2\beta_3(n-1)\Sigma_1(0)\right] \ , \end{aligned}$$

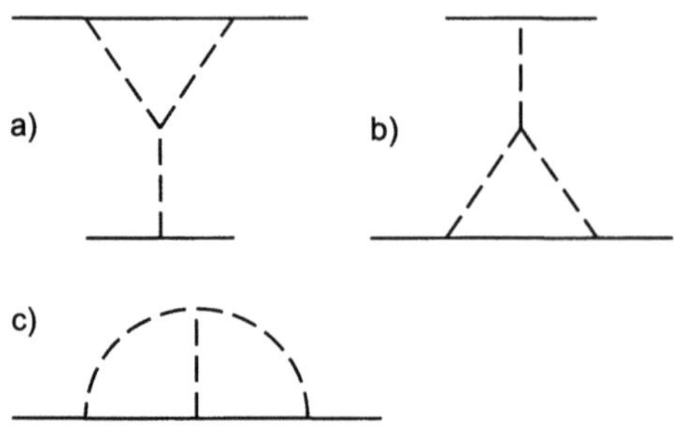

Fig. 10.1. (**a, b**) Diagrams of first order in β_3, leading to an additive renormalization of the two-body interaction. (**c**) Diagram contributing to the additive renormalization of the chemical potential

where the first term gives the leading contribution of the two-body interaction, and we used the notation β_2 to replace β_e. Again using Eq. (A 4.13) we find

$$G_c^{(0,2)}(\mathbf{0},\ldots,\mathbf{0};n,n) = -\frac{\Omega}{(4\pi\ell^2)^{d/2}}(n-1)^2\Big[\beta_2 + 2\beta_3\zeta\left(\frac{d}{2}\right) - \beta_3 n^{-1+\varepsilon/2}\frac{4}{\varepsilon(1-\varepsilon/2)} + O(n^{-2})\Big] \quad . \tag{10.15}$$

The leading three-body contribution $\sim \beta_3\zeta\left(\frac{d}{2}\right)$ results from summation over small fractions of the chains, contained in the triangular substructure of diagrams 10 a,b. It clearly can be absorbed into a redefinition of the two-body coupling: $\beta_2 + 2\beta_3\zeta\left(\frac{d}{2}\right) \to \beta_e$. The remaining contribution conforms to our expectations, being of relative order $n_c^{(3)}(n)/n_c^{(2)}(n) \sim n^{-1+\varepsilon/2} \sim n/R_0^d$ compared to the leading 'two-body' term.

Inserting the three-body interaction into the single chain Greensfunction (Fig. 10.1c), we find that it also generates a one-body term. Subtracting the appropriate one-body and two-body interactions we can construct an additively renormalized three-body interaction, which conforms to the estimate (10.1).

This analysis generalizes to all higher interactions. $\beta_\mu(\mathbf{r}_{j_1},\ldots,\mathbf{r}_{j_\mu})$ generates interactions $\beta_{\mu'}(\mathbf{r}_{j_1},\ldots,\mathbf{r}_{j_{\mu'}})$ for all $1 \le \mu' < \mu$, and after subtraction of all these parts the effect of the subtracted interaction is correctly given by our simple estimate.

To summarize, in constructing our model in principle we should start from the highest interaction $\beta_\mu, \mu = \mu_0$ included, subtracting it by all terms $\mu < \mu_0$ so that the remainder generates corrections obeying the estimate (10.1). The subtracted terms are to be absorbed into a redefinition of the $\beta_\mu, \mu < \mu_0$. We then proceed the same way with β_{μ_0-1}, etc. We finally reach a two-body interaction β_e, which differs from β_2 by contributions extracted from all $\beta_\mu, \mu > 2$. This in turn has to be subtracted by a one-body contribution, which is absorbed into the segment chemical potential.

For a full discussion we must include momentum dependent interactions. For instance, starting from a two-body potential of finite range we can expand the Fourier transform $v_2(\mathbf{q}^2)$ as

$$v_2(\mathbf{q}^2) = \beta_2 + (\mathbf{q}\ell)^2 \beta_{2,1} + \frac{((\mathbf{q}\ell)^2)^2}{2} \beta_{2,2} + \dots \quad . \tag{10.16}$$

Writing

$$(\mathbf{q}\ell)^2 = \frac{(\mathbf{q}R_0)^2}{n}$$

and recalling that we are interested in macroscopic scales $q^{-1} \sim r \sim R_0$, we naively would conclude that each factor of $\mathbf{q}^2$ suppresses the interaction by a factor of $1/n$. Again this is true only after appropriate subtractions. Furthermore we should consider higher order perturbation theory in all the couplings. Clearly the matter becomes extremely involved. No general discussion is available.

What should we learn from this argument? First of all our model, simple as it looks, is highly nontrivial. β_e is not just a rough parameterization of a microscopic two-body interaction, but summarizes the leading effect of all many-body interactions within small fractions of the chains. For $n \to \infty$ these fractions can be taken arbitrarily small but still involve infinitely many segments. The additive renormalization $\beta_2 \to \beta_e$ is no 'short chain effect', in the sense that it would involve only a fixed finite number of neighboring segments. Thus β_e is best viewed as a parameterization of the interaction among whole strands of the chain, short in relation to the total chain, but arbitrarily large on a microscopic scale. From this point of view simple microscopic models of the excluded volume lack physical relevance, except if we can show that many-body interactions are negligibly small on the microscopic scale. We must be aware, however, that the elimination of microscopic degrees of freedom sketched in Chap. 2 automatically generates many-body terms.

As a second message we note that all available evidence suggests that we can additively renormalize the many-body interactions, so that their effect becomes negligible for long chains in the Θ-region $z \lesssim 1$. The restriction to the region of small z reflects the fact that all our arguments above are based on simple perturbation theory. With this qualification and with the additional observation that for $d = 3$ three-body interactions may play some role this suggests that universality in the sense of the two-parameter scheme is valid close to the Θ- temperature.

We finally note that this discussion gains additional importance with respect to the continuous chain limit. In Chap. 7 we have shown that we can construct the continuous chain model only after an additive renormalization, which essentially extracts a one-body part from the two-body potential. If we want to include many-body interactions, we can take the continuous chain limit only after we carried through the additional additive renormalizations sketched here. They are thus an essential step in the very definition of the continuous chain model. A particularly lucid discussion may be found in [Dup88].

10.2 The general structure of the renormalization group

10.2.1 Dilatations

As has been noted in Sect. 8.1 the renormalization group is closely connected to the group of spatial dilatations. We therefore first discuss some general properties of the dilatation group, translating the results to physics in the next sections.

The basic transformation to be considered is a dilatation of the elementary length scale ℓ:

$$T_\lambda[\ell] = \frac{\ell}{\lambda} \quad , \tag{10.17}$$

Dilatations continuously depend on the scale parameter $\lambda > 0$ and form a very simple group with respect to multiplication:

$$T_{\lambda'\lambda}[\ell] = T_{\lambda'}\left[T_\lambda[\ell]\right] \quad . \tag{10.18}$$

Any one-parameter set of transformations obeying this rule under successive applications can be said to be a representation of the dilatation group. Thus relation (8.5) identifies the RG mapping constructed in Sect. 8.1 as a representation of dilatations in the space of models parameterized by $\{\ell, n, \beta_e\}$. We now work out some properties of the dilatation group represented in some general space with coordinates $\vec{Y} \equiv \{Y_j\} \equiv \{Y_1, Y_2, Y_3, \ldots\}$. This space for instance may be the space of all Hamiltonians, the coordinates Y_j being the coupling constants.

The simplest representations take the form

$$T_\lambda[Y_j] = \lambda^{D_j} Y_j \quad , \quad j = 1, 2, 3, \ldots \tag{10.19}$$

where the exponents D_j are arbitrary real numbers. Such representations are called 'linear', the transformed coordinates being linear functions of the original ones. In RG theory the Y_j are known as (linear) 'scaling fields' of scaling dimensions D_j, a notation of purely historical origin.

An important concept is the notion of 'fixed points'. A point $\vec{Y}^* \equiv \{Y_1^*, Y_2^*, Y_3^*, \ldots\}$ in parameter space is said to be a fixed point under the representation T_λ if it obeys

$$T_\lambda \left[\vec{Y}^*\right] \equiv \vec{Y}^* \tag{10.20}$$

for all λ. Clearly a linear representation for finite values of the scaling fields has the single fixed point $Y_j \equiv 0, j = 1, 2, \ldots$.

We now consider some quantity $Q = Q\{Y_j\}$ invariant under the linear representation T_λ:

$$Q\{Y_j\} \equiv Q\Big\{T_\lambda[Y_j]\Big\} \equiv Q\Big\{\lambda^{D_j} Y_j\Big\} \quad . \tag{10.21}$$

This is a special case of Euler's homogeneous functions, which more generally are defined to obey the transformation law

$$E\{Y_j\} = \lambda^{D_E} E\Big\{\lambda^{D_j} Y_j\Big\} \quad . \tag{10.22}$$

Homogeneous functions obey linear scaling laws, where the attribute 'linear' refers to the underlying linear representation. These scaling laws result from choosing λ to eliminate one variable ($Y_1 > 0$, for definiteness):

$$\lambda = const. \;\; Y_1^{-1/D_1} \quad . \tag{10.23}$$

As a result we find

$$E\{Y_1, Y_2, Y_3, \ldots\} = Y_1^{-D_E/D_1} \bar{E}\Big\{Y_2 Y_1^{-D_2/D_1}, Y_3 Y_1^{-D_3/D_1}, \ldots\Big\} \quad , \tag{10.24}$$

where $\bar{E}\{\ldots\}$ is the scaling function and the $Y_j Y_1^{-D_j/D_1}$ are the scaling variables. Thus a homogeneous function, a priori depending on F variables, nontrivially depends only on $F - 1$ combinations of its variables.

We now consider the structure of the general 'nonlinear' representations of the dilatation group, where the transformed quantities are nonlinear functions of the original ones:

$$\vec{Y}(\lambda) = T_\lambda^{(n\ell)}[\vec{Y}] = \Big\{f_j(\lambda; \vec{Y})\Big\}_{j=1,2,\ldots} \quad . \tag{10.25}$$

The group structure under a change of λ forces $\vec{Y}(\lambda)$ to change according to a first order ordinary differential equation. To show this we write

$$\begin{aligned} T_{\lambda+d\lambda}^{(n\ell)} &= T_{(1+\frac{d\lambda}{\lambda})\lambda}^{(n\ell)} = T_{(1+\frac{d\lambda}{\lambda})}^{(n\ell)} \; T_\lambda^{(n\ell)} \\ &= (1 + d\ln\lambda \; \hat{T}) T_\lambda^{(n\ell)} \quad , \end{aligned} \tag{10.26}$$

where we used the obvious relation $T^{(n\ell)}_{\lambda=1} \equiv 1$ and assumed the representation to be analytic in λ.

$$\hat{T} = \frac{d\,T^{(n\ell)}_{\lambda}}{d\,\ln\,\lambda}\Big|_{\lambda=1}$$

is the 'generator' of the representation. Acting on $\vec{Y}$, Eq. (10.26) yields

$$\vec{Y}(\lambda + d\lambda) - \vec{Y}(\lambda) = d\ln\lambda\ \hat{T}[\vec{Y}(\lambda)] \quad , \tag{10.27}$$

or in terms of components Y_j:

$$\frac{d\,Y_j(\lambda)}{d\,\ln\,\lambda} = \hat{T}_j[\{Y_i(\lambda)\}];\ \ j = 1, 2, 3, \ldots \quad . \tag{10.28}$$

These equations define a flow in parameter space.

Fixed points $\vec{Y}^*$ (Eq. (10.20)) obey

$$\hat{T}_j[\vec{Y}^*] = 0;\ \ j = 1, 2, 3, \ldots \quad . \tag{10.29}$$

Close to a fixed point the nonlinear representation reduces to a linear one. Introducing

$$\delta Y_j = Y_j - Y_j^* \quad , \tag{10.30}$$

we may write

$$\frac{d\,\delta\,Y_j}{d\,\ln\,\lambda} = \sum_i D_{ji}\delta Y_i + O(\delta Y^2) \quad , \tag{10.31}$$

where the coefficients

$$D_{ji} = D_{ji}[\vec{Y}^*] = \frac{\partial}{\partial Y_i}\Big|_{\vec{Y}^*}\ \hat{T}_j[\vec{Y}]$$

depend on the fixed point. If the matrix $D = (D_{ji})$ can be diagonalized by a linear transformation: $D \to (D_j\ \delta_{ij})$, we recover a representation of the form (10.19), the scaling fields being linear combinations of the δY_j. In the more general case, where the diagonalization is not possible, the transformation law (10.19) is corrected by powers of $\ln\lambda$, a situation not considered further here. Furthermore, if D_{ji} can be diagonalized, we may construct a (nonlinear) coordinate transformation $Y_j \to Y_j^{(n\ell)}\{Y_i\}$ introducing 'nonlinear scaling fields' $Y_i^{(n\ell)}$, which under dilatations globally transform according to Eq. (10.19). These nonlinear scaling fields are a set of curvilinear coordinates in parameter space, following the flow lines of the nonlinear representation.

Any quantity invariant under a nonlinear representation

$$Q\{Y_j\} = Q\{Y_j(\lambda)\} \tag{10.32}$$

obeys nonlinear scaling laws, which again are derived by choosing λ to eliminate the explicit dependence on one variable:

$$\lambda = \lambda(Y_1); \quad Y_1(\lambda(Y_1)) = 1 \tag{10.33}$$

$$Q\{Y_1, Y_2, Y_3, \ldots\} = \bar{Q}\{Y_2(\lambda(Y_1)), Y_3(\lambda(Y_1)), \ldots\} \ . \tag{10.34}$$

Thus quite generally a dilatation invariant variable, which a priori depends on F scaling fields, can be reduced to a scaling function depending on $F-1$ scaling variables.

The transformation properties of $Q\{Y_j\}$ are often given in terms of 'Ward identities' which are found by taking the derivative of Eq. (10.32):

$$0 = \lambda \frac{d}{d\lambda} Q\{Y_j(\lambda)\} = \sum_j \frac{d\,Y_j(\lambda)}{d\,\ln\,\lambda} \frac{\partial}{\partial\,Y_j} Q \Big|_{\{Y_i(\lambda)\}} \ .$$

Equation (10.28) yields

$$\sum_j \hat{T}_j[\vec{Y}] \frac{\partial}{\partial\,Y_j} Q[\vec{Y}] = 0 \quad , \tag{10.35}$$

which is the general form of the Ward identity.

10.2.2 Comparison to our previous results

Let us now embed the renormalization group, constructed in Chap. 8, into this general framework. As mentioned above, relation (8.5) shows that the RG we are searching for must be a nonlinear representation of the group of dilatations in the space of parameters $\{\ell, n, \beta_e\}$. These are the microscopic parameters of the model, and the representation shall leave macroscopic observables invariant. Furthermore we want the representation to show a nontrivial fixed point. In Sect. 8.2 we have constructed such a representation based on first order perturbation theory. The invariance constraint is obeyed within deviations of order $n_0^{-1+\varepsilon/2}$, $n_0 = n(\lambda = 1)$. Equations (8.38), (8.42) give the parameter flow under this nonlinear representation in the standard form (10.28).

With the explicit form (8.1), (8.38), (8.42) of the mapping the Ward identities (10.35) read

$$\left[-\ell\frac{\partial}{\partial \ell} + W(\beta_e)\frac{\partial}{\partial \beta_e} + \frac{1}{\nu(\beta_e)} n \frac{\partial}{\partial\,n}\right] Q(\ell, n, \beta_e) = O(n_0^{-1+\varepsilon/2}) \ . \tag{10.36}$$

These equations are known as the *renormalization group equations*. They are clearly equivalent to the integrated form

$$Q(\ell_0, n_0, \beta_{e0}) = Q(\ell(\lambda), n(\lambda), \beta_e(\lambda))(1 + O(n_0^{-1+\varepsilon/2})) \ , \tag{10.37}$$

where $\ell(\lambda) = \ell_0/\lambda$ (cf. Eq. (8.1)).

This representation shows two fixed points. We first consider the trivial, or 'Gaussian', fixed point $1/n = 0, \beta_e = 0$. The associated linear representation $T_\lambda^{(0)}$ is of the form (10.19) with $D_{1/n}^{(0)} = -2, D_{\beta_e}^{(0)} = -\varepsilon$, resulting immediately from Eq. (8.29). These are the 'canonical dimensions' of $1/n$ or β_e, respectively. ($D_\ell = -1$ holds by definition of λ.) Showing negative scaling dimensions, both $1/n$ or β_e for $\lambda \to 0$ are driven away from the Gaussian fixed point: $1/n(\lambda) = \lambda^{-2}/n_0; \beta_e(\lambda) = \lambda^{-\varepsilon}\beta_{e,0}$. It should be noted that the continuous chain model is constructed to be rigorously invariant under $T_\lambda^{(0)}$ (cf. Eq. (8.4)). We however pay the price that physical observables, which necessarily involve a microscopic segment size $\ell > 0$, are invariant only up to corrections $\sim \beta_e n^0$. Therefore the naive scaling laws predicting the crossover from Θ- to excluded volume conditions to be a function of the scaling variable $z = \beta_e n^{\varepsilon/2}$, only hold close to the Θ-point: $\beta_e \ll 1$, as pointed out in Chap. 7.

The nontrivial fixed point $1/n^* = 0, \beta_e = \beta_e^*$ induces another linear representation T_λ^*. To find the corresponding scaling dimensions, we write, following Eq. (10.30)

$$\begin{aligned} \delta\beta_e(\lambda) &= \beta_e(\lambda) - \beta_e^* \\ \delta\left(\frac{1}{n(\lambda)}\right) &= \frac{1}{n(\lambda)} - \frac{1}{n^*} \equiv \frac{1}{n(\lambda)} \ , \end{aligned}$$

and we use the RG flow equations to find

$$\begin{aligned} \frac{d\delta\beta_e(\lambda)}{d\ln\lambda} &= \left.\frac{\partial W}{\partial\beta_e}\right|_{\beta_e^*} \delta\beta_e(\lambda) + O(\delta\beta_e^2) \\ \frac{d\delta(1/n(\lambda))}{d\ln\lambda} &= -\frac{1}{\nu(\beta_e^*)}\,\delta\left(1/n(\lambda)\right) + O(\delta\beta_e) \ . \end{aligned}$$

Thus

$$\begin{aligned} D_{1/n}^* &= -\frac{1}{\nu} \\ D_{\delta\beta_e}^* &= \left.\frac{\partial W(\beta_e)}{\partial\beta_e}\right|_{\beta_e^*} \equiv \omega = \varepsilon + O(\varepsilon^2) \ . \end{aligned} \tag{10.38}$$

The difference $D_{1/n}^* - D_{1/n}^{(0)} = 2 - 1/\nu$ is known as the 'anomalous dimension' of $1/n$. The scaling dimension ω of $\delta\beta_e$ is the 'correction to scaling' exponent. Since $\omega > 0$ ($d < 4$) the nontrivial fixed point is repulsive only in direction $1/n$, but attracts all the region $\beta_e > 0$.

The representation T_λ^* underlies the scaling laws in the excluded volume limit, discussed in Chap. 9. These laws are of the general form (10.24), but following standard practice, we assumed β_e to be precisely at the fixed point β_e^*. Corrections due to small deviations $\beta_e - \beta_e^*$ in the theory of critical phenomena are known as 'corrections to scaling', a somewhat ambiguous notation since these corrections have nothing to do with higher interactions or with a breakdown of the linear representation T_λ^* valid in the neighborhood of the fixed point.

To show the general contents of the linear representation T_λ^* we consider once more the end-to-end distance of an isolated chain, essentially rephrasing the calculation of Eqs. (8.13)–(8.16) or (8.54) ff. Ignoring $1/n$ corrections the nonlinear representation yields

$$R_e(\ell, n, \beta_e) = \ell/\lambda \, \bar{R}_e(n(\lambda), \beta_e(\lambda)) \ .$$

Close to the excluded volume fixed point it reduces to the linear representation T_λ^* resulting in

$$\begin{aligned} n(\lambda) &= const \ n \, \lambda^{1/\nu} \\ \beta_e(\lambda) - \beta_e^* &= const \ \lambda^{\omega} \ . \end{aligned} \tag{10.39}$$

Choosing

$$n(\lambda) = const \ \ \rightarrow \lambda = const \ n^{-\nu} \ , \tag{10.40}$$

we find

$$\begin{aligned} R_e &= const \ \ell \, n^\nu \bar{R}_e(const, \beta_e(\lambda)) \\ &= const \ \ell \, n^\nu \left(1 + const \ n^{-\nu\omega} + O(n^{-2\nu\omega})\right) \ . \end{aligned} \tag{10.41}$$

To find the last line we expanded $\bar{R}_e(const, \beta_e(\lambda))$ about β_e^*, neglecting terms of order $(\beta_e(\lambda) - \beta_e^*)^2$. This is consistent with the linearization (10.31) of $T_\lambda^{(n\ell)}$. Of the result (10.41) only the leading power law $R_e \sim n^\nu$ is kept in standard scaling theories, the term $\sim n^{-\nu\omega}$ being a 'correction to scaling'.

10.2.3 The general structure of the renormalization group

The renormalization group deals with *critical systems*, i.e. systems for which the correlation length diverges and which therefore show important fluctuations on all length scales. Consider for instance a normal fluid. Measuring the radial distribution function, i.e. the correlations among the positions of the molecules, we generically find that a given molecule influences only its immediate neighborhood. It creates a small density fluctuation which extends over a correlation length ξ of a few Ångströms. However, driving this system close to the critical point of its fluid-gas phase transition, we observe that ξ increases tremendously. This is signalled by the phenomenon of critical opalescence, i.e. the strong scattering of normal light, which shows that there exist density fluctuations over distances of the order of the wavelength of light: $\xi \gtrsim 10^3$ Å. Furthermore, the strong scattering of radiation of smaller wavelength shows that there exist fluctuations on all scales $r \lesssim \xi$. Together with the divergence of ξ at the critical point this is the defining property of a critical system.

According to this definition dilute solutions of long macromolecules are critical. The role of the correlation length is played by the radius of gyration $R_g \sim N^\nu \to \infty, N \to \infty$, and by virtue of the chain structure a polymer coil shows density fluctuations on all scales $r \lesssim R_g$. Indeed, a blob of size r is just a correlated fluctuation of the density.

Now for a critical system straightforward perturbation theory breaks down, since it cannot deal with a macroscopically large number of strongly interacting degrees of freedom. This typically is signaled by the divergence of the expansion parameter ($z = \beta_e N^{\varepsilon/2} \to \infty$ in the polymer problem). We can however base our theory on the very same feature that ruins naive perturbation theory: there exist strong fluctuations on all scales. In particular at the critical point $\xi = \infty$ there exists no distinguished scale governing the macroscopic physics. A critical system looks the same at each magnification, provided the resolution of our experiment is not high enough to detect the microstructure. We therefore should be able to describe the same macroscopic physics by a set of Hamiltonians $\mathcal{H}_\lambda$, defined on the continuously varying microscopic scale $\ell' = \ell/\lambda$, $0 < \lambda \leq 1$. Instead of solving the problem by brute force in a single step we might try to construct the mapping $\mathcal{H} \equiv \mathcal{H}_1 \to \mathcal{H}_\lambda$, which clearly is a representation of the dilatation group in the space of Hamiltonians. Such a representation is called a renormalization group, $\mathcal{H}_\lambda$ being known as the renormalized Hamiltonian. From the general discussion of dilatations we know that it is sufficient to construct the generator $d\mathcal{H}_\lambda/d\ln\lambda|_{\lambda=1}$, which is a manageable problem, since it involves only the elimination of the few degrees of freedom governing the physics in the infinitesimal range $\ell \leq r < \ell(1-d\lambda)$. Integrating the flow equations up to $\lambda = \ell/\xi$ we then hopefully can find a Hamiltonian $\mathcal{H}_{\ell/\xi}$ governing the physics on macroscopic scales, where all effects of scales $r < \xi$ are absorbed into a renormalization of the couplings. This is the general RG scenario.

We now discuss the basic features of the RG flow. This amounts to giving a physical interpretation of our general discussion of the dilatation group, and it explains qualitatively the characteristic features like universality, power laws, and scaling observed in a critical system. For a given class of systems (fluids, uniaxial ferromagnets, polymer solutions, etc.) we envisage the space of all Hamiltonians to be parameterized by a microscopic scale ℓ and dimensionless couplings $v_j, j = 1, 2, 3, \ldots$. The Hamiltonian is then represented by a point $(\ell, \{v_j\})$, corresponding to $\{Y_j\}$ in the analysis of Sect. 10.2.1. The RG induces a flow $(\ell/\lambda, \{v_j(\lambda)\})$ in coupling constant space. Playing a special role, ℓ will not be considered as a coordinate of that space. It is rather replaced by the scale parameter λ of the flow.

By definition, criticality is reached if the correlation length diverges:

$$0 = \xi^{-1} = \ell^{-1}\bar{\xi}^{-1}\{v_j\} \quad . \tag{10.42}$$

This relation defines the *critical manifold* (often called critical surface), which is an (infinite dimensional) submanifold of parameter space. Since the macroscopic scale ξ is invariant under renormalization, $\bar{\xi}$ scales as

$$\lambda\bar{\xi}^{-1}\{v_j(\lambda)\} = \bar{\xi}^{-1}\{v_j(1)\} \ , \tag{10.43}$$

which shows that the critical surface $\bar{\xi}^{-1} \equiv 0$ is invariant under the RG: critical systems are mapped onto critical systems. Furthermore for a given class of systems generically the critical behavior is found to be independent of microscopic details. This *universality* suggests that the RG for $\lambda \to 0$ maps all of – or a large part of – the critical manifold onto a single *fixed point Hamiltonian* $\mathcal{H}^* \hat{=} \{v_j^*\}$, that governs the critical behavior. We then may linearize the RG about the fixed point to construct the *linear scaling fields.*

Let us assume for simplicity that the linearized RG is of the form (10.19), so that asymptotically the scaling fields w_j, constructed as linear combinations of the $v_k - v_k^*$, obey

$$\frac{d\, w_j}{d \ln \lambda} = D_j w_j \quad . \tag{10.44}$$

Fields of fixed point dimension $D_j > 0$ for $\lambda \to 0$ are driven into the fixed point and therefore are called *irrelevant.* $D_j < 0$ identifies a *relevant* field, being driven away from the fixed point. There also may occur *marginal* fields: $D_j = 0$. For these we have to go beyond the linear approximation to decide whether they are *marginally irrelevant* (driven into the fixed point) or *marginally relevant* (driven away from the fixed point). Marginal fields are one source of logarithmic corrections in scaling laws. To illustrate this, let us consider the simplest case, assuming that the field $w > 0$ to leading order transforms as

$$\frac{d\, w}{d \ln \lambda} = w^2 \quad .$$

Thus

$$w(\lambda) \sim \frac{1}{|\ln \lambda|} \to 0 \quad , \quad \lambda \to 0 \quad ,$$

the parameter $w > 0$ being marginally irrelevant. As mentioned before another source of logarithmic corrections is found if the matrix D_{ij}, defining the linearized RG according to Eq. (10.31), cannot be diagonalized.

Leaving the fixed point in the direction of an irrelevant field, we stay within the critical surface. Going in the direction of a relevant field we leave the region of attraction of the fixed point, in general leaving the critical surface. In physical applications the number of relevant fields must be small. Criticality must be reached by adjusting only a few parameters. For instance, for a ferromagnet the magnetic field must vanish and the temperature must approach the Curie temperature. For the liquid-gas critical point, temperature and chemical potential have to be adjusted. For polymer solutions the critical conditions are $1/N \to 0, \ell^d c \to 0$. It is these parameters that correspond to the relevant directions in parameter space. In all these examples we thus have two relevant scaling fields, which is the generic case.

So far we considered the structure of coupling constant space in the vicinity of a fixed point, as induced by the linearized RG. We can now globally define flow lines $\vec{w}_{n\ell}^{(j_0)}(\lambda) \equiv \{w_j^{(j_0)}(\lambda)\}$, distinguished by the fact that they approach the fixed point from direction w_{j_0}. These lines define a set of curvilinear coordinates in Hamiltonian space reflecting the global structure of the RG. Measuring the distance along those lines in terms of a variable $w_{n\ell}^{(j_0)}$ obeying

$$\frac{d\, w_{n\ell}^{(j_0)}}{d \ln \lambda} = D_{j_0} w_{n\ell}^{(j_0)} \quad , \tag{10.45}$$

we arrive at the nonlinear scaling fields. For marginal fields the definition has to be changed appropriately.

We are now in the position to sketch the RG flow globally. A schematic picture of the flow in the critical manifold is shown in Fig. 10.2, where the manifold is approximated as a plane parameterized by $w_{n\ell}^{(i_0)}, w_{n\ell}^{(i_1)}$. Assuming that $w_{n\ell}^{(i_0)}$ is the irrelevant field of smallest fixed point dimension we note that

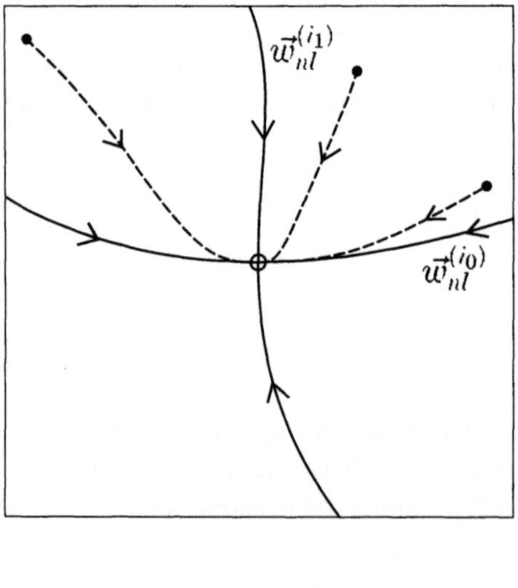

Fig. 10.2. Schematic flow diagram on the critical surface close to a fixed point. *Full lines* show the flow of nonlinear scaling fields, where $D_{i_1} > D_{i_0} > 0$. That relation is signalled by the *broken lines* which illustrate that the flow from a generic point on the critical surface approaches the fixed point from direction $\vec{w}_{n\ell}^{(i_0)}$. The tangents of $\vec{w}_{n\ell}^{(i_0)}, \vec{w}_{n\ell}^{(i_1)}$ in the fixed points define the linear scaling fields, which here arbitrarily have been chosen perpendicular to each other

a generic point on the critical manifold for $\lambda \to 0$ is driven close to the flow line $\vec{w}_{n\ell}^{(i_0)}$, approaching the fixed point from direction $w^{(i_0)}$. The exceptional case where the scaling field of the next smallest dimension governs the flow into the fixed point presupposes that the original critical Hamiltonian, if parameterized by the nonlinear scaling fields, shows $w_{n\ell}^{(i_0)}(\lambda = 1) = 0$. Thus $w_{n\ell}^{(i_0)}$ plays a special role, inducing the leading corrections to the fixed point behavior which generally are addressed as *corrections to scaling.* Depending only on the single additional parameter $w_{n\ell}^{(i_0)}(1)$, these corrections have some degree of universality. Corrections due to the other irrelevant fields are known as '*nonuniversal corrections*'.

The flow outside the critical manifold is sketched in Fig. 10.3, where we schematically represent the critical manifold by $\vec{w}_{n\ell}^{(i_0)}$. Asymptotically the flow is driven in the direction of the flow line of that relevant field $w_{n\ell}^{(r)}$, which is the field with negative dimension of largest absolute value contained in the parameterization of the original Hamiltonian. Now consider some typical experiment, for instance changing the temperature of a ferromagnet at zero magnetic field: $h \equiv 0$. (In that case we deal only with a single relevant field $\sim T - T_c$.) The original Hamiltonian $\mathcal{H}_{\lambda=1}(T)$ will trace out a physical path in parameter space, intersecting the critical manifold at the critical temperature T_c. For T close to T_c, $\mathcal{H}_\lambda(T)$ under the RG will be driven into the fixed point region, where we can replace the RG by its linearized form. In that region the observables thus reduce to homogeneous functions of the parameter λ, and fixing λ so that the relevant field $\sim T - T_c$ takes some fixed small value independent of T we derive the scaling laws. (See Eq. (10.21) ff.) In an equivalent, but more elegant formulation of the argument we assume the Hamiltonian to be parameterized by nonlinear scaling fields obeying Eq. (10.44). With such a set of basic coordinates the homogeneity (10.21) of

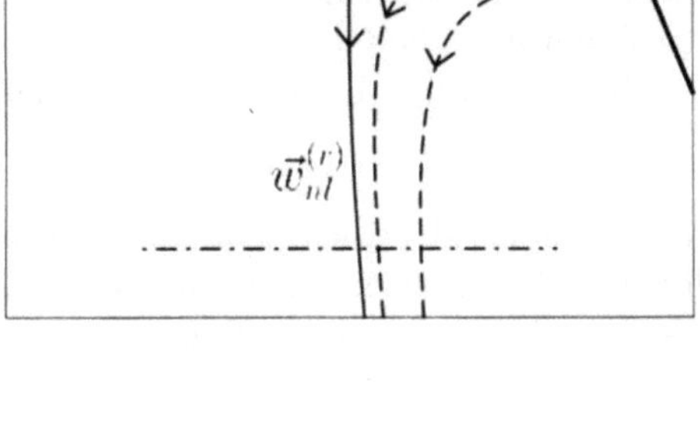

Fig. 10.3. RG flow illustrated with the example of a single relevant field $\vec{w}_{n\ell}^{(r)}$ (essentially parameterizing $T - T_c$) and a single irrelevant field $\vec{w}_{n\ell}^{(i_0)}$, representing the critical surface. The *shaded ellipsoid* shall indicate the region where the linearized RG is appropriate. The fat line represents the path in parameter space followed by a physical system upon changing T, for instance. Typical renormalization group trajectories are shown by the *broken lines.* The *dot-dashed lines* represent the uncritical manifolds far from criticality, where perturbation theory holds

an RG-invariant observable holds rigorously:

$$Q\{w_{n\ell}^{(j)}(1)\} \equiv Q\{w_{n\ell}^{(j)}(\lambda)\} \equiv Q\{\lambda^{D_j} w_{n\ell}^{(j)}(1)\} \quad . \tag{10.46}$$

Choosing $\lambda^{D_1} w_{n\ell}^{(1)}(1) = 1$, we quite generally find the relation

$$Q\{w_{n\ell}^{(j)}(1)\} \equiv \bar{Q}\Big\{w_{n\ell}^{(j)}(1)/(w_{n\ell}^{(1)}(1))^{D_j/D_1}\Big\}_{j=2,3,\dots} \quad . \tag{10.47}$$

To derive the physical scaling laws we note that the $w_{n\ell}^{(j)}(1)$, parameterizing the physical path, should be analytic functions of the physical parameters (T, h, or $N, \ell^d c$, etc), the relevant fields vanishing at the critical manifold. Again using the magnetic example we may thus assume $w_{n\ell}^{(1)}(1) \sim h$, $w_{n\ell}^{(2)}(1) \sim T - T_c$, valid to leading order close to the critical manifold. Thus the first scaling variable takes the form $h/|T-T_c|^{D_2/D_1}$. The neglect of higher order terms in the expansion of $w_{n\ell}^{(j)}(1)$ about $h = 0, T = T_c$ is equivalent to the use of the linearized RG in the first argument.

To calculate the scaling functions we now follow the flow up to some fixed surface chosen far from the critical manifold, such that $\ell/\lambda \sim \xi$. The resulting 'macroscopic' Hamiltonian allows for a perturbative treatment, all critical effects being absorbed into the renormalization of the couplings. On the purely formal level the precise choice of this *uncritical manifold*, where the scaling functions are to be evaluated, is of no importance. This is guaranteed by the scale invariance of the theory, evaluated to arbitrarily high order of perturbation theory. In practice, however, we can evaluate the scaling functions only to low order approximation. Thus the precise choice of the uncritical manifold is an important part of the approximation scheme, which will be of much concern to us later.

Up to now we tacitly assumed that the single fixed point $\{v_j^*\}$ dominates the critical manifold. This is not the full story. In the applications of interest there are at least two fixed points located on the critical manifold. Besides $\{v_j^*\}$ there is a fixed point $\{v_j^0\}$ at which an additional coordinate $v^{(i_0)}$ is relevant, the other fields in the critical surface staying irrelevant. The attractive manifold of $\{v_j^0\}$ separates the critical surface into a part attracted to $\{v_j^*\}$ and a part where the flow either is governed by another fixed point or runs away to infinity. Generally the fixed point $\{v_j^0\}$ is 'trivial', describing the noninteracting theory which is the starting point of simple perturbation theory. For our polymer problem it is the Gaussian fixed point $1/N = 0, \ell^d c = 0, \beta_e = 0$, the excluded volume coupling being the additional relevant coordinate.

So far we have considered the RG for a general critical system. Figure 10.3, in particular, sketches the flow for a typical phase transition problem, where the physical path truly intersects the critical manifold. This manifold separates the regions corresponding to the two phases, and under the RG the system is driven into different directions on the two sides. For polymer solutions the picture is somewhat simpler. The critical manifold $1/N = 0$, $\ell^d c = 0$ bounds the physical region $1/N > 0$, $\ell^d c > 0$. It does not separate two phases, only parameter space on one side of the critical manifold being physical. The Gaussian fixed point $1/N = 0, \ell^d c = 0, \beta_e = 0$ describes a Θ-chain. The region $\beta_e > 0$ of the critical surface is driven into the excluded volume fixed point. For $\beta_e < 0$ we find a runaway. Within our model the chain collapses under the influence of the attractive forces into a region of the size of the elementary segment ℓ_0. Physically this collapse is blocked by a hard core of the two-body potential or by repulsive three-body forces, but to incorporate such features in a consistent RG calculation at present exceeds our capabilities.

This structure of the critical manifold is valid for $d < 4$. For $d \to 4$ the nontrivial fixed point merges into the Gaussian fixed point. For $d \geq 4$ only the Gaussian fixed point is physical, governing the system for all $\beta_e \geq 0$. In RG language this is the reason for the triviality of the results for $d > 4$, pointed out in Chap. 6.

Here we should also stress another feature of the polymer problem. In standard critical phenomena only the relevant parameters are controlled easily, for instance by changing the magnetic field or the temperature. The irrelevant parameters are determined by the material at hand, with little chance of systematic variation. If we want to control more parameters, we must turn to more complicated systems like multicomponent solutions. For polymer solutions, however, besides the relevant parameters $1/N, \ell^d c$, we can control the temperature. This strongly affects the excluded volume coupling which in turn rules the leading irrelevant scaling field. In contrast to many other critical phenomena we thus can very easily explore a three dimensional range of physical Hamiltonians. From a theoretical point of view this is one source of fascination in the polymer problem.

We now have described the general scenario identifying the RG as a representation of the dilatations in the space of all Hamiltonians. To make this idea work, we have to carry out two steps:

(i) we have to construct the generator of infinitesimal dilations,

and

(ii) we have to integrate the resulting flow equations.

The first step involves the elimination of only an infinitesimally small fraction of all the degrees of freedom, which is a manageable problem. Depending on the precise prescription of the elimination process, several RG generators have been discussed in the literature. The second step necessarily involves heavy approximations, since it is impossible to integrate the nonlinear flow equations in an infinitely dimensional space. A good truncation is essential, and it is here where we give up the idea that the renormalized Hamiltonian $\mathcal{H}_\lambda$ on scale $r > \ell_0/\lambda$ *rigorously* describes the same physics as the microscopic Hamiltonian. The most drastic – but also most successful – truncation restricts the space to those fields relevant at the Gaussian fixed point (i.e. $1/N, \ell^d c, \beta_e$ for the polymer problem). This leads to the field theoretic renormalization group, which we now discuss.

10.3 Field theoretic renormalization group and universality

Field theory works with an extremely restricted parameter space, keeping only the linear scaling fields relevant at the Gaussian fixed point. Within that 'field theoretic manifold' we cannot find an exact representation of the dilatation group of the kind discussed in the previous section. Rather we have to be content with a mapping leaving physical observables invariant up to irrelevant corrections. In terms of polymer physics we therefore are restricted to the region of long chains $N \gg 1$ and small concentrations $\ell^d c \ll 1$. In trying to do better we would have to enlarge the parameter space by further (irrelevant) fields corresponding to three-body forces, chain stiffness, etc.

Within its restricted parameter space the field theoretic RG can be constructed to all orders in powers of a renormalized coupling u replacing the excluded volume constant β_e. u parameterizes the critical surface. Perturbation theory can be resummed to yield a most precise RG mapping. Besides the Gaussian fixed point $u^{(0)} = 0$ this mapping shows an excluded volume fixed point $u^* > 0 (d < 4)$. Within the field theoretic manifold we thus have an *approximate* representation of the dilatation group consistent with the general structure outlined in the previous section.

There remains the question of whether all further interaction parameters are irrelevant and die out faster than all terms included in the field theoretic model. Proving this globally we would prove universality in the sense that the field theoretic model adequately describes the critical behavior. Field theory cannot attack this question globally, but only locally via expansion about the field theoretic model. We present the results in the language of polymer physics, first discussing the neighborhood of the Gaussian fixed point. Here the results of Sect. 10.1 show that the appropriately subtracted coupling $\beta_{\mu,\sigma}$, representing a μ-body interaction with σ factors of type $(\mathbf{q}\ell)$ (see Eq. (10.16)) is a linear scaling field of dimension

$$D^{(0)}_{\mu,\sigma} = \mu(d-2) + \sigma - d \ . \tag{10.48}$$

Indeed to leading order the interaction energy due to $\beta_{\mu,0}$ is $W_\mu \sim \beta_{\mu,0} n_c^{(\mu)}(n)$, and using Eq. (10.1) and the invariance of W_μ under rescaling we find

$$\beta_{\mu,0}(\lambda) = \beta_{\mu,0}\left(\frac{n}{n(\lambda)}\right)^{\mu-(\mu-1)d/2} = \beta_{\mu,0}\lambda^{\mu(d-2)-d} \ .$$

Note that $n(\lambda) = \lambda^2 n$ holds at the Gaussian fixed point (see Sect. 10.2.2). An additional factor of $(\mathbf{q}\ell)$ induces an additional factor of λ. In field theoretic language the argument amounts to naive power counting presented in Sect. 7.2. We thus find that for $d > 3$ all higher couplings ($\mu \geq 3$ or $\mu = 2, \sigma > 0$) are irrelevant at the Gaussian fixed point. In $d = 3$ the three-body coupling $\mu = 3, \sigma = 0$ is marginal and in fact turns out to be marginally irrelevant. In the present language this just rephrases an argument given several times before.

Let us now turn to the nontrivial fixed point. For $d \to 4$ it merges into the Gaussian fixed point. Analyticity in the 'parameter' d of the RG flow presupposed by the ε-expansion then guarantees that the scaling fields at u^* as well as their dimensions $D^*_{\mu,\sigma}$ for $d < 4$ smoothly develop from their equivalent at the Gaussian fixed point, $d = 4$. Indeed it has recently been shown [KWP93] that

$$D^*_{\mu,\sigma}(4-\varepsilon) = D^{(0)}_{\mu,\sigma}(d=4) + a_{\mu,\sigma}\varepsilon + O(\varepsilon^2) \ ,$$

where $a_{\mu,\sigma} \geq 0$ for all μ, σ. This is a strong hint that all higher interactions, including the three-body interaction, stay irrelevant at the nontrivial fixed point also in $d = 3$, their dimensions being larger than the dimension of the renormalized two-body coupling u. In summary, we have reasons to believe that the field theoretic RG conforms to the general picture, establishing universality at least for systems close to the field theoretic manifold.

We close this section with a remark on the central limit theorem. It states that interactions, confined to segments closely spaced along the chain, do not affect the large scale Gaussian behavior of an otherwise noninteracting chain. In RG language it states that all such interactions are irrelevant globally. The field theoretic analysis extends this by stating that all higher interactions are irrelevant at least locally, the two-body coupling driving the system into a new non-Gaussian fixed point.

Notes

The serious problems encountered if we aim at a 'microscopically realistic' calculation of the excluded volume forces have clearly been noted in the literature. See, for instance, [DLR96].

The general renormalization group as a representation of dilatations in the space of all Hamiltonians has been introduced by Wilson (see [WK74] for a review). Much of the abstract structure has been worked out by Wegner, as reviewed in [Weg76]. See also [Ma76].

The field theoretic RG has been discussed in numereous books or reviews. We in particular refer to [ZJ89, Ami84, BGZJ76]. Special attention to the role of irrelevant corrections has been given in [ZJ89, Jug82, KWP93, KS94].

Part III

Field Theoretic Formulation of the Theory

We present here a most efficient form of the theory. It is based on 'renormalized perturbation theory', which results from taking a kind of continuous chain limit. However, this limit is taken in a much more refined way than in the 'naive' approach discussed in Chap. 7. The resulting renormalized theory preserves the effect of the leading cut off corrections and exists also in $d = 4$. All nonuniversal corrections due to higher interactions etc. are eliminated. The microscopic segment size ℓ is replaced by a scale ℓ_R, which can be chosen arbitrarily. The renormalization group equations give the change of the parameters of the renormalized model under a change of ℓ_R. The calculation of macroscopic observables then proceeds in two well separated steps. We first use the RG to map the system from the critical region to an uncritical manifold, where the renormalized perturbation theory is well behaved. In practice this amounts to choosing ℓ_R to be of the order of the smallest macroscopic length scale of interest. We then use renormalized perturbation theory to calculate the universal scaling functions. All the calculation proceeds within the renormalized formalism, with no 'matching' to unrenormalized theory involved.

This separation of the approach into the two steps (i) construction of the RG mapping, and (ii) calculation of the scaling functions is of fundamental significance. It turns out that a good approximation for the RG mapping (step i), including precise values of the exponents, is most important. Fortunately there exists a special renormalization scheme, called 'minimal subtraction', where a very precise form of the RG mapping is known. It results from pushing perturbation theory to high orders. The calculation of the scaling functions (step ii) is restricted to low order perturbation theory, however. Fortunately this seems to be sufficient, since the scaling functions do not show much structure.

In this part we first explain in general terms the construction of renormalized perturbation theory. We show how the RG results from the arbitrariness of ℓ_R and establish the general scaling form (Chap. 11). We then turn to the specific technique of minimal subtraction and show how to calculate the scaling functions (Chap. 12). The RG mapping, used in the sequel, is presented and discussed in Chap. 13. We finally (Chap. 14) illustrate the theory with an evaluation of the tree approximation.

The theory as presented here has been called the 'field theoretic approach', since it relies on powerful results derived in quantum field theory or in statistical field theory of critical phenomena. This aspect, however, will not be important to us. Yet a reader acquainted with critical phenomena will easily see close connections and hopefully may take some interest in the construction of crossover scaling functions as advocated here.

11. Renormalized Perturbation Theory and Field-Theoretic Renormalization Group

The version of the renormalization group constructed in Chap. 8 shows some principle disadvantage: calculating cross-over phenomena by matching to 'bare' (i.e. unrenormalized) perturbation theory, we introduce model-dependent terms masking the expected universal behavior of the scaling functions. Can we construct a better version of the theory, clearly distinguishing universal from model-dependent features?

Such a formulation indeed exists and is known as *'renormalized perturbation theory'*. The construction uses a modified form of the continuous chain limit. As explained in Chap. 7, naively the continuous chain limit $\ell \to 0$ is taken with $R_0^2 = \ell^2 n, z = \beta_e n^{\varepsilon/2}$ fixed. In results like (4.16 ii):

$$\frac{1}{2d} R_e^2 = R_0^2 \left[1 + z \left(\frac{2}{\varepsilon} - \frac{2}{2+\varepsilon} + \left(\alpha_1 - \frac{2}{\varepsilon} \right) (\ell / R_0)^\varepsilon + O\left(\frac{\ell^2}{R_0^2} \right) \right) \right]$$

we thereby suppress the nonuniversal correction terms $\sim \alpha_1 - 2/\varepsilon$, which reflect the discrete microstructure. Furthermore we have the technical benefit of replacing summations by integrals, which are much easier to handle. However, we also lose those terms subleading in n which according to Sect. 8.2 are important in the construction of the renormalization group. Furthermore, the result does not exist in four dimensions and therefore is not compatible with the idea of expanding in powers of $\varepsilon = 4 - d$. This expansion, however, was essential in constructing a consistent perturbative scheme. Finally, as has been stressed in Chap. 7, the continuous chain model is a valid model of a real polymer only close to the Θ-temperature. It thus a priori cannot be applied to systems like polystyrene-toluene, which for all accessible temperatures are far from Θ-conditions.

We can overcome all of these problems by taking the continuous chain limit in a more sophisticated way. We introduce a 'renormalized' interaction constant u and a 'renormalized' chain length n_R by the formal relations

$$\beta_e \ell^{-\varepsilon} = u \ell_R^{-\varepsilon} Z_u(u, \ell/\ell_R) \qquad (11.1\ \text{i})$$

$$n\ell^2 = n_R \ell_R^2 Z_n(u, \ell/\ell_R) \quad , \qquad (11.1\ \text{ii})$$

where ℓ_R is some arbitrary length scale taking care of the spatial dimensions in the renormalized theory. The 'renormalization factors' Z_u, Z_n are independent of chain length and are defined via their Taylor series expansion in powers of u. *The coefficients in these series are chosen so as to absorb all the leading microstructure corrections occurring in the expansions of the scaling functions, reordered according to powers of u.* In the scaling functions only corrections of relative order n^{-1} are left, which are eliminated by taking the continuous chain limit $\ell \to 0$, but keeping $\underline{u, n_R, \ell_R}$ fixed. The resulting *'renormalized perturbation theory'* is purged from microstructure effects and reproduces the experimentally observable scaling functions up to corrections roughly of order n^{-1}, which are negligibly small in all the excluded volume regime. This is to be contrasted to the naive continuous chain limit, which results from setting $Z_u = Z_n = 1$, and reproduces the results of the discrete chain model up to corrections of order n^0, only. Furthermore the renormalized expansion exists also in four dimensions and thus is compatible with the ε-expansion. All these statements hold to arbitrary order in u, which is the highly nontrivial content of the *Theorem of Renormalizability.*

Note that the concept of *renormalization*, as sketched here, is independent of the *renormalization group*, even though Eqs. (11.1) are most similar to Eqs. (8.1), which expressed the mapping under a change of the microscopic length scale. But here the primary goal is to suppress model dependent features, not to change the segment size. Of course the two aspects are deeply related, since the microstructure-independent renormalized theory is automatically found to be dilatation invariant. Note further that upon renormalization the important leading corrections to the continuous chain limit are not lost, but effectively are saved in the dependence on ℓ/ℓ_R of the renormalization factors. The limit $\ell \to 0$ is taken in the renormalized theory, not in the renormalization factors! This point is important conceptually and must be stressed here, even though in later chapters for technical reasons we will take the limit $\ell \to 0$ also in the Z-factors. This method, which essentially starts from the naive continuous chain limit, has to be justified by quite a delicate discussion.

With respect to the general discussion of the previous chapter renormalization keeps all the relevant parameters $\left(n_R \hat{=} n, \ell_R^d c_p n_R \hat{=} \ell^d c\right)$ and the most important irrelevant one $(u \hat{=} \beta_e)$, but via the limit $\ell \to 0$ suppresses all the other irrelevant parameters describing many-body interactions. It thus maps the bare Hamiltonian $\mathcal{H}$ found somewhere in parameter space onto a renormalized Hamiltonian $\mathcal{H}_R$ found on a field theoretic manifold parameterized by $n_R, \ell_R^d c_p, u$. $\mathcal{H}$ and $\mathcal{H}_R$ differ only by irrelevant interaction terms which induce corrections of order $1/n$ in the critical limit. It must be clearly stressed that this step of renormalization in fact changes the physical observables, but only by negligible terms.

Having constructed renormalized perturbation theory, in a further step we have to determine the renormalized parameters u, n_R as functions of $\beta_e, n, \ell/\ell_R$. As it stands, Eq. (11.1) cannot be used for that purpose, since the Taylor expansion of the renormalization factors suffers from all the drawbacks of the unrenormalized cluster expansion. A priori, in strict ε-expansion it orders according to powers of $u \cdot \ln(\ell/\ell_R)$. Now the renormalized perturbation theory for the scaling functions proceeds in powers of $un_R^{\varepsilon/2}$ corresponding to the parameter z of the naive continous chain limit. To make these expansions well behaved we have to choose $n_R \sim 1$, which via Eq. (11.1 ii) implies $(\ell_R/\ell)^2 \sim n$. The expansion of the Z-factors therefore effectively orders according to powers of $u \cdot \ln(n)$, breaking down in the excluded volume limit of long chains. (Note that the choice $n_R \sim 1$ replaces the matching condition $n' \sim 1$ of Sect. 8.3). Renormalizing the theory we do not seem to have solved the problem. Rather we have shifted it to the renormalization factors.

This problem can be circumvented, however. It is here, where the renormalization group comes to our help. Equation (11.1) defines a one-parameter set of renormalized theories, all being equivalent to the given unrenormalized theory. This set is parameterized by the arbitrary length scale ℓ_R, which can be viewed as the size of an effective segment. Taking the logarithmic derivative with respect to ℓ_R, keeping all unrenormalized parameters fixed, from Eqs. (11.1) we derive flow equations which give the change of the renormalized coupling u and of the renormalized chain length n_R under an infinitesimal change of the effective segment size, i.e. we derive the RG equations. As in Chap. 8, the functions occurring in the flow equations have a valid expansion in powers of u and ε, independent of ℓ/ℓ_R. We can thus determine $u(\ell/\ell_R)$ etc. by integrating the flow equations evaluated to some finite order in u and ε, starting from some initial conditions ($u(1)$, etc.). In that way we circumvent all problems related to a direct perturbative determination of the renormalization factors for $(\ell_R/\ell)^2 \sim n$. We can then exploit the freedom in choosing ℓ_R to eliminate one renormalized variable, taking $n_R = 1$, for instance. (In later chapters a more refined choice will be advocated.) The remaining renormalized variables become the 'scaling variables', and the renormalized expressions for the physical observables become the scaling functions.

The scaling variables depend on the initial conditions, i.e. on the renormalized theory defined on scale $\ell_R = \ell$, which is of microscopic order. Therefore the initial values $u(1)$ etc. should be analytic functions of the bare couplings, which indeed might be calculable by exploiting Eqs. (11.1) for $\ell_R = \ell$. However that step is unnecessary if not futile. We have learned in Sect. 10.1 that the microscopic excluded volume strength β_e is not simply related to the physical pair-potential among segments, but is just some effective parameter analytic in T. Thus the relation among $u(1)$ and β_e is of no interest, the same statement holding for the value of $Z_n(u(1), 1)$. Without losing information we may use the initial conditions in the integration of the RG flow equations as our effective temperature- and chemistry-dependent parameters. *Once we have established the renormalized theory we need never refer to the underlying bare model*, which after all is a quite artificial object. It is the renormalized theory, that contains all the universal physics we want to describe.

This insight can hardly be overemphasized. Our theory is constructed to deal with the critical limit $N \to \infty, \ell^d c \to 0$. Renormalization yields a consistent theory of the universal behavior, whereas the bare model is no microscopically valid representation of any realistic system. The renormalized theory essentially depends on two parameters $(u(1), Z_n(u(1), 1))$ only. Thus our theory is a generalized two-parameter theory, also holding outside the Θ-region. But we pay with being unable to construct a microscopic theory of these parameters. Furthermore, the theory cannot simply be extended into the oligomer region or to higher concentration, where the correction terms suppressed by renormalization will play an important role.

The theory as sketched here is quite general. It establishes and exploits scale invariance of the polymer problem. In the present chapter we concentrate on these general aspects, presenting in precise terms all the qualitative results of the field theoretic RG. For this purpose we never need to specify explicitly the choice of the renormalization factors Z_n, Z_u. This choice is relevant only for concrete perturbative calculations of the scaling functions, a topic taken up in the next chapters. Here we only illustrate the basic ideas by renormalizing the first order results for R_e^2, A_2 (Sect. 11.1), thereby showing how the Z-factors absorb the microstructure dependence. We are then prepared to formulate the Theorem of Renormalizability (Sect. 11.2). We will not prove that theorem; this would need a book of its own. Only in Appendix A 11.1 we will point out some ideas underlying the proof. With the help of the basic theorem we can prove the existence of the renormalization group flow equations (Sect. 11.3). Integrating these equations we find the global RG flow (Sect. 11.4). The integration constants define the nonuniversal parameters of the two parameter theory (Sect. 11.5). We can then derive scaling laws both in the dilute limit (Sect. 11.6) and for finite overlap (Sect. 11.7). We close this chapter with some remarks on universality and on the mathematical status of the theory (Sect. 11.8).

11.1 Illustration of renormalized perturbation theory

In Chap. 8 we have constructed the renormalization group, starting from bare perturbation theory for the discrete chain model. This expansion involves nonuniversal microstructure corrections which we will now absorb into renormalization factors, introduced via a redefinition of the interaction constant and the chain length. According to Eq. (11.1) we write

$$\beta_e \ell^{-\varepsilon} = u \ell_R^{-\varepsilon} Z_u(u, \ell/\ell_R) \ ,$$
$$n\ell^2 = n_R \ell_R^2 Z_n(u, \ell/\ell_R) \ ,$$

where ℓ_R is an arbitrary length scale. The renormalized coupling u and the renormalized chain length n_R are well defined only after the renormalization factors Z_u, Z_n have been specified. It will turn out that we have some freedom in choosing these factors. However we insist here on three features:

(i) the Z-factors must be independent of n, so that n_R is proportional to n,

(ii) they must have a power series expansion of the form

$$Z_y = \sum_{k=0}^{\infty} A_y^{(k)}(\ell/\ell_R) u^k \quad ; \ y = u, n \ , \tag{11.2}$$

(iii) they should be finite for all $d \leq 4$.

(I should note that condition (iii) will be weakend in the next chapter.)

With the ansatz (11.2) we reorder bare perturbation theory in powers of u. For simplicity taking $A_y^{(0)} = 1$, from Eq. (4.16 ii) we find

$$\frac{R_e^2}{2d} = n_R \ell_R^2 \Big\{ 1 + u \Big[\frac{2}{\varepsilon} \left(n_R^{\varepsilon/2} - \left(\frac{\ell}{\ell_R} \right)^{\varepsilon} \right) - \frac{2}{2+\varepsilon}\, n_R^{\varepsilon/2} + \left(\frac{\ell}{\ell_R} \right)^{\varepsilon} \alpha_1(\varepsilon) + A_n^{(1)}(\ell/\ell_R) + O\left(\frac{\ell^2}{n_R \ell_R^2} \right)^{1-\varepsilon/2} \Big] + O(u^2) \Big\} \ . \tag{11.3}$$

If we now choose

$$A_n^{(1)}(\ell/\ell_R) = \frac{2}{\varepsilon} \left[\left(\frac{\ell}{\ell_R} \right)^{\varepsilon} - 1 \right] - \left(\frac{\ell}{\ell_R} \right)^{\varepsilon} \alpha_1(\varepsilon) \ , \tag{11.4}$$

the expansion takes the simple form

$$\frac{R_e^2}{2d} = n_R \ell_R^2 \Big\{ 1 + u \Big[\frac{2}{\varepsilon} \left(n_R^{\varepsilon/2} - 1 \right) - \frac{2}{2+\varepsilon}\, n_R^{\varepsilon/2} + O\left(\frac{\ell^2}{n_R \ell_R^2} \right)^{1-\varepsilon/2} \Big] + O(u^2) \Big\} \ . \tag{11.5}$$

This is just a reordered expansion for the given microscopic model. In Eq. (11.5) we now take the limit $\ell \to 0$ with u, n_R, ℓ_R fixed, which amounts to dropping the terms of order $(\ell^2/(\ell_R^2 n_R))^{1-\varepsilon/2}$. Since $n_R \ell_R^2$ is of order $R_e^2 \sim n\ell^2$, these terms truly are negligible. They arise from the terms of order $n^{-1+\varepsilon/2}$ in the bare expansion. The important corrections of order n^0 are absorbed into the renormalization factors. The renormalized expansion itself is purged from microstructure-dependent terms, like $\alpha_1(\varepsilon)$. In addition, the renormalized expansion shows another important feature. In contrast to the naive continuous chain limit it also exists in the limit $\varepsilon \to 0$. This is most important, since by virtue of $u^* \sim \varepsilon$ our expansion in principle proceeds in powers of ε. To first order of the ε-expansion we find from Eq. (11.5)

$$\frac{R_e^2}{2d} = n_R \ell_R^2 \Big\{ 1 + u(\ln n_R - 1) + O(u^2, u\varepsilon) \Big\} \ . \tag{11.6}$$

We now repeat these steps, starting from Eq. (4.32) for A_2. Choosing

$$A_u^{(1)}(\ell/\ell_R) = \frac{8}{\varepsilon}\left(1 - \left(\frac{\ell}{\ell_R}\right)^{\varepsilon}\right) + \left(\frac{\ell}{\ell_R}\right)^{\varepsilon} (\alpha_2(\varepsilon) + 2\,\alpha_1(\varepsilon)) \ , \tag{11.7}$$

we find

$$\frac{A_2}{(4\pi)^{d/2}} = u\, n_R^2 \ell_R^d \Big\{ 1 - u \Big[\frac{4}{\varepsilon}\,(n_R^{\varepsilon/2} - 1) - a(\varepsilon)\, n_R^{\varepsilon/2} + O\left(\left(\frac{\ell^2}{n_R \ell_R^2}\right)^{1-\varepsilon/2}\right) \Big]$$
$$+ O(u^2) \Big\} \ , \tag{11.8}$$

or for $\varepsilon \to 0, \ell \to 0$ as above,

$$\frac{A_2}{(4\pi)^{d/2} u\, n_R^2\, \ell_R^d} = 1 - u\,[2 \ln n_R - a(0)] + O(u^2, u\varepsilon) \ . \tag{11.9}$$

The constant $a(\varepsilon)$ is given in Eq. (A 4.16). Again the renormalized expression is finite in four dimensions and does not refer to a specific microstructure.

This calculation may look like a silly exercise. Introducing two arbitrary power series Z_u, Z_n we can of course give the expansions for R_e^2, A_2 almost any form we like. The results (11.6), (11.9) are by no means unique. We, for instance, may replace $A_n^{(1)}$ by $A_n^{(1)'} = A_n^{(1)} + c(\varepsilon)$, thus effectively changing the definition of n_R. As a result, our renormalized expressions are modified by the arbitrary constant $c(\varepsilon)$. A priori the renormalized expansion does not seem to have much meaning.

The power of renormalization becomes apparent if we consider more observables or higher orders of the expansion. Calculating more observables like the radius of gyration or the scattering functions we make the gratifying observation that all the annoying microstructure effects in any observable are cancelled by the same choice of the renormalization factors: *a redefinition of the coupling and of the chain length is sufficient to eliminate all microstructure effects in first order calculations.* (This holds only for properly normalized quantities. The general case is considered in the next section.)

Furthermore this result generalizes to arbitrary order of renormalized perturbation theory. The renormalized expansion therefore yields microstructure-independent scaling functions reproducing the physical observables up to negligible corrections.

This result is of central importance, and we formulate it in precise terms in a separate section.

11.2 The theorem of renormalizability

The general result of renormalization is best expressed in terms of the cumulants $G_c^{(L,M)}$ introduced in Sect. 4.3. (Recall that $G_{c,m_1\dots m_L}^{(L,M)}$ is given by the set of all connected diagrams contributing to the Greensfunction $G_{m_1\dots m_L}^{(L,M)}$, which describes a system of M chains with insertions of the local segment densities of chains $m_1 \dots m_L$.) We reorder the perturbation expansion, introducing the renormalized coupling u and the renormalized chain length n_R via relations (11.1). We furthermore need to introduce an additional renormalization factor $Z(u, \ell/\ell_R)$ as well as the chemical potential per segment $\mu_s^*(\beta_e)$ of an isolated infinitely long chain. (The role of μ_s^* is to take care of the one-body interaction generated by the two-body interaction, as discussed in Sects. 7.2 or 10.1.) All the renormalization factors Z_u, Z_n, Z, by definition allow for a power series expansion in terms of u and are independent of the chain lengths. We now define the renormalized cumulant G_{cR} by the relation

$$\exp\left(\mu_s^*(\beta_e)\sum_{m=1}^{M} n_m\right) G_{c,m_1\dots m_L}^{(L,M)}(\mathbf{q}_1,\dots,\mathbf{q}_L;\mathbf{p}_1,\dots,\mathbf{p}_{2M};n_1,\dots,n_M;\beta_e,\ell)$$
$$= \frac{\Omega}{\ell^d}(4\pi)^{-Md/2}\left(\frac{\ell}{\ell_R}\right)^{d-2L-dM} \left(Z\left(u,\ell/\ell_R\right)\right)^M \left(Z_n\left(u,\ell/\ell_R\right)\right)^{L-M}$$
$$G_{cR,m_1\dots m_L}^{(L,M)}(\bar{\mathbf{q}}_1,\dots,\bar{\mathbf{q}}_L;\bar{\mathbf{p}}_1,\dots,\bar{\mathbf{p}}_{2M};n_{1R},\dots,n_{MR};u,\ell/\ell_R)\ , \tag{11.10}$$

where

$$\bar{\mathbf{q}}_j = \mathbf{q}_j\ell_R \quad ;\bar{\mathbf{p}}_j = \mathbf{p}_j\ell_R \quad . \tag{11.11}$$

The precise form of Eq. (11.10) results from the field theoretic analysis sketched in Appendix A 11.1.

The central result of the renormalization approach is the

Theorem of Renormalizability.

The renormalization factors can be chosen such that the limit $\ell \to 0$ for $G_{cR}^{(L,M)}$ exists for all dimensions $d \le 4$, the corrections being of canonical order $\ell^2/(n_R\ell_R^2)$:

$$G^{(L,M)}_{cR,m_1\dots m_L}(\bar{\mathbf{q}}_1,\dots,\bar{\mathbf{p}}_{2M};n_{1R},\dots,n_{MR};u,\ell/\ell_R)$$
$$= G^{(L,M)}_{cR,m_1\dots m_L}(\bar{\mathbf{q}}_1,\dots,\bar{\mathbf{p}}_{2M};n_{1R},\dots,n_{MR};u)\left(1+O\left(\frac{\ell^2}{n_R\ell_R^2}\right)\right). \tag{11.12}$$

This result is valid in perturbation theory to all orders in a simultaneous expansion in u and ε.

The notion of 'canonical order' needs some explanation. A priori in μ-th order of perturbation theory in u the leading neglected term is found to be proportional to

$$u^\mu \frac{\ell^2}{\ell_R^2 n_R}\left(\frac{(\ell_R^2 n_R/\ell^2)^{\varepsilon/2}-1}{\varepsilon}\right)^\mu .$$

Expanding in ε we find a correction of order

$$u^\mu \ell^2/(\ell_R^2 n_R)\left[\ln(\ell^2 n_R/\ell^2)\right]^\mu .$$

Clearly the term $\ell^2/(\ell_R^2 n_R)$ dominates and suppresses the corrections. The notion of canonical order refers to that leading behavior.

Let us very clearly express the important and nontrivial content of the Theorem of Renormalizability: going to higher orders we in μ-th order the unrenormalized expansion find microstructure-dependent corrections of type $\beta_e^\mu n^{\mu'\varepsilon/2}$ for all $0 \le \mu' < \mu$. *All such terms are eliminated by renormalization, even though the renormalization factors are independent of n!* In the renormalized cumulants only correction terms of canonical order $\ell^2/(n_R\ell_R^2)$ survive. Since $n_R\ell_R^2 \sim R_g^2$ is of macroscopic order, these corrections can be safely neglected, at least in the ε-expansion. *Renormalization is a partial resummation of perturbation theory, that absorbs the microstructure dependence.*

We now might wonder whether this result, being heavily based on the expansion about $d = 4$, guarantees that the corrections are still negligible at the physical dimension $d = 3$. In principle we could imagine the logarithms in the correction terms to sum up to some power overwhelming the prefactor $(\ell^2/\ell_R^2 n_R)$. It is here where we invoke the general RG-structure discussed in the previous chapter. The correction terms correspond to the irrelevant perturbations of our model. As explained in Sect. 10.3, we have reasons to believe that these stay irrelevant in the excluded volume region also for $d = 3$. We therefore believe that such a dangerous conspiracy of the logarithmic terms does not occur.

In Eq. (11.12) the limit $\ell \to 0$ suppresses the nonuniversal features, and the theorem thus solves the problem of separating universal and nonuniversal aspects. It, however, contains much more. We can interpret Eqs. (11.10),

(11.12) as a mapping from the physical bare model ℓ, β_e, n to a renormalized model ℓ_R, u, n_R, where ℓ_R can be chosen freely. This is exactly the essence of the renormalization group, since the arbitrariness of ℓ_R just means scale invariance. We will elaborate on this aspect in the next sections, showing that the theorem of renormalizability implies the RG flow equations and generalized scaling laws.

It may be appropriate to add a remark on the proof of the basic theorem. In fact this proof has never been formulated directly for the polymer system. Rather, one first carries through a Laplace transform with respect to the chain length variables n_m to arrive at an equivalent field theory. (See Chap. 7.) In quantum physics such theories describe systems of interacting elementary particles. In this context the problem of renormalization emerges most naturally. The microscopic length scale ℓ corresponds to the size of the particles. Now, in a relativistically invariant theory elementary particles have to be point-like, which forces us to take the limit $\ell \to 0$. Furthermore, the theory naturally is defined in $d = 4$ (three space- and one time-dimension). Like the naive continuous chain limit the limit $\ell \to 0$ in naive perturbation theory for $d = 4$ leads to singularities, which are to be eliminated by renormalization. Historically the development of the proof of renormalizability has started in the late 1940s and was eventually completed around 1970. Even for our problem, which in some respects is the simplest nontrivial field theory, the presentation of the proof needs a full one-semester course. In Appendix A 11.1 we therefore only discuss some structural aspects relevant for renormalizability and show how the field theoretic results lead to Eqs. (11.10), (11.12).

11.3 Renormalization group equations

The theorem of renormalizability can be read in two ways. With the renormalized theory taken to be fixed, it implies the existence of a one-parameter class, parameterized by ℓ, of bare theories, all equivalent to the given renormalized theory and thus equivalent to each other. This aspect is related to universality: a whole class of microscopic models yields the same scaling functions. In the next chapter we will use this aspect to get rid of the technical complications of the discrete chain model. We can however also interpret the theorem as establishing the existence of a one-parameter class of renormalized theories, all equivalent to a given bare theory. This class is parameterized by the length scale ℓ_R or the scaling parameter

$$\lambda = \frac{\ell}{\ell_R} \quad , \tag{11.13}$$

equivalently. In this form the theorem establishes the scale invariance of the renormalized theory. The RG flow equations refer to this latter interpretation. They give the change of the renormalized theory under an infinitesimal dilatation of the basic length scale ℓ_R.

To find the RG equations we take the derivative with respect to ℓ_R of Eq. (11.10) with Eq. (11.12) inserted, keeping all unrenormalized variables $\beta_e, \ell, n_m, \mathbf{q}_j, \mathbf{p}_j$ fixed. A little calculation yields

$$0 = \Big\{2L + d(M-1) + M\ell_R \frac{d}{d\ell_R} \ln Z + (L-M)\,\ell_R \frac{d}{d\ell_R} \ln Z_n + \sum_{j=1}^{L} \bar{\mathbf{q}}_j \nabla_{\bar{\mathbf{q}}_j} + \sum_{j=1}^{2M} \bar{\mathbf{p}}_j \nabla_{\bar{\mathbf{p}}_j} + \ell_R \frac{du}{d\ell_R}\frac{\partial}{\partial u} + \sum_{m=1}^{M} \ell_R \frac{dn_{mR}}{d\ell_R} \frac{\partial}{\partial n_{mR}}\Big\} \cdot G^{(L,M)}_{cR,m_1\ldots m_L}(\bar{\mathbf{q}}_1, \ldots, \bar{\mathbf{p}}_{2M}; n_{1R}, \ldots, n_{MR}; u) \ , \quad (11.14)$$

where

$$\bar{\mathbf{q}}_j \nabla_{\bar{\mathbf{q}}_j} \equiv \sum_{\alpha=1}^{d} \bar{\mathbf{q}}_{j,\alpha} \frac{\partial}{\partial \bar{\mathbf{q}}_{j,\alpha}} \quad , \text{ etc} \quad . \quad (11.15)$$

From relation (11.1 ii) among n and n_R we find

$$\ell_R \frac{dn_{mR}}{d\ell_R} = -n_{mR}\left(2 + \ell_R \frac{d}{d\ell_R} \ln Z_n\right) \quad . \quad (11.16)$$

Recall that the derivatives

$$\ell_R \frac{d}{d\ell_R} \equiv -\lambda \frac{d}{d\lambda} \quad (11.17)$$

are understood keeping all bare parameters fixed. We now introduce the notation

$$-\ell_R \frac{d}{d\ell_R} u = \lambda \frac{du}{d\lambda} = W(u) \quad (11.18\ \text{i})$$

$$\lambda \frac{d}{d\lambda} \ln Z_n = 2 - \frac{1}{\nu(u)} \quad (11.18\ \text{ii})$$

$$\lambda \frac{d}{d\lambda} \ln Z = \eta(u) \quad . \quad (11.18\ \text{iii})$$

Below we will show that these derivatives indeed depend on u only. With these definitions Eq. (11.14) can be written as

$$0 = \Big\{\sum_{j=1}^{L} \bar{\mathbf{q}}_j \nabla_{\bar{\mathbf{q}}_j} + \sum_{j=1}^{2M} \bar{\mathbf{p}}_j \nabla_{\bar{\mathbf{p}}_j} - W(u)\frac{\partial}{\partial u} - \frac{1}{\nu(u)}\left(M - L + \sum_{m=1}^{M} n_{mR} \frac{\partial}{\partial n_{mR}}\right) + M(d + 2 - \eta(u)) - d\Big\} G^{(L,M)}_{cR,m_1\ldots m_L}(\bar{\mathbf{q}}_1, \ldots, \bar{\mathbf{p}}_{2M}; n_{1R}, \ldots, n_{MR}; u) \quad . \quad (11.19)$$

This result is known as *renormalization group equation* for the (L, M)-cumulant. It is clearly just a Ward identity of the dilatation group (see Sect. 10.2.1).

Equations (11.18) are called the RG flow equations, giving the flow in parameter space.

The only, but essential, use we make here of Eq. (11.19) is to extract the properties of the functions $W(u)$ etc. Note that by construction these functions are independent of (L, M), of momenta, and of chain lengths. Writing down Eq. (11.19) for three different choices of (L, M) we therefore have a system of three equations linear in $W, 1/\nu, \eta$, which allows us to express these functions in terms of renormalized cumulants. Three important results immediately follow.

(i) W, ν, η can depend only on renormalized variables. Since they are independent of momenta and chain length, they depend on u only.

(ii) The functions $W(u)$, etc., exist for all $d \leq 4$ and can be calculated in terms of an expansion in powers of u and ε. They are independent of microstructure effects.

(iii) Integration of Eq. (11.18) can be used to determine the Z-factors. By virtue of statement (ii) the result is independent of microstructure effects, up to integration constants, which are treated as adjustable parameters containing all the dependence on the physical parameters.

Note that in this way we determine expressions for the Z-factors valid for $\varepsilon \geq 0$ and all $\lambda = \ell/\ell_R$. This is to be contrasted to the expansion (11.2), which breaks down for $\lambda \to 0, \varepsilon \to 0$ (cf. Eq. 11.4).

11.4 Global form of the RG flow

Starting from the general form (11.18) we add the fixed point assumption:

the function $W(u)$ shows a nontrivial zero $u = u^*$. For $d < 4$, u^* is positive, vanishing as $u^* \sim \varepsilon$ for $d \to 4$.

This assumption is fulfilled for all realizations of the renormalization group of interest here. Furthermore, $u = 0$ is a trivial fixed point, since the expansion of $W(u)$ starts as

$$W(u) = -\varepsilon\, u + O(u^2) \quad . \tag{11.20}$$

This equation is completely general, following directly from Eqs. (11.1 i), (11.18 i) together with the ansatz (11.2). It is useful to introduce a normalized coupling

$$f = \frac{u}{u^*} \quad . \tag{11.21}$$

In the sequel most of our results will be written in terms of f.

We first consider the flow of the coupling, written as

$$\lambda \frac{d}{d\lambda} f = W_f(f) \quad . \tag{11.22}$$

Note that $W_f(f) \equiv W(u^* f)/u^*$. Separation of variables yields

$$\ln \lambda = \int_{f_0}^{f(\lambda)} \frac{dx}{W_f(x)} \quad , \tag{11.23}$$

where $f_0 = f(1)$. We introduce the derivative of $W_f(f)$ at the fixed point $f = 1$:

$$\frac{d}{dx} W_f(x) \mid_{x=1} = \omega \quad . \tag{11.24}$$

Using this definition as well as Eq. (11.20) we write $1/W_f(x)$ in the form

$$\frac{1}{W_f(x)} = -\frac{1}{\varepsilon\, x} + \frac{1}{\omega(x-1)} + p_f(x) \quad , \tag{11.25}$$

where $p_f(x)$ is some function which we assume to be regular for $0 \leq x$. With this form the integral (11.23) yields

$$\begin{aligned} \ln \lambda = &-\frac{1}{\varepsilon} \ln \frac{f(\lambda)}{f_0} + \frac{1}{\omega} \ln \frac{1 - f(\lambda)}{1 - f_0} \\ &+ P_f(f(\lambda)) - P_f(f_0) \quad . \end{aligned} \tag{11.26}$$

Here $P_f(f)$ is the integral of $p_f(f)$, the integration constant being fixed as $P_f(0) = 0$. Exponentiation yields

$$\ell_R = \frac{\ell}{\lambda} = f^{1/\varepsilon} |1 - f|^{-\frac{1}{\omega}} \exp(-P_f(f)) s_\ell \quad , \tag{11.27}$$

where $f = f(\lambda)$. The parameter

$$s_\ell = \ell f_0^{-1/\varepsilon} |1 - f_0|^{1/\omega} e^{P_f(f_0)} \tag{11.28}$$

absorbs all f_0-dependence. Equation (11.27) is an implicit equation for $f = f(\lambda, s_\ell)$, where s_ℓ replaces the starting value f_0. This equation effectively determines Z_u.

We next consider Z_n. Written in terms of f, Eq. (11.18 ii) reads

$$\lambda \frac{d}{d\lambda} \ln Z_n(f(\lambda)) = 2 - \frac{1}{\nu_f(f(\lambda))} \quad ,$$

or, equivalently

$$\frac{d}{df}\ln Z_n(f) = \frac{2 - \frac{1}{\nu_f(f)}}{W_f(f)} \; . \tag{11.29}$$

At the fixed points we have $\nu_f(0) = 1/2$ or $\nu_f(1) = \nu$, respectively. This suggests

$$2 - \frac{1}{\nu_f(f)} = f\left[2 - \frac{1}{\nu} + (1-f)p_n(f)\right] \; , \tag{11.30}$$

where $p_n(f)$ is some regular function. Using the form (11.25), (11.30) in integrating Eq. (11.29) we find

$$\ln\frac{Z_n(f)}{Z_n(f_0)} = \frac{1}{\omega}\left(2 - \frac{1}{\nu}\right)\ln\frac{1-f}{1-f_0} + P_n(f) - P_n(f_0) \; , \tag{11.31}$$

where $P_n(x)$ collects regular terms and can be specified to obey $P_n(0) = 0$. We thus find

$$Z_n(f) = Z_n(f_0)\Big|1 - f_0\Big|^{\frac{1}{\omega}(\frac{1}{\nu}-2)} e^{-P_n(f_0)} \cdot \Big|1 - f\Big|^{\frac{1}{\omega}(2-\frac{1}{\nu})} e^{P_n(f)} \; . \tag{11.32}$$

We now substitute this result into Eq. (11.1 ii), eliminating ℓ_R with the help of Eq. (11.27). A little calculation, also using the definition (11.28) of s_ℓ, yields

$$n_R = f^{-2/\varepsilon}\Big|1 - f\Big|^{\frac{1}{\omega\nu}} \exp\left[2P_f(f) - P_n(f)\right] s_n\, n \tag{11.33}$$

$$s_n = Z_n^{-1}(f_0) f_0^{2/\varepsilon}\Big|1 - f_0\Big|^{-\frac{1}{\omega\nu}} \exp\left[P_n(f_0) - 2P_f(f_0)\right] \; . \tag{11.34}$$

The remaining factor Z is determined in the same way, starting from

$$\frac{d}{df}\ln Z(f) = \frac{\eta_f(f)}{W_f(f)} \; , \tag{11.35}$$

where $\eta_f(0) = 0, \eta_f(1) = \eta$. Writing

$$\eta_f(f) = f(\eta + (1-f)p_z(f)) \; ,$$

we find

$$Z(f) = Z(f_0)\left|\frac{1-f}{1-f_0}\right|^{\eta/\omega} e^{P_z(f) - P_z(f_0)} \; , \tag{11.36}$$

where $P_z(\ldots)$ again collects regular terms.

We should note that the integration yields two branches of the global RG map: $0 \le f < 1$ or $1 < f$, respectively, depending on $f_0 \lessgtr 1$. This two branched structure is consistent with the general discussion of the RG (Sect. 10.2.3), showing that the nontrivial fixed point ($f = 1$) can be approached generically from two different directions (see Fig. 10.2). This structure will show up in all the scaling functions.

11.5 The nonuniversal parameters and two parameter theory

The quantities $f_0, Z_n(f_0), Z(f_0)$ have been introduced as initial conditions in the integration of the flow equations, fixed at $\ell_R = \ell$. It might thus be tempting to identify $u^* f_0$ with β_e. This, however, is not correct. The coupling β_e of the bare theory is not identical to the coupling $u^* f_0$ of the renormalized theory at $\ell_R = \ell$. Equations (11.1), (11.2), (11.7), for instance, yield

$$\beta_e = u^* f_0 \left[1 + (\alpha_2 + 2\alpha_1) u^* f_0 + O(f_0^2)\right] \quad ,$$

illustrating that f_0 will depend on the details of the underlying model. All we can say is that generally f_0 should take the form

$$f_0 = \beta_e \hat{f}(\beta_e) \quad , \tag{11.37}$$

where $\hat{f}(\beta_e)$ can be expanded in powers of β_e. Since β_e contains the temperature dependence of our theory, this result implies that $f_0 = f_0(T)$ is a regular function of T, vanishing at $T = \Theta$. Similar results hold for $Z_n(f_0), Z(f_0)$.

In the equations for n_R or f we have collected the f_0-dependence into the parameters s_n or s_ℓ. The connection to the more familiar formulation of the two-parameter theory is established by introducing

$$\begin{aligned}
\tilde{z} &= (s_n\, n)^{\varepsilon/2} = \tilde{v}(T) n^{\varepsilon/2} \\
\tilde{R}_0^2 &= s_\ell^2 s_n\, n = \tilde{\ell}^2(T) n \\
\tilde{v}(T) &= f_0 |1 - f_0|^{-\frac{\varepsilon}{2\nu\omega}} Z_n^{-\varepsilon/2}(f_0) \exp\left[\frac{\varepsilon}{2} P_n(f_0) - \varepsilon P_f(f_0)\right] \\
\tilde{\ell}(T) &= \ell |1 - f_0|^{\frac{2\nu-1}{2\nu\omega}} Z_n^{-1/2}(f_0) \exp\left[\frac{1}{2} P_n(f_0)\right] \quad .
\end{aligned} \tag{11.38}$$

In terms of these parameters the RG mapping (11.27), (11.33) reads

$$n_R^{\varepsilon/2} = f^{-1} \Big|1 - f\Big|^{\frac{\varepsilon}{2\omega\nu}} \exp\left[\varepsilon P_f(f) - \frac{\varepsilon}{2} P_n(f)\right] \tilde{z} \tag{11.39}$$

$$\ell_R^2 n_R = \Big|1 - f\Big|^{\frac{1}{\omega}\left(\frac{1}{\nu} - 2\right)} \exp\left[-P_n(f)\right] \tilde{R}_0^2 \quad . \tag{11.40}$$

This shows that the variables of the renormalized theory depend on temperature, chain length, and segment size via the two parameters $\tilde{z}, \tilde{R}_0^2$. For $T \to \Theta$, i.e. $f_0 \sim T - \Theta \to 0$, we find

$$\begin{aligned}
\tilde{z} &= v_T \frac{T - \Theta}{\Theta} \left(1 + O\left(\frac{T - \Theta}{\Theta}\right)\right) n^{\varepsilon/2} \\
\tilde{R}_0^2 &= \ell^2 \left(1 + O\left(\frac{T - \Theta}{\Theta}\right)\right) n \quad ,
\end{aligned} \tag{11.41}$$

where we used $Z_n(0) = 1, P_n(0) = P_f(0) = 0$. For $T \to \Theta$ the parameters $\tilde{z}, \tilde{R}_0^2$ thus tend to the variables of the naive two-parameter scheme

$$\begin{aligned} \tilde{z} &\to const\ z \\ & \qquad\qquad T \to \Theta \\ \tilde{R}_0^2 &\to R_0^2 \quad . \end{aligned} \tag{11.42}$$

This is consistent with our previous discussion (see Chap. 7), where we stressed that the continuous chain model, underlying naive two-parameter theory, is a valid representation of a general polymer model only close to the Θ-point.

In typical normalized observables all the prefactors in Eq. (11.10), i.e. $\exp(\mu_{\mathfrak{s}}^* \Sigma n_m)$, $\Omega/\ell^d, Z^M$, etc., drop out. Only the two parameters $\tilde{z}, \tilde{R}_0^2$ (or $\tilde{v}(T), \tilde{\ell}(T)$, equivalently) occurring in the RG mapping (11.39), (11.40) survive. RG theory therefore establishes a generalized two-parameter theory, valid also outside the Θ-limit. It depends on two regular functions $\tilde{v}(T), \tilde{\ell}(T)$, which are not determined by the theory. We have stressed in the introductory section that this does not imply that upon renormalizing the theory we have lost information. Even if we were able to calculate functions like $\hat{f}(\beta_e)$ (Eq. (11.37)) for a given microscopic model in a reliable way, we would be confronted with the problem of calculating the temperature dependence of β_e. As we stressed in Chap. 10, β_e is a complicated object collecting contributions from all higher terms in a realistic description of the macromolecule. Thus $\beta_e = \beta_e(T)$ cannot be calculated, and the parameterization of the temperature dependence in terms of the two unknown but regular functions $\tilde{v}(T), \tilde{\ell}(T)$ is the best we can get. From this point of view all attempts to estimate the strength of the excluded volume from a specific microscopic model are idle. This of course does not exclude the possibility that by studying a lot of chemically different systems we might be able to find empirical rules for the determination of $\tilde{v}(T), \tilde{\ell}(T)$. Even with this more modest goal in mind we have however to note that the parameters $\tilde{v}(T), \tilde{\ell}(T)$ or s_ℓ, s_n, equivalently, not only depend on the underlying bare theory, but also on the specific realization of the renormalization group used, i.e. on the specific choice of the renormalization factors. We comment on this in Sect. 11.8. We should note furthermore that the two-branched structure of the RG mapping for any given $\tilde{z}$ implies the existence of two couplings ($f > 1$ or $f < 1$, respectively) solving Eq. (11.39). In applying the theory therefore for a given system we first have to decide whether we are on the weak coupling ($f < 1$) or the strong coupling ($f > 1$) branch.

The representation (11.39), (11.40) of the RG mapping, introducing two parameters $\tilde{z}, \tilde{R}_0^2$, is adequate outside the excluded volume limit. In the excluded volume limit $u \to u^*$, i.e. $f \to 1$, the two parameters combine into a single parameter. The limiting form of the RG mapping is best derived by eliminating $|1-f|$ from Eqs. (11.27), (11.33). We find

$$\ell_R n_R^\nu = B\, n^\nu \quad , \tag{11.43}$$

where

$$B = s_\ell s_n^\nu \exp\left((2\nu - 1)P_f(1) - \nu P_n(1)\right) \tag{11.44}$$

plays the role of the microscopic length scale in the excluded volume limit. Clearly $B = B(T)$ can also be expressed in terms of $\tilde{\ell}(T), \tilde{v}(T)$:

$$B(T) \sim \tilde{\ell}(T)\tilde{v}(T)^{\frac{1}{\varepsilon}(2\nu-1)} \quad . \tag{11.45}$$

11.6 Scaling laws in the dilute limit

Having discussed the general structure of the RG mapping we can derive the general scaling laws. These best are formulated for properly normalized quantities, for which all explicit prefactors relating unrenormalized to renormalized cumulants drop out. As an example we consider the endpoint distribution function $P(\mathbf{p}, n)$. Equations (4.5), (4.20) yield

$$P(\mathbf{p}, n) = \frac{G^{(1)}(\mathbf{p}, -\mathbf{p}; n)}{G^{(1)}(\mathbf{0}, \mathbf{0}; n)} \quad . \tag{11.46}$$

By virtue of Eq. (11.10) this is identical to a ratio of renormalized cumulants, and invoking the theorem of renormalizability we find

$$P(\mathbf{p}, n) = \frac{G_R^{(0,1)}(\bar{\mathbf{p}}, -\bar{\mathbf{p}}; n_R; f)}{G_R^{(0,1)}(\mathbf{0}, \mathbf{0}; n_R; f)} \equiv P_R(\bar{\mathbf{p}}, n_R, f) \quad . \tag{11.47}$$

To find the general scaling law we now fix the renormalized length scale by the condition

$$n_R = 1 \quad , \tag{11.48}$$

which in view of Eqs. (11.39), (11.40) yields

$$\begin{aligned} f &= f_\lessgtr(\tilde{z}) \\ \ell_R &= \tilde{R}_0 L(f) \quad . \end{aligned} \tag{11.49}$$

We thus find the scaling law in the form

$$P(\mathbf{p}, n) = \tilde{P}_\lessgtr(\mathbf{p}\tilde{R}_0, \tilde{z}) \quad . \tag{11.50}$$

Clearly corresponding laws hold for all properly normalized quantities.

To summarize: *the renormalization group proves two parameter scaling. The two parameters $\tilde{R}_0, \tilde{z}$ however show a more complicated temperature dependence than assumed in the naive two-parameter scheme. The latter is correct only close to the Θ-point. Furthermore the scaling functions take two different forms, representing the weak or the strong coupling branch.*

In Eqs. (11.49), (11.50) we indicated the branch by the index $(<,>)$. In the sequel we will omit that index, the two-branched structure being understood throughout.

In the excluded volume limit $f = 1$ we find the linear scaling laws. Equations (11.43), (11.48) yield

$$\ell_R = B\, n^{\nu} = const\; R_g \quad . \tag{11.51}$$

Thus

$$P(\mathbf{p}, n) \underset{f=1}{=} P^*(\mathbf{p}R_g) \quad , \tag{11.52}$$

which is the expected scaling law. Quite generally we note that, since for $n_R = 1 = f$ the only parameter left is the length scale $\ell_R \sim R_g$, the linear scaling laws can be derived by simple dimensional analysis in terms of R_g. The renormalization group thus proves the hypothesis underlying scaling theory in the excluded volume limit. (Cf. Chap. 9.)

More examples of scaling laws for specific quantities will be presented in later chapters. Here we want to extend the discussion in two respects: We consider not normalized quantities or quantities depending on more than one chain length.

As an example of the first type we consider the single chain partition function. Equations (4.19), (11.10) yield

$$\mathcal{Z}^{(1)}(n) = \frac{\Omega}{(4\pi\ell^2)^{d/2}} \frac{Z}{Z_n} e^{-\mu_s^*(\beta_e)n} G_{cR}^{(0,1)}(\mathbf{0}, \mathbf{0}, n_R, f) \quad . \tag{11.53}$$

Restricting ourselves to the excluded volume limit we note that the factor $G_{cR}^{(0,1)}$ reduces to a constant by virtue of $n_R = 1; f = 1$. Equations (11.27), (11.32), (11.36) yield

$$\frac{Z}{Z_n} \sim \left(\frac{\ell_R}{\ell}\right)^{2-\frac{1}{\nu}-\eta} \quad . \tag{11.54}$$

Using Eq. (11.51) we thus find

$$\begin{aligned} \mathcal{Z}^{(1)}(n) &\underset{n\to\infty}{=} \frac{\Omega}{(4\pi\ell^2)^{d/2}} \left(\frac{\ell_R}{\ell}\right)^{2-\frac{1}{\nu}-\eta} e^{-\mu_s^*(\beta_e)n} \cdot const \\ &= const\, n^{\gamma-1} e^{-\mu_s^*(\beta_e)n} \quad , \end{aligned} \tag{11.55}$$

where the exponent γ is defined as

$$\gamma = \nu(2-\eta) \quad . \tag{11.56}$$

Note, that in Eq. (11.55) only the exponent $\gamma-1$ is independent of microstructure.

Our discussion is easily extended to a system containing chains of different length. We note that ratios of chain length variables are invariant under renormalization. Indeed Eq. (11.1 ii) yields

$$\frac{n_m}{n_{m'}} = \frac{n_{mR}}{n_{m'R}} \quad . \tag{11.57}$$

It is therefore suggestive to introduce these ratios as scaling variables. To give an example we consider the second virial coefficient A_2. Generalized to two chains of different lengths, Eq. (4.28) reads

$$A_2(n_1,n_2) = -\Omega \frac{G_c^{(0,2)}(\mathbf{0},\mathbf{0},\mathbf{0},\mathbf{0};n_1,n_2)}{G^{(0,1)}(\mathbf{0},\mathbf{0};n_1)G^{(0,1)}(\mathbf{0},\mathbf{0};n_2)} \quad . \tag{11.58}$$

Renormalization yields

$$\begin{aligned} A_2(n_1,n_2) &= -\ell_R^d \frac{G_{cR}^{(0,2)}(\mathbf{0},\mathbf{0},\mathbf{0},\mathbf{0};n_{1R},n_{2R};f)}{G_R^{(0,1)}(\mathbf{0},\mathbf{0};n_{1R};f)G_R^{(0,1)}(\mathbf{0},\mathbf{0};n_{2R};f)} \\ &= \ell_R^d A_{2R}\left(n_{1R},\frac{n_2}{n_1},f\right) \quad . \end{aligned} \tag{11.59}$$

Using the condition $n_{1R}=1$ we thus find

$$A_2(n_1,n_2) = \tilde{R}_{01}^d \bar{A}_{2R}\left(\frac{n_2}{n_1},\tilde{z}_1\right) \quad . \tag{11.60}$$

A more symmetric form of this scaling law will be presented later (Chap. 16).

Finally we should explain the notion of 'universal ratios'. Quite generally these ratios are combinations of observable quantities, depending only on f. At the fixed points they therefore reduce to pure numbers, being universal just like the exponents. The prototype of such a quantity is the ratio R_g/R_e discussed in the context of Eq. (8.19). From each scaling function we can construct an infinity of such ratios. Consider, for instance, the expansion of P_R in terms of $\bar{\mathbf{p}} = \mathbf{p}\ell_R$:

$$P_R(\bar{\mathbf{p}},n_R,f) = 1 + b_2\bar{\mathbf{p}}^2 + b_4\bar{\mathbf{p}}^4 + b_6\bar{\mathbf{p}}^6 + \ldots \quad . \tag{11.61}$$

Once n_R has been fixed, for instance, by imposing the condition $n_R = 1$, the coefficients $b_2, b_4, \ldots$ depend only on f and thus take well defined values for $f=1$. Now it is easily checked that the ratios $b_{2m}/(b_2)^m$ can be expressed in terms of the coefficients of the Taylor expansion of $P(\mathbf{p},n)$ in powers of p^2:

$$P(\mathbf{p},n) = 1 + \hat{b}_2\mathbf{p}^2 + \hat{b}_4\mathbf{p}^4 + \hat{b}_6\mathbf{p}^6 + \ldots \tag{11.62}$$

$$\frac{b_{2m}}{(b_2)^m} \equiv \frac{\hat{b}_{2m}}{(\hat{b}_2)^m} \quad . \tag{11.63}$$

Thus these ratios are observable quantities, taking a well defined numerical value in the excluded volume limit. This is the definition of a universal ratio.

11.7 Renormalization and scaling at finite overlap

11.7.1 Renormalized expressions

We first consider the grand potential $\Pi[\mu_p]$, which according to the Linked Cluster Theorem can be written as

$$\Pi[\mu_p] = \lim_{\Omega\to\infty} \frac{1}{\Omega} \sum_{M=1}^{\infty} \frac{1}{M!} \sum_{n_1 \dots n_M = 1} \prod_{m=1}^{M} e^{\mu_p(n_m)} G_c^{(0,M)}(\mathbf{0},\dots,\mathbf{0}; n_1,\dots,n_M) \; . \tag{11.64}$$

(See Sect. 5.1.) Invoking the theorem of renormalizability we write

$$\ell_R^d \Pi[\mu_p] = \sum_{M=1}^{\infty} \frac{1}{M!} \sum_{n_1 \dots n_M = 1} \prod_{m=1}^{M} \left(e^{\mu_p(n_m) - \mu_s^* n_m} \frac{\lambda^{-d}}{(4\pi)^{d/2}} \frac{Z}{Z_n} \right) \cdot G_{cR}^{(0,M)}(\mathbf{0},\dots,\mathbf{0}; n_{1R},\dots,n_{MR}; f) \; . \tag{11.65}$$

Here, up to corrections of order $1/n$, we may replace the summation over n_m by integration:

$$\sum_{n_m} \longrightarrow \int_0^\infty dn_m = \lambda^{-2} Z_n \int_0^\infty dn_{mR} \; , \tag{11.66}$$

and we define the renormalized chemical potential as

$$e^{\mu_R(n_R)} = \frac{\lambda^{-d-2}}{(4\pi)^{d/2}} Z e^{\mu_p(n) - \mu_s^* n} \tag{11.67}$$

As a result, Eq. (11.65) takes the form

$$\Pi[\mu_p] = \ell_R^{-d} \Pi_R([\mu_R], f) \tag{11.68}$$

$$\Pi_R([\mu_R], f) = \sum_{M=1}^{\infty} \frac{1}{M!} \int dn_{1R} \dots dn_{MR} \prod_{m=1}^{M} e^{\mu_R(n_{mR})} G_{cR}^{(0,M)}(\mathbf{0},\dots,\mathbf{0}; n_{1R},\dots,n_{MR}; u), \tag{11.69}$$

which expresses the grand potential in terms of renormalized quantities.

We will mainly be interested in results expressed in terms of concentrations instead of the chemical potential. We thus renormalize Eq. (5.4):

$$\begin{aligned} c_p(n) &= \frac{\partial}{\partial \mu_p(n)} \Pi[\mu_p] \\ &= \lim_{\Omega\to\infty} \frac{1}{\Omega} \sum_{M=0}^{\infty} \frac{1}{M!} \sum_{n_1 \dots n_M = 1} e^{\mu_p(n)} \prod_{m=1}^{M} e^{\mu_p(n_m)} \\ &\cdot G_c^{(0,M+1)}(\mathbf{0},\dots,\mathbf{0}; n, n_1, \dots, n_M) \ . \end{aligned} \tag{11.70}$$

Following the same steps as above we find

$$\ell_R^d c_p(n) = \frac{\lambda^2}{Z_n} c_{pR}(n_R) \tag{11.71}$$

$$\begin{aligned} c_{pR}(n_R) = e^{\mu_R(n_R)} \sum_{M=0}^{\infty} \frac{1}{M!} \int_0^\infty dn_{1R} \dots dn_{MR} \prod_{m=1}^{M} e^{\mu_R(n_{mR})} \\ G_{cR}^{(0,M+1)}(\mathbf{0},\dots,\mathbf{0}; n_R, n_{1R}, \dots, n_{mR}; f) \ . \end{aligned} \tag{11.72}$$

Upon summation over n, Eq. (11.71) yields

$$\ell_R^d c_p = \frac{\lambda^2}{Z_n} \sum_n c_{pR}(n_R) = \int_0^\infty dn_R c_{pR}(n_R) = c_{pR} \quad , \tag{11.73}$$

where we used Eq. (11.66). In terms of the reduced chain length distribution $p(y)$ we find (cf. Sect. 5.3)

$$\begin{aligned} c_{pR}(n_R) &= \frac{Z_n}{\lambda^2} \ell_R^d c_p \frac{1}{N} p\left(\frac{n}{N}\right) \\ &= c_{pR} \frac{1}{N_R} p\left(\frac{n_R}{N_R}\right) \quad , \end{aligned} \tag{11.74}$$

where N_R is the renormalized number averaged chain length. $p(y)$ is expressed completely in terms of renormalized quantities, and also the normalization (5.38) $\int dy\, p(y) = 1 = \int dy\, y\, p(y)$ is RG invariant. Thus it is an RG invariant quantity itself, a rather obvious fact, since ratios of chain length variables are RG invariant. Equation (11.72) can be solved formally for $\mu_R(n_R)$, thus allowing us to replace $\mu_R(n_R)$ by $c_{pR}(n_R)$ or $c_{pR}, N_R, p(y)$, equivalently. Consequently the equation of state can be written as

$$\frac{1}{k_B T} \hat{\Pi} = \ell_R^{-d} \Pi_R(c_{pR}, N_R, f, [p(y)]) \ . \tag{11.75}$$

This derivation is easily extended to all correlation functions of interest. Any grand-canonical cumulant can be expressed by summing appropriate M-chain cumulants. We then use the theorem of renormalizability, and we eliminate $\mu_R(n_R)$ in favor of $c_{pR}(n_R)$. The result is most easily formulated for quantities obeying RG invariant normalization conditions, where it amounts to replacing $\ell^d c_p$ by $c_{pR}, \mathbf{q}\ell$ by $\bar{\mathbf{q}} = \mathbf{q}\ell_R, N$ by N_R, and β_e by $u^* f$. We in particular define (cf. Eq. (9.4))

$$J_{a(c)}(\mathbf{q}) = \frac{1}{c_p N^2} I_{d,a(c)}(\mathbf{q}) \quad . \tag{11.76}$$

By virtue of the sum rule (5.24) J_a is normalized according to

$$J_a(\mathbf{0}) = \tilde{p}_2 \quad , \tag{11.77}$$

where $\tilde{p}_2 = N_w/N$ (cf. Eq. (5.42 i)). The same normalization condition holds for $J_c(\mathbf{q})$ in the single chain limit ($c_p \to 0$), where $J_c(\mathbf{q}) \equiv J_a(\mathbf{q})$. From the above, the form of the renormalized expressions follows:

$$J_{a(c)}(\mathbf{q}) = J_{a(c)R}(\bar{\mathbf{q}}, c_{pR}, N_R, f, [p]) \quad . \tag{11.78}$$

The corresponding results are easily established for endpoint correlations.

11.7.2 Scaling laws

For a simple derivation of the scaling laws we again choose

$$N_R = 1 \quad , \tag{11.79}$$

immediately noting that this choice is adequate for a general discussion, but not optimal for perturbative calculations in strongly overlapping systems. From Eqs. (11.49), (11.73) we find

$$\bar{\mathbf{q}} = \mathbf{q}\tilde{R}_0 L(f) \tag{11.80}$$
$$f = f(\tilde{z}) \tag{11.81}$$
$$c_{pR} = \tilde{R}_0^d c_p L^d(f) \quad , \tag{11.82}$$

where $\tilde{R}_0, \tilde{z}$ are defined in terms of the average chain length N. We introduce the overlap variable

$$\tilde{s} = \tilde{R}_0^d c_p \tag{11.83}$$

to take care of the concentration. All normalized scaling functions will depend on the interaction parameter $f = f(\tilde{z})$, overlap $\tilde{s}$, polydispersity $p(y)$, and momenta $\mathbf{q}\tilde{R}_0$, and we will always find two branches due to the two-branched structure of the relation $f \leftrightarrow \tilde{z}$.

Exemplifying the general structure with the osmotic pressure, we note that $\hat{\Pi}/(k_B T c_p)$ obeys an RG invariant normalization condition:

$$\lim_{c_p \to 0} \frac{\hat{\Pi}}{k_B T c_p} = 1 \quad . \tag{11.84}$$

This is van't Hoff's law. We can thus write

$$\frac{\hat{\Pi}}{k_B T c_p} = 1 + \mathcal{P}(\tilde{s}, f(\tilde{z}), [p(y)]) \quad , \tag{11.85}$$

where the thermodynamic scaling function $\mathcal{P}$ from Eq. (11.75) is found as

$$\mathcal{P} = \frac{\Pi_R}{c_{pR}} - 1 \quad . \tag{11.86}$$

Equation (11.85) is the nonlinear scaling law. The linear scaling law results at the fixed point $f = 1$. As is easily derived with the help of Eq. (11.43) it takes the form discussed in Sect. 9.1: $\mathcal{P}$ becomes a function of $s = c_p R_g^d$ and $p(y)$.

11.8 A comment on the universality of the scaling functions and the status of our results

Here some remarks on the universality of the scaling functions seem appropriate. Up to now we have not specified the renormalization factors, and indeed even staying within the field theoretic scheme we can use different prescriptions leading to differing renormalized theories, which in turn yield different expressions for the scaling functions. How can this be compatible with universality? The question is resolved by noting that scaling functions, calculated to *all orders* in different renormalized theories, can be transformed into each other by readjusting the nonuniversal parameters. This easily follows from the theorem of renormalizability. To show this we start from some bare theory and renormalize it by two different schemes, distinguished by superscripts (1) or (2). Since the bare theory is fixed, Eqs. (11.1), (11.10) imply a relation among the renormalized theories.

$$u^{(2)} = \frac{Z_u^{(1)}}{Z_u^{(2)}}\, u^{(1)} \tag{11.87 i}$$

$$n_R^{(2)} = \frac{Z_n^{(1)}}{Z_n^{(2)}}\, n_R^{(1)} \tag{11.87 ii}$$

$$G_{cR}^{(L,M)(2)} = \left(\frac{Z^{(1)}}{Z^{(2)}}\right)^{M} \left(\frac{Z_n^{(1)}}{Z_n^{(2)}}\right)^{L-M} G_{cR}^{(L,M)(1)} \quad . \tag{11.87 iii}$$

Since all renormalized quantities exist for $d \leq 4$ this implies that the ratios $Z_u^{(1)}/Z_u^{(2)}$ etc. have a finite limit for $d = 4$. Using the power series representation of the Z-factors we can in principle solve Eq. (11.87 i) to find $u^{(2)}$ as power series in $u^{(1)}$. Substituting the result into Eqs. (11.87 ii), (11.87 iii) we find $n_R^{(2)}, G_{cR}^{(L,M)(2)}$ expressed in terms of the renormalization scheme (1). Thus different renormalization schemes are related by a so-called finite renormalization, which amounts to a redefinition of the renormalized parameters, or to a different parameterization of the field theoretic manifold, equivalently.

This 'finite renormalization' has two consequences. First, the nonuniversal parameters depend on both the microscopic system and the renormalized theory chosen. They thus have no direct microscopic meaning. Physical information is contained in the relative change upon changing the chemical microstructure or temperature, but not in the absolute values. Second, on a more technical level, numerical results of *finite order* calculations will differ for different renormalization schemes. This is a principle problem, unavoidable in low order calculations of scaling functions. Unambiguous results are found only for quantities not involving the nonuniversal constants, like exponents or critical ratios, or normalized scaling functions expressed in terms of RG-invariant variables. The function $P^*(\mathbf{p}R_G)$ (Eq. (11.52)) is an example. For such quantities the ε-expansion is unique. This aspect will be discussed further in Sect. 12.4.

It should now be clear that two-parameter scaling in terms of $\tilde{z}, \tilde{R}_0$ or the linear scaling laws found in the excluded volume limit are unavoidable consequences of the renormalization group or dilatation symmetry, being based on the theorem of renormalizability. Since this has been proven to all orders of perturbation theory these results must hold as long as perturbation theory makes any sense. Now even the problem of convergence of the perturbation expansion can be analyzed to some extent. As a first result one finds that the theory does not converge in the proper sense. The basic argument is simple. Assume that the segment interaction is attractive instead of being repulsive. As a result the chain will minimize its free energy by maximizing its interaction energy at the expense of losing entropy. It will collapse into a dense globule, and for long enough chains this collapse will occur for arbitrarily small attractive interaction. Mathematically this implies that the Greensfunctions show an essential singularity for $u \to 0, n_R \to \infty$, which ruins the convergence of perturbation theory. In the general RG framework of Sect. 10.2 this amounts to the runaway of a negative coupling.

Using field theoretic methods this divergence has been analyzed to some extent. The analysis strongly suggests that the expansion is in fact an asymptotic expansion in the mathematical sense. Using the powerful method of Borel transformation one can resum the expansion to get a unique finite result. We can thus state that the theory is well defined and all the qualitative results discussed above are firmly established – at least to the satisfaction of a theoretical physicist. Of course, from the point of view of a mathematician the argument is not at all rigorous.

A 11. Appendix

A 11.1 Remarks on renormalizability

General considerations of renormalizability are best carried through within the field theoretic formulation. We recall from Sect. 7.2 that a Laplace transform with respect to the chain length variables eliminates the segment summations. In the resulting field theoretic formulation a propagator line of momentum $\mathbf{k}$, being part of the m-th polymer line, yields a factor (cf. Eq. (7.17))

$$G_{0F}(\mathbf{k}, \tau_m) = \frac{e^{-\sigma(\mathbf{k}^2+\tau_m)}}{\mathbf{k}^2 + \tau_m} \quad ,$$

where τ_m is the Laplace variable conjugate to $R_{0,m}^2$. For simplicity we started from the cut off continuous chain model, where $\sigma \sim \ell^2$.

We now want to extract the leading microstructure dependence of the theory, i.e. the leading dependence on the cut off σ. We therefore analyze the ultraviolet properties of the theory for $d = 4$, noting that our approach is based on an expansion in $\varepsilon = 4 - d$. As has been pointed out in Sect. 7.2 it is the virtue of the field theoretic formulation that we only have to consider the one-line-irreducible (1-ℓ-i) vertex functions $\tilde{\Gamma}_F^{(L,M)}$, defined by the sum of all 1-ℓ-i graphs with L density insertions and $2M$ legs. For $d = 4$ the superficial degree of divergence of a diagram contributing to $\tilde{\Gamma}_F^{(L,M)}$ depends only on (L, M), but is independent of the order of perturbation theory. Equation (7.29) yields

$$\delta = 4 - 2M - 2L; \quad d = 4 \quad . \tag{A 11.1}$$

Recall that δ specifies the leading cut off dependence of a diagram as $\sigma^{-\delta/2}$. As Eq. (A 11.1) shows, we have $\delta \leq -2$ for $L + M \geq 3$. So the u.v. behavior of such vertex functions induces only negligible microstructure corrections of order $\sigma \sim \ell^2$. The dangerous vertex functions, which would diverge for $\sigma \to 0$, can be classified as $(L, M) = (0, 1)$ [Fig. 11.1a], $(L, M) = (0, 2)$ [Fig. 11.1b], $(L, M) = (1, 1)$ [Fig. 11.1c]. Now consider for instance the one loop diagram included in Fig. 11.1b, which contributes to $\tilde{\Gamma}_F^{(0,2)}$. The analytical expression reads

$$I = \tilde{u}_0^2 \int_k G_{0F}(\mathbf{k}, \tau_1) G_{0F}(\mathbf{k} + \mathbf{p}, \tau_2) \quad ,$$

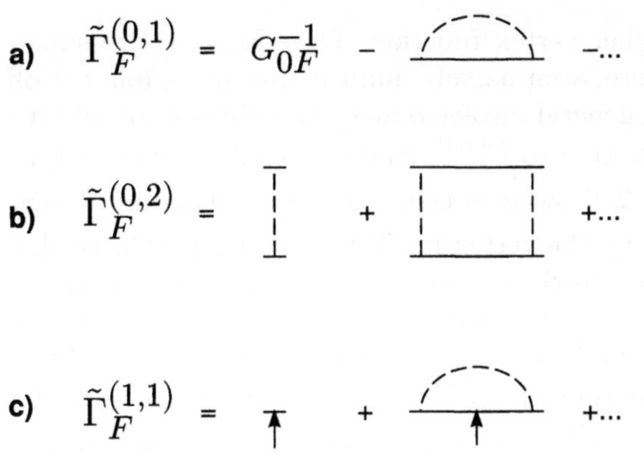

Fig. 11.1a–c. Superficially divergent vertex functions

where **p** stands for the external momenta. We now add and subtract an integral, which is independent of τ_j, **p** and shows the same u.v. behavior:

$$I = \tilde{u}_0^2 \int_k e^{-2\sigma k^2} \left(\frac{1 + O(\sigma)}{(k^2 + \tau_1)((\mathbf{k} + \mathbf{p})^2 + \tau_2)} - \frac{1}{(k^2 + \ell_R^{-2})^2} \right)$$
$$+ \tilde{u}_0^2 \int_k \frac{e^{-2\sigma k^2}}{(k^2 + \ell_R^{-2})^2} \quad .$$

The first term exists for $\sigma \to 0$, with corrections of order $\sigma \ln(\sigma \ell_R^{-2})$. The second integral yields a contribution of order $\ln(\sigma \ell_R^{-2})$. Since it is independent of **p**, τ_j, it can be absorbed into a redefinition of the vertex, i.e. into the first diagram of Fig. 11.1b. In standard field theoretic notation we write

$$\tilde{u}_0 = u \ell_R^{-\varepsilon} \frac{Z_4(u, \sigma/\ell_R^2)}{Z^2(u, \sigma/\ell_R^2)} \qquad \text{(A 11.2)}$$

and absorb the leading cut off dependence into the renormalization factor. We clearly have some freedom in defining the subtraction term, which gives rise to different, but equivalent, renormalization prescriptions. We further note that in the main text we use the notation

$$\frac{Z_4}{Z^2} = Z_u \quad . \qquad \text{(A 11.3)}$$

In a similar way the leading cut off dependence of $\tilde{\Gamma}_F^{(1,1)}$ (Fig. 11.1c) can be absorbed into a renormalization factor Z_2/Z multiplying the density insertion itself. Finally consider structures of type $(L, M) = (0, 1)$ (Fig. 11.1a), which for $d = 4$ show $\delta = 2$. Mass subtraction reduces δ to $\delta = 0$, which still corresponds to logarithmic singularities. These singularities can be absorbed into a redefinition of $\hat{\tau}_m = \tau_m - \tau_c$ (Eq. (7.25))

$$\hat{\tau}_m = \ell_R^{-2} \frac{Z_2(u, \sigma/\ell^2)}{Z(u, \sigma/\ell_2)} t_m \qquad \text{(A 11.4)}$$

and an additional factor of Z for each polymer line.

As pointed out above, higher vertex functions ($L+M \geq 3$) are superficially convergent and therefore, seen naively, induce only negligible cut off corrections. They, however, in general involve dangerous subdiagrams of type $\tilde{\Gamma}_F^{(L,M)}$, $L+M<3$. A contribution to $\tilde{\Gamma}_F^{(0,3)}$, showing a subdiagram of type $\tilde{\Gamma}_F^{(0,2)}$, is presented in Fig. 11.2. Now it is clear that a subdiagram of type $\tilde{\Gamma}_F^{(0,2)}$ can always be replaced by the vertex itself, the resulting reduced diagram being a valid contribution to the vertex function considered. Since the leading cut off dependence of the subdiagram is independent of its embedding into a larger diagram, it can be absorbed into the vertex of the reduced diagram, exactly like in the analysis of $\tilde{\Gamma}_F^{(0,4)}$. The same argument works for subdiagrams of type $\tilde{\Gamma}_F^{(1,1)}$ or $\tilde{\Gamma}_F^{(0,2)}$. The explicit calculation, showing that this scheme works in low orders, can be found in standard text books. (See the 'Notes' section for references.) The proof to all orders is most complicated.

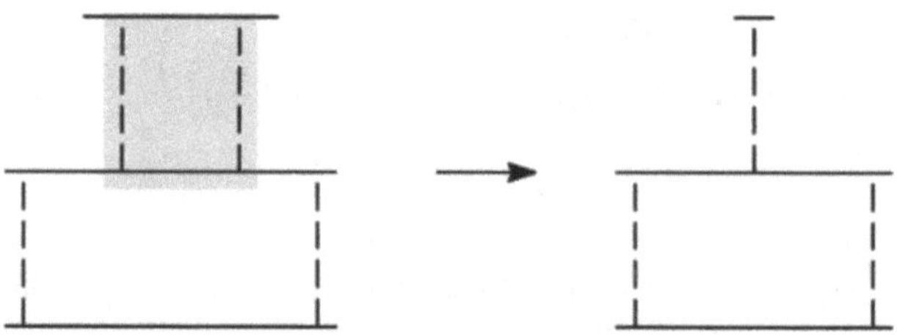

Fig. 11.2. A contribution to $\tilde{\Gamma}_F^{(0,3)}$ containing a divergent subdiagram of type $\tilde{\Gamma}_F^{(0,2)}$, indicated by the *grey box*. The divergence contributes to the renormalization of the upper vertex in the diagram *on the r.h.s.*

Collecting all factors we find the field theoretic result corresponding to Eqs. (11.10), (11.12):

$$\begin{aligned} &\tilde{G}^{(L,M)}_{FC,m_1\dots m_2}(\mathbf{q}_1,\dots,\mathbf{q}_L;\mathbf{p}_1,\dots,\mathbf{p}_{2M};\tau_1,\dots,\tau_M,u_o,\sigma) \\ &= \ell_R^{2L+M(d+2)-d}\left(\frac{Z_2}{Z}\right)^{-L} Z^M \\ &\cdot\, G^{(L,M)}_{FR,m_1\dots m_2}(\bar{\mathbf{q}}_1,\dots,\bar{\mathbf{q}}_L;\bar{\mathbf{p}}_1,\dots,\bar{\mathbf{p}}_{2M};t_1,\dots,t_M,u)\left(1+O\left(\left(\frac{\ell}{\ell_R}\right)^2\right)\right). \end{aligned} \tag{A 11.5}$$

Here $G_{FR}^{(L,M)}$ is the dimensionless renormalized cumulant, finite for $d=4$. Inverting now the multiple Laplace transform we immediately find the result (11.10), (11.12). Note that in the inverse Laplace transform we have to renormalize the integration variable τ_m which yields

$$\int_{-i\infty}^{i\infty} d\tau_m e^{\tau_m R^2_{o,m}} \ldots = e^{\tau_c \ell^2 n_m} \ell_R^{-2} \frac{Z_2}{Z} \int_{-i\infty}^{i\infty} dt_n \exp\left[t_m \left(\frac{\ell}{\ell_R}\right)^2 \frac{Z_2}{Z} n_m\right] \cdots \tag{A 11.6}$$

This on the one hand produces additional renormalization factors and on the other hand identifies Z_n as

$$Z_n = \frac{Z}{Z_2} \quad . \tag{A 11.7}$$

(Cf. Eq. (11.1 ii)), so that the renormalized exponential takes the form $\exp(t_m n_{mR})$. Furthermore we have to identify $\tau_c \ell^2 = -\mu_s^*$.

We finally note that this analysis resulted in an explicit identification of the renormalization factors Z_u, Z_n, Z of polymer theory with renormalization factors of field theory. We can thus take over higher order results for the renormalization factors and the resulting RG flow, established in field theory. This is a most useful result since a good representation of the RG mapping is crucial in the application of the theory, but higher order calculations are most complicated.

Notes

Field theoretic renormalization in low orders is discussed in many books (see [ZJ89, Ami84]). [Smi91], p. 262, sketches the history of the subject and gives references to the full proof.

Field theory typically deals with a model containing a single mass variable $\tau \hat{=} \tau_m$. In terms of polymer physics this yields a nearly exponential chain length distribution, slowly varying with overlap and excluded volume interaction. (See, for instance [Sch92].) The original work of de Gennes [dG72] and Des Cloizeaux [Clo75] exploited the renormalizability of this field theoretic ensemble. (We encountered this ensemble before in Appendix A 7.1.) Extension to problems involving a few chains of definite lengths and thus a few variables τ_m is due to Burch and Moore [BM76a, BM76b]. The generalization to an arbitrary number of τ_m variables, necessary for a treatment of arbitrary polydispersity, was presented in [SW80], starting from the discrete chain model. Except for the replacement of segment summations by integrals the presentation here follows this work.

The convergence of perturbation theory is discussed in [ZJ89], where the method of Borel transformation and resummation is also explained.

12. Dimensional Regularization and Minimal Subtraction

The results of the previous chapter are completely general. They are valid for any field theoretic renormalization scheme, i.e. independent of the specific choice of the renormalization factors. For quantitative calculations we of course have to specify the Z-factors, and as pointed out in Sect. 11.1, we have some freedom there. We will use the scheme of 'dimensional regularization' and 'minimal subtraction'. This scheme is most efficient for actual calculations, but its underlying basis is a little bit delicate. It needs some careful explanation.

As formulated in the previous chapter, renormalization proceeds by first introducing u, n_R according to Eq. (11.1), thereby absorbing the leading microstructure dependence into the renormalization factors. In a second step the limit $\ell \to 0$ is taken in the renormalized Greensfunctions to eliminate the remaining nonuniversal corrections. This limit, however, is not taken in the bare theory or in the renormalization factors. The first step is technically complicated since we have to evaluate the summations occurring in the discrete chain model. Equivalently, working with the continuous chain model we have to keep the finite cut-off. Now it turns out that the two steps can be inverted. First we can take the naive continuous chain limit for $d < 4$, thus converting the summations into integrals without cut-off. We may then renormalize the continuous chain model to arrive at the same renormalized theory as with the original approach.

This statement should immediately meet with opposition. We have stressed before that the continuous chain model has a number of shortcomings.

(i) It is finite only for $d < 4$, but does not exist for $d = 4$. This suggests problems with the ε-expansion.

(ii) The subleading terms important for renormalization are lost.

(iii) It is valid only infinitesimally close to the Θ-point.

How then can it be used to construct the renormalized theory, finite for $d \leq 4$ and valid for all temperatures?

Concerning the first two points the answer is that in some sense they compensate each other. Working with the continuous chain model the purpose of renormalization is not to eliminate the microstructure dependence, which effectively is suppressed by the limit $\ell \to 0$, but to make the theory finite for $d = 4$. It turns out that all divergencies occurring for $d \to 4$ in the (additively renormalized) continuous chain model can be absorbed into renormalization factors Z, Z_u, Z_n. The renormalized theory exists for $d \leq 4$. It furthermore turns out that the divergences of the continuous chain model carry the same information as the leading cut-off corrections for a discrete chain.

Concerning the third point we note that the renormalized theory, expressed in terms of u, n_R, ℓ_R, does not know whether the underlying bare theory is close to the Θ-point or elsewhere in the region $\beta_e > 0$. We in fact can construct the same renormalized theory starting either from the continuous chain model or from the discrete chain model. We may thus exploit the technical benefits of the continuous chain limit to construct the renormalized theory, which is then interpreted in terms of the discrete chain model, valid also for T not close to Θ. This only amounts to giving up the simple temperature dependence (11.41) of the nonuniversal parameters, which is generally valid in the continuous chain model, only.

The step of first taking the limit $\ell \to 0$ for $d < 4$ to arrive at a finite unrenormalized theory without cut-off is known as *dimensional regularization*, in contrast to the 'cut-off regularization' of the discrete chain model. The breakdown of the dimensionally regularized theory as $d \to 4$ shows up in pole terms $\sim (1/\varepsilon)^k$. These terms have to be absorbed into the renormalization factors. The scheme of *minimal subtraction* is distinguished in that it extracts only the pole terms but no regular parts. It turns out that in that scheme the functions $W(u), 1/\nu(u), \eta(u)$ occurring in the RG flow equations (11.18) show some marvelous property: Except for the trivial contribution $-\varepsilon\, u$ to $W(u)$ (Eq. (11.20)) they are strictly independent of dimension. We can thus use them without any problem at physical dimension $d = 3$, even though they are derived by ε-expansion valid close to $d = 4$. Furthermore it has been shown that the scaling functions in minimal subtraction can be calculated directly for $d = 3$. We thus end up with a calculational scheme making no explicit use of expanding about $d = 4$.

In this chapter we first show that the continuous chain model is renormalizable by taking the naive continuous chain limit of the theorem of renormalizability. We then argue that we can construct renormalization schemes for the continuous or the discrete chain models, equivalent in the sense that they yield the same renormalized theory (Sect. 12.1). In Sect. 12.2 we establish the general properties of the flow equations which result from minimal subtraction. We then are in the position to fully describe the scheme which we use in calculating the scaling functions (Sect. 12.3). We will later evaluate the scaling functions directly in $d = 3$, without recourse to the ε-expansion. Virtues and drawbacks of the two methods are discussed in Sect. 12.4.

This chapter is concerned with important but technical aspects of the actual calculations. A reader more interested in the results may go on directly to the next chapter, where the explicit form of the RG mapping is presented.

12.1 Equivalence of discrete or continuous chains

12.1.1 Renormalizability of the continuous chain model

For $d < 4$ in the theorem of renormalizability we can take the naive continuous chain limit, addressed here as NCL for shortness.

NCL: We take $\ell \to 0$; keeping $\tilde{u}_0 = (4\pi)^{d/2}\, \beta_e \ell^{-\varepsilon}, R_0^2 = n\ell^2$ fixed.

We know that the additively renormalized bare theory (the left hand side in Eq. (11.10)) exists in that limit. Also the Z-factors attain finite limits $Z(u), Z_u(u), Z_n(u)$. Indeed they are constructed as power series in u, the coefficients for $\ell > 0, \varepsilon > 0$ taking the form of polynomials in $(\ell/\ell_R)^{\varepsilon}$. Thus no problem results from setting $\ell = 0$. Since the renormalized theory is finite for $d = 4$, whereas the bare continuous chain model diverges for $d \to 4$, showing poles in ε, also the Z-factors must diverge for $d \to 4$. In the NCL we can therefore formulate the theorem of renormalizability as follows:

The renormalization factors can be chosen to absorb all the pole terms of the dimensionally regularized bare theory to yield a renormalized theory finite for $d \leq 4$.

We again illustrate this with R_e^2 or A_2, just taking the NCL in our calculation of Sect. 11.1. We find (cf. Eqs. (7.11), (7.12))

$$\frac{R_e^2}{2d} = R_0^2 \left[1 + \frac{4z}{\varepsilon(2+\varepsilon)} + O(z^2) \right] \tag{12.1}$$

$$\frac{A_2}{(4\pi)^{d/2}} = R_0^d z \left[1 - \left(\frac{4}{\varepsilon} - a(\varepsilon) \right) z + O(z^2) \right] , \tag{12.2}$$

where $z = \beta_e n^{\varepsilon/2} = (4\pi)^{-d/2}\, \tilde{u}_0 R_0^{\varepsilon}$. Equations (11.1) reduce to

$$(4\pi)^{-d/2}\,\tilde{u}_0 = u\ell_R^{-\varepsilon} Z_u(u) \tag{12.3}$$

$$R_0^2 = n_R \ell_R^2 Z_n(u) \quad . \tag{12.4}$$

Using a power series ansatz analogous to (11.2) with $A_y^{(0)} = 1$ we find

$$\frac{R_e^2}{2d} = n_R\,\ell_R^2\Big\{1 + u\left[\left(\frac{2}{\varepsilon} - \frac{2}{2+\varepsilon}\right) n_R^{\varepsilon/2} + A_n^{(1)}\right] + O(u^2)\Big\} \tag{12.5}$$

$$\frac{A_2}{(4\pi)^{d/2}} = u\,n_R^2\,\ell_R^d\Big\{1 - u\left[\left(\frac{4}{\varepsilon} - a(\varepsilon)\right) n_R^{\varepsilon/2} - A_u^{(1)} - 2A_n^{(1)}\right] + O(u^2)\Big\} \quad . \tag{12.6}$$

Now choosing

$$A_n^{(1)} = -\frac{2}{\varepsilon} \tag{12.7}$$

$$A_u^{(1)} = \frac{8}{\varepsilon} \quad , \tag{12.8}$$

which results from taking $\ell = 0$ in expressions (11.4), (11.7) we cancel all the pole terms and find renormalized results identical to Eqs. (11.5), (11.8) and finite for $\varepsilon = 0$ (cf. Eqs. (11.6), (11.9)). Indeed, all this calculation just amounts to setting $\ell = 0$ in Eqs. (11.3)–(11.8).

12.1.2 Existence of the equivalent continuous chain model

In Chap. 7 we have shown that the bare discrete chain or continuous chain models are naively equivalent only close to the Θ-point. We thus might wonder whether the equivalence of the two models, shown above to one loop order, can hold generally. We thus have to show that starting from these different bare theories we nevertheless can construct identical renormalized theories. We consider the renormalized continuous chain limit (RCL), used in the theorem of renormalizability.

RCL: We take $\ell \to 0$, keeping ℓ_R, n_R, u fixed.

We thus move along a set of bare theories, equivalent in the sense that they yield the same renormalized theory. To relate this way of taking the limit $\ell \to 0$ to the NCL from Eqs. (11.27), (11.33) we note that the set of all equivalent bare theories is defined by fixed values of the parameter s_ℓ and of the combination $s_n\,n$. Now assuming a starting value $f_0 < 1$ we note from Eq. (11.28) that $s_\ell = const$ implies the asymptotic behavior

$$f_0 \sim \ell^{\varepsilon} \to 0, \quad \ell \to 0 \quad . \tag{12.9}$$

For $f_0 \to 0$ the relation $f_0 \sim \beta_e$ holds (cf. Eq. (11.37)), so that asymptotically the set of equivalent theories obeys

$$\beta_e \ell^{-\varepsilon} \sim const \sim \tilde{u}_0 \quad . \tag{12.10 i}$$

In the same region Eq. (11.34) for s_n, if combined with Eq. (12.9), implies

$$const = s_n \, n \sim \ell^2 n = R_0^2 \quad . \tag{12.10 ii}$$

Thus asymptotically the NCL and the RCL, if applied to the bare functions for $d < 4$, are identical. In other words, for $f_0 < 1$ the set of equivalent theories for $\ell \to 0$ reaches the Θ-region, where the continuous chain model and the discrete chain model coincide. The same renormalized theory can therefore be constructed from both models.

Two questions remain.

(i) Does this construction of the equivalent continuous chain model imply a restriction on the renormalization schemes allowed?

(ii) Does the restriction to $f_0 < 1$ pose a serious problem?

Concerning the first question we note that the result of any renormalization scheme based on the continuous chain model via a finite renormalization can be mapped on the renormalized theory derived from the discrete chain model, and vice versa. After renormalization the models are completely equivalent.

We now turn to the restriction $f_0 < 1$, which in view of Eq. (11.26) implies $f < 1$, i.e. $u < u^*$. A priori this is a serious point of concern. Equation (11.28) shows that starting from $f_0 > 1$ (which implies $f > 1$) and decreasing ℓ for s_ℓ fixed we always stay with $f_0 > 1$. In fact, ignoring $P_f(f_0)$ and taking into account that the exponent ω is found to be smaller than ε we would predict that f_0 diverges. Thus for a starting value $f_0 > 1$ the continuous chain limit $f_0 \sim \ell^\varepsilon \to 0$ cannot be reached. The strong coupling branch does not end in a Θ-point, and the discrete chain model is not equivalent to a well defined continuous chain model. Noting that decreasing the segment size increases the interaction we rather may speculate that the model wants to approximate some stiff microstructure, not the infinitely flexible microstructure of the continuous chain. Since the Gaussian model cannot describe a stiff chain, the model eventually breaks down. This could be signaled by the increase of f_0 beyond the region where our perturbative constructions can be trusted.

The problem, however, does not ruin our construction of the renormalized theory, and it does not keep us from using the results in some region $u = fu^* > u^*$. Concerning the RG flow we note that we will use the special scheme of 'minimal subtraction', where the flow equations depend on d only trivially and cannot show a singularity at the dimension-dependent fixed point u^*. Thus our RG mapping also holds in some region $u > u^*$. In calculating the scaling functions we first use the RG to map our system on an uncritical system of short chains or higher concentration. (See the discussion related to Fig. 10.3.) Such a system should also show no sign of singularities related to u^*, which are the signature of a critical system.

We thus take the following attitude. For technical reasons we calculate the renormalized theory starting from the continuous chain model. By equivalence of the bare theories for $f_0 < 1$ we know that we can derive the same theory from the discrete chain model. Since the renormalized theory in the way we construct it should show no singularity at u^*, we can use it for $u > u^*$. This region however can be interpreted only in terms of the discrete chain model.

12.2 Minimal subtraction

Starting from the continuous chain model as the dimensionally regularized theory we write the renormalization factors as

$$Z_u = \frac{1}{2} b_u(\varepsilon) \left(1 + \sum_{k=1}^{\infty} A_u^{(k)} u^k \right) \tag{12.11 i}$$

$$Z_n = 1 + \sum_{k=1}^{\infty} A_n^{(k)} u^k \tag{12.11 ii}$$

$$Z = 1 + \sum_{k=1}^{\infty} A^{(k)} u^k \quad . \tag{12.11 iii}$$

We introduced a parameter $b_u(\varepsilon)$, which is assumed to be a smooth function obeying

$$b_u(\varepsilon) = 1 + b_1 \varepsilon + O(\varepsilon^2) \quad . \tag{12.12}$$

A change of b_u just amounts to a special finite renormalization. b_u is kept here to illustrate later the effect of such a renormalization (see Sect. 12.4). However in the next chapter we will find that b_u can be absorbed rigorously into the parameters fixing the uncritical manifold where the scaling functions are to be evaluated. Without losing generality we could therefore set $b_u(\varepsilon) \equiv 1$ from the outset. Also the factor 1/2 in Eq. (12.11 i) has no deeper meaning and is included only with regard to previous calculations.

In the dimensionally regularized theory the coefficients $A_y^{(k)}$ are functions only of ε, which have to be chosen to cancel the singularities of the bare theory occurring for $\varepsilon \to 0$. This goal can be reached with the ansatz

$$A_y^{(k)} = \sum_{k'=1}^{k} a_y^{(k,k')} \varepsilon^{-k'} \quad , \tag{12.13}$$

which allows only for singular terms. This ansatz is known as 'minimal subtraction of dimensional poles', and *it can be proven that the very requirement that the* $G_{cR}^{(L,M)}$ *exist in* $d = 4$ *uniquely determines all the* $a_y^{(k,k')}$. Any other prescription for choosing the Z-factors just adds regular terms to the expansion (12.13).

We now discuss the resulting properties of the RG flow equations. Writing Eq. (12.3) as

$$\ln\left((4\pi)^{-d/2}\tilde{u}_0\right) = -\varepsilon \ln \ell_R + \ln\, u + \ln Z_u(u)$$

and taking the derivative $\ell_R\, d/d\ell_R, \tilde{u}_0$ fixed, we find

$$0 = -\varepsilon + \frac{1}{u}\ell_R \frac{d\,u}{d\,\ell_R} + \ell_R \frac{d\,u}{d\,\ell_R}\frac{\partial}{\partial\,u} \ln Z_u(u) \quad .$$

We now introduce $W(u) = -\ell_R\, du/d\ell_R$ (Eq. (11.18 i)) to find the expression

$$W(u) = \frac{-\varepsilon\, u}{1 + u\frac{\partial}{\partial\,u} \ln Z_u(u)} \quad . \tag{12.14}$$

As pointed out in Sect. 11.3 the renormalizability of the theory guarantees that $W(u)$ can be expanded in powers of u, the coefficients of this expansion staying finite for $\varepsilon \to 0$. By virtue of Eq. (12.14) all terms in this expansion higher than the leading linear term have the form

$$\varepsilon\, u^k\, W_k, \quad k \geq 2 \quad ,$$

where W_k is a linear combination of products of the $A_u^{(k')}, k' \leq k$. Thus W_k contains only pole terms. Since the coefficient $\varepsilon\, W_k$ is finite for $\varepsilon \to 0$, all higher order poles have to cancel, leaving only a first-order pole

$$W_k = \frac{w_k}{\varepsilon} \quad .$$

As a result we find that $W(u)$ has the structure

$$W(u) = -\varepsilon\, u + \hat{W}(u) \quad , \tag{12.15}$$

where

$$\hat{W}(u) = \sum_{k=2}^{\infty} w_k u^k$$

is strictly independent of ε. Since $\hat{W}(u)$ can be calculated for $\varepsilon = 0$, in view of Eq. (12.12) it must also be independent of the parameter $b_u(\varepsilon)$.

The argument is easily extended to show that the functions $1/\nu(u)$ (Eq. (11.18 ii)) or $\eta(u)$ (Eq. (11.18 iii)) are also independent of ε. Consider for instance $1/\nu(u)$. Equations (11.18), (12.11 ii) yield

$$\begin{aligned} \frac{1}{\nu(u)} &= 2 - \lambda \frac{du}{d\lambda} \frac{\partial}{\partial u} \ln Z_n \\ &= 2 + (\varepsilon\, u - \hat{W}(u)) \frac{\partial}{\partial u} \ln\left(1 + \Sigma A_n^{(k)} u^{(k)}\right) \quad . \end{aligned}$$

In the expansion in powers of u again all higher order poles have to cancel, so that there survive only those simple poles multiplied by $\varepsilon\, u$. As a result the power series for $1/\nu(u)$ is independent of ε.

These results show that the minimal subtraction scheme eliminates ambiguities inherent in an extrapolation of ε-expansion results to physical dimension $d = 3$. For the flow equations no extrapolation is necessary. Furthermore they are strictly independent of the parameter $b_u(\varepsilon)$.

12.3 Final form of renormalized perturbation theory

We now are in the position to formulate precisely our scheme of renormalized perturbation theory.

12.3.1 General procedure

Starting from the discrete chain model we carry through the following steps.

(i) We write down the formal expressions resulting from the diagrams to the order considered to find a power series in β_e.

(ii) We extract appropriate factors of $\exp(\mu_s^*(\beta_e)n)$, combining in the result the summations such that the naive continuous chain limit exists.

(iii) We take the naive continuous chain limit, which just amounts to replacing summations by integrals:

$$\sum_{j=j_<}^{j_>} \longrightarrow \int_{j_<}^{j_>} \quad .$$

(iv) In the result we introduce renormalized variables according to

$$\beta_e = \lambda^{\varepsilon} u Z_u(u) \tag{12.16 i}$$

$$n = \lambda^{-2} n_R Z_n(u) \tag{12.16 ii}$$

$$\ell = \lambda \ell_R \tag{12.16 iii}$$

and we multiply by appropriate renormalization factors, reordering the expression according to powers of u. The renormalized expression results.

Aiming at the ε-expansion we

(v) expand the result in powers of ε, counting u as order ε.

As Dohm and co-workers [SD89] have shown however, we may also

(v') evaluate the u-expansion directly in three dimensions.

We will generally choose this latter approach.

We should note that except for the free energy all observables can be normalized appropriately, so that the factors $\exp(\mu_s^*(\beta_e)n)$ as well as the explicit Z-factors drop out. Furthermore it is not necessary to introduce explicitly the intermediate quantities $\tilde{u}_0, R_0^2$ occurring in the NCL.

We have formulated the method starting from the cluster expansion valid in the dilute limit. The loop expansion is renormalized by exactly the same steps. The only new feature is the occurrence of the parameter $W = W(N)$ (cf. Eq. (5.1)), which under these steps transforms as

$$\begin{aligned} W &= (4\pi)^{d/2}\, \beta_e \ell^d c_p N^2 \\ &= (4\pi)^{d/2}\, \lambda^{\varepsilon} u Z_u \ell^d c_p \lambda^{-4} N_R^2 Z_n^2 \\ &= (4\pi)^{d/2}\, u c_{pR} N_R^2 Z_u Z_n^2 \\ &= W_R Z_u Z_n^2 \quad . \end{aligned} \tag{12.17}$$

Here $c_{pR} = \ell_R^d c_p$ (Eq. (11.73)), and we should recall that N denotes the number averaged chain length. N_R is defined by Eq. (12.16 ii) with n replaced by N.

In the following chapters we will present results of first order renormalized perturbation theory (one loop). To this order we need the following expressions

$$\mu_s^*(\beta_e) = \beta_e \sum_{j=1}^{\infty} j^{-d/2} + O(\beta_e^2) \tag{12.18}$$

$$Z_u(u) = \frac{b_u(\varepsilon)}{2}\left(1 + \frac{4}{\varepsilon}\,u + O(u^2)\right) \tag{12.19 i}$$

$$Z_n(u) = 1 - \frac{1}{\varepsilon}\,u + O(u^2) \tag{12.19 ii}$$

$$Z(u) = 1 + O(u^2) \ . \tag{12.19 iii}$$

The additive renormalization (12.18) has been derived previously (cf. Sects. 5.4.3 or 10.1). The modification of the terms $\sim u/\varepsilon$ in Eqs. (12.19) as compared to Eqs. (12.7), (12.8) results from the additional factor 1/2 included in our present definition of u (Eq. 12.11 i).

If we want to evaluate the results in strict ε-expansion we need in addition the ε-expansion of u^*, which is found as

$$u^* = \frac{\varepsilon}{4} + \frac{21}{128}\,\varepsilon^2 + O(\varepsilon^3) \ . \tag{12.20}$$

Note that u^*, defined as nontrivial zero of $W(u)$, is necessarily independent of $b_u(\varepsilon)$, a feature which may seem somewhat surprising in view of the definition of u (Eqs. (12.16 i), (12.11 i)).

12.3.2 A simple example

We again exemplify the general method with a calculation of the endpoint distribution of a single chain. As shown in Eq. (11.47) this quantity is renormalized as

$$P(\mathbf{p}, n) = P_R(\bar{\mathbf{p}}, n_R, f) = \frac{G_R^{(0,1)}(\bar{\mathbf{p}}, -\bar{\mathbf{p}}; n_R; f)}{G_R^{(0,1)}(\mathbf{0}, \mathbf{0}; n_R; f)} \ , \tag{12.21}$$

where both the chemical potential $\mu_s^*(\beta_e)$ and the renormalization prefactor Z/Z_n drop out. To first-order the unrenormalized single chain Greensfunction $G_c^{(0,1)}$ has been calculated in Sect. 4.1. Steps (i) and (ii) immediately yield

$$P(\mathbf{p}, n) = e^{-p^2\ell^2 n}\Big\{1 - \beta_e \sum_{0<j_1<j_2<n} (j_2 - j_1)^{-d/2}\left(e^{p^2\ell^2(j_2-j_1)} - 1\right) + O(\beta_e^2)\Big\}.$$

Step (iii) together with the substitution $j_i = n\, s_i$ yields the dimensionally regularized result

$$P(\mathbf{p}, n) = e^{-p^2\ell^2 n}\Big\{1 - \beta_e n^{\varepsilon/2} \int_{0<s_1<s_2<1} (s_2 - s_1)^{-d/2} \left(e^{p^2\ell^2 n(s_2-s_1)} - 1\right) + O(\beta_e^2)\Big\} \ . \tag{12.22}$$

We now turn to renormalization (step iv). Equations (12.16), (12.19 ii) yield

$$p^2\ell^2 n = \bar{\mathbf{p}}^2 n_R \left(1 - \frac{u}{\varepsilon} + O(u^2)\right) \quad , \tag{12.23}$$

where we should recall the notation $\bar{\mathbf{p}} = \mathbf{p}\ell_R$ (Eq. (11.11)). The factor $\beta_e n^{\varepsilon/2}$ to the order considered can be written as

$$\beta_e\, n^{\varepsilon/2} = \frac{b_u}{2}\, u n_R^{\varepsilon/2}(1 + O(u)) \ . \tag{12.24}$$

Carrying out step (iv) we find

$$P_R(\bar{\mathbf{p}}, n_R, f) = e^{-\bar{\mathbf{p}}^2 n_R}\Big\{1 - \frac{b_u}{2}\, u n_R^{\varepsilon/2} \int_{0<s_1<s_2<1} (s_2 - s_1)^{-d/2} \left(e^{\bar{\mathbf{p}}^2 n_R (s_2 - s_1)} - 1\right) + \bar{\mathbf{p}}^2 n_R \frac{u}{\varepsilon} + O(u^2)\Big\} \quad , \tag{12.25}$$

which is the renormalized result which may be evaluated directly for $d = 3$ ($\varepsilon = 1$).

Aiming at the ε-expansion we still have to analyze the integral. The analysis is simple. We write

$$\begin{aligned}
&\int\limits_{0<s_1<s_2<1} (s_2 - s_1)^{-d/2} \left(e^{(s_2-s_1)x} - 1\right) \\
&= x \int\limits_{0<s_1<s_2<1} (s_2 - s_1)^{1-d/2} \\
&+ \int\limits_{0<s_1<s_2<1} (s_2 - s_1)^{-d/2} \left(e^{(s_2-s_1)x} - 1 - (s_2 - s_1)x\right) \ .
\end{aligned}$$

The second integral is finite for $\varepsilon = 0$. The first integral is easily evaluated to yield

$$\int\limits_{0<s_1<s_2<1} (s_2 - s_1)^{-d/2} \left(e^{(s_2-s_1)x} - 1\right) = \frac{4}{\varepsilon(2+\varepsilon)} x + I_1^{(\varepsilon)}(x) \quad ,$$

where $I_1^{(\varepsilon)}$ stands for the second integral. Substituting this result into Eq. (12.25) we find

$$\begin{aligned}
P_R(\bar{\mathbf{p}}, n_R, f) = e^{-\bar{\mathbf{p}}^2 n_R}\Big\{&1 + \bar{\mathbf{p}}^2 n_R \frac{u}{\varepsilon}\left(1 - b_u(\varepsilon)\frac{n_R^{\varepsilon/2}}{1 + \varepsilon/2}\right) \\
&- \frac{b_u(\varepsilon)}{2}\, u\, n_R^{\varepsilon/2} I_1^{(\varepsilon)}(\bar{\mathbf{p}}^2 n_R) + O(u^2)\Big\} \ .
\end{aligned} \tag{12.26}$$

We now expand in ε, using Eq. (12.12): $b_u(\varepsilon) = 1 + \varepsilon\, b_1 + O(\varepsilon^2)$. The result takes the form

$$P_R(\bar{\mathbf{p}}, n_R, f) = e^{-\bar{\mathbf{p}}^2 n_R}\Big\{1 - \bar{\mathbf{p}}^2 n_R \frac{\varepsilon}{8} f\,(\ln n_R + 2b_1 - 1) - \frac{\varepsilon}{8} f I_1^{(0)}(\bar{\mathbf{p}}^2 n_R) + O(\varepsilon^2)\Big\} \ , \tag{12.27}$$

where

$$I_1^{(0)}(x) = \int_0^1 ds(1-s)s^{-2}(e^{sx} - 1 - sx) \ . \tag{12.28}$$

We used $u = u^* f$, inserting the value (12.20) of u^*.

In later chapters we will see a few more examples of explicit calculations. Specifically we will present the calculation of the single-chain density correlation function in Appendix A 15.2 and the calculation of the osmotic pressure in Appendix A 17.1.

12.4 Perturbation theory in three dimensions vs ε-expansion

In the construction of the RG, dimension $d = 4$ plays a special role as upper critical dimension of the theory. This for instance shows up in the estimate of the nonuniversal corrections to the theorem of renormalizability, or in the feature that the nontrivial fixed point u^* merges with the Gaussian fixed point for $d \to 4$. It naturally leads to the ε-expansion. However, the RG mapping constructed in minimal subtraction only trivially depends on ε. Also results of renormalized perturbation theory do not necessarily ask for further expansion in ε. Equation (12.25) gives an example. We should thus consider the practical implications of the ε-expansion in some more detail.

12.4.1 Virtues and disadvantages of the ε-expansion

Consider the ε-expansion (12.27) of the renormalized end-to-end distribution. It contains the constant b_1 and thus depends on our renormalization scheme. This dependence can be eliminated by replacing the chain length, which is a microscopic parameter, by the end-to-end distance

$$R_e^2 = -\Delta_p|_0 \ln P(\mathbf{p}, n) = -2d\ell_R^2 \frac{\partial}{\partial \bar{\mathbf{p}}^2}|_0 \ln P_R(\bar{\mathbf{p}}, n_R, f) \ . \tag{12.29}$$

Equation (12.27) yields

$$R_e^2 = 2d\ell_R^2 n_R \left[1 + \frac{\varepsilon}{8} f\,(\ln n_R + 2b_1 - 1) + O(\varepsilon^2)\right] \ . \tag{12.30}$$

Thus

$$\bar{\mathbf{p}}^2 n_R = \mathbf{p}^2 \frac{R_e^2}{2d}\left[1 - \frac{\varepsilon}{8} f\left(\ln n_R + 2b_1 - 1\right) + O(\varepsilon^2)\right] \quad . \tag{12.31}$$

We substitute this result into Eq. (12.27) to find

$$P(\mathbf{p}, n) = P_R(\mathbf{p}, n_R, f) = e^{-\frac{\mathbf{p}^2 R_e^2}{2d}}\left[1 - \frac{\varepsilon}{8} f I_1^{(0)}\left(\frac{\mathbf{p}^2 R_e^2}{2d}\right) + O(\varepsilon^2)\right] \quad , \tag{12.32}$$

a result independent of $b_u(\varepsilon)$.

However, the result still depends on $f = u/u^*$, which is not an observable quantity. We can eliminate this parameter in favor of the interpenetration ratio

$$\psi = \left(\frac{d}{12\pi}\right)^{d/2} \frac{A_2}{R_g^d} \quad . \tag{12.33}$$

This quantity will be analyzed in detail in Sect. 15.4. The ε-expansion result reads

$$\psi = \frac{\varepsilon}{8} f(1 + O(\varepsilon)) \quad .$$

Thus Eq. (12.32) can be written as

$$P(\mathbf{p}, n) = e^{-\frac{\mathbf{p}^2 R_e^2}{2d}}\left[1 - \psi I_1^{(0)}\left(\frac{\mathbf{p}^2 R_e^2}{2d}\right) + O(\varepsilon^2)\right] \quad , \tag{12.34}$$

a fully universal result containing only observable variables.

This calculation illustrates a general feature: we may write down scaling laws for normalized quantities in terms of scaled momenta $\mathbf{q}R_g$ (or $\mathbf{q}R_e$, equivalently), scaled concentrations $c_p R_g^3$, and ψ replacing the coupling. Such relations involve only physically observable macroscopic quantities. They must have a uniquely defined ε-expansion, where $\psi = O(\varepsilon)$ acts as an expansion parameter. The result is necessarily independent of any conventions of the renormalization scheme. Not even the form of the RG flow equations matters. Furthermore, in establishing such results, we never have to invoke a condition like $n_R = 1$. These are the great virtues of consistent ε-expansion.

We may ask why we do not work exclusively within that scheme, which gives seemingly unambiguous results. The answer is that the above statement – correct as it is – shows the matter in too rosy a light. Consider again Eqs. (12.33), (12.31). We have chosen to keep some factors of d, the reason being that these factors come in 'naturally' in the definition of the variables R_e^2, ψ. This clearly is some prejudice. Expanding out these factors we construct a perfectly legitimate ε-expansion result. In physical dimension

$d = 3(\varepsilon = 1)$ the two approximations, both being consistent first order ε results, yield quantitatively different predictions. The matter becomes worse if we note the great freedom we have in choosing the physical variables. For instance the factor $\left(\frac{d}{12\pi}\right)^{d/2}$ in the definition (12.33) of ψ has purely historical reasons. We could omit it, replacing ψ by a different interpenetration ratio. We could also decide to expand $\ln P(\mathbf{p}, n)$ in place of $P(\mathbf{p}, n)$. Again the first order results, extrapolated to $\varepsilon = 1$, would be different quantitatively. Such ambiguities could be avoided only by pushing the ε-expansion to high orders and using appropriate resummation methods to extract the scaling functions from these asymptotic expansions. Unfortunately this is a forbidding program, except for a few most simple quantities $(R_e^2(c_p = 0), R_g^2(c_p = 0), A_2)$. In general we can calculate only low order (one loop) approximations and thus have to live with these ambiguities.

There are more problems. The result (12.34), for instance, yields no information on chain-length or temperature dependence, which is hidden in R_e^2 or ψ. To extract it, we have to write down RG flow equations for these variables. This results in the so-called 'direct renormalization' scheme [Clo81], which however has not been pushed to the same high order as the flow equations derived for u, N_R in minimal subtraction.

A more subtle problem occurs for quantities involving several characteristic length scales. Consider for instance the density correlation function in the limit of large momenta: $(\mathbf{q}R_g)^2 \gg 1$ where $1/q$ defines a length scale of interest, which is much smaller than R_g. In the excluded volume limit simple scaling considerations (cf. Sect. 9.1, Eq. (9.20)) suggest

$$J_a(\mathbf{q}R_g) \underset{|\mathbf{q}R_g| \to \infty}{\sim} |\mathbf{q}R_g|^{-1/\nu} \ . \tag{12.35}$$

Strict ε-expansion cannot yield this result. It expands the exponent according to $1/\nu = 2 - \varepsilon/4 + O(\varepsilon^2)$ (cf. Eq. (8.34)) and yields a power series in $\ln(\mathbf{q}^2 R_g^2)$:

$$J_a(\mathbf{q}R_g) \sim (\mathbf{q}R_g)^{-2} \left(1 + \frac{\varepsilon}{8} \ln(\mathbf{q}^2 R_g^2) + O(\varepsilon^2)\right) \ .$$

To yield reasonable results we have to resum such power series, thus giving up the idea of strict ε-expansion. If then we insist on finding a form of $J_a(\mathbf{q}R_g)$ smoothly interpolating among the limits of small or large momenta, we again have a large amount of freedom in constructing this form. We only note that Eq. (12.35) asymptotically is equivalent to

$$J_a(\mathbf{q}R_g) \sim \left(a + (\mathbf{q}R_g)^2\right)^{-\frac{1}{2\nu}} \ ,$$

where the arbitrary constant $a > 0$ determines the region where $J_a(\mathbf{q}R_g)$ crosses over from analytic behavior for small $\mathbf{q}$ to the asymptotic power law.

Similar problems are abundant as soon as we leave the region of small momenta and isolated chains. As a final example we consider the semidilute limit. Using the unrenormalized loop expansion in Sect. 5.4.3 we have calculated the first order correction to $\mu_p(n)$. We found a correction of order $c^{-\varepsilon/2}$, where c is the segment concentration. The form of this term is due to screening and has nothing to do with the critical behavior treated by renormalization and ε-expansion. It thus should not be expanded in powers of ε. We can trace it back to the occurrence of the size of the concentration blobs as an additional length scale.

To summarize, strict ε-expansion a priori seems to yield unambiguous results. Closer inspection, however, reveals that in low order calculations considerable ambiguity is hidden in the definition of the physical observables used as variables or chosen to calculate. What is worse, the ε-expansion does not incorporate relevant physical ideas predicting the behavior outside the small momentum range or beyond the dilute limit. In particular, it does not give a reasonable form for crossover scaling functions. On the other hand, it can be used to calculate well-defined critical ratios, which are a function of dimensionality only. Even then, however, the precise definition of the ratio matters.

On a deeper level we observe that the ε-expansion does not properly respect the structure of the theory. As discussed before (see Sects. 8.3 or 10.2, in particular), we should first use the RG to map the system to an uncritical manifold, which for dilute systems is determined by $N_R = O(1)$, for instance. In a second step the scaling functions on the uncritical manifold can be calculated by renormalized perturbation theory. These two steps are well separated conceptually. The exponents, for instance, are properties of the RG, independent of specific scaling function. Strict ε-expansion mixes these two steps since it simultaneously expands exponents and scaling functions. The expected scaling structure then has to be put in by hand at the end of the calculation.

12.4.2 Direct evaluation in three dimensions

Avoiding the ε-expansion, we can work directly with renormalized perturbation theory in powers of u, evaluated in $d = 3$ dimensions. We still have to decide what quantities to expand, and it clearly will make some difference whether we for instance use the result (12.25) for $P_R(\bar{\mathbf{p}}, n_R, f)$ or rather expand $\ln P_R(\bar{\mathbf{p}}, n_R, f)$. Having chosen some basic set of quantities, however, we consistently evaluate observables of interest without further expansion. We for instance calculate R_g^2, A_2 perturbatively, constructing ψ from Eq. (12.33) without further manipulation. Clearly then some trivial change in the definition of ψ has no deeper consequences.

However, the ambiguity in defining the expansion parameter, which in the ε-expansion is related to the precise choice of ψ, is not eliminated. Rather it is made very explicit in our choice of $b_u(1)$. Also the other ambiguities, concerning the crossover behavior in momenta or overlap, not only remain but are clearly brought out and condensed into a few theoretical parameters. As recalled above, in evaluating the scaling functions we, following the general RG philosophy, map the theory into a parameter region far from the criticality, i.e. into an uncritical manifold where our expansion is well behaved (see Fig. 10.3). Again the precise choice of that manifold would not matter if we were able to calculate up to high orders, but with the low order expressions available it will influence the quantitative results. We will see in the next chapter that the choice of the uncritical manifold introduces three parameters, which fix the crossover from Θ- to excluded volume conditions, from small to large momenta, or from the dilute to the semidilute limit, respectively. $b_u(1)$, however, can be absorbed into these parameters. *Once the parameters are fixed, the RG mapping generates smooth crossover functions, and produces nontrivial asymptotic behavior without any further manipulations.* The asymptotic power law (12.35), for instance, is found automatically even in a zero order calculation. We will illustrate this remarkable power of the RG in Chap. 14 with an evaluation of the zero loop approximation. Furthermore the renormalized expansion in u also preserves the structure of screening in the semidilute limit.

To summarize, using not the ε-expansion but renormalized perturbation theory in powers of u we make optimal use of the renormalization group, exhibiting and preserving the physical structure suggested by intuitive arguments. The method does not eliminate the arbitrariness inherent in low order calculations, but has the virtue of condensing it into a few well controllable parameters. The precise choice of these parameters and thus of the uncritical manifold will be discussed in Sect. 13.2 and Appendix A 13.1.

We must however add a word of caution. The method as sketched here has been established to deal with the small momentum behavior of the scaling functions. For large momenta we may run into problems in that the *subleading* momentum dependence is not treated correctly. On the quantitative level however this does not seriously invalidate our results. We will discuss this aspect further in Sect. 15.2. See also Sect. 13.2.3.

We should not close this section without touching upon some delicate technical point. A priori even for a discrete chain the critical chemical potential μ_s^*, defined as chemical potential per segment of an infinitely long chain, does not exist outside the ε-expansion. We already encountered this problem in Sect. 7.2, where we found that τ_c, proportional to μ_s^*, for $d < 4$ in naive perturbation theory suffers from infrared singularities. That problem has been considered in the field theoretic framework. The results, expressed in polymer language, show that the u-expansion is not invalidated, as long as we consider quantities which do not involve μ_s^* explicitly. Almost all the quantities of interest to us are of this type. Only in the equation of state relating $\mu_p(n)$ and $c_p(n)$ does an explicit contribution $\mu_s^* n$ occur. But even here the problem shows up in higher orders only and can be eliminated by taking an appropriate derivative of $\mu_p(n)/n$. We therefore do not consider this problem further.

Notes

In field theoretic context the method of dimensional regularization and minimal subtraction has been proposed in [tHV72]. To the level considered here it is discussed in standard textbooks [ZJ89, Ami84]. There the explicit calculation of the Z-factors can also be found. The method has been applied directly to the polymer system in the continuous chain limit in [Dup86a], where different versions of the approach are compared.

The evaluation of the scaling functions directly in $d = 3$ has been advocated in [SD89, SD90], where the method has been shown to work for small momenta. Also the treatment of the additive 'mass renormalization' (corresponding to the introduction of μ_s^*) has been carefully considered there.

Concerning the application of field theory for $f > 1$ there have been some discussions in the literature (see [BB90, BB94, LMS95, Sch94]). This extension of the theory was initiated by the work of [Nic91]. The present formulation was advocated and compared to Monte Carlo or physical data in [Sch94, KS96, GSS97].

13. Quantitative Form of the Renormalization Group Mapping

In quantitative calculations we use the renormalization group to map the system from its critical region to an uncritical parameter region, where low order perturbation theory can be expected to yield good results. We thus need two ingredients: a good representation of the renormalization group and a sensible condition fixing the uncritical region.

A most accurate form of the renormalization group mapping has been constructed in field theory, using results of the ε-expansion carried to order ε^5 and additional information on the asymptotically high orders of the expansion. We will present the results as specialized to polymer theory in Sect. 13.1. It is this form of the mapping which we will use in all further analysis.

To provide the second ingredient, the choice of the uncritical manifold poses a delicate problem. It amounts to fixing the renormalized segment size ℓ_R and thus the scale parameter $\lambda = \ell/\ell_R$ in terms of the important macroscopic length scales. Clearly the very necessity of such a choice intrinsically is tied to the approximate nature of our calculation: the full renormalized functions by construction of the RG mapping are invariant under a change of ℓ_R, except for overall Z-factors. However, since we can calculate the scaling functions only approximately, we break strict scale invariance and the choice of ℓ_R matters. Here the physical insight provided by the blob concept turns out to be most helpful. In semidilute solutions we expect to find a blob size $R_B^* \ll R_g$ (cf. Sect. 9.1), so that nontrivial interaction effects are suppressed for $r \gg R_B^*$. Since renormalization shall deal with strong interaction effects on scale ℓ_R, we choose ℓ_R to interpolate smoothly among $\ell_R \sim R_g$ (dilute limit) and $\ell_R \sim R_B^*$ (semidilute limit). More generally, we choose ℓ_R to be of the order of the smallest macroscopic length scale in the system. This is in keeping with the general interpretation of the RG as constructing an effective Hamiltonian valid for phenomena on scales $r \gtrsim \ell_R$ (Sect. 10.2.3).

However, such physical considerations do not guarantee that the resulting choice of ℓ_R really works. As is clear from the above, ultimately any prescription for choosing the uncritical surface must be judged on the basis of perturbation theory. Higher order corrections must stay reasonably small, the correct qualitative behavior being embodied into the lowest order (tree) approximation. In the next chapter we will show that our choice fulfills the latter condition. Indeed, the tree approximation reproduces the results of scaling theory. A general discussion of the higher order corrections has not been given. However, some detailed analysis suggests that screening and thus the crossover from dilute to semidilute solutions is consistently treated within our approach at least for $(qR_B^*)^2 \lesssim 1$. For large q the situation is less clear.

The condition, which fixes ℓ_R, necessarily involves some numerical coefficients. To determine these we have to search for regions where the physical quantities are fairly insensitive to the precise choice of these parameter values. The existence of such a region is a necessary condition for the intrinsic consistency of our approach, reflecting the fact that physical quantities, if evaluated rigorously, are independent of these parameters. It is thus most pleasing that we indeed can find such a region, where furthermore the predictions for typical universal critical ratios like the interpenetration ratio $\varPsi$ are found to be close to experimental values. This strongly supports our approach. We discuss the choice of the uncritical manifold in Sect. 13.2, where we also give one example for the determination of a theoretical parameter. The determination of the other two parameters is discussed in Appendix A 13.2, and in Appendix A 13.1 it is shown that the parameter $b_u(\varepsilon)$ can be absorbed into the other parameters of the theory. This completes the construction of the renormalized theory to the order of one loop.

With respect to the value of the renormalized coupling the general parameter range decomposes into different regions: the weak coupling region $f < 1$ describes the crossover from Θ- to excluded volume conditions. There is also the strong coupling region $f > 1$, not connected to Θ-conditions. These two regions are separated by the excluded volume fixed point $f = 1$. In Sect. 13.3 we discuss these three regimes, specializing the RG mapping to various limits of interests. In the weak coupling region we establish the connection to the Daoud-Jannink diagram (cf. Sect. 9.2). In the strong coupling region only some restricted part of the RG flow is of relevance, due to the role of nonuniversal corrections. This will be explained in some detail.

13.1 The renormalization group mapping in three dimensions

The construction of the quantitative RG mapping [SD89] starts from the perturbation expansion of $W(u)$ evaluated to order u^6 or from the expansions for $\eta(u)$ or $2 - 1/\nu(u)$ evaluated to order u^5. These expansions are only asymptotic, but they can be resumed, using the Borel method. Within the accuracy of the calculation the results for $d = 3$ are reproduced by the analytical expressions

$$W_f(f) = -f(1-f)\frac{1+.824f}{1+1.280f} \tag{13.1}$$

$$2 - \frac{1}{\nu_f(f)} = .364f - .166f^2 + .174f^3 - .073f^4 \tag{13.2}$$

$$\eta_f(f) = .033f^2 - .002f^3 \quad , \tag{13.3}$$

where $f = u/u^*$ (Eq. (11.21)). The nontrivial fixed point is found at

$$u^* = .364 \quad . \tag{13.4}$$

Note, that the coupling u of [SD89] corresponds to $u/8$ in our notation, and $W_f(f) = \beta_u(u,1)/u^*; 2 - 1/\nu_f(f) = f_2(u), \eta_f(f) = -f_3(u)$; ([SD89], Eqs. (5.14)–(5.16)). Furthermore we have slightly corrected the parameterization of $2-1/\nu_f(f)$ so as to find the value $\nu = \nu_f(1) = .588$ instead of the value .59. The necessary changes are within the accuracy of the parameterization. They are motivated by independent higher order calculations of the exponents, both within the ε-expansion and within the u-expansion in $d = 3$. (See, for instance, [GZJ85].) The best theoretical value of ν resulting from that work is $\nu = 0.588$. Real experiments are not accurate enough to check this value. Experimentally it is even hard to discriminate between $\nu \approx 0.59$ and the Flory value $\nu = 0.6$. However, recent Monte Carlo simulations [LMS95], using extremely long chains, yield $\nu = 0.5877 \pm 0.0006$, in excellent agreement with the field theoretic result.

Rigorous integration of the resulting flow equations (cf. Sect. 11.4) yields

$$\lambda = \frac{f_0}{f}\left(\frac{1-f}{1-f_0}\right)^{1/\omega}\left(\frac{1+.824f}{1+.824f_0}\right)^{-.250} \tag{13.5}$$

$$Z_n(f) = Z_n(f_0)\left(\frac{1-f}{1-f_0}\right)^{(2-1/\nu)/\omega}\left(\frac{1+.824f}{1+.824f_0}\right)^{.289}\frac{F_n(f)}{F_n(f_0)} \tag{13.6}$$

$$Z(f) = Z(f_0)\left(\frac{1-f}{1-f_0}\right)^{\eta/\omega}\left(\frac{1+.824f}{1+.824f_0}\right)^{-.013}\frac{F_z(f)}{F_z(f_0)} \quad , \tag{13.7}$$

where

$$F_n(f) = \exp(-.228f + .103f^2 - .038f^3) \tag{13.8}$$

$$F_z(f) = \exp(.049f - .0016f^2) \quad . \tag{13.9}$$

The exponents ν, ω, η take the values

$$\nu = 0.588; \quad \omega = 0.80; \quad \eta = 0.03 \quad . \tag{13.10}$$

The factor $(1+.824f)^{.289}F_n(f)$ in the interval $0 \le f \le 1.5$ can be parameterized as

$$(1+.824f)^{-.289}F_n^{-1}(f) \longrightarrow H(f) = 1 - .005f - .028f^2 + .022f^3 \quad . \tag{13.11}$$

Our final expression for the RG mapping therefore reads

RG Mapping ($d = 3$)

$$\ell_R = \frac{\ell}{\lambda} = f\Big|1-f\Big|^{-1/\omega} H_u(f) s_\ell \tag{13.12 i}$$

$$n_R = \lambda^2 Z_n^{-1} n = f^{-2}\Big|1-f\Big|^{1/\nu\omega} H_n(f) s_n n \quad , \tag{13.12 ii}$$

where

$$H_u(f) = (1+.824f)^{0.25} \tag{13.13}$$

$$H_n(f) = \frac{H(f)}{H_u^2(f)} \quad . \tag{13.14}$$

The nonuniversal constants s_ℓ, s_n can be expressed in terms of the starting value $f_0 = f(\lambda = 1)$ as

$$s_\ell = f_0^{-1}\Big|1-f_0\Big|^{1/\omega} H_u^{-1}(f_0)\ell \tag{13.15 i}$$

$$s_n = f_0^2\Big|1-f_0\Big|^{-1/\nu\omega} H_n^{-1}(f_0) Z_n^{-1}(f_0) \quad . \tag{13.15 ii}$$

13.2 Choice of the renormalized length scale

13.2.1 Qualitative considerations

It is clear a priori that it does not make sense to choose $\ell_R > R_g$, since then the coil would be smaller than an effective segment. To leading order of bare perturbation theory R_g is given by the expression for a noninteracting chain: $R_g^2 \approx \ell^2 N$ (Eq. (3.32); $d = 3$), where we neglect polydispersity effects. Renormalizing this expression we find

$$R_g^2 \approx \ell_R^2 N_R \quad ,$$

suggesting the choice

$$N_R \approx 1 \tag{13.16}$$

which implicitly fixes ℓ_R as $\ell_R \approx R_g$.

This is a good choice as long as R_g is the only length scale of interest, i.e. as long as we consider isolated chains in the limit of small momenta $qR_g \lesssim 1$. In the semidilute region the blob size R_B^* replaces R_g in defining the scale of the typical excluded volume correlations, the concentration blobs interacting but weakly with each other. This suggests to identify the effective segment with a blob: $\ell_R \approx R_B^*$. Now a quantitative measure of the blob size is provided by the screening length ξ_E introduced in Sect. 5.4.2. Using Eqs. (12.11 i), (12.16) to renormalize the lowest order expression (5.63) we find

$$\xi_E^2 = \frac{\ell_R^2}{2fc_R} \quad ,$$

where we introduced the renormalized segment concentration

$$c_R = \tilde{u} c_p \ell_R^d N_R \tag{13.17}$$

$$\tilde{u} = (4\pi)^{d/2} \frac{u^*}{2} b_u \quad . \tag{13.18}$$

This suggests the choice

$$2fc_R \approx 1 \quad . \tag{13.19}$$

Here the factor of the coupling f has a good physical meaning. Clearly the screening effect underlying the blob concept needs interacting chains. For strictly noninteracting chains there is no semidilute region. Thus $R_B^* \sim \xi_E$ depends on f, a feature ignored in Sect. 9.1, where we only considered the excluded volume limit $f = 1$.

Finally, momentum defines a length scale $r \sim 1/q$, and evaluating scattering functions for large momenta, i.e. looking for correlations deep inside a blob, we should choose

$$\ell_R \approx \frac{1}{q}$$

or

$$(\ell_R \mathbf{q})^2 \equiv \bar{\mathbf{q}}^2 \sim 1 \quad , \tag{13.20}$$

equivalently.

Equations (13.16), (13.19), or (13.20) are adequate in limiting situations, where the length considered is much shorter than the other macroscopic length scales. To treat the crossover we fix ℓ_R by a relation smoothly interpolating among these limits. We choose

$$\frac{\bar{\mathbf{q}}^2}{q_0^2} + \frac{n_0}{N_R} + f\,\frac{c_R}{c_0} = 1 \quad , \tag{13.21}$$

where the numerical constants q_0^2, n_0, c_0 take into account that our qualitative considerations fix ℓ_R only up to constants. Once we settled upon definite values of these constants this equation determines the uncritical manifold, where the scaling functions are to be evaluated. It fixes ℓ_R to be of the order of the smallest macroscopic length of interest. This is easily seen by using the appropriate expressions for R_g^2, ξ_E^2 to write Eq. (13.21) in the form

$$\frac{\mathbf{q}^2}{q_0^2} + \frac{n_0}{R_g^2} + \frac{1}{2c_0\xi_E^2} \approx \frac{1}{\ell_R^2} \quad .$$

The parameters q_0, n_0, c_0 determine the relative weight of the different lengths in the determination of ℓ_R and thus determine the crossover regions among the different limits. The choice (13.21) is in complete accord with the general RG ideas expounded in Sect. 10.2.3. Being guided by the blob concept in identifying the relevant lengths it incorporates the physical ideas of the scaling approach into RG theory. In Chap. 14 we will find that it guarantees that the correct scaling behavior is found in renormalized tree approximation, i.e. zero order perturbation theory incorporates the relevant physics. This, of course, is an essential property of a reasonable expansion.

13.2.2 Determination of the theoretical parameters

The theory involves four parameters related to the use of low order perturbation theory:

$$b_u(1), \;\; n_0, \;\; c_0 \;\; q_0^2 \quad .$$

One of these parameters can be fixed arbitrarily. This is due to the two-parameter nature of the theory which guarantees that the renormalized variables occur only in the combination $uN_R^{\varepsilon/2}, \bar{\mathbf{q}}^2 N_R, c_{pR} N_R^{d/2}$, corresponding to $z, \mathbf{q}^2 R_0^2, s_0$ in the unrenormalized continuous chain model (see Sect. 7.1). The formal argument is given in Appendix A 13.1. We exploit this freedom to choose $b_u(1) = 1$. The remaining parameters can be seen to govern the crossover in temperature (n_0), overlap (c_0), or momentum (q_0). We exemplify the choice of these parameters with n_0.

The choice of n_0 governs ℓ_R for small momenta in the dilute limit where Eq. (13.21) reduces to $N_R = n_0$. We thus should consider some universal ratio, defined for $c \to 0, q \to 0$, which must be most sensitive to N_R. The interpenetration ratio ψ is a good candidate, since it depends on N_R even in zero-loop approximation. From Sect. 15.4, Eq. (15.54) we take the one-loop order result

$$\psi^* = 0.182 n_0^{1/2} \frac{1.728 - 0.521 n_0^{1/2}}{\left(0.636 + 0.232 n_0^{1/2}\right)^{3/2}} \quad .$$

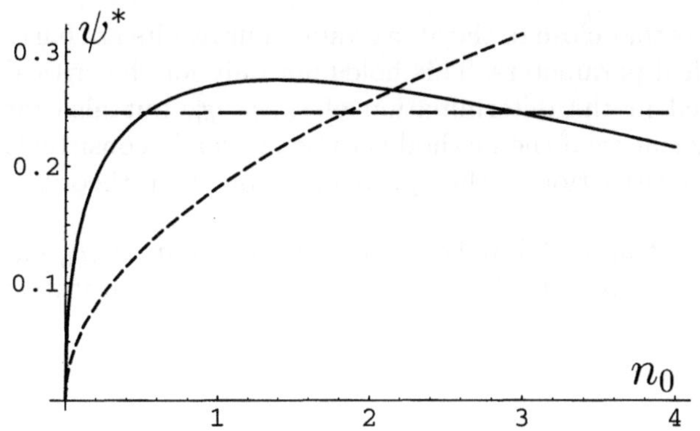

Fig. 13.1. Fixed point value of the interpenetration ratio as function of the parameter n_0. *Full line*: one-loop result; *short dashes*: zero-loop order; *long dashes*: $\psi^* = 0.245$

valid for $d = 3, f = 1$ in a monodisperse system. In Fig. 13.1 we have plotted ψ^* as function of n_0, together with the zero-loop approximation $\psi^* = 0.182 n_0^{1/2}$. It is found that the one-loop approximation shows a maximum $\psi^* \approx .275$ for $n_0 \approx 1.4$ and therefore becomes insensitive to n_0 in a neighborhood of that value. We should choose n_0 in that region since we know that the exact expression for ψ^* is independent of n_0. It thus is very pleasing to note that the experimental value of ψ^* is not far from the maximum found here. A careful analysis [Cot80] of a set of physical experiments yields

$$\psi^*_{exp} = 0.245 \pm 0.005 \quad ,$$

a value which has also been supported by Monte Carlo experiments: $\psi^*_{MC} = 0.247$ [Nic91]. Fitting to this value we may choose either $n_0 = 0.53$ or $n_0 = 2.95$. It turns out that these values essentially yield identical results for scaling functions like the temperature dependent swelling factor of a chain. Polydispersity effects, however, are strongly magnified for larger n_0, to the extent where the results look unreasonable. We therefore choose $n_0 = .53$.

A similar determination of c_0 or q_0 based on appropriate critical ratios is presented in Appendix A 13.2. Summarizing the results we choose the theoretical parameters as

$$\begin{aligned} b_u(1) &= 1 \\ n_0 &= 0.53 \\ c_0 &= 1.2 \\ q_0^2 &= 50 \quad . \end{aligned} \tag{13.22}$$

The resulting value of $\tilde{u}$ (Eq. (13.18)) is $\tilde{u} = 8.107$. It must be stressed that the parameter values (13.22) are adjusted to a one-loop evaluation of the scaling functions within the present formulation of renormalized perturbation theory in $d = 3$ and would have to be readjusted if for instance we would work to two-loop order.

We should further stress that close to the above values our results are fairly insensitive to the theoretical parameters. This holds not only for the critical ratios explicitly considered in the determination of c_0, n_0, q_0^2, but also for others. This is a necessary feature if the method is to be internally consistent. It strongly supports our construction of the approximate one-loop theory.

In the sequel we will put $b_u = 1$ but keep the parameters n_0, c_0, q_0^2 explicit in general expressions, inserting the numerical values in quantitative evaluations only.

13.2.3 Some warning

The analysis of the next chapters shows that our way to deal with the crossover works for most observables of interest in normal binary solutions, at least for small momenta. Such quantities are governed by a single scale showing nontrivial behavior, viz. the blob size ξ_E, reducing to R_g in the dilute limit. Adjusting ℓ_R we can deal with this single scale. There are more complicated problems governed by more than one nontrivial scale. In favorite cases they can be dealt with by including the additional scale into the choice of the uncritical manifold (Eq. (13.21)). (See the discussion of $A_2(n_1, n_2)$ in Sect. 16.4.) However, this does not always work. An example is the behavior of a long chain N_1 in a semidilute solution of shorter chains $N \ll N_1$. Here two scales $\xi_E(N)$ and $R_g(N_1)$ are important, and both show nontrivial power law behavior. In that case no good crossover form reproducing the expected behavior in all limits can be found with the RG presented here. We either have to resort to ad hoc exponentiating strict ε-expansion results or we have to adjust the renormalization method itself to the problem at hand, taking the renormalization factors to depend on the ratio of the two scales [AG78]. The latter clearly would be the better strategy, but it would involve an extreme amount of labor to establish an appropriate and accurate form of the RG mapping. We essentially would have to repeat the high order calculation leading to our mapping 13.1-13.3 in a more complicated situation. Restricting ourselves to the simplest and most important situation of normal binary solutions we do not consider such more complicated phenomena.

In binary solutions similar problems occur for correlation functions evaluated at large momenta. From field theory it is known that correlation functions in the regime $q\xi_E \gg 1$ can be expanded in a series of anomalous powers $(q\xi_E)^{-x_i}$. Only the leading term of that series is reproduced correctly by our method. It is found, however, that the subleading terms of the large-q expansion are not important quantitatively, which justifies our approach at least on the level of a one-loop calculation.

The general procedure to identify such dangerous cases consists in a careful analysis of perturbation theory to find whether singularities survive, which are not cured by our choice of the uncritical manifold.

13.3 The RG mapping in different regions of parameter space

13.3.1 General form ($d = 3$)

In general a solution will contain a distribution of chain lengths, characterized by the reduced chain length distribution $p(y)$ (Eq. (5.36)) and the number averaged chain length N. As pointed out in Sect. 11.6, $p(y)$ is invariant under the renormalization group. The other physical variables characterizing the solution can be expressed in terms of N. From Eqs. (11.38), (11.83) we recall the definitions

$$\begin{aligned} \tilde{z} &= (s_n N)^{1/2} = \tilde{v} N^{1/2} \\ \tilde{R}_0^2 &= s_\ell^2 s_n N = \tilde{\ell}^2 N \\ \tilde{s} &= c_p \tilde{R}_0^3 \quad . \end{aligned} \tag{13.23}$$

From Eqs. (13.12), (13.17) the relation to the renormalized variables $N_R, c_R, \bar{\mathbf{q}}$ takes the form

$$\begin{aligned} \tilde{z} &= f|1-f|^{-1/2\nu\omega} H_n^{-1/2}(f) N_R^{1/2} \\ \mathbf{q}^2 \tilde{R}_0^2 &= |1-f|^{(2-\frac{1}{\nu})/\omega} H^{-1}(f) N_R \bar{\mathbf{q}}^2 \\ \tilde{u}\tilde{s} &= |1-f|^{(2-\frac{1}{\nu})\frac{3}{2\omega}} H^{-3/2}(f) N_R^{1/2} c_R \quad . \end{aligned} \tag{13.24}$$

This mapping is to be combined with Eqs. (13.21), (13.22) fixing the uncritical manifold. A compact form results by introducing an intermediate variable w via

$$\frac{n_0}{N_R} = w^2 \left(1 - \frac{\bar{\mathbf{q}}^2}{q_0^2}\right) \quad . \tag{13.25}$$

Equation (13.21) takes the form

$$f \frac{c_R}{c_0} = \left(1 - \frac{\bar{\mathbf{q}}^2}{q_0^2}\right)(1 - w^2) \quad . \tag{13.26}$$

w controls the overlap. The dilute limit corresponds to $w = 1$, the semidilute limit is reached for $w = 0$. All the scaling functions can be expressed in terms of w, f, and momentum variables of type $\bar{\mathbf{q}}^2$. The RG mapping can be rewritten as

$$\frac{\tilde{u}}{c_0 n_0}\tilde{s}\tilde{z} = |1-f|^{\frac{1}{\omega}(3-\frac{2}{\nu})} H_u(f) H^{-2}(f) \frac{1-w^2}{w^2} \tag{13.27 i}$$

$$n_0^{-1/2}\tilde{z} = f|1-f|^{\frac{-1}{2\nu\omega}} H_n^{-1/2}(f) w^{-1} \left(1-\frac{\bar{\mathbf{q}}^2}{q_0^2}\right)^{-1/2} \tag{13.27 ii}$$

$$\frac{1}{n_0}\mathbf{q}^2\tilde{R}_0^2 = |1-f|^{(2-\frac{1}{\nu})\frac{1}{\omega}} H^{-1}(f) \frac{\bar{\mathbf{q}}^2}{1-\frac{\bar{\mathbf{q}}^2}{q_0^2}} w^{-2} \ , \tag{13.27 iii}$$

where

$$\frac{1}{\omega}\left(3-\frac{2}{\nu}\right) = -0.502; \quad \frac{1}{2\nu\omega} = 1.063; \quad \left(2-\frac{1}{\nu}\right)\frac{1}{\omega} = .374 \ . \tag{13.28}$$

These values result from Eq. (13.10). These equations map the physical manifold, parameterized by $\tilde{z}, \tilde{s}, \mathbf{q}^2 R_0^2$ onto the uncritical manifold, parameterized by $f, w, \bar{\mathbf{q}}^2$. Equations (13.27 i, iii) can be solved for w or $\bar{\mathbf{q}}^2$, respectively. Substituting the result into Eq. (13.27 ii) we find an implicit equation for f as function of $\tilde{z}, \tilde{s}, \mathbf{q}^2\tilde{R}_0^2$.

13.3.2 Special limits

We will often be concerned with experimental results taken in special limiting parameter regions. We therefore collect here the relevant forms of the mapping (13.27).

a) Dilute limit: $\tilde{s} \to 0$
Clearly this limit is reached for $w = 1$. The mapping reduces to

$$n_0^{-1/2}\tilde{z} = f|1-f|^{-\frac{1}{2\nu\omega}} H_n^{-1/2}(f) \left(1-\frac{\bar{\mathbf{q}}^2}{q_0^2}\right)^{-1/2} \tag{13.29 i}$$

$$\frac{1}{n_0}\tilde{R}_0^2 = |1-f|^{(2-\frac{1}{\nu})\frac{1}{\omega}} H^{-1}(f) \ell_R^2 \left(1-\frac{\bar{\mathbf{q}}^2}{q_0^2}\right)^{-1} \ . \tag{13.29 ii}$$

b) Semidilute limit: $\tilde{s} \to \infty$
This limit is reached for $w \to 0$. Using Eq. (13.27 ii) to eliminate w we find

$$\frac{\tilde{u}}{c_0}\frac{\tilde{s}}{\tilde{z}} = f^{-2}|1-f|^{\frac{1}{\omega}(3-\frac{1}{\nu})} H_u^{-1}(f) H^{-1}(f) \left(1-\frac{\bar{\mathbf{q}}^2}{q_0^2}\right) \tag{13.30 i}$$

$$\frac{\tilde{R}_0^2}{\tilde{z}^2} = f^{-2}|1-f|^{2/\omega} H_u^{-2}(f) \ell_R^2 \ . \tag{13.30 ii}$$

Since

$$\frac{\tilde{s}}{\tilde{z}} = s_\ell^3 c_p s_n N = s_\ell^3 s_n c \tag{13.31}$$

$$\frac{\tilde{R}_0^2}{\tilde{z}^2} = s_\ell^2 \ , \tag{13.32}$$

this shows that in the semidilute limit the renormalized parameters are functions of monomer concentration c, of $\bar{\mathbf{q}}^2$ and of temperature, but independent of N. Up to possible trivial factors of N this property will be transfered to all scaling functions. We furthermore should note that $w \to 0$ implies $N_R \to \infty$ (cf. Eq. (13.25)). Due to screening this does not lead to singularities in the loop expansion, as was demonstrated in Sect. 5.4.3 within the framework of unrenormalized perturbation theory.

c) Θ-region: $f \to 0$
The mapping reduces to

$$\frac{\tilde{u}}{c_0 n_0}\tilde{s}\tilde{z} = (1 + 0.718 f + O(f^2))\frac{1-w^2}{w^2} \tag{13.33 i}$$

$$n_0^{-1/2}\tilde{z} = f(1+O(f))w^{-1}\left(1 - \frac{\bar{\mathbf{q}}^2}{q_0^2}\right)^{-1/2} \tag{13.33 ii}$$

$$n_0^{-1}\tilde{R}_0^2 = (1 - 0.369 f + O(f^2))\frac{\ell_R^2}{1 - \frac{\bar{\mathbf{q}}^2}{q_0^2}} w^{-2} \ . \tag{13.33 iii}$$

d) Excluded volume limit: $f \to 1$
Equation (13.27 ii) yields

$$|1-f|^{1/\omega} = \tilde{z}^{-2\nu} H_n^{-\nu}(1) w^{-2\nu}\left(1 - \frac{\bar{\mathbf{q}}^2}{q_0^2}\right)^{-\nu} n_0^\nu \ . \tag{13.34}$$

Substituting this result into Eqs. (13.27 i, iii) we find

$$\frac{n_0}{c_0}\tilde{u}\hat{s} = (1-w^2)w^{2-6\nu}\left(1 - \frac{\bar{\mathbf{q}}^2}{q_0^2}\right)^{2-3\nu} \tag{13.35 i}$$

$$B^2 N^{2\nu} = w^{-4\nu}\ell_R^2\left(1 - \frac{\bar{\mathbf{q}}^2}{q_0^2}\right)^{-2\nu} \ , \tag{13.35 ii}$$

where

$$\hat{s} = (H_n^\nu(1) H_u(1) n_0^{-\nu})^3 \tilde{s}\tilde{z}^{6\nu-3} = c_p N^{3\nu} B^3 , \tag{13.36}$$

and

$$B = H_n^\nu(1) H_u(1) n_0^{-\nu} s_\ell\, s_n^\nu = .968 s_\ell s_n^\nu n_0^{-\nu} \ . \tag{13.37}$$

Using the result (15.14,15.15) for the radius of gyration of an isolated chain (see Sect. 15.1), we find that $\hat{s}$ is related to the geometrical overlap $s = R_g^3 c_p$ via

$$\hat{s} = 3.584 \ s \ \ (n_0 = .53) \ .$$

Inserting the geometrical factor $4\pi/3$ for the volume of a sphere we find that $\hat{s} \sim 1$ corresponds to a fictitious system filled with excluded volume coils:

$$\frac{4}{3}\,\pi\, s = \frac{4}{3}\,\pi\, c_p R_g^3 = 1.169\,\hat{s} \quad .$$

In other words, $\hat{s}$ fairly quantitatively measures the average number of chains interpenetrating a given coil, in the excluded volume limit.

13.3.3 Crossover diagram for $f \leq 1$

Specializing to $\mathbf{q} = 0$ we now discuss the crossover diagram of Daoud and Jannink (Sect. 9.2) in the light of the RG mapping. This diagram structures the weak coupling region $f < 1$. We first consider the equivalent diagram on the uncritical manifold. For $q = 0$ the renormalized variables $w = (n_0/N_R)^{1/2}$ and f in the weak coupling region range over the square

$$\begin{aligned} 0 &\leq w \leq 1 \\ 0 &\leq f \leq 1 \quad , \end{aligned}$$

as shown in Fig. 13.2. The edges of the square represent the asymptotic limits discussed above. The corners correspond to the single chain excluded volume limit (A), single chain Θ-limit (B), semidilute Θ-limit (C), or semidilute excluded volume limit (D), respectively. Shaded areas correspond to limiting regions, where simple power laws are to be expected.

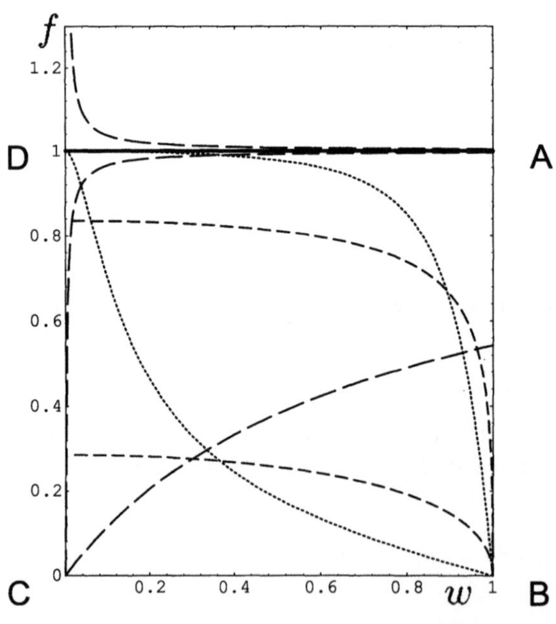

Fig. 13.2. Crossover diagram in the $w-f$ plane (uncritical surface). The *shaded parts* indicate regions dominated by asymptotic power law behavior. Weak coupling region ($f < 1$): *Long dashes*: lines of fixed $\tilde{z}$ ($= 300; 1$, from above). *Short dashes*: lines of fixed $\tilde{s}/\tilde{z}$ ($= 0.01; 1$, from above). *Dotted*: lines of fixed $\tilde{s} = (0.02; 0.8$, from above). In the strong coupling region a line of fixed $\tilde{z} = 300$ is shown

It is of interest to trace out paths in the $w - f$ diagram corresponding to typical experiments.

(Type 1) <u>$N, T > \Theta$ fixed; c varies</u>

This amounts to increasing the overlap for fixed $\tilde{z}$. If we are in the excluded volume limit $f = 1$, with increasing $\hat{s}$ we go along the upper edge from point A to point D. For any $f < 1$ the path at $w = 1$ (i.e. $\tilde{s} = 0$) starts out almost horizontal, but eventually bends down to reach $w = 0$ (i.e. $\tilde{s} = \infty$) in point C, representing the semidilute Θ-limit. The semidilute excluded volume limit (D) can be reached only for very small concentrations of very long chains. For $|1 - f| \ll 1$ we might see a double crossover A-region $\rightarrow$ D-region $\rightarrow$ C.

(Type 2) <u>$c > 0, T > \Theta$ fixed; N varies</u>

Both $\tilde{s}, \tilde{z}$ vary, but $\tilde{s}/\tilde{z}$ is constant. For $c = 0 = \tilde{s}$ (dilute limit) we would go along the right edge from B (small N) to A (large N). In the general case we go from B towards the semidilute limit, reaching $w = 0$ (i.e. $\tilde{s} = \infty$), at some $f = f(\tilde{s}/\tilde{z}) < 1$ (cf. Eq. (13.30 i)). A double crossover B $\rightarrow$ A-region $\rightarrow$ D-region is possible for very small concentrations.

(Type 3) <u>$c > 0, N$ fixed; T varies</u>

With the assumptions of the naive two-parameter theory this amounts to varying $\tilde{z}$, keeping $\tilde{s}$ fixed. We go from the dilute Θ-limit B to the semidilute excluded volume limit D. Depending on the value of $\tilde{s}$ we can pass close by the dilute excluded volume limit A or the semidilute Θ-limit C. For more general temperature dependence both $\tilde{z}$ and $\tilde{s}$ will vary. The curves will still start in B, but the exact shape will depend on details of the chemical system. They in fact might cross the line $f = 1$ and escape into the strong coupling region.

To construct the equivalent diagram in physical parameter space we use Eqs. (13.27 i, ii) ($q = 0$) to map all the curves into the $\tilde{s} - \tilde{z}$ plane. Figure 13.3 shows the result in a log-log plot. Clearly now the curves of type (1)–(3) simply become straight lines. For $w \gtrsim 0.9$ or $w \lesssim 0.1$ we are in the dilute or the semidilute region, respectively. For $0.1 \lesssim w \lesssim 0.9$ we expect to see crossover behavior. The line $c = c^*$ of the Daoud-Jannink diagram roughly corresponds to $w \approx 0.5$. It is easily checked that for fixed w and $f \rightarrow 1$ the RG mapping yields $\tilde{s} \sim \tilde{z}^{3-6\nu}$, so that c^* behaves as

$$c^* \sim \tilde{s}(f, w = .5) N^{-1/2} \sim t^{3-6\nu} N^{1-3\nu} \quad ,$$

where we used $\tilde{z} \sim t N^{1/2}$, valid for $t = (T - \Theta)/\Theta \ll 1$. This result coincides with Eq. (9.27). From Fig. 13.3, this asymptotic behavior of c^* is found to be

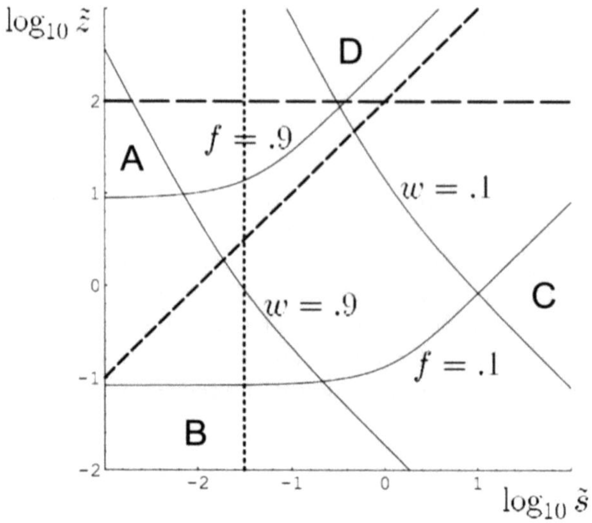

Fig. 13.3. Crossover diagram in the $\tilde{s} - \tilde{z}$ plane (logarithmic scale). *Shaded areas* roughly correspond to power law regions. A: dilute excluded volume region; B: dilute θ-region; C: semidilute θ-region; D: semidilute excluded volume region. Also drawn are lines $\tilde{z} = 100$ (*long dashes*); $\tilde{s}/\tilde{z} = 0.01$ (*short dashes*); $\tilde{s} = 0.03$ (*dotted*)

valid for $\tilde{z} \gtrsim 20$. In the opposite limit, w fixed, $f \to 0$, we find $\tilde{s} \sim \tilde{z}^{-1}$. Close to the Θ-point the semidilute region is thus shifted to

$$c > \frac{const}{t\,N} \quad .$$

In Fig. 13.3 we also plotted lines $f = 0.9, f = 0.1$, which bound the excluded volume or the Θ-region, respectively. In the dilute limit $w \to 1$ these lines reduce to $\tilde{z} = const$. In the semidilute limit they behave as $\tilde{z} \sim \tilde{s}$, corresponding to

$$\tilde{c} \sim t \quad .$$

Accidentally this relation agrees with Eq. (9.33), defining the tricritical region. This coincidence happens only in three dimensions. Indeed, our crossover diagrams ignore the three-body forces which induce an additional structure in the Θ-region. This structure involves the strength of the three-body interaction as a new scale and therefore is not fixed relative to the curves shown here.

13.3.4 The strong coupling region

We now turn to the domain $f > 1$. Clearly Eqs. (13.27) are still a valid representation of the RG mapping, and the scaling variables still take the form (13.23). However the strong coupling region is not connected to a Θ-region, and $s_n^{1/2} = \tilde{v}(T)$, $\tilde{\ell} = \tilde{\ell}(T)$ are unknown functions of T. In contrast, in the weak coupling regime $f < 1$ we know their functional form in the vicinity of the Θ-temperature. Curves like the line N, T fixed, c varying, included in Fig. 13.2 run off to some unknown limit.

A priori the strong coupling region of our theory is bounded by the condition that low order renormalized perturbation theory should be trusted. This will be found to restrict f to about $f \lesssim 1.5$. A much more stringent restriction, however, is imposed by the nonuniversal corrections. Let us assume for simplicity that we are concerned with small momentum behavior in a dilute solution ($c \to 0, q \to 0$). As we have discussed in Chap. 10, higher interactions neglected in our model in the excluded volume region introduce corrections of order $\sim 1/N$. In the weak coupling region we can suppress these corrections by increasing the chain length, still keeping $\tilde{z} < \infty$ or $f < 1$ fixed by appropriately decreasing T towards Θ. Thus the whole crossover $0 \leq f \leq 1$ is observable by analyzing data for long chains close to the Θ-temperature. The nonuniversal corrections strictly vanish in the limit $T \to \Theta$, $N \sim (T - \Theta)^{-2}$, $\tilde{v}(T) \sim T - \Theta$, where the naive two-parameter theory is recovered. Now in the strong coupling regime we cannot expect that $\tilde{v}(T)$ changes drastically with T. We thus cannot control $\tilde{z}$ independent of N. Since for $\tilde{v}(T)$ fixed, f with increasing N tends towards the fixed point $f = 1$, we have to admit that in the strong coupling region only some neighborhood of the fixed point yields universal results.

The RG mapping (13.27) allows for a more precise statement. Solving Eq. (11.27 ii) for $w = 1, \bar{\mathbf{q}} = 0$ in the limit $f \to 1$ we find

$$f - 1 = const \left(\tilde{v}(T) N^{1/2}\right)^{-2\nu\omega} + O\left(N^{-2\nu\omega}\right) \quad . \tag{13.38}$$

Since $\nu\omega = .472$ we conclude that the leading corrections to fixed point behavior in the strong coupling regime as described by our theory are proportional to $N^{-.472}$ and thus dominate the nonuniversal corrections $\sim N^{-1}$ not taken into account. But already the next term in an expansion about $f = 1$: $(1 - f)^2 \sim N^{-.944}$ will in general be mixed with the nonuniversal corrections. In the terminology of general critical phenomena (Sect. 10.2) we can trust only the 'corrections to scaling'. Analysis of typical scaling functions suggests that this restricts the applicability of the theory in the strong coupling regime to $f - 1 \lesssim .2$, a restriction valid also for semidilute systems or larger momenta.

Clearly this discussion does not exclude the possibility of finding special chemical systems where the universal strong coupling regime happens to be larger, but we have no systematic handle to enforce that situation. Previous work therefore concentrated on the weak coupling region, the extension to strong coupling being considered only recently. The experimental phenomenology seems to be consistent with this discussion. Experimentally either the systems are in a 'Θ-solvent', so that the Θ-temperature can be reached and the scaling functions of the weak coupling regime are observed, or the system is in a 'good solvent', always close to the excluded volume limit.

A 13. Appendix

A 13.1 Elimination of $b_u(1)$

We can scale away $b_u(1)$, exploiting the fact that the renormalized theory is a two-parameter scheme. Recall that we may start from the naive continuous chain limit, where all observables are expressed in terms of the variables $z, \mathbf{q}^2 R_0^2, s_0$. Renormalizing these variables according to Eqs. (12.16) we find the variables of renormalized perturbation theory:

$$\begin{aligned} z = \beta_e N^{1/2} &\to f u^* b_u N_R^{1/2} \\ \mathbf{q}^2 R_0^2 = \mathbf{q}^2 \ell^2 N &\to \mathbf{q}^2 \ell_R^2 N_R \\ s_0 = c_p R_0^3 &\to c_p \ell_R^3 N_R^{3/2} \ . \end{aligned} \tag{A 13.1}$$

Note that the Z-factors only introduce some additional dependence on f. We now define

$$\begin{aligned} N_R' &= b_u^2 N_R \\ \ell_R' &= b_u^{-1} \ell_R \end{aligned} \tag{A 13.2}$$

to write the renormalized variables as $f u^* N_R'^{1/2}, \mathbf{q}^2 \ell_R'^2 N_R', c_p \ell_R'^3 N_R'^{3/2}$, where the explicit factor b_u has dropped out. We furthermore rewrite c_R (Eq. (13.17)) as

$$\begin{aligned} c_R &= (4\pi)^{3/2} \frac{u^*}{2} b_u c_p \ell_R^3 N_R \\ &= b_u^2 c_R' \end{aligned} \tag{A 13.3}$$

$$c_R' = (4\pi)^{3/2} \frac{u^*}{2} c_p \ell_R'^3 N_R' \ . \tag{A 13.4}$$

In terms of these variables our condition fixing the uncritical manifold takes the form

$$1 = \frac{\mathbf{q}^2 \ell_R'^2}{q_0^2 b_u^{-2}} + \frac{b_u^2 n_0}{N_R'} + f \frac{c_R'}{c_0 b_u^{-2}} \ , \tag{A 13.5}$$

showing that the remaining dependence on b_u can be absorbed into q_0^2, n_0, c_0. We thus can take $b_u = 1$ from the outset.

A 13.2 Parameter dependence of renormalized perturbation theory

Here we analyze to which extent the choice of the parameters n_0, c_0, q_0 in Eq. (13.21) fixing the uncritical manifold

$$1 = \frac{\bar{\mathbf{q}}^2}{q_0^2} + \frac{n_0}{N_R} + f\,\frac{c_R}{c_0}$$

influences our results, and we determine the values to be used in quantitative evaluations. In the main text we have determined a value $n_0 = 0.53$ from an analysis of the interpenetration ratio. We still have to determine c_0 and q_0^2.

A 13.2.1 Choice of q_0. The parameter q_0 determines the momentum region where the observables cross over from being governed by scales $r \sim R_g$ or $r \sim R_B^*$ to being dominated by $r \sim 1/q \ll R_g, R_B^*$. The radius of gyration R_g is defined by the small momentum behavior of the (normalized) density correlations of an isolated chain

$$J_a^{-1}(\mathbf{q}) = 1 + \frac{1}{3}\mathbf{q}^2 R_g^2 + O(q^4) \ . \tag{A 13.6}$$

To determine q_0 we consider the large momentum behavior. In the excluded volume limit the scaling theory predicts (cf. Eq. (9.20))

$$J_a^{-1}(\mathbf{q}) \underset{q\to\infty}{\longrightarrow} \mathcal{R}_{\infty/2}\left(R_g^2\mathbf{q}^2\right)^{\frac{1}{2\nu}} \ . \tag{A 13.7}$$

This result is also born out by the renormalized perturbation theory as will be shown in Chap. 15. $\mathcal{R}_{\infty/2}$ is a universal ratio connecting the limits $q \to 0$ or $q \to \infty$. This is clearly exhibited by using Eqs. (A 13.7), (A 13.6) to write

$$\mathcal{R}_{\infty/2} = \left[3\frac{\partial}{\partial q^2}\Big|_0 J_a^{-1}(\mathbf{q})\right]^{-\frac{1}{2\nu}} \lim_{q\to\infty} q^{-1/\nu} J_a^{-1}(\mathbf{q}) \ . \tag{A 13.8}$$

By construction $\mathcal{R}_{\infty/2}$ is particularly sensitive to the choice of q_0. A first-order calculation, evaluated in three dimensions and using $n_0 = .53$ yields

$$\mathcal{R}_{\infty/2} = q_0^{2-1/\nu}\left(0.348 + \frac{1.108}{q_0}\right) \ . \tag{A 13.9}$$

(With a little calculation this result can be extracted from Appendix A 15.3.) In Fig. 13.4 we plot this expression as function of q_0^2. As q_0^2 increases we find that $\mathcal{R}_{\infty/2}$ approaches a plateau region, where it is fairly insensitive to q_0^2. We choose the value $q_0^2 = 50$ right in the plateau region. It yields $\mathcal{R}_{\infty/2} = .91$, in good agreement with the value $\mathcal{R}_{\infty/2} = .926$ suggested on the basis of the ε-expansion (see Sect. 15.2.3).

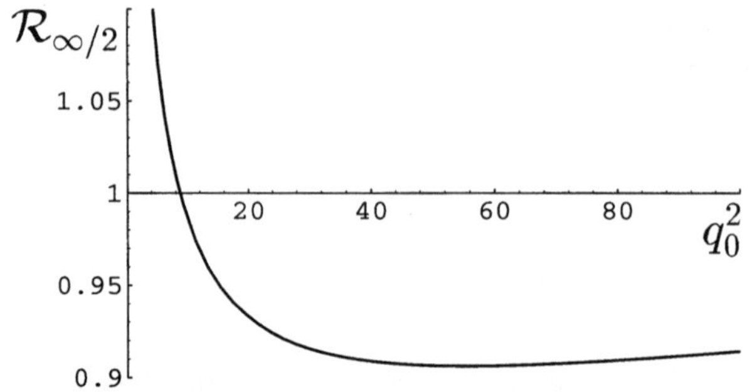

Fig. 13.4. The ratio $\mathcal{R}_{\infty/2}(f = 1)$ as function of the parameter q_0^2, evaluated in one-loop approximation

A 13.2.2 Choice of c_0. The parameter c_0 determines the crossover among the dilute or semidilute regimes. It is thus related to the overlap concentration c^* introduced in the blob model. We may determine it from an analysis of the osmotic pressure in the excluded volume limit. The scaling law reads

$$\frac{\hat{\Pi}}{k_B T} = c_p \left(1 + \mathcal{P}(\hat{s})\right) \quad , \tag{A 13.10}$$

where $\hat{s}$ (Eq. (13.36)) is the overlap variable in the excluded volume limit. For $\hat{s} \to 0$ the virial expansion holds:

$$\mathcal{P}(\hat{s}) = \frac{A_2}{2}\, \hat{s} + O(\hat{s}^2) \quad . \tag{A 13.11}$$

In the semidilute limit renormalized perturbation theory verifies the law (cf. Eq. (9.14))

$$\mathcal{P}(\hat{s}) \underset{\hat{s}\to\infty}{\longrightarrow} \hat{s}^{\frac{1}{\nu d - 1}} \quad . \tag{A 13.12}$$

We can construct a universal ratio combining the limits (A 13.11), (A 13.12), this ratio being sensitive to the choice of c_0:

$$\mathcal{A}_{\infty/2} = \left[\lim_{s\to\infty} s^{-1/(\nu d-1)} \mathcal{P}(s)\right]^{\nu d - 1} \left[\lim_{s\to 0} s^{-1} \mathcal{P}(s)\right]^{-1} \tag{A 13.13}$$

For $d = 3$ the first-order calculation of Chap. 17 yields

$$\mathcal{A}_{\infty/2} = 1.541 c_0^{3\nu-2} \left(1 - \frac{.352}{c_0^{1/2}}\right)^{3\nu-1} \quad . \tag{A 13.14}$$

This result is plotted in Fig. 13.5. After a rapid initial increase the variation becomes quite smooth, the function taking a maximum around $c_0 = 1, \mathcal{A}_{\infty/2} \approx 1.1$. This value of $\mathcal{A}_{\infty/2}$ is consistent with experiment. In the sequel we choose $c_0 = 1.2$, close to the maximum.

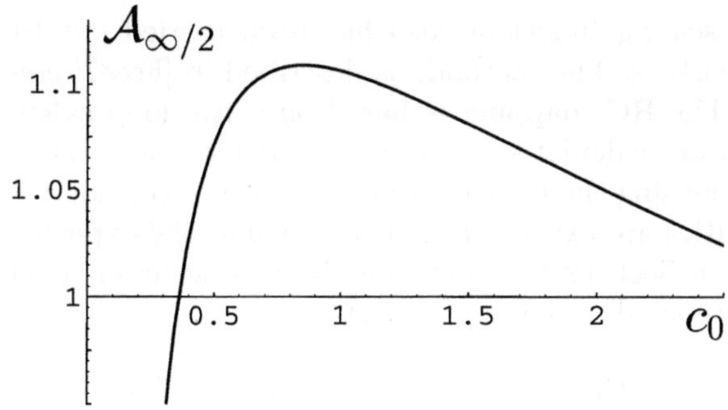

Fig. 13.5. The ratio $\mathcal{A}_{\infty/2}(f=1)$ as function of the parameter c_0, evaluated in one-loop approximation

Notes

Almost all the work on scaling functions in polymer physics is on the level of a first-order (one-loop) calculation. Only for R_e^2 or R_g^2 has the ε-expansion been pushed to second order [BM85], and higher order calculations within unrenormalized two-parameter theory have been published for R_e^2, R_g^2, or A_2 [MN87, Nic91]. The exponents, of course, have been calculated to high orders. Such high order expansions have to be resumed, using additional information on the asymptotic structure of perturbation theory. The relevant technique of 'Borel summation' is explained in [ZJ89], where references to higher order calculations are also given. This work essentially concentrates on a precise determination of the exponents.

It is the distinguishing feature of the work of Dohm et al. [SD89] that it applies these methods to a determination of the flow equations, not just the exponents. It thus provides us with a most precise form of the RG flow. It should be noted, however, that the resulting form of the flow equations is very close to an RG mapping suggested on the basis of two-loop order results [Sch84]. Quite generally the results for exponents or the RG mapping show a remarkable stability against inclusion of additional higher order terms.

For low order calculations of the scaling functions a variety of implementations of the RG have been used. The present formulation has grown out of the work [Sch84]. The basic philosophy is the same, but in this earlier work the renormalization scheme was based on field theoretic 'renormalization conditions'. This amounts to using a non-minimally subtracted theory, where the Z-factors are determined by imposing specific values to certain renormalized field theoretic vertex functions. The renormalized coupling, for instance, is defined as the value of $\Gamma_{F,R}^{(0,2)}(\mathbf{q}_1, \mathbf{q}_2, \mathbf{q}_3, \mathbf{q}_4)$ at some special momenta of order $q \sim \ell_R^{-1}$, this implicitly fixing Z_u.

Much work on polymer scaling functions also has been carried out by Freed, Ohno, Ohta, and coworkers. The method, as described in [Fre87] employs minimal subtraction. The RG mapping is based on a two-loop calculation, but scaling functions are calculated to order of one-loop in strict ε-expansion. The problem of choosing the uncritical manifold is not considered. Instead logarithmic singularities are exponentiated to arrive at the expected power laws. As pointed out in Sect. 12.4.1, such singularities occur in strict ε-expansion outside the dilute small momentum limit.

A parameter representation of the temperature crossover similar to that presented here first has been worked out in [Eld81]. An early treatment of crossover in momentum, similar to that discussed here, may be found in [Nel76].

In all this earlier work the dependence of low order renormalized theory on parameters like n_0, c_0, q_0 has not been considered. Rather some 'reasonable' choice has been made a priori.

Des Cloizeaux advocates a non-minimally subtracted renormalization scheme, in which renormalization conditions are set directly in terms of polymer observables [Clo81, CJ90]. It goes by the name of 'direct renormalization'. The renormalized coupling constant, for instance, in essence is identified with the interpenetration ratio, and the renormalized length scale is identified with the end-to-end distance R_e. Section 12.4.1 gives a typical result (12.34). In some sense this is the most physical approach. It cannot however deal with crossover phenomena, except by using strict ε-expansion. According to our general considerations the choice $\ell_R \sim R_e$ (or $\ell_R \sim R_g$, equivalently) is not adequate for semidilute systems or higher momenta. In these regions the ε-expansion is exponentiated ad hoc, the full crossover functions not being considered explicitly. A crossover calculation in that scheme, except for the difference in basic variables would be quite similar to the method of Freed and coworkers.

14. The Renormalized Tree Approximation: a Simple Illustration of the Theory

In this chapter we illustrate the power of the theory with a discussion of the tree approximation, which yields most simple expressions for the scaling functions. We thus can study the effect of renormalization and RG flow without struggling with the technical complexities of higher loop orders. We will see the results of scaling theory and blob model emerging naturally from our form of the renormalized theory, and we will encounter the characteristic features of crossover scaling functions. Clearly we should not expect that the tree approximation is very accurate quantitatively – though it performs not too bad. (In [BL86] a set of critical ratios was evaluated in renormalized tree approximation, with reasonable success.) However, almost all the qualitative aspects of the theory can be studied here.

An essential aspect, we want to demonstrate, is the fundamental importance of a (qualitatively) correct choice of the uncritical manifold. It is by this choice that our theory generates smooth crossover functions, which interpolate among asymptotic power law behavior as expected from scaling theory. It is most important that the correct behavior is found even in lowest order approximation. Otherwise higher order corrections, trying to reconstruct the correct asymptotics, must blow up. Then a one loop calculation cannot be reliable quantitatively.

In its unrenormalized form the tree approximation has been worked out in Sect. 5.4.1. We now renormalize these expressions according to the steps described in Sect. 12.3.1. Since to lowest order the Z-factors obey

$$Z_n = 1 + O(u), \quad Z_u = \frac{1}{2}(1 + O(u)) \quad , \tag{14.1}$$

($b_u = 1$), renormalization here just amounts to replacing the unrenormalized variables by their renormalized counterparts. The results are then to be evaluated with the full RG mapping discussed in the previous chapter, a mapping which is based on a high order calculation. Contrary to what one might believe at first, there is no inconsistency in combining such a sophisticated mapping with low order results for the scaling functions. In our theory the construction of the RG, which maps the system on the uncritical manifold, is a step well separated from the calculation of the scaling functions within the uncritical manifold. The two steps interfere only in the choice of the theoretical parameters n_0, c_0, q_0. This choice, however, does not influence the form of the RG.

In the following we will discuss the most important experimental quantities: the density autocorrelations (Sect. 14.1), the osmotic pressure (Sect. 14.2), or the density cumulant (Sect. 14.3). Since this chapter is meant as an illustration only, we ignore polydispersity, assuming a monodisperse system. We furthermore specialize to $d = 3$; and we do not readjust the theoretical parameters n_0, c_0, q_0 to a zero loop calculation. We consider only the weak coupling branch, since we want to compare to scaling predictions, which are not available for the strong coupling situation outside the excluded volume limit.

14.1 Density autocorrelations

14.1.1 General expression

The correlation function $I_{d,a}(\mathbf{q})$ of the segment density within a given chain in the solution has been defined in Eqs. (5.10), (5.17). The corresponding scaling function $J_a(\mathbf{q})$ has been introduced in Eq. (9.4):

$$J_a(\mathbf{q}) = \frac{1}{cN} I_{d,a}(\mathbf{q}) \quad . \tag{14.2}$$

Comparing to Eq. (5.54) we find the tree approximation in unrenormalized form:

$$J_a(\mathbf{q}) = D(N\mathbf{q}^2\ell^2) \quad . \tag{14.3}$$

(Recall that we consider monodisperse systems.) Using relations (12.16), (14.1) we find

$$N\mathbf{q}^2\ell^2 = \bar{\mathbf{q}}^2 N_R(1 + O(u)) \quad . \tag{14.4}$$

Thus

$$J_a(\mathbf{q}) = D(\bar{\mathbf{q}}^2 N_R) \tag{14.5}$$

is the renormalized tree approximation.

14.1.2 Radius of gyration

The radius of gyration is found as

$$\begin{aligned} R_g^2 &= -\frac{1}{2}\Delta_q\Big|_0 \ln I_{d,a}(\mathbf{q}) \\ &= \ell_R^2 N_R \quad , \end{aligned} \tag{14.6}$$

(recall $d = 3$). Though this expression is most simple, the RG mapping induces a lot of structure. For $\mathbf{q} = 0$ the general form (13.27) reduces to

$$\frac{\tilde{u}}{c_0 n_0}\tilde{s}\tilde{z} = \left|1-f\right|^{\frac{1}{\omega}\left(3-\frac{2}{\nu}\right)} H_u(f) H^{-2}(f) \frac{1-w^2}{w^2} \tag{14.7 i}$$

$$n_0^{-1/2}\tilde{z} = f\left|1-f\right|^{-\frac{1}{2\nu\omega}} H_n^{-1/2}(f) w^{-1} \tag{14.7 ii}$$

$$\frac{1}{n_0}\tilde{R}_0^2 = \left|1-f\right|^{\left(2-\frac{1}{\nu}\right)\frac{1}{\omega}} H^{-1}(f)\ell_R^2 w^{-2} \quad , \tag{14.7 iii}$$

where $w^{-2} = N_R/n_0$ (cf. Eq. (13.25)). Thus Eq. (14.6) equivalently can be written as

$$\frac{R_g^2}{\tilde{R}_0^2} = \left|1-f\right|^{\left(\frac{1}{\nu}-2\right)\frac{1}{\omega}} H(f) \quad , \tag{14.8}$$

showing that in the Θ-limit

$$R_g^2 \underset{f\to 0}{\longrightarrow} \tilde{R}_0^2 = \tilde{\ell}^2 N \tag{14.9}$$

is independent of concentration. (Recall that we ignore three-body interactions.)

We now first consider the dilute limit $w = 1$, i.e.

$$N_R = n_0 \ . \tag{14.10}$$

Equation (14.7 ii) yields f as function of $\tilde{z}$. For $f \to 1$, in particular, we find (cf. Eq. (13.29))

$$\left|1-f\right|^{\frac{1}{\omega}} \sim \tilde{z}^{-2\nu} \quad , \tag{14.11}$$

so that

$$R_g^2 \sim \tilde{R}_0^2 \tilde{z}^{4\nu-2} \quad , \tag{14.12}$$

which is the power law expected in the dilute excluded volume limit (cf. the scaling result 9.25). For general $f, 0 \le f \le 1$, i.e. general $\tilde{z}$ in the weak coupling region, Eqs. (14.8), ((14.7 ii); $w = 1$) describe the crossover among Θ- and excluded volume-limits, in close analogy to the argument given in Sect. 8.3 for the swelling of R_e^2. We must note only one important difference. In the present calculation without any problem we can set $N_R = n_0 = O(1)$, since all the corrections of order $1/N$ are rigorously suppressed in the renormalized theory.

Let us now turn to the concentration crossover in the excluded volume limit. Equations (13.35) yield

$$\frac{n_0}{c_0}\tilde{u}\hat{s} = (1-w^2)w^{2-6\nu}$$

$$\ell_R^2 = w^{4\nu} B^2 N^{2\nu} \quad . \tag{14.13}$$

Using the form (14.6) of R_g^2 we thus find

$$\begin{aligned} R_g^2 &= B^2 N^{2\nu} w^{4\nu-2} n_0 \\ &= n_0^{2\nu} B^2 \left(\frac{N}{N_R}\right)^{2\nu} N_R \ . \end{aligned} \tag{14.14}$$

If compared to Eq. (9.17) the second equation shows that N/N_R is to be identified with the number of segments n^* per concentration blob, whereas $N_R = N/n^*$ is the number of blobs per chain. The first equation shows that we find a smooth crossover from the dilute limit $w \to 1$ to the semidilute limit $w \to 0$. In the latter limit Eqs. (14.13), (14.14) yield the expected power law

$$\begin{aligned} R_g^2 &\sim B^2 N^{2\nu} \hat{s}^{\frac{2\nu-1}{1-3\nu}} \\ &\sim B^2 N \left(B^3 c\right)^{\frac{2\nu-1}{1-3\nu}} \ , \end{aligned} \tag{14.15}$$

where the last expression results from Eq. (13.36), $\hat{s} = B^3 c N^{3\nu-1}$.

Using Eq. (14.8) together with the full form (14.7) of the mapping we can easily map out the double crossover from the dilute to the semidilute excluded volume regimes and further to the semidilute Θ-regime. Figure 14.1a shows an example, where we plotted $R_g^2/\tilde{R}_0^2$ as function of the concentration as measured by the overlap $\tilde{s} = c_p \tilde{R}_0^3$, at fixed chain length and temperature, i.e. fixed $\tilde{z}$, $\tilde{R}_0^2$. The chain is taken to be very long, so that for isolated chains $c \to 0$ we are close to the excluded volume limit. Working within the weak coupling regime we run along a curve of type 1 in the crossover diagram of Fig. 13.2, which for the present value $\tilde{z} = 1000$ closely passes by the semidilute excluded volume limit. In the doubly logarithmic plot (14.1) we see a horizontal line corresponding to the dilute excluded volume limit Eq. (14.12) (point A in Fig. 13.2), which bends over to a straight line governed by the law (14.15) (semidilute excluded volume limit, point D) and finally crosses over to a horizontal line ($R_g^2 = \tilde{R}_0^2$) representing the semidilute Θ-limit Eq. (14.9) (point C).

Such doubly logarithmic plots are a common tool in visualizing crossover functions. They however give a strongly distorted picture of the physical behavior. Consider for instance the crossover from the dilute to the semidilute excluded volume behavior, which for our example takes place in the range $10^{-4} \lesssim \tilde{s} \lesssim 0.1$. Figure 14.1b shows this crossover in a direct plot. Clearly the 'crossover' just consists in cutting off the singularity exhibited by the semidilute power law (14.15) for $\tilde{s} \to 0$. There is no 'dilute regime' in the sense $R_g^2 \approx const$, but a steep decrease of R_g^2 starts directly at $\tilde{s} = 0$.

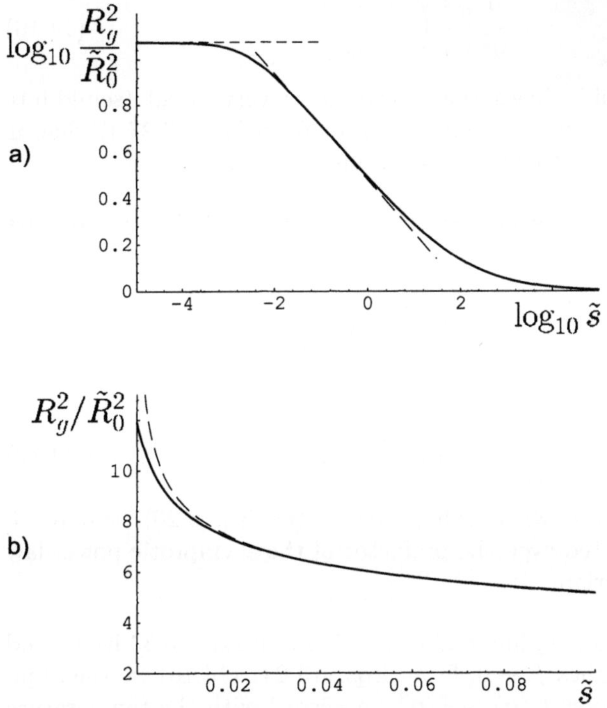

Fig. 14.1. (**a**) $\log_{10} R_g^2/\tilde{R}_0^2$ as function of $\log_{10} \tilde{s}$ for $\tilde{z} = 10^3$. Also included are the dilute limit (*short dashes*) and the semidilute excluded volume limit (*long dashes*) (Eq. (14.15)). (**b**) $R_g^2/\tilde{R}_0^2$ as function of $\tilde{s}$, showing the gradual crossover from dilute to semidilute behavior. The semidilute power law (Eq. (14.15)) is represented by the *dashed line*

14.1.3 Momentum dependence

The simple expression (14.5) could lead to the erroneous conclusion that the momentum dependence of the autocorrelation function in tree approximation follows the Debye function, reproducing the result for a noninteracting chain. This conclusion is false, since the uncritical surface and thus both N_R and $\bar{\mathbf{q}}^2$ nontrivially depend on $\mathbf{q}^2$. The effects show up in the region of large momenta: $\mathbf{q}^2\tilde{R}_0^2 \gg 1$.

Equation (13.24) shows that $\mathbf{q}^2\tilde{R}_0^2 \gg 1$ for f bounded necessarily implies $\bar{\mathbf{q}}^2 N_R \gg 1$, so that we can replace the Debye function by the asymptotic expression (5.51 ii):

$$J_a(\mathbf{q}) \to \frac{2}{\bar{\mathbf{q}}^2 N_R}$$

$$= \frac{2}{\bar{\mathbf{q}}^2} \frac{w^2}{n_0} \left(1 - \frac{\bar{\mathbf{q}}^2}{q_0^2}\right) \quad . \tag{14.16}$$

Furthermore Eq. (13.27 iii) shows that either w is very small (semidilute region) or $\bar{\mathbf{q}}^2 \to q_0^2$. Keeping $\tilde{s}, \tilde{z}$ fixed we find from Eq. (13.27 i) that w cannot vanish. Thus the limit $\bar{\mathbf{q}}^2 \tilde{R}_0^2 \to \infty$ implies $\bar{\mathbf{q}}^2 \to q_0^2$.

We now first consider the excluded volume limit $f = 1$. We can rewrite Eq. (13.35 ii) to find

$$w^2 \left(1 - \frac{\bar{\mathbf{q}}^2}{q_0^2}\right) \underset{\bar{\mathbf{q}}^2 \to q_0^2}{\longrightarrow} \left(\frac{\mathbf{q}^2}{q_0^2} B^2 N^{2\nu}\right)^{-\frac{1}{2\nu}} \quad .$$

Thus

$$J_a(\mathbf{q}) \to \frac{2}{q_0^2 N_0} \left(\frac{\mathbf{q}^2}{q_0^2} B^2\right)^{-\frac{1}{2\nu}} N^{-1} \quad . \tag{14.17}$$

This result is in full accord with scaling theory (cf. Eq. (9.20), and recall $I_{d,a} = c_p N^2 J_a$.) As expected even the prefactor of the asymptotic power law is independent of the overlap.

Now consider the weak coupling region $f < 1$. From Eq. (13.33 ii) we find that $\bar{\mathbf{q}}^2 \to q_0^2, \tilde{z}$ fixed, implies $f \to 0$. From Eqs. (13.24), (14.16) we then find the trivial limiting behavior $J_a(\mathbf{q}) \sim 1/\mathbf{q}^2$, in accord with the temperature blob model. For a system close to the excluded volume limit for large $\mathbf{q}$ we first may find some region described by the law (14.17), the trivial behavior developing only for extreme values of the momentum (see Fig. 14.2).

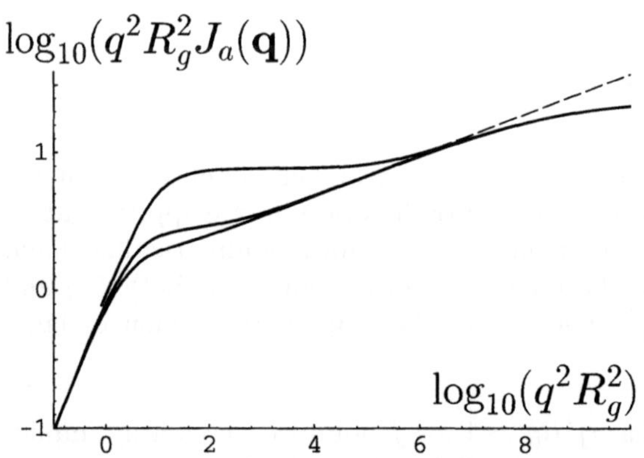

Fig. 14.2. $\log_{10}(q^2 R_g^2 J_a(\mathbf{q}))$ as function of $\log_{10}(q^2 R_g^2)$ for $\tilde{z} = 10^3$. R_g here is the radius of an isolated chain. The curves are for values $\tilde{s} = 10^{-4}, 10^{-2}, 1$ (from below). Note that for that value of $\tilde{z}$ the excluded volume overlap $\hat{s}$ is found as $\hat{s} \approx 100\tilde{s}$. The *broken line* gives the large momentum behavior in the excluded volume limit

In the dilute region $w \lesssim 1$ the autocorrelation function directly crosses over from Debye-type behavior on scales $\mathbf{q}^2 R_g^2 \lesssim 1$ to the large momentum behavior as discussed above. For semidilute systems $w \ll 1$, there develops an intermediate regime, governed by the asymptotics of the Debye function,

$J_a(\mathbf{q}) \sim D(\mathbf{q}^2 R_g^2) \sim 1/\mathbf{q}^2$. This regime is bounded by $\mathbf{q}^2\xi_E^2 \sim 1$, where ξ_E is the screening length. We thus find a rich behavior of $J_a(\mathbf{q})$. The primary scale is ξ_E. For $\mathbf{q}^2\xi_E^2 \lesssim 1$, i.e. length scales $r \gtrsim \xi_E$ we find Debye-type behavior, crossing over from the small momentum region $\mathbf{q}^2 R_g^2 \lesssim 1$ to an intermediate power law $J_a \sim 1/\mathbf{q}^2$, valid for $R_g^{-2} \ll \mathbf{q}^2 \lesssim \xi_E^{-2}$. For $\mathbf{q}^2\xi_E^2 \gg 1$, i.e. $r \ll \xi_E$ we look inside the concentration blob. We then may find a range of excluded volume behavior, followed by trivial behavior for very large $\mathbf{q}$. All these predictions of scaling theory here follow from our RG mapping applied to the trivial renormalized tree approximation.

Numerically the crossover function for $J_a(\mathbf{q})$ is to be calculated from Eq. (14.5) with the help of the full RG mapping (13.27). Figure 14.2 shows a doubly logarithmic plot of $\mathbf{q}^2 R_g^2 J_a(\mathbf{q})$ vs $\mathbf{q}^2 R_g^2$, where the isolated chain radius is used to set the scale of $\mathbf{q}$. A fixed large value $\tilde{z} = 1000$ is used, but the overlap varies from a very dilute system ($\tilde{s} = 10^{-4}$) to a semidilute system ($\tilde{s} = 1$). All curves coincide for large $\mathbf{q}$, this region testing the correlations inside a blob. At the highest values $\mathbf{q}^2 R_g^2$ shown, we see the beginning of the crossover towards trivial Θ-point behavior $J_a \sim \mathbf{q}^{-2}$. The corresponding plateau $\mathbf{q}^2 J_a = const$, expected in that plot, is reached only for values $\mathbf{q}^2 R_g^2 \sim 10^{14}$, far beyond the range shown here. In the dilute limit we see that the expected power law $J_a \sim q^{-1/\nu}$ holds to a good approximation over several decades $50 \lesssim q^2 R_g^2 \lesssim 10^6$. With increasing overlap from the small momentum side of that range there develops a plateau corresponding to $J_a \sim q^{-2}$. This is the intermediate region $1 \ll \mathbf{q}^2 R_g^2 \ll R_g^2/\xi_E^2$.

14.2 Osmotic pressure

In terms of the thermodynamic scaling function $\mathcal{P}(\tilde{s}, f)$ the osmotic pressure is written as (cf. Eq. (11.85))

$$\frac{\hat{\Pi}}{k_B T c_p} = 1 + \mathcal{P}(\tilde{s}, f) \quad . \tag{14.18}$$

The unrenormalized tree approximation (5.58) yields

$$\begin{aligned} \mathcal{P}(\tilde{s}, f) &= \frac{u_0}{2} c_p N^2 \\ &= \frac{1}{2}(4\pi)^{d/2} \beta_e \ell^d c_p N^2 \quad . \end{aligned} \tag{14.19}$$

Renormalizing with the help of Eqs. (12.16) we find

$$u_0 c_p N^2 = \tilde{u} f \ell_R^d c_p N_R^2 = f c_R N_R \tag{14.20}$$

(cf. the definitions (13.18), (3.17) of $\tilde{u}, c_R$). Thus

$$\mathcal{P}(\tilde{s}, f) = \frac{f}{2} c_R N_R \quad . \tag{14.21}$$

Again the renormalized or unrenormalized expressions show identical structure. Equations (13.25), (13.26) ($\bar{\mathbf{q}} \equiv 0$) yield a simple expression in terms of the variable w:

$$\mathcal{P}(\tilde{s}, f) = \frac{c_0 n_0}{2} \frac{1 - w^2}{w^2} \quad . \tag{14.22}$$

By virtue of Eq. (14.7 i) this can also be written as

$$\mathcal{P}(\tilde{s}, f) = \frac{\tilde{u}}{2} \tilde{s} \tilde{z} \Big| 1 - f \Big|^{\frac{1}{\omega}\left(\frac{2}{\nu} - 3\right)} H^2(f) H_u^{-1}(f) \quad . \tag{14.23}$$

In the Θ-limit $f \to 0$ we find, using the expressions (13.23) for $\tilde{z}, \tilde{s}$

$$\mathcal{P}(\tilde{s}, f) \to \frac{\tilde{u}}{2} \tilde{s} \tilde{z} = \frac{\tilde{u}}{2} \, \tilde{v} \tilde{\ell}^3 c_p N^2 \quad , \tag{14.24}$$

which up to some change in notation is the unrenormalized result. In the dilute limit we introduce the second virial coefficient $\tilde{A}_2$ as

$$\mathcal{P}(\tilde{s}, f) = \frac{1}{2} \tilde{s} \tilde{A}_2 + O(\tilde{s}^2) \quad . \tag{14.25}$$

Equation (14.23) yields

$$\tilde{A}_2 = \tilde{u} \tilde{z} \Big| 1 - f \Big|^{\frac{1}{\omega}\left(\frac{2}{\nu} - 3\right)} H^2(f) H_u^{-1}(f) \quad , \tag{14.26}$$

where $f = f(\tilde{z})$ is to be taken from Eq. (14.7 ii), $w = 1$. In the Θ-limit we find

$$\tilde{A}_2 = \tilde{u} \tilde{z} \quad . \tag{14.27}$$

In the excluded volume limit $f \to 1$ we can use Eq. (14.11) to find

$$\tilde{A}_2 \sim \tilde{z}^{6\nu - 3} \sim N^{3(\nu - 1/2)} \quad , \tag{14.28}$$

which in view of $\tilde{s} \sim c_p N^{3/2}$ yields the expected power law for A_2:

$$A_2 \sim N^{3\nu} \quad .$$

In the semidilute limit Eqs. (13.30)–(13.32) show that f is a function of c and T, only. Since the same holds for $c_p \tilde{s} \tilde{z} \sim c_p^2 N^2$ we find the expected result: $\hat{\Pi}/k_B T$ is independent of N. In the semidilute excluded volume limit, in particular, it is easily checked from Eqs. (14.13), (14.22), $w \to 0$, that we recover des Cloizeaux's power law $\hat{\Pi}/k_B T \sim c^{\frac{3\nu}{3\nu - 1}}$ (cf. Eq. (9.14)).

Again we can easily calculate the full crossover. As an example Fig. 14.3 shows the scaling function $\mathcal{P}/\hat{s}$ as function of $\hat{s}$ in the excluded volume limit. In unrenormalized tree approximation this ratio would be a constant proportional to the second virial coefficient. In renormalized theory we see a pronounced variation which rapidly approaches the asymptotic power law.

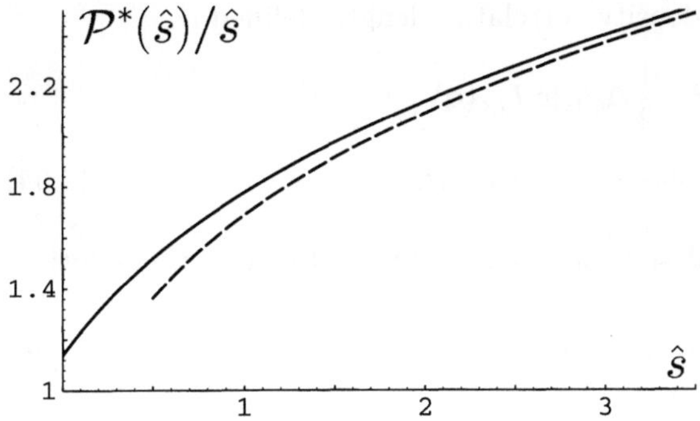

Fig. 14.3. The thermodynamic scaling function in the excluded volume limit $\mathcal{P}^*(\hat{s})$, divided by the excluded volume overlap $\hat{s}$. The *broken line* gives the asymptotic power law $\mathcal{P}^*/\hat{s} \sim \hat{s}^{\frac{1}{3\nu-1}-1}$

14.3 Density cumulant

Equation (5.56) yields the scaling function $J_c(\mathbf{q})$ of the density cumulant in unrenormalized tree approximation:

$$J_c(\mathbf{q}) = \frac{1}{cN} I_{d,c}(\mathbf{q}) = \frac{D(N\mathbf{q}^2\ell^2)}{1 + u_0 cND(N\mathbf{q}^2\ell^2)} \quad . \tag{14.29}$$

This expression is renormalized with Eqs. (14.4), (14.20):

$$J_c(\mathbf{q}) = \frac{D(\bar{\mathbf{q}}^2 N_R)}{1 + f c_R N_R D(\bar{\mathbf{q}}^2 N_R)} \quad . \tag{14.30}$$

For large momenta $J_d(\mathbf{q})$ reduces to $J_a(\mathbf{q})$. We thus only consider here the limit of small $\mathbf{q}$, leaving the discussion of the momentum crossover to Chap. 19.

For a monodisperse system $I_{d,c}(0)$ obeys the compressibility sum rule

$$\frac{1}{cN} I_{d,c}(0) = \left[\frac{\partial}{\partial c_p} \frac{\hat{\Pi}}{k_B T} \right]^{-1} \quad , \tag{14.31}$$

which will be derived in Sect. 19.1. In terms of renormalized scaling functions it reads

$$J_c^{-1}(0) = 1 + \mathcal{P} + c_R \frac{\partial}{\partial c_R} \mathcal{P} \quad . \tag{14.32}$$

It is easily checked from Eq. (14.21) that the tree approximation

$$J_c^{-1}(0) = 1 + f c_R N_R \tag{14.33}$$

obeys this sum rule. So its behavior is equivalent to that of $\mathcal{P}$ and needs no further discussion.

Of special interest is the density correlation length defined as

$$\begin{aligned}\xi_d^2 &= -\frac{1}{2}\Delta_q|_0 \ln I_{d,c}(\mathbf{q}) \\ &= 3\ell_R^2 \frac{\partial}{\partial\bar{\mathbf{q}}^2}|_0 \ln J_c^{-1}(\mathbf{q}) \quad . \end{aligned} \tag{14.34}$$

It is the range of the typical segment density fluctuations in the solution. Equation (14.30) yields

$$\xi_d^2 = \frac{\ell_R^2 N_R}{1 + f c_R N_R} \quad . \tag{14.35}$$

With respect to our choice (13.21) of the uncritical manifold it is of interest to rewrite Eq. (14.35) as

$$\frac{\ell_R^2}{\xi_d^2} = \frac{1}{N_R} + f c_R \quad . \tag{14.36}$$

A glance to Eq. (13.21) shows that for $\bar{\mathbf{q}}^2 \ll q_0^2$ we fix the renormalized segment size ℓ_R to be of the order of ξ_d. This is in full accord with the general discussion of renormalization, since the density correlation length is the smallest macroscopic length of interest in the small momentum region.

In the dilute limit ξ_d of course equals the radius of gyration. We thus consider the semidilute limit, where ξ_d is of the order of the screening length. (As will be discussed in Chap. 19, it is not strictly identical to the screening length.) Equation (14.35) reduces to

$$\begin{aligned}\xi_d^2 &= \frac{\ell_R^2}{f c_R} \\ &= \frac{\ell_R^2}{c_0} \\ &= \frac{\tilde{R}_0^2}{c_0}\left(\frac{f}{\tilde{z}}\right)^2 \left|1 - f\right|^{-2/\omega} H_u^2(f) \quad . \end{aligned} \tag{14.37}$$

Here we used Eq. (13.30 ii). Again this expression only depends on c or T. In the (semidilute) Θ-limit Eqs. (13.33 i, ii) imply

$$\left(\frac{f}{\tilde{z}}\right)^2 = \frac{w^2}{n_0} = \frac{c_0}{\tilde{u}\tilde{z}\tilde{s}} \quad , \tag{14.38}$$

so that

$$\xi_d^2 = \frac{\tilde{R}_0^2}{\tilde{u}\tilde{z}\tilde{s}} = \frac{\tilde{\ell}^2}{\tilde{u}\tilde{v}\tilde{\ell}^3 c} \quad , \tag{14.39}$$

in keeping with the unrenormalized result. The factor $\tilde{v}$ in the denominator might give rise to concern, since it vanishes for $T \to \Theta\left(\tilde{z} = \tilde{v}N^{\frac{1}{2}} \to 0\right)$. The resulting divergence of ξ_d^2, however, is unphysical, since Eq. (14.39) only holds in the semidilute regime $w \ll 1$. Thus in the Θ-regime $f \to 0$ Eqs. (14.37), (14.38) imply $\xi_d^2 \ll \tilde{R}_0^2$. On the other hand, taking $T \to \Theta$ at fixed concentration we leave the semidilute regime and reach the dilute Θ-limit where $\xi_d^2 = R_g^2 = \tilde{R}_0^2$. (Compare Fig. 13.2.) Of course three body effects will become important, which destroy this simple result.

In the semidilute excluded volume limit we may use Eqs. (13.35) to find

$$\begin{aligned} \ell_R^2 &= \omega^{4\nu} B^2 N^{2\nu} \\ &= \left(\frac{N_0}{c_0}\tilde{u}\hat{s}\right)^{\frac{2\nu}{1-3\nu}} B^2 N^{2\nu} \\ &= \left(\frac{N_0}{c_0}\tilde{u}B^3 c\right)^{\frac{2\nu}{1-3\nu}} B^2 \ . \end{aligned} \tag{14.40}$$

Thus

$$\xi_d^2 \sim B^2 \left(B^3 c\right)^{-\frac{2\nu}{3\nu-1}} \tag{14.41}$$

obeys the power law expected for the blob size (Eq. (9.8)).

Figure 14.4 shows $\xi_d^2/\tilde{R}_0^2$ for $\tilde{s}$ fixed as function of $\tilde{z}$, which describes a crossover driven by changing the temperature for c, N fixed. $\tilde{s}$ was chosen large ($\tilde{s} = 10$), so that a large part of the curve is in the semidilute regime. Still it does not properly reach the semidilute Θ-limit before bending over towards excluded volume behavior. The corresponding path in the $w - f$ diagram starts at the dilute Θ-limit (point B) passes by the semidilute Θ-limit (point C) and ends for large $\tilde{z}$ in the semidilute excluded volume limit (point D).

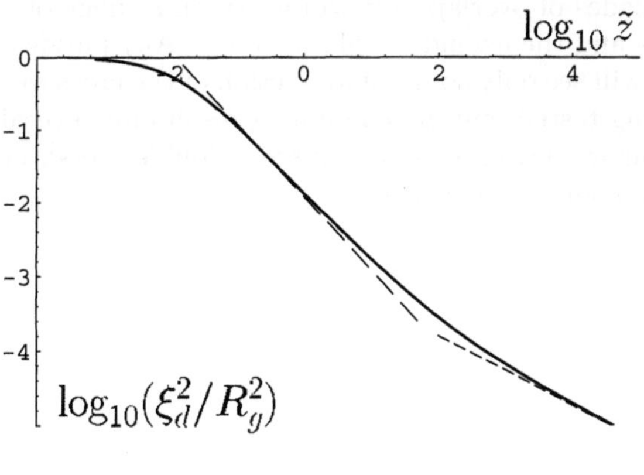

Fig. 14.4. $\log_{10} \xi_d^2/\tilde{R}_0^2$ as function of $\log_{10} \tilde{z}$ for $\tilde{s} = 10$, fixed. The *dashed lines* represent the asymptotic behavior in the semidilute Θ-limit (*long dashes*) or the semidilute excluded volume limit (*short dashes*)

14.4 Summary of important aspects

All the results for the scaling functions expressed in terms of $N_R, c_R, f, \bar{\mathbf{q}}^2$, as written down here, look very simple. They result from simply replacing unrenormalized by renormalized variables in the tree approximation. Still our expressions yield highly nontrivial behavior, reproducing the results of scaling theory and giving crossover functions among the various asymptotic scaling limits. This is due to the use of the RG together with a well chosen uncritical manifold. It is by this latter choice that we establish the connection to scaling theory.

All the results shown here are correct qualitatively, and also on the quantitative level the scaling functions will not be changed too much by the one-loop corrections. This leads us to another point to be stressed: all scaling functions are extremely smooth, showing no pronounced structure. The crossover regions are extremely broad, also depending on the quantity considered. Thus any definition of a c^*, where the behavior changes from dilute to semidilute, or of a q^*, where the anomalous power law $J_{c(a)} \sim q^{-1/\nu}$ sets in, is plagued by large ambiguities. Such concepts are often used in the evaluation of experiments, but their only valid role is to set a scale for the definition of asymptotic limits ($c/c^* \ll 1$: dilute, $c/c^* \gg 1$: semidilute, for instance). Otherwise they should be abandoned. This in particular becomes clear if we do not use doubly logarithmic plots (cf. Fig. 14.1b).

The very broad crossover has important implications for an experimental test of the theory. Indeed, to illustrate clearly the effects, in the figures of this chapter we have driven the parameters into extreme regions. No experiment, for instance, will cover six decades of overlap, nor will we reach a range of $\tilde{z}$ sufficient to cover both the Θ- and the excluded volume limits. Any physical experiment or any simulation will see only some small section of the crossover function, the full function being tested only by combining results for several chemical systems. In particular we have no chance to see a double crossover as typically exhibited here, in a single experiment.

Part IV

One Loop Approximation: Detailed Results and Comparison to Experiment

This part is concerned with the evaluation of the renormalized theory to one-loop approximation. All experimentally relevant quantities will be considered, and special consideration will be given in the light of our results to an examination of heuristic methods of data analysis. In addition we will consider quantities like the internal correlations within an isolated chain. Though such quantities in practice are accessible only to computer experiments, they are of great interest in exhibiting the fractal nature of the chain configuration.

The full analytic results are typically complicated and will be presented in appendices. The main text concentrates on the results evaluated for $d = 3$. For some of the more important experimental quantities we will also construct simple fit formulae, which numerically reproduce the theoretical results.

With regard to a comparison with experiment we note here the conventions for physical variables. All unrenormalized chain lengths are measured as monomer number per chain, i.e. N is the number averaged polymerization index. Concentrations c_p, c are measured in polymer number or monomer number per unit volume, respectively. So in relating to typical experimental definitions some factors of Avogadro's number or of the molecular weight of a monomer have to be inserted. Furthermore all temperatures will be measured on the absolute scale ($^\circ$K).

15. Dilute Limit: Density Correlations, Virial Coefficients, and Interpenetration Ratio

Properties of isolated chains are easily accessible in computer experiments, but are hard to measure in a real solution. Nevertheless a large amount of labor has been spent to extract single chain properties by extrapolating experimental data to infinite dilution. Partly this may be due to the fact that properties involving a single chain or a few chains only are the easiest to analyze theoretically, but they also are of practical interest: they yield unambiguous information on such important quantities as the average chain length or the polydispersity of the sample.

In this chapter we consider those few-chain quantities which are accessible in real experiments. Specifically we will discuss the density correlation function of an isolated chain as well as related quantities like the swelling factor. Among the properties of pairs of chains we will consider the second virial coefficient or the interpenetration ratio. For all these quantities we will also critically examine the experimental situation assuming, however, that these few-chain properties indeed have been measured. We will not consider here the process of extrapolating to zero concentration. Rather this task will be taken up in later chapters, where we discuss solutions of finite overlap.

In computer experiments we can measure additional quantities. In particular much work has been devoted to a determination of the average end-to-end distance. But we can also measure more delicate quantities, like internal correlations within a polymer coil or the interaction of two coils as function of the center-of-mass distance. Such quantities will be considered in the next chapter.

The density correlation function $I_d(\mathbf{q}, n)$ of an isolated chain is of primary importance and will be discussed in great detail. The Taylor expansion for small q yields the radius of gyration (Sect. 15.1). We can equivalently consider the swelling factor $\alpha_g^2 = R_g^2(T,n)/R_g^2(\Theta,n)$, which is traditionally used to describe the crossover from the Θ-point to the excluded volume limit. Strictly speaking, α_g is not a universal ratio, universality holding only within the realm of naive two-parameter theory. By comparison to experiment, however, we find the two-parameter assumption to be justified for many systems explicitly analyzed. Indeed, the theoretical result agrees nicely with the experimental data. Besides the crossover towards Θ, i.e. the weak coupling region, we will also consider the strong coupling region $f > 1$, where, however, good experimental data are rare.

In discussing the $\mathbf{q}$-dependence of I_d (Sect. 15.2) we have to distinguish the regions $qR_g \lesssim 1$ or $qR_g \gg 1$. For small q, I_d strongly depends on polydispersity, and to a good approximation is given by the polydispersity-dependent form of the Debye function. For large q, I_d becomes independent of polydispersity, but deviates from the Debye function. Whereas the latter asymptotically vanishes proportional to q^{-2}, the former in the excluded volume limit behaves as $q^{-1/\nu}$, in agreement with the simple scaling arguments of Sect. 9.1 (Eq. (9.20)). In Sect. 14.1.3 we encountered that behavior even in renormalized tree approximation. It seems to be compatible with experiment.

Another quantity of great interest is the second virial coefficient A_2, which yields information on a pair of mutually interacting chains (Sect. 15.3). A_2 can be derived from the virial expansions of either the osmotic pressure or the density cumulant $I_{d,c}(\mathbf{q} = 0)$. These two definitions are identical for a strictly monodisperse solution, but differ somewhat for general polydispersity. Here again we will discuss polydispersity effects and the crossover from Θ-to excluded volume conditions, as well as the strong coupling regime.

Next we will analyze the interpenetration ratio $\psi \sim A_2/R_g^d$ (Sect. 15.4). This quantity can be interpreted as measuring the interchain interaction, and indeed its crossover behavior is a fairly direct mapping of the flow of the renormalized coupling constant. This makes it a most attractive quantity for a detailed theoretical analysis. However, ψ is found to be strongly dependent not only on excluded volume strength but also on polydispersity. Furthermore, the predicted polydispersity-dependence is sensitive to our choice of n_0, a feature which will also be discussed here.

Having worked out the theoretical results we turn to a comparison with experiment. This is the topic of Sect. 15.5, where we also discuss heuristic methods of data analysis.

Some more material is presented in appendices. In Appendix A 15.1 we derive formal expressions for $I_d(\mathbf{q})$ or A_2, valid for arbitrary polydispersity. Appendix A 15.2 presents the one-loop calculation for $I_d(\mathbf{q})$. This is meant as another illustration of the perturbative techniques. Finally in Appendix A 15.3 we summarize perturbative results for general d, and we present the ε-expansion for some of the more important critical ratios.

15.1 The radius of gyration

15.1.1 General expressions

The density correlation function of a single chain is defined as (cf. Sect. 3.2)

$$I_d(\mathbf{q}, n) = \sum_{j_1 \neq j_2} \left\langle e^{i\mathbf{q}(\mathbf{r}_{j_2} - \mathbf{r}_{j_1})} \right\rangle \quad . \tag{15.1}$$

According to Eq. (A 15.12) for general chain length distribution we have to consider the function

$$I_d(\mathbf{q}, [P]) = \sum_n P(n) I_d(q, n) \quad , \tag{15.2}$$

where $P(n) = \frac{1}{N}\, p(n/N)$ is the chain length distribution. (Cf. Sect. 5.3.) $I_d(\mathbf{q}, [P])$ for $\mathbf{q} \to 0$ obeys the sum rule (cf. Eq. (5.24))

$$I_d(0, [P]) = \tilde{p}_2 N^2 = N N_w \quad . \tag{15.3}$$

Following Eq. (11.76) we define the normalized scaling functions

$$J_d(\mathbf{q}, n) = \frac{1}{n^2} I_d(\mathbf{q}, n) \tag{15.4}$$

or

$$\begin{aligned} J_d(\mathbf{q}, [p]) &= \frac{1}{N^2} I_d(\mathbf{q}, [P]) \\ &= \int_0^\infty dy\, p(y) y^2 J_d(\mathbf{q}, yN) \quad , \end{aligned} \tag{15.5}$$

respectively.

For a polydisperse system the radius of gyration is defined as

$$\begin{aligned} R_g^2[p] &= -\frac{1}{2} \Delta_q \ln I_d(\mathbf{q}, [P]) \Big|_{\mathbf{q}=0} \\ &= -\frac{1}{2} \Delta_q \ln J_d(\mathbf{q}, [p]) \Big|_{\mathbf{q}=0} \quad . \end{aligned} \tag{15.6}$$

Using Eqs. (15.3), (15.5) we can easily express $R_g^2[p]$ in terms of the radius of gyration $R_g^2(n)$ of an isolated chain of length n:

$$R_g^2[p] = \int_0^\infty dy\, \frac{p(y)}{\tilde{p}_2}\, y^2 R_g^2(yN) \quad . \tag{15.7}$$

The resulting one-loop expression is given in Appendix A 15.3.1, Eq. (A 15.30). For $d = 3, N_R = n_0$ the result reads

$$R_g^2[p] = \frac{\tilde{p}_3}{\tilde{p}_2} \ell_R^2 n_0 \left(1 + \left(\frac{67}{105} n_0^{1/2} - 1 \right) u^* f + f n_0^{1/2} c_g[p] \right) \quad , \tag{15.8}$$

where

$$c_g[p] = -\frac{67}{105} u^* \left(1 - \int_0^\infty dy \, y^{7/2} \frac{p(y)}{\tilde{p}_3} \right) \tag{15.9}$$

is a polydispersity correction, which vanishes for the monodisperse ensemble. For the set of Schultz distributions (5.43) $c_g[p]$ can be expressed in terms of Γ-functions. It turns out that the result in the relevant range $1 \le \tilde{p}_2 \le 2$ can be parameterized as

$$c_g(\tilde{p}_2) = 0.288 \, (\tilde{p}_2 - 1) - 0.070 \, (\tilde{p}_2 - 1)^2 \quad , \tag{15.10}$$

the error being less than 0.002 in absolute value.

We introduce the theoretical swelling factor

$$\bar{\alpha}_g^2(f, [p]) = \frac{\tilde{p}_2}{\tilde{p}_3} \frac{R_g^2[p]}{\tilde{R}_0^2} \quad . \tag{15.11}$$

We should stress that $\bar{\alpha}_g^2$ is not identical to the experimental swelling factor $\alpha_g^2 = R_g^2(T)/R_g^2(\Theta)$, due to a possible temperature dependence of the effective segment size $\tilde{\ell}$ and thus of $\tilde{R}_0$. Now using Eq. (13.29 ii) ($\bar{\mathbf{q}}^2 = 0$) to eliminate ℓ_R^2 and setting $n_0 = 0.53$ from Eqs. (15.8), (15.11) we find

$$\bar{\alpha}_g^2(f, [p]) = \left| 1 - f \right|^{\frac{1}{\omega}\left(\frac{1}{\nu} - 2\right)} H(f)(1 - 0.195 f + 0.728 c_g[p] f) \quad , \tag{15.12}$$

$$\frac{1}{\omega} \left(\frac{1}{\nu} - 2 \right) = -0.374 \quad .$$

f is related to $\tilde{z} = \tilde{v} N^{1/2}$ by Eq. (13.29 i):

$$\tilde{z} = n_0^{1/2} f \left| 1 - f \right|^{-\frac{1}{2\nu\omega}} H_n^{-1/2}(f) \quad , \tag{15.13}$$

$$\frac{1}{2\nu\omega} = 1.063 \quad .$$

We now evaluate these results in various regions of f.

15.1.2 Excluded volume limit

For $f = 1$ we use Eq. (13.35 ii), $w = 1, \bar{\mathbf{q}}^2 = 0$ to express ℓ_R^2 in terms of $B^2 N^{2\nu}$. Equation (15.8) yields

$$R_g^2[p] = a_g[p] B^2 N_w^{2\nu} \tag{15.14}$$

$$a_g[p] = \tilde{p}_3 \tilde{p}_2^{-1-2\nu}(0.427 + 0.386 c_g[p]) \quad , \tag{15.15}$$

where we expressed R_g^2 in terms of the weight average chain length N_w. Any experimental determination of R_g^2 also yields N_w via the sum rule (15.3).

Equation (15.14) constitutes the well known power law for the radius of gyration. The coefficient $a_g B^2$ depends on the microscopic structure, but ratios $a_g[p_1]/a_g[p_2]$ for two different but fixed chain length distributions are universal. This allows for an interesting check. From Eqs. (15.7), (15.14) we derive the exact relation

$$\frac{a_g[p]}{a_g[m]} = \tilde{p}_2^{-1-2\nu} \int_0^\infty dy\, p(y) y^{2+2\nu} \quad , \tag{15.16}$$

where $a_g[m]$ denotes the monodisperse result. The one-loop approximation (15.15) yields

$$\frac{a_g[p]}{a_g[m]} = \tilde{p}_3 \tilde{p}_2^{-1-2\nu}(1 + 0.904 c_g[p]) \quad . \tag{15.17}$$

Figure 15.1 compares these two expressions evaluated for Schultz distributions $1 \leq \tilde{p}_2 \leq 2$. We also included the zero loop result $a_g[p]/a_g[m] = \tilde{p}_3/\tilde{p}_2^{1+2\nu}$. It is found that the one-loop approximation is not too bad. It somewhat underestimates the polydispersity dependence, but even for $\tilde{p}_2 = 2$ the error is smaller than 5%. (We note that the exact relation (15.16) holds only in the excluded volume limit. In the crossover regime we have only the loop expansion available.)

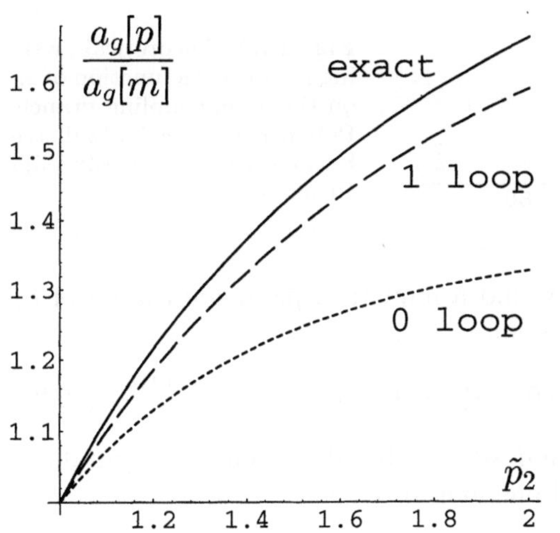

Fig. 15.1. Polydispersity dependence (Schultz distributions) of the coefficient of the asymptotic power law for the radius of gyration

15.1.3 Weak coupling region

In the weak coupling region $f < 1$, $\bar{\alpha}_g^2$ (Eq. (15.12)) varies from

$$\bar{\alpha}_g^2 = 1 + (0.239 + c_g[p])\tilde{z} + O(\tilde{z}^2), \quad \tilde{z} \ll 1 \tag{15.18}$$

to

$$\bar{\alpha}_g^2 = (0.843 + 0.762 c_g[p])\tilde{z}^{4\nu-2}, \quad \tilde{z} \to \infty \tag{15.19}$$
$$4\nu - 2 = 0.352 \quad .$$

The latter expression represents the excluded volume limit. Figure 15.2 shows $\bar{\alpha}_g^2$ as function of

$$\tilde{z}_w = \tilde{p}_2^{1/2}\tilde{z} \tag{15.20}$$

for a set of Schultz distributions, as found by evaluating Eqs. (15.12), (15.13). A small polydispersity effect is visible. With increasing $\tilde{p}_2$, $\bar{\alpha}_g^2$ increases for fixed $\tilde{z}_w$. Using $\tilde{z}_w$ in place of $\tilde{z}$, however, we have absorbed most of the polydispersity dependence.

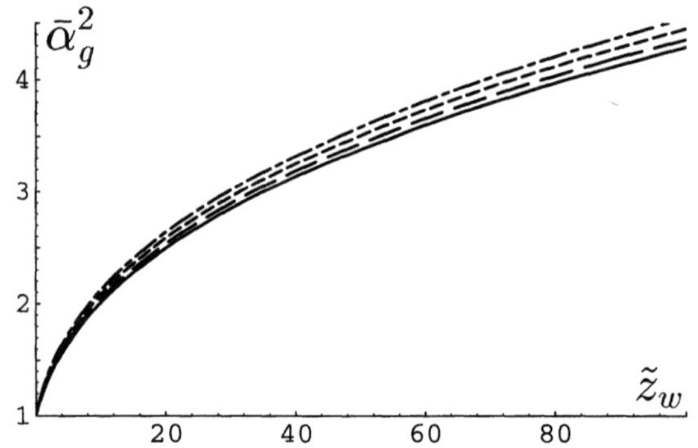

Fig. 15.2. Theoretical swelling factor $\bar{\alpha}_g^2$ as function of $\tilde{z}_w$ on the weak coupling branch. Polydispersity: Schultz distributions with $\tilde{p}_2 = 1; 1.2; 1.5; 2$ (from below)

It turns out that we can easily find a numerical parameterization of $\bar{\alpha}_g^2$ as function of $\tilde{z}$. The simple expression

$$\bar{\alpha}_g^2 = \left[1 + (1.32 + 5.263 c_g[p])\tilde{z} + 0.378(1 + 0.904 c_g[p])^{5.682}\tilde{z}^2\right]^{0.176} \tag{15.21}$$

reproduces our results for all $\tilde{p}_2$ and all $\tilde{z}$ within deviations less than 0.2%. It is constructed to match the full one-loop result for $\tilde{z} \to \infty$.

Again we can carry through a meaningful test. Shanes and Nickel [SN90] have calculated $\bar{\alpha}_g^2$ for a monodisperse system in the continuous chain model to order z^4, using this result as input in a direct renormalization scheme. Their final result is parameterized as

$$\bar{\alpha}^2_{g,SN} = (1 + 1.48\tilde{z} + 0.378\tilde{z}^2)^{0.176} \quad , \tag{15.22}$$

where we adjusted the nonuniversal scale of $\tilde{z}$ such that the asymptotic behavior (15.19) is reproduced. For all $\tilde{z}$ the deviation from Eq. (15.21), $c_g[p] = 0$ is less than 1.2%. The main difference is that Eq. (15.22) starts out with a somewhat larger slope:

$$\bar{\alpha}^2_{g,SN} = 1 + 0.260\tilde{z} + O(\tilde{z}^2) \ .$$

The good overall agreement among the two calculations again strongly supports the belief that our one-loop theory is accurate within corrections of the order of 1% for reasonably narrow chain length distributions.

15.1.4 Strong coupling region

Equations (15.12), (15.13) also hold in the strong coupling region $1 < f \lesssim 1.5$, where the upper bound of f reflects my prejudice on the range of validity of low order renormalized perturbation theory. Recall, however, that the usual interpretation of $\tilde{z}, \tilde{R}_0$ in terms of quantities measured in a Θ-region is not even approximately valid, since the strong coupling branch is not connected to a Θ-point. Rather $\tilde{z}$ just is a measure of $N^{1/2}$, whereas $\tilde{R}_0$ is the radius of a fictitious noninteracting reference chain.

Numerically evaluating $\bar{\alpha}^2_g$ we find that for the monodisperse system we can again construct a simple fit formula

$$\bar{\alpha}^2_g = 0.843\tilde{z}^{0.352}\left[1 - 0.54\tilde{z}^{-0.941}\left(1 + \frac{0.44}{\tilde{z}}\right)\right] \quad . \tag{15.23}$$

For $1 < f \lesssim 1.5$, corresponding to $\tilde{z} \gtrsim 3$, this deviates from the true numerical result by less than 0.1%. The form (15.23) is suggested by the discussion of the corrections to scaling (Sect. 10.2.2), which shows that the leading corrections to the dominant behavior $\tilde{z}^{4\nu-2}$ are of relative order $\tilde{z}^{-2\nu\omega}$, $2\nu\omega = 0.941$. It turns out that Eq. (15.23) also gives a good fit for general polydispersity, provided $\tilde{z}$ is interpreted as $\tilde{z}_w$ and the prefactor $0.843\tilde{z}^{0.352}$ is replaced by the appropriate asymptotic behavior (15.19).

A plot of $\bar{\alpha}^2_g$ as function of $\tilde{z}_w$ in the strong coupling branch looks most similar to the weak coupling result (Fig. 15.2) and therefore is not reproduced here. Rather for a monodisperse system Fig. 15.3a compares the three branches of the swelling factor. In this plot the fixed point branch is given by the asymptotic behavior (15.19) (broken line). The weak coupling branch lies above and the strong coupling branch lies below the fixed point branch. Asymptotically both branches converge towards the fixed point branch. The approach is fairly slow, the leading additive correcting being proportional to

$\tilde{z}^{-0.589}$ (cf. Eq. (15.23)). Figure 15.3b shows the same results in terms of the effective exponent

$$\nu_{eff} = \frac{1}{2} + \frac{1}{4}\frac{\partial \ln \bar{\alpha}_g^2}{\partial \ln \tilde{z}} \quad .$$

It illustrates that ν_{eff} rapidly approaches $\nu = 0.588$ up to corrections of order $|\nu_{eff} - \nu| \approx 0.01$, but the final convergence is very slow.

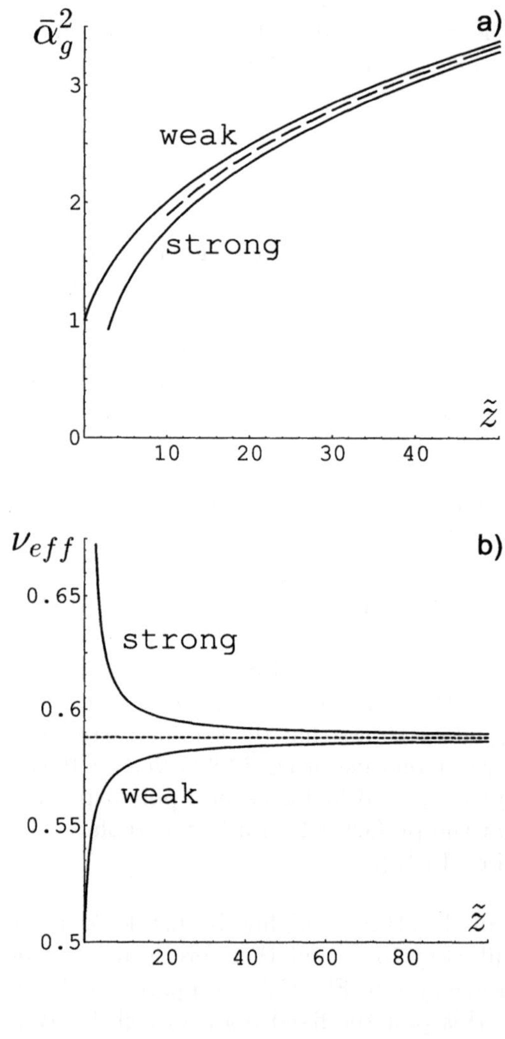

Fig. 15.3. (**a**) The three branches of $\bar{\alpha}_g^2$ for a monodisperse system. *Broken line*: excluded volume branch; *upper (lower) line*: weak (strong) coupling branch. (**b**) Effective exponent ν_{eff} derived from $\bar{\alpha}_g^2$. *Broken line*: fixed point value; *upper (lower) line*: strong (weak) coupling branch

15.2 Momentum dependence of the scattering form factor

15.2.1 General expressions

In tree approximation $J_d(\mathbf{q}, [p])$ is given by the (polydispersity dependent) Debye function:

$$J_d(\mathbf{q}, [p]) = D_p(Q_R) \quad ,$$

where

$$Q_R = \bar{\mathbf{q}}^2 N_R \quad . \tag{15.24}$$

From the discussion of Sect. 14.1 we know that this form qualitatively reproduces the scaling results. This suggests to write the one-loop approximation in a form where the corrections affect the argument of the Debye function. We thus write

$$J_d(\mathbf{q}, [p]) = D_p(\hat{Q}) \quad , \tag{15.25}$$

where $\hat{Q}$ is given in Appendix A 15.2, Eq. (A 15.26). For $d = 3$ we find

$$\hat{Q} = Q_R(1 - u^* f) - \frac{u^*}{2} f N_R^{1/2} \frac{\Delta J_d(Q_R, [p])}{D_p'(Q_R)} \quad . \tag{15.26}$$

$\Delta J_d(Q_R, [p])$ is to be calculated from Eqs. (A 15.20), (A 15.23). For the monodisperse ensemble the result within 1% deviation can be parameterized as

$$\Delta J_d(Q, [m]) = \frac{0.425Q}{1 + 0.400Q + 0.071Q^{3/2} + 0.044Q^2 + 0.0191Q^{2.5}} \quad . \tag{15.27}$$

This expression strictly reproduces the one-loop result in the limits $Q \to 0$ or $Q \to \infty$. For the exponential ensemble, i.e. the Schultz distribution with $\sigma = 1$ (cf. Eq. (5.43)), ΔJ_d can be evaluated analytically to yield

$$\Delta J_d(Q, [e]) = \frac{4\sqrt{\pi}}{(1+Q)^2} \left\{ \frac{1}{Q^2} \left[(1+Q)^{1/2} - 1 - \frac{Q}{2} \right] - 1 + \frac{1-2Q}{8+2Q} \right.$$
$$\left. + (1+Q) \frac{2}{\sqrt{Q}} \operatorname{arctg} \frac{\sqrt{Q}}{2} \right\} \quad . \tag{15.28}$$

In explicit evaluations we will consider these two cases, which span the relevant range of polydispersity.

The form (15.25), (15.26) has the virtue that the effective momentum variable $\hat{Q}$ changes smoothly with Q_R. For the monodisperse ensemble in the excluded volume limit, for instance, $\hat{Q}/Q_R$ varies from $0.580, Q_R \to 0$ to $0.922, Q_R \to \infty$. Note that in the limit of large momenta the one-loop term just yields a finite correction to the coefficient of the leading Q_R-behavior. The dangerous factor $N_R^{1/2}$ is cancelled rigorously. (See Sect. 15.2.3 for a more explicit discussion.)

15.2.2 The region of small momenta

For small momenta $J_d(\mathbf{q}, [p])$ may be expanded according to

$$J_d(\mathbf{q}, [p]) = \tilde{p}_2 - \bar{R}_2[p] Q_R + \bar{R}_4[p] Q_R^2 + O(Q_R^3) \ . \tag{15.29}$$

From $\bar{R}_4, \bar{R}_2$ we can build a ratio

$$\begin{aligned} \mathcal{R}_{4/2}[p] &= \frac{\bar{R}_4[p]}{\bar{R}_2^2[p]} \\ &= \frac{1}{2} \frac{\partial^2}{(\partial q^2)^2} J_d \left(\frac{\partial}{\partial q^2} J_d \right)^{-2} \Bigg|_{q=0} \ , \end{aligned} \tag{15.30}$$

which in the excluded volume limit is a universal function of polydispersity. For the monodisperse system we find, directly evaluating Eqs. (A 15.28), (A 15.29) for $\varepsilon = 1$

$$\mathcal{R}_{4/2}[m] = 0.673 \quad \text{(direct evaluation)} \ ,$$

whereas the ε-expansion (Eq. (A 15.32)) yields

$$\mathcal{R}_{4/2}[m] = 0.706 \quad (\varepsilon\text{-expansion}) \ .$$

The tree approximation yields

$$\mathcal{R}_{4/2}[m] = 3/4 \ \text{(zero loop)} \ .$$

These results may be compared to Monte Carlo data of Bishop and Saltiel [BS91], who find $\mathcal{R}_{4/2}[m] = 0.694$. Clearly the agreement is very good.

This result implies that $J_d(q, [p])$ for the monodisperse system for small q^2 falls below the Debye approximation. Formulated equivalently in terms of the more commonly considered inverse scattering functions, J_d^{-1} for small q^2 is curved upwards stronger than the Debye function. However, this effect sensitively depends on polydispersity, as may be seen by comparing to the results for the exponential ensemble. There we find

$$\mathcal{R}_{4/2}[e] = 0.510 \quad \text{(direct evaluation)} \ ,$$

to be compared to

$$\mathcal{R}_{4/2}[e] = 1/2 \ . \quad \text{(zero-loop)} \ .$$

Compared to the monodisperse system the one-loop effect is not only much smaller but is also of the opposite sign.

The ratio $\mathcal{R}_{4/2}$ has first been calculated in [WS81]. Pushing the expansion (15.29) of $J_d(\mathbf{q}, [p])$ to order Q_R^m we can clearly calculate a coefficient $\bar{R}_{2k}[p]$ and associated ratios like $\mathcal{R}_{2k/2} = \bar{R}_{2k}/\bar{R}_2^k$. For the monodisperse system results to all orders Q_R^m in one-loop approximation are given in [Dup86b]. We do not further pursue this subject here for two reasons. First, within our method we would have to take into account that N_R depends on q^2 via the choice of the uncritical manifold. Due to the large value $q_0^2 = 50$ this effect can be neglected for the low order terms of the expansion, but must be taken into account in high orders. This just being a technical complication, the second observation is more important. Numerically evaluating our full form for $J_d(\mathbf{q}, [p])$ (see Sect. 15.2.4 for the details) we find that our result as function of $q^2 R_g^2[p]$ over a fairly large range closely follows the Debye approximation $D_p(\tilde{p}_2 q^2 R_g^2/\tilde{p}_3)$. This is illustrated in Fig. 15.4 with J_d^{-1} evaluated in the excluded volume limit for monodisperse or exponential chain length distributions. Even the one-loop effect on $\bar{R}_4$ can hardly be seen. For the monodisperse system the Debye approximation is excellent for $q^2 R_g^2[m] \lesssim 20$, whereas for the exponential distribution the deviations start around $q^2 R_g^2[e] \approx 10$. In this situation the Taylor series in powers of Q_R is of little use. Indeed, the individual terms of the Taylor series have little meaning. The origin lies in the fact that the asymptotic behavior of the scaling functions closely matches the behavior for small argument. For instance, J_d^{-1} varies from

$$J_d^{-1} = \tilde{p}_2^{-1} \left(1 + \frac{q^2}{3} R_g^2\right) \quad , \quad \text{small } q^2$$

to

$$J_d^{-1} = const\, q^{1/\nu} = const\, q^{1.70}, \quad q^2 \to \infty \quad .$$

Thus there can be no region where a q^4-term alone is significant, but all terms $q^{2m}, m \geq 2$, conspire to generate the asymptotic power law. This is nicely illustrated with the results of [Dup86b] where it is found that the $\bar{R}_{2k}$ alternate in sign, the $O(\varepsilon)$ contribution increasing roughly proportional to k. Similar features will be found later (Sect. 17.3) for the virial expansion.

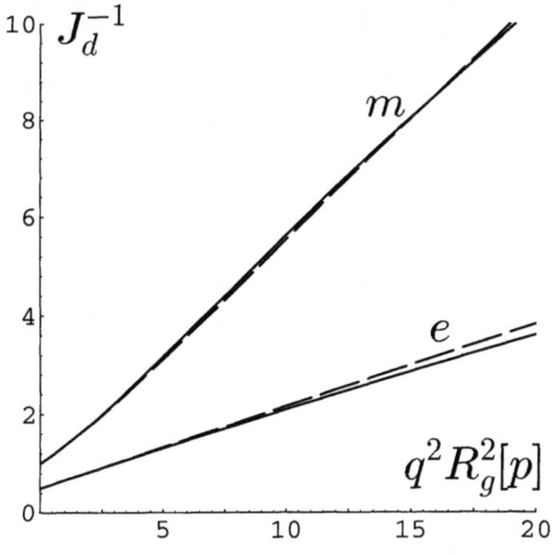

Fig. 15.4. $J_d^{-1}(\mathbf{q}, [p])$ as function of $\mathbf{q}^2 R_g^2[p]$ in the excluded volume limit. *Upper lines*: monodisperse ensemble; *lower lines*: exponential ensemble. *Broken lines*: Debye approximation; *full lines*: one loop results

15.2.3 The limit of large momenta

For large values of Q_R Eqs. (15.25), (15.26) reduce to

$$J_d^{-1}(\mathbf{q},[p]) = \frac{1}{2}Q_R\left[1+u^*f\left(\frac{\pi^{3/2}}{(\bar{\mathbf{q}}^2)^{1/2}}-1\right)\right] \quad . \tag{15.31}$$

Recall that from the simple scaling theory (Sect. 9.1) we expect the scattering at large momenta to be due to short subchains, not testing the global structure of the coil. This is consistent with the fact that $J_d(\mathbf{q},[p])$ becomes independent of N_R or of the chain length distribution.

In Eq. (15.31), or more generally Eq. (A 15.33), the prefactor $N_R^{\varepsilon/2}$ multiplying the loop integral is cancelled by a contribution $Q_R^{-\varepsilon/2} = (\bar{\mathbf{q}}^2 N_R)^{-\varepsilon/2}$ extracted from ΔJ_d. As a result, the correction stays finite in the limit $q^2 \to \infty$, corresponding to $\bar{\mathbf{q}}^2 \to q_0^2$. This supports our choice of the renormalized length scale, which thus at least for J_d and to one loop order suppresses any singular contribution in perturbation theory. More precisely, the first-order correction contains a contribution $(\bar{\mathbf{q}}^{-\varepsilon}-1)/\varepsilon$. Now the commonly employed choice $N_R = const$ implies $\bar{\mathbf{q}}^2 \to \infty$ for $q^2 \to \infty$. In ε-expansion we thus would find a divergent contribution. It is this divergence which is suppressed by using the appropriate crossover form of ℓ_R.

According to Eqs. (15.24), (13.25) Q_R is to be expressed in terms of $\bar{\mathbf{q}}^2$ as

$$Q_R = \frac{\bar{\mathbf{q}}^2 n_0}{1-\bar{\mathbf{q}}^2/q_0^2} \quad . \tag{15.32}$$

Restricting ourselves to the excluded volume limit we furthermore note the relation (cf. Eq. (13.35 ii), $w=1$)

$$\left(1-\frac{\bar{\mathbf{q}}^2}{q_0^2}\right)^{-1} = \left(\frac{\mathbf{q}^2 B^2 N^{2\nu}}{\bar{\mathbf{q}}^2}\right)^{\frac{1}{2\nu}} \quad , \tag{15.33}$$

and we recall that $Q_R \to \infty$ implies $\bar{\mathbf{q}}^2 \to q_0^2$ (cf. Eq. (15.32)). Using these results together with the values $u^* = 0.364, n_0 = 0.53, q_0^2 = 50$ we find

$$J_d^{-1}(\mathbf{q},[p]) \underset{q^2R_g^2 \gg 1}{\longrightarrow} 0.439\left(q^2B^2N^{2\nu}\right)^{\frac{1}{2\nu}} \quad . \tag{15.34}$$

Due to the occurrence of the length parameter B this expression is not universal. A universal result is found by measuring q on the scale of the (monodisperse) radius of gyration $R_g[m]$. We thus define a critical ratio

$$\mathcal{R}_{\infty/2} = \lim_{q\to\infty}\left[\left(\frac{3}{d}R_g^2[m]q^2\right)^{-\frac{1}{2\nu}} J_d^{-1}(\mathbf{q},[p])\right] \quad . \tag{15.35}$$

In contrast to $\mathcal{R}_{4/2}$ etc. this ratio is a very important quantity since it combines the limits $q^2 \to 0, q^2 \to \infty$ and thus determines the overall shape of J_d^{-1}. Equations (15.14), (15.15), (15.34) yield

$$\mathcal{R}_{\infty/2} = 0.905 \quad \text{(direct evaluation)} \quad .$$

The ε-expansion (Eq. (A 15.35)) yields [Wit82]

$$\mathcal{R}_{\infty/2} = 0.782(\varepsilon = 1) \quad .$$

These results deviate strongly from the tree approximation

$$\mathcal{R}_{\infty/2} = \frac{1}{2} \quad \text{(zero loop)} \quad .$$

Duplantier [Dup86b] calculates a ratio, where $\frac{3}{d}R_g^2[m]$ is replaced by the end-to-end distance: $\frac{1}{2d}R_e^2[m]$. To find his best estimate he uses some kind of direct evaluation in $d = 3$, improved by information on a one-dimensional system. His result may be converted into an expression for $\mathcal{R}_{\infty/2}$ to yield $\mathcal{R}_{\infty/2} = 0.926$, remarkably close to our estimate. We however should recall that we used this ratio to determine q_0^2 (see Appendix A 13.2.1), searching for a region where it is insensitive to q_0^2. It is pleasing that in this way we found an estimate completely consistent with the quite different approach of [Dup86b], but we must still be aware of the ambiguities inherent in estimating a ratio which turns out to be far from its zero-loop value. This is also a good example for the inherent ambiguities of the ε-expansion. Expanding $\mathcal{R}_{\infty/2}^{-1}$ and numerically inverting the result from Eq. (A 15.35) we find a value $\mathcal{R}_{\infty/2} = 1.15$ instead of 0.78.

So far we have discussed the leading asymptotic behavior $J_d(\mathbf{q}) \sim q^{-1/\nu}$. We might also be interested in extracting the first corrections to these asymptotics. This, however, turns out to be a very complicated task. Quite generally, asymptotic expansions of critical correlation functions are known to proceed in a series of anomalous powers:

$$J_d(\mathbf{q}) \sim g_0 q^{-x_0} + g_1 q^{-x_1} + g_2 q^{-x_2} + \ldots; \quad x_0 = 1/\nu$$

where the subleading powers are definitely larger than x_0 ($x_1 = x_0 + 2$ in $d = 4$), but are not simply related to x_0. They are predicted by a field theoretic technique, known as 'short distance expansion' (see, for instance, [ZJ89], [Eis93]). Via the appropriate choice of ℓ_R our method deals correctly with the leading power $\sim q^{-1/\nu}$, but it does not reproduce the subleading terms of the short distance expansion. This, however, is no serious problem, since, as with the small q expansion, here also all subleading terms have to conspire to provide the correct analytical behavior for $q \to 0$. Thus these subleading terms have not much relevance individually, and their combined effect should be adequately given by our crossover calculation.

15.2.4 Crossover in momentum and excluded volume strength

We now evaluate Eqs. (15.25), (15.26) numerically, using Eq. (13.25), $w = 1$:

$$N_R = n_0 \left(1 - \frac{\bar{\mathbf{q}}^2}{q_0^2}\right)^{-1}$$

together with Eqs. (13.27), which relate $f, \bar{\mathbf{q}}^2$ to $\tilde{z}, \mathbf{q}^2 \tilde{R}_0^2$:

$$n_0^{-1/2} \tilde{z} = f \left|1 - f\right|^{-\frac{1}{2\nu\omega}} H_n^{-1/2}(f) \left(1 - \frac{\bar{\mathbf{q}}^2}{q_0^2}\right)^{-1/2}$$

$$\frac{1}{n_0} \mathbf{q}^2 \tilde{R}_0^2 = \left|1 - f\right|^{\left(2-\frac{1}{\nu}\right)\frac{1}{\omega}} H^{-1}(f) \bar{\mathbf{q}}^2 \left(1 - \frac{\bar{\mathbf{q}}^2}{q_0^2}\right)^{-1} .$$

For the excluded volume branch we use Eq. (13.35 ii):

$$\mathbf{q}^2 B^2 N^{2\nu} = \bar{\mathbf{q}}^2 \left(1 - \frac{\bar{\mathbf{q}}^2}{q_0^2}\right)^{-2\nu} .$$

We plot the results as function of $\mathbf{q}^2 R_g^2[m]$, where (cf. Eqs. (15.14), (15.15), (15.11))

$$\mathbf{q}^2 R_g^2[m] = \begin{cases} 0.427 \ \ \mathbf{q}^2 B^2 N^{2\nu} & (f = 1) \\ \mathbf{q}^2 \tilde{R}_0^2 \ \ \bar{\alpha}_g^2(f, [m]) & (f \neq 1) \end{cases} .$$

For $\bar{\alpha}_g^2$ we use the parameterizations (15.21), (15.23), respectively.

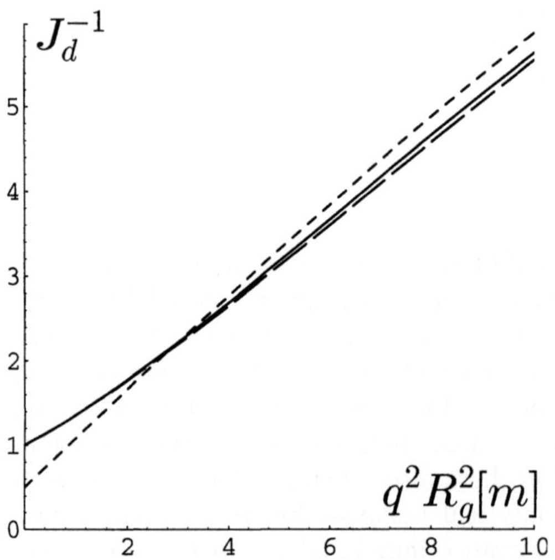

Fig. 15.5. Polydispersity dependence of $J_d^{-1}(\mathbf{q}, [p])$ in the excluded volume limit. *Full line*: monodisperse; *short dashes*: exponential distribution $\tilde{p}_2 = 2$; *long dashes*: Debye approximation, monodisperse

The choice of the 'monodisperse' variable $\mathbf{q}^2 R_g^2[m]$ in place of $\mathbf{q}^2 R_g^2[p]$ has the virtue to suppress polydispersity effects. This is illustrated in Fig. 15.5 which compares $J_d^{-1}(\mathbf{q}, [p])$ for the monodisperse and the equilibrium ensemble in the excluded volume limit. Polydispersity becomes important only for small momenta $q^2 R_g^2[m] \lesssim 2$. For larger momenta even the fairly broad exponential ensemble stays close to the monodisperse result. Figure 15.5 should be compared to Fig. 15.4, where we plotted J_d^{-1} as function of $q^2 R_g^2[p], p = m, e$. Since polydispersity effects die out so rapidly, in the sequel we restrict ourselves to the monodisperse ensemble.

Figure 15.6 shows $J_d^{-1}(\mathbf{q}, [m])$ evaluated in the excluded volume limit (full line) or for the value $\tilde{z} = 10$ in both the weak coupling or strong coupling regimes (long dashes). Also shown are the Debye approximation (short dashes) or the asymptotic power law (dotted line). For $q^2 R_g^2 \gtrsim 20$ the excluded volume behavior leaves the Debye approximation and bends over towards the asymptotic behavior. The full line runs more or less parallel to the asymptote, due to subleading terms which die out very slowly. Indeed, the leading corrections to the asymptotic behavior $q^{1/\nu}$ of J_d^{-1} are expected to behave as $q^{1/\nu - x}, x \approx 2$, so that they yield an almost constant contribution. The lines for $\tilde{z} = 10$ split off from the excluded volume limit towards larger (weak coupling) or smaller (strong coupling) values. With decreasing $\tilde{z}$ this effect becomes more pronounced. We finally note that plotting J_d in place of J_d^{-1} suppresses the shift of the large $-q$ behavior. Such a plot closely follows the asymptotics even for $q^2 R_g^2 \gtrsim 10$.

It is again instructive to calculate the effective exponent

$$\nu_{eff} = \left[\frac{\partial \ln J_d^{-1}(q)}{\partial \ln q} \right]^{-1} ,$$

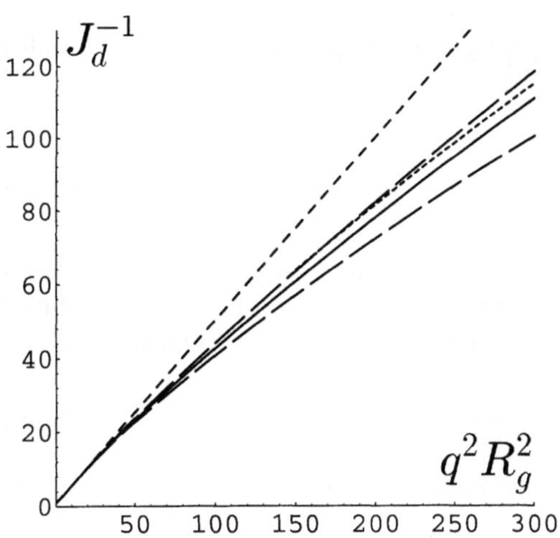

Fig. 15.6. $J_d^{-1}(\mathbf{q}, [m])$ as function of $q^2 R_g^2[m]$ to one-loop approximation. *Full line*: excluded volume limit; *long dashes*: $\tilde{z} = 10$, the *lower curve* is in the strong coupling regime; *short dashes*: Debye approximation; *dotted line*: asymptotic power law

which gives the inverse slope of a doubly logarithmic plot. Figure 15.7 shows the result in the excluded volume limit or for $\tilde{z} = 100$ or $\tilde{z} = 10$ in both the weak or strong coupling regions. Clearly, if we do not hit the excluded volume limit precisely, the effective exponent will run off to larger values (strong coupling) or towards $\nu = 0.5$ (weak coupling). In some intermediate range of $q^2 R_g^2$ it is varying only slowly, so that an experiment over some restricted momentum range can be described by an effective power law. We note that the weak coupling behavior is consistent with the qualitative scaling considerations based on 'temperature blobs' (Sect. 9.3).

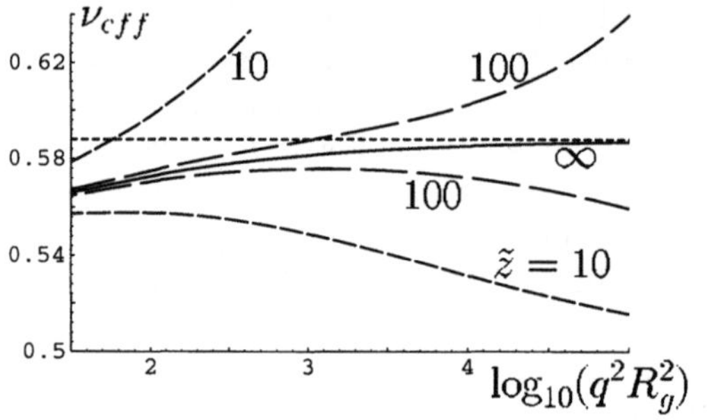

Fig. 15.7. Effective momentum dependent exponent ν_{eff} as function of $\log_{10}(\mathbf{q}^2 R_g^2[m])$. Values of $\tilde{z}$ are indicated at the curves. The *upper two curves* are in the strong coupling regime

15.3 The second virial coefficients

15.3.1 General results

The second virial coefficient A_2 can be defined in terms of the virial expansions of either the osmotic pressure

$$\frac{\hat{\Pi}}{k_B T} = c_p + \frac{1}{2} A_2^{\Pi} c_p^2 + O(c_p^3) \tag{15.36}$$

or the forward scattering intensity

$$c I_{d,c}^{-1}(\mathbf{q} = \mathbf{0}, c_p, N) = \frac{1}{N_w} + \frac{A_2^S}{N_w^2} c + O(c^2) \quad . \tag{15.37}$$

Recall that $c = N c_p$ is the monomer concentration and that we measure all concentrations as particle number per volume. As defined here, A_2^S is related to the standard definition given, for instance, in [Yam71] by the equation

$$A_2^S = 2 \frac{M_w^2}{N_A} A_2 \quad , \tag{15.38}$$

where M_w is the weight average molecular weight and N_A is Avogadro's number.

As shown in Appendix A 15.1 both A_2^Π and A_2^S can be expressed in terms of the same basic quantity

$$A_2(n_1, n_2) = -\Omega \frac{G_c^{(0,2)}(\mathbf{0}, \mathbf{0}, \mathbf{0}, \mathbf{0}; n_1, n_2)}{G^{(0,1)}(\mathbf{0}, \mathbf{0}; n_1) G^{(0,1)}(\mathbf{0}, \mathbf{0}; n_2)} \quad , \tag{15.39}$$

which is the virial coefficient of two chains of prescribed lengths:

$$A_2^\Pi = \sum_{n_1, n_2} P(n_1) P(n_2) A_2(n_1, n_2) \tag{15.40}$$

$$A_2^S = \sum_{n_1, n_2} \frac{n_1\, n_2}{N^2} P(n_1) P(n_2) A_2(n_1, n_2) \quad . \tag{15.41}$$

For a monodisperse system A_2^S and A_2^Π are identical, but they differ for general polydispersity. Here we will concentrate on the directly observable quantities A_2^Π, A_2^S, postponing a discussion of $A_2(n_1, n_2)$ to the next chapter.

The one-loop expressions for A_2^Π, A_2^S are given in Appendix A 15.3.3. Using Eq. (13.29) to eliminate ℓ_R in three dimensions we find

$$(4\pi)^{-3/2} \frac{A_2^\Pi}{\tilde{R}_0^3} = a_2(f, c_A^\Pi) \tag{15.42 i}$$

$$(4\pi)^{-3/2} \frac{A_2^S}{(\tilde{p}_2)^2 \tilde{R}_0^3} = a_2(f, c_A^S) \quad , \tag{15.42 ii}$$

where $(n_0 = 0.53)$

$$\begin{aligned} a_2(f, c_A) &= \left(\frac{\ell_R}{\tilde{R}_0}\right)^3 \bar{a}_2(f, c_A) \\ &= 0.1325 \Big|1 - f\Big|^{\frac{3}{\omega}\left(\frac{1}{2\nu} - 1\right)} f H^{3/2}(f)(1 + 0.348 f + 0.728\, c_A f) \\ \frac{3}{\omega}\left(\frac{1}{2\nu} - 1\right) &= -0.561 \quad . \end{aligned} \tag{15.43}$$

The polydispersity corrections take the form

$$\begin{aligned} c_A^\Pi = -0.3883 \Big[5 \int_0^\infty dy\, p(y) y^{3/2} + 2 \int_0^\infty dy\, p(y) y^{5/2} \\ - \int_0^\infty dy_1 dy_2 p(y_1) p(y_2)(y_1 + y_2)^{5/2} + 4\sqrt{2} - 7 \Big] \end{aligned} \tag{15.44 i}$$

or

$$\begin{aligned} c_A^S = -0.3883 \Big[5 \int_0^\infty dy \frac{p(y)}{\tilde{p}_2} y^{5/2} + 2 \int_0^\infty dy \frac{p(y)}{(\tilde{p}_2)^2}\, y^{7/2} \\ - \int_0^\infty dy_1 dy_2 \frac{p(y_1) p(y_2)}{(\tilde{p}'')^2} y_1 y_2 (y_1 + y_2)^{5/2} + 4\sqrt{2} - 7 \Big] \quad . \end{aligned} \tag{15.44 ii}$$

For the set of Schultz distributions (Eq. (5.43)) c_A^{Π}, c_A^{S} can be expressed in terms of Γ-functions. In the relevant range $1 \le \tilde{p}_2 \le 2$ the result can be parameterized as

$$c_A^{\Pi}(\tilde{p}_2) = -0.125\ (\tilde{p}_2 - 1) \tag{15.45 i}$$

$$c_A^{S}(\tilde{p}_2) = -0.385\ (\tilde{p}_2 - 1) + 0.107\ (\tilde{p}_2 - 1)^2 - 0.027\ (\tilde{p}_2 - 1)^3\ , \tag{15.45 ii}$$

the errors being less than 1.0% or 0.3%, respectively.

15.3.2 Excluded volume limit

In the excluded volume limit we substitute the result (13.35 ii) for ℓ_R, $(w = 1, \bar{\mathbf{q}}^2 = 0)$, into Eqs. (A 15.38) to find

$$(4\pi)^{-3/2}\, A_2^{\Pi *} = (0.0689 + 0.0372 c_A^{\Pi}) B^3 N^{3\nu} \tag{15.46 i}$$

$$(4\pi)^{-3/2}\, A_2^{S *} = (0.0689 + 0.0372 c_A^{S}) \tilde{p}_2^2 B^3 N^{3\nu}\ . \tag{15.46 ii}$$

The light scattering virial coefficient for $1 \le \tilde{p}_2 \lesssim 1.6$ to a good approximation can be represented as

$$(4\pi)^{-3/2}\, A_2^{S *} = 0.0689 B^3 N_w^{3\nu}\ . \tag{15.47}$$

15.3.3 Weak coupling region

We evaluate Eqs. (15.42) (15.43), using $f = f(\tilde{z})$ as given by Eq. (15.13). For the osmotic virial coefficient plotted as $(4\pi)^{-3/2}\, A_2^{\Pi} / \tilde{R}_0^3$ vs $\tilde{z}$ we see some effect of polydispersity (Fig. 15.8). For the light scattering virial coefficient this effect is suppressed almost completely, if we plot $(4\pi)^{-3/2}\, A_2^{S} / \tilde{R}_{0,w}^3$ against $\tilde{z}_w = \tilde{p}_2^{1/2} \tilde{z}$. Here

$$\tilde{R}_{0,w}^2 = \tilde{p}_2 \tilde{R}_0^2 = \tilde{\ell}^2 N_w\ . \tag{15.48}$$

The plot therefore coincides with the appropriate reinterpretation of the full line in Fig. 15.8.

We can find a simple parameterization of these results. It turns out that the function $a_2(f(\tilde{z}), c_A)$, Eq. (15.43), within 0.5% deviation can be represented as

$$\tilde{a}_2(\tilde{z}, c_A) =$$

$$0.182\tilde{z} \left[1 + (2.15 - 4.24\, c_A)\tilde{z} + \frac{0.82\tilde{z}^2}{(1 + 0.54\, c_A)^{4.24}} \right]^{-0.236}\ , \tag{15.49}$$

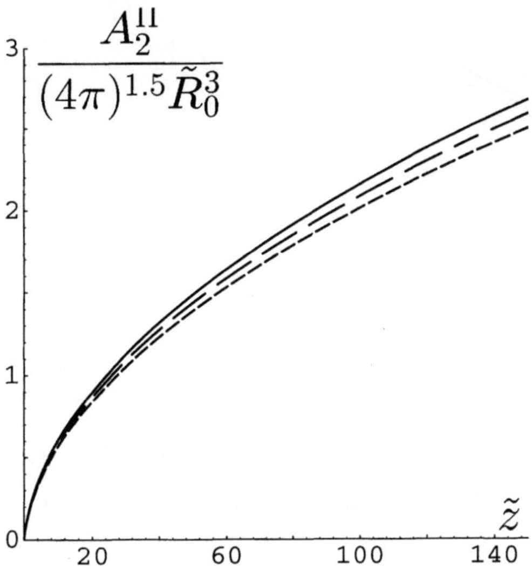

Fig. 15.8. Reduced second virial coefficient as function of $\tilde{z}$. *Full line*: monodisperse; *long dashes*: Schultz distribution $\tilde{p}_2 = 1.5$; *short dashes*: exponential ensemble ($\tilde{p}_2 = 2$)

where again the full one-loop result is reproduced for $\tilde{z} \to 0$ or $\tilde{z} \to \infty$. In view of the insensitivity with respect to polydispersity this implies for A_2^S the parameterization

$$(4\pi)^{-3/2} \frac{A_2^S}{\tilde{R}_{0,w}^3} = \frac{0.182\tilde{z}_w}{(1 + 2.15\tilde{z}_w + 0.82\tilde{z}_w^2)^{0.236}} , \tag{15.50}$$

valid to a good approximation for normal chain length distributions.

15.3.4 Strong coupling region

Qualitatively the behavior of the virial coefficients in the strong or weak coupling region looks quite similar, except that in the strong coupling region we are restricted to $\tilde{z} \gtrsim 3$. $A_2^{II}/\tilde{R}_0^2$ as function of $\tilde{z}$ again shows some polydispersity dependence, whereas $A_2^S/\tilde{R}_{0,w}^3$ as function of $\tilde{z}_w$ is essentially independent of polydispersity. A simple parameterization reads

$$(4\pi)^{-3/2} \frac{A_2^S}{\tilde{R}_{0,w}^3} = 0.191\tilde{z}_w^{0.528} \left[1 + 0.528\tilde{z}_w^{-0.941}\left(1 + \frac{2.18}{\tilde{z}_w}\right)\right] . \tag{15.51}$$

For the monodisperse case it is accurate up to errors of 0.5%.

For a monodisperse system ($A_2^{II} \equiv A_2^S \equiv A_2$) Fig. 15.9 shows the three branches of $(4\pi)^{-3/2} A_2/\tilde{R}_0^3$. The strong coupling branch lies above and the weak coupling branch lies below the excluded volume power law, respectively. Again the approach towards the asymptotic power law is quite slow.

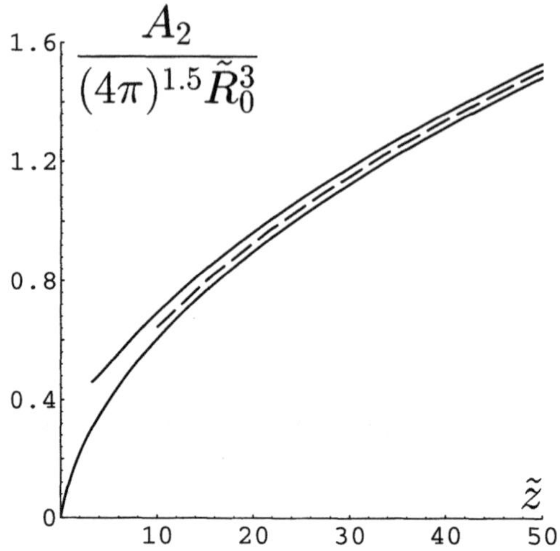

Fig. 15.9. Reduced second virial coefficient (monodisperse) as function of $\tilde{z}$. *Broken line*: asymptotic power law; *upper (lower) full line*: strong (weak) coupling branch

15.4 The interpenetration ratio

Combining the second virial coefficient and the radius of gyration we construct the interpenetration ratio

$$\psi^{S(\Pi)} = \left(\frac{d}{12\pi}\right)^{d/2} \frac{A_2^{S(\Pi)}}{R_g^d} \quad . \tag{15.52}$$

Roughly speaking, A_2 measures the volume which a given chain effectively excludes for all the others. Thus the dimensionless ratio ψ is proportional to the fraction of the coil volume excluded. To get a feeling for this quantity we may evaluate it for hard spheres of radius $R = \sqrt{\frac{5}{3}} R_g$ to find ($d = 3$)

$$\psi_{hardsphere} = (4\pi)^{-3/2} \frac{4}{3}\pi(2R)^3/R_g^3 \approx 1.6 \quad .$$

The interpenetration ratio has received much attention. For fixed polydispersity it is a universal function of the theoretical swelling factor $\bar{\alpha}_g^2$, reducing to a universal number ψ^* in the excluded volume limit. This results from the fact that both ψ and $\bar{\alpha}_g^2$ are well defined functions of $\tilde{z}$, so that we can eliminate this intermediate microstructure dependent variable. As a result we of course find two branches of the relation $\psi \Leftrightarrow \bar{\alpha}_g^2$, corresponding to strong or weak coupling, respectively.

In the sequel we consider $\psi = \psi^S$, only. Results for ψ^Π are similar qualitatively and are easily derived from the expressions given above.

15.4.1 Excluded volume limit

The ε-expansion of ψ^* can be constructed from the results given in Appendix A 15.3. In particular, for the monodisperse ensemble from Eqs. (A 15.31), (A 15.41) we find

$$\begin{aligned}\psi^* &= \frac{\varepsilon}{8}\left(1+\frac{77}{96}\,\varepsilon+\frac{\varepsilon}{2}\,\ln\,2+O(\varepsilon^2)\right)\\ &= 0.269 \quad (\varepsilon=1)\ .\end{aligned} \tag{15.53}$$

We note that the first order correction is extremely large. Higher orders have not been calculated.

Evaluating Eq. (15.52) directly in three dimensions we find, keeping n_0 explicit,

$$\psi^* = (\tilde{p}_2)^2\left(\frac{\tilde{p}_2}{\tilde{p}_3}\right)^{3/2} 0.182 n_0^{1/2}\frac{1.728-0.521n_0^{1/2}+c_A^S n_0^{1/2}}{\left(0.636+0.232n_0^{1/2}+c_g n_0^{1/2}\right)^{3/2}}\ . \tag{15.54}$$

For the monodisperse case $\tilde{p}_2 = 1 = \tilde{p}_3, c_A^S = 0 = c_g$ we have used this expression to fix n_0 in Sect. 13.2.2. Thus the value $\psi^*(n_0 = 0.53) = 0.247$ is not to be considered as a prediction in our formulation of the theory, except for the consistency condition that it should be found in the region where ψ^* is not very sensitive to n_0.

ψ^* turns out to depend strongly on polydispersity. Figure 15.10 shows the result (15.54), $n_0 = 0.53$, for Schultz distributions, where we also included the result predicted from the zero-loop behavior $\psi^* \sim (\tilde{p}_2)^2\,(\tilde{p}_2/\tilde{p}_3)^{3/2}$. We find

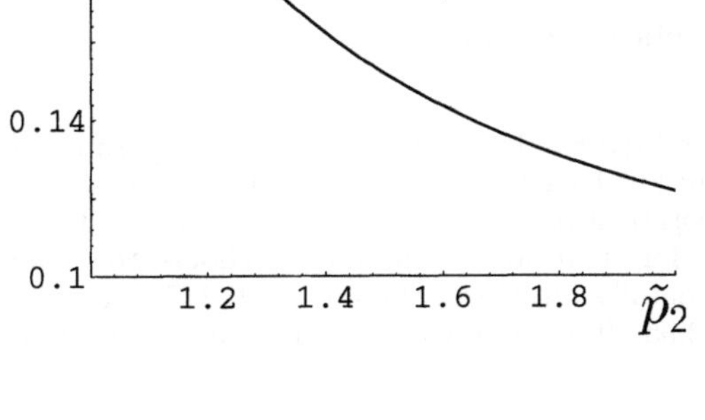

Fig. 15.10. ψ^* as function of $\tilde{p}_2$ for Schultz distributions. *Full line*: one-loop; *broken line*: polydispersity correction evaluated to zero-loop order

a decrease with increasing polydispersity, the effect being strongly enhanced by the one-loop effects. It is due to the strong increase with polydispersity of R_g (cf. Eq. (15.15) and Fig. 15.1). As a result, a meaningful experimental determination of ψ^* needs samples of well characterized polydispersity.

15.4.2 Crossover behavior

It is found that ψ/ψ^* as function of $\tilde{z}_w$ is almost independent of polydispersity. Figure 15.11 shows both the weak coupling ($\psi/\psi^* < 1$) and the strong coupling branches ($\psi/\psi^* > 1$), as well as the fixed point branch $\psi = \psi^*$. The plot is for the monodisperse ensemble, but even the exponential ensemble ($\tilde{p}_2 = 2$) yields results which on this scale could hardly be distinguished from the monodisperse results. We again see the characteristic features of all these crossover plots: ψ/ψ^* with increasing $\tilde{z}_w$ fairly rapidly approaches values close to its limiting value, the final convergence to the limit, however, being quite slow. This holds for both the weak or strong coupling branches.

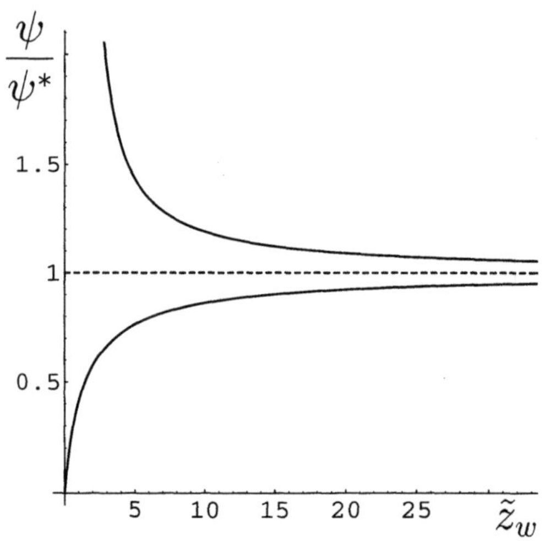

Fig. 15.11. The three branches of ψ/ψ^* as function of $\tilde{z}_w$

15.4.3 Some comment on the choice of n_0

We found here a consistent set of one-loop results for the important experimental observables in the dilute limit. They are based on a direct evaluation in $d = 3$, avoiding the ε-expansion. Of the theoretical parameters fixing the uncritical manifold, only n_0 shows up in the results for small momenta, and we choose the value $n_0 = 0.53$. Adjusting to the experimentally measured value $\psi^* = 0.247$ (monodisperse) we could also choose $n_0 = 2.95$. It turns out that this makes almost no difference for scaling functions of monodisperse systems. We only have to readjust the nonuniversal scale of $\tilde{z}$ to find results for $\tilde{\alpha}_g^2, A_2^{\pi(S)}, \psi$ which are indistinguishable for the two choices of n_0 on the scale of our figures. Only for $J_d^{-1}(\mathbf{q})$ can we detect some difference, the results for $n_0 = 2.95$ and $(\mathbf{q}R_g)^2 \gtrsim 20$ being about 5% larger than those shown in Fig. 15.6. This remarkable stability of the quantitative predictions strongly supports the validity of our method.

The situation changes somewhat if we consider the polydispersity dependence. It turns out that the choice $n_0 = 2.95$ dramatically magnifies all polydispersity effects. In going from $\tilde{p}_2 = 1$ to $\tilde{p}_2 = 2$, ψ^*, in particular, is found to decrease by a factor of 5 instead of a factor of 2 as exhibited for $n_0 = 0.53$ in Fig. 15.10. This is due to the factor of $n_0^{1/2}$ multiplying all polydispersity corrections. It is these unreasonably strong polydispersity effects which led us to favor $n_0 = 0.53$. Still this strong sensitivity of the polydispersity effects indicates that they cannot be trusted with the same confidence as the crossover results for small polydispersity. Fortunately, however, polydispersity effects typically are restricted to the region of small overlap and small momenta. Furthermore, they are also not very well known on the experimental side.

15.5 Comparison to experiment

15.5.1 Some remarks on data analysis

With the only exception of A_2^{Π}, all quantities considered in this chapter are typically measured by light scattering. In the present section we therefore concentrate on scattering experiments, which yield the (inverse) single chain form factor as

$$\lim_{c\to 0} \frac{c}{I_{d,c}(\mathbf{q})} = \frac{1}{N\, J_d(\mathbf{q})} \quad . \tag{15.55}$$

The extrapolation to zero concentration of course has problems of its own. We will discuss it in later chapters, after we have calculated the scattering function for finite overlap. Furthermore the measured intensity – the 'Rayleigh ratio' – is related to $I_{d,c}$ by some nonuniversal prefactor of microscopic origin. For this and other experimental problems we refer to standard literature [Kra87]. We here assume that $N/J_d(\mathbf{q})$ has indeed been accurately determined.

In a scattering experiment $I_{d,c}(\mathbf{q})$ is measured as function of the scattering angle ϑ, which is related to $\mathbf{q}$ as

$$\mathbf{q}^2 = \left(\frac{4\pi}{\lambda_0}\, sin\, \frac{\vartheta}{2}\right)^2 \quad . \tag{15.56}$$

Here λ_0 is the wavelength in the solution of the radiation used. Some order of magnitude estimates are appropriate. For light scattering a typical wave length is $\lambda_0 \approx 3000\,\text{Å}$, and in typical experiments ϑ covers some range of order $20^o \lesssim \vartheta \lesssim 160^o$. With these numbers we can estimate the range of $\mathbf{q}^2$:

$$q_{min}^2 \leq \mathbf{q}^2 \leq q_{max}^2$$
$$q_{min}^2 \approx 5 \cdot 10^{-7}\,\text{Å}^{-2}; \quad q_{max}^2 \approx 2 \cdot 10^{-5}\,\text{Å}^{-2} \quad .$$

In the present context $\mathbf{q}^2 R_g^2[m]$ is the relevant dimensionless variable. Using typical values of $R_g^2[m]$ as measured for polystyrene in cyclohexane ($T = \Theta$): $R_g^2[m] \approx 8N\,[\text{Å}^2]$ or in benzene (good solvent): $R_g^2[m] \approx 4N^{2\nu}\,[\text{Å}^2]$ we find the results of Table 15.1. The extrapolation to $\mathbf{q} = 0$ should thus pose no problem for $N \lesssim 50{,}000$ in a Θ-solvent, or for $N \lesssim 20{,}000$ in a good solvent. For longer chains a direct determination of R_g^2, being defined in terms of the slope of $N/J_d(\mathbf{q})$ for $\mathbf{q} \to 0$, becomes ambiguous. On the other hand, only for long chains can we reach the region $q^2 R_g^2 \gtrsim 20$, where in good solvents deviations from the Debye approximation are to be expected.

Table 15.1. Typical experimental range of the scaling variable $q^2 R_g^2$

	N	$q_{min}^2 R_g^2[m]$	$q_{max}^2 R_g^2[m]$	$R_g[m][\text{Å}]$
Θ-solvent	1000	0.004	0.16	90
	10000	0.04	1.6	280
	100000	0.4	16	900
good solvent	1000	0.006	0.26	116
	10000	0.1	4	450
	100000	1.5	60	1740

Neutron scattering uses very small wavelengths: $\lambda_0 \approx 5\,\text{Å}$. Large values of $\mathbf{q}^2 R_g^2$ are easily reached, but clearly the method is not appropriate for covering the region of small $\mathbf{q}^2 R_g^2$.

We now turn to the question how to determine R_g^2 for chains so long that the data do not allow for a straightforward extrapolation to $\mathbf{q} = 0$. In that case we clearly need some model of $N/J_d(\mathbf{q})$, valid in the region of small $\mathbf{q}^2 R_g^2$. We discuss here two approaches, proposed by Berry [Ber66] or Fujita [Fuj70], respectively.

Berry's method is based on the observation that the inverse (monodisperse) Debye function $D^{-1}(x)$ for $x \lesssim 6$ shows pronounced curvature (see Fig. 15.5), which however is greatly reduced by plotting $D^{-1/2}(x)$ in place of $D^{-1}(x)$. Thus Berry's approximation consists in fitting the data in the range of small momenta to the expression

$$\left(\frac{1}{N\,J_d(\mathbf{q})}\right)^{1/2} = \frac{1}{N_w^{(B)1/2}}\left(1 + \frac{q^2 R_g^{(B)2}}{6}\right) \quad , \tag{15.57}$$

where the superscript (B) stands for Berry's square root plot. Now according to our theory $J_d(\mathbf{q})$ in the range $\mathbf{q}^2 R_g^2 \lesssim 10$ is well represented by the Debye function, generalized to arbitrary polydispersity:

$$J_d(\mathbf{q}, [p]) = D_p\left(\frac{\tilde{p}_2}{\tilde{p}_3}(\mathbf{q}R_g[p])^2\right) \quad . \tag{15.58}$$

We thus can use this expression as 'exact' input to test the results of a square root plot based on some prescribed range of $q^2 R_g^2$. Figure 15.12 shows the results in terms of correction factors $N_w^{(B)}/N_w, R_g^{(B)2}/R_g^2$, which have been determined by fixing $N_w^{(B)}, R_g^{(B)2}$ (Eq. (15.57)) in terms of two 'exact data points' $(NJ_d(q_1), q_1 R_g^2)$, $(NJ_d(q_2), q_2 R_g^2)$ calculated from Eq. (15.58) $(q_1 R_g^2 = q^2 R_g^2 - 0.15,\ q_2 R_g^2 = q^2 R_g^2 + 0.15)$. We find that both N_w and R_g^2 are systematically underestimated. The errors are quite small, however, as long as the experiment covers a region $q^2 R_g^2 \lesssim 1.5$. From Fig. 15.12 we then find deviations reaching about 1% for N_w or 4% for R_g^2. We thus might conclude that we can safely use the square root plot except for extremely long chains. This is true, however, only for very narrow chain length distributions. Even a small polydispersity makes a large effect. As is obvious from Fig. 15.5 an increase of polydispersity suppresses the initial curvature of $J_d^{-1}(\mathbf{q})$. The effect on the square root plot is quite dramatic, as is illustrated in Fig. 15.12 by including effective parameters extracted for small polydispersity $N_w/N - 1 = \tilde{p}_2 - 1 = 0.1$. For R_g^2 in particular, the estimated value rapidly falls below 0.9 of the true value. Since experimentally polydispersity in most cases is not well known, this is a source of sizeable systematic errors, which are hard to estimate.

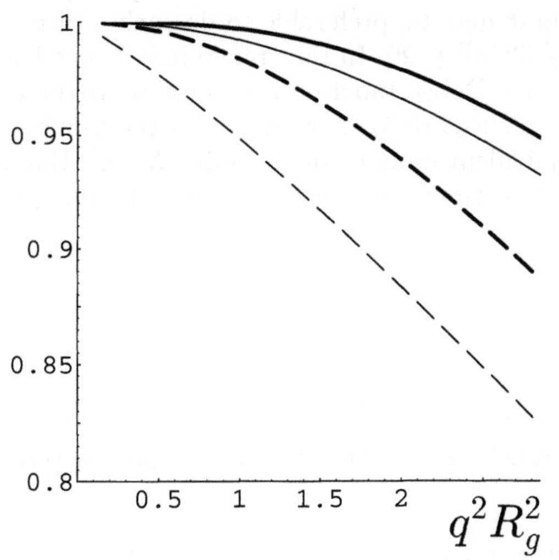

Fig. 15.12. Effective parameters $N_w^{(B)}/N_w$ (*full lines*) or $R_g^{(B)2}/R_g^2$ (*broken lines*). Monodisperse ensemble: *fat lines*, Schultz distribution $\tilde{p}_2 = 1.1$; *thin lines*

For very long chains Fujita [Fuj70] suggested another method of data analysis. It is also based on the observation that $J_d(\mathbf{q},[m])$ for not too large $\mathbf{q}^2 R_g^2[m]$ closely follows the Debye function. Within the Debye approximation, however, it is easily checked that a plot of

$$\frac{1}{Y_F} = \lim_{c\to 0} \frac{c}{I_{d,c}(\mathbf{q}^2)} = \frac{1}{N D(\mathbf{q}^2 R_g^2)}$$

against

$$Z_F = \left(q^4 Y_F(q^2)\right)^{-1} \int_0^{q^2} dx\, x^2 Y_F(x)$$

is strictly linear:

$$\frac{1}{Y_F} = \frac{R_g^2}{N} Z_F + \frac{1}{N} \quad .$$

Fitting the integrated data to a linear relation $1/Y_F \leftrightarrow Z_F$ we thus can immediately extract N and R_g^2.

From Fig. 15.4 we conclude that this should be a good procedure for chains of arbitrary length, as long as the range of data is restricted to $q^2 R_g^2 \lesssim 20$. Again, however, this method is sensitive to polydispersity. For $\tilde{p}_2 > 1$ the relation $1/Y_F \leftrightarrow Z_F$ develops some curvature, and a linear approximation based on some intermediate range of $q^2 R_g^2$ underestimates both N_w and R_g^2. The effects are quite similar to those shown for the square root plot.

Fujita's method is equivalent to directly fitting to the (monodisperse) Debye function. In view of our results it may be preferable to do such a direct fit for momenta in the range $3 \lesssim q^2 R_g^2[m] \lesssim 20$. In that range polydispersity effects are weak and the monodisperse Debye function is a good approximation irrespective of polydispersity (cf. Fig. 15.5). This fit yields the number average chain length N and the equivalent monodisperse radius $R_g^2[m]$. Data for $q^2 R_g^2[m] \lesssim 3$ may then be used to estimate N_w and the true radius $R_g^2[p]$.

15.5.2 α_g^2, A_2, ψ: Θ-solvents

Our results for the theoretical swelling factor $\bar{\alpha}_g^2$ or the second virial coefficients A_2^S on the weak coupling branch are summarized by the parameterizations

$$\bar{\alpha}_g^2 = \left[1 + (1.32 + 5.263 c_g)\tilde{z} + 0.378\,(1 + 0.904 c_g)^{5.682}\, \tilde{z}^2\right]^{0.176} \quad , \qquad \text{(Eq. 15.21)}$$

$$(4\pi)^{-3/2}\,\frac{A_2^S}{\tilde{R}_{0,w}^3} = 0.182\tilde{z}_w\left(1+2.15\tilde{z}_w+0.82\tilde{z}_w^2\right)^{-0.236}\,, \qquad \text{(Eq. 15.50)}$$

$$c_g = 0.288\,(\tilde{p}_2-1) - 0.070\,(\tilde{p}_2-1)^2\,. \qquad \text{(Eq. 15.10)}$$

The interpenetration ratio ψ^S (Eq. (15.52)) can be expressed in terms of $\bar{\alpha}_g^2, A_2^S$ as

$$\psi^S = (4\pi)^{-3/2}\,\frac{A_2^S}{\tilde{R}_{0,w}^3}\left(\frac{\tilde{R}_{0,w}}{R_g}\right)^3 \tag{15.59}$$

$$\left(\frac{\tilde{R}_{0,w}}{R_g}\right)^3 = \left(\frac{\tilde{p}_2^2}{\tilde{p}_3}\right)^{3/2}\bar{\alpha}_g^{-3} = \left(2-\frac{1}{\tilde{p}_2}\right)^{-3/2}\bar{\alpha}_g^{-3}\,, \tag{15.60}$$

where the last relation holds for Schultz distributions. Since most experiments of interest use light scattering, we do not consider A_2^{Π} or ψ^{Π} here. We recall the relations

$$\tilde{z} = \tilde{v}(T)N^{1/2}; \quad \tilde{R}_0^2 = \tilde{\ell}^2(T)N\,,$$
$$\tilde{z}_w = \tilde{p}_2^{1/2}\tilde{z} = \tilde{v}(T)N_w^{1/2}; \quad \tilde{R}_{0,w}^2 = \tilde{p}_2\tilde{R}_0^2 = \tilde{\ell}^2(T)N_w\,.$$

As has been stressed before (see Sect. 11.5), the temperature dependence of $\tilde{v}(T), \tilde{\ell}(T)$ is not predicted by our theory. In particular, the experimental swelling factor

$$\alpha_g^2 = \frac{R_g^2(T)}{R_g^2(\Theta)} \tag{15.61}$$

may differ from $\bar{\alpha}_g^2$. Indeed, writing

$$R_\Theta^2 = \frac{\tilde{p}_3}{\tilde{p}_2}\tilde{R}_0^2(T=\Theta) = \frac{\tilde{p}_3}{\tilde{p}_2}\tilde{\ell}^2(\Theta)N\,, \tag{15.62}$$

we find from Eqs. (15.11), (15.61)

$$\alpha_g^2 = \frac{\tilde{\ell}^2(T)}{\tilde{\ell}^2(\Theta)}\bar{\alpha}_g^2\,. \tag{15.63}$$

It is only close to the Θ-temperature, that the approximations of naive two parameter theory

$$\tilde{\ell}^2(T) = \tilde{\ell}^2(\Theta), \quad \tilde{v}(T) = v_T t, \quad t = \frac{T-\Theta}{\Theta}\,. \tag{15.64}$$

can be expected to hold.

Figures 15.13–15.15 show results for $\bar{\alpha}_g^2$ and $(4\pi)^{-3/2}A_2^S/\tilde{R}_{0,w}^3$ as function of $\tilde{z}_w$, or ψ^S as function of $\bar{\alpha}_g^2$, respectively. For these plots we selected systems where the temperature dependence of the parameters $\tilde{\ell}, \tilde{v}(T)/t$ seems to be weak. Specifically the parameter values chosen are

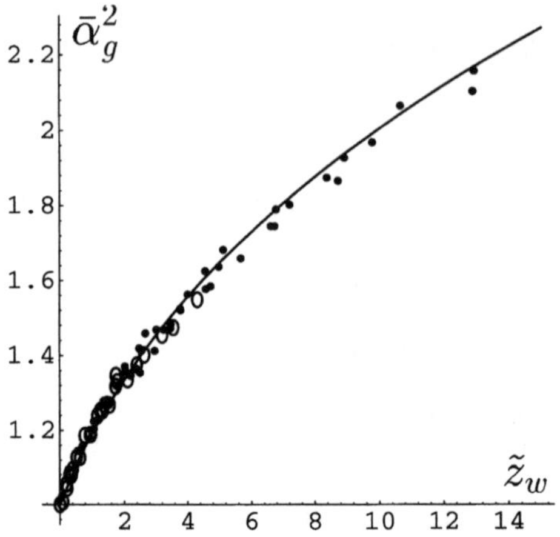

Fig. 15.13. $\bar{\alpha}_g^2$ as function of $\tilde{z}_w$, weak coupling. Data: polystyrene-cyclohexane (•) [MEF78], [MF81], [YFTY71], P-isobutylene-isoamylisovalerate (∘) [MNF72]

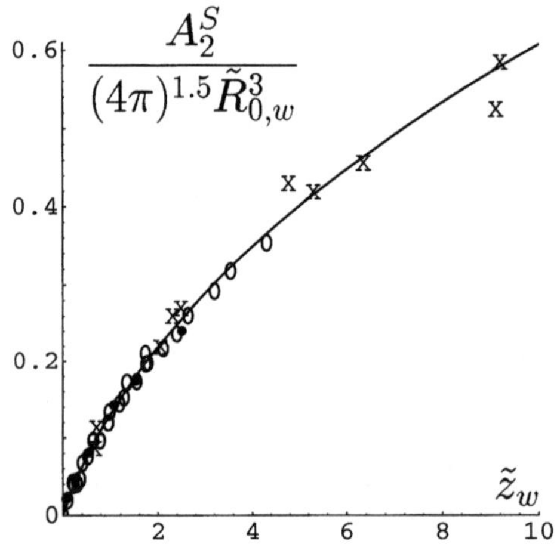

Fig. 15.14. $(4\pi)^{-3/2}A_2^S/\tilde{R}_{0,w}^3$ as function of $\tilde{z}_w$, weak coupling, symbols as in Fig. 15.13. Additional data on polystyrene-transdecaline (X) [FFKH74]

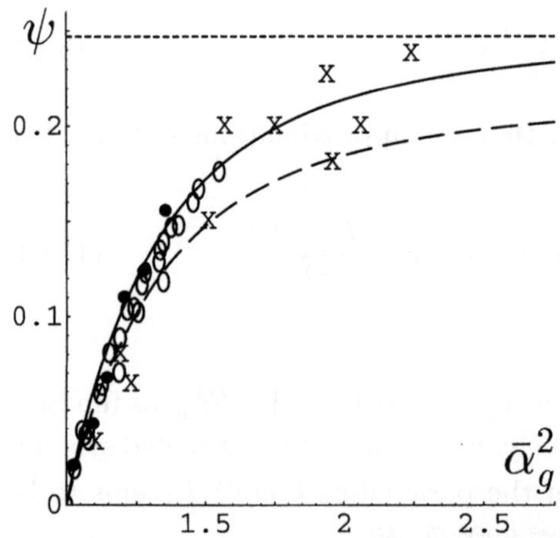

Fig. 15.15. ψ as function of $\bar{\alpha}_g^2$, weak coupling, symbols as in Fig. 15.14. The *full line* gives the one-loop result for a monodisperse system; *broken line*: $\tilde{p}_2 = 1.1$. The *dotted line* gives ψ^*

polystyrene-cyclohexane

$$\tilde{v}(T) = 0.35t; \quad \tilde{\ell} = 3.0\,\text{Å} \quad ; \quad 0 \le t \lesssim 0.067$$

P-isobutylene-isoamyl isovalerate

$$\tilde{v}(T) = 0.13t(1 - 0.7t); \quad \tilde{\ell} = 2.30\,\text{Å} \quad ; \quad 0 \le t \lesssim 0.13$$

($\tilde{v}(T) = 0.13t$ would also yield an acceptable fit)

polystyrene-transdecaline

$$\tilde{v}(T) = 0.43t; \quad \tilde{\ell} = 2.9\,\text{Å} \quad ; \quad 0 \le t \lesssim 0.12 \quad .$$

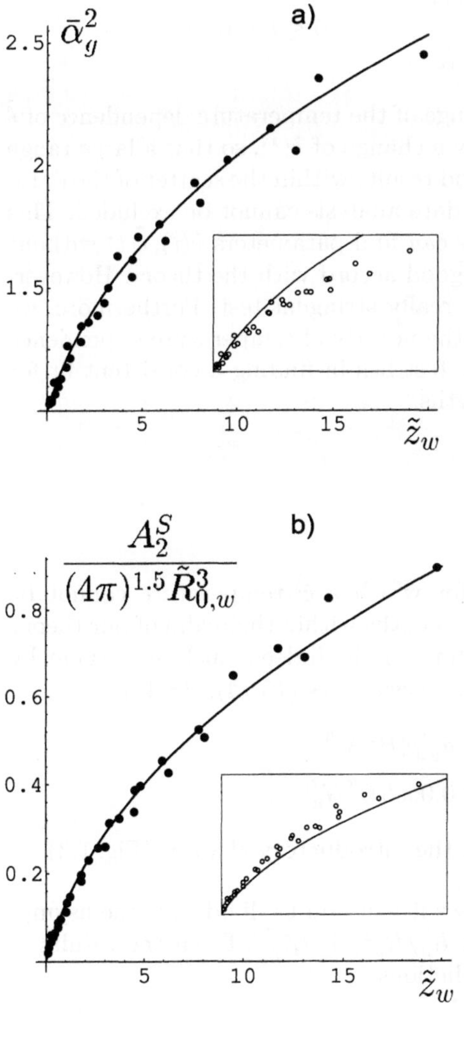

Fig. 15.16a, b. System P-chloroprene-transdecaline [NKTF69]: (**a**) $\bar{\alpha}_g^2$ as function of $\tilde{z}_w$; (**b**) $(4\pi)^{-3/2} A_2^S / \tilde{R}_{0,w}^3$ as function of $\tilde{z}_w$. The curves are calculated with $\tilde{\ell}(T), \tilde{v}(T)$ given in the text. The *inserts* give a fit with the ansatz $\tilde{v}(t) = v_T t, \tilde{\ell} = const$

For each system results for several chain lengths are included, but not distinguished in the figures. Clearly the data are consistent with the theory.

There are other systems, where a nontrivial temperature variation of the parameters is clearly visible. An example is shown in Fig. 15.16, with data for three different chain lengths ($N_w \approx 6600, 9800; 18{,}800$) of P-chloroprene in transdecaline. No reasonable fit with $\tilde{\ell} = const, \tilde{v}(T)/t = v_T = const$ is possible. In particular, fixing $\tilde{v}_T$ by the slope of α_g^2 for $\tilde{z}_w \to 0$, we see clear deviations for larger $\tilde{z}_w$ (inserts in Figs. 15.16). The good fit shown is found for the choice

P-chloroprene-transdecaline

$$\begin{aligned} \tilde{v}(T) &= 0.38t(1 + 3t + 25t^2) \\ \tilde{\ell} &= 2.65(1 - 2t)^{1/2}\,\text{Å} \quad . \end{aligned} \qquad 0 \leq t \lesssim 0.17$$

However, it must be noted, that a change of the temperature dependence of $\tilde{\ell}$ can be compensated to some extent by a change of $\tilde{v}(t)$, so that a large range of parameterizations yields equally good results within the scatter of the data. Furthermore systematic errors due to data analysis cannot be excluded. This example is quite typical. We generally can find parameters $\tilde{v}(t), \tilde{\ell}(t)$ without any problem which bring the data in good accord with the theory. However, the data are not precise enough for a really stringent test. Furthermore, we want to recall and stress that, due to the nontrivial temperature dependence of $\tilde{\ell}, \tilde{v}$ allowed by the theory, the true test lies in finding a consistent fit for both α_g^2 and A_2^S for several chain lengths.

15.5.3 α_g^2, A_2, ψ: Good Solvents

There are polymer solvent systems for which a Θ-temperature cannot be reached. We then expect that for chain lengths within the realm of our theory we are fairly close to the excluded volume limit. Indeed, such data typically have been described by the asymptotic power laws (15.14), (15.47):

$$\begin{aligned} R_g^{*2} &= a_g[p] B^2 N_w^{2\nu} \\ (4\pi)^{-3/2} A_2^{*S} &= 0.0689 B^3 N_w^{3\nu} \quad . \end{aligned}$$

Typical examples have been shown in the introductory chapter (Fig. 1.3).

To detect deviations from these laws it is useful to divide out the asymptotic behavior, thus concentrating on $R_g^2/R_g^{*2}, A_2^S/A_2^{S*}$. From the results of the previous sections we take the predictions

$$\frac{A_2^S}{A_2^{S*}} = \begin{cases} 1 + 0.528\tilde{z}_w^{-0.941}\left(1 + \dfrac{2.18}{\tilde{z}_w}\right) & \text{strong coupling} \\ \left(1 + \dfrac{2.622}{\tilde{z}_w} + \dfrac{1.219}{\tilde{z}_w^2}\right)^{-0.236} & \text{weak coupling} \end{cases} \tag{15.65}$$

$$\frac{R_g^2}{R_g^{*2}} = \begin{cases} 1 - 0.54\tilde{z}_w^{-0.941}\left(1 + \dfrac{0.44}{\tilde{z}_w}\right) & \text{strong coupling} \\ \left(1 + \dfrac{3.492}{\tilde{z}_w} + \dfrac{2.646}{\tilde{z}_w^2}\right)^{0.176} & \text{weak coupling} \end{cases} \tag{15.66}$$

where we employed the numerical parameterizations of the theoretical results. Now it turns out that for most experiments published so far the statistical scatter is too large to identify any systematic trend of the data so reduced. This in particular is true for 'moderately good' solvents on the weak coupling branch, where systematic and precise measurements covering a large range of chain lengths are missing. Thus most experiments only allow for a verification of the asymptotic power laws, together with an estimate of ψ^*. The situation is slightly better for the strong coupling branch, such 'very good solvent' systems comprising two of the most thoroughly examined examples: polystyrene in benzene or in toluene. In particular the latter system has recently been analyzed in much detail by the group of Yamakawa [AEYY93, YAE93, EAY93]. Whereas the data for R_g^2/R_g^{*2} scatter too much to allow for firm conclusions, the data for A_2^S/A_2^{S*} show a clear trend. They are plotted in Fig. 15.17, where we included data for $N_w \gtrsim 100$ only. The parameter values are

$$\text{PS-Toluene}, T = 15^oC : B = 3.07\,\text{Å}; \tilde{v}(T) = 0.35 \quad .$$

The parameter B is determined with high precision ($\Delta B \approx \pm.03$), but $\tilde{v}(T)$, which sets the scale of $\tilde{z}_w$, cannot be determined with high confidence. Still this plot clearly shows that this system has to be described by the strong coupling branch.

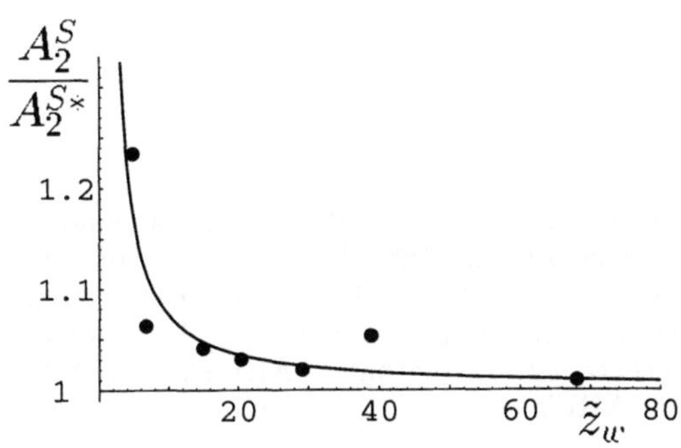

Fig. 15.17. A_2^S/A_2^{S*} as function of $\tilde{z}_w$; strong coupling. Data: polystyrene-toluene [AEYY93]

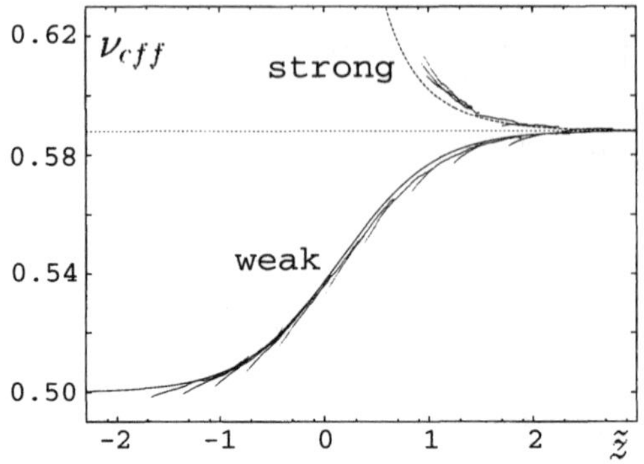

Fig. 15.18. $\nu_{eff} = \partial \ln R_g / \partial \ln N$ as function of $\tilde{z}$. Monte Carlo data [GSS97] for the Domb-Joyce model on a cubic lattice. Shown are dense sets of data points, each representing ν_{eff} with varying N for a given excluded volume strength. The *smooth* curves are the theoretical results

The situation as described here is typical for systems close to the excluded volume limit. An extensive analysis of physical solution data may be found in [KS96]. To improve the situation, extensive computer experiments have been carried through [GSS97], where both the chain length and the excluded vol-

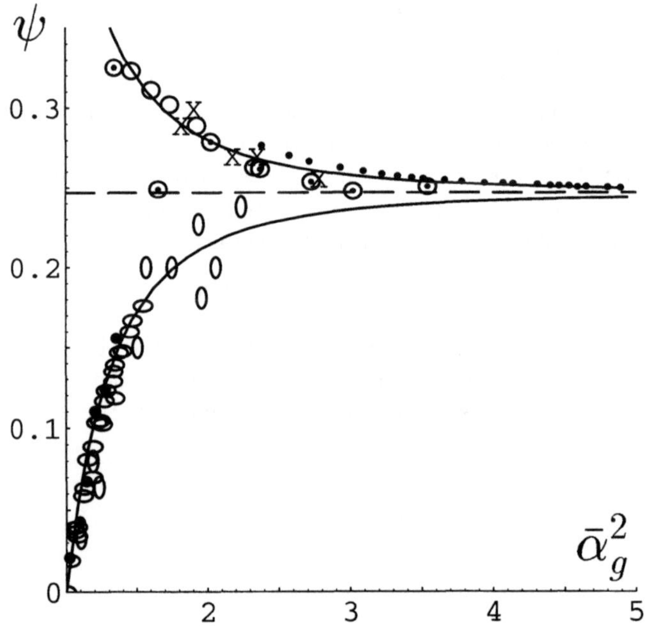

Fig. 15.19. ψ as function of $\bar{\alpha}_g^2$. The theoretical curves are calculated for a monodisperse ensemble. Data, weak coupling (lower branch): (●) Polystyrene-cyclohexane [YFTY71]; (▢) P-isobutylene-isoamyl valerate [MNF72]; (**0**) Polystyrene-transdecaline [FFKH74]. Data, strong coupling (upper branch): (⊙) Polystyrene-toluene [AEYY93]; (X) P-methylmethacrylate-nitroethane [AHEY94, AEY94]; (◯) P-methylmethacrylate-chloroform [AHEY94, AEY94]; (•) self-avoiding walks on a simple cubic lattice [LMS95]

ume strength have been varied systematically over a large range. Figure 15.18 shows the effective exponent ν_{eff} of the radius of gyration (cf. Sect. 15.14 and Fig. 15.3b), found in that work. Clearly, after some initial transients due to short chains ($N \lesssim 100$), the data merge on a two-branched scaling curve, which agrees nicely with our theoretical result.

Finally we consider the interpenetration function. Figure 15.19 shows the theoretical results together with typical data, most of which having been used in our previous analysis. We also included data for two other 'very good' solvent systems (PMMA in nitroethane or chloroform) as well as recent Monte Carlo data for self-avoiding walks on the simple cubic lattice. Clearly the structure of the two branches, separated by the fixed point value $\psi = \psi^*$, explains the data well.

More precise data would be most welcome, in particular for 'moderately good' solvents, i.e. for systems close to the excluded volume limit, but on the weak coupling branch. Nevertheless we think that this analysis solves the problem of the 'nonuniform' approach of ψ to the fixed point value ψ^* [FN85, Fuj90, KAEY95].

15.5.4 Momentum dependence of the scattering function

For $\mathbf{q}^2 R_g^2 \lesssim 20$, $J_d(\mathbf{q})$ is very close to the Debye function, a feature which has been observed in many experiments. Figure 15.20 illustrates this fact with data for the system polystyrene-toluene, compared to $J_d^{-1}(\mathbf{q})$ as calculated in Sect. 15.2. Recall that for $q^2 R_g^2[m] \lesssim 20$ our theoretical result essentially reproduces the Debye function.

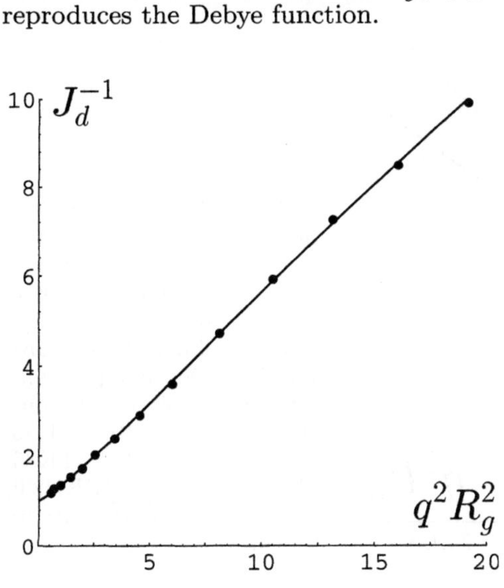

Fig. 15.20. $J_d^{-1}(\mathbf{q})$ as function of $q^2 R_g^2$. Theory: monodisperse one-loop result. Data: polystyrene-toluene [TK86, UUTK78]

For large momenta, $\mathbf{q}^2 R_g^2 \gg 1$, scaling theory or the renormalization group predict (cf. Eq. (15.34))

$$J_d(\mathbf{q}) \sim q^{-\frac{1}{\nu}} \quad ,$$

this power law being valid in the excluded volume limit. As shown in Fig. 15.7, however, it needs extremely large values $q^2 R_g^2 \gtrsim 10^3$ to reach the asymptotic exponent within 1% deviation, and any violation of the excluded volume limit will show up as a pronounced deviation from the asymptotic curve. Still there have been attempts to measure ν in terms of the asymptotic behavior (15.34) via neutron scattering [CDF+74]. The system was a polystyrene of $N \sim 10^4$ dissolved in the good solvent CS_2. Since the large momentum behavior is independent of concentration, a solution of moderate concentration was used. Within the range $250 \lesssim (qR_g)^2 \lesssim 6000$ the data show power law behavior with $\nu_{eff} = 0.595 \pm 0.02$. This is perfectly consistent with Fig. 15.7. However it must be noted that the momenta $q \gtrsim 3\,10^{-2}\,\text{Å}^{-1}$ used in that experiment test structures on scales less than about 50 Å, and so microstructure effects cannot be ignored. Indeed, Rawiso et al. [RDP87] repeated the experiment with chains deuterated in different places of the monomer unit and found a pronounced microstructure effect. Thus the above quoted result seems somewhat fortuitous.

For the system polystyrene-toluene there exist extensive light scattering experiments covering the intermediate momentum range for very long chains. In Fig. 15.21 typical results are compared to the theoretical prediction for the universal scaling function J_d^{-1} as function of $q^2 R_g^2$ in the excluded volume

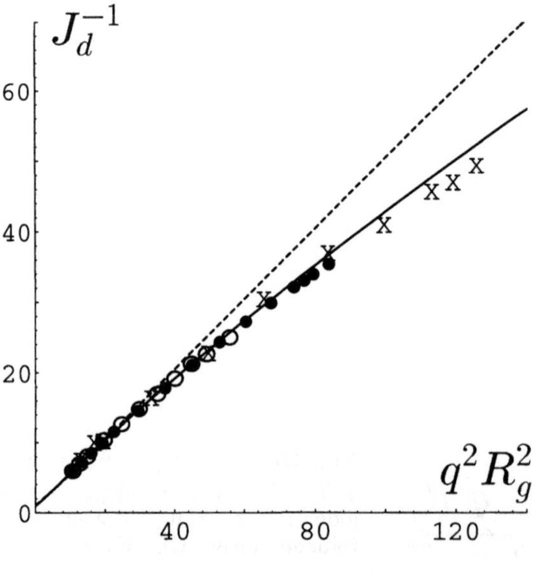

Fig. 15.21. $J_d^{-1}(\mathbf{q})$ as function of $q^2 R_g^2$ in the intermediate momentum range. *Full line*: one-loop approximation, monodisperse, excluded volume limit. *Broken line*: Debye-approximation. Data: polystyrene-toluene (∘) $N \approx 10^5$ [TK86] (•) $N \approx 1.5 \cdot 10^5$ [TK86] (X) $N \approx 2 \times 10^5$ [NIKN83]; only data for $q^2 R_g^2 > 10$ are shown

limit. The agreement is remarkable, in particular if we recall that no nonuniversal fit parameter is involved. Closer inspection suggests a violation of the universal excluded volume limit for the largest momenta measured for each sample. This violation qualitatively agrees with that expected for a system on the strong coupling branch (cf. Sect. 15.2.4 and Fig. 15.6), which is consistent with our findings for A_2^S, R_g^2 for that system (cf. Sect. 15.5.3). A quantitative description, however, needs unreasonably small values of $\tilde{v}(T)$, not compatible with the results for $q^2 R_g^2 \ll 1$. The analysis shows that for all sets of data the deviation is found for $q^{-1} \lesssim 250\,\text{Å}$, irrespective of the chain length. At such q-values the scattering intensity could well pick up small contributions from microscopic length scales, which suggests that we might see the onset of nonuniversal microstructure effects.

We finally note a recent simulation of the single chain form factor of polyethylene [DLR96], where an effective power law $J_d(\mathbf{q}) \sim q^{-1/\nu}$ has been observed. The asymptotic ratio is found as $\mathcal{R}_{\infty/2} \approx 0.83$. This result is perfectly consistent with the theoretical estimate $\mathcal{R}_{\infty/2} \approx 0.905$, if we take into account that in the intermediate q-range $J_d^{-1}(\mathbf{q})$ is found below the true asymptote (cf. Fig. 15.6).

A 15. Appendix

A 15.1 Formal results on polydispersity dependence

We first generalize the analysis of Appendix A 4.1 to chains of different lengths. Recall from Sect. 5.1 the representation of the grand potential $\Pi[\mu_p]$ in terms of M-chains cumulants:

$$\begin{aligned} \Omega\, \Pi[\mu_p] &= \ln \mathcal{Z}[\mu_p] \\ &= \sum_n e^{\mu_p(n)} \mathcal{Z}^{(1)}(n) \\ &\quad + \frac{1}{2} \sum_{n_1 n_2} e^{\mu_p(n_1)+\mu_p(n_2)} \mathcal{Z}_c^{(2)}(n_1, n_2) + \dots\,, \end{aligned} \tag{A 15.1}$$

where

$$\begin{aligned} \mathcal{Z}_c^{(2)}(n_1, n_2) &= \mathcal{Z}^{(2)}(n_1, n_2) - \mathcal{Z}^{(1)}(n_1)\mathcal{Z}^{(1)}(n_2) \\ &\equiv G_c^{(0,2)}(\mathbf{0}, \mathbf{0}, \mathbf{0}, \mathbf{0}; n_1, n_2) \quad . \end{aligned}$$

The concentration of chains of length n_1 is derived as (see Eq. (5.4))

$$c_p(n_1) = \frac{1}{\Omega} \frac{\partial}{\partial \mu_p(n_1)} \ln \mathcal{Z}[\mu_p] \quad .$$

Inserting Eq. (A 15.1) and solving for $\mu_p(n)$ we find

$$e^{\mu_p(n_1)} = \frac{\Omega\, c_p(n_1)}{\mathcal{Z}^{(1)}(n_1)} \left[1 + \sum_{n_2} c_p(n_2) A_2(n_1, n_2) + \ldots \right] \quad , \tag{A 15.2}$$

where

$$\begin{aligned} A_2(n_1, n_2) &= -\Omega \frac{\mathcal{Z}_c^{(2)}(n_1, n_2)}{\mathcal{Z}^{(1)}(n_1)\mathcal{Z}^{(1)}(n_2)} \\ &= -\Omega \frac{G_c^{(0,2)}(\mathbf{0},\mathbf{0},\mathbf{0},\mathbf{0}; n_1, n_2)}{G^{(0,1)}(\mathbf{0},\mathbf{0}; n_1) G^{(0,1)}(\mathbf{0},\mathbf{0}; n_2)} \quad . \end{aligned} \tag{A 15.3}$$

Inserting this result into Eq. (A 15.1) and using Eq. (5.35):

$$c_p(n) = c_p P(n) \quad ,$$

we find the first terms of the virial expansion of the osmotic pressure

$$\frac{\hat{\Pi}}{k_B T} = c_p + \frac{1}{2} A_2^{\Pi} c_p^2 + O(c_p^3) \quad , \tag{A 15.4}$$

$$A_2^{\Pi} = \sum_{n_1 n_2} P(n_1) P(n_2) A_2(n_1, n_2) \quad . \tag{A 15.5}$$

We may repeat the calculation for the density correlation function (5.17)

$$I_{d,c}(\mathbf{q}, c_p, N) = \int_{\Omega} d^3 r e^{i\mathbf{q}\mathbf{r}} \left\langle (\rho(\mathbf{r}) - c)(\rho(\mathbf{0}) - c) \right\rangle \quad .$$

Evaluating the Fourier transform, we find

$$I_{d,c}(\mathbf{q}) = \frac{1}{\Omega} \left\langle \sum_{m_1 m_2} \sum_{j_1 j_2} e^{i\mathbf{q}\left(\mathbf{r}_{j_1}^{(m_1)} - \mathbf{r}_{j_2}^{(m_2)}\right)} \right\rangle - \Omega c^2 \delta_{\mathbf{q}\mathbf{0}} \, . \tag{A 15.6}$$

The expectation value can be written as

$$\begin{aligned} &\langle \ldots \rangle = \\ &\frac{1}{\mathcal{Z}} \sum_{m=1}^{\infty} \frac{1}{(m-1)!} \sum_{n_1 \ldots n_m} \prod_j e^{\mu_p(n_j)} G_{11}^{(2,m)}(\mathbf{q}, -\mathbf{q}; \mathbf{0}, \ldots, \mathbf{0}; n_1, \ldots, n_m) + \\ &\frac{1}{\mathcal{Z}} \sum_{m=2}^{\infty} \frac{1}{(m-2)!} \sum_{n_1 \ldots n_m} \prod e^{\mu_p(n_j)} G_{12}^{(2,m)}(\mathbf{q}, -\mathbf{q}; \mathbf{0}, \ldots, \mathbf{0}; n_1, \ldots, n_m) \quad , \end{aligned} \tag{A 15.7}$$

where in the first part both density insertions are in the same chain (1), whereas they are in different chains (1,2) in the second part. Expanding as above, we find after a little calculation

$$
\begin{aligned}
I_{d,c}(\mathbf{q}, c_p, N) = & \\
& \sum_{n_1} \frac{c_p(n_1)}{\mathcal{Z}^{(1)}(n_1)} G_{11}^{(2,1)}(\mathbf{q}, -\mathbf{q}; \mathbf{0}, \mathbf{0}; n_1) \left(1 + \sum_{n_2} c_p(n_2) A_2(n_1, n_2)\right) \\
& + \Omega \sum_{n_1 n_2} \frac{c_p(n_1) c_p(n_2)}{\mathcal{Z}^{(1)}(n_1) \mathcal{Z}^{(1)}(n_2)} \Big(G_{c11}^{(2,2)}(\mathbf{q}, -\mathbf{q}; \mathbf{0}, \mathbf{0}, \mathbf{0}, \mathbf{0}; n_1, n_2) \\
& + G_{c12}^{(2,2)}(\mathbf{q}, -\mathbf{q}; \mathbf{0}, \mathbf{0}, \mathbf{0}, \mathbf{0}; n_1, n_2) \Big) + O(c_p^3) \quad . \qquad \text{(A 15.8)}
\end{aligned}
$$

The term $\Omega c^2 \delta_{\mathbf{q},\mathbf{0}}$ in Eq. (A 15.6) is eliminated by introducing the cumulants.

Using the rigorous relations

$$
\begin{aligned}
G_{11}^{(2,1)}(\mathbf{0}, \mathbf{0}, \mathbf{0}, \mathbf{0}; n_1) &= n_1^2 \mathcal{Z}^{(1)}(n_1) \\
G_{c11}^{(2,2)}(\mathbf{0}, \ldots, \mathbf{0}; n_1, n_2) &= n_1^2 \mathcal{Z}_c^{(2)}(n_1, n_2) \\
G_{c12}^{(2,2)}(\mathbf{0}, \ldots, \mathbf{0}; n_1, n_2) &= n_1 n_2 \mathcal{Z}_c^{(2)}(n_1, n_2) \quad , \qquad \text{(A 15.9)}
\end{aligned}
$$

we find

$$
\begin{aligned}
I_{d,c}(\mathbf{0}, c_p, N) = c_p \sum_n n^2 P(n) - c_p^2 \sum_{n_1 n_2} n_1 n_2 P(n_1) P(n_2) A_2(n_1, n_2) \\
+ O(c_p^3) \quad , \qquad \text{(A 15.10)}
\end{aligned}
$$

which yields the second virial coefficient from light scattering

$$
A_2^S = \sum_{n_1 n_2} \frac{n_1 n_2}{N^2} P(n_1) P(n_2) A_2(n_1, n_2) \quad . \qquad \text{(A 15.11)}
$$

Note that as a byproduct we find from Eq. (A 15.8)

$$
\frac{1}{c_p} I_{d,c}(\mathbf{q}, c_p, N) = \sum_n P(n) I_d(\mathbf{q}, n) + O(c_p) \quad , \qquad \text{(A 15.12)}
$$

where $I_d(\mathbf{q}, n)$ is the scattering function of an isolated chain of length n.

A 15.2 One-Loop calculation of I_d (q)

We calculate here the normalized scaling function

$$
\begin{aligned}
J_d(\mathbf{q}, n) &= \frac{1}{n^2} I_d(\mathbf{q}, n) \\
&= \frac{2}{n^2} \sum_{j_1 < j_2} \left\langle e^{i\mathbf{q}(\vec{r}_{j_2} - \vec{r}_{j_1})} \right\rangle \quad , \qquad \text{(A 15.13)}
\end{aligned}
$$

following [Dup86b]. The expectation value in Eq. (A 15.13) gives the correlations among specific segments j_1, j_2. We may introduce the corresponding Greensfunction

$$G_{j_1 j_2}(\mathbf{q}, n) = \int D[r] \exp\left(-\frac{1}{4\ell^2}\sum_{j=1}^{n}(\vec{r}_j - \vec{r}_{j-1})^2\right)$$
$$\cdot\prod{}' \left[1 - (4\pi\ell^2)^{d/2}\beta_e \delta^d(\vec{r}_{i_2} - \vec{r}_{i_1})\right] \exp(iq(\vec{r}_{j_2} - \vec{r}_{j_1})) \quad , \tag{A 15.14}$$

which for zero momentum reduces to the partition function. We can thus write

$$\left\langle e^{i\mathbf{q}(\vec{r}_{j_2} - \vec{r}_{j_1})}\right\rangle = \frac{G_{j_1 j_2}(\mathbf{q}, n)}{G_{j_1 j_2}(0, n)} \quad . \tag{A 15.15}$$

The diagrammatic expansion of $G_{j_1 j_2}(\mathbf{q}, n)$ is shown in Fig. 15.22. These diagrams are to be evaluated according to the rules of Sect. 4.2, except that the summation over the segments j_1, j_2 attached to the density vertices has to be omitted. Furthermore the prefactor $\Omega(4\pi\ell^2)^{-d/2}$ is common to all contributions and thus cancels in the ratio (A 15.15). Ignoring it from the outset we note that diagram (0) just yields $G_0(q, j_2 - j_1)$. Denoting by $-\beta_e I_\nu$ the contribution of diagram $(\nu), \nu = 1, \ldots 6$, we find

$$\frac{G_{j_1 j_2}(\mathbf{q}, n)}{G_{j_1 j_2}(0, n)} = G_0(\mathbf{q}, j_2 - j_1) - \beta_e \sum_{\nu=1}^{6}\left[I_\nu(\mathbf{q}) - I_\nu(\mathbf{0}) G_0(\mathbf{q}, j_2 - j_1)\right]$$
$$+ O(\beta_e^2) \quad . \tag{A 15.16}$$

It is easily checked that the contributions $\nu = 1, 2$ identically vanish. The further evaluation is exemplified with the contribution $\nu = 4$.

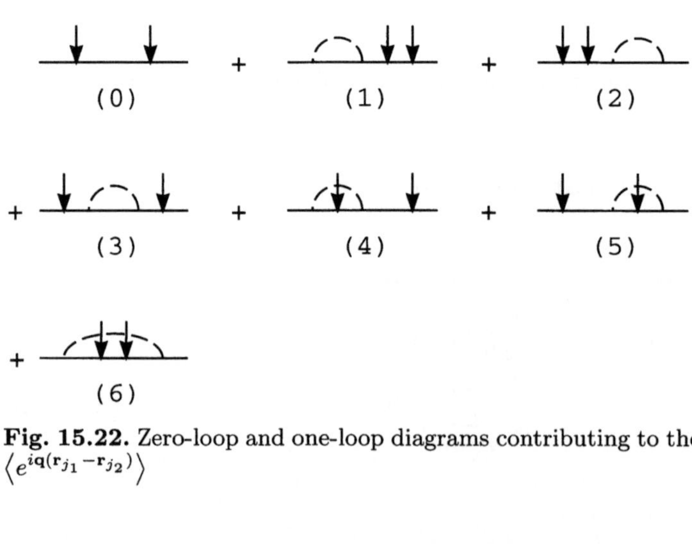

Fig. 15.22. Zero-loop and one-loop diagrams contributing to the expectation value $\left\langle e^{i\mathbf{q}(\mathbf{r}_{j_1} - \mathbf{r}_{j_2})}\right\rangle$

$$I_4(\mathbf{q}) - I_4(\mathbf{0})G_0(\mathbf{q}, j_2 - j_1) =$$
$$(4\pi\ell^2)^{d/2} \sum_{j_3=1}^{j_1-1} \sum_{j_4=j_1+1}^{j_2-1} \Big\{ \int_k \exp\Big[-\vec{k}^2\ell^2(j_1-j_3) - (\vec{k}+\mathbf{q})^2\ell^2(j_4-j_1)\Big]$$
$$\exp\left[-\mathbf{q}^2\ell^2(j_2-j_4)\right] - \int_k \exp\left[-\vec{k}^2\ell^2(j_4-j_3) - \mathbf{q}^2\ell^2(j_2-j_1)\right] \Big\}$$
$$= \sum_{j_3=1}^{j_1-1} \sum_{j_4=j_1+1}^{j_2-1} e^{-q^2\ell^2(j_2-j_1)}(j_4-j_3)^{-d/2}\left[\exp\left(q^2\ell^2\frac{(j_4-j_1)^2}{j_4-j_3}\right)-1\right] .$$

We used the results of Appendix A 3.3 to carry through the k-integrals. We now take the continuous chain limit, replacing the summations by integrations over $s_{\mu_p} = j_{\mu_p}/n$. With the notation

$$Q = q^2\ell^2 n \tag{A 15.17}$$

we find

$$I_4(\mathbf{q}) - I_4(\mathbf{0})G_0(\mathbf{q}, j_2 - j_1) =$$
$$n^{\varepsilon/2}\int_0^{s_1} ds_3 \int_{s_1}^{s_2} ds_4 e^{-Q(s_2-s_1)}(s_4-s_3)^{-d/2}\left[\exp\left(Q\frac{(s_4-s_1)^2}{s_4-s_3}\right)-1\right] .$$

For $J_d(\mathbf{q}, n)$ this yields the contribution

$$J_4 = -\beta_e \frac{2}{n^2}\sum_{j_1<j_2}\left(I_4(\mathbf{q}) - I_4(\mathbf{0})G_0(\mathbf{q}, j_2-j_1)\right)$$
$$= -2\beta_e n^{\varepsilon/2}\int_{0<s_3<s_1<s_4<s_2<1} e^{-Q(s_2-s_1)}(s_4-s_3)^{-d/2}$$
$$\left[\exp\left(Q\frac{(s_4-s_1)^2}{s_4-s_3}\right)-1\right] .$$

We now introduce difference variables

$$x_1 = s_1 - s_3, \quad x_2 = s_4 - s_1, \quad x_3 = s_2 - s_4$$

and integrate over s_3 to find

$$J_4 = -2\beta_e n^{\varepsilon/2}\int_0^1 dx_1 \int_0^{1-x_1} dx_2 \int_0^{1-x_1-x_2} dx_3$$
$$(1-x_1-x_2-x_3)(x_2+x_1)^{-d/2}e^{-Q(x_3+x_2)}\left[\exp\left(Q\frac{x_2^2}{x_2+x_1}\right)-1\right] .$$

The x_3-integral can be carried out to yield a factor of the Debye function

$$J_4 = -\beta_e n^{\varepsilon/2}\int_0^1 dx_1 \int_0^{1-x_1} dx_2 (1-x_1-x_2)^2(x_1+x_2)^{-d/2}e^{-Qx_2}$$
$$D(Q(1-x_1-x_2))\left[\exp\left(Q\frac{x_2^2}{x_2+x_1}\right)-1\right] .$$

We finally change variables according to

$$x_1 = (1-\alpha)x \quad , x_2 = \alpha x \quad ,$$

the integration measure transforming with the functional determinant

$$\begin{vmatrix} \frac{dx_1}{dx} & \frac{dx_1}{d\alpha} \\ \frac{dx_2}{dx} & \frac{dx_2}{d\alpha} \end{vmatrix} = x \quad .$$

This yields our final expression

$$J_4 = -\beta_e n^{\varepsilon/2} \int_0^1 dx x^{-1+\varepsilon/2} (1-x)^2 D(Q(1-x)) \cdot \left[F(Qx) - \frac{1}{Qx} \left(1 - e^{-Qx}\right) \right] \quad ,$$

where

$$F(Qx) = \int_0^1 d\alpha e^{-Qx\alpha(1-\alpha)} \tag{A 15.18}$$

can be expressed in terms of the error function.

By reflection symmetry of the chain, diagrams 4 or 5 yield identical contributions to $J_d(\mathbf{q}, n)$. The remaining terms can be evaluated following the lines given above. We find

$$J_d(\mathbf{q}, n) = D(Q) - \beta_e n^{\varepsilon/2} \Delta J_d(Q) \quad , \tag{A 15.19}$$

$$\begin{aligned} \Delta J_d(Q) = & -\frac{2}{\varepsilon} Q D^{'}(Q) + \frac{Q^2}{2} D^{'}(Q) \int_0^1 dx \, x^{\varepsilon/2} D(Qx) \\ & + \int_0^1 dx \, x^{-2+\varepsilon/2} (e^{-Qx} - 1) \left[(1-x)^3 D^{'}(Q(1-x)) - D^{'}(Q) \right] \\ & + 2 \int_0^1 dx \, x^{-1+\varepsilon/2} (1-x)^2 D(Q(1-x)) \left[F(Qx) - \frac{1}{Qx} \left(1 - e^{-Qx}\right) \right] \\ & + \int_0^1 dx x^{\varepsilon/2} (1-x) \left[F(Qx) - D(Qx) \right] \quad . \end{aligned} \tag{A 15.20}$$

This is the unrenormalized, but dimensionally regularized expression. In Eq. (A 15.20) the last contribution is due to diagram (6), the last but one contribution is due to diagrams (4) and (5), the remaining terms are due to diagram (3). Only this latter diagram yields a singular term of order $1/\varepsilon$.

To renormalize according to Eqs. (12.16), (12.19) we write

$$Q = \mathbf{q}^2 \ell^2 n = \bar{\mathbf{q}}^2 n_R Z_n = \bar{\mathbf{q}}^2 n_R \left(1 - \frac{u}{\varepsilon} + O(u^2) \right)$$

$$\beta_e n^{\varepsilon/2} = u n_R^{\varepsilon/2} Z_u Z_n^{\varepsilon/2} = \frac{u}{2} n_R^{\varepsilon/2} + O(u^2) \, ,$$

where we used $b_u(\varepsilon) = 1$. Substituting these results into Eqs. (A 15.19), (A 15.20) and ordering according to powers of u we find

$$\begin{aligned} J_d(\mathbf{q}, n) &= J_{dR}(\mathbf{q}, n_R, u) \\ &= D(Q_R) - \frac{u}{\varepsilon} Q_R D'(Q_R) - \frac{u}{2} n_R^{\varepsilon/2} \Delta J_d(Q_R) \quad . \end{aligned} \tag{A 15.21}$$

We now use Eqs. (A 15.12), (A 15.13) to average over the chain length distribution:

$$\begin{aligned} J_d(\mathbf{q}, [p]) &= \int_0^\infty dy\, p(y) y^2 J_d(\mathbf{q}, yN) \\ &= D_p(Q_R) - \frac{u}{\varepsilon} Q_R D_p'(Q_R) - \frac{u}{2} N_R^{\varepsilon/2} \Delta J_d(Q_R, [p]) \end{aligned} \tag{A 15.22}$$

$$\Delta J_d(Q_R, [p]) = \int_0^\infty dy\, p(y) y^{2+\varepsilon/2} \Delta J_d(y Q_R) \tag{A 15.23}$$

$$Q_R = \bar{\mathbf{q}}^2 N_R \quad . \tag{A 15.24}$$

In final form we write

$$J_d(\mathbf{q}, [p]) = D_p(\hat{Q}) \tag{A 15.25}$$

$$\hat{Q} = Q_R \left(1 - \frac{u}{\varepsilon}\right) - \frac{u}{2} N_R^{\varepsilon/2} \frac{\Delta J_d(Q_R, [p])}{D_p'(Q_R)} \quad . \tag{A 15.26}$$

A 15.3 Further perturbative results

A 15.3.1 Small momentum behavior of $J_d\ (\mathbf{q}, n)$. Expanding in powers of Q_R we find from Eqs. (A 15.21), (A 15.20)

$$J_d(\mathbf{q}, n) = 1 - Q_R \bar{R}_2 + Q_R^2 \bar{R}_4 + O(\bar{q}^6) \tag{A 15.27}$$

$$\bar{R}_2 = \frac{1}{3} \left[1 + \frac{u^*}{\varepsilon} f \left(n_R^{\varepsilon/2} - 1 \right) - \frac{u^*}{2} f n_R^{\varepsilon/2} \left(\frac{2}{2+\varepsilon} + \frac{1}{4+\varepsilon} - \frac{1}{6+\varepsilon} \right) \right] \tag{A 15.28}$$

$$\begin{aligned} \bar{R}_4 = \frac{1}{12} \Bigg[& 1 + 2 \frac{u^*}{\varepsilon} f \left(n_R^{\varepsilon/2} - 1 \right) \\ & - \frac{u^*}{2} f\, n_R^{\varepsilon/2} \left(\frac{20}{6+3\varepsilon} - \frac{16}{20+5\varepsilon} + \frac{4}{30+5\varepsilon} - \frac{4}{120+15\varepsilon} \right) \Bigg] \quad . \end{aligned} \tag{A 15.29}$$

For general polydispersity the radius of gyration is found as (cf. Eq. (15.7))

$$R_g^2 = \frac{d}{3}\frac{\tilde{p}_3}{\tilde{p}_2}\ell_R^2 N_R \Big[1 - \frac{u^*}{\varepsilon} f + \frac{u^*}{2} f N_R^{\varepsilon/2}\left(\frac{2}{\varepsilon} - \frac{2}{2+\varepsilon} - \frac{1}{4+\varepsilon} + \frac{1}{6+\varepsilon}\right)$$
$$\cdot \frac{1}{\tilde{p}_3}\int_0^\infty dy\, p(y) y^{3+\varepsilon/2}\Big] \quad . \tag{A 15.30}$$

For a monodisperse system the ε-expansion reads

$$R_g^2 = \frac{d}{3}\ell_R^2 n_R \left[1 + \frac{\varepsilon}{8} f \left(\ln n_R - \frac{13}{12}\right)\right] \quad . \tag{A 15.31}$$

From $\bar{R}_4, \bar{R}_2$ we can construct the critical ratio $\mathcal{R}_{4/2}$ (cf. Eq. (15.30)). The ε-expansion reads

$$\mathcal{R}_{4/2} = \frac{3}{4}\left(1 - \frac{7}{120}\,\varepsilon\, f\right) \quad . \tag{A 15.32}$$

A 15.3.2 Large momentum behavior of J_d (q, n). For $Q_R \to \infty$ a somewhat tedious analysis yields

$$J_d^{-1}(\mathbf{q}, n) \to \frac{\hat{Q}}{2}$$
$$\to \frac{Q_R}{2}\left[1 + \frac{u^*}{\varepsilon} f \left(\frac{\Gamma(1+\varepsilon/2)\Gamma^2(1-\varepsilon/2)}{\Gamma(2-\varepsilon)}(1+\varepsilon)\left(\bar{q}^2\right)^{-\varepsilon/2} - 1\right)\right] \quad . \tag{A 15.33}$$

As discussed in Sect. 15.2.3 we may construct a critical ratio connecting the large or small momentum limits:

$$\mathcal{R}_{\infty/2} = \lim_{q\to\infty}\left[\left(\frac{3}{d} R_g^2 q^2\right)^{-\frac{1}{2\nu}} J_d^{-1}(\mathbf{q}, n)\right] \quad . \tag{A 15.34}$$

Using Eqs. (A 15.31), (A 15.33) together with $1/2\nu = 1 - \varepsilon/8$ for a monodisperse system we find in strict ε-expansion [Wit82]

$$\mathcal{R}_{\infty/2} = \frac{1}{2}\left(1 + \frac{\varepsilon}{8} f \left(\frac{61}{12} - \gamma_{Eu}\right)\right) \quad , \tag{A 15.35}$$

where $\gamma_{Eu} = .5772$ is Euler's number.

A 15.3.3 The second virial coefficients. The basic quantity is the second virial coefficient $A_2(n_1, n_2)$ for two chains of lengths n_1, n_2, respectively (cf. Eq. (A 15.3)). To first order the renormalized loop expansion yields

$$(4\pi)^{-d/2} A_2(n_1, n_2) = \frac{u^*}{2} f\, n_{R1}\, n_{R2} \ell_R^d$$
$$\left[1 + \frac{2}{\varepsilon} u^* f \Big(1 - \frac{n_{R1}^{\varepsilon/2} + n_{R2}^{\varepsilon/2}}{1-\varepsilon^2/4} - \frac{n_{R1}^{2+\varepsilon/2} + n_{R2}^{2+\varepsilon/2} - (n_{R1}+n_{R2})^{2+\varepsilon/2}}{(1-\varepsilon^2/4)(2+\varepsilon/2) n_{R1} n_{R2}}\Big)\right] \quad . \tag{A 15.36}$$

Strict ε-expansion results in

$$\left(4\pi\ell_R^2\right)^{-d/2} A_2(n_1,n_2) = \frac{u^*}{2}\, f\, n_{R1}\, n_{R2}$$
$$\left[1 - \frac{\varepsilon}{8}\, f\Big(1 + 2\ln(n_{R1}\, n_{R2}) + \frac{n_{R1}}{n_{R2}}\ \ln\, n_{R1} + \frac{n_{R2}}{n_{R1}}\ \ln\, n_{R2}\right.$$
$$\left. - \frac{\left(n_{R1}+n_{R2}\right)^2}{n_{R1}\, n_{R2}}\ \ln(n_{R1}+n_{R2})\Big)\right] \quad , \qquad \text{(A 15.37)}$$

where the ε-expansion (12.20) of u^* may be inserted.

Equations (A 15.5), (A 15.11), (A 15.36) yield for the osmotic or light scattering virial coefficients

$$\left(4\pi\ell_R^2\right)^{-d/2} A_2^{\Pi} = \bar{a}_2(f, c_A^{\Pi}) \qquad \text{(A 15.38 i)}$$

$$\left(4\pi\ell_R^2\right)^{-d/2} \frac{A_2^S}{\tilde{p}_2^2} = \bar{a}_2(f, c_A^S) \quad , \qquad \text{(A 15.38 ii)}$$

where

$$\bar{a}_2(f,c) = \frac{u^*}{2}\, f\, N_R^2\Big[1 + \frac{2}{\varepsilon}u^*\, f\left(1 - N_R^{\varepsilon/2}\frac{6+\varepsilon-2^{2+\varepsilon/2}}{(1-\varepsilon^2/4)(2+\varepsilon/2)}\right)$$
$$+ f\, N_R^{\varepsilon/2}\, c\Big] \quad . \qquad \text{(A 15.39)}$$

The polydispersity corrections $c_{A_2}^{\Pi}, c_{A_2}^S$ are defined as

$$c_A^{\Pi} = -\frac{2u^*}{\varepsilon(1-\varepsilon^2/4)(2+\varepsilon/2)}\Big[(4+\varepsilon)\int_0^\infty dy\, p(y) y^{1+\varepsilon/2}$$
$$+ 2\int_0^\infty dy\, p(y) y^{2+\varepsilon/2} - \int_0^\infty dy_1\, dy_2 p(y_1)p(y_2)\left(y_1+y_2\right)^{2+\varepsilon/2}$$
$$- 6 - \varepsilon + 2^{2+\varepsilon/2}\Big] \quad , \qquad \text{(A 15.40 i)}$$

$$c_A^S =$$
$$-\frac{2u^*}{\varepsilon(1-\varepsilon^2/4)(2+\varepsilon/2)}\Big[(4+\varepsilon)\int_0^\infty dy\, \frac{p(y)}{\tilde{p}_2} y^{2+\varepsilon/2} + 2\int_0^\infty dy \frac{p(y)}{(\tilde{p}_2)^2} y^{3+\varepsilon/2}$$
$$- \int_0^\infty dy_1\, dy_2 \frac{p(y_1)p(y_2)}{(\tilde{p}_2)^2}\, y_1 y_2 \left(y_1+y_2\right)^{2+\varepsilon/2} - 6 - \varepsilon + 2^{2+\varepsilon/2}\Big] \quad .$$
$$\text{(A 15.40 ii)}$$

They vanish for the monodisperse ensemble.

We finally note the ε-expansion for the monodisperse system

$$\bar{a}_2(f,0) = \frac{\varepsilon}{8}\, f\, N_R^2\left[1 + \frac{21}{32}\varepsilon - \frac{\varepsilon}{4}\, f\left(\ln N_R + \frac{1}{2} - 2\ln 2\right)\right] \quad . \qquad \text{(A 15.41)}$$

Notes

Calculations of $\bar{\alpha}_g^2, A_2, \psi$ are abundant in the literature, most work being concerned with monodisperse systems. The older work is based mainly on naive two parameter theory. Low orders of the expansions in $z = \beta_e n^{1/2} (d = 3)$ have been calculated, and the results were extrapolated to the excluded volume limit $z \to \infty$ using fairly heuristic arguments. These approaches have been carefully reviewed in Yamakawa's book.

After the discovery of the renormalization group again many different calculations of $\bar{\alpha}_g^2, A_2, \psi$ have been published. Almost all that work is restricted to the order of one-loop, and of course the results are equivalent within strict ε-expansion framework. Also the numerical results in $d = 3$ turn out to be in good agreement. We therefore do not review all that work, but we restrict ourself to noting some small subset of publications, which to my feeling contributed important steps to our present day understanding of the subject.

The identification of ψ^* as a critical ratio is due to Witten and Schäfer [WS78]. Previous calculations, not based on the RG, generally predicted that ψ diverges in the excluded volume limit. Elderfield [Eld80] was the first to present a calculation of the crossover from Θ- to excluded volume conditions using a running coupling constant $u = u(\lambda)$. A discussion of this crossover for many observable quantities has then been published by Oono and Freed [OF82], noting that a unified treatment greatly enhances the significance of a comparison to experiment. This work was based on a low order representation of the RG combined with an exponentiated form of the ε-expansion for the scaling functions. A first unified discussion similar to the present approach may be found in [Sch84]. To finish this short selection we mention some work of Duplantier [Dup86a, Dup86c], showing how to renormalize consistently on the basis of the z-expansion of naive two-parameter theory.

In the excluded volume limit a calculation of $J_d^{-1}(\mathbf{q}, [n])$ has been first published by Ohta et al. [OOF81a], who use an exponentiated form of the ε-expansion. Also Duplantier [Dup86b] has evaluated his results in closed form for all $\mathbf{q}$. The results are most similar to those shown here, both on the qualitative and on the quantitative level. The subleading terms in the large q expansion of $J_d^{-1}(\mathbf{q})$ have recently been discussed in [JBJ95].

Up to 1991 all work exclusively considered the weak coupling branch or the excluded volume limit. The reason has been clearly stated in [Sch84]: only in a neighborhood of $T = \Theta$ can we systematically study scaling in chain length *and* temperature. Thus like in naive two-parameter theory, variations of $T - \Theta$ or N were assumed essentially to be equivalent. The existence of the strong coupling branch was observed only recently in an important contribution by Nickel [Nic91], in an attempt to understand most precise

Monte Carlo data. It may be somewhat surprising that a qualitative feature of that importance could have been overlooked for so many years. Now, on the theoretical side there was a strong prejudice against using RG theory for $u > u^*$, which we have discussed in Sect. 12.1. But also on the experimental side there was no clear evidence that our picture was incomplete, as can be understood by a glance to the experimental situation. Most experiments on dilute systems can be grouped into two classes. The first class aims at a verification of two-parameter theory. It thus works with a Θ-solvent close to $T = \Theta$ and can adequately be explained by RG results on the weak coupling branch for fairly small values of $\tilde{z}$. The second class, using 'good' or 'moderately good' solvents, was concerned with the excluded volume limit, verifying the fixed point behavior as established by the RG. Now, as shown by Figs. 15.3a or 15.9, if we are missing good data in the intermediate region of $\tilde{z}$, we can easily draw a smooth curve connecting data on the weak coupling branch for $\tilde{z}$ small to data on the strong coupling branch for $\tilde{z}$ large. An example of such an interpolation is the often used Domb-Barrett equation for α_g^2 [DB76, LBD78], which connects perturbative results of naive two-parameter theory (i.e. weak coupling) to Monte Carlo results for long self-avoiding walks on a lattice. The latter now have been identified [Sch94] as being described by the strong coupling branch. Some inconsistency of an interpretation of experimental data in terms of a simple unique dependence on z had been noted in the literature [FN85, Fuj90], but in view of the freedom in adjusting the nonuniversal scale of z, of unknown polydispersity effects, or other experimental problems this observation generally was not considered to signal a serious breakdown of the theory. Only recently in a series of papers Yamakawa and coworkers [Yam92, AEYY93, YAE93, EAY93, KAEY95] have reconsidered this problem in some detail. They suggested an interpretation of the apparent violations of the two-parameter scheme in terms of chain stiffness. It must however be noted that this explanation is has been critizised in [KS96]. In my view only the two-branched structure of the RG gives a consistent explanation.

16. Dilute Limit: Details on the Internal Structure of Isolated Coils

The present chapter deals with the internal structure of a polymer coil. This is easily accessible only to computer experiments, and from the point of view of physical experiments it might be felt that we belabor fairly academic questions. However, a good understanding of the internal structure may help our intuition also in interpreting physical experiments, correcting the oversimplified picture put forward in classical work. Furthermore it turns out that the structure imposed by dilatation invariance is most fascinating, offering polymer physics a distinguished place in modern statistical mechanics. In this chapter we review analytical and numerical results, which show details of this nontrivial internal structure.

Traditional 'Flory-type' theories view the polymer coil as a cloud of monomers characterized by some smooth segment density profile. At the beginning of Chap. 10 we criticized this picture, since it naively yields wrong estimates for the number of interaction points. A related feature will be stressed here: the smoothed density model is basically wrong, simply because the chain structure creates huge variations of the local segment density. This holds even for a noninteracting Gaussian chain, as is easily demonstrated. Consider some sphere of radius $L, \ell \ll L \ll R_g(n)$, placed somewhere inside a coil of radius R_g, chain length n. According to the smoothed density model such a sphere would contain $\hat{n}(L) \sim nL^d/R_g^d \sim (L/\ell)^d n^{1-d/2}$ monomers, more or less independent of its position within the coil. However, taking into account chain connectedness we find $\hat{n}(L) \sim (L/\ell)^2$ independent of n, provided the chain passes through the sphere. This estimate simply expresses the fact that it needs $(L/\ell)^2$ monomers to span a distance L. On the other hand, if the chain does not meet the sphere, then clearly $\hat{n}(L) = 0$. For $d > 2$, empty spheres necessarily exist, since otherwise the molecular weight of the chain would scale as R_g^d instead of R_g^2. Thus even for a Gaussian chain there exist fluctuations of local segment density, ranging from $\hat{n}(L)/L^d \sim L^{2-d}/\ell^2$ to zero. For long chains these fluctuations are huge compared to the average density $n/R_g^d \sim n^{1-d/2} \to 0$. Furthermore, due to scale invariance this structure must persist on all intermediate scales L. Clearly a smoothed density model is completely unrealistic, since it ignores the most basic feature of a polymer coil, the underlying chain.

Strongly inhomogeneous scale invariant objects can often be described as 'fractals', a concept which has become quite popular in condensed matter physics. The basic notion is the fractal dimension, defined by extending the above argument. Consider some object embedded in d-dimensional space, and cover it by d-dimensional spheres of radius L. Assume that the covering needs $n(L)$ spheres. Now repeat this Gedanken experiment with spheres of radius L'. If in some large range $L_{min} \ll L, L' \ll L_{max}$ we find the law $n(L)/n(L') = \left(L'/L\right)^{d_c}$, d_c fixed, then the object is a 'fractal' of fractal dimension d_c. By construction, d_c can be any positive real number, not restricted to integers. To show that d_c indeed is some generalization of the standard Euclidian dimension, we consider a continuous body of standard dimension d_1, embedded in $d \geq d_1$ dimensional space. With Ω_1 being the d_1-dimensional volume of the body, a covering needs $n(L) \approx const\ \Omega_1/L^{d_1}$ spheres. Thus $n(L)/n(L') = \left(L'/L\right)^{d_1}$, and $d_c = d_1$. (We should note here that there exist different ways to define a fractal dimension. d_c, as introduced here, is known as 'capacity' dimension, closely related to, but not necessarily identical to the famous Hausdorff dimension.)

Turning to a Gaussian coil of chain length n, we note that a covering needs $n(L) \sim n/\hat{n}(L) \sim n\,(\ell/L)^2$ spheres of size L. Thus $d_c = 2$: a Gaussian coil intrinsically is a two-dimensional object, which however by no means implies that its configuration is flat. Embedded in $d > 2$ dimensional space, the coil extends over a d-dimensional volume, of which it essentially fills only a two-dimensional part. It thus shows a very loose structure.

The concept of fractal dimension offers another way to understand the role of $d = 4$. Placing two objects of fractal dimensions $d_c^{(1)}$, $d_c^{(2)}$ and linear extension R into the same d-dimensional volume of extension R, they will generically overlap in a manifold of fractal dimension $d_c' = d_c^{(1)} + d_c^{(2)} - d$, provided $d_c' \geq 0$. (Otherwise the intersection is empty.) For two Gaussian coils: $d_c^{(1)} = d_c^{(2)} = 2$, we thus find $d = 4$ as borderline dimension. For $d > 4$ two Gaussian coils randomly placed in the same volume will not overlap and thus any excluded volume interaction among the coils is ineffective. We note that even though the underlying picture is quite different, we in this respect find the same results as with the smoothed density model (cf. also Chap. 6).

What is the fractal dimension of a self-repelling chain in the excluded volume limit? This is easily taken from the renormalization group. Indeed, in the present context Eqs. (8.1), (8.12), which give the number of effective segments of size $\ell_R = \ell/\lambda$: $n(\lambda) = \lambda^{1/\nu}\, n$, can be read as stating that it needs $n(L) \sim n(\ell/L)^{1/\nu}$ spheres of size L to cover the chain. Thus $d_c = 1/\nu$, which endows the exponent ν with a geometrical meaning. Specifically in three dimensions we find $d_c \approx 1.7$, which means that a self-repelling chain fills the volume even less effectively than a Gaussian chain.

How can we test the fractal nature of a polymer coil? Most results of the previous chapter are inadequate for this purpose. They are compatible both with a fractal or a homogeneous density distribution, since they refer to the chain as a whole. Only the large momentum behavior of the scattering function gives some hint: taking the intensity $I(q, N) \sim N(\ell q)^{-1/\nu}, qR_g \gg 1$, as measure of the number of monomer pairs of spatial distances $r \lesssim L \sim q^{-1}$, we conclude that a sphere of size L centered at any given segment contains of order of $(L/\ell)^{1/\nu}$ monomers. This is in agreement with fractal behavior but inconsistent with a Flory-type smoothed density model. More direct evidence can be taken from an analysis of either the internal correlations or the interaction among polymer coils of different size.

Consider first internal correlations, i.e. correlations among two specific segments $j_1 < j_2$. Scale invariance suggests that such correlations closely resemble those among the ends of a chain of length $j_2 - j_1$. We for instance expect internal distances to scale like end-to-end distances:

$$\mathcal{R}^2_{i/e}(j_1, j_2 - j_1, n - j_2) = \frac{\left\langle (\mathbf{r}_{j_1} - \mathbf{r}_{j_2})^2 \right\rangle}{R^2_e(j_2 - j_1)} \sim 1 \ . \tag{16.1}$$

This implies $\left\langle (\mathbf{r}_{j_1} - \mathbf{r}_{j_2})^2 \right\rangle \sim (j_2 - j_1)^{2\nu}$, in agreement with the above interpretation of the large momentum behavior of $I_d(\mathbf{q}, n)$. That power law is indeed borne out both by analytical calculations and Monte Carlo simulations. This then proves the scale invariance or self-similarity of the chain: internal distances obey the same power law as the size of the total chain. (Of course in all this discussion we assume to be in the excluded volume limit.)

A more detailed analysis shows that the ratio $\mathcal{R}^2_{i/e}$ is weakly dependent on the position of segments j_1, j_2 within the chain: it is larger for segments in the center of the chain. This can be understood as an effect of the dangling ends which exert some entropic strain on the piece (j_1, j_2). It is consistent with the finding that the probability of close encounters of interior segments is reduced compared to that of the chain ends. To be more precise, the distribution $\hat{P}(\mathbf{r}; j_1, j_2 - j_1, n - j_2)$ of the distance $\mathbf{r}_{j_2} - \mathbf{r}_{j_1} = \mathbf{r}$ for small $\mathbf{r}$ vanishes as

$$\hat{P}(\mathbf{r}; j_1, j_2 - j_1, n - j_2) \sim |\mathbf{r}|^{\vartheta} \ , r^2 / \left\langle (r_{j_1} - r_{j_2})^2 \right\rangle \ll 1 \ . \tag{16.2}$$

The exponent ϑ governing this 'correlation hole' is larger for interior segments than for endpoints. The very existence of the correlation hole is an interaction effect, which qualitatively distinguishes the excluded volume chain from a Gaussian chain.

The study of internal correlations stresses the aspect of scale invariance of the internal structure. The concept of fractal dimension is most directly tested by placing a small test chain n_2 at position $\mathbf{R}$ inside a large coil n_1, and calculating the free energy excess $U(\mathbf{R}, n_1, n_2)$ relative to the two isolated coils. It is found that U is a function of $R/R_g^{(1)}, n_2/n_1$, and u only, *finite even in the limit of infinitely long chains.* If the position $\mathbf{R}$ is chosen at random, then U vanishes in the limit $n_2/n_1 \to 0$. If $\mathbf{R}$ is restricted to lie on chain n_1, then in the same limit U takes some finite value close to $U(\mathbf{R}, n_2, n_2)$. All these results are supported by Monte Carlo simulations. They clearly rule out any smoothed density model, directly demonstrating the existence of large fluctuations in the local segment density. More precisely they show that any randomly chosen small volume in the large chain is almost certainly empty. Finite segment density is concentrated in a vanishing fraction of the coil volume.

To stress even more the great difference between the two concepts, we estimate $U(\mathbf{R}, n_1, n_2)$ according to the smoothed density model. With $\rho(\mathbf{r}, n)$ being the segment density with respect to the center of mass for a chain of length n, we find in a Flory-type mean-field approximation

$$\begin{aligned} U(\mathbf{R}, n_1, n_2) &\sim \beta_e\, \ell^d \int d^d r \rho(\mathbf{r}, n_1) \rho(\mathbf{r} - \mathbf{R}, n_2) \\ &\sim \beta_e\, \ell^d \frac{n_1}{\left(R_g^{(1)}\right)^d} \frac{n_2}{\left(R_g^{(2)}\right)^d} \left(R_g^{(2)}\right)^d \tilde{U}\left(\frac{\mathbf{R}}{R_g^{(2)}}, \frac{n_2}{n_1}\right) \quad , \end{aligned}$$

where the factor $\tilde{U}(\ldots)$ depends on the exact form of the density distribution. Thus

$$U(\mathbf{R}, n_1, n_2) \sim \beta_e n_1^{2-\nu d} \frac{n_2}{n_1} \tilde{U}\left(\frac{\mathbf{R}}{R_g^{(2)}}, \frac{n_2}{n_1}\right)$$

diverges in the limit $n_1, n_2 \to \infty, n_2/n_1$ fixed: $U \sim n_1^{2-\nu d} \sim n_1^{0.24}, (d = 3)$. Since $\exp(-U(\mathbf{R}, n_1, n_2))$ is proportional to the probability of finding the second chain at position $\mathbf{R}$ within the first chain, this implies that large coils rapidly become inpenetrable, irrespective of $n_2/n_1 > 0$. Thus large coils should behave like hard spheres, an idea first put forward by Flory and often invoked since then. This is in sharp contrast to the RG result, which implies that coils can interpenetrate irrespective of n_1, n_2: U stays finite in the above limit. Indeed, for $n_2/n_1 \ll 1$ the small coil can even diffuse through the large coil without being seriously disturbed, simply because it rarely meets strands of the large coil. This is a most striking conceptual difference arising from RG theory as compared to the traditional approach.

Unfortunately this aspect hardly can be tested in a real experiment. A principle possibility has been pointed out by Khokhlov [Kho81] [GKK82]. We might imagine preparing polymers with some reactive group placed in the center of the chain. Reactions among such macromolecules in solution according to the hard sphere concept are virtually impossible. They are suppressed exponentially with increasing chain length! RG theory yields a qualitatively different result. Allowing for interpenetration it yields a reaction rate which decreases only like some power of N. This power is determined by the probability of close contacts among interior segments of two different chains, which follows a law of type (16.2). Some less direct evidence can be taken from a study of the second virial coefficient $A_2(n_1, n_2)$ of two chains of very different lengths. Allowing for interpenetration, the fractal nature yields a much smaller value than predicted by a hard sphere model based on the smoothed density concept.

We now review further results presented in this chapter. The internal correlation function $\hat{P}(\mathbf{r}; j_1, j_2 - j_1, n - j_2)$ can be used to evaluate the 'hydrodynamic radius' R_h defined as

$$R_h^{-1} = \frac{1}{n^2} \sum_{j_1 \neq j_2} \left\langle |\mathbf{r}_{j_2} - \mathbf{r}_{j_1}|^{-1} \right\rangle \quad . \tag{16.3}$$

R_h plays an important role in certain approximate treatments of dynamical problems (see, for instance, Chap. 6 of [Yam71]), where it governs the chain length dependence of the diffusion coefficient or of the excess viscosity in the solution. Being a static length, R_h in the excluded volume limit should be proportional to R_g or R_e, equivalently: $R_h \sim R_g \sim R_e \sim n^\nu$. This is not borne out experimentally. The effective exponent ν_h extracted from R_h is typically found to be smaller than ν. This has (erroneously) been explained as a cross-over effect. Assume that the solvent is good enough to yield $R_e \sim n^\nu$ for $n > n_t^* \gg 1$, but not so good that the swelling is complete for shorter chains: $R_e(n) \sim n^{\nu_{eff}(n)}, \nu_{eff}(n) < \nu$ for $n < n_t^*$. (We assume to be on the weak coupling branch.) With $R_{j_1 j_2}^2 = \left\langle (\mathbf{r}_{j_2} - \mathbf{r}_{j_1})^2 \right\rangle \sim R_e^2(|j_2 - j_1|)$ we then expect small parts of the chain to be less swollen than the total chain. Now R_h, as compared to R_g or R_e, gives much stronger weight to the small parts of the chain. Thus it is well conceivable that ν_h has not yet reached its asymptotic value ν, even though R_g or R_e shows the limiting power law. This idea forms the basis of the 'temperature blob' model, briefly mentioned in Sect. 9.3. It is widely believed to explain the discrepancy among R_h or R_g, R_e. Having an explicit expression for $\hat{P}(\mathbf{r}; j_1, j_2 - j_1, n - j_2)$ we can check this explanation. It is found to be wrong. The excess swelling due to the dangling ends essentially compensates for the incomplete swelling of short subchains, leading to $R_h \sim R_e$ also in the cross-over regime. The experimental finding for R_h cannot be explained within the present theory. It may be due to chain stiffness, which influences R_h much stronger than R_g or R_e – as do all microstructure effects.

Besides internal correlations and interpenetrating coils in this chapter we also discuss features of the average segment density distribution. Both the total segment density distribution and the distribution of specific segments relative to the center of mass have been calculated within the RG formalism. The total segment density distribution is not very different from that in a Gaussian coil, but a definite difference is found for specific segments. Whereas in a Gaussian coil the distribution of each segment relative to the center of mass is itself Gaussian, having its maximum at the center of mass $\mathbf{R} = 0$, in an excluded volume coil the position of the maximum may occur at $|\mathbf{R}| \neq 0$, depending on the position of the segment along the chain. The mean squared distance from the center of mass, if measured in units of R_g^2, still is close to that in a Gaussian chain, but fluctuations around the average position are suppressed. This shows that self-repulsion makes the chain more rigid.

We finally review results on the shape of a coil. Being a fluctuating object, the shape of a Gaussian coil on the average is not spherical but in first approximation can be described as a prolate ellipsoid. Quantities characterizing the shape of this ellipsoid have been calculated both for a Gaussian or a self-avoiding chain. It is found that the excluded volume interaction makes the coil more aspherical and prolate.

To summarize, the analysis reviewed here identifies the polymer coil as a cigar-shaped object, not as soft as a Gaussian coil but essentially empty: it is a fractal, filling only a part of dimension $d_c = 1/\nu \approx 1.7$ of its three dimensional volume. The fractal nature implies that any structure self-similarly is repeated on all scales $\ell \ll r \ll R_g$. In particular the coil configuration shows holes of all sizes, so that a small coil can slip through a large coil without being disturbed. Clearly this picture implies a dramatic change of concepts, as compared to traditional smoothed density ideas.

In the following sections we present the results in more detail. We start with a discussion of endpoint correlations (Sect. 16.1), which, though not revealing internal structure, can be viewed as a limiting case of internal correlations. Internal correlations are discussed in Sect. 16.2, also including an analysis of the hydrodynamic radius and the temperature blob model. Section 16.3 deals with density distributions. We finally consider interpenetrating coils, discussing $A_2(n_1, n_2)$ in Sect. 16.4 and presenting results on the free energy excess $U(\mathbf{R}, n_1, n_2)$ in Sect. 16.5. Since most quantities considered here are not measurable in real solutions, we do not rederive the results within the present framework of renormalized perturbation theory in three dimensions. Mostly restricting the discussion to the excluded volume limit, we rather analyze ε-expansion results as found in the literature, referring to the original work for the derivations. The more complicated formulae are collected in Appendix A 16.1. A formal expression for R_h is derived in Appendix A 16.2. In Appendix A 16.3 we briefly discuss the entropy and the internal energy of a single chain, and in Appendix A 16.4 in the light of our results we consider the celebrated Flory argument, which results in $\nu = 3/5$ for $d = 3$.

16.1 Endpoint correlations

The endpoint correlation function

$$P(\mathbf{p}, n) = \langle \exp\left(i\mathbf{p}(\mathbf{r}_n - \mathbf{r}_0)\right)\rangle \tag{16.4}$$

has been used to exemplify the structure of the theory in previous chapters. From Sect. 11.6 we know that it is renormalized according to the relation

$$P(\mathbf{p}, n) = P_R(\bar{\mathbf{p}}, n_R, f) \quad . \tag{16.5}$$

The resulting nonlinear scaling law takes the form (cf. Eq. (11.50))

$$P(\mathbf{p}, n) = \tilde{P}(\mathbf{p}\tilde{R}_0, \tilde{z}) \quad .$$

From Appendix A 16.1, Eqs. (A 16.1)–(A 16.7) we find

$$\begin{aligned} P_R\left(\bar{\mathbf{p}}; n_R, f\right) = e^{-\bar{\mathbf{p}}^2 n_R}\Big\{1 - \frac{\varepsilon}{8} f\Big[\bar{\mathbf{p}}^2 n_R(\ln n_R + 1) + 1 \\ - e^{\bar{\mathbf{p}}^2 n_R} + \left(\bar{\mathbf{p}}^2 n_R - 1\right)\hat{E}\left(\bar{\mathbf{p}}^2 n_R\right)\Big] + O(\varepsilon^2)\Big\} \quad , \end{aligned} \tag{16.6}$$

where

$$\hat{E}(x) = \int_0^x \frac{dz}{z}\left(e^z - 1\right) \quad . \tag{16.7}$$

A little calculation shows that this result is identical to Eqs. (12.27), (12.28). (Recall that we take $b_1 = 0$.)

16.1.1 End-to-end distance

Consider first the expansion for small $\bar{\mathbf{p}}$.

$$P_R\left(\bar{\mathbf{p}}; n_R, u\right) = 1 - \bar{\mathbf{p}}^2 \bar{R}_e^2 + O(\bar{\mathbf{p}}^4) \tag{16.8}$$

$$\bar{R}_e^2 = n_R\left[1 + \frac{\varepsilon}{8} f\left(\ln n_R - 1\right) + O(\varepsilon^2)\right] \quad . \tag{16.9}$$

The mean squared end-to-end distance takes the form

$$R_e^2(n) = 2d\ell_R^2 \bar{R}_e^2 \quad . \tag{16.10}$$

From R_e^2 and R_g^2 (Eq. (A 15.31)) we can construct a critical ratio

$$\mathcal{R}_{g/e}^2 = \frac{R_g^2(n)}{R_e^2(n)} = \frac{1}{6}\left(1 - \frac{\varepsilon}{96} f + O(\varepsilon^2)\right) \quad . \tag{16.11}$$

This result shows that to first order the interaction effect on $\mathcal{R}_{g/e}^2$ is extremely small. Evaluated in the excluded volume limit and at three dimensions: $\varepsilon = 1$,

Eq. (16.11) yields $6\mathcal{R}^2_{g/e} \approx 0.99$. Monte Carlo simulations yield $6\mathcal{R}^2_{g/e} \approx 0.959$ [LMS95], a somewhat larger but still quite small deviation from the Gaussian chain result. (An extensive collection of MC results for $\mathcal{R}^2_{g/e}$ can be found in [WKG96].) Now, this is one of the rare examples where a two-loop calculation has been presented [BM85]. In the excluded volume limit $f = 1$ the result reads

$$\mathcal{R}^2_{g/e} = \frac{1}{6}\left(1 - \frac{\varepsilon}{96} - 0.0306\varepsilon^2 + O(\varepsilon^3)\right) \quad , \tag{16.12}$$

giving $6\mathcal{R}^2_{g/e} = 0.959$, in excellent agreement with the Monte Carlo data.

Since R_e^2 is one of the prime quantities determined in Monte Carlo simulations, we also discuss here the crossover behavior within the frame of our theory, working directly in $d = 3$. Our result for R_e^2, analogous to Eq. (15.8) for R_g^2 follows from Eq. (12.25) and reads for a monodisperse system

$$\frac{1}{2d} R_e^2 = \ell_R^2 n_0 \left[1 + u^* f \left(\frac{2}{3} n_0^{1/2} - 1\right)\right] \quad . \tag{16.13}$$

It yields a theoretical swelling factor

$$\bar{\alpha}_e^2 = \frac{R_e^2}{2d\tilde{R}_0^2} = \left|1 - f\right|^{\frac{1}{\omega}\left(\frac{1}{\nu} - 2\right)} H(f)(1 - 0.187f) \quad , \tag{16.14}$$

($n_0 = 0.53$). From Eqs. (15.8), (16.13) we can calculate the critical ratio $\mathcal{R}^2_{g/e}$, which turns out to coincide numerically with the $O(\varepsilon)$ result. So in this respect the $d = 3$ formulation brings about no improvement.

We now turn to the crossover of the swelling factor, using Eq. (15.13) to relate f and $\tilde{z}$. It again turns out that we can find simple parameterizations in the style of Eqs. (15.21), (15.23):

$$\bar{\alpha}_e^2 = \left(1 + 1.40\tilde{z} + 0.40\tilde{z}^2\right)^{0.176}$$
$$0 \le \tilde{z} < \infty \qquad \text{weak coupling} \; , \tag{16.15 i}$$

$$\bar{\alpha}_e^2 = 0.851\tilde{z}^{0.352}\left[1 - 0.530\tilde{z}^{-0.941}\left(1 + \frac{0.424}{\tilde{z}}\right)\right]$$
$$3 \le \tilde{z} \le \infty \qquad \text{strong coupling} \; . \tag{16.15 ii}$$

These parameterizations reproduce our full results within deviations less than 0.3%. For the weak coupling branch there again exist very precise higher order calculations [MN87], which have been analyzed with renormalization group techniques [Nic91, CCJ85]. After a rescaling of $\tilde{z}$ Eq. (16.15 i) reproduces these results within deviations less than 0.2% for $0 \le \tilde{z} \le 200$.

In Sect. 15.1.3 we found a similar good agreement for our calculation of $\bar{\alpha}_g^2$ with high order results. A slight inconsistency, however, must be noted. As pointed out above, the renormalized first-order calculation falls short of reproducing the correct result for $R^2_{g/e}$ by about 3%. In comparing $\bar{\alpha}^2_g, \bar{\alpha}^2_e$ to higher order calculations this shows up in the occurrence of a mismatch in the rescaling of $\tilde{z}$, the factors differing by about 8%. Since in principle the nonuniversal factor relating $\tilde{z}$ to the z-variable used in the higher order calculations is unique, this illustrates the limitations of the one-loop theory. In analyzing Monte Carlo results for R^2_e or R^2_g the same inconsistency shows up, of course. Results for other quantities like the second virial coefficient are consistent with the nonuniversal scale extracted for R_g. This reflects the fact that our theory by construction reproduces the correct value of ψ^*.

An extensive comparison of results equivalent to Eqs. (16.15) to Monte Carlo data has been published in [Sch94]. (In this work a slightly different RG mapping, yielding $\nu = 0.59$, has been used and the conventions concerning $b_u(1)$, n_0 are different. This explains the numerical difference in the parameterizations of $\bar{\alpha}^2_e$. In the relevant range of $\tilde{z}$, however, the results are completely equivalent, provided $\tilde{z}$ is rescaled appropriately.) Full agreement among theory and simulation data has been established. In particular the two branches are clearly exhibited. Again this implies the breakdown of the Domb-Barrett approach [DB76], which interpolates among the weak or strong coupling branches. Figure 16.1 illustrates the quality of the fit. In Fig. 16.1a we compare to Monte Carlo data taken for a hard-sphere chain in continuous space. In this model spheres of radius R_0 are connected by rods fixing the distance of successive spheres to ℓ_0 (center-to-center). The ratio $\delta_0 = 2R_0/\ell_0 \leq 1$ governs the strength of the excluded volume, and according to [KBB81] the excluded volume limit $u = u^*$ is attained for $\delta_0 \approx 0.5$. Figure 16.1a shows R^2_e divided by the asymptotic power law for $\delta_0 = 0.99; 0.75$ (strong coupling) or $\delta_0 = 0.25$ (weak coupling). It nicely illustrates the two branches. Figure 16.1b shows data for self-avoiding walks on the simple-cubic lattice, covering the huge range $100 \leq n \leq 80{,}000$ and thus a huge range of $\tilde{z}$. Self-avoiding walks on typical lattices are represented by the strong coupling branch, as is clearly exhibited. A further very detailed test of the two-branched structure may be found in recent Monte Carlo work [GSS97]. Again the results nicely fit with the theory.

16.1.2 The shape of the end-to-end correlation function

The end-to-end correlation function in real space

$$\hat{P}(\mathbf{r}, n) = \int_p e^{-i\mathbf{p}\mathbf{r}} P(\mathbf{p}, n) \tag{16.16}$$

has been considered in detail in the literature, the work concentrating on the excluded volume limit. In particular, Fisher [Fis66] and McKenzie and Moore [KM71] have established the behavior for large $|\mathbf{r}|$:

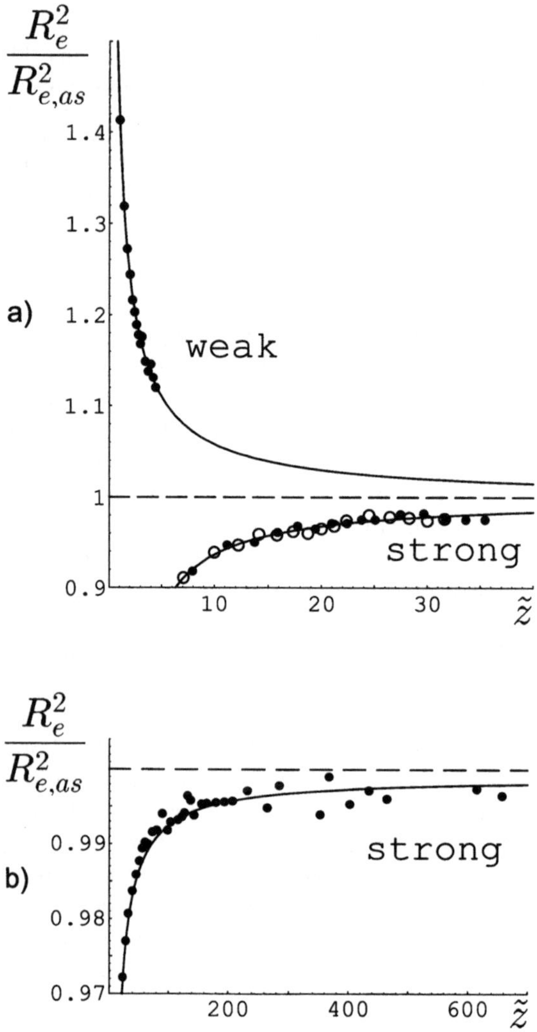

Fig. 16.1a, b. $R_e^2/R_{e,as}^2$ as function of $\tilde{z}$, where $R_{e,as}^2 = 6\tilde{\ell}^2\tilde{v}(T)^{4\nu-2}0.851n^{2\nu}$. Curves represent the theory: (**a**) lower branch: strong coupling, upper branch: weak coupling. MC-data for the hard sphere model [BMB91], upper branch: $\delta_0 = 0.25$; lower branch: $\bullet$ $\delta_0 = 0.75$; $\circ$ $\delta_0 = 0.99$; (**b**) data: self avoiding walks on the simple cubic lattice [LMS95]. Note the different scales of the figures

$$
\begin{aligned}
R_e^d\hat{P}(\mathbf{r},n) \underset{|\mathbf{r}|>>R_e}{\longrightarrow} a_1\left|\frac{\mathbf{r}}{R_e}\right|^{\frac{2+d-2\gamma}{2(1-\nu)}-d} \exp\left(-a_2\left|\frac{\mathbf{r}}{R_e}\right|^{\frac{1}{1-\nu}}\right) \\
\sim \left|\frac{\mathbf{r}}{R_e}\right|^{0.257} \exp\left(-a_2\left|\frac{\mathbf{r}}{R_e}\right|^{2.427}\right), \quad (d=3) \ , \qquad (16.17)
\end{aligned}
$$

where a_1, a_2 are (positive) constants. The exponent γ has been introduced in Eq. (11.56) in the context of the scaling of the single chain partition function. In $d = 3$ it is found to take the value $\gamma = 1.158$. (See Appendix A 16.3.)

For small $|\mathbf{r}|$ Des Cloizeaux [Clo74] has derived the power law

$$R_e^d \hat{P}(\mathbf{r}, n) \underset{|\mathbf{r}|<<R_e}{\longrightarrow} a_3 \left|\frac{\mathbf{r}}{R_e}\right|^{\frac{\gamma-1}{\nu}} \sim \left|\frac{\mathbf{r}}{R_e}\right|^{0.269} (d=3) \ , \tag{16.18}$$

where a_3 is another (positive) constant. This proves the existence of a correlation hole: close encounters of the chain ends are suppressed. Both the short distance and the large distance results strongly differ from the Gaussian behavior of a noninteracting chain.

The exponent of the prefactor in Eq. (16.17) is sufficiently close to the power governing the small distance behavior (16.18) to suggest a simple interpolation, commonly known as des Cloizeaux's scaling form:

$$\hat{P}(\mathbf{r}, n) = \frac{\tilde{a}_1}{R_e^3} \left|\frac{\mathbf{r}}{R_e}\right|^{0.269} \exp\left(-\tilde{a}_2 \left|\frac{\mathbf{r}}{R_e}\right|^{2.427}\right), \quad d=3 \ . \tag{16.19}$$

The normalization of $\hat{P}$ and the definition of the end-to-end distance R_e fix the two constants as $\tilde{a}_1 = 0.299; \tilde{a}_2 = 1.269$. With these values $\hat{P}(r/R_e) = R_e^3 \hat{P}(\mathbf{r}, n)$ is plotted in Fig. 16.2 and compared to Gaussian behavior. The correlation hole is clearly exhibited. The predicted behavior of $\hat{P}(\mathbf{r}, n)$ has been verified in Monte Carlo simulations [Bau81, BC91, BCRF91]. In particular a recent extensive simulation [Val96] shows that the deviations from the simple expression (16.19) are very small.

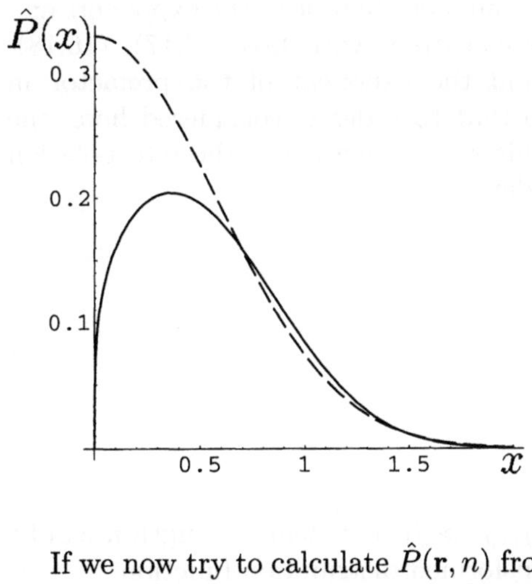

Fig. 16.2. $\hat{P}(x) = R_e^3 \tilde{P}(\mathbf{r}, n), x = r/R_e$, as given by Eq. (16.19) (*full line*). The *broken line* represents the Gaussian approximation

If we now try to calculate $\hat{P}(\mathbf{r}, n)$ from Eq. (16.16), using the one-loop approximation (16.6) for $P_R(\bar{\mathbf{p}}, n_R, f)$, we apparently run into problems. From the behavior for small $|\mathbf{r}|$, Eq. (16.18), we expect to see a large momentum behavior

$$\begin{aligned} P(\mathbf{p}, n) \underset{|\mathbf{p}R_e| \gg 1}{\sim} \left|\mathbf{p}R_e\right|^{-d+\frac{1-\gamma}{\nu}} &\sim \left|pR_e\right|^{-4+\frac{3\varepsilon}{4}+O(\varepsilon^2)} \\ \sim \quad \left|\mathbf{p}R_e\right|^{-4} &\left(1 + \frac{3}{4}\varepsilon \ \ln|\mathbf{p}R_e| + O(\varepsilon^2)\right) \ . \end{aligned}$$

Evaluating Eq. (16.6) for large $\bar{\mathbf{p}}^2 n_R$ we find

$$P_R\left(\bar{\mathbf{p}}, n_R, f\right) = e^{-\bar{\mathbf{p}}^2 n_R} - \frac{\varepsilon}{8} f \frac{1}{(\bar{\mathbf{p}}^2 n_R)^2}\left(1 + O\left(\frac{1}{\bar{\mathbf{p}}^2}\right)\right) \ , \tag{16.20}$$

and thus no sign of the logarithmic $O(\varepsilon)$-correction is seen. This reflects the fact that the Gaussian zero-order term cannot approximate the leading large momentum behavior, which is due to the correlation hole. The leading behavior $\sim p^{-4}$ therefore is an interaction effect and first shows up in one-loop order. Thus a momentum space calculation is not adapted to give the behavior of the endpoint correlations outside the small momentum limit. However, analytically calculating the Fourier transform of the ε-expansion (16.6) and ad hoc exponentiating the result, Duplantier [Dup86b] has derived the expression

$$\begin{aligned} \hat{P}(\mathbf{r}, n) \quad &= \quad \left(\frac{2\pi}{d} R_e^2\right)^{-d/2} \left(1 + \frac{\varepsilon}{8}(\gamma_{Eu} - 2)\right) \left(\frac{d}{2}\frac{r^2}{R_e^2}\right)^{\varepsilon/8} \\ &\exp\left[-\left(\frac{d}{2}\frac{r^2}{R_e^2}\right)^{1+\varepsilon/8} \left(1 + \frac{\varepsilon}{8}(\gamma_{Eu} - 2)\right)\right] \ , \end{aligned} \tag{16.21}$$

where $\gamma_{Eu} = 0.5772\ldots$ denotes Euler's number. To order ε the exponents and thus the total expression (16.21) are consistent with Eqs. (16.17), (16.18). Note that $(\gamma - 1)/\nu$ (Eq. (16.18)) and the exponent of the prefactor in Eq. (16.17) differ only in order ε^2, so that to order ε, considered here, the powers are the same. Furthermore, this result shows that the interpolation form (16.19) is correct to one loop order.

16.2 Internal correlations

16.2.1 General results

Spatial correlations among segments $j_1, j_2 > j_1$ in a chain of length n can be calculated in terms of the Fourier transformed distribution function

$$P(\mathbf{p}; n_1, n_2, n_3) = \langle \exp\left(i\mathbf{p}\left(\mathbf{r}_{j_2} - \mathbf{r}_{j_1}\right)\right)\rangle \ , \tag{16.22}$$

where we use the notation

$$n_1 = j_1; \; n_2 = j_2 - j_1; \; n_3 = n - j_2 \ .$$

Since $P(\ldots)$ obeys a renormalization invariant normalization condition: $P\left(\mathbf{p}^2=0;n_1,n_2,n_3\right) = 1$, the nonlinear scaling law takes the form (cf. Sect. 11.6)

$$\begin{aligned} P\left(\mathbf{p};n_1,n_2,n_3\right) &= P_R\left(\bar{\mathbf{p}};n_{2R},\frac{n_1}{n_2},\frac{n_3}{n_2};f\right) \\ &= \tilde{P}\left(\mathbf{p}R_e(n_2),\frac{n_1}{n_2},\frac{n_3}{n_2};\tilde{z}\right) \quad , \end{aligned} \tag{16.23}$$

where we identified ℓ_R with the end-to-end distance of a chain of length n_2. At the fixed point $\tilde{P}(\ldots)$ becomes a fully universal function.

We will be mainly concerned here with moments of the distribution function. Specifically the mean squared distance among segments j_1, j_2 is found as

$$\begin{aligned} R_i^2(n_1,n_2,n_3) &= \left\langle (\mathbf{r}_{j_2}-\mathbf{r}_{j_1})^2\right\rangle \\ &= -2dR_e^2(n_2)\frac{\partial}{\partial x^2}\bigg|_0 \tilde{P}\left(x,\frac{n_1}{n_2},\frac{n_3}{n_2};\tilde{z}\right) \quad . \end{aligned} \tag{16.24}$$

The ratio

$$\mathcal{R}_{i/e}^2(n_1,n_2,n_3) = \frac{R_i^2(n_1,n_2,n_3)}{R_e^2(n_2)} \tag{16.25}$$

gives the excess swelling due to the embedding of the piece n_2 into the larger chain. For $f=1$ it reduces to a fully universal function of the RG invariant ratios $n_1/n_2, n_3/n_2$:

$$\mathcal{R}_{i/e}^2 \to -2d\frac{\partial}{\partial x^2}\bigg|_0 \tilde{P}\left(x,\frac{n_1}{n_2},\frac{n_3}{n_2},\infty\right) \quad .$$

This demonstrates that interior parts indeed swell according to the same power law as the total chain, as expected for a scale invariant object. With respect to the fractal nature of the coil this is a most important result. It is essential that the result depends only on the ratios $n_1/n_2, n_3/n_2$, but not on n_1, n_3 individually. The latter dependence would introduce a separate scale for the end effects and thus would violate scale invariance.

We also are interested in the inverse moment

$$R_h^{-1}(n_1,n_2,n_3) = \left\langle |\mathbf{r}_{j_2}-\mathbf{r}_{j_1}|^{-1}\right\rangle \quad , \tag{16.26}$$

which upon summation over j_1, j_2 yields the hydrodynamic radius (cf. Eq. (16.3)). To relate R_h^{-1} to $P(\ldots)$ it needs some algebra presented in Appendix A 16.2. In three dimensions we find

$$R_h^{-1}(n_1,n_2,n_3) = \frac{1}{\pi R_e(n_2)}\int_0^\infty \frac{dx}{\sqrt{x}}\tilde{P}\left(\sqrt{x};\frac{n_1}{n_2},\frac{n_3}{n_2};\tilde{z}\right) \quad . \tag{16.27}$$

In analogy to Eq. (16.25) we define a ratio

$$\mathcal{R}^2_{hi/he}(n_1, n_2, n_3) = \frac{R^2_h(n_1, n_2, n_3)}{R^2_h(0, n_2, 0)} \quad , \tag{16.28}$$

which at the fixed point again is a universal function of $n_1/n_2, n_3/n_2$.

All these quantities have been calculated within the ε-expansion [Gri83, Dup85, Dup86b, SB86]. The basic result for $P_R\left(\bar{\mathbf{p}}; n_{2R}, \frac{n_1}{n_2}, \frac{n_3}{n_2}; f\right)$ is lengthy and is presented in Appendix A 16.1 (Eqs. (A 16.1)–(A 16.5)).

16.2.2 Excess swelling of internal correlations

Equation (A 16.8) yields for the ratio $\mathcal{R}^2_{i/e}$

$$\mathcal{R}^2_{i/e}(n_1, n_2, n_3) = 1 + \frac{\varepsilon}{8} f \left[\frac{1}{2}\left(\frac{n_2}{n} - 1\right) + \frac{n_1}{n_2} \ln\left(1 + \frac{n_2}{n_1}\right) + \frac{n_3}{n_2} \ln\left(1 + \frac{n_2}{n_3}\right) \right] + O(\varepsilon^2) \ . \tag{16.29}$$

This expression takes its maximum value for $n_1/n_2 \to \infty, n_3/n_2 \to \infty$

$$\mathcal{R}^2_{i/e}(\infty, n_2, \infty) - 1 = \frac{3}{16}\varepsilon f + O(\varepsilon^2) \ . \tag{16.30 i}$$

For a piece starting at the end of an infinitely long chain we find

$$\mathcal{R}^2_{i/e}(0, n_2, \infty) - 1 = \frac{1}{16}\varepsilon f + O(\varepsilon^2) \ . \tag{16.30 ii}$$

This inhomogeneous swelling of subchains is an end effect, which however does not die out in the limit of an infinitely long chain. Rather swelling stays inhomogeneous over finite fractions of the chain. As stressed above, this is due to scale invariance. Figure 16.3a, b shows $\mathcal{R}^2_{i/e}(n_1, n_2, n_3) - 1$ for the central piece $n_1 = n_3 = (n - n_2)/2$ or an end piece $n_1 = 0, n_3 = n - n_2$ in the excluded volume limit $f = 1$, together with Monte Carlo data. The data scatter somewhat but support the analytical results. Recent simulations [WKG96] are consistent with the data shown here.

We now consider the inverse moment. For endpoint correlations we find from Eq. (A 16.9)

$$\begin{aligned} \mathcal{R}_{e/he} &= \left\langle |\mathbf{r}_0 - \mathbf{r}_n|^{-1} \right\rangle R_e(n) \\ &= \sqrt{\frac{6}{\pi}} \left(1 - \frac{\varepsilon}{8} f (1 - \ln 2) + O(\varepsilon^2)\right) \ . \end{aligned} \tag{16.31}$$

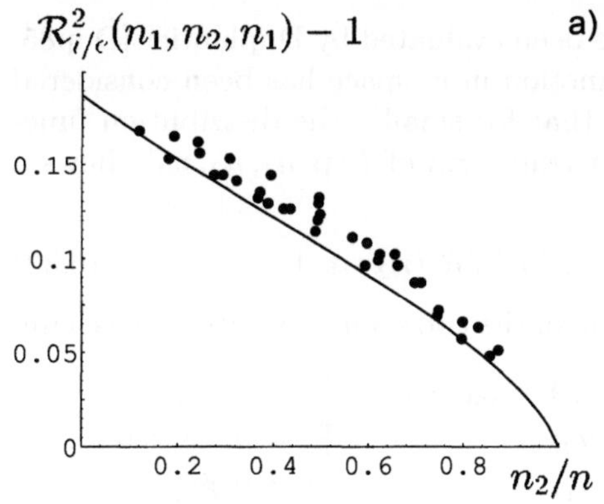

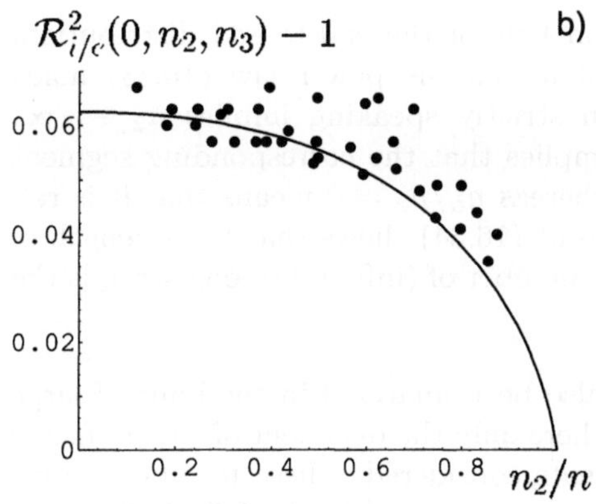

Fig. 16.3. (**a**) $\mathcal{R}^2_{i/e}\left(\frac{n-n_2}{2}, n_2, \frac{n-n_2}{2}\right) - 1$ as function of n_2/n in the excluded volume limit. The data are taken for the hard bead chain, $\delta_0 = 0.53$ (from [SB86]). (**b**) Same as (**a**) for $\mathcal{R}^2_{i/e}(0, n_2, n - n_2) - 1$

Again the first order correction is quite small, resulting in the estimate $\sqrt{\pi/6}\,\mathcal{R}_{e/he} \approx 0.96$ valid for $f = 1, \varepsilon = 1$. Monte Carlo simulations [SB86] in the excluded volume limit yield a result of the same order of magnitude: $\sqrt{\pi/6}\,\mathcal{R}_{e/he} \approx 0.92$. As found before, first order ε-results underestimate somewhat the small correction.

The analytical expression for $\mathcal{R}_{hi/he}(n_1, n_2, n_3)$, giving the excess swelling of a subchain in terms of the inverse moments, is somewhat complicated. The full results can be found in the appendix (Eqs. (A 16.10), (A 16.11)). Here we only note the special limits

$$\mathcal{R}^2_{hi/he}(\infty, n_2, \infty) - 1 = \frac{3}{4}\varepsilon f(2\ln 2 - 1) + O(\varepsilon^2) \qquad (16.32\text{ i})$$

$$\mathcal{R}^2_{hi/he}(0, n_2, \infty) - 1 = \frac{\varepsilon}{4} f(2\ln 2 - 1) + O(\varepsilon^2)\ , \qquad (16.32\text{ ii})$$

which should be compared to Eqs. (16.30). It is found that the effect of excess swelling is stronger for $R_h(n_1, n_2, n_3)$ than for $R_i(n_1, n_2, n_3)$. Further evaluation of $\mathcal{R}^2_{hi/he}$ yields results closely resembling Fig. 16.3.

Higher moments of $(\mathbf{r}_{j_1} - \mathbf{r}_{j_2})$ have been evaluated by Duplantier [Dup85, Dup86b]. Also the full distribution function in $\mathbf{r}$- space has been considered [Clo80b, OO81, Dup86b]. It is found that for small $\mathbf{r}$ the distribution function $\hat{P}(\mathbf{r}; n_1, n_2, n_3)$ defined as Fourier-transform of $P(\mathbf{p}; n_1, n_2, n_3)$ shows a correlation hole:

$$\hat{P}(\mathbf{r}; n_1, n_2, n_3) \sim |\mathbf{r}|^{\vartheta} \quad , \mid \mathbf{r} \mid / R_e(n_2) \ll 1 \quad . \tag{16.33}$$

The exponent ϑ depends on the position of the two segments within the chain:

$$\vartheta = \begin{cases} \vartheta_0 = (\gamma - 1)/\nu & = \varepsilon/4 + \frac{9\varepsilon^2}{128} + \ldots ; \quad \frac{n_1}{n_2} = 0 = \frac{n_3}{n_2} \\ \vartheta_1 = \varepsilon/2 - \frac{3}{64}\varepsilon^2 + \ldots & > \vartheta_0 ; \quad \frac{n_1}{n_2} = 0 ; \frac{n_3}{n_2} \neq 0 \\ \vartheta_2 = \varepsilon - \frac{15}{32}\varepsilon^2 + \ldots & > \vartheta_1 ; \quad \frac{n_1}{n_2} \neq 0 \neq \frac{n_3}{n_2} \end{cases} \quad . \tag{16.34}$$

Note that ϑ_0 describes the correlation hole of the end-to-end distribution, as discussed in Sect. 16.1.2. By construction the power law (16.33) holds in the excluded volume limit, which strictly speaking implies $n_2 \rightarrow \infty$. $n_\alpha / n_2 > 0, (\alpha = 1 \text{ or } 3)$, therefore implies that the corresponding segment is infinitely far from the chain end whereas $n_\alpha / n_2 = 0$ means that it is relatively close to the chain end. The result (16.34) shows that the strength of the correlation hole increases with the number of (infinitely) long strands the segments considered carry along.

The distribution function $\hat{P}$ has also been analyzed in the limit of large $\mid \mathbf{r} \mid$. A law of type (16.17) is found, where only the exponent of the prefactor depends on the position of the segments considered. These results to some extent have been verified by computer experiments [Bau81, RFBC93].

Summarizing this discussion we may note that the interaction also leads to a swelling of the internal parts of the chain. Consistent with scale invariance, the dominant swelling behavior of an internal part of length n_2 is proportional to the swelling of a chain of total length $n_2 : R_i(n_1, n_2, n_3) \sim R_e(n_2) \sim n_2^{\nu}$, but in addition there is some swelling contained in the prefactor of this power law. This excess swelling depends on the position of the piece n_2 within the chain and is related to the correlation hole, i.e. to the effect that the interaction suppresses close encounters of the segments. Since this repulsive effect is stronger for the interior parts of the chain, it leads to a stronger swelling of such pieces. Also the inverse moment weights small distances stronger and therefore is more affected.

16.2.3 Temperature blobs and hydrodynamic radius

The excess swelling is of relevance for an understanding of the hydrodynamic radius. To appreciate its implications we need to discuss the 'temperature blob' model, which we briefly mentioned before in our discussion of scaling theory. In the excluded volume limit $\tilde{z} \to \infty$ but close to the Θ-temperature $t = (T-\Theta)/\Theta \ll 1$, scaling theory predicts a specific temperature dependence of the single chain radius: $R_g(n,t) \sim t^{(2\nu-1)/\varepsilon} n^{\nu}$. As explained in Sect. 9.3 this can be interpreted by modeling the chain as a swollen configuration of n/n_t^* temperature blobs of length $n_t^* \sim t^{-2/\varepsilon}$, which internally are not swollen. It is tempting to use this model to explain the difference among the experimental findings for R_h or R_g. As mentioned in the introductory section, in terms of effective exponents

$$\nu_{eff}^{(h,g)} = \frac{d \ln R_{h,g}}{d \ln n}$$

most experiments find values $\nu_{eff}^h < \nu_{eff}^g$. In contrast to R_g, the radius R_h hardly ever reaches the excluded volume limit, where $\nu_{eff}^h = \nu_{eff}^g = \nu$ holds. Since R_h as compared to R_g weights small subchains $n \lesssim n_t^*$ more strongly, it is suggestive to interpret this finding as a crossover effect. For R_h the contribution from the unswollen temperature blobs dies out so slowly that for the accessible chain lengths the effective exponent ν_{eff}^h is still in the crossover regime from Θ- to excluded volume conditions.

This basic idea of the temperature blob model is contained in the following approximate equations for R_g or R_h:

$$R_g^2 = \frac{1}{n^2} \sum_{j_1<j_2} \left\langle (\mathbf{r}_{j_2} - \mathbf{r}_{j_1})^2 \right\rangle \approx \frac{1}{n^2} \sum_{j_1<j_2} R_e^2(j_2 - j_1) \qquad (16.35\ \mathrm{i})$$

$$R_h^{-1} = \frac{2}{n^2} \sum_{j_1<j_2} \left\langle |\mathbf{r}_{j_2} - \mathbf{r}_{j_1}|^{-1} \right\rangle \approx \frac{2}{n^2} \sum_{j_1<j_2} \left\langle (\mathbf{r}_{j_1} - \mathbf{r}_{j_2})^2 \right\rangle^{-\frac{1}{2}}$$

$$\approx \frac{2}{n^2} \sum_{j_1<j_2} R_e^{-1}(j_2 - j_1) \ . \qquad (16.35\ \mathrm{ii})$$

Note that this involves two approximations: we ignore the excess swelling of internal parts, and in Eq. (16.35 ii) we furthermore replace the inverse moment by the inverse square root of the second moment. To proceed, we need a model for $R_e(n)$. In the simplest implementation of the temperature blob model we may use the ansatz

$$R_e^2(n) = \begin{cases} \ell^2 n, & n \le n_t^* \\ \ell^2 n_t^* \left(\dfrac{n}{n_t^*} \right)^{2\nu} & n \ge n_t^* \end{cases} , \qquad (16.36)$$

which takes the temperature blob concept literally. More realistic models of $R_e(n)$, smearing out the behavior (16.36) have also been employed. As a result the temperature blob model indeed seems to explain the observed discrepancy among ν_{eff}^h and ν_{eff}^g.

With the full distribution function of the internal distances at hand, we can check this explanation. It is found to be just wrong. The detailed analysis [SB86] is lengthy, and here we only review the essential results. The first (rigorous) equalities in Eqs. (16.35) relate R_g^2, R_h^{-1} to the moments discussed in the previous section. We may evaluate these equations fixing the renormalized length scale by the relation $n_{2R} = 1$. This incorporates the idea of the blob model by focussing on the swelling of the piece $j_2 - j_1 = n_2$ as the basic effect. For a system on the weak coupling branch it implies that the effective coupling constant f depends on n_2, decreasing with decreasing n_2 as expected. Now evaluating the segment summations using *zero order* results for the moments we find results as given by the blob model. Since in this approximation the moments equal the corresponding moments for a chain of length n_2, it is equivalent in spirit to using the approximate equalities in (16.35). However, including the *first order* terms and thus the excess swelling we find that the difference among ν_{eff}^g and ν_{eff}^h never exceeds $\nu_{eff}^g - \nu_{eff}^h \lesssim 0.01$, which is below the experimental accuracy. The excess swelling thus counterbalances the influence of crossover among Θ- or excluded volume limits, incorporated in the temperature blob model. This is easily understood since, as pointed out in the previous section, the excess swelling necessarily is stronger for $R_h(n_1, n_{21}, n_3)$ than for $R_i(n_1, n_{21}, n_3)$.

In the context of this discussion the recent discovery of the strong coupling branch is also of interest. Experimentally a strong deviation $\nu_{eff}^h < \nu_{eff}^g$ has been found for the system polystyrene-benzene, which is likely to be on the strong coupling branch. Also Monte Carlo simulations for self-avoiding walks, which clearly are on the strong coupling branch, show similar effects. For these systems neither a Θ-temperature nor temperature blobs exist and the effective exponent ν_{eff} *increases* with decreasing chain length. (See Fig. 15.3b.) A calculation in the style of the blob model would thus suggest $\nu_{eff}^h > \nu_{eff}^g$, in sharp contrast to the data. A closer inspection of the Monte Carlo data suggests that the experimental finding $\nu_{eff}^h < \nu_{eff}^g$ is a microstructure effect, reflecting some stiffness of the chains on short scales. It is thus not contained in the present two-parameter model.

The temperature blob model has also been used to derive predictions for the single chain scattering function. Clearly such attempts suffer from the same shortcoming, neglecting the important excess swelling of the internal structure. To summarize, the temperature blob model should be abandoned.

16.3 Density distributions

16.3.1 Distribution function for specific segments

Introducing the center-of-mass of the polymer chain

$$\mathbf{R}_{cm} = \frac{1}{n} \sum_{j_1=1}^{n} \mathbf{r}_{j_1} \quad ,$$

we define the (Fourier transformed) distribution function of the relative position of segment j as

$$\rho_{cm}(\mathbf{q}, j, n) = \langle \exp\left[i\mathbf{q}\left(\mathbf{r}_j - \mathbf{R}_{cm}\right)\right]\rangle \quad . \tag{16.37}$$

This function is renormalizable, and the center of mass does not introduce any new renormalization factor. This is easily proven by considering the expansion of $\rho_{cm}(\mathbf{q}, j, n)$ in powers of q^2. All the coefficients of that expansion can be expressed in terms of segment density correlation functions with their known renormalization properties. Since $\rho_{cm}(\mathbf{q}, j, n)$ is normalized, the general scaling law reads

$$\begin{aligned} \rho_{cm}(\mathbf{q}, j, n) &= \rho_{cm,R}\left(\bar{\mathbf{q}}, \frac{j}{n}, n_R, f\right) \\ &= \tilde{\rho}_{cm}\left(\mathbf{q}^2 R_g^2, \frac{j}{n}, \tilde{z}\right) \quad . \end{aligned} \tag{16.38}$$

In the excluded volume limit, ρ_{cm} becomes a fully universal function of $q^2 R_g^2$ and j/n.

$\rho_{cm,R}$ has been calculated to first order in ε [SK88]. The lengthy result is given in Appendix A 16.1.2, Eqs. (A 16.12), (A 16.13). Here we only consider the expansion for small $\mathbf{q}$:

$$\begin{aligned} \rho_{cm,R}\left(\bar{\mathbf{q}}, \frac{j}{n}, n_R, f\right) &= 1 - \bar{\mathbf{q}}^2 \bar{R}_2\left(\frac{j}{n}, n_R, f\right) \\ &+ \bar{R}_4\left(\frac{j}{n}, n_R, f\right)\left(\bar{R}_2\left(\frac{j}{n}, n_R, f\right)\bar{\mathbf{q}}^2\right)^2 + O\left(\left(\bar{\mathbf{q}}^2\right)^3\right) . \end{aligned} \tag{16.39}$$

Combining $\ell_R^2 \bar{R}_2(j/n, n_R, f)$ with R_g^2, we can construct a universal ratio $\left\langle \left(\mathbf{r}_j - \mathbf{R}_{cm}\right)^2 \right\rangle / R_g^2$. $\bar{R}_4$ is a universal ratio by itself:

$$\left(2+\frac{4}{d}\right)\bar{R}_4\left(\frac{j}{n},n_R,f\right)-1=\frac{\left\langle\left[(\mathbf{r}_j-\mathbf{R}_{cm})^2-\left\langle(\mathbf{r}_j-\mathbf{R}_{cm})^2\right\rangle\right]^2\right\rangle}{\left\langle(\mathbf{r}_j-\mathbf{R}_{cm})^2\right\rangle^2}\ . \tag{16.40}$$

The results for these ratios are given in the appendix, Eqs. (A 16.14), (A 16.15).

Evaluating $\left\langle(\mathbf{r}_j-\mathbf{R}_{cm})^2\right\rangle/R_g^2$ as function of $\bar{j}=j/n$ we find that the first order correction is remarkably small, this ratio within deviations less than 5% being given by the leading Gaussian approximation to $\rho_{cm,R}$. Thus the mean squared distance of a specific segment from the center of mass essentially is given by the 'swollen Gaussian approximation', which replaces the self-repelling coil by a Gaussian coil modified only by replacing the parameter $\ell^2 n$ by the correct end-to-end distance. This is consistent with and may be the origin of the finding that ratios like R_e/R_g or R_e/R_h are fairly insensitive to the interaction.

Turning to the ratio (16.40) we find a completely different result. For a Gaussian chain this ratio takes the value $2/d$, independent of j/n. For the self-repelling chain in the excluded volume limit at $d=3$ it is smaller by about 10–20% and strongly depends on $\bar{j}$. It reaches ~ 0.5 for $\bar{j}\sim 0.2$. Since this ratio measures the fluctuations of $(\mathbf{r}_j-\mathbf{R}_{cm})^2$ about its average value this implies that the interaction makes the chain stiffer, forcing the segments to stay closer to their average position. This is also very clearly brought out by calculating the full distribution function of $\mathbf{r}_j-\mathbf{R}_{cm}$. Numerically evaluating the inverse Fourier-transform of $\rho_{cm,R}\left(\bar{\mathbf{q}}^2,j/n,n_R,u^*\right)$ (Eq. (A 16.12)), for segments not too close to the chain center we find a shallow maximum for $|\mathbf{r}_j-\mathbf{R}_{cm}|/R_g>0$. In contrast, for a noninteracting Gaussian chain the distribution function for any segment j is a Gaussian centered at $\mathbf{r}_j=\mathbf{R}_g$. The interaction therefore suppresses fluctuations of the segments towards the center of the chain, but leaves the average position essentially unchanged, except for the overall swelling. Detailed plots of these results may be found in [SK88].

16.3.2 Distribution function of the total segment density

Summing over all segments we construct the total segment density distribution function

$$\rho_{cm}(\mathbf{q}^2,n)=\frac{1}{n}\sum_{j=1}^{n}\rho_{cm}(\mathbf{q}^2,j,n)\quad , \tag{16.41}$$

which again can be expressed in terms of renormalized variables without any nonuniversal prefactor:

$$\begin{aligned}\rho_{cm}(\mathbf{q}^2, n) &= \tilde{\rho}_{cm}\left(\mathbf{q}^2 R_g^2(n), \tilde{z}\right) \\ &= \int_0^1 dx \tilde{\rho}_{cm}\left(\mathbf{q}^2 R_g^2(n), x, \tilde{z}\right) \quad . \end{aligned} \tag{16.42}$$

With the form of $\rho_{cm,R}$ given in the appendix we may numerically evaluate that integral. As expected, the main effect of the interaction consists in a reduction of the density in the center. Compared to the swollen Gaussian approximation the central peak is reduced by about 20%. As should be clear from the previous section this is not due to a shift in the average position of the segments, as one might suspect at first. Rather fluctuations towards the center, which build up the large central density in a Gaussian coil, are suppressed. We again refer to [SK88] for the details.

16.3.3 The shape of a polymer coil

The ensemble of polymer configurations is isotropic and does not distinguish any direction. Expectation values therefore are rotationally invariant. In particular, two-point correlations, which necessarily depend on a single vector-type variable $\mathbf{q}$, in fact can depend on the invariant $\mathbf{q}^2$, only. They thus contain no information on the average shape of the coil. For such information to survive the averaging process, it needs tensor type quantities depending on at least two independent directions. The simplest quantity to be considered is the shape tensor $Q_{\alpha\beta}$. For a given configuration $\{\mathbf{r}_j\} = \{\mathbf{r}_0, \mathbf{r}_1, \ldots, \mathbf{r}_n\}$ of the chain, $Q_{\alpha\beta}$ is defined as

$$Q_{\alpha\beta}\{\mathbf{r}_j\} = \frac{1}{2(n+1)^2} \sum_{i,j=0}^{n} (r_{i\alpha} - r_{j\alpha})(r_{i\beta} - r_{j\beta}), \ \alpha, \beta = 1, \ldots, d \ . \tag{16.43}$$

It is related to the inertia tensor $I_{\alpha\beta}$:

$$I_{\alpha\beta}\{\mathbf{r}_j\} = \delta_{\alpha\beta} \, \mathrm{tr}Q\{\mathbf{r}_j\} - Q_{\alpha\beta}\{\mathbf{r}_j\} \quad .$$

Here

$$\mathrm{tr}Q = \sum_{\alpha=1}^{d} Q_{\alpha\alpha} \tag{16.44}$$

denotes the trace. Note that

$$\mathrm{tr}Q\{\mathbf{r}_j\} = R_g^2\{\mathbf{r}_j\}$$

is the radius of gyration of the configuration considered.

By virtue of rotational invariance, the expectation value of Q reduces to

$$\langle Q_{\alpha\beta}\{\mathbf{r}_i\}\rangle = \delta_{\alpha\beta} \frac{1}{d} \langle \mathrm{tr}Q\{\mathbf{r}_j\}\rangle = \delta_{\alpha\beta} \frac{R_g^2}{d} \ , \tag{16.45}$$

and thus yields no information on the shape of the coil. Such information is contained in the eigenvalues $\lambda^{(\alpha)}\{\mathbf{r}_j\}$ of $Q\{\mathbf{r}_j\}$. If exactly one $\lambda^{(\alpha)}\{\mathbf{r}_j\}$ is larger than the average $\bar{\lambda}\{\mathbf{r}_j\} = R_g^2\{\mathbf{r}_j\}/d$, then the coil configuration is prolate. In simulations the configurational average of the ordered eigenvalues $\lambda^{(\alpha)}\{\mathbf{r}_j\} \geq \lambda^{(\alpha')}\{\mathbf{r}_j\}, \alpha < \alpha'$ has been determined. The results are collected in Table 16.1. It is found that even a non-interacting chain is strongly prolate. Its shape is that of a fairly thin asymmetric ellipsoid. Self-repulsion increases somewhat the deviations from spherical shape. This is easily understood since an increase of asymmetry increases the low-density surface region at the expense of the high-density bulk region, thus decreasing the total interaction energy.

Table 16.1. $\lambda^{(\alpha)}/\bar{\lambda}$ from Monte Carlo simulations a) [SS71], [Sol71]; b) [JEK92]

α	1	2	3
Gaussian[a)]	0.76	0.176	0.065
self-avoiding[b)]	0.782	0.163	0.055

The eigenvalues of $Q\{\mathbf{r}_j\}$ are not easily accessible to analytic calculations. (An approximate calculation has been presented in [MH81].) Information can be gained by calculating invariants of Q. In particular the 'asphericity' has been considered, which is defined as

$$\mathcal{A}_d\{\mathbf{r}_j\} = \frac{d}{d-1} \frac{\mathrm{tr}\hat{Q}^2\{\mathbf{r}_i\}}{(\mathrm{tr}Q\{\mathbf{r}_j\})^2} \quad . \tag{16.46}$$

Here $\hat{Q}$ is the traceless tensor

$$\hat{Q}_{\alpha\beta}\{\mathbf{r}_j\} = Q_{\alpha\beta}\{\mathbf{r}_j\} - \delta_{\alpha\beta}\bar{\lambda}\{\mathbf{r}_j\} \quad . \tag{16.47}$$

$\mathcal{A}_d$ can be expressed in terms of the eigenvalues $\lambda^{(\alpha)}$.

$$\mathcal{A}_d\{\mathbf{r}_j\} = \frac{1}{d(d-1)} \sum_{\alpha=1}^{d} \left(\frac{\lambda^{(\alpha)}\{\mathbf{r}_j\}}{\bar{\lambda}\{\mathbf{r}_j\}} - 1 \right)^2 \quad . \tag{16.48}$$

It varies from $\mathcal{A}_d\{\mathbf{r}_j\} = 0$ for spherical shape to $\mathcal{A}_d\{\mathbf{r}_j\} = 1$ for a rod (prolate limit). For a 'pancake' (oblate limit) it takes the value $1/(d-1)^2$. $\mathcal{A}_d$ cannot discriminate between prolate or oblate shapes. Thus a second invariant must be considered, which is based on the determinant of $\hat{Q}\{\mathbf{r}_j\}$:

$$\mathcal{S}_d\{\mathbf{r}_j\} = \frac{\det \hat{Q}\{\mathbf{r}_j\}}{\left(\frac{1}{d}\mathrm{tr}Q\{\mathbf{r}_j\}\right)^d} = \prod_{\alpha=1}^{d} \left(\frac{\lambda^{(\alpha)}\{\mathbf{r}_j\}}{\bar{\lambda}\{\mathbf{r}_j\}} - 1 \right) \quad . \tag{16.49}$$

For $d = 3, \mathcal{S}_d\{\mathbf{r}_j\}$ is positive for prolate or negative for oblate configurations.

The evaluation of the configurational average of $\mathcal{A}_d\{\mathbf{r}_j\}$ or $\mathcal{S}_d\{\mathbf{r}_j\}$ even for a Gaussian chain is not completely trivial. Averaging ratios of fluctuating quantities is difficult. A simple bypass consists in replacing $\langle\mathcal{A}_d\{\mathbf{r}_j\}\rangle$ or $\langle\mathcal{S}_d\{\mathbf{r}_j\}\rangle$ by ratios of averages [RG86, AN86]:

$$\hat{\mathcal{A}}_d = \frac{d}{d-1}\frac{\left\langle \mathrm{tr}\hat{Q}^2\{\mathbf{r}_j\}\right\rangle}{\left\langle (\mathrm{tr}Q\{\mathbf{r}_j\})^2\right\rangle} \tag{16.50 i}$$

$$\hat{\mathcal{S}}_d = d^d\frac{\left\langle \det \hat{Q}\{\mathbf{r}_j\}\right\rangle}{\left\langle (\mathrm{tr}Q\{\mathbf{r}_j\})^d\right\rangle} \quad . \tag{16.50 ii}$$

These expressions can be evaluated by standard methods. However, Diehl and Eisenriegler [DE89] devised a trick to deal with the ratio as originally defined. They write

$$(\mathrm{tr}Q)^{-m} = \int_0^\infty dy \frac{y^{m-1}}{(m-1)!} e^{-y\,\mathrm{tr}Q} \quad ,$$

thus representing for instance $\langle\mathcal{A}_d\{\mathbf{r}_j\}\rangle = \mathcal{A}_d$ as

$$\mathcal{A}_d = \frac{d}{d-1}\int_0^\infty dy\, y \left\langle \mathrm{tr}\hat{Q}^2\{\mathbf{r}_j\}e^{-y\,\mathrm{tr}\,Q\{\mathbf{r}_j\}}\right\rangle \quad . \tag{16.51}$$

As a result, the evaluation of the expectation value for a Gaussian chain reduces to Gaussian integrals.

Analyzing these ratios with the help of the renormalization group one finds that upon renormalization they reduce to functions of the renormalized coupling only, taking universal values at the excluded volume fixed point. These values have been calculated [JEK92] to first order in ε. The results are in reasonable agreement with Monte Carlo data. Table 16.2 collects the results for $d = 3$, together with those for Gaussian chains. All results confirm that the polymer coil is of rather elongated prolate shape, the corrections due to the interaction being relatively small. The large difference among $\hat{\mathcal{A}}_3$ or $\mathcal{A}_3$ ($\hat{\mathcal{S}}_3$ or $\mathcal{S}_3$) is remarkable. It suggests that the distribution of $\mathcal{A}_3\{\mathbf{r}_j\}$ or $\mathcal{S}_3\{\mathbf{r}_j\}$ is fairly broad, a feature verified by computer experiments [Jag92]. An extensive study of the distributions of the asphericity and the eigenvalues, including also correlations among the eigenvalues, has recently been published

Table 16.2. Asphericity and normalized determinants for three dimensional coils. Data taken from Table IV of [JEK92]

	$\hat{\mathcal{A}}_3$	$\mathcal{A}_3$	$\hat{\mathcal{S}}_3$	$\mathcal{S}_3$
Gaussian	0.526	0.394	0.887	0.475
self-avoiding	0.535	0.416	0.893	–

[Sci96]. The broadness of the distribution stresses the fact that an excluded volume coil, though being somewhat more rigid than a Gaussian coil, is still a strongly fluctuating object.

16.4 The second virial coefficient of two coils of different size

16.4.1 Analytical results

For two chains of lengths n_1, n_2 the second virial coefficient $A_2(n_1, n_2)$ has been defined in Sect. 15.3, Eq. (15.39):

$$A_2(n_1, n_2) = -\Omega \frac{G_c^{(0,2)}(\mathbf{0}, \mathbf{0}, \mathbf{0}, \mathbf{0}; n_1, n_2)}{G^{(0,1)}(\mathbf{0}, \mathbf{0}; n_1) G^{(0,1)}(\mathbf{0}, \mathbf{0}; n_2)} \quad .$$

From this definition we conclude that it renormalizes according to

$$A_2(n_1, n_2) = \ell_R^d A_{2R}(n_{1R}, n_{2R}, f) \quad . \tag{16.52}$$

The one-loop result has been given in Appendix A 15.3.3, Eq. (A 15.36). In strict ε-expansion it can be written as

$$(4\pi)^{-\frac{d}{2}} A_2(n_1, n_2) = \frac{u^*}{2} f \ell_R^d n_{1R} n_{2R} \Big\{ 1 + \frac{\varepsilon}{8} f \Big[2 \ln \left(\frac{1}{n_{2R}} + \frac{1}{n_{1R}} \right) - 1 + \frac{1}{\bar{n}} \ln(1 + \bar{n}) + \bar{n} \ln \left(1 + \frac{1}{\bar{n}} \right) \Big] + O(\varepsilon^2) \Big\} \quad , \tag{16.53}$$

where

$$\bar{n} = \frac{n_{2R}}{n_{1R}} \equiv \frac{n_2}{n_1} \quad . \tag{16.54}$$

By symmetry we can restrict ourselves to $\bar{n} \lesssim 1$.

Naively fixing ℓ_R by the mean size of the chains: $N_R = \frac{1}{2}(n_{1R} + n_{2R}) = 1$, in the limit $\bar{n} \to 0$ we encounter singularities due to the contribution $\ln(1/n_{2R} + 1/n_{1R})$. Such singularities are a typical feature of problems involving two important length scales of very different size, which in the present case are the radii of the two coils. The singularities here signal the occurrence of power laws governing the $\bar{n}$-dependence for a system of widely differing chain lengths. In favorable cases like the present one, the power laws can be extracted by adjusting the definition of ℓ_R. We recall from Sect. 13.2 that ℓ_R should be chosen to be of order of the smallest macroscopic length in the system. For the present problem this implies $\ell_R \sim R_g(n_2)$ for $\bar{n} \to 0$. We thus impose the condition

$$1 = \frac{1}{n_{1R}} + \frac{1}{n_{2R}} \quad , \tag{16.55}$$

which for $\bar{n} = 0$ reduces to $n_{2R} = 1$. (Being concerned with strict ε-expansion here we do not introduce the parameter n_0, which is relevant for direct calculations in $d = 3$.) Using this condition together with the RG mapping in the excluded volume limit $f = 1$ we find

$$\ell_R = B' \left(\frac{1}{n_1} + \frac{1}{n_2} \right)^{-\nu}$$

$$B'^{-d} (4\pi)^{-\frac{d}{2}} \frac{2}{u^*} A_2(n_1, n_2) = (n_1 + n_2)^{\nu d} \left(\frac{\bar{n}}{(1+\bar{n})^2} \right)^{\nu d - 1}$$
$$\cdot \left\{ 1 + \frac{\varepsilon}{8} \left[\frac{1}{\bar{n}} \ln(1+\bar{n}) + \bar{n} \ln\left(1 + \frac{1}{\bar{n}}\right) - 1 \right] + O(\varepsilon^2) \right\} \quad . \tag{16.56}$$

Taking in this expression the limit $\bar{n} \to 0$ we find

$$A_2(n_1, n_2) = const\, n_1^{\nu d} \bar{n}^{\nu d - 1} = const\, n_1 n_2^{\nu d - 1} \quad , \tag{16.57}$$

which implies

$$\frac{A_2(n_1, n_2)}{A_2(n_1, n_1)} \sim \bar{n}^{\nu d - 1} \to 0 \quad .$$

Thus the second virial coefficient of two chains of widely different lengths $n_2 \ll n_1$ becomes arbitrarily small compared to the second virial coefficient $A_2(n_1, n_1) \sim n_1^{\nu d}$ of the longer species. This finding is easily understood. We consider the long chain $n_1 \gg n_2$ to be composed of n_1/n_2 blobs of size n_2. Only configurations in which the small chain overlaps such a blob will contribute to $A_2(n_1, n_2)$, the contribution being proportional to $A_2(n_2, n_2) \sim n_2^{\nu d}$. Taking the number of these n_2-blobs into account $A_2(n_1, n_2)$ should behave as $(n_1/n_2) n_2^{\nu d}$, which explains (16.57). This simple argument has first been given in a talk by De Gennes, as refered to in [WP82]. Note that this argument considers the chain as a coiled tube of radius $\sim n_2^{\nu}$ rather than as homogeneously filled sphere of radius $\sim n_1^{\nu}$. It thus stresses the strong inhomogeneity of the local density within the sphere. We should compare this prediction to a hard sphere model resulting from a smoothed density ansatz. Since hard spheres cannot overlap, such a model yields

$$A_2(n_1, n_2) = const\, (R_g(n_1) + R_g(n_2))^d$$
$$\sim (n_1 + n_2)^{\nu d} \left(\frac{1 + \bar{n}^{\nu}}{(1+\bar{n})^{\nu}} \right)^d \underset{\bar{n} \to 0}{\longrightarrow} n_1^{\nu d} \quad , \tag{16.58}$$

or

$$\frac{A_2(n_1, n_2)}{A_2(n_1, n_1)} \to const > 0 \quad .$$

In a smoothed density model even for arbitrarily small $\bar{n}$ the virial coefficients $A_2(n_1, n_2)$ and $A_2(n_2, n_2)$ are of the same order of magnitude.

16.4.2 The experimental situation

In principle $A_2(n_1, n_2)$ can be determined by measuring the second virial coefficient in 'bimodal' solutions containing two sharp polymer fractions of length n_1, n_2. For such a system from Eq. (A 15.11), which relates the light scattering virial coefficient A_2^S to $A_2(n_1, n_2)$, we easily derive the result

$$A_2^S = w_1^2 A_2(n_1, n_2) + 2w_1 w_2 A_2(n_1, n_2) + w_2^2 A_2(n_2, n_2) \quad . \tag{16.59}$$

Here w_j is the weight fraction, expressed in terms of the segment concentration c_j of species j as

$$w_j = \frac{c_j}{c_1 + c_2} \quad .$$

Varying w_1/w_2 we thus should be able to extract $A_2(n_1, n_2)$. Relating our result to standard experimental definitions we must keep track of a factor $N_w^{-2} = (w_1 n_1 + w_2 n_2)^{-2}$ multiplying A_2^S (cf. Eq. (15.38)), and we replace w_j by a variable $\xi_j = w_j n_j / (w_1 n_1 + w_2 n_2)$. Figure 16.4 shows our results for $N_w^{-2} A_2^S / \left(n_1^{-2} A_2(n_1, n_1)\right)$. It is found that the RG result fairly smoothly interpolates among the values of the pure fractions. In contrast, the hard sphere model predicts a maximum at some intermediate composition, an effect which increases with decreasing n_2/n_1. This is a characteristic feature of all Flory-type smoothed density models. Quite similar results are found for the osmotic virial coefficient.

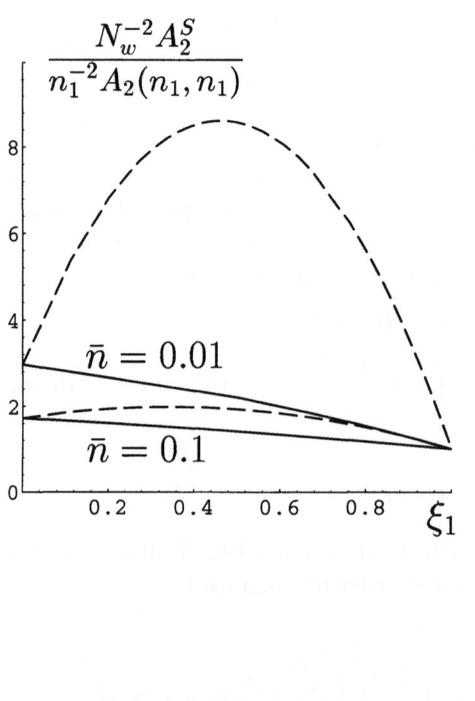

Fig. 16.4. $N_w^{-2} A_2^S / (n_1^{-2} A_2(n_1, n_1))$ as function of ξ_1. *Lower curves:* $n_2/n_1 = 0.1$; *upper curves:* $n_2/n_1 = 0.01$. *Full lines:* RG result; *broken lines:* hard spheres (Eq. (16.58))

Over the years there has been considerable effort to measure $A_2(n_1, n_2)$ and to verify the maximum of A_2 predicted by traditional theories. Very high precision is needed in such an experiment. The outcome has been inconclusive. Quite often a maximum has been found, the result not being compatible with RG theory. However, in such experiments the shorter chain length typically was in the range $n_2 \lesssim 500$, and thus nonuniversal $1/n_2$-corrections may influence the results. Closer examination of data for longer chains shows that the data analysis may suffer from systematic errors, the maximum being a spurious effect. A reanalysis of older osmotic pressure data [Sch82] found the data to be consistent with RG theory. Also the fairly recent light scattering data of [SNF87], [LMS+92] are more in favor of RG theory than of a smoothed density ansatz. In a plot like Fig. 16.5 they show an almost straight variation among the pure limits, with only very small curvature. A detailed quantitative comparison with the theoretical prediction (16.56) does not seem to be appropriate, however. In all cases the data for the pure fractions obey the expected power law behavior for A_2^S only with deviations of the order of 10%. This might indicate a violation of the excluded volume conditions in particular for the shorter chains (cf. Fig. 15.17), but might also be due to problems of data analysis. We thus conclude that the final experimental test of RG theory for $A_2(n_1, n_2)$ still has to be done.

16.5 The free energy excess of interpenetrating coils

In the present section we consider two functions $U^{(j)}(\mathbf{r}), j = 1, 2$, defined as

$$\exp\left(-U^{(j)}(\mathbf{r})\right) = \frac{\Omega^2}{Z(n_1)Z(n_2)} \int D\left(\mathbf{r}^{(1)}, \mathbf{r}^{(2)}\right) \delta^d(\mathbf{R}_j)\delta^d\left(\mathbf{R}_{cm}^{(2)} - \mathbf{r}\right) \exp\left(-V(1,2)\right) \quad , \tag{16.60}$$

where

$$\begin{aligned} \mathbf{R}_1 &\equiv \mathbf{R}_{cm}^{(1)} \\ \mathbf{R}_2 &\equiv \mathbf{r}_{n_1/2}^{(1)} \quad . \end{aligned} \tag{16.61}$$

$V(1,2)$ is the potential energy of two chains, as defined in Eq. (2.3). Clearly $U^{(j)}(\mathbf{r})$ is the excess free energy of two coils of mutual distance $\mathbf{r}$, where the distance is measured in terms of the centers of mass ($j = 1$) or of the center of mass of chain 2 and the central segment of chain 1 ($j = 2$). For $\mathbf{r} \gg R_g(n_1), R_g(n_2)$, the multiple integral in Eq. (16.60) factorizes into the contributions of the two chains, and as a result the r.h.s. of Eq. (16.60) tends to one: the free energy excess clearly vanishes if the two chains are far apart. This suggests to introduce the pair correlation functions

$$A_2^{(j)}(\mathbf{r}, n_1, n_2) = 1 - e^{-U^{(j)}(\mathbf{r})} \quad . \tag{16.62}$$

It is easily checked that the integral over $\mathbf{r}$ of $A_2^{(j)}(\mathbf{r}, n_1, n_2)$ yields the second virial coefficient $A_2(n_1, n_2)$:

$$\int d^d r A_2^{(j)}(\mathbf{r}, n_1, n_2) = A_2(n_1, n_2) \quad ,$$

this sum rule being valid both for $j = 1$ or $j = 2$. Equation (16.52) therefore directly yields the renormalization of $A_2^{(j)}(\mathbf{r}, n_1, n_2)$:

$$A_2^{(j)}(\mathbf{r}, n_1, n_2) = A_{2R}^{(j)}\left(\frac{\mathbf{r}}{\ell_R}, n_{1R}, n_{2R}, f\right) \quad . \tag{16.63}$$

Identifying ℓ_R with some appropriate radius R we find $A_2^{(j)}$ as a universal function

$$A_2^{(j)}(\mathbf{r}, n_1, n_2) = \tilde{A}_2^{(j)}\left(\frac{\mathbf{r}}{R}, \frac{n_2}{n_1}, \tilde{z}\right) \quad , \tag{16.64}$$

a result, which by virtue of Eq. (16.62) leads to the corresponding result for $U^{(j)}(\mathbf{r})$:

$$U^{(j)}(\mathbf{r}) = \tilde{U}^{(j)}\left(\frac{\mathbf{r}}{R}, \frac{n_1}{n_2}, \tilde{z}\right) \quad . \tag{16.65}$$

For both measures of the distance among the two chains the free energy excess is a function of $\mathbf{r}/R, n_1/n_2, \tilde{z}$, only. *It does not depend on the absolute size of a coil other than by an implicit dependence via* $\tilde{z}$. This result truly is remarkable as becomes clear if we consider the excluded volume limit and place the two chains on top of each other: $\mathbf{r} = 0$. Then $U^{(j)}(\mathbf{0}), \tilde{z} \to \infty$, is a universal function of n_1/n_2, independent of the absolute size of the chains: it is not changed by increasing both chains at the same rate. As has been stressed in the introductory section such a behavior rules out any smoothed density model. Indeed, the two fractals must arrange such that they interpenetrate without effectively overlapping with each other. This, in fact, has recently been verified by a direct calculation of the average number of the interaction points of two coils [MS98]. In the excluded volume limit this number is found to vanish as $n^{2-\nu(d+\vartheta_2)} \sim n^{-0.24}(d = 3)$, where ϑ_2 is the correlation hole exponent introduced in Sect. 16.2.2.

A quantitative analysis to one-loop order [KSB89] supports these qualitative considerations. Let us first consider $\tilde{U}^{(1)}(\mathbf{0}, \bar{n}, \tilde{z} = \infty)$. Here the centres of mass of two excluded volume chains coincide. For $\bar{n} \ll 1$ the small chain n_2 almost certainly will sit in a hole of the large chain n_1, and $\tilde{U}^{(1)}(0, \bar{n}, \infty)$ should vanish. Indeed,

$$\tilde{U}^{(1)}(\mathbf{0}, \bar{n}, \infty) \to const\ \bar{n}^{\nu d-1}, \bar{n} \to 0 \quad , \tag{16.66}$$

in full accord with the discussion of $A_2(n_1, n_2)$ in the previous section. Since for the small probe chain the center of mass of the large coil is just equivalent to an arbitrarily chosen point within the coil volume, this shows that generically for $\bar{n} \ll 1$ the interaction among the two coils vanishes, and the small coil can diffuse freely through the volume 'occupied' by the large coil.

Now we place the second chain on top of the central segment of the first chain, thus considering $\tilde{U}^{(2)}(\mathbf{0}, \bar{n}, \infty)$. For $\bar{n} \to 0$ this function does not vanish but saturates at a value close to $\tilde{U}^{(2)}(0, 1, \infty)$. Again this is in full accord with the blob picture for $A_2(n_1, n_2)$. Taken together, the result for $\tilde{U}^{(1)}(\mathbf{0}, \bar{n} \to 0, \infty)$ and $\tilde{U}^{(2)}(\mathbf{0}, \bar{n} \to 0, \infty)$ demonstrate the strong inhomogeneity of the density distribution within a fractal.

Indeed, this result can also be found from $\tilde{U}^{(2)}(\mathbf{0}, \bar{n}, \infty)$ alone. Figure 16.5 shows $\tilde{U}^{(2)}(0, \bar{n}, \infty)$ as function of $\bar{n}$. For $\bar{n} \to 0$ it demonstrates saturation as discussed above. For $\bar{n} \to \infty$, however, the small chain n_1 sits at the center of mass of the large chain n_2. Thus the law (16.65) applies, resulting in a linear decrease $\log \tilde{U}^{(2)} \to const - (3\nu - 1) \log \bar{n}$ for large $\log \bar{n}$.

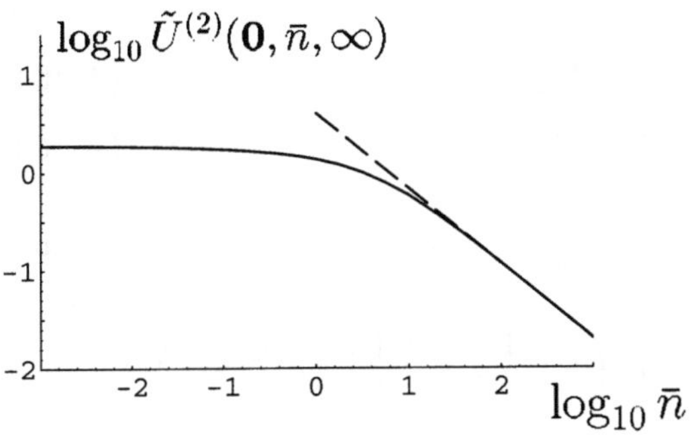

Fig. 16.5. $\log_{10} \tilde{U}^{(2)}(\mathbf{r} = 0, \bar{n}, \tilde{z} = \infty)$ as function of $\log_{10}(\bar{n})$. The *broken line* corresponds to the asymptotic power law (Eq. (16.66))

It is interesting to note also a quantitative result for the excess free energy. A value $\exp\left[-\tilde{U}^{(1)}(\mathbf{r} = 0, \bar{n} = 1, \tilde{z} = \infty)\right] \approx 0.2$ is found. Compared to two well separated chains the Boltzmann factor $e^{-\tilde{U}^{(1)}}$ therefore decreases only by a factor of 5 even for infinitely long chains of the same size. Monte Carlo simulations yield $e^{-\tilde{U}^{(1)}} \approx 0.1$, but for the chain lengths used the value still increases with increasing $n_1 = n_2$. The asymptotic limit was not yet reached. All this demonstrates that also on the quantitative level chains interpenetrate strongly. For more details, including quantitative results for the $\mathbf{r}$-dependence of $\tilde{U}^{(j)}$, we refer to [KSB89].

A 16. Appendix

A 16.1 ε-Expansion results

A 16.1.1 Internal correlations. Except for the summations over j_1, j_2 the internal correlation function $P(\mathbf{p}, n_1, n_2, n_3)$ is identical to the density correlation function of an isolated chain, which has been discussed in the previous chapter. Closely following the calculation outlined in Appendix A 15.2 to first order in the ε-expansion we find the following renormalized result:

$$\begin{aligned} & P_R\left(\bar{\mathbf{p}}, n_{2R}, \frac{n_1}{n_2}, \frac{n_3}{n_2}; f\right) \\ & = e^{-P}\Big\{1 + \frac{\varepsilon}{8} f\Big[e^P - 3 - (1 + ln\, n_{2R})P \\ & + \frac{3}{P}\left(e^P - 1\right) + \frac{\bar{n}}{P_R}\left(e^{P/\bar{n}} - 1\right) - (2 + P)\hat{E}(P) \\ & - \hat{E}\left(\frac{P}{\bar{n}}\right) + \hat{H}\left(P, \frac{n_1}{n_2}\right) + \hat{H}\left(P, \frac{n_3}{n_2}\right)\Big] + O(\varepsilon^2)\Big\} \quad , \end{aligned} \tag{A 16.1}$$

where

$$P = \bar{p}^2 n_{2R} \tag{A 16.2}$$

$$\bar{n} = 1 + \frac{n_1 + n_3}{n_2} \tag{A 16.3}$$

$$\hat{E}(x) = \int_0^x \frac{dz}{z}\left(e^z - 1\right) \tag{A 16.4}$$

$$\begin{aligned} \hat{H}(P, y) = {} & \hat{E}\left(\frac{P}{1+y}\right) - \ln\left(1 + \frac{1}{y}\right) - \frac{1+y}{P}\left(e^{\frac{P}{1+y}} - 1\right) \\ & + \frac{1}{yP}\int_0^{1/y} \frac{dz}{z^2}\left(e^{Py\frac{z^2}{1+z}} - 1\right) \quad . \end{aligned} \tag{A 16.5}$$

With regard to the analysis of end-to-end correlations we note $\hat{H}(P, y)$ for $y \to 0$:

$$\hat{H}(P, 0) = 1 + 2\hat{E}(P) - \frac{2}{P}\left(e^P - 1\right) \quad . \tag{A 16.6}$$

Clearly the renormalized end-to-end distance distribution is found as

$$\mathcal{P}_R(\bar{\mathbf{p}}, n_R; f) = \mathcal{P}_R(\bar{\mathbf{p}}, n_R, 0, 0; f) \quad . \tag{A 16.7}$$

The mean squared distance of segments j_1, j_2 can be calculated as

$$R_i^2(n_1, n_2, n_3) = 2d\ell_R^2 \frac{\partial}{\partial \bar{p}^2}\Big|_0 P_R\left(\bar{\mathbf{p}}, n_{2R}, \frac{n_1}{n_2}, \frac{n_3}{n_2}\right) \quad .$$

It turns out that in this expression all integrals can be evaluated analytically. We find

$$R_i^2(n_1,n_2,n_3) = 2d\ell_R^2 n_{2R}\Big[1+\frac{\varepsilon}{8}f\Big(\ln n_{2R}-1$$
$$-\frac{1}{2}\frac{n_1+n_3}{n_1+n_2+n_3}+\frac{n_1}{n_2}\ln\left(1+\frac{n_2}{n_1}\right)+\frac{n_3}{n_2}\ln\left(1+\frac{n_2}{n_3}\right)\Big)+O(\varepsilon^2)\Big]. \tag{A 16.8}$$

The inverse moment is to be calculated according to Eq. (16.27). Note that we evaluate the momentum integral in three dimensions, but use the ε-expansion for $\tilde{P}\left(\sqrt{x},\frac{n_1}{n_2},\frac{n_3}{n_2};\tilde{z}\right)$. For the end-to-end distance we find

$$\begin{aligned}(R_{he}(n))^{-2} &= \left\langle |\mathbf{r}_0-\mathbf{r}_n|^{-1}\right\rangle^2\\ &= \frac{1}{\pi\ell_R^2 n_R}\left[1-\frac{\varepsilon}{8}f\,(\ln n_R+1-2\ln 2)+O(\varepsilon^2)\right].\end{aligned} \tag{A 16.9}$$

A complicated result is found for the internal distances

$$\begin{aligned}&(R_h(n_1,n_2,n_3))^{-2} = \left\langle |\mathbf{r}_{j_2}-\mathbf{r}_{j_1}|^{-1}\right\rangle^2\\ &= \frac{1}{\pi\ell_R^2 n_{2R}}\Big[1-\frac{\varepsilon}{8}f\Big(\ln n_{2R}+5+6\ln 2\\ &+4\bar{n}^{1/2}(\bar{n}-1)^{1/2}+2\ln\bar{n}-4\ln\left(\bar{n}^{1/2}+(\bar{n}-1)^{1/2}\right)-2H(\frac{n_1}{n_2})\\ &-2H(\frac{n_3}{n_2})\Big)+O(\varepsilon^2)\Big]\end{aligned} \tag{A 16.10}$$

$$\begin{aligned}H(y) &= \frac{2y^{1/2}}{(1+y)^{1/2}}+2y^{1/2}(1+y)^{1/2}+\ln y-2\ln\left(y^{1/2}+(1+y)^{1/2}\right)\\ &+2\int_0^1 dz\,\left(y+z-z^2\right)^{-1/2}(y+z)^{-3/2}\left(y+\frac{z}{2}\right).\end{aligned} \tag{A 16.11}$$

A 16.1.2 Density distributions. The basic result for $\rho_{cm,R}\left(\bar{\mathbf{q}},\frac{j}{n},n_R,f\right)$ reads [SK88]:

$$\begin{aligned}\rho_{cm,R}(\bar{\mathbf{q}},\bar{j},n_R,f) &= \exp\left\{-\frac{\bar{q}^2 n_R}{3}\left(\bar{j}^3+(1-\bar{j})^3\right)\right\}\\ &\cdot\left\{1-\frac{\varepsilon}{8}f\left(J_1(\bar{\mathbf{q}}^2 n_R,\bar{j})+J_1(\bar{q}^2 n_R,1-\bar{j})+J_2(\bar{\mathbf{q}}^2 n_R,\bar{j})\right)+O(\varepsilon^2)\right\}\end{aligned} \tag{A 16.12}$$

$$
\begin{aligned}
J_1(Q,\bar{\jmath}) = & \frac{\bar{\jmath}^3}{3} Q\left(\ln \bar{\jmath}+\frac{2}{3}\right)-\frac{1}{3}\int_0^1 \frac{dz}{z}\left(e^{Q\bar{\jmath}^3 z/4}-1\right) \\
& +\frac{1}{3}\int_0^1 \frac{dz}{z}\left[\frac{4-8z+3z^2}{z(2-z)^2}\left(e^{Q\bar{\jmath}^3 z\left(1-\frac{z}{2}\right)^2}-1\right)-Q\bar{\jmath}^3\right]
\end{aligned}
\qquad \text{(A 16.13 i)}
$$

$$
\begin{aligned}
& J_2(Q,\bar{\jmath}) \\
& = \int_{0<z_1<\bar{\jmath}<z_2<1} (z_2-z_1)^{-2}\Big\{\exp\Big[-Q\Big(\frac{z_1^3}{3}+\frac{(z_2-z_1)^3}{12}+\frac{(1-z_2)^3}{3} \\
& +(z_2-\bar{\jmath})(\bar{\jmath}-z_1)\left(\frac{1}{z_2-z_1}-1\right)-\frac{1}{3}+\bar{\jmath}-\bar{\jmath}^2\Big)\Big]-1\Big\} \ .
\end{aligned}
\qquad \text{(A 16.13 ii)}
$$

The result has been rewritten in terms of the present renormalization conventions. Expanding for small $\mathbf{q}$ we can carry out the integrals and calculate universal ratios

$$
\begin{aligned}
\frac{\left\langle(\mathbf{r}_j-\mathbf{R}_{cm})^2\right\rangle}{R_g^2} & = 2-6\bar{\jmath}(1-\bar{\jmath})-\frac{\varepsilon}{4} f\left[\frac{1}{8}+\frac{25}{4}\bar{\jmath}(1-\bar{\jmath})\right. \\
& \left.+3\bar{\jmath}\left(1-\frac{\bar{\jmath}}{2}\right)\ln \bar{\jmath}+3(1-\bar{\jmath})\left(1-\frac{1-\bar{\jmath}}{2}\right)\ln(1-\bar{\jmath})\right]+O(\varepsilon^2)
\end{aligned}
\qquad \text{(A 16.14)}
$$

$$
\begin{aligned}
& \frac{\left\langle\left[(\mathbf{r}_j-\mathbf{R}_{cm})^2-\langle(\mathbf{r}_j-\mathbf{R}_{cm})^2\rangle\right]^2\right\rangle^2}{\langle(\mathbf{r}_j-\mathbf{R}_{cm})^2\rangle^2} = \frac{2}{d}-9\left(1+\frac{2}{d}\right)\frac{\varepsilon}{4} f \\
& \cdot(1-3\bar{\jmath}(1-\bar{\jmath}))^{-2}\left[\frac{31}{960}+\frac{23}{20}\bar{\jmath}(1-\bar{\jmath})+\frac{29}{40}\bar{\jmath}^2(1-\bar{\jmath})^2\right. \\
& \left.+\left(1+\frac{\bar{\jmath}^2}{4}-\frac{\bar{\jmath}^3}{10}\right)\bar{\jmath}^2\ln \bar{\jmath}+\left(1+\frac{(1-\bar{\jmath})^2}{4}-\frac{(1-\bar{\jmath})^3}{10}\right)(1-\bar{\jmath}])^2\ln(1-\bar{\jmath})\right] \\
& +O(\varepsilon^2) \ .
\end{aligned}
\qquad \text{(A 16.15)}
$$

Here $\bar{\jmath}=j/n$, and in Eq. (A 16.15) we kept some explicit dimensions d, which are due to the algebra of the $\mathbf{q}$-expansion.

A 16.2 An expression for the hydrodynamic radius

We start from the expression

$$
R_h^{-1}(n_1,n_2,n_3)=\int d^d r \frac{1}{|\mathbf{r}|}\hat{P}(\mathbf{r};n_1,n_2,n_3) \quad , \qquad \text{(A 16.16)}
$$

where

$$
\hat{P}(\mathbf{r};n_1,n_2,n_3)=\left\langle\delta^d\left(\mathbf{r}-\mathbf{r}_{n_2+n_1}+\mathbf{r}_{n_1}\right)\right\rangle \quad . \qquad \text{(A 16.17)}
$$

We may express $\hat{P}(\mathbf{r},\ldots)$ in terms of its Fourier transform $P(\mathbf{p};n_1,n_2,n_3)$ (Eq. (16.22)) and interchange integrals to find

$$R_h^{-1}(n_1,n_2,n_3) = \lim_{\eta\to 0_+} \int_{\mathbf{p}} P(\mathbf{p};n_1,n_2,n_3) \int d^d r \frac{e^{i\mathbf{pr}-\eta|\mathbf{r}|}}{|\mathbf{r}|} \quad . \tag{A 16.18}$$

Here $\eta > 0$ serves as a large distance regulator in the $\mathbf{r}$-integral. We easily find

$$\lim_{\eta\to 0} \int d^d r \frac{e^{i\mathbf{pr}-\eta|\mathbf{r}|}}{|\mathbf{r}|} = 2\pi^{d/2} \frac{\Gamma(d-1)}{\Gamma(d/2)} |\mathbf{p}|^{1-d} \quad .$$

Substituting this result into Eq. (A 16.18) and integrating over the direction of $\mathbf{p}$ we find

$$\begin{aligned} R_h^{-1}(n_1,n_2,n_3) &= 2^{2-d} \frac{\Gamma(d-1)}{\Gamma^2(d/2)} \int_0^\infty dp\, P(p,n_1,n_2,n_3) \\ &= 2^{1-d} \frac{\Gamma(d-1)}{\Gamma^2(d/2)} \frac{1}{R_e(n_2)} \int_0^\infty dx\, x^{-1/2} \tilde{p}\left(\sqrt{x}, \frac{n_1}{n_2}, \frac{n_3}{n_2}; \tilde{z}\right) \quad , \end{aligned} \tag{A 16.19}$$

where in the last line we used Eq. (16.23). Equation (16.27) follows by specializing to $d = 3$.

A 16.3 Free energy, internal energy, or entropy of a single chain

In dimensional regularization the single-chain partion function $\mathcal{Z}(n)$ is found as

$$\mathcal{Z}^{(1)}(n) = \frac{\Omega}{(4\pi\ell^2)^{d/2}} e^{-\mu_s^* n} \left[1 + \beta_e n^{\varepsilon/2}\left(\frac{2}{\varepsilon} + \frac{1}{1-\varepsilon/2}\right) + O(\beta_e^2)\right] \quad . \tag{A 16.20}$$

According to the theorem of renormalizability (Eqs. (11.10), (11.12)) $\mathcal{Z}(n)$ takes the general renormalized form (cf. Eq. (11.53))

$$\begin{aligned} \mathcal{Z}^{(1)}(n) &\equiv G^{(1)}(\mathbf{0},\mathbf{0};n) \\ &= \frac{\Omega}{(4\pi\ell^2)^{d/2}} e^{-\mu_s^* n} \frac{Z}{Z_n} G_R^{(1)}(\mathbf{0},\mathbf{0};n_R,u) \quad . \end{aligned} \tag{A 16.21}$$

Comparing Eqs. (A 16.20), (A 16.21) we can read off $Z/Z_n G_R^{(1)}$ and express the result in terms of renormalized variables, following Sect. 12.3.1. We find

$$G_R^{(1)}(\mathbf{0},\mathbf{0};n_R,u) = 1 + \frac{u}{\varepsilon}\left(n_R^{\varepsilon/2} - 1\right) + \frac{u\, n_R^{\varepsilon/2}}{2-\varepsilon} + O(u^2) \quad . \tag{A 16.22}$$

This result may be evaluated either by ε-expansion or directly in $d = 3$. With our conventions ($n_R = n_0 = 0.53; u = u^* f, u^* = 0.364$) the latter method yields

$$G_R^{(1)}(\mathbf{0}, \mathbf{0}; n_0, u^* f) = 1 + 0.166 f \quad , \tag{A 16.23}$$

where f as function of $\tilde{z} = v n^{1/2}$ follows from Eq. (15.13). The partition function follows by substituting this result into Eq. (A 16.21) and using the expressions (13.6), (13.7) for the Z-factors. We express the result in terms of an excess free energy density defined as

$$\Delta F = -\ln\left(\frac{(4\pi\ell^2)^{d/2}}{\Omega}\mathcal{Z}(n)\right) \quad . \tag{A 16.24}$$

We find

$$\Delta F = \mu_s^* \, n + H_F(f_0) + \frac{\gamma - 1}{\nu\omega} \ln |1 - f| + F_1(f) \tag{A 16.25}$$

$$F_1(f) = -\ln\left[(1 + 0.824 f)^{-0.013} F_z(f) H(f)(1 + 0.166 f)\right] \quad . \tag{A 16.26}$$

Here $\gamma = \nu(2 - \eta)$ (Eq. (11.56)), and the function $H_F(f_0)$ collects all f_0-dependent factors. The functions $F_z(f), H(f)$ are due to the Z-factors and have been given in Sect. 13.1.

At the fixed point $f \to 1$ Eq. (A 16.25) together with the relation among f and $\tilde{z}$ yields

$$\Delta F = \mu_s^* \, n - (\gamma - 1) \ln \, n + const \quad , \tag{A 16.27}$$

in agreement with Eq. (11.55). Quite generally we find that ΔF is dominated by the nonuniversal contribution $\mu_s^* \, n$ and shows an additional nonuniversal constant $H_F(f_0)$. Only the remaining part yields a universal dependence on $\tilde{z}$.

This universal crossover dependence may conveniently be extracted by taking the logarithmic derivative with respect to n. We thus define an effective exponent

$$\begin{aligned} \gamma_{eff} &= 1 - \frac{\partial}{\partial \ln n}(\Delta F - \mu_s^* n) \\ &= 1 - \frac{1}{2} \frac{\partial f}{\partial \ln \tilde{z}} \frac{\partial}{\partial f}\left[\frac{\gamma - 1}{\nu\omega} \ln |1 - f| + F_1(f)\right] \quad . \end{aligned} \tag{A 16.28}$$

The derivative $\partial f / d \ln \tilde{z}$ has to be taken from Eq. (15.13). On the weak coupling branch this exponent varies from $\gamma_{eff}(\tilde{z} = 0) = 1$ at the Θ-point to $\gamma_{eff} = \gamma \approx 1.158$ in the excluded volume limit. Figure 16.6 shows the two branches of $\gamma_{eff}(\tilde{z})$, resulting from numerical evaluation of Eq. (A 16.28). The data are from a recent simulation of the Domb-Joyce model [GSS97]. The nonuniversal parameter $\tilde{v}(T)$ contained in $\tilde{z}$ has been adjusted to data

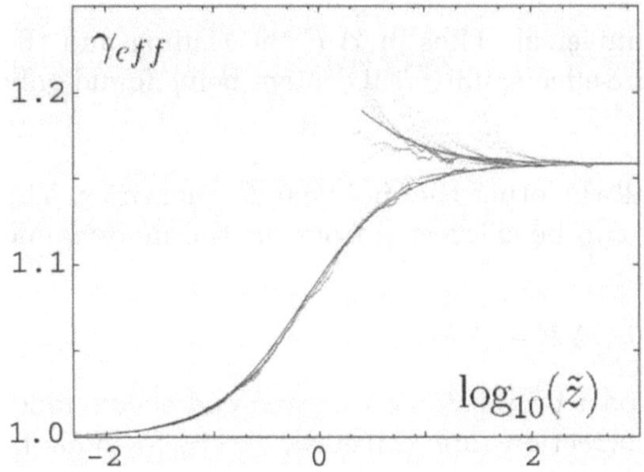

Fig. 16.6. γ_{eff} (Eq. (A 16.28)) as function of $\log_{10}\tilde{z}$. Data from [GSS97]. The *smooth curves*, almost covered by the data, represent the theory

for R_e^2 for the same model, so that this plot involves no adjustable parameter. The figure shows essentially dense sets of data as function of n, for several values of the excluded volume strength. Clearly the agreement with the theory is excellent.

From ΔF we may derive the excess of the internal energy density as

$$\Delta E = T \frac{\partial}{\partial T} \Delta F \quad , \tag{A 16.29}$$

where we have to note that both μ_s^* and f_0 depend on T, and f depends on T via its starting value f_0. We furthermore recall that we absorbed factors $1/k_B T$ into the thermodynamic quantities. Thus

$$\begin{aligned} \Delta E = {} & T \frac{\partial \mu_s^*}{\partial T} n + T \frac{\partial}{\partial T} H_F(f_0) \\ & + T \frac{\partial f}{\partial T} \frac{\partial}{\partial f} \left(\frac{\gamma - 1}{\nu\omega} \ln |1 - f| + F_1(f) \right) \quad . \end{aligned}$$

The derivative $T \partial f / \partial T$ may be taken from Eq. (15.13), written as

$$\tilde{v}(T) n^{1/2} = f \Big|1 - f\Big|^{-\frac{1}{2\nu\omega}} H_n^{-1/2}(f) \quad .$$

Substituting the result into the expression for ΔE we find after a little calculation

$$\begin{aligned} \Delta E = {} & T \frac{\partial \mu_s^*}{\partial T} n + T \frac{\partial}{\partial T} H_F(f_0) \\ & - 2 \frac{T}{\tilde{v}} \frac{\partial \tilde{v}}{\partial T} \, \frac{(\gamma - 1) f - \nu\omega f |1 - f| \partial F_1 / \partial f}{f + 2\nu\omega |1 - f| - \nu\omega \, f |1 - f| \partial \ln \, H_n / \partial f} \quad . \end{aligned} \tag{A 16.30}$$

In the excluded volume $f \to 1$ this expression reduces to

$$\Delta E = T \frac{\partial \mu_s^*}{\partial T} n + const + const \, n^{-\nu\omega} \quad , \tag{A 16.31}$$

where only the exponent $\nu\omega$ is universal. Thus in $\Delta\,E$ the leading and the subleading term are microstructure effects, universal content being found only in the next correction.

This structure is typical for all the other thermodynamic quantities. The entropy excess $\Delta\,S$, for instance, can be calculated from the thermodynamic relation

$$T\,\Delta\,S = \Delta\,E - \Delta\,F \quad .$$

We again encounter a dominant part proportional to n and end effects independent of n. Only subleading corrections are universal, except for nonuniversal prefactors. In such quantities therefore the theoretically interesting universal features are masked by nonuniversal effects.

We finally note that ΔE essentially counts the number of two segment contacts in a self-repelling chain. This results from the observation that in the not renormalized model the temperature dependence is inherent in $\beta_e = \beta_e(T-\Theta)$, so that the derivative $\partial/\partial T$ is equivalent to $\partial/\partial\beta_e$. This allows for a simple interpretation of the terms in Eq. (A 16.31). The leading contribution $\sim \partial\mu_s^*/\partial T$ is due to back-folding of short parts of the chain, short on the scale of the total chain length. The constant contribution is an end effect, and only the contribution $\sim n^{-\nu\omega} \approx n^{-0.47}$ counts contacts of segments spaced a finite fraction of the chain length. Thus the number of these nontrivial contacts vanishes in the critical limit of an infinitely long chain. The crossing properties of self-repelling walks have recently been analyzed in more detail in [MS98], where the crossover behavior of ΔE on the weak coupling branch may also be found.

A 16.4 The Flory argument

Flory's celebrated argument [Flo49], [FF51] is based on postulating a form of the free energy $F(R,n)$ of a chain of length n, constrained to have radius R. Aiming only at a calculation of ν and not of the crossover function for the swelling factor, we use a somewhat simplified ansatz which reads

$$F(R,n) = \frac{R^2}{\ell^2 n} + (4\pi)^{\frac{d}{2}}\,\beta_e \ell^d \frac{n^2}{R^d} \quad . \tag{A 16.32}$$

A more general discussion may be found in [Yam71].

In Eq. (A 16.32) the second term is a mean field estimate for the interaction energy of the chain, based on the smoothed density model, i.e. on the estimate (6.2) for the number of two-segment contacts. The first term, up to a constant independent of R, gives the free energy of a Gaussian chain of end-to-end distance R, calculated as $F_0 = -\ln \hat{P}_0(R, n)$ (cf. Eq. (3.8)). Minimizing $F(R, n)$ with respect to R we find

$$0 = 2\frac{R}{\ell^2 n} - d\,(4\pi)^{\frac{d}{2}}\,\beta_e \ell^d \frac{n^2}{R^{d+1}} \quad , \tag{A 16.33}$$

which yields

$$R(n) \sim n^\nu \tag{A 16.34}$$

$$\nu = \frac{3}{d+2} \quad . \tag{A 16.35}$$

The result of this simple argument is remarkable. The value of ν is known to be exact for $d = 1, 2, 4$, and for $d = 3$ it is an excellent approximation, experimentally hardly distinguishable from the best RG value $\nu = 0.588$, which has been established by an immense amount of work. Only the most recent Monte Carlo simulations unambiguously rule out Flory's value. In view of this success, Flory-type arguments are quite popular, having been applied to related problems like branched polymers. Typically the results are quite reasonable.

Despite its great success the Flory argument has resisted any attempt to give a convincing justification. On the contrary, the ansatz (A 16.32) plainly is wrong, as has been stressed by Des Cloizeaux [Clo76]. In fact, as should be clear from the discussion of this chapter, the very idea of modeling a polymer coil as a cloud of monomers, underlying the estimate of the interaction energy, is misleading, ignoring the all-important chain structure. Also the first term in Eq. (A 16.32) is inadequate. As we have discussed in the beginning of Chap. 6, the typical random walk configurations, underlying this estimate, are strongly suppressed by the interaction. In summary, neither the entropy nor the internal energy of the coil are adequately represented in the Flory ansatz, as is also obvious from the results of the previous section.

To put the criticism on a more quantitative basis we may evaluate the free energy excess as calculated from Eqs. (A 16.32), (A 16.33):

$$\begin{aligned} \Delta F_{Flory} &= F(R(n), n) \\ &= \left(1 + \frac{2}{d}\right) \frac{R^2(n)}{\ell^2 n} \\ &= \left(1 + \frac{2}{d}\right) \left(\frac{d}{2}(4\pi)^{d/2} \beta_e n^{\varepsilon/2}\right)^{\frac{2}{d+2}} \quad . \end{aligned} \tag{A 16.36}$$

This clearly has no similarity to Eq. (A 16.27), which is an exact result of the RG in the excluded volume limit, amply verified by Monte Carlo simulations.

There have been many attempts to derive the Flory expression for ν on an internally consistent basis. However, in general these attempts at some point introduce a mean field type assumption, ignoring internal correlations of the coil. Such assumptions cannot properly be controlled and cannot be embedded in a consistent theory. This is not surprising since the nontrivial dilatation symmetry of a polymer coil is not compatible with mean field assumptions.

Notes

The end-to-end distance has been calculated in numerous publications. The most extensive work [MN87] is based on a field theoretic calculation pushed up to six-loop order, which is resumed by Padé-Borel methods or used in a renormalization procedure [CCJ85]. A low order analysis of the crossover, similar in spirit to that given here, was presented in [FD88].

The ratio R_g^2/R_e^2 to order ε has first been calculated in [WS78]. This work used the field theoretic ensemble characterized by a fixed segment chemical potential. Transformation to fixed chain length then employed the precisely known values of the critical exponents. As a result the estimate $6R_{g/e}^2 = 0.975$ is found, somewhat closer to the simulation results than the direct $O(\varepsilon)$ expression. Using higher order results of the z-expansion in $d = 3$, Nickel [Nic91] finds a result which essentially is identical to the second-order ε result quoted here, (Eq. (16.12)).

The first calculation of the end-to-end distribution function $\hat{P}(\mathbf{r}, n)$ directly in $\mathbf{r}$-space can be found in [OOF81b]. It is based on the ε-expansion to $O(\varepsilon)$ and presents an exponentiated form which differs somewhat from Eq. (16.21).

The temperature blob model has independently been suggested in [WC79], [AH79], using the ansatz (16.36). Subsequently more refined crossover forms for $R_e^2(n)$ have been employed [FSW80], [ABA81], with considerable success. Though this approach had immediately been criticized on the basis of Monte Carlo data [CS80], it enjoyed great popularity. Our discussion follows [SB86], where more details on the crossover behavior can also be found.

Analytical work on the density distribution relative to the center of mass is due to [SK88]. In that work further results and plots are also found. Simulation results supporting the analytical work can be found in [Zif82].

The most extensive work on the shape of a polymer coil has been published in [JEK92], to which we refer for details and further references.

There are numerous calculations of $A_2(n_1, n_2)$ to be found in the literature, both using RG methods or models of smoothed density type. Whereas the former essentially agree with the theory as presented here the latter yield results close to the hard sphere model. A compilation of older theories may be found in [SNF87], where the serious problems of data analysis are also clearly discussed.

Our discussion of the free energy excess follows [KSB89], where the $\mathbf{r}$-dependence of $U^{(j)}(\mathbf{r})$ has also been discussed. Furthermore the effect of interpenetration on chain radii and segment densities has been calculated, all one-loop results being supported by Monte Carlo data. A similar free energy excess, where the distance of the chains has been fixed in terms of the chain ends, has been analyzed in [WP82]. Using this definition of $\mathbf{r}$ the results are distorted somewhat by the correlation hole.

The internal energy or entropy of a single chain in the excluded volume limit have been discussed in [Sch76],[Clo76]. For a more detailed discussion of the free energy excess or the internal energy outside the excluded volume limit see [MS98]. An analysis of the entropy excess in the crossover region has been presented in [Eld81].

17. Thermodynamic Quantities for Finite Overlap

This chapter deals with the thermodynamic properties of the solution. As quantity of central importance we identify the thermodynamic scaling function $\mathcal{P}$, defined in terms of the osmotic pressure as

$$\frac{\hat{\Pi}}{k_B T} = c_p \left[1 + \tilde{\mathcal{P}}\left(\tilde{s}, \tilde{z}, [p]\right)\right] \quad . \tag{17.1}$$

Here $\tilde{s} = c_p \tilde{R}_0^d$ is the overlap variable introduced in Eq. (13.23). Clearly $\tilde{\mathcal{P}}$ vanishes for $\tilde{s} \to 0$ and thus measures the deviation from ideal solution behavior.

It turns out that the concentration dependence of all thermodynamic quantities can be expressed in terms of $\tilde{\mathcal{P}}$. To reconstruct the full thermodynamics we only need to consider in addition some quantity defined in the dilute limit, like the free energy of a single chain. As a consequence, knowledge of the scaling function $\tilde{\mathcal{P}}$ essentially amounts to having full control over the thermodynamics. We will illustrate this by expressing the chemical potential or the free energy in terms of $\tilde{\mathcal{P}}$.

The only 'thermodynamic' quantity deserving a separate treatment is the forward scattering intensity $I_{d,c}(\mathbf{q} = 0)$. Only for a monodisperse system is it simply related to the osmotic compressibility and thus to $\tilde{\mathcal{P}}$, but for general polydispersity the relation is less direct. We therefore will calculate $I_{d,c}(\mathbf{0})$ separately.

For small overlap both $\tilde{\mathcal{P}}$ and $I_{d,c}(\mathbf{0})$ allow for a virial expansion. We will however find this expansion to be not very useful beyond its first two terms. In the excluded volume limit $\tilde{\mathcal{P}}$ or $I_{d,c}(\mathbf{0})$ with increasing overlap $s = c_p R_g^d$ tend to nontrivial power laws

$$\tilde{\mathcal{P}} \sim s^{1/(d\nu - 1)} \sim \frac{c_p N^2}{I_{d,c}(\mathbf{0})} \quad ,$$

and the crossover towards that asymptotic behavior starts for quite small overlap and is not well represented by the virial expansion. The situation here is quite similar to that found for the momentum dependence of the single chain form factor (Sect. 15.2.2), where the Taylor expansion in powers of q^2 also does not lead very far.

The asymptotic large overlap behavior reflects screening. It is strictly independent of polydispersity. Indeed polydispersity effects only show up for small overlap and rapidly die out in the crossover region.

In comparing to experiments we must recall that the theory applies to small concentrations only. Empirically we find that it should not be used for concentrations larger than about 10 wt%. For such low concentrations only the osmotic pressure or the scattering intensity can be determined with good precision. Other experiments like a determination of the heat of dilution need higher concentrations to yield a significant effect. We again discuss standard methods of data analysis, demonstrating that the virial expansion is inadequate in general. For longer chains most data are taken in the crossover region, where neither the virial expansion nor asymptotic power laws apply. However, data compare well to the full crossover functions. Furthermore, the familiar square root plot for not too long chains ($N \lesssim 10^4$) yields reliable values of N and of the second virial coefficient, provided for $I_{d,c}(\mathbf{0})$ the extrapolation to $\mathbf{q} = 0$ does not introduce systematic errors.

This chapter is organized as follows. In Sect. 17.1 we use general thermodynamic relations to establish the prominent role of the thermodynamic scaling function. One-loop results for $\tilde{\mathcal{P}}$ or $I_{d,c}(\mathbf{0})$ are presented in Sect. 17.2. Section 17.3 is devoted to the virial expansion, including a detailed discussion of the third virial coefficient, which is the highest term which possibly is of some use. The general crossover in overlap and excluded volume strength is discussed in Sect. 17.4, and Sect. 17.5 is concerned with the experimental situation. In Appendix A 17.1 we sketch the perturbative calculation of $\tilde{\mathcal{P}}$, and the general one loop results for both $\tilde{\mathcal{P}}$ and $I_{d,c}(\mathbf{0})$ are collected in Appendix A 17.2. A short discussion of the chemical potential or the free energy is given in Appendix A 17.3. In the present theory $\tilde{\mathcal{P}}/\tilde{s}$ replaces the Flory-Huggins parameter χ familiar from traditional mean field theories, and the relation among $\tilde{\mathcal{P}}$ and χ is analyzed in Appendix A 17.4.

17.1 Thermodynamic scaling function and general thermodynamic relations

The general scaling form of the osmotic pressure has been derived in Sect. 11.7.2. We briefly recall the argument.

In the limit of infinite dilution the osmotic pressure $\hat{\Pi}$ obeys van't Hoff's law:

$$\frac{\hat{\Pi}}{k_B T} = c_p \quad ,$$

which provides a renormalization invariant normalization condition: $\hat{\Pi}/(k_B T\, c_p) \to 1, c_p \to 0$. Thus $\hat{\Pi}/(k_B T\, c_p)$ must be a universal function of the scaling variables $\tilde{z} = (s_n N)^{\varepsilon/2}, \tilde{s} = c_p\, \tilde{R}_0^d$. In addition it depends on polydispersity. We write this function in the form of Eq. (17.1)

$$\frac{\hat{\Pi}}{k_B T c_p} = 1 + \tilde{\mathcal{P}}\left(\tilde{s}, \tilde{z}, [p]\right) \quad ,$$

thus introducing the thermodynamic scaling function $\mathcal{P}$ to take into account interaction effects among the macromolecules. In the sequel we will again suppress factors of $1/k_B T$: $\hat{\Pi}/k_B T \to \Pi$.

The thermodynamic scaling function governs the concentration dependence of all thermodynamic quantities. This is the consequence of a Gibbs-Duhem relation. For any reversible and isothermal change of the system the variation $d\Pi$ of the osmotic pressure is related to the variation $d\mu_p(n)$ of the chemical potential as

$$d\Pi = c_p \sum_n P(n) d\mu_p(n) \quad , \tag{17.2}$$

where $P(n)$ is the chain length distribution (Sect. 5.3). Applying this relation to an infinitesimal change of c_p keeping $P(n)$ fixed, we find

$$\frac{\partial}{\partial c_p} \Pi = c_p \frac{\partial \bar{\mu}}{\partial c_p} \quad , \tag{17.3}$$

where

$$\bar{\mu}(c_p) = \sum_n P(n)\, \mu_p(n) \tag{17.4}$$

is the polydispersity averaged chemical potential. We now insert Eq. (17.1) to rewrite this result as

$$\frac{\partial}{\partial c_p}\left[\bar{\mu} - \tilde{\mathcal{P}} - \ln c_p\right] = \frac{1}{c_p} \tilde{\mathcal{P}} \quad . \tag{17.5}$$

Upon integration we find

$$\begin{aligned} \bar{\mu}(c_p) &= \ln \ell^d c_p + \bar{\mu}_0 + \tilde{\mathcal{P}}(\tilde{s}, \tilde{z}, [p]) \\ &+ \int_0^{\tilde{s}} \frac{dt}{t} \tilde{\mathcal{P}}(t, \tilde{z}, [p]) \quad , \end{aligned} \tag{17.6}$$

where

$$\bar{\mu}_0 = \lim_{c_p \to 0} \left[\bar{\mu}(c_p) - \ln \ell^d c_p\right] \tag{17.7}$$

is an integration constant to be determined in the dilute limit. Note that $\lim_{t\to 0} \tilde{\mathcal{P}}(t,\dots)/t$ exists and essentially is given by the second virial coefficient (cf. Eq. (17.25)). Equation (17.6) shows that the nontrivial concentration dependence of $\bar{\mu}(c_p)$ is fully determined by the thermodynamic scaling function.

From Π and $\bar{\mu}(c_p)$ we can construct the free energy excess per unit volume

$$\Delta F = c_p \bar{\mu}(c_p) - \Pi \ . \tag{17.8}$$

Equations (17.1), (17.6) yield

$$\frac{1}{c_p} \Delta F = \ln \ell^d c_p + \bar{\mu}_0 - 1 + \int_0^{\tilde{s}} \frac{dt}{t} \tilde{\mathcal{P}}(t, \tilde{z}, [p]) \ . \tag{17.9}$$

Since the free energy density expressed as function of $c_p, T, P(n)$ determines all thermodynamic quantities this proves the result announced above: *the concentration dependence of all thermodynamic quantities is determined by the thermodynamic scaling function.*

We now consider the forward scattering intensity defined as (cf. Sect. 5.1.4)

$$I_{d,c}(\mathbf{0}) = \frac{1}{\Omega} \int d^d r \, d^d r' \left\langle (\hat{\rho}(\mathbf{r}) - c)(\hat{\rho}(\mathbf{r}') - c) \right\rangle \ , \tag{17.10}$$

where $c = c_p N$ is the segment concentration and $\hat{\rho}(\mathbf{r})$ is the local segment density. We may write

$$\int d^d r \, \hat{\rho}(\mathbf{r}) = \sum_n n \, M(n) \ , \tag{17.11}$$

where $M(n)$ is the number of polymer molecules of length n in the solution. We must note that this quantity fluctuates in our grand canonical ensemble, the average value being

$$\left\langle \sum n \, M(n) \right\rangle = \Omega \, c \ . \tag{17.12}$$

Using Eqs. (17.11), (17.12) we rewrite Eq. (17.10) as

$$\begin{aligned} I_{d,c}(\mathbf{0}) &= \frac{1}{\Omega} \left\langle \left(\sum_n nM(n) - \Omega c \right)^2 \right\rangle \\ &= \frac{1}{\Omega} \sum_{n,n'} nn' \left(\left\langle M(n) M(n') \right\rangle - \left\langle M(n) \right\rangle \left\langle M(n') \right\rangle \right) \ . \end{aligned}$$

These expectation values can be expressed in terms of the grand potential $\Pi[\mu]$ (see Sect. 5.1):

$$
\begin{aligned}
I_{d,c}(\mathbf{0}) &= \sum_{nn'} nn' \frac{\partial}{\partial \mu(n)} \frac{\partial}{\partial \mu(n')} \Pi \\
&= \sum_{nn'} nn' \frac{\partial c_p(n')}{\partial \mu(n)} \quad ,
\end{aligned}
\tag{17.13}
$$

This is the general sum rule, expressing the forward scattering intensity in terms of thermodynamic quantities. It holds for arbitrary polydispersity and can be rewritten in terms of the matrix of second derivatives of the free energy, $\partial^2 \Delta F / \partial c_p(n)\, \partial c_p(n')$.

In the dilute limit $c_p \to 0$ we can use relation (A 15.2):

$$
c_p(n) = \frac{1}{\Omega}\, \mathcal{Z}^{(1)}(n) e^{\mu_p(n)} (1 + O(c_p))
$$

to find

$$
\frac{\partial c_p(n')}{\partial \mu_p(n)} = \delta_{n'n} c_p(n)(1 + O(c_p)) \quad .
$$

This yields

$$
I_{d,c}(\mathbf{0}) \underset{c_p \to 0}{=} \sum_n n^2 c_p(n) = N_w c \quad , \tag{17.14}
$$

a relation valid for all polydispersity.

The general result (17.13) is simplified if we consider a monodisperse ensemble of chains of length N:

$$
\begin{aligned}
e^{\mu_p(n)} &= \delta_{nN} e^{\mu} \\
c_p(n) &= c_p \delta_{nN}
\end{aligned}
$$

Equation (17.13) reduces to

$$
I_{d,c}(\mathbf{0}) = N^2 \frac{\partial c_p}{\partial \mu} \quad .
$$

Taking Π as an explicit function of $c_p(\mu)$: $\Pi(\mu) = \Pi(c_p(\mu))$, we find

$$
c_p = \frac{\partial}{\partial \mu} \Pi = \frac{\partial c_p}{\partial \mu} \frac{\partial}{\partial c_p} \Pi \quad .
$$

This allows us to express $\partial c_p / \partial \mu$ in terms of the compressibility $\left(\frac{\partial}{\partial c_p} \Pi\right)^{-1}$

$$
I_{d,c}(\mathbf{0}) = c \left[\frac{\partial}{\partial c} \Pi \right]^{-1} \quad , \tag{17.15}
$$

which is the well-known compressibility sum rule valid for a monodisperse system. Polydispersity corrections to $I_{d,c}(0)$ or $\frac{\partial}{\partial c_p} \Pi$ obviously differ.

Physically the sum rule stresses that the scattering is due to density fluctuations. If the polymer system were incompressible, fluctuations of the density could not occur. On the theoretical side the sum rule poses some constraint on the approximation schemes. Any consistent approximation should preserve the sum rule. We will find that the loop expansion conforms to that criterion.

We in general will consider a scaling function $J_c(\mathbf{q})$ defined as

$$J_c(\mathbf{q}) = \frac{I_{d,c}(\mathbf{q})}{c\,N} \quad . \tag{17.16}$$

The monodisperse sum rule takes the form

$$J_c^{-1}(0,[m]) = 1 + \frac{\partial}{\partial\,\tilde{s}}\left\{\tilde{s}\,\tilde{\mathcal{P}}(\tilde{s},\tilde{z},[m])\right\} \quad , \tag{17.17}$$

where $[m]$ stands for 'monodisperse'.

17.2 General one-loop results

Renormalization yields the basic result (cf. Sect. 11.7)

$$\frac{1}{c_p}\,\Pi = 1 + \mathcal{P}_R(c_{pR}, N_R, f, [p]) \quad .$$

As has been recalled above, in view of the RG mapping $\mathcal{P}_R$ takes the scaling form

$$\mathcal{P}_R(c_{pR}, N_R, f, [p]) \rightarrow \tilde{\mathcal{P}}(\tilde{s},\tilde{z},[p]) \quad .$$

For $f \rightarrow 1$ it reduces to

$$\tilde{\mathcal{P}}(\tilde{s},\tilde{z},[p]) \underset{f\rightarrow 1}{\longrightarrow} \mathcal{P}^*(\hat{s},[p]) \quad ,$$

where $\hat{s} = B^3 c_p N^{3\nu}$ (Eq. (13.36)). In the excluded volume limit we thus recover Des Cloizeaux's scaling law

$$\Pi = c_p\left[1 + \mathcal{P}^*(\hat{s},[p])\right] \quad . \tag{17.18}$$

Results of the same structure hold for the normalized forward scattering intensity $J_c(\mathbf{0})$ (Eq. (17.16)). In view of Eq. (17.14) this function obeys the RG-invariant condition

$$J_c(\mathbf{0}) \underset{c\rightarrow 0}{=} \tilde{p}_2 \tag{17.19}$$

and therefore it is renormalized according to

$$J_c(\mathbf{0}) = J_{c,R}(\mathbf{0}; c_{pR}, N_R, f, [p]) \quad .$$

As above we find the scaling laws

$$\begin{aligned} J_{c,R}(\mathbf{0}; c_{pR}, N_R, f, [p]) &\longrightarrow \tilde{J}_c(\mathbf{0}; \tilde{s}, \tilde{z}, [p]) \\ &\underset{f \to 1}{\longrightarrow} J_c^*(\mathbf{0}, \hat{s}, [p]) \quad . \end{aligned} \tag{17.20}$$

For general d the one-loop results for $\mathcal{P}_R$ or $J_{c,R}$ are given in Appendix A 17.2. We quote here only the results for $d = 3$, in an appropriately parameterized form. From Eqs. (A 17.6)–(A 17.8), (A 17.10) we find

$$\begin{aligned} \mathcal{P}_R(c_{pR}, N_R, f, [p]) &= \frac{1}{2}(1 + 2u^* f)W_R \\ -\frac{\pi^{1/2}}{3} u^* f N_R^{1/2} &\left[(1 + 2W_R)^{1/2}\left(1 - \frac{1}{W_R}\right) + \frac{1}{W_R}\right](1 + \Delta\mathcal{P}_1(W_R, [p])) , \end{aligned} \tag{17.21}$$

where

$$W_R = f\, c_R\, N_R \quad . \tag{17.22}$$

Here $\Delta\mathcal{P}_1$ is a polydispersity correction, vanishing for an exponential chain length distribution. The parameterized form of $\Delta\mathcal{P}_1$ adequate for Schultz distributions, is given in Eq. (A 17.11). For the exponential ensemble Eq. (17.21) is based on the analytical evaluation of the one-loop integral (A 17.7).

A similar expression for $J_{c,R}$ is found from Eqs. (A 17.12), (A 17.16):

$$\begin{aligned} J_{c,R}^{-1}(\mathbf{0}; c_{pR}, N_R, f, [p]) &= \frac{1}{\tilde{p}_2} + (1 + 2u^* f)W_R \\ -\frac{\pi^{1/2}}{8} u^* \frac{f N_R^{1/2}}{W_R} &\left[\frac{(1 - 4W_R)^2}{2(1 + 2W_R)^{1/2}} + \frac{1}{2} + \left(4 - \frac{1}{W_R}\right)\left((1 + 2W_R)^{1/2} - 1\right)\right] \\ \cdot [1 + \Delta J_1(W_R, [p])] \quad . \end{aligned} \tag{17.23}$$

The polydispersity correction $\Delta J_1(W_R, [p])$ is given in Eq. (A 17.17). It again by construction vanishes for the exponential ensemble. All the parameterizations quoted here reproduce the full numerical results for Schultz-type distributions within deviations less than 0.3%.

In the previous section we have established the compressibility sum rule (17.17). In terms of $J_{c,R}, \mathcal{P}_R$ it reads

$$\begin{aligned} &J_{c,R}^{-1}(\mathbf{0}; c_{pR}, N_R, f, [m]) \\ &= 1 + \frac{\partial}{\partial c_{pR}}\left[c_{pR}\mathcal{P}_R(c_{pR}, N_R, f, [m])\right] \quad . \end{aligned} \tag{17.24}$$

From the results given in the appendix it is easily checked that the one loop results obey the sum rule, provided we keep f, N_R fixed while taking the derivative.

Therefore the final physical results also obey the sum rule, provided we choose the renormalized length scale ℓ_R to be independent of c_p. However, in order to describe the concentration crossover we have chosen ℓ_R to be concentration dependent, imposing the condition (13.21) ($\mathbf{q} \equiv 0$):

$$\frac{n_0}{N_R} + f\,\frac{c_R}{c_0} = 1 \quad .$$

This induces some concentration dependence into f or N_R. Therefore our final results evaluated for $d = 3$ do not rigorously respect the sum rule. The effect is largest in the semidilute limit, where it is found that $J_c^{-1} - 1$ differs from $\partial(\tilde{s}\mathcal{P}(\tilde{s}))/\partial\,\tilde{s}$ by about 3%. In view of the experimental accuracy this inconsistency seems perfectly tolerable, however.

17.3 The virial expansion

17.3.1 Definitions

In our notation the virial expansion of the thermodynamic scaling function takes the form

$$\mathcal{P} = \frac{1}{2} A_2^{\Pi}\, c_p + \frac{1}{6} A_3^{\Pi}\, c_p^2 + \frac{1}{24} A_4^{\Pi}\, c_p^3 + \dots \quad . \tag{17.25}$$

The sum rule (17.17) suggests to write the expansion of $J_{d,c}(\mathbf{0})$ as

$$J_c^{-1}(\mathbf{0}) = \frac{1}{\tilde{p}_2} + \frac{A_2^S}{\tilde{p}_2^2}\, c_p + \frac{A_3^S}{2\tilde{p}_2^3}\, c_p^2 + \frac{A_4^S}{6\tilde{p}_2^4}\, c_p^3 + \dots \quad . \tag{17.26}$$

On the one hand, this definition guarantees the identity

$$A_k^{\Pi}[m] \equiv A_k^S[m], \quad k = 2, 3, \dots \quad . \tag{17.27}$$

On the other hand it yields a simple relation of A_k^S to the experimentally observed virial coefficients A_k. The latter typically are defined by the expansion

$$const \cdot \frac{c_w}{I_{d,c}(\mathbf{0})} = \frac{1}{M_w} + 2A_2 c_w + 3A_3 c_w^2 + 4A_4 c_w^3 + \dots \quad , \tag{17.28}$$

where M_w is the weight average molecular weight and c_w is the weight concentration of polymer [measured in g/cm^3]. As shown in Eq. (15.38) A_2^S is expressed totally in terms of M_w, A_2, as observed in the scattering experiment: $A_2^S = 2\frac{M_w^2}{N_A} A_2$. Similar relations exist for $A_k^S, k > 2$. In particular

$$\frac{A_3^S}{\left(A_2^S\right)^2} = \frac{3}{2}\frac{A_3}{A_2^2 M_w} \quad . \tag{17.29}$$

The second virial coefficients have been discussed in detail in Sect. 15.3. Also the third virial coefficient is of some interest. However, before discussing theoretical results for the ratio (17.29) we should examine the range of concentrations where the virial expansion may be useful.

17.3.2 General features of the virial expansion

Consider the thermodynamic scaling function, specialized to the exponential ensemble $\Delta\mathcal{P}_1 \equiv 0$ at the excluded volume fixed point $f = 1$. We rewrite the virial expansion in terms of $\hat{s} = B^3 c_p N^{3\nu}$ as

$$\mathcal{P}^*(\hat{s}, [e]) = \sum_{k=2}^{\infty} \frac{a_k^{*\Pi}}{k!} \hat{s}^{k-1} \quad , \tag{17.30}$$

where clearly

$$a_k^{*\Pi}(B^3 N^{3\nu})^{k-1} \equiv A_k^{\Pi} \quad .$$

In this form the virial coefficients $a_k^{*\Pi}$ are universal numbers. The one-loop approximation is derived from Eq. (17.21) together with the RG mapping (cf. Sect. 13.3)

$$\begin{aligned} N_R &= \frac{n_0}{w^2} \\ W_R &= c_0 n_0 \left(\frac{1}{w^2} - 1\right), \quad (f \equiv 1) \\ \frac{n_0}{c_0}\tilde{u}\hat{s} &= (1 - w^2) w^{2-6\nu} \quad . \end{aligned} \tag{17.31}$$

(Recall Eq. (13.18): $\tilde{u} = (4\pi)^{3/2} \frac{u^*}{2} = 8.1075$.) A little calculation, preferably with the help of computer algebra, yields

$$\begin{aligned} \mathcal{P}^*(\hat{s}, [e]) = {} & 1.433\hat{s} + 1.065\hat{s}^2 - .648\hat{s}^3 \\ & + .518\hat{s}^4 - .506\hat{s}^5 + 1.075\hat{s}^6 \\ & - 3.852\hat{s}^7 + 13.39\hat{s}^8 - 41.62\hat{s}^9 + \ldots \quad , \end{aligned}$$

where we used $n_0 = 0.53, c_0 = 1.2$. The alternating character of this series is easily understood by observing that for large $\hat{s}$ it must match the asymptotic behavior

$$\mathcal{P}^* \sim \hat{s}^{1/(3\nu-1)} = \hat{s}^{1.309} \quad .$$

All terms beyond the first one have to conspire to approximate that power law. This expansion is useful only for very small $\hat{s}$. Figure 17.1 shows approximations based on the first terms of the expansion. The linear term $a_2^{*\pi}\hat{s}/2$

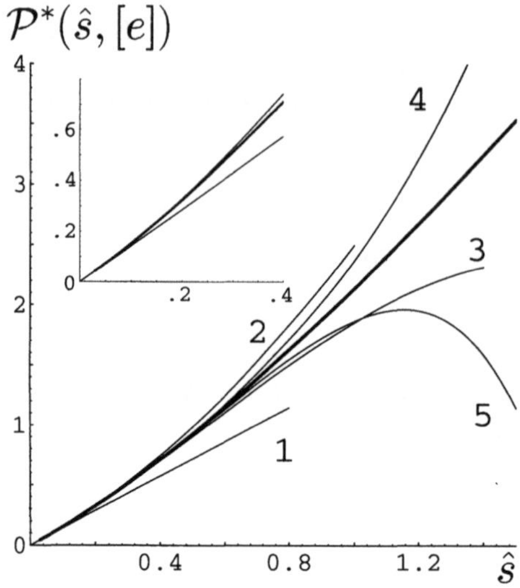

Fig. 17.1. Thermodynamic scaling function for the exponential ensemble $\mathcal{P}^*(\hat{s}, [e])$ (*thick line*). The *thin lines* represent low orders of the virial expansion. The *curves* are identified by the order of the highest term included. The *insert* shows the region $\hat{s} \leq 0.4$, including the approximations of order $\hat{s}$ or $\hat{s}^2$

yields a reasonable approximation for $\hat{s} \lesssim 0.16$, only. Including the third virial coefficient we increase the region of validity to about $\hat{s} \lesssim 0.4$. Higher terms improve the approximation only very slowly. Indeed, the expansion breaks down before we reach $\hat{s} \approx 1$. Since $\frac{4\pi}{3} c_p R_g^3 = 1.169\hat{s}$ (cf. Sect. 13.3.2) this implies that the virial expansion does not extend very far into the crossover region, where the chains start overlapping.

This finding is characteristic of all virial expansions encountered in the present context. Quite generally the leading terms fairly smoothly match the limiting semidilute behavior, and higher terms of the expansion have little meaning individually. The useful range of the expansions barely reaches the crossover region, where for longer chains typically most of the data are taken. Recall that we found the same situation in our discussion of momentum crossover. The implications of this result for the analysis of experimental data will be discussed in Sect. 17.5.

17.3.3 The third virial coefficient

In the previous section we found that the third virial coefficient may still have some relevance individually. We therefore discuss here A_3, or rather the combination (17.29):

$$\mathcal{A}_{3/2}^{S(\Pi)} = \frac{2}{3} \frac{A_3^{S(\Pi)}}{\left[A_2^{S(\Pi)}\right]^2} \quad . \tag{17.32}$$

At the excluded volume fixed point this ratio is a universal function of polydispersity. We first consider this fixed point behavior. (Note that in the literature $\mathcal{A}^S_{3/2}$ is commonly called g.)

Using pure ε-expansion for a monodisperse system a straightforward but lengthy calculation based on Eq. (A 17.6) yields (see also [Clo82])

$$\begin{aligned}\mathcal{A}^{*S}_{3/2} = \mathcal{A}^{*\Pi}_{3/2} &= \varepsilon\left(8\,\ln\,2 - \frac{9}{2}\,\ln\,3 - \frac{1}{6}\right)\\ &= 0.435\ (\varepsilon = 1)\ .\end{aligned}$$

This result is independent of our choice of n_0, c_0, as it should be. Direct evaluation of $\mathcal{A}^*_{3/2}$ in $d = 3$ yields a result that weakly depends on these parameters. With our values $n_0 = 0.53, c_0 = 1, 2$, we find

$$\begin{aligned}&\mathcal{A}^{*S}_{3/2} = \mathcal{A}^{*\Pi}_{3/2} = 0.310\quad\text{(monodisperse)}\\ &\left.\begin{aligned}\mathcal{A}^{*S}_{3/2} &= 0.361\\ \mathcal{A}^{*\Pi}_{3/2} &= 0.519\end{aligned}\right\}\ \text{(exponential ensemble)}\ .\end{aligned}\qquad(17.33)$$

Thus for light scattering, in particular, the residual dependence of $\mathcal{A}^*_{3/2}$ on polydispersity is weak. The main polydispersity effect is taken into account by the explicit factors of $\tilde{p}_2$ in Eq. (17.26). For the set of Schultz distributions the polydispersity dependence of $\mathcal{A}^{*S}_{3/2}$ can be parameterized as

$$\mathcal{A}^{*S}_{3/2}[p] = 0.675 - 0.695\tilde{p}_2 + 0.390\tilde{p}_2^2 - 0.061\tilde{p}_2^3\quad , \qquad (17.34)$$

showing a shallow minimum $\mathcal{A}^{*S}_{3/2}(\tilde{p}_2 \approx 1.2) \approx 0.30$. We finally note that for the monodisperse system Douglas and Freed [DF85] give a value $\mathcal{A}^*_{3/2}[m] = 0.277$. Their result is based on a combination of unrenormalized two-parameter theory in $d = 3$ with the ε-expansion, which is a somewhat unsystematic approach.

We now discuss the ratio $\mathcal{A}^S_{3/2}$ as function of the interaction parameter $\tilde{z}$, restricting ourselves to the monodisperse ensemble. Since f implicitly depends on c_p via the choice of ℓ_R, an analytical evaluation is cumbersome. It is much easier to determine A^S_2, A^S_3 by numerical differentiation of the scaling function $\tilde{J}_c^{-1}(\mathbf{0}; \tilde{s}, \tilde{z}, [m])$. The result is shown in Fig. 17.2. We see the typical two-branched structure. On the weak coupling branch $\mathcal{A}^S_{3/2}$ vanishes for $\tilde{z} \to 0$, whereas on the strong coupling branch $\mathcal{A}^S_{3/2}$ rapidly increases with decreasing $\tilde{z}$. Note that close to the Θ-point our result clearly will break down due to three-body forces.

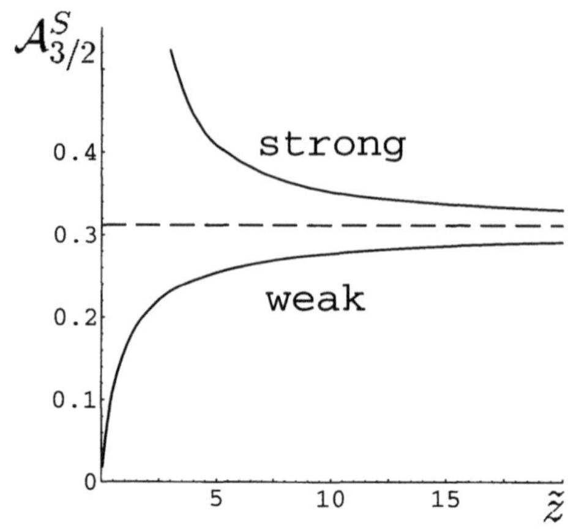

Fig. 17.2. $\mathcal{A}^S_{3/2}[m]$ as function of $\tilde{z}$, showing the two branched structure of temperature crossover

17.4 Evaluation for larger overlap

17.4.1 The semidilute limit

Scaling arguments suggest that the osmotic pressure in the semidilute excluded volume limit behaves like a power of the concentration:

$$\Pi \sim c^{3\nu/(3\nu-1)} \sim c^{2.309} \quad . \tag{17.35}$$

As pointed out in Sect. 9.1 this law follows from the scaling law (17.18) by assuming that Π becomes independent of the chain length. In terms of the blob picture Π essentially is given by the concentration of blobs.

For the thermodynamic scaling function Eq. (17.35) implies the behavior

$$\mathcal{P}^*(\hat{s}, [p]) \underset{\hat{s}\to\infty}{\longrightarrow} A^*_\infty \hat{s}^{1/(3\nu-1)} \quad . \tag{17.36}$$

By virtue of the sum rule (17.15) equivalent results are found for $I^{-1}_{d,c}(\mathbf{0})$ or $J^{-1}_{d,c}(\mathbf{0})$. Equations (17.17) (17.36), in particular, imply

$$J^{*-1}_c(\mathbf{0}; \hat{s}, [p]) \underset{\hat{s}\to\infty}{\longrightarrow} \frac{3\nu}{3\nu-1} A^*_\infty \hat{s}^{1/(3\nu-1)} \quad . \tag{17.37}$$

Recall that the semidilute behavior is expected to be independent of polydispersity.

Our one-loop results allow us to check these predictions. The semidilute limit quite generally is found as (cf. Sect. 13.3)

$$\begin{aligned} f\, c_R &\to c_0 \\ N_R &= \frac{n_0}{w^2} \to \infty \end{aligned} \tag{17.38}$$

i.e.

$$W_R = f\, c_R N_R = \frac{c_0 n_0}{w^2} \to \infty \quad . \tag{17.39}$$

With these results Eqs. (17.21) or (17.23) reduce to

$$\mathcal{P}_R = \frac{0.318}{w^2}(1 + 0.173 f) \tag{17.40}$$

$$J_{c,R}^{-1} = \frac{0.636}{w^2}(1 + 0.312 f) \quad , \tag{17.41}$$

where we inserted our standard values of c_0, n_0. As expected, these results are independent of polydispersity. The polydispersity corrections $\Delta\mathcal{P}_1$ or ΔJ_1 vanish for $W_R \to \infty$. Equations (17.40), (17.41) differ from the zero-loop results only by f-dependent correction factors, which stay finite in the semidilute limit $w \to 0$. As explained in Sect. 5.4.3, this is a consequence of screening which guarantees that higher order terms do not spoil the large overlap behavior as found in tree approximation.

In the semidilute excluded volume limit w is related to the overlap variable $\hat{s}$ by Eq. (17.31), $w \to 0$:

$$\frac{n_0}{c_0}\, \tilde{u}\hat{s} = w^{2-6\nu} \quad .$$

Inserting this result we find from Eqs. (17.40), (17.41), $f = 1$,

$$\begin{aligned} \mathcal{P}^*(\hat{s}, [p]) &\to 1.981 \hat{s}^{1.309} \\ J_c^{*-1}(\mathbf{0}; \hat{s}, [p]) &\to 4.432 \hat{s}^{1.309} \quad , \end{aligned} \tag{17.42}$$

where the exponent of $\hat{s}$ is $1/(3\nu - 1)$. The ratio of the prefactors takes the value $J^{*-1}/\mathcal{P}^* \to 2.237$, to be compared to $3\nu/(3\nu - 1) = 2.309$ resulting from the sum rule (cf. Eq. (17.37)). This violation of the sum rule is a prize we pay for our use of a concentration dependent renormalized length scale, which effectively coincides with the size of the concentration blob. Being very small, this violation clearly is not serious. In fitting Eqs. (17.42) to experiment it amounts to an uncertainty of about 1% of the nonuniversal scale B contained in $\hat{s} = B^3 c_p N^{3\nu}$.

Combining the small and large overlap behaviors we can construct a universal ratio

$$\mathcal{A}^{\Pi}_{\infty/2}[p] = \lim_{\hat{s}\to\infty} \left[\hat{s}^{-1/(3\nu-1)} \mathcal{P}^*(\hat{s}, [p])\right]^{3\nu-1} \lim_{\hat{s}\to 0} \left[\hat{s}^{-1}\mathcal{P}^*(\hat{s}, [p])\right]^{-1} \quad . \tag{17.43}$$

It characterizes the overall shape of $\mathcal{P}^*(\hat{s}, [p])$ and as such is of considerable interest. Evaluating $\mathcal{A}^{\Pi}_{\infty/2}[p]$ for the monodisperse ensemble we find

$$\mathcal{A}^{\Pi}_{\infty/2}[m] \approx 1.098 \quad (d = 3, \text{direct evaluation}) \quad .$$

We should recall that we used this ratio to fix the value of c_0 (see Appendix A 13.2.2). In strict ε-expansion we find

$$\mathcal{A}^{\Pi}_{\infty/2}[m] = 1 + \frac{\varepsilon}{4}\gamma_{Eu} + O(\varepsilon^2) \approx 1.144 \, (\varepsilon = 1) \quad .$$

An estimate of this ratio can also be extracted from [CN82], where ε-expansion supplemented by information for $d = 1$ is used to find $\mathcal{A}^{\Pi}_{\infty/2}[m] \approx 1.087$. Clearly this supports our result derived by direct evaluation in $d = 3$. For the special equilibrium chain length distribution, which is close to the exponential ensemble, $\mathcal{A}^{\Pi}_{\infty/2}$ has been calculated to order ε^2 [KSW81]:

$$\mathcal{A}^{\Pi}_{\infty/2}[e] = 1 + 0.22\varepsilon + 0.06\varepsilon^2 + O(\varepsilon^3) \quad .$$

The somewhat larger value reflects the decrease of the second virial coefficient as polydispersity increases (see Sect. 15.3.2). The smallness of the $O(\varepsilon^2)$-correction supports our one-loop evaluation of the theory.

An equivalent ratio can be constructed from $J_c^{-1}(\mathbf{0})$:

$$\begin{aligned} \mathcal{A}^{S}_{\infty/2}[p] = & \lim_{\hat{s}\to\infty} \left[\hat{s}^{-1/(3\nu-1)} J_c^{*-1}(\mathbf{0}; \hat{s}, [p])\right]^{3\nu-1} \\ & \cdot \lim_{\hat{s}\to 0} \left[\hat{s}^{-1}\left(J_c^{*-1}(\mathbf{0}; \hat{s}, [p]) - \tilde{p}_2^{-1}\right)\right]^{-1} \quad . \end{aligned} \tag{17.44}$$

In view of Eqs. (17.25), (17.26), (17.36), (17.37) for a monodisperse system it obeys the relation

$$\mathcal{A}^{S}_{\infty/2}[m] = \frac{1}{2}\left(\frac{3\nu}{3\nu - 1}\right)^{3\nu-1} \mathcal{A}^{\Pi}_{\infty/2}[m] \quad . \tag{17.45}$$

With the result for $\mathcal{A}^{\Pi}_{\infty/2}[m]$ this yields

$$\mathcal{A}^{S}_{\infty/2}[m] \approx 1.040 \quad .$$

So far we considered the semidilute *excluded volume* limit. We of course can evaluate Eqs. (17.40), (17.41) also away from the fixed point $f = 1$. In the semidilute limit the RG mapping reduces to (cf. Eqs. (13.27 ii), (13.30 i))

$$w^{-1} = f^{-1}\left|1 - f\right|^{\frac{1}{2\nu\omega}} H_n^{1/2}(f)\frac{\tilde{z}}{n_0^{1/2}} \tag{17.46}$$

$$\frac{\tilde{u}}{c_0}\frac{\tilde{s}}{\tilde{z}} = f^{-2}\left|1 - f\right|^{(3-\frac{1}{\nu})\frac{1}{\omega}} H_u^{-1}(f) H^{-1}(f) \quad . \tag{17.47}$$

The final results are best expressed in terms of Π or $1/I_{d,c}(\mathbf{0})$. A little calculation yields

$$\Pi = 4.054 c_p \tilde{z}\tilde{s} |1-f|^{0.502} H^2(f) H_u^{-1}(f)(1+0.173f) \tag{17.48}$$

or

$$\frac{1}{I_{d,c}(\mathbf{0})} = 8.108 \tilde{\ell}^3 \tilde{v} |1-f|^{0.502} H^2(f) H_u^{-1}(f)(1+0.312f) \quad . \tag{17.49}$$

In the last equation we used (cf. Eq. (13.23))

$$\tilde{z}\tilde{s} = \tilde{\ell}^3 \, \tilde{v} \, c \, N \quad . \tag{17.50}$$

In view of Eq. (17.50) the prefactor $c_p \tilde{z}\tilde{s}$ in Eq. (17.48) reproduces the trivial behavior expected in mean-field theory: $\Pi \sim \tilde{\ell}^3 \tilde{v} c^2$, where $\tilde{\ell}^3 \tilde{v}$ plays the role of the (dimensional) excluded volume parameter. The corresponding prefactor in (17.49) then follows directly from the sum rule (17.17). This trivial behavior is corrected by f-dependent factors, where f depends only on $\tilde{s}/\tilde{z} = c\tilde{\ell}^3/\tilde{v}$. In the semidilute limit both Π and $I_{d,c}(\mathbf{0})$ therefore are independent of chain length and polydispersity, as expected. We should recall, however, that to reach the semidilute limit in the Θ-region of the weak coupling branch it needs extremely long chains: $N \sim (T-\Theta)^{-1}$ (cf. Sect. 13.3.3). Since precise experiments systematically varying T in the semidilute limit are missing, we do not plot out the results (17.48), (17.49).

17.4.2 The concentration crossover

The crossover both in temperature and concentration is found by evaluating Eqs. (17.21), (17.23) with the RG-mapping as given by Eq. (13.27), $\mathbf{q} \equiv 0$. Since the behavior of $\tilde{\mathcal{P}}(\tilde{s}, \tilde{z}, [p])$ or $\tilde{J}_c^{-1}(\mathbf{0}; \tilde{s}, \tilde{z}, [p])$ is closely parallel, we illustrate here the results mainly with a discussion of the thermodynamic scaling function.

Consider first the weak coupling branch $f < 1$, not too close to the excluded volume limit. From the virial expansion we know the behavior at small overlap:

$$\frac{1}{\tilde{s}} \tilde{\mathcal{P}}(\tilde{s}, \tilde{z}, [p]) \underset{\tilde{s} \to 0}{\longrightarrow} \frac{A_2^{\Pi}}{2\tilde{R}_0^3} \quad .$$

With increasing overlap $\tilde{s}$, but fixed $\tilde{z}$, $\tilde{\mathcal{P}}/\tilde{s}$ increases. However, the coupling f decreases, since the system moves along a path of type 1 in the crossover diagram of Fig. 13.2. It tends to the semidilute Θ-limit (Fig. 13.2, point C), where $f = 0$. From Eq. (17.48) we therefore find

$$\frac{1}{\tilde{s}} \tilde{\mathcal{P}}(\tilde{s}, \tilde{z}, [p]) \underset{\tilde{s} \to \infty}{\longrightarrow} 4.054 \tilde{z} \quad .$$

Thus in both limits $\tilde{\mathcal{P}}/\tilde{s}$ tends to finite nonvanishing values. These limiting values increase with increasing $\tilde{z}$.

We should note that the large overlap limit $\tilde{s} \to \infty = \text{const}$, described here, is not likely to be of experimental relevance. To reach it we have to increase the segment concentration c, keeping $T > \Theta, N \gg 1$ fixed. With increasing c, however, we will enter the region where three-body forces become important. These lead to a further increase of $\tilde{\mathcal{P}}/\tilde{s}$. Only for extremely long chains can we expect to find an intermediate region of small enough c but large enough $\tilde{s}$ so that some leveling off of $\tilde{\mathcal{P}}/\tilde{s}$ might be observed.

We now consider the neighborhood of the excluded volume limit. Here it is preferable to switch to the variable $\hat{s} = B^3 c_p N^{3\nu}$ which is related to $\tilde{s}, \tilde{z}$ via

$$\hat{s} = 2.776 \tilde{s} \tilde{z}^{3(2\nu-1)} \quad , \tag{17.51}$$

an equation easily derived from Sect. 13.3.2. Figure 17.3 shows $\tilde{\mathcal{P}}/\hat{s}$ as function of $\hat{s}$ for the excluded volume limit and for large values of $\tilde{z}$ both in the weak or strong coupling regimes. With increasing overlap the weak or strong coupling curves leave the excluded volume curve in opposite directions. In some intermediate range of $\hat{s}$ they approximate power laws with some effective $\tilde{z}$-dependent exponent. It is interesting to note that on the weak coupling side this exponent is smaller than the asymptotic excluded volume value $1/(3\nu - 1)$, corresponding to a value $\nu_{eff} > \nu$. The strong coupling side yields $\nu_{eff} < \nu$. This trend is opposite to that observed for instance for R_g^2 in the dilute limit (cf. Fig. 15.3b). Thus we cannot even approximately take care of small deviations from the excluded volume limit by introducing a single effective exponent $\nu_{eff}(\tilde{z})$ for all quantities.

The explicit results shown here are given for the monodisperse ensemble. Polydispersity effects are small even in the dilute limit and rapidly die out

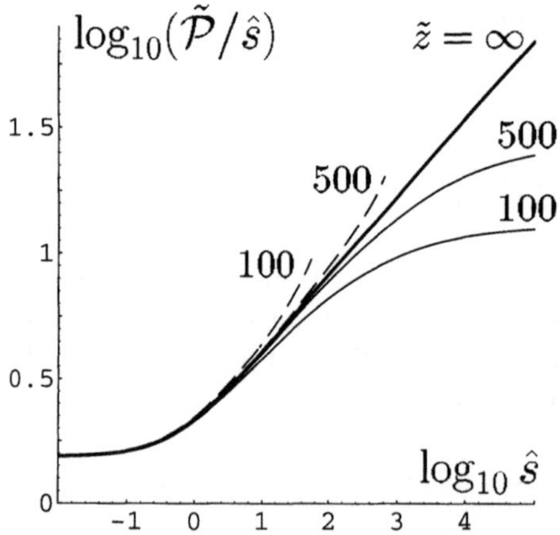

Fig. 17.3. $\log_{10}[\mathcal{P}(\tilde{s}, \tilde{z}, [m])/\hat{s}]$ as function of $\log_{10} \hat{s}$. *Thick line*: excluded volume limit; *thin lines*: weak coupling branch; *broken lines*: strong coupling branch. Values of $\tilde{z}$ are given with the curves

with increasing overlap. For fairly sharp distributions $\tilde{p}_2 \lesssim 1.1$ the deviation of $\mathcal{P}^*(\hat{s}, [p])$ from the monodisperse result never exceeds 1%. For the forward scattering intensity it is found that $J_c^{*-1}(\mathbf{0}, \hat{s}, [p]) - 1/\tilde{p}_2$ is almost independent of polydispersity. In other words, the correction ΔJ_1 plays no serious role.

For a comparison to experiment it is useful to provide simple parameterizations of the scaling functions, valid for monodisperse systems in the excluded volume limit. It turns out that the expressions

$$\mathcal{P}^*(\hat{s}, [m]) = 1.536\hat{s} \left(\frac{1 + 2.509\hat{s} + 1.360\hat{s}^2}{1 + 0.596\hat{s}} \right)^{0.309} \tag{17.52}$$

$$J_c^{*-1}(\hat{s}, [m]) - 1 = 3.073\hat{s} \left(\frac{1 + 3.80\hat{s} + 5.67\hat{s}^2}{1 + 1.73\hat{s}} \right)^{0.309} \tag{17.53}$$

reproduce our results within 0.3% deviation. The latter expression also yields an excellent fit for polydisperse systems, provided we replace J_c^{*-1} by $\tilde{p}_2 J_c^{*-1}$ and $\hat{s}$ by

$$\hat{s}_w = B^3 c N_w^{3\nu - 1} = \tilde{p}_2^{0.764} \hat{s} \quad .$$

Note that this implies the relation

$$\frac{c N_w}{I_{d,c}(0)} \approx J_c^{*-1}(\hat{s}_w, [m]) \quad .$$

17.5 Comparison to experiment

17.5.1 The dilute range

In Chap. 15 we have discussed the second virial coefficients of the osmotic pressure or the forward scattering intensity, taking the attitude that these important quantities can in practice be determined by taking data for sufficiently small concentrations. The results of the present chapter allow us to check this assumption. We thus consider the following questions.

(i) What typically are the limitations of the dilute range, where a virial expansion can be trusted?

(ii) How reliable are typical methods used to extract the chain length or the second virial coefficient?

(iii) Can we determine higher virial coefficients?

In our present analysis we must assume that for $J_c(\mathbf{q})$ the extrapolation towards $\mathbf{q} = 0$ does not introduce systematic errors. As discussed in Sect. 15.5.1 this implies that the data are taken in the range $q^2 R_g^2 \lesssim 1$. We furthermore found that the extrapolation $\mathbf{q} \to 0$ is quite sensitive to polydispersity effects. The simultaneous extrapolation in $\mathbf{q}$ and c will be considered in Sect. 19.5.

The answer to our first question may be taken from Sect. 17.3, where we found that close to the excluded volume limit we need many terms of the virial expansion to go beyond an overlap $\hat{s} \approx 0.4$. For a good solvent system we thus may define the dilute range by the inequality

$$\hat{s} \lesssim 0.4 \quad .$$

Going to moderate solvents or Θ-solvents we should recall from Sect. 13.3 that the 'dilute' range in principle is determined not by the geometrical overlap ($\hat{s}$ or $\tilde{s}$) alone, but also by the renormalized coupling constant. It thus increases if we approach the Θ-temperature, as has been illustrated in Fig. 13.3. It is limited for $T = \Theta$ by three-body forces. In the following discussion we will consider only systems close to the excluded volume limit.

We may translate the restriction $\hat{s} \lesssim 0.4$ into a condition for the weight concentration c_w. In our notation

$$\hat{s} = c_p B^3 N^{3\nu} = c B^3 N^{3\nu - 1} \quad ,$$

where c is measured as monomer number per volume. Taking as typical values $B = 3\,\text{Å}$ and a molecular weight of the monomer of 100 g/mol we find

$$\hat{s} \approx 0.16\, N^{0.764}\, c_w \quad ,$$

where the experimental concentration c_w is measured in g/cm^3. Thus the dilute range is estimated as

$$c_w \lesssim 2.5\, N^{-0.764} [g/cm^3] \quad .$$

This yields, for instance,

$$c_w \lesssim \begin{cases} 1.3\ 10^{-2} & ; \ N = 10^3 \\ 2.2\ 10^{-3} & ; \ N = 10^4 \\ 3.8\ 10^{-4} & ; \ N = 10^5 \quad . \end{cases}$$

Osmotic measurements typically use concentrations $c_w \gtrsim 2 \cdot 10^{-3} [g/cm^3]$, so that the dilute range is reached only for fairly short chains ($N \lesssim 3000$). Light scattering can be carried to much lower concentrations: $c_w \gtrsim 10^{-5} g/cm^3$, though typically the lowest concentrations are somewhat larger ($c_w \gtrsim 10^{-4}$ g/cm^3). As a result, experiments often operate close to the boundary of the dilute range, at least for good solvent systems. Therefore a careful examination of questions (ii) or (iii) formulated above is appropriate.

We consider scattering experiments which yield data in a restricted range of overlap, $0 < \hat{s}_1 \leq \hat{s} \leq \hat{s}_2$, and we take our one-loop result for $J_c^{-1}(\mathbf{0})$ as ideal input data. We first fit the 'data' by an effective virial expansion of the form

$$\frac{1}{N} J_c^{-1}(\mathbf{0}) = \frac{c}{I_{d,c}(\mathbf{0})} = \frac{1}{N_{w,eff}} + 2A_{2,eff}c + 3A_{3,eff}c^2 \tag{17.54}$$

comparing the effective parameters to those resulting from the exact expansion of the one-loop scaling function. Typical results for a monodisperse system ($\tilde{p}_2 = 1$) or a system of large polydispersity ($\tilde{p}_2 = 1.5$) in the excluded volume limit are shown in Table 17.1. We observe the following behavior.

(i) The chain length in principle is overestimated, the effect however being negligibly small, if sufficient data for $\hat{s} \lesssim 0.3$ are included.

(ii) Also the second virial coefficient is overestimated. Compared to N_w the error is larger, but still stays on the level of a few percent for a range $\hat{s} \lesssim 0.3$.

(iii) The effective third virial coefficient has little to do with the true value of A_3. It is definitely smaller, even for the range of smallest overlap considered.

(iv) The effects are fairly insensitive to polydispersity.

Table 17.1. Correction factors for effective virial coefficients

$\tilde{p}_2 = 1$	Virial expansion			Square root plot	
Overlap range	$\frac{N_{w,eff}}{N_w}$	$\frac{A_{2,eff}}{A_2}$	$\frac{A_{3,eff}}{A_3}$	$\frac{N_{w,eff}}{N_w}$	$\frac{A_{2,eff}}{A_2}$
$0.012 \leq \hat{s} \leq 0.088$	1.0001	1.001	0.95	0.9996	0.994
$0.05 \leq \hat{s} \leq 0.15$	1.0006	1.007	0.90	0.998	0.987
$0.1 \leq \hat{s} \leq 0.3$	1.005	1.03	0.80	0.99	0.97
$0.2 \leq \hat{s} \leq 0.6$	1.03	1.08	0.65	0.95	0.95
$0.2 \leq \hat{s} \leq 1.0$	1.08	1.14	0.55	0.90	0.94
$\tilde{p}_2 = 1.5$					
$0.012 \leq \hat{s} \leq 0.088$	1.0001	1.002	0.94	0.999	0.990
$0.05 \leq \hat{s} \leq 0.15$	1.001	1.01	0.88	0.995	0.98
$0.1 \leq \hat{s} \leq 0.3$	1.01	1.04	0.75	0.978	0.96

Concerning the third virial coefficient it must be noted that the situation is actually worse than indicated in Table 17.1. For $\hat{s} \lesssim 0.1$ the A_3-term in Eq. (17.54) contributes less than 2% of the total. The scatter of experimental data will not allow for a reliable determination of A_3 from that range. We need to go up to $\hat{s} \approx .3$, if the A_3-term is to make up for 10% of $J_c^{-1}(\mathbf{0})$, resulting in a 20% underestimation of A_3. This shows that the effect of the third virial coefficient can hardly be separated from the influence of the higher terms of the virial expansion. A correct extraction of A_3 from an evaluation based on the direct virial expansion is virtually impossible. Similar results follow from an analysis of the osmotic pressure. Taken together, these results imply that the ratio $\mathcal{A}_{3/2}$ (Eq. (17.29)) is likely to be seriously underestimated.

The same problem plagues the often used 'Bawn plot' [BFK50]. It consists in analyzing

$$\left(\frac{c_2}{I_{d,c}}\bigg|_{c=c_2} - \frac{c_1}{I_{d,c}}\bigg|_{c=c_1} \right) \frac{1}{c_2 - c_1}$$

as function of $c_1 + c_2$. In terms of $J_c(\mathbf{0})$ it amounts to the approximation

$$\frac{\delta\, J_c^{*-1}}{\delta\, \hat{s}} = \frac{J_c^{*-1}(0)\Big|_{\hat{s}_2} - J_c^{*-1}(\mathbf{0})\Big|_{\hat{s}_1}}{\hat{s}_2 - \hat{s}_1} = a_{2,eff} + \frac{1}{2} a_{3,eff}(\hat{s}_2 + \hat{s}_1) \quad . \qquad (17.55)$$

Being based on the virial expansion, such an analysis essentially reproduces the results of Table 17.1, despite the somewhat different weighting of the data points implied by taking all different pairs $(\hat{s}_2, \hat{s}_1)$ into account. The virtue of this plot lies in the fact that it involves only two-fit parameters, and that it should allow to identify deviations due to higher virial coefficients more easily. Such deviations in principle lead to some downward curvature of the plot. Furthermore, data points for different $(\hat{s}_2, \hat{s}_1)$ but fixed $\hat{s}_1 + \hat{s}_2$ in principle do not fall on top of each other. Observations of a systematic splitting of the points also would indicate the importance of higher terms in the virial expansion. However, both these effects turn out to be not very strong. This is illustrated in Fig. 17.4 with theoretical 'data' taken in the interval $0.05 \leq \hat{s} \leq 0.15$. Clearly the Bawn plot looks perfect, despite resulting in errors as given in Table 17.1. Since in using this method we need to take a ratio of differences, which largely magnifies the statistical scatter of real data, we in analyzing such data would have no chance to identify the small systematic deviations of the points from the straight line. It should be noted that this figure closely resembles typical Bawn plots found in the literature.

If we do not ask for information on the third virial coefficient we may use a 'square root plot', evaluating the data according to the ansatz

$$[N J_c(\mathbf{0})]^{-1/2} = \left[\frac{c}{I_{d,c}(\mathbf{0})} \right]^{1/2} = \frac{1}{N_{w,eff}^{1/2}} + A_{2,eff} N_{w,eff}^{1/2}\, c \quad . \qquad (17.56)$$

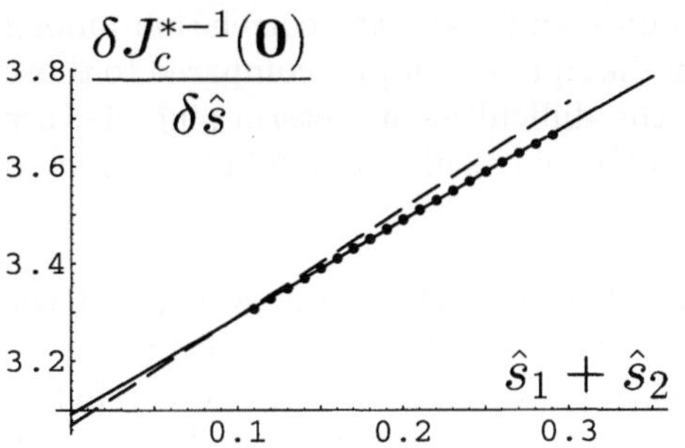

Fig. 17.4. Bawn-plot for 'theoretical data' taken in the interval $0.05 \leq \hat{s} \leq 0.15$ (*points*), using the one-loop result for $J_c^*(\mathbf{0})$ in the monodisperse ensemble. The *full line* is the best fit using Eq. (17.55). The *broken line* derives from Eq. (17.55) with the exact virial coefficients inserted

This method is based on the empirical observation that $\sqrt{J_c^{*-1}}$ seems to depend linearly on $\hat{s}$ in a fairly large interval. It is supported by our one-loop calculation, as is illustrated in Fig. 17.5. Up to $\hat{s} \approx .5$ the linear approximation deviates from $J_c^{*-1/2}$ by less than 2%. This is to be compared to the linear approximation to J_c^{*-1}, which reaches 2% deviation for $\hat{s} \approx .1$. This feature of the square root plot may be understood by noting that it effectively yields a ratio $\mathcal{A}_{3/2} = 1/3$ that compares favorably to the theoretical estimate (17.33) for $\mathcal{A}^*_{3/2}[m]$.

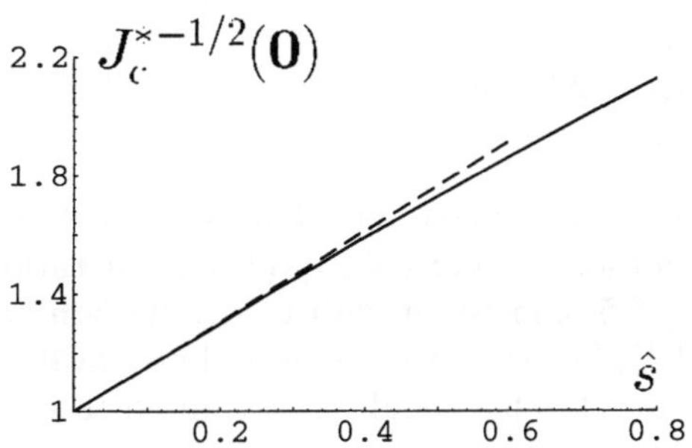

Fig. 17.5. $J_c^{*-1/2}(\mathbf{0}, [m])$ as function of $\hat{s}$. The *broken line* gives the initial slope

Evaluating theoretical data as above, but now using the square root plot, we find the results given in the last two columns of Table 17.1. We note that now we underestimate both N_w and A_2, a feature obvious from inspecting Fig. 17.5. The results for N_w are somewhat worse than those found from the virial expansion, but the estimates for A_2 resulting from a range of larger $\hat{s}$-values are improved.

The present discussion is concerned with the forward scattering intensity. For the osmotic pressure a similar discussion has been presented in [Sch82], with equivalent results. Even though we have used our one-loop results as 'ideal data' in the analysis, we should note that the qualitative results do not depend on the details of the scaling functions. Rather they reflect the alternating character of the virial expansion, which asymptotically has to match a power $\hat{s}^x, 1 < x < 2$. The main results, viz. the correlation among the errors, the tendency of the errors in the square root plot compared to those of the virial expansion analysis, and the difficulties in determining A_3, are therefore expected to hold even if the one-loop calculation would be corrected by higher order terms.

To summarize, this analysis shows that both the chain length and the second virial coefficient can be reliably determined from typical data, provided in a scattering experiment the extrapolation towards $\mathbf{q} = 0$ is reliable. Furthermore, since the errors in the square root plot or the virial expansion go in opposite directions, presumably very precise values could be found by comparing results from both methods of data analysis. The third virial coefficient, on the other hand, is necessarily subject to large systematic errors, which can be estimated only on the basis of a definite theoretical model.

Turning now to a quantitative comparison with data we recall from Sect. 15.5 that good agreement has been found among theory and experiment for the second virial coefficient. It is thus reassuring to observe that A_2 can indeed be measured to good precision. Attempts to measure A_3 are rare. Only recently a series of light scattering experiments has been published [NNT91], [NAK+92], covering a large range of chain lengths in good solvent systems:

polyisobutylene-cyclohexane, $465 \lesssim N_w \lesssim 2.5 \cdot 10^5$;
polystyrene-benzene, $99 \lesssim N_w \lesssim 2 \cdot 10^5$.

The experiments cover the dilute range as identified here ($0.03 \lesssim \hat{s}_w \lesssim 0.3$). The outcome is somewhat surprising. For long chains ($N_w \gtrsim 30{,}000$) the ratio $\mathcal{A}^S_{3/2}$ levels off at a value $0.4 \lesssim \mathcal{A}^S_{3/2} \lesssim 0.5$, consistent with the prediction of first-order ε-expansion, but somewhat higher than the value of direct evaluation in three dimensions ($\mathcal{A}^{*S}_{3/2}[m] \approx 0.31$). This could be attributed to the low order of our calculation, but it must be kept in mind that according to our analysis the Bawn-plot, used in the evaluation of the data, should *underestimate* $\mathcal{A}^S_{3/2}$ by about 20% for the overlap range of the experiment. What is worse, for both systems $\mathcal{A}^S_{3/2}$ is found to approach its asymptotic value from below, increasing from $\mathcal{A}^S_{3/2} \approx .3$, $N_w \approx 1000$, to the above quoted value. For the PS-benzene system this tendency has been verified by scattering experiments of Li et al. [LWXM95]. Both systems, however, are to be described by the strong coupling branch. This follows from the observation that for both systems A^S_2 approaches its asymptotic power law from above. Indeed, the analysis of the A^S_2 data for the PS-benzene system, given in [KS96] shows reasonably good agreement with the strong coupling branch of our theory. For a strong coupling system, however, $\mathcal{A}^S_{3/2}$ should approach its asymptotic value from above, as shown in Fig. 17.2. We thus conclude that the data are not compatible with any two-parameter type theory, a conclusion also drawn by the authors.

To analyze the situation further we may compare the original data for $I_{d,c}(\mathbf{0})$ to our theory[1]. Figure 17.6 shows the results for the system polystyrene-benzene. To bring out the effects, we plot $(\tilde{p}_2 J_c^{-1}(\mathbf{0}) - 1)/\hat{s}_w$ as function of $\hat{s}_w = B^3 c N_w^{3\nu-1}$. (In a direct plot $J_c^{-1}(\mathbf{0})$ vs $\hat{s}_w$ deviations can hardly be detected.) In Fig. 17.6 for clarity we have shifted the results of the different probes by constant amounts, the chain length decreasing from bottom to top. We used parameters $B = 3.13\,\text{Å}$, $\tilde{v} = 0.4$, which are in good accord with other measurements on this system. For $N_w \lesssim 500$ we expect to see nonuniversal corrections. We therefore omitted the data for the two shortest samples ($N_w < 200$). For all the other probes we expect in principle our theory to apply. In fitting the data we allowed for a readjustment of N_w. Compared to the literature values the change is always less than 1%, except for the longest chain, where we increased N_w by about 4%. As is clear from Fig. 17.6, in several cases we find definite deviations among theory and experiment. It must be noted that quantitatively the deviations are not large. They amount at most to 6% in the present plot, resulting in about 3% deviation in $c/I_{d,c}^{-1}(\mathbf{0})$. Still the effect seems to be outside the statistical scatter of the data. However, inspecting the data for the different samples it is hard to detect any systematics in these deviations. A similar picture is found for the system polyisobutylene-cyclohexane.

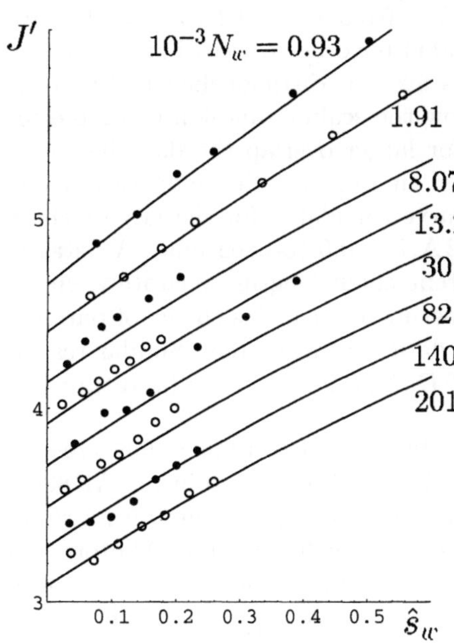

Fig. 17.6. $J' = (\tilde{p}_2 J_c^{-1}(\mathbf{0}) - 1)/\hat{s}_w$ as function of $\hat{s}_w$. Data [NNT91]: polystyrene-benzene of different chain lengths, shifted by constant amounts (+0.2)

[1] I am indebted to Dr. Norisuye for sending me a compilation of the data.

Norisuye et al. [NNA93] try to explain the experimental findings by including chain stiffness and three-body interactions. Their theory builds upon the weak coupling branch and uses ideas of [Yam92], which have been criticized in [KS96]. Therefore this interpretation might be doubted. Indeed we might point to the extrapolation $\mathbf{q} \to 0$, which at least for the longer chains could be a source of systematic errors (see Sect. 19.5.1).

All problems of extrapolation would be absent in a determination of A_3 via Monte Carlo simulations. Bruns [Bru97] has recently presented such data for self-avoiding walks on three-dimensional lattices. The data are consistent with the theory in that they seem to approach an asymptotic value $\mathcal{A}^*_{3/2} \approx 0.3$, coming from above as expected for such strong coupling systems. However, the chains are much too short ($n \leq 65$) to exclude the possibility that $\mathcal{A}_{3/2}(n)$ for $n \to \infty$ might increase again, as found in the experimental data. In summary we feel that the precise determination of A_3 still poses some problems.

17.5.2 The concentration crossover

Experiments covering a large range of overlap aim at testing the asymptotic power laws (17.35)–(17.37). We first consider the osmotic pressure. For the system P-(α-methylstyrene)-toluene results of an extensive study [NKKN81] are shown in Fig. 17.7. The overlap ranges from $\hat{s} \approx 0.02$ to $\hat{s} \approx 25$. Five samples of chain lengths $600 \lesssim N \lesssim 16{,}000$ have been used, but in view of the scatter of the data different lengths are not distinguished in Fig. 17.7. For smaller overlap the data closely follow the scaling function predicted for the excluded volume limit (full line). For larger overlap the data lie somewhat above that curve. As is illustrated in Fig. 17.3 this indicates strong coupling behavior, and we included a curve calculated for the largest chain length, using parameter values $B = 3.02\,\text{Å}$, $\tilde{v} = 0.5$ (broken line). A detailed examination of the data, treating different chain lengths separately, shows good agreement between theory and experiment with this choice of parameters, which is also consistent with other data for this system. Similar results prevail for other experimental studies of $\hat{\Pi}/k_BT$ found in the literature.

We now turn to the forward scattering intensity. Figure 17.8 shows results for P-(α-methylstyrene)-toluene [NKKN81] or polystyrene-toluene [AD82] [2]. We again find good agreement between theory and experiment, with a slight tendency of the data to lie above the excluded volume curve. The original evaluation of the polystyrene-toluene data in terms of the asymptotic power law (17.37): $I^{-1}_{c,d}(\mathbf{0}) \sim c^{1/(3\nu_{eff}-1)}$ resulted in a value $\nu_{eff} \approx 0.575$. From our

[2] I am indebted to Dr. McDonnell for providing me with a compilation of their data.

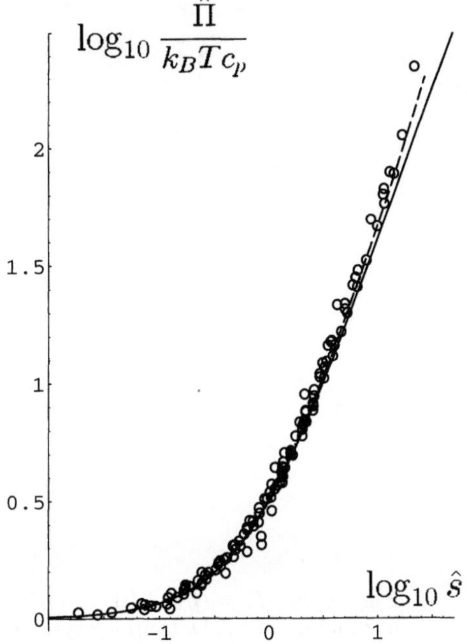

Fig. 17.7. $\log_{10} \frac{\hat{\Pi}}{k_B T c_p}$ as function of $\log_{10} \hat{s}$. Data [NKKN81]: P-(α-methylstyrene)-toluene. Theory in the excluded volume limit: *full curve*. Fit to the strong coupling branch: *broken curve*

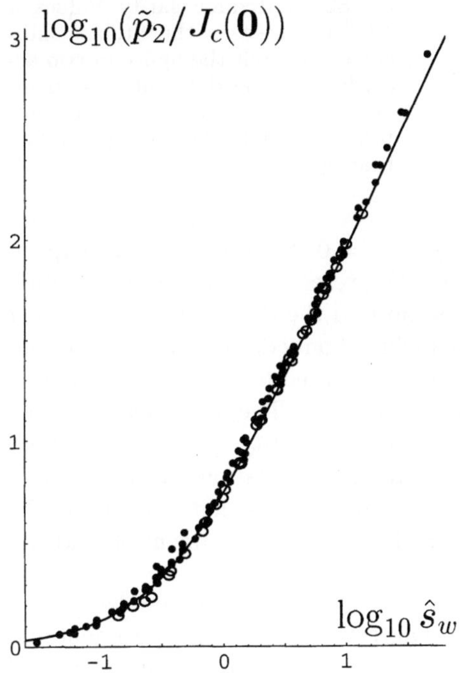

Fig. 17.8. $\log_{10}(\tilde{p}_2/J_c(\mathbf{0}))$ as function of $\log_{10} \hat{s}_w$. *Circles*: P-(α-methylstyrene)-toluene [NKKN81]. *Points*: polystyrene-toluene [AD82]. The theoretical curve for the excluded volume limit is included

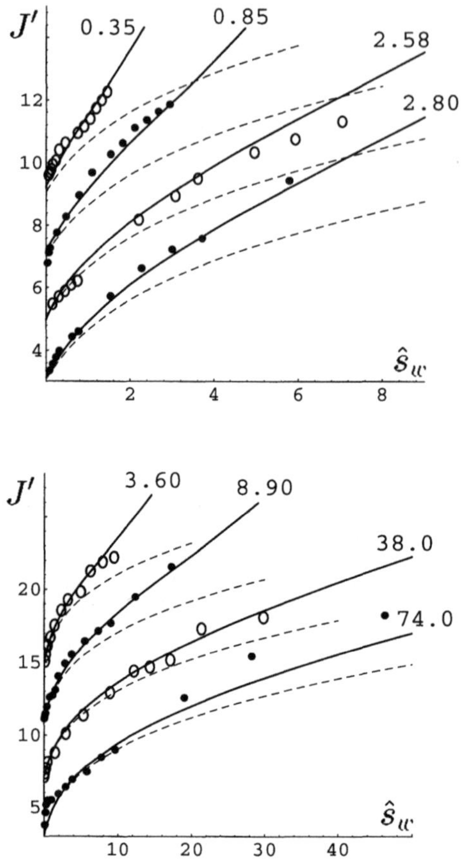

Fig. 17.9. $J' = (\tilde{p}_2/J_c(\mathbf{0}) - 1)/\hat{s}_w$ as function of $\hat{s}_w$. Data: polystyrene-toluene for different chain lengths, [AD82], shifted for clarity. Values of $10^{-3}N_w$, used in the fit, are indicated. We split the figure in two according to the different $\hat{s}_w$-ranges. *Broken lines*: excluded volume limit; *full lines*: fit to the strong coupling branch

analysis it is clear that such a value $\nu_{eff} < \nu = 0.588$ in the present experiment indicates strong coupling behavior. We present a more detailed evaluation of the data in Fig. 17.9, where we separated the data according to their chain lengths ($350 \lesssim N_w \lesssim 74{,}000$). We included only data for weight concentrations less than 10%. The full curves are calculated for the strong coupling branch, using parameter values $B = 3.15\,\text{Å}$, $\tilde{v} = 0.7$. The fit is most satisfactory, in particular if we recall that the plotted quantity $(\tilde{p}_2 J_c^{-1}(\mathbf{0}) - 1)/\hat{s}_w$ is most sensitive to small errors. The systematic deviations from excluded volume behavior are nicely explained. We finally note that except for the longest chain the extrapolation $\mathbf{q} \to 0$ in the present experiment should not be dangerous.

In summary, we have found very good agreement between experiment and theory for thermodynamic quantities in a large range of overlap. For good solvent systems some small deviations from the excluded volume limit are identified as strong coupling behavior. Only the third virial coefficient poses problems, but in view of the good overall agreement and of the difficulties in a determination of A_3 discussed above, we believe that the origin of these problems should be sought in the experimental evaluation rather than in the theory.

We have been concerned here with testing the theory in good solvent systems. Of course, the crossover towards Θ-conditions is also of great interest. Unfortunately extensive experiments covering this crossover outside the dilute region are rare. Typically for a given system only a single quantity is measured for one or two chain lengths. We are aware of only one more extensive set of measurements, which will be discussed in Sect. 19.5.

A 17. Appendix

A 17.1 Perturbative calculation of $\mathcal{P}$

The loop expansion for Π has been derived in Appendix A 5.2. It formally reads (cf. Eq. (5.31))

$$\Pi = K^{(0,0)} + \frac{u_0}{2} c^2 \quad , \tag{A 17.1}$$

where $K^{(0,0)}$ denotes the set of all vertex irreducible graphs contributing to Π. The series of one-loop diagrams is shown in Fig. 17.10. The symmetry factor σ_m of a one-loop diagram of m chains equals $\sigma_m = 1/(2m)$, where the factor $1/2$ is due to right-left symmetry and the factor $1/m$ takes into account cyclic permutations of the chains. Resuming the series we find the contribution

$$-\frac{1}{2}\int_k \ln\left(1 + u_0 c N D_p(k^2\ell^2 N)\right) \quad . \tag{A 17.2}$$

The zero-loop diagram yields

$$\frac{1}{2}\sum_n \frac{2}{(4\pi\ell^2)^{d/2}}\, e^{\mu_p(n) - u_0 c n} \quad . \tag{A 17.3}$$

The equation of state relating $\mu_p(n)$ to $c_p(n)$ to one-loop order of unrenormalized perturbation theory has been derived in Sect. 5.4.3, where we pointed out that we have to take care of an additive renormalization in the single-chain contribution. The very same problem occurs with the single-chain one-loop contribution to $K^{(0,0)}$. Indeed comparing Figs. 5.8, 17.10 we find that in calculating $\mu_p(n)$ (or $K^{(0,1)}$, equivalently) or $K^{(0,0)}$ we deal with diagrams of identical structure. It turns out that in calculating $K^{(0,0)}$ the additive renormalization cancels. So instead of introducing the additive renormalization

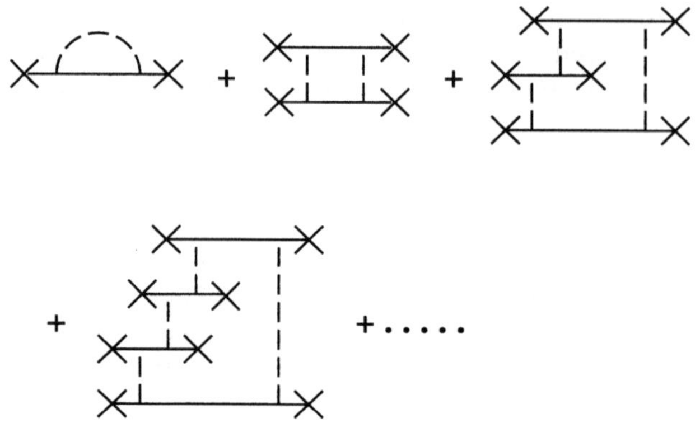

Fig. 17.10. Series of one-loop diagrams contributing to $K^{(0,0)}$

in (A 17.2) we may simply ignore it also in the calculation of $\mu_p(n)$, using Eqs. (5.66), (3.22), (5.60) to write

$$\frac{1}{(4\pi\ell^2)^{d/2}} e^{\mu_p(n)-u_0cn} = c_p(n)\left[1+u_0\,\frac{n^2}{2}\int_k \frac{D(k^2\ell^2 n)}{1+u_0cND_p(k^2\ell^2N)}\right] \quad . \tag{A 17.4}$$

As a result we find

$$\frac{1}{c_p}\Pi = 1+\mathcal{P}$$

$$\mathcal{P} = \frac{u_0}{2}\,cN + \frac{1}{2c_p}\,\frac{(4\pi)^{-d/2}}{\Gamma\left(\frac{d}{2}\right)}\int_0^\infty dx\, x^{1-\varepsilon/2}\left[\frac{u_0cND_p(x\ell^2N)}{1+u_0cND_p(x\ell^2N)} - \ln\left(1+u_0cND_p(x\ell^2N)\right)\right] \quad , \tag{A 17.5}$$

which is the dimensionally regularized expression. Note that in expanding the integrand with respect to c the single-chain contribution (linear term) drops out. No term needing additive renormalization survives.

A 17.2 One-loop results for $\mathcal{P}$ or $I_{d,c}(0)$

A 17.2.1 The thermodynamic scaling function. Since Π/c_p obeys a renormalization invariant normalization, the renormalization of $\mathcal{P}$ proceeds by replacing all unrenormalized parameters by their renormalized counterparts (see Sect. 12.3.1). A simple calculation yields

$$\mathcal{P}_R = \frac{1}{2} W_R \left[1 + \frac{2}{\varepsilon} f u^* + \frac{1}{d\Gamma\left(\frac{d}{2}\right)} f u^* N_R^{\varepsilon/2} \int_0^\infty dx\, x^{2-\varepsilon/2} \frac{D_p'(x) D_p(x)}{\left(1 + W_R D_p(x)\right)^2} \right] , \tag{A 17.6}$$

where $W_R = f c_R N_R$. We simplified the integral by partial integration.

The loop integral

$$\mathcal{P}_1(W_R, [p]) = \int_0^\infty dx\, x^{2-\varepsilon/2} \frac{D_p'(x) D_p(x)}{(1 + W_R D_p(x))^2} \tag{A 17.7}$$

deserves some consideration. For the exponential chain length distribution the Debye-function takes the form

$$D_p(x) = \frac{2}{1+x} \quad ,$$

and $\mathcal{P}_1(W_R, [e])$ can be evaluated analytically:

$$\mathcal{P}_1(W_R, [e]) = -\frac{2\pi}{\sin \frac{\varepsilon}{2}\pi} \frac{1}{W_R} \left[(1 + 2W_R)^{1-\varepsilon/2} \left(1 - \frac{\varepsilon}{2} - \frac{1}{2W_R} \right) + \frac{1}{2W_R} \right] . \tag{A 17.8}$$

For general chain length distributions only the limit $W_R \to \infty$, relevant for the large overlap regime, is available analytically:

$$\begin{aligned} \mathcal{P}_1(W_R, [p]) \underset{W_R \to \infty}{\longrightarrow} & -4 \int_0^\infty dx \frac{x^{1-\varepsilon/2}}{(x + 2W_R)^2} \\ & = -\frac{4\pi}{\sin \frac{\varepsilon}{2}\pi} \left(1 - \frac{\varepsilon}{2} \right) (2W_R)^{-\varepsilon/2} . \end{aligned} \tag{A 17.9}$$

Note that this limit is independent of polydispersity. Note further, that the simple form of that result is a consequence of screening. The evaluation of the loop integral (A 17.7) closely follows the analysis of the one-loop correction to the chemical potential, as given in Sect. 5.4.3.

For general values W_R we write

$$\mathcal{P}_1(W_R, [p]) = \mathcal{P}_1(W_R, [e]) \left[1 + \Delta\mathcal{P}_1(W_R, [p]) \right] \quad , \tag{A 17.10}$$

where the polydispersity correction $\Delta\mathcal{P}_1$ is to be evaluated numerically. A simple parameterization may be found, valid for Schultz-distributions in physical dimension $d = 3$:

$$\Delta\mathcal{P}_1(W_R, [p]) = \frac{0.192(\tilde{p}_2 - 2)}{1 + (0.86 + 0.36\tilde{p}_2) W_R} \quad . \tag{A 17.11}$$

This form reproduces the numerical result for $\Delta\mathcal{P}_1(W_R, [p])$ $(d = 3)$ within deviations less than .004 in absolute value.

A 17.2.2 One-loop results for $J_c(\mathbf{0})$. The calculation of the full density correlation function will be sketched in Chap. 19. Here we only give the expression for the renormalized forward scattering intensity. To one-loop order of renormalized perturbation theory we find

$$J_{c,R}^{-1}(\mathbf{0}; c_{pR}, N_R, f, [p]) = \tilde{p}_2^{-1} + W_R\Big(1 + \frac{2}{\varepsilon} f u^* - \frac{1}{\Gamma(\frac{d}{2})} f u^* N_R^{\varepsilon/2} J_1(W_R, [p])\Big) \qquad \text{(A 17.12)}$$

$$J_1(W_R, [p]) = \frac{1}{(\tilde{p}_2)^2} \int_0^\infty dx\, x^{-3-\varepsilon/2} \left(\frac{\tilde{p}'(x) + 1 - \tilde{p}_2 x}{1 + W_R D_p(x)}\right)^2 . \qquad \text{(A 17.13)}$$

For the monodisperse ensemble the relation $\tilde{p}'(x) = -\tilde{p}(x)(= -e^{-x})$ holds, and it is easily checked that Eqs. (A 17.6), (A 17.12) obey the sum rule (17.17), where the derivative is to be taken with N_R, f fixed.

For the exponential ensemble the loop integral J_1 can again be evaluated analytically

$$J_1(W_R, [e]) = \frac{\pi}{4 \sin \frac{\varepsilon}{2}\pi} \frac{1}{W_R^2} \Big[\frac{1 - \varepsilon/2}{4}\left(1 + \frac{(4W_R - 1)^2}{(1 + 2W_R)^{\varepsilon/2}}\right) + \left(1 - \frac{1}{4W_R}\right)\left((1 + 2W_R)^{1-\varepsilon/2} - 1\right)\Big] . \qquad \text{(A 17.14)}$$

In the semidilute limit we find

$$J_1(W_R, [p]) \underset{W_R \to \infty}{\longrightarrow} \frac{\pi}{\sin \frac{\varepsilon}{2}\pi} \left(1 - \frac{\varepsilon}{2}\right) (2W_R)^{-\varepsilon/2} , \qquad \text{(A 17.15)}$$

independent of polydispersity.

For general polydispersity $J_1(W_R, [p])$ is to be evaluated numerically. We write

$$J_1(W_R, [p]) = J_1(W_R, [e]) \left[1 + \Delta J_1(W_R, [p])\right] . \qquad \text{(A 17.16)}$$

ΔJ_1 can be parameterized as

$$\Delta J_1(W_R, [p]) = -\frac{0.369 + 1.210\, c_A^S[p]}{[1 + (2.40\,\tilde{p}_2 - 0.21)W_R + (1.2\,\tilde{p}_2 - 0.2)^2 W_R^2]^{1/2}} , \qquad \text{(A 17.17)}$$

where $c_A^S[p]$ denotes the polydispersity correction occurring for the second virial coefficient:

$$c_A^S[p] = -0.385(\tilde{p}_2 - 1) + 0.107(\tilde{p}_2 - 1)^2 - 0.027(\tilde{p}_2 - 1)^3$$

(cf. Eq. (15.45 ii)). Equations (A 17.16), (A 17.17) reproduce the numerical result for $J_1(W_R, [p])$ within deviations less than 0.3%.

A 17.3 Chemical potential and free energy excess

A 17.3.1 The chemical potential. The results for the average chemical potential

$$\bar{\mu} = \sum_n P(n)\,\mu_p(n)$$

could be derived by renormalizing the expression for $\mu_p(n)$ given in Sect. 5.4.3. However here we base the discussion on Eq. (17.6) derived from the Gibbs-Duhem relation.

We first need an expression for the integration constant $\bar{\mu}_0$ (Eq. (17.7)). This constant being defined in the dilute limit, we may exploit the single chain result (see Appendix A 15.1)

$$e^{\mu_p(n)} = \frac{\Omega c_p(n)}{\mathcal{Z}^{(1)}(n)} \quad .$$

Restricting ourselves to the excluded volume limit we may use the exact expression (A 16.25):

$$\frac{\ell^d}{\Omega}\mathcal{Z}^{(1)}(n) = const\ n^{\gamma-1}e^{-\mu_s^* n}$$

to find

$$\bar{\mu}_0 = \mu_s^* N + \sum_n P(n)\ln\left(n^{1-\gamma}P(n)\right) + const \quad . \tag{A 17.18}$$

As a result Eq. (17.6) yields

$$\bar{\mu}(c_p) = \mu_s^* N + \sum_n P(n)\ln\left(n^{1-\gamma}P(n)\right) + \ln \ell^d c_p$$
$$+ const + \bar{\mu}_1^*(\hat{s},[p]) \tag{A 17.19}$$

$$\bar{\mu}_1^*(\hat{s},[p]) = \mathcal{P}^*(\hat{s},[p]) + \int_0^{\hat{s}} \frac{dt}{t}\mathcal{P}^*(t,[p]) \quad . \tag{A 17.20}$$

In the dilute limit we immediately find from Eq. (17.30)

$$\bar{\mu}_1^* = a_2^{*\Pi}\,\hat{s} + O(\hat{s}^2) \quad . \tag{A 17.21}$$

In the semidilute limit Eq. (17.36) yields

$$\bar{\mu}_1^* \to 3\nu A_\infty^* \hat{s}^{1/(3\nu-1)} \quad . \tag{A 17.22}$$

We thus find for the average chemical potential per segment

$$\lim_{N\to\infty} \frac{\bar{\mu}(c_p)}{N} = \mu_s^* + 3.49\left(B^3 c\right)^{1.309} \quad , \tag{A 17.23}$$

where we inserted $A_\infty^* = 1.98$ (Eq. (17.42)). This result is valid in the semidilute excluded volume limit, independent of polydispersity. The full concentration or temperature dependence of $\bar{\mu}$ closely resembles that of $\mathcal{P}$ and can be discussed in the same style.

A 17.3.2 Free energy excess and related quantities. We consider here the excluded volume limit of Eq. (17.9):

$$\frac{1}{c_p}\Delta F = \ln \ell^3 c_p + \bar{\mu}_0 - 1 + \int_0^{\hat{s}} \frac{dt}{t}\, \mathcal{P}^*(t,[p]) \ . \qquad \text{(A 17.24)}$$

In the semidilute limit this expression again reduces to a function of the segment concentration only:

$$\Delta F = c\left[\mu_s^* + 1.51\left(B^3 c\right)^{1.309}\right] \ . \qquad \text{(A 17.25)}$$

Other important quantities are the entropy excess per unit volume

$$\Delta\,\hat{S} = -\frac{\partial \Delta \hat{F}}{\partial T}$$

or the excess of the internal energy

$$\begin{aligned}\Delta\,\hat{E} &= \Delta\,\hat{F} + T\,\Delta\,\hat{S} \\ &= -T^2\,\frac{\partial}{\partial\,T}\,\frac{\Delta\,\hat{F}}{T} \ .\end{aligned}$$

(Recall $\Delta\,\hat{F} = k_B T\,\Delta\,F$, etc.) In the dilute limit such quantities have been discussed in Appendix A 16.3. In the semidilute excluded volume limit the results are of the structure (A 17.25). To give an example

$$\frac{\Delta\,\hat{E}}{k_B T} = -c\left[T\,\frac{\partial \mu_s^*}{\partial\,T} + 5.93\left(T\,\frac{\partial}{\partial\,T}\ln B\right)\left(B^3 c\right)^{1.309}\right] \ .$$

Thus the power law $c^{1/(3\nu-1)} = c^{1.309}$ is characteristic to the semidilute excluded volume regime for all thermodynamic quantities. The temperature and concentration crossover of all these quantities qualitatively is the same as that of $\mathcal{P}$.

A 17.4 Relation to the Flory-Huggins Parameter

Flory and Huggins independently presented a mean field theory for the thermodynamics of polymer solutions. Their approach was meant to hold for concentrated solutions, but has also been used to analyze data in the semidilute or even the dilute regime.

If specialized to small concentrations the results of Flory-Huggins theory in our notation read

$$\frac{\hat{\Pi}}{k_B T c_p} = 1 + \left(\frac{1}{2} - \chi\right)\Omega_0 c N \qquad \text{(A 17.26)}$$

$$\bar{\mu}_{FH}(c_p) = \ln(\Omega_0 c_p) + \mu_0 + (1 - 2\chi)\Omega_0 cN \tag{A 17.27}$$

$$\frac{\Delta \hat{F}_{FH}}{k_B T c_p} = \ln(\Omega_0 c_p) + \mu_0 - 1 + \left(\frac{1}{2} - \chi\right) \Omega_0 cN \tag{A 17.28}$$

$$\mu_0 = 1 + \ln N + (\chi - 1)N \ . \tag{A 17.29}$$

Here Ω_0 denotes the volume of a monomer and χ is the Flory-Huggins interaction parameter, which depends on temperature, but in the original version of the theory was assumed to be independent of chain length or concentration. It is easily checked that under this condition Eqs. (A 17.26)–(A 17.29) obey the Gibbs-Duhem relation as well as the other thermodynamic relations among $\Delta\hat{F}, \bar{\mu}$, or $\hat{\Pi}$. If χ is allowed to depend on concentration, different χ-parameters $\chi^{\Pi}, \chi^{\mu}, \chi^{F}$ have to be introduced such that the thermodynamic relations are fulfilled.

A glance to Sect. 5.4 reveals that the Flory-Huggins approximation is identical to the unrenormalized tree approximation, provided we identify $\Omega_0 = \ell^3, (1 - 2\chi)\Omega_0 = u_0$, and we replace μ_0 by $3/2 \ln(4\pi)$. The latter replacement takes into account a difference in the normalization of the single chain partition function. Giving in principle the effect of mixing polymer and solvent, Flory-Huggins theory normalizes the partition function with the partition function of a polymer chain in the melt. $\bar{\mu}_{FH}$ or $\Delta\hat{F}_{FH}$ therefore coincide with $\bar{\mu}, \Delta\hat{F}$ as discussed here up to an additive contribution $\mu_0 - 3/2 \ln(4\pi)$.

Comparing Eqs. (A 17.26)–(A 17.29) to the results of the renormalized formalism we find expressions for the effective Flory-Huggins parameters:

$$\frac{1}{2} - \chi^{\Pi} = \frac{1}{\ell^3 cN} \mathcal{P}_R(c_{pR}, N_R, [p], f) \tag{A 17.30}$$

$$\frac{1}{2} - \chi^{\mu} = \frac{1}{2\ell^3 cN} \left[\bar{\mu}_R(c_{pR}, N_R, [p], f) - \bar{\mu}_0 - \ln(\ell^3 c_p)\right] \tag{A 17.31}$$

$$\chi^F = 2\chi^{\mu} - \chi^{\Pi} \ . \tag{A 17.32}$$

Clearly the relation of the Flory-Huggins parameters to the universal scaling functions is messed up by nonuniversal factors needed for transverting c, N into appropriate scaling variables. Specifically, in the excluded volume limit we find

$$\frac{1}{2} - \chi^{\Pi} = \frac{B^3}{\ell^3} N^{3\nu-2} \frac{\mathcal{P}^*}{\hat{s}} \underset{\hat{s}\to\infty}{\longrightarrow} A^*_{\infty} \frac{B^3}{\ell^3} \left(B^3 c\right)^{\frac{1}{3\nu-1}-1} \tag{A 17.33}$$

$$\frac{1}{2} - \chi^{\mu} = \frac{B^3}{\ell^3} N^{3\nu-2} \frac{\bar{\mu}(c_R) - \mu_0 - \ln \ell^3 c_p}{2\hat{s}} \underset{\hat{s}\to\infty}{\longrightarrow} \frac{3}{2}\nu A^*_{\infty} \frac{B^3}{\ell^3} \left(B^3 c\right)^{\frac{1}{3\nu-1}-1} \tag{A 17.34}$$

In the semidilute limit the effective parameters $\frac{1}{2} - \chi^{\Pi}, \frac{1}{2} - \chi^{\mu}$ therefore differ by a factor $3\nu/2 = .882$.

Equations (A 17.33), (A 17.34) show that in the strict excluded volume limit $N \to \infty, \hat{s}$ fixed, the Flory Huggins parameters tend to 1/2, the correction being of order $N^{3\nu-2}$. This is interesting, since according to the traditional interpretation $\chi \to \frac{1}{2}$ implies the vanishing of the interaction and thus the Θ-limit. It shows that a naive interpretation of data in terms of χ-parameters may lead to wrong conclusions.

To summarize, RG theory identifies the χ-parameters as nonuniversal chain-length, concentration- and temperature-dependent quantities. Analyzing experiments at small concentrations with the help of χ-parameters just amounts to a reparameterization of the data, which furthermore may yield misleading results.

We finally note some further aspect of interest. We may naively extend RG theory beyond its region of applicability by increasing the concentration beyond $c \sim 10\%$ in volume. For weak coupling we then formally run along a path of type 1 in the crossover diagram of Fig. 13.2, eventually ending in the semidilute Θ-fixed-point. As a result we recover naive Flory-Huggins theory. The osmotic pressure, for instance, takes the form

$$\frac{\hat{\Pi}}{k_B T c_p} = 1 + \frac{\tilde{u}}{2}\,\tilde{s}\tilde{z} = \frac{1}{2}\,\tilde{u} v_T \frac{T-\Theta}{\Theta} \cdot \tilde{\ell}^3 c N \quad ,$$

leading to the identification

$$(1-2\chi)\Omega_0 = \tilde{u} v_T \frac{T-\Theta}{\Theta} \tilde{\ell}^3 \quad .$$

It should thus be possible to set up a heuristic theory interpolating between RG theory for small concentrations and Flory Huggins theory for large concentrations. Clearly in such a theory $\tilde{\ell}^3 c$ should be replaced by the appropriate volume fraction and three-body forces should be included in some empirical way. No theoretical effort in that direction seems to have been made, but a good discussion from the experimental side may be found in [NHKF84].

Notes

For the osmotic pressure the scaling law in the excluded volume limit was first derived by Des Cloizeaux [Clo75], this work laying the foundation of RG theory for solutions of finite overlap. The scaling of the scattering function has been established in [SW77]. That early work used the field theoretic equilibrium ensemble with its broad and concentration dependent chain length distribution. The generalization to arbitrary polydispersity was established in [SW80]. The first calculation of the thermodynamic scaling function is due to [KSW81], this work going to two-loop order in the field theoretic ensemble. Subsequently there have been published the results of one-loop calculations for arbitrary polydispersity [OO82],[Sch82],[ON83], [Sch84],[NO85]. In technical details that work differs from the present approach, but it clearly is equivalent in the sense of the ε-expansion. Also the numerical results for $d = 3$ are not too different. ε-Expansions for critical ratios like $A^*_{3/2}$ or $\mathcal{R}_\mathcal{P}$ have been given in [CN82]. A detailed analysis of osmotic pressure data may be found in [Sch82]. An extensive discussion of the entropy may be found in [Eld81].

18. Density Autocorrelations at Finite Overlap

The density autocorrelation function $I_{d,a}(\mathbf{q})$ measures the conformation of a probe chain in a solution containing an arbitrary concentration of other chains. In the dilute limit it of course coincides with the density correlation function of a single chain, as analyzed in Chap. 15. In the semidilute limit it reflects the basic features of the concentration blob model. As has been discussed in Sect. 9.1 we may view the chain as a sequence of concentration blobs of typical spatial extension ξ_E. This blob size or screening length measures the range of nontrivial excluded volume correlations in the solution. Roughly speaking, in the excluded volume limit two segments j_1, j_2 found within spatial distance $|\mathbf{r}_{j_2} - \mathbf{r}_{j_1}| < \xi_E$, most likely are part of the same chain. ξ_E depends on concentration and temperature, but is independent of the chain length. It thus defines a new scale, characteristic of the semidilute limit and small compared to the radius of gyration R_g. According to the blob model correlations on scale $r \gg \xi_E$ are Gaussian, since the interaction among the blobs is screened. Pairs of segments deep inside a blob, however, feel the full interaction and show excluded volume correlations like in an isolated chain.

Our calculations support and sharpen these ideas. Our results are best expressed in terms of the normalized scaling function

$$J_a(\mathbf{q}) = \frac{I_{d,a}(\mathbf{q})}{Nc} \quad ,$$

which in the dilute limit reduces to $J_d(\mathbf{q})$ introduced in Sect. 15.1. We find as follows.

(i) In the semidilute limit $J_a(\mathbf{q})$ for momenta $0 \le qR_g \ll R_g/\xi_E$ reduces to the Debye function:

$$J_a(\mathbf{q}) \to D_p\left(\frac{\tilde{p}_2}{\tilde{p}_3}\, q^2 R_g^2\right) \quad ,$$

where R_g is the proper chain-length-, concentration-, and temperature-dependent radius of the probe chain.

(ii) For momenta $q\xi_E \gg 1$ and for *all overlap* $J_a(\mathbf{q})$ reduces to the single chain correlation function $J_d(\mathbf{q})$.

(iii) In the intermediate range $1 \ll qR_g \lesssim R_g/\xi_E$ there exist small but characteristic deviations from the Debye function, which reflect the residual interaction among the blobs.

Findings (i) or (ii) verify the blob model. They are expected to be rigorous. The intermediate range (iii) of course is not handled adequately by simple blob considerations.

Another important quantity to be considered here is the radius of gyration of the probe chain. For simplicity restricting the present discussion to the excluded volume limit we again first recall the blob model result (Sect. 9.1): In a semidilute system R_g behaves as

$$R_g^2 \sim c^{\frac{1-2\nu}{3\nu-1}} N \ .$$

The crossover from the dilute limit $R_g^2 \sim N^{2\nu}$ is driven by increasing the segment concentration c, and it has been speculated that R_g^2 stays close to its isolated chain value, up to a concentration c^* where the chains start overlapping. There a rapid decrease should set in. Our calculation verifies the asymptotic law, but does not support that picture of the crossover. Rather, as shown in Fig. 18.1, R_g^2 rapidly decreases with increasing c, without any sign of an initial plateau. In fact, the slope $\partial R_g^2/\partial c$ takes its maximum value at $c = 0$. Furthermore the crossover proceeds most smoothly and does not allow for a sensible definition of a critical concentration c^*, where a qualitative change from dilute to semidilute behavior should take place. A similar result has been noted in the previous chapter for the concentration crossover of thermodynamic quantities. We thus again find that the blob model is stretched too far, if taken to predict a quantitative form of the crossover scaling functions.

As is clear from the above discussion, the density autocorrelation function directly checks the blob picture and thus is of much interest for our understanding of solutions at finite concentration. Unfortunately an experimental determination meets with considerable difficulties, which in fact are

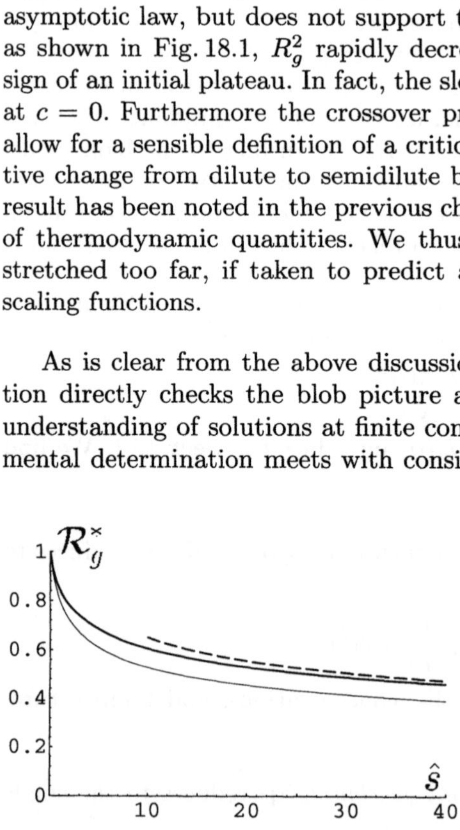

Fig. 18.1. $\mathcal{R}_g^*(\hat{s})$ as function of $\hat{s}$ for the monodisperse (*thick line*) or the exponential (*thin line*) ensemble. For a monodisperse system the asymptotic power law is given by the *broken line*

a matter of principle. To measure $I_{d,a}(\mathbf{q})$ we have to label a few chains in the solution, and this labeling is not allowed to change any chemical or physical properties of the chain except for its scattering cross section. We may use isotopic labeling, i.e. a solute consisting of a mixture of deuterated or protonated chains. Deuteration strongly changes the neutron scattering cross section, and comparing the scattering signal from solutes of different relative composition we may hope to extract the autocorrelation function. In this way for instance the conformation of a chain in the melt has been determined. Now, deuteration clearly does not change the chemical properties, but it somewhat influences the excluded volume interaction, as can be seen from the fact that it slightly changes the Θ-temperature. A more detailed analysis needs the framework of the theory of ternary polymer solutions containing two different polymer species in a common solvent (see Chap. 22). There it is found that $I_{d,a}(\mathbf{q})$ can be reliably measured only in the excluded volume limit, where any difference in the microscopic interactions is suppressed by renormalization. Outside that limit $I_{d,a}(\mathbf{q})$ is not directly accessible in real experiments, and even the method of isotopic labeling in general will not allow for an experimental check of the crossover towards Θ-conditions. In this chapter we therefore restrict our comparison with experiments to good solvent systems. We find reasonable agreement among the theory and the (very few) experimental data.

This chapter is organized as follows. In Sect. 18.1 we recall the definition and the scaling properties of the autocorrelation function. Explicit one-loop evaluations of R_g^2 or of $I_{d,a}(\mathbf{q})$ are the topic of Sects. 18.2 or 18.3, respectively. The experimental situation is discussed in Sect. 18.4. The complicated and lengthy one-loop results in their full analytic form are summarized in the appendix.

18.1 Definition and scaling properties of $I_{d,a}(\mathbf{q})$

We define $I_{d,a}(\mathbf{q})$ as the sum of the autocorrelations of all individual chains in the solution (cf. Sect. 5.1.4):

$$I_{d,a}(\mathbf{q}) = \frac{1}{\Omega} \int d^d r \, d^d r' \, e^{i\mathbf{q}(\mathbf{r}-\mathbf{r}')} \left\langle \sum_m \hat{\rho}_m(\mathbf{r}) \hat{\rho}_m(\mathbf{r}') \right\rangle . \tag{18.1}$$

Recall that $\hat{\rho}_m(\mathbf{r})$ denotes the local segment density of the m-th chain. The definition (18.1) sums over all the chains in the solution, but a little formal manipulation shows that $I_{d,a}/c_p$ indeed gives the scattering from a single marked chain, averaged over the chain length distribution $P(n)$.

$I_{d,a}(\mathbf{q})$ obeys the sum rule

$$I_{d,a}(\mathbf{0}) = \tilde{p}_2 c_p N^2 \quad , \tag{18.2}$$

valid for all concentrations (cf. Eq. (5.24)). We therefore introduce the normalized correlation function

$$J_a(\mathbf{q}) = \frac{I_{d,a}(\mathbf{q})}{c_p N^2} \quad , \tag{18.3}$$

which renormalizes without any renormalization prefactors:

$$J_a(\mathbf{q}) = J_{aR}(\bar{\mathbf{q}}, c_{pR}, N_R, f; [p]) \tag{18.4}$$

(cf. Sect. 11.7.1).

Exploiting the general form of the RG-mapping we can immediately write down the scaling law

$$J_a(\mathbf{q}) = \tilde{J}_a(q^2 \tilde{R}_0^2, \tilde{s}, \tilde{z}; [p]) \quad , \tag{18.5}$$

where (cf. Sect. 13.3.1) $\tilde{R}_0 = \tilde{\ell} N^{1/2}, \tilde{s} = c_p \tilde{R}_0^3, \tilde{z} = \tilde{v} N^{1/2}$. In the excluded volume limit $\tilde{z} \to \infty$ the scaling law reduces to

$$J_a(\mathbf{q}) = J_a^*(q^2 R_g^2(N,0), \hat{s}; [p]) \quad . \tag{18.6}$$

Here and in the sequel $R_g^2(N,0)$ denotes the radius of gyration of the chain at zero concentration in the excluded volume limit.

18.2 The radius of gyration

From the definition

$$R_g^2 = -\frac{1}{2} \Delta_q \ln I_{d,a}(\mathbf{q}) \Big|_{q=0} \tag{18.7}$$

we immediately find the result of Eq. (A 18.7), which in three dimensions reads

$$\begin{aligned} R_g^2 &= \frac{\tilde{p}_3}{\tilde{p}_2} \ell_R^2 N_R \left[1 - u^* f \left(1 + \frac{1}{2} N_R^{1/2} \mathcal{M}_a(0, W_R) \right) \right] \\ W_R &= f c_R N_R \quad . \end{aligned} \tag{18.8}$$

The one-loop correction $\mathcal{M}_a(0, W_R)$ is given by Eq. (A 18.8). Explicit expressions for the exponential or the monodisperse ensemble are given in Eqs. (A 18.11) or (A 18.13), respectively.

We first consider the excluded volume limit $f = 1$, which results in the scaling form

$$\frac{R_g^2(N,\hat{s})}{R_g^2(N,0)} = \mathcal{R}_g^*(\hat{s}) \quad . \tag{18.9}$$

To one-loop order Eq. (18.8) yields

$$\mathcal{R}_g^*(\hat{s}) = w^{4\nu-2}\frac{0.636 - 0.182N_R^{1/2}\mathcal{M}_a(0,W_R)}{0.636 - 0.182n_0^{1/2}\mathcal{M}_a(0,0)} \quad . \tag{18.10}$$

Recall the relations (cf. Eqs. (17.31) or Sect. 13.3.2, for instance)

$$N_R = \frac{n_0}{w^2}; \quad W_R = c_0 n_0\left(\frac{1}{w^2}-1\right)$$
$$\frac{n_0}{c_0}\tilde{u}\hat{s} = w^{2-6\nu}(1-w^2) \quad ,$$

valid for $q = 0$. Recall further that $\hat{s} = B^3 c N^{3\nu}$ up to a numerical constant equals the geometric overlap in the solution: $\frac{4\pi}{3}c_p R_g^3(N,0) = 1.169\hat{s}$ (monodisperse R_g), so that $\mathcal{R}_g^*(\hat{s})$ is a fully universal observable function depending on overlap and polydispersity.

From Eqs. (18.10), (A 18.14) we easily can determine the virial expansion of $\mathcal{R}_g^*(\hat{s})$.

$$\mathcal{R}_g^*(\hat{s}) = \begin{cases} 1 - 0.331\hat{s} + 0.447\hat{s}^2 + O(\hat{s}^3); & [m] \\ 1 - 0.380\hat{s} + 0.384\hat{s}^2 + O(\hat{s}^3); & [e] \end{cases} \quad , \tag{18.11}$$

where $[m]$ or $[e]$ stand for the monodisperse or the exponential ensemble, respectively. Irrespective of polydispersity $\mathcal{R}_g^*(\hat{s})$ shows its maximal slope for $\hat{s} \to 0$, so that $R_g^2(N,\hat{s})$ starts decreasing rapidly with increasing overlap, with no indication of a plateau. In the semidilute limit Eq. (A 18.12) yields the asymptotic behavior

$$\mathcal{R}_g^*(\hat{s}) \underset{\hat{s}\to\infty}{\longrightarrow} \hat{s}^{-\frac{2\nu-1}{3\nu-1}} \cdot \begin{cases} 1.103 & [m] \\ 0.922 & [e] \end{cases} \quad . \tag{18.12}$$

The structure of that limit is somewhat more explicit, if we consider $R_g^2(N,\hat{s})$ directly. Equation (18.8) yields

$$R_g^2(N,\hat{s}) = \frac{\tilde{p}_3}{\tilde{p}_2}B^2 N \cdot 0.471\left(B^3 c\right)^{-\frac{2\nu-1}{3\nu-1}} \quad , \tag{18.13}$$

a result as expected from the blob model. Polydispersity- and chain length-dependence are identical to those for a noninteracting chain, the interaction essentially introducing a power of the concentration.

Using Eqs. (A 18.11) or (A 18.13) we can determine the crossover function $\mathcal{R}_g^*(\hat{s})$ for all overlap. Figure 18.1 shows the result. As expected we see a monotonous decrease, without any special structure.

So far we have considered the excluded volume limit. Outside that limit we may write the scaling law as

$$\frac{R_g^2}{\tilde{R}_0^2} = \tilde{\mathcal{R}}_g(\tilde{s}, \tilde{z}; [p]) \quad . \tag{18.14}$$

Restricting ourselves to a monodisperse system we find from Eq. (18.8) together with the RG-mapping

$$\begin{aligned} \tilde{\mathcal{R}}_g(\tilde{s}, \tilde{z}; [m]) &= \Big|1-f\Big|^{\frac{1}{\omega}\left(\frac{1}{\nu}-2\right)} H(f) \left[1 - u^* f \left(1 + \frac{n_0^{1/2}}{2w} \mathcal{M}_a(0, W_R)\right)\right] \\ \frac{\tilde{u}}{c_0 n_0^{1/2}}\, \tilde{s} &= f^{-1}\Big|1-f\Big|^{\frac{3}{\omega}\left(1-\frac{1}{2\nu}\right)} H^{-3/2}(f) \frac{1-w^2}{w^2} \\ \frac{\tilde{z}}{n_0^{1/2}} &= f\Big|1-f\Big|^{-\frac{1}{2\nu\omega}} H_n^{-1/2}(f) w^{-1} \quad . \end{aligned} \tag{18.15}$$

(cf. Sect. 13.3.1). It is easily checked that $\tilde{\mathcal{R}}_g(\ldots)$ in the semidilute limit $w \to 0$ becomes a function of $\tilde{z}/\tilde{s} = \tilde{v}(T)(\tilde{\ell}^3(t)c)^{-1}$, only. This limit is approached fairly rapidly with increasing $\tilde{s}$. The concentration dependence at fixed T, N is shown in Fig. 18.2. Note that $\hat{s}$ here is calculated according to Eq. (17.51): $\hat{s} = 2.776\tilde{s}\tilde{z}^{3(2\nu-1)}$.

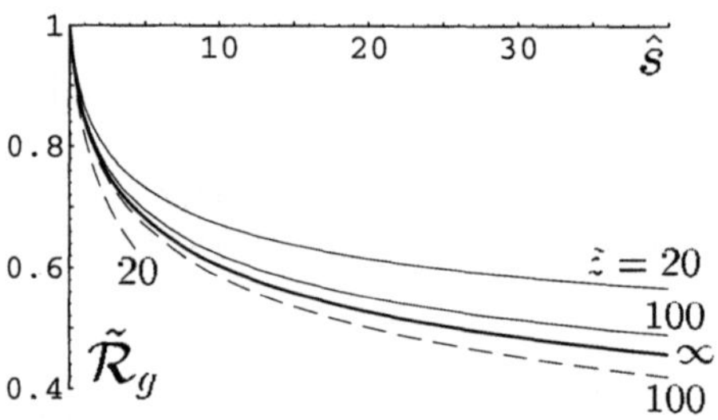

Fig. 18.2. $\tilde{\mathcal{R}}_g(\tilde{s}, \tilde{z}, [m])$ as function of $\hat{s}$ for several values of $\tilde{z}$ both for weak (*thin lines*) and strong (*broken lines*) coupling. The *thick line* represents the excluded volume limit

18.3 Momentum dependence

Expecting the form of $J_a(\mathbf{q})$ to be roughly given by the (polydispersity dependent) Debye function we follow our treatment of the density correlations of a single chain in writing

$$J_a(\mathbf{q}) = D_p(\hat{Q}) \quad . \tag{18.16}$$

All the dependence on momentum, overlap, etc. is absorbed into the function

$$\hat{Q} = \hat{Q}(\bar{\mathbf{q}}^2, c_{pR}, N_R, f; [p]) \quad .$$

To one loop-order Eq. (A 18.5), $d = 3$, yields

$$\hat{Q} = Q_R \left[1 - u^* f \left(1 + \frac{1}{2} N_R^{1/2} \mathcal{M}_a(Q_R, W_R)\right)\right] , \tag{18.17}$$

where $Q_R = \bar{\mathbf{q}}^2 N_R$, and the one-loop correction $\mathcal{M}_a(\ldots)$ is given in Eq. (A 18.6).

The dilute limit of $J_a(\mathbf{q})$ in detail has been discussed in Chap. 15. It was found to follow closely the Debye function up to momenta of order $q^2 R_g^2(N,0) \approx 20$ (excluded volume limit). For larger momenta $J_a(\mathbf{q})$ decreases less rapidly than the Debye function, and in the excluded volume limit it approaches a power law $J_a(\mathbf{q}) \sim \left[q^2 R_g^2(N,0)\right]^{-\frac{1}{2\nu}}$. Of course deviations from the Debye approximation decrease if we approach the Θ-temperature.

18.3.1 The semidilute limit

In the semidilute limit the solution besides R_g shows the blob size or screening length ξ_E as another important length scale. A precise definition of ξ_E will be given in the next chapter. (See Sect. 19.3.) For the present purpose it is sufficient to invoke the blob picture to note that ξ_E, giving the range of pronounced excluded volume correlations, is independent of N but depends on concentration and temperature. Specifically, in the excluded volume limit

$$\xi_E^2 \sim B^2 \left(B^3 c\right)^{-\frac{2\nu}{3\nu-1}} .$$

Since the semidilute limit can be interpreted as the limit $N \to \infty, B^3 c \ll 1$ fixed, it implies $\xi_E \ll R_g$. More precisely

$$\xi_E^2 / R_g^2 \sim \left(B^3 c\right)^{-\frac{1}{3\nu-1}} \frac{1}{N} \to 0 \tag{18.18}$$

(cf. Eq. (18.13)). Due to the occurrence of ξ_E as a second scale, $J_a(\mathbf{q})$ shows a richer structure, which we now discuss.

We first keep $q^2 R_g^2$ fixed, while taking the semidilute limit. We therefore concentrate on length scales large compared to ξ_E. In renormalized variables this amounts to the limit $N_R \to \infty, Q_R$ fixed. From the results of the appendix it is easily found that in this limit the one-loop correction $\mathcal{M}_a(Q_R, W_R)$ tends to $\mathcal{M}_a(0, W_R)$. As a consequence, the function $\hat{Q}$ (Eqs. (18.17), (A 18.5)) tends to $\frac{3}{d}\tilde{p}_2/\tilde{p}_3 q^2 R_g^2(N, \tilde{s}, \tilde{z}; [p])$ (cf. Eq. (A 18.7)). This proves the result announced in the introductory section:

$$\lim_{\tilde{s}\to\infty,\, q^2 R_g^2 = const.} J_a(\mathbf{q}) = D_p\left(\tilde{p}_2/\tilde{p}_3 q^2 R_g^2(N, \tilde{s}, \tilde{z}; [p])\right) . \tag{18.19}$$

Note that this result holds strictly for arbitrary, but finite, values of $q^2 R_g^2$ in a system of infinite overlap. Taking the semidilute limit in this special way we shift all the deviations from the Debye functions towards $q^2 R_g^2 = \infty$. The Debye function is thus approached not only for momenta of order R_g^{-1}, but also the 'large momentum' tail $D_p(\hat{Q}) \to 2/\hat{Q}$ should be observable, for any strength of the excluded volume.

We now concentrate on length scales of order ξ_E, taking the semidilute limit with fixed $q^2 \xi_E^2$. In terms of renormalized variables this amounts to the limit $N_R \to \infty, \bar{\mathbf{q}}^2$ fixed. The one-loop correction $\mathcal{M}_a(Q_R, W)$ for $d = 3$ can be evaluated analytically and yields the result of Eq. (A 18.17). For simplicity we restrict the evaluation to the excluded volume limit. The relevant form of the RG-mapping is given in Sect. 13.3.2, Eqs. (13.34), (13.35), where we have to take the semidilute limit $w \to 0$. We furthermore from the next chapter take the quantitative result for ξ_E:

$$\xi_E^2 = 0.639 B^2 \left(\frac{n_0}{c_0} \tilde{u} B^3 c \right)^{-\frac{2\nu}{3\nu-1}} . \tag{18.20}$$

After a little calculation we find as final result

$$\begin{aligned} \hat{s}^{-\frac{1}{3\nu-1}} J_a^{-1} = \frac{n_0}{2} \left(\frac{n_0}{c_0} \tilde{u} \right)^{\frac{1}{3\nu-1}} \bar{\mathbf{q}}^2 \left(1 - \frac{\bar{\mathbf{q}}^2}{q_0^2} \right)^{-\frac{1}{3\nu-1}} \cdot \Bigg\{ 1 - u^* + \frac{u^*}{2} \\ \left(\frac{\pi}{2c_0} \right)^{1/2} \left(1 - \frac{\bar{\mathbf{q}}^2}{q_0^2} \right)^{-1/2} \Bigg[\frac{2}{Q_1} - 2 + \pi Q_1^{1/2} + \left(2 - \frac{1}{Q_1} - Q_1 \right) Q_1^{-1/2} \\ \cdot \arccos \left(\frac{1 - Q_1}{1 + Q_1} \right) \Bigg] \Bigg\} , \end{aligned} \tag{18.21}$$

where

$$Q_1 = \frac{\bar{\mathbf{q}}^2}{2c_0 (1 - \bar{\mathbf{q}}^2 / q_0^2)} .$$

The intermediate variable $\bar{\mathbf{q}}$ is related to $q\xi_E$ via

$$(q\xi_E)^2 = 0.639 \bar{\mathbf{q}}^2 \left(1 - \frac{\bar{\mathbf{q}}^2}{q_0^2} \right)^{-\frac{2\nu}{3\nu-1}} . \tag{18.22}$$

Note that in terms of the measured function $I_{d,a}(\mathbf{q})$ this result reads

$$I_{d,a}(q) = const\, c^{\frac{3\nu-2}{3\nu-1}} F_a(q^2 \xi_E^2) ,$$

where the function F_a is independent of chain length and polydispersity. In full accord with the blob model the correlations on scale ξ_E are independent of the overall size of the chains. In taking this form of the semidilute limit we have shrunk to zero the polydispersity and chain length dependent range $q^2 R_g^2 < \infty$ considered above.

The result (18.21) shows some peculiar behavior in the limit of small $q^2\xi_E^2$. Upon expanding we find

$$\hat{s}^{-\frac{1}{3\nu-1}} J_a^{-1}(\mathbf{q}) = 2.623 q^2\xi_E^2 + 1.163(q^2\xi_E^2)^{3/2} + O(q^4\xi_E^4) \quad . \tag{18.23}$$

In contrast to standard behavior this result is not analytic in q^2. Rather we generally expect it to order according to products of powers of q^2 and powers of $(q^2)^{1/2} = (q^2)^{1-\varepsilon/2}$. This is a typical effect of the residual interaction among the blobs. These odd powers of $(q^2)^{1/2}$ have the same origin and are the direct counterpart of the powers of $c^{1/2} = c^{1-\varepsilon/2}$, that occur in the semidilute limit for $q = 0$ (see Sect. 5.4.3). Recall that even in the semidilute limit the screened interaction vanishes only on scales of order $R_g \gg \xi_E$.

In the limit of large momenta $q^2\xi_E^2 \gg 1$ our result reduces to

$$\hat{s}^{-\frac{1}{3\nu-1}} J_a^{-1}(\mathbf{q}) = 3.413 \left(q^2\xi_E^2\right)^{\frac{1}{2\nu}} \quad . \tag{18.24 i}$$

Using Eq. (18.20) we easily find that this is identical to the result (15.34) found for $q^2R_g^2 \gg 1$ in the dilute excluded volume limit:

$$J_a^{-1}(\mathbf{q}) = 0.439 \left(q^2B^2N^{2\nu}\right)^{\frac{1}{2\nu}} \quad . \tag{18.24 ii}$$

Indeed, from the large momentum behavior of the one-loop correction, Eq. (A 18.16) it is easily checked that $J_a^{-1}(\mathbf{q})$ for $q \to \infty$ becomes independent of overlap or polydispersity, the scattering being due to parts of a single chain deep inside a blob. Thus Eq. (18.24 ii) is valid in the excluded volume limit independent of the overlap, which only determines the momentum range where this behavior is attained.

The combination $\hat{s}^{-\frac{1}{3\nu-1}} J_a^{-1}(\mathbf{q})/(2.632q^2\xi_E^2)$ in the semidilute excluded volumed limit is plotted in Fig. 18.3 as function of $q^2\xi_E^2$. We divided by the behavior for $q \to 0$ (cf. Eq. (18.23)) to show clearly the effect of the q^3-term. It leads to a rapid increase and severely restricts the region, where the approximation $J_a^{-1} \sim q^2\xi_E^2$ is accurate within a few percent. The overshooting, however, is not dramatic, since J_a^{-1} for $q^2\xi_E^2 \gtrsim 1$ bends over towards the asymptotic power law (18.24). In the present plot this yields the asymptotic behavior $J_a^{-1}/(q^2\xi_E^2) \sim \left(q^2\xi_E^2\right)^{\frac{1}{2\nu}-1} \sim \left(q^2\xi_E^2\right)^{-0.150}$. Recall that in this plot the region $q^2R_g^2 < \infty$ is shrunk to zero. The square root singularity, seen here, will be washed out on scale $q^2 \sim R_g^{-2}$.

18.3.2 General overlap

The above discussion of the semidilute limit with fixed $q^2R_g^2$ might suggest that with increasing overlap $J_a(\mathbf{q})$ approaches the Debye function by simply

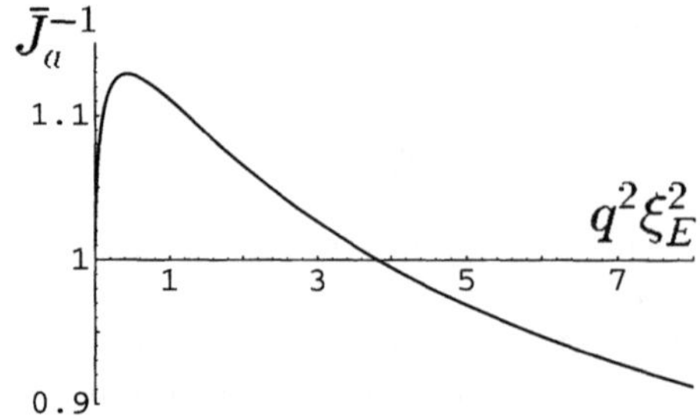

Fig. 18.3. $\bar{J}_a^{-1} = \hat{s}^{\frac{1}{3\nu-1}} J_a^{-1}(\mathbf{q}) / (2.632 q^2 \xi_E^2)$ as function of $q^2 \xi_E^2$ in the semidilute excluded volume limit

shifting to higher values the point where it bends over towards the asymptotic behavior. However the situation is more complicated. Figure 18.4 shows $D(q^2 R_g^2)/J_a(\mathbf{q})$ as function of $q^2 R_g^2$ for different values of $\hat{s}$. For dilute systems ($\hat{s} = 0.01$) in the range of small $q^2 R_g^2$ we observe the very small overshooting already discussed in Chap. 15 for isolated chains. With increasing overlap this bump increases, becoming more important even for small $q^2 R_g^2$. For $\hat{s} = 1$, for instance, it is a quite pronounced feature in all the range $q^2 R_g^2 \lesssim 50$. Only if we approach the semidilute limit the bump moves to higher q-values, and

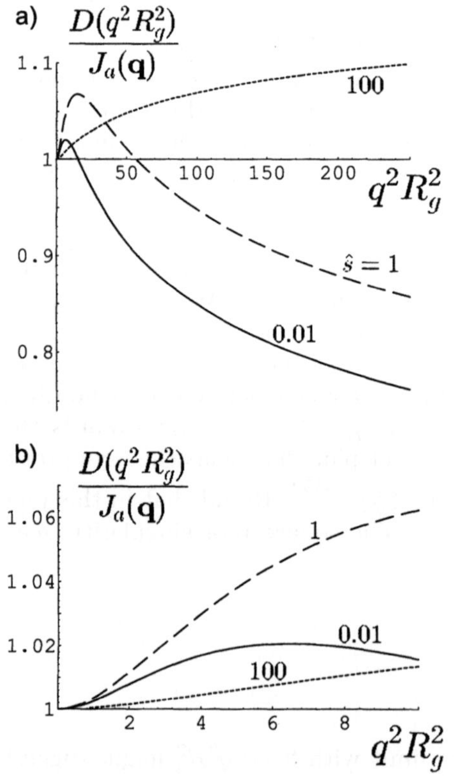

Fig. 18.4. (**a**) $D(q^2 R_g^2) J_a^{-1}(\mathbf{q})$ (monodisperse) as function of $(qR_g)^2$ in the excluded volume limit for values $\hat{s} = 0.01, 1, 100$. (**b**) Magnification of the small q region

$J_a^{-1}(\mathbf{q})$ very slowly on scale $q^2R_g^2$ approaches $D^{-1}(q^2R_g^2)$ from above. This peculiar behavior can be seen as a precursor of the term $(q^2\xi_E^2)^{3/2}$ found in the semidilute limit on scale ξ_E.

So far we have considered the excluded volume limit. Outside that limit for larger momenta we find deviations very much like those discussed in the single chain limit (cf. Fig. 15.6). Indeed, for systems not too far from the excluded volume limit the scaling function J_a as calculated in the excluded volume limit is an excellent approximation up to fairly large values $q^2R_g^2 \gtrsim 20$, provided the correct $R_g = R_g(N, \tilde{s}, \tilde{z})$ is taken. We therefore do not consider this matter further.

18.4 The experimental situation

To explain the method of isotopic labeling we first consider a normal binary solution of polymer concentration c_p. The total density correlation function

$$I_{d,c}(\mathbf{q}) = \frac{1}{\Omega}\int d^dr\, d^dr'\, e^{i\mathbf{q}(\mathbf{r}-\mathbf{r}')} \cdot \langle(\sum_m \rho_m(\mathbf{r}) - c)(\sum_{m'} \rho_{m'}(\mathbf{r}') - c)\rangle \tag{18.25}$$

can be decomposed into a single chain part $I^{(1)}(\mathbf{q})$ and the contribution $I^{(2)}(\mathbf{q})$ involving the segment densities of two different chains:

$$I_{d,c}(\mathbf{q}) = c_p I^{(1)}(\mathbf{q}) + c_p^2 I^{(2)}(\mathbf{q}) \ . \tag{18.26}$$

Clearly $I^{(1)}, I^{(2)}$ besides chain lengths etc. will depend on c_p: Equation (18.26) is an exact decomposition of the product of local densities in Eq. (18.25), not a virial expansion. Indeed, $c_p I^{(1)}(\mathbf{q}) \equiv I_{d,a}(\mathbf{q})$ is just the density autocorrelation function. $c_p^2 I^{(2)}(\mathbf{q})$ gives the interchain correlation. If we denote by ν_i the scattering increment of a polymer segment relative to the solvent, the total scattering intensity reads

$$I(q) = \nu_i^2 I_{d,c}(\mathbf{q}) \ .$$

We now extend this consideration to a solution containing two polymer species A, B, which differ in their scattering increments

$$\nu_i \rightarrow \nu_i^{(A)}, \nu_i^{(B)} \quad ,$$

but otherwise are completely identical,

$$
\begin{aligned}
I_A^{(1)} &\equiv I_B^{(1)} \equiv I^{(1)} \\
I_{AA}^{(2)} &\equiv I_{BB}^{(2)} \equiv I_{AB}^{(2)} \equiv I^{(2)} \ .
\end{aligned}
\tag{18.27}
$$

These functions clearly depend on the total concentration

$$
c_p = c_p^{(A)} + c_p^{(B)} \ , \tag{18.28}
$$

but are independent of the relative composition

$$
x = \frac{c_p^{(A)}}{c_p} \ . \tag{18.29}
$$

This feature is essential. It results from, and precisely expresses, the assumed identity of the two species. For such a solution the total scattering intensity reads

$$
\begin{aligned}
I(\mathbf{q}) &= \nu_i^{(A)2}\left(xc_pI^{(1)} + x^2c_p^2I^{(2)}\right) \\
&+ \nu_i^{(B)2}\left((1-x)c_pI^{(1)} + (1-x)^2c_p^2I^{(2)}\right) \\
&+ 2\nu_i^{(A)}\nu_i^{(B)}x(1-x)c_p^2I^{(2)} \\
&= \left(x\nu_i^{(A)2} + (1-x)\nu_i^{(B)2}\right)c_pI^{(1)} + \left(x\nu_i^{(A)} + (1-x)\nu_i^{(B)}\right)^2 c_p^2I^{(2)} \ .
\end{aligned}
\tag{18.30 i}
$$

Using Eq. (18.26) to eliminate $I^{(2)}$ in favor of the total scattering function $I_{d,c}(\mathbf{q})$ of a binary system, we can also write the result as

$$
\begin{aligned}
I(\mathbf{q}) &= \left(x\nu_i^{(A)} + (1-x)\nu_i^{(B)}\right)^2 I_{d,c}(\mathbf{q}) \\
&+ (\nu_i^{(A)} - \nu_i^{(B)})^2x(1-x)c_pI^{(1)}(\mathbf{q}) \ .
\end{aligned}
\tag{18.30 ii}
$$

Varying x, but keeping c_p fixed, we can now extract $I^{(1)}, I^{(2)}$ separately. Several methods have been employed.

(i) 'Dilute A-system'
Data are taken for a set of values $x, x \ll 1$, and the results are extrapolated to determine $I^{(1)}$ as

$$
(\nu_i^{(A)} - \nu_i^{(B)})^2c_pI^{(1)} = \lim_{x\to 0}\frac{1}{x}\left[I(q) - \nu_i^{(B)}(\nu_i^{(B)} + 2x(\nu_i^{(A)} - \nu_i^{(B)}))I_{d,c}(\mathbf{q})\right] \ . \tag{18.31}
$$

Note, that $I_{d,c}(\mathbf{q})$ is independent of the species and can be determined in a binary solution ($x = 0$, for instance). This method essentially treats the system as a dilute solution of A-chains, in an effective solvent comprising the B-chains.

(ii) 'Compensation method'
Under favorable conditions $\nu_i^{(A)}, \nu_i^{(B)}$ are of different sign, so that we can choose a composition $x = x_0$ such that the coefficient of $I_{d,c}(\mathbf{q})$ or $I^{(2)}(\mathbf{q})$ in Eqs. (18.30) vanishes. $I^{(1)}$ is determined directly, without extrapolation.

(iii) 'High concentration method'
$I(\mathbf{q})$ is measured for two convenient values of x to determine $I^{(1)}, I^{(2)}$ (or $I_{d,c}(\mathbf{q})$) by solving Eqs. (18.30).

All these methods have been used in the literature.

Clearly these methods hinge upon the identities (18.27). As we discussed in the introductory section, these identities in general do not even hold for chemically identical chains prepared from deuterated or protonated monomers. Rather we must invoke renormalization to argue that in a good solvent on large scales any difference existing on the microscopic level is suppressed, provided the system is close enough to the excluded volume limit. We must note, however, that upon increasing the overlap a ternary solution quite often separates into an A-rich and a B-rich phase. Only if the solution stays homogeneous in a parameter region close enough to the excluded volume fixed point, we will be able to extract $I_{d,a}(\mathbf{q})$ reliably.

Fortunately the high concentration method allows us to check the internal consistency of the analysis. Using more than two compositions x we can extract $I^{(1)}, I^{(2)}$ from different pairs of measurements. If Eqs. (18.27) hold, the results must be identical within the accuracy of the experiment.

We are aware of only two extensive sets of experiments relevant for the present discussion. In both cases the neutron scattering cross section was determined for isotopically labeled mixtures of polystyrene in good solvents. (carbon-disulfide or toluene, respectively). In the first set of experiments [DCF+75] the power law $I_{d,a}(\mathbf{q}) \sim q^{1/\nu}$ was found both in dilute and in semidilute solutions, as expected. Furthermore R_g could be extracted directly from $I_{d,a}(\mathbf{q})$ measured in the range $qR_g \lesssim 1$. Here $I_{d,a}(\mathbf{q})$ was determined by extrapolating the data to $x \to 0$. Also the normal dilute limit was considered, allowing for a determination of $R_g(N,0)$. We therefore can plot the data as $R_g^2(N,\hat{s})/R_g^2(N,0)$ vs $\hat{s}$, this plot involving no fit parameter and being universal, provided the system has reached the excluded volume limit. The result is shown in Fig. 18.5. Clearly the agreement among theory and data is quite satisfactory. We should note that the statistical error of the data is estimated to be quite large, being of the order of 10%.

In the second set of experiments [KBWU85],[UBK88] the data analysis proceeded according to methods (ii) or (iii) described above. Care was taken to check the validity of the approach by comparing results from different

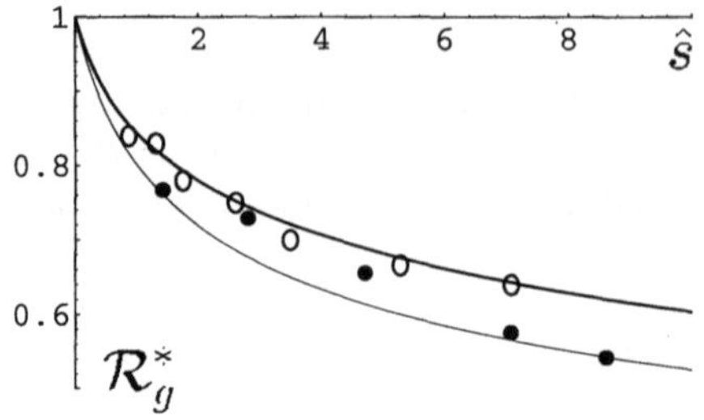

Fig. 18.5. $\mathcal{R}_g^*(\hat{s}) = R_g^2(N,\hat{s})/R_g^2(N,0)$ as function of $\hat{s}$. *Thick line*: monodisperse ensemble; *thin line*: exponential ensemble. Data: polystyrene-CS_2 [DCF+75]: *points*; polystyrene-toluene [KBWU85]: *circles*

compositions. The momentum range of the experiment, however, did not cover the region $qR_g < 1$, and the radius of gyration was determined by fitting the data to the Debye-approximation for $J_a(\mathbf{q})$ in the momentum range $1 \lesssim q^2R_g^2 \lesssim 30$. The resulting values of R_g nicely fit the excluded volume curve for $R_g^2(N,\hat{s})/R_g^2(N,0)$. (See Fig. 18.5.)

Still this latter result is somewhat misleading. First we must note that $R_g(N,0)$ was not measured. According to our evaluation the data cover a range $0.6 \lesssim \hat{s} \lesssim 7$. Thus our analysis via the estimate for $R_g^2(N,0)$ involves the nonuniversal fit parameter B, which for deuterated toluene may differ somewhat from the previously established value refering to normal toluene. Indeed, in our analysis it comes out to be about 10% larger. We should also note that for polystyrene in normal toluene indications of strong coupling behavior are found, so that for the fairly short chains ($N \approx 1000$) and high concentrations ($2\% \lesssim c \lesssim 16\%$) used here we would also expect to see deviations from the excluded volume limit. The most serious point of concern, however, relates to the way of data analysis. As is clear from Fig. 18.4, in the momentum and overlap range measured here our theory predicts definite deviations from the Debye approximation. These deviations, however, can hardly be detected without using the theoretical result, since in the relevant range our one-loop result can be brought close to the Debye function by adjusting R_g. The effective R_g would come out larger than the radius of gyration as given by our theory, the correction factor depending on the overlap. This is illustrated in Fig. 18.6, where the data of [KBWU85], are adjusted to our theory or to the Debye function, respectively. Both fits seem equally good, though a close inspection shows that the somewhat larger curvature predicted by our calculation may yield a slightly better fit. The resulting R_g-values differ by about 5%: $R_g = 109.8$ Å (Debye Model) $R_g = 104.5$ Å (one-loop calculation). Similar results are found for other samples, where $I_{d,c}(\mathbf{q})$ data are given. Taking into account the systematic trend of these corrections which would somewhat increase the slope of the data points in Fig. 18.5, we also find that these data indicate strong coupling behavior of the present system. Unfortunately a more detailed analysis is not too meaningful since additional information allowing for an independent determination of the nonuniversal parameters is missing.

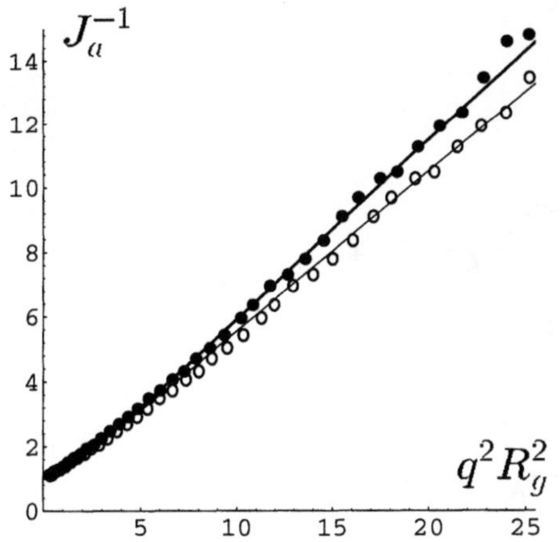

Fig. 18.6. Fit of data of [KBWU85], sample 'SS 12.5 AO', to the Debye function (*thin line, circles*) or the one-loop approximation (*thick line, points*)

We finally comment on attempts to measure the temperature crossover of R_g in semidilute systems. For isotopically labeled polystyrene in cyclohexane such a measurement has been carried through by Cotton and coworkers [CNB+76]. The data were interpreted to verify the simple scaling predictions discussed in Sect. 9.2. They however are inconsistent both with the crossover form as derived here and with results on R_g in dilute systems. A detailed analysis of the problem has been presented in [Sch93]. The clue lies in the observation that deuteration changes somewhat the Θ-temperature. It thus changes the excluded volume strength. The effect on the parameter $v_T = \tilde{v}(T)/t, t = (T - \Theta)/\Theta$, was found to be surprisingly large, but recently has been verified by new measurements [WXW+97] which for deuterated samples yield a v_T value about twice that for normal polystyrene. Thus a solution containing a mixture of deuterated and hydrogenated polymers close to $T = \Theta$ cannot be treated as a binary system. Within the framework of ternary solution theory a consistent interpretation of the data can be given. We conclude that the temperature crossover of R_g in binary solutions of finite overlap is not accessible experimentally.

A 18. Appendix

A 18.1 Full one-loop results

The diagrams contributing to $I_{d,a}(\mathbf{q})$ to one-loop order are shown in Fig. 18.7. The evaluation is straightforward, but tedious, and we only present the renormalized result:

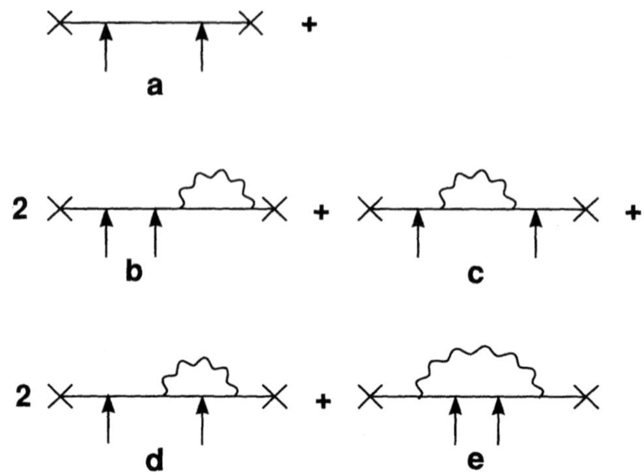

Fig. 18.7a–e. Diagrams for $I_{d,a}(\mathbf{q})$ to one-loop order

$$
\begin{aligned}
&J_{aR}(\bar{\mathbf{q}}, c_{pR}, N_R, u; [p]) = D_p(Q_R) - \frac{u^*}{\varepsilon} f\, Q_R D_p^{'}(Q_R) \\
&+\frac{u^* f}{2\Gamma(\frac{d}{2})} N_R^{\varepsilon/2} \int_0^\infty dy\, y^{-\varepsilon/2} \frac{D_p(y) - D_p(Q_R)}{Q_R - y} \left(1 + \frac{2}{\rho(y)}\right) \\
&+\frac{u^* f}{2\Gamma(\frac{d}{2})} W_R\, D^{'}(Q_R) N_R^{\varepsilon/2} \int_0^\infty dy\, \frac{y^{-\varepsilon/2}}{\rho(y)} D_p(y) \\
&-\frac{u^* f}{2\Gamma(\frac{d}{2})} N_R^{\varepsilon/2} Q_R \int_0^\infty dy\, y^{-\varepsilon/2} H_1(y, Q_R) \\
&-\frac{u^* f}{2\Gamma(\frac{d}{2})} N_R^{\varepsilon/2} \int_0^\infty dy\, \frac{y^{-1-\varepsilon/2}}{\rho(y)} \\
&\left[D_p(Q_R) - \frac{2}{Q_R^2}(\tilde{p}(y+Q_R) - \tilde{p}(y) - Q_R \tilde{p}^{'}(y))\right] \\
&+\frac{u^* f}{2}(4\pi)^{d/2} W_R\, N_R^{\varepsilon/2} \int_{\hat{k}} \frac{D_p(K)}{\rho(K)} H_1(K_1, Q_R) \\
&-u^* f (4\pi)^{d/2} N_R^{\varepsilon/2} \int_{\hat{k}} \frac{1}{\rho(K)} H_2(K, K_1, Q_R) \\
&-\frac{u^* f}{2}(4\pi)^{d/2} N_R^{\varepsilon/2} \int_{\hat{k}} \frac{1}{\rho(K)} H_1(K_1, K)\ .
\end{aligned}
\qquad \text{(A 18.1)}
$$

We introduced the notation

$$
\begin{aligned}
Q_R &= \bar{\mathbf{q}}^2 N_R; \quad W_R = f c_R N_R\ ; \\
K &= k^2 N_R, \quad K_1 = (\mathbf{k} + \bar{\mathbf{q}})^2 N_R; \quad \hat{\mathbf{k}} = \mathbf{k} N_R^{1/2}\ ,
\end{aligned}
$$

as well as the functions

$$\rho(y) = 1 + W_R D_p(y) \tag{A 18.2}$$

$$H_1(y_1, y_2) = \frac{1}{(y_1 - y_2)^2}\left[D_p(y_1) - D_p(y_2) - (y_1 - y_2)D_p'(y_2)\right] \tag{A 18.3}$$

$$H_2(y_1, y_2, y_3) = \frac{D_p(y_1)}{(y_2 - y_1)(y_3 - y_1)} + \frac{D_p(y_2)}{(y_3 - y_2)(y_1 - y_2)} + \frac{D_p(y_3)}{(y_1 - y_3)(y_2 - y_3)} . \tag{A 18.4}$$

In Eq. (A 18.1) the last three terms are due to diagrams c,d,e, respectively. Diagrams a,b together with the single chain contribution to c yield the remaining terms, where we have to take care of the fact that the transformation from $e^{\mu_p(n)}$ to $c_p(n)$ in the first diagram yields additional one-loop terms.

Introducing the intermediate momentum type quantity $\hat{Q}$ via

$$J_{aR}(\bar{\mathbf{q}}, c_{pR}, N_R, u; [p]) = D_p(\hat{Q}) \quad ,$$

we write our result in the final form

$$\hat{Q} = Q_R\left[1 - \frac{u^*}{\varepsilon} f - \frac{u^*}{2} f\, N_R^{\varepsilon/2} \mathcal{M}_a(Q_R, W_R)\right] \quad , \tag{A 18.5}$$

$$\begin{aligned} \mathcal{M}_a(Q_R, W_R) = {} & \left[\Gamma(\tfrac{d}{2}) Q_R D_p'(Q_R)\right]^{-1} \cdot \int_0^\infty dy\, y^{-\varepsilon/2} \\ & \cdot \Big[\frac{D_p(Q_R) - D_p(y)}{Q_R - y}\left(1 + \frac{2}{\rho(y)}\right) - W_R D_p'(Q_R) \frac{D_p(y)}{\rho(y)} + \frac{D_p(Q_R)}{y\,\rho(y)} \\ & - \frac{2}{y\,\rho(y) Q_R^2}(\tilde{p}(y + Q_R) - \tilde{p}(y) - Q_R \tilde{p}'(y)) + Q_R H_1(y, Q_R)\Big] \\ & + \frac{(4\pi)^{d/2}}{Q_R D_p'(Q_R)} \int_{\hat{k}} \frac{1}{\rho(K)} \Big[H_1(K_1, K) + 2H_2(K, K_1, Q_R) \\ & - W_R D_p(K) H_1(K_1, Q_R)\Big] \quad . \end{aligned} \tag{A 18.6}$$

The radius of gyration is found as

$$R_g^2 = \frac{d}{3}\frac{\tilde{p}_3}{\tilde{p}_2}\, \ell_R^2 N_R \left[1 - \frac{u^*}{\varepsilon} f - \frac{u^*}{2} f\, N_R^{\varepsilon/2} \mathcal{M}_a(0, W_R)\right] \quad . \tag{A 18.7}$$

$\mathcal{M}_a(0, W_R)$ can be written as

$$\mathcal{M}_a(0, W_R) = \frac{-6}{d\Gamma(d/2)}\frac{1}{\tilde{p}_3} \int_0^\infty dy\, \frac{y^{-\varepsilon/2}}{\rho(y)} C_p^{(0)}(y) \tag{A 18.8}$$

$$C_p^{(0)}(y) = \frac{1}{6}\frac{d^4\tilde{p}(y)}{d\,y^4} - \frac{4}{y^2} T_1[\tilde{p}''(y)] + \frac{16}{y^3} T_2[\tilde{p}'(y)] - \frac{20}{y^4} T_3[\tilde{p}(y)] \;, \tag{A 18.9}$$

where the T_j are 'Taylor operators' defined as

$$T_j[g(y)] = g(y) - \sum_{i=0}^{j} \frac{y^i}{i!}\frac{\partial^i}{\partial y^i}\Big|_0 g(y) \quad . \tag{A 18.10}$$

A 18.2 Evaluation and parameterization in three dimensions

Specializing to $d = 3$ we can for some cases evaluate the integrals of the previous section analytically. This in particular holds true for the exponential ensemble $\tilde{p}(y) = (1+y)^{-1}$. For general polydispersity certain limits, where the result becomes independent of polydispersity, can be treated.

A 18.2.1 Radius of gyration. For the exponential ensemble evaluation of Eq. (A 18.8) ($d = 3$) yields

$$\mathcal{M}_a^{[e]}(0, W_R) = -\frac{\sqrt{\pi}}{3}\left[2\left(2 - \frac{1}{2W_R^2}\right)^2 (1 + 2W_R)^{-1/2} - \frac{11}{4W_R} + \frac{13}{4W_R^2} + \frac{1}{2W_R^3} - \frac{1}{2W_R^4}\right] . \tag{A 18.11}$$

For general polydispersity only the semidilute limit $W_R \to \infty$ is accessible to an analytic calculation. Indeed, this limit is independent of polydispersity, and we find

$$\mathcal{M}_a(0, W_R) = -\frac{4}{3}\sqrt{2\pi} W_R^{-1/2} + O\left(\frac{1}{W_R}\right) , \tag{A 18.12}$$

consistent with Eq. (A 18.11).

For a monodisperse system we have evaluated $\mathcal{M}_a(0, W)(d = 3)$ numerically. The result can be parameterized as

$$\mathcal{M}_a^{[m]}(0, W_R) = -\frac{4}{3}\sqrt{\pi}(0.2 + 0.5W_R)^{-1/2}\frac{1.775 + W_R}{0.945 + W_R} + \frac{8.650}{1 + 1.979W_R} , \tag{A 18.13}$$

where as usual the analytic form chosen has no deeper meaning. The error of that parameterization is less than 0.6%.

We finally note the results in the small overlap regime:

$$\begin{aligned} \mathcal{M}_a^{[m]}(0, W_R) &= -1.276 + 0.205W_R - 0.0733W_R^2 + O(W_R^3) \\ \mathcal{M}_a^{[e]}(0, W_R) &= -2.474 + 1.145W_R - 1.014W_R^2 + O(W_R^3) \end{aligned} . \tag{A 18.14}$$

Polydispersity thus somewhat affects the small overlap regime, but comparing the results (A 18.11), (A 18.13) one finds that the difference among $\mathcal{M}_a^{[m]}, \mathcal{M}_a^{[e]}$ rapidly decreases with increasing W_R, reaching the level of 20% for $W_R \approx 18$.

A 18.2.2 Momentum dependence. For the exponential ensemble $\mathcal{M}_a(Q,W)$ can again be evaluated analytically:

$$\mathcal{M}_a^{[e]}(Q,W) = -\frac{\sqrt{\pi}}{Q}\Big\{2\frac{2W-1-Q}{(1+2W)^{1/2}} - 2 + \frac{1+Q}{W}$$
$$\left[1-\frac{1}{(1+2W)^{1/2}} + \frac{(1+2W)^{-1/2}-(1+Q)^{-1/2}}{2W-Q} + \frac{1-(1+Q)^{-1/2}}{Q}\right]\Big\}$$
$$-\frac{\sqrt{\pi}}{2W}Q^{-3/2}\Big[(1+Q)^2 \arccos\frac{4-Q}{4+Q}$$
$$-(1+Q-2W)^2 \arccos\frac{4W^2+4Q\sqrt{1+2W}-Q^2}{4W^2+4Q(1+W)+Q^2}\Big] \quad . \tag{A 18.15}$$

In general, however, only certain limits, where $\mathcal{M}_a(Q,W)$ becomes independent of polydispersity, can be treated analytically. In these limits the integrals in Eq. (A 18.6) are dominated by the range of large momenta (i.e. large y or large $\hat{k}^2$) so that the reduced chain length distribution drops out. Specifically we find in the large momentum limit $Q \to \infty$, W fixed

$$\mathcal{M}_a(Q,W) = -2\pi^{3/2}Q^{-1/2} + O\left(\frac{1}{Q}\right) \quad , \tag{A 18.16}$$

independent of W or polydispersity. Note that this result coincides with the large momentum behavior of the one-loop correction in the dilute limit, as it can be extracted from Eq. (A 15.33).

We can also find an analytical result in the semidilute limit

$$Q \to \infty, \quad W \to \infty$$
$$Q_1 = \frac{Q}{2W} \quad \text{fixed} \quad .$$

$$W^{1/2}\mathcal{M}_a(Q,W) = \mathcal{M}_a^{[sd]}(Q_1) = -\sqrt{\frac{\pi}{2}}\Big[\frac{2}{Q_1} - 2 + \pi Q_1^{1/2}$$
$$+\left(2-\frac{1}{Q_1}-Q_1\right)Q_1^{-1/2}\arccos\left(\frac{1-Q_1}{1+Q_1}\right)\Big] \quad . \tag{A 18.17}$$

Again this holds for all polydispersities. It is easily checked that for $Q_1 \to \infty$ Eq. (A 18.17) reduces to Eq. (A 18.16), whereas it yields Eq. (A 18.12) for $Q_1 \to 0$.

In the dilute limit $\mathcal{M}_a(Q,0)$ of course coincides with the single-chain one-loop correction as discussed in Appendix A 15.2.

We may finally ask for an explicit parameterization of $\mathcal{M}_a(Q,W)$ in the monodisperse ensemble. Insisting on getting the correct asymptotic results in the large momentum or small momentum limits as well as in the semidilute limit, we construct a parameterization on the basis of the function $\mathcal{M}_a^{[sd]}(Q_1)$ (Eq. (A 18.17)):

$$\mathcal{M}_a^{[m]}(Q,W) = -\left[\mathcal{M}_a^{[m]}(0,W)^{-2} + \frac{Q}{4\pi^3}\right]^{-1/2} \cdot \left\{1 + \left(1 - \frac{0.2}{(1+0.17W)^2}\right)\left[\hat{\mathcal{M}}(Q_{eff}(Q,W)) - 1\right]\right\} \tag{A 18.18}$$

$$\hat{\mathcal{M}}(X) = -\left(\frac{9}{32\pi} + \frac{X}{2\pi^3}\right)^{1/2} \mathcal{M}_a^{[sd]}(X) \tag{A 18.19}$$

$$Q_{eff}(Q,W) = \left[b_1(W) + \frac{Q^2}{(b_2(W)+2W)^2}\right]^{1/2} - b_1(W)^{1/2} \tag{A 18.20}$$

$$b_1(W) = \frac{1.55 + 0.85W}{2 + 0.4W + 0.34W^2 + 0.018W^3}$$
$$b_2(W) = 12\,\frac{1+0.11W}{1+0.06W} \quad . \tag{A 18.21}$$

This form reproduces the numerical results for $\mathcal{M}_a^{[m]}(Q,W)$ within less than 2% deviation.

Notes

The overlap dependence of R_g has been discussed in [ON83], [Fre83],[Sch84]. Whereas the first two references use an exponentiated ε-expansion, the last reference is similar in spirit to the method used here.

The momentum dependence of $I_{d,a}(\mathbf{q})$ has first been analyzed in [NO85], the results being qualitatively similar to those presented here, except that the use of the ε-expansion destroys the full structure of the semidilute limit (cf. Eq. (18.23)).

19. The Density Cumulant

The density cumulant $\hat{I}_{d,c}(\mathbf{r}-\mathbf{r}')$ describes the correlations among local segment densities $\hat{\rho}(\mathbf{r}), \hat{\rho}(\mathbf{r}')$ in the solution. The Fourier transform $I_{d,c}(\mathbf{q})$ is measured in any standard scattering experiment. In the dilute limit $I_{d,c}(\mathbf{q})/c_p$ tends to the density correlation function $I_d(\mathbf{q}, n)$ of an isolated chain. In analyzing $I_{d,c}(\mathbf{q})$ for larger overlap, we again have to distinguish three different regions of momenta.

(i) $0 \lesssim qR_g \lesssim 1$

For very small momenta $I_{d,c}(\mathbf{q})$ can be expanded as

$$I_{d,c}^{-1}(\mathbf{q}) = I_{d,c}^{-1}(\mathbf{0}) \left[1 + \frac{1}{3}\mathbf{q}^2 \xi_d^2 + O(q^4)\right] \quad . \tag{19.1}$$

This expansion defines the density correlation length ξ_d. The leading term $I_{d,c}^{-1}(\mathbf{0})$ is related to the osmotic pressure by the compressibility sum rule and has been discussed in Chap. 17 in the context of thermodynamics. The density correlation length ξ_d in the dilute limit reduces to the radius of gyration of an isolated chain. With increasing overlap ξ_d/R_g rapidly decreases, and in the semidilute limit ξ_d becomes independent of the chain length and essentially proportional to the screening length ξ_E. Specifically in the semidilute excluded volume limit it behaves as

$$\xi_d \sim N^\nu \hat{s}^{-\frac{\nu}{3\nu-1}} \sim c^{-\frac{\nu}{3\nu-1}} \quad .$$

This is consistent with the blob concept, since by its very definition the blob gives the range of the density correlations in the solution. However, ξ_d should not rigorously be identified with the size of a blob, since even in the semidilute limit it depends on polydispersity. In contrast, the blob concept intrinsically is independent of the chain length distribution. This leads us to distinguish between ξ_d and the screening length or blob size ξ_E. The polydispersity dependence of ξ_d reflects the fact that it is extracted from momenta of order $q \lesssim R_g^{-1}$, corresponding to length scales large compared to the blob size.

To extract the blob size and thus the screening length we have to consider the intermediate momentum range

(ii) $\xi_d/R_g \ll q\xi_d \lesssim 1$.

The analysis acquires a precise meaning in the semidilute limit, where $\xi_d/R_g \to 0$. Taking this limit with $q\xi_d$ fixed, i.e. concentrating on length scales $r \sim \xi_d \ll R_g$, we suppress all polydispersity and chain length dependence. For $q\xi_d \ll 1$ the result can be expanded as

$$I_{d,c}^{-1}(\mathbf{q}) = I_{d,c}^{-1}(\mathbf{0}) \left(1 + \frac{1}{3}\mathbf{q}^2 \xi_E^2 + \dots \right) \quad , \tag{19.2}$$

which defines the screening length. ξ_E is thus constructed by first taking the semidilute limit *keeping q fixed on scale ξ_d^{-1}*, and then expanding with respect to q. Inverting these steps we find the semidilute limit of ξ_d, which in general differs from ξ_E. These two lengths refer to different momentum regions, which are well separated for a semidilute system. ξ_E is found to obey the same power law as ξ_d:

$$\xi_E \sim c^{-\frac{\nu}{3\nu-1}} \quad ,$$

but with a polydispersity independent prefactor. We note however that the leading term $I_{d,c}(\mathbf{0})$ is independent of interchanging the limits $q \to 0$, $\tilde{s} \to \infty$.

We finally note that in the large momentum region,

(iii) $1 \ll q\xi_d$,

scattering is dominated by pairs of segments deep inside a blob, which typically are found on the same chain. Thus

$$I_{d,c}(\mathbf{q}) \underset{q\,\xi_d \to \infty}{\longrightarrow} I_{d,a}(\mathbf{q}) \quad , \tag{19.3}$$

and in the excluded volume limit it develops the characteristic power law

$$I_{d,c}(\mathbf{q}) \sim q^{-\frac{1}{\nu}} \quad ,$$

irrespective of the overlap.

These general results support the qualitative blob considerations. However, as for $I_{d,a}(\mathbf{q})$, on the quantitative level we find some interesting structure not covered by the blob model. Due to the residual interaction of the blobs both $I_{d,a}^{-1}(\mathbf{q})$ and $I_{d,c}^{-1}(\mathbf{q})$ in the semidilute limit develop terms of order q^3 and thus yield singular curvature for $q^2 \to 0$ on scale ξ_E^2. As a result both correlation functions are strongly influenced by the one-loop terms and rapidly deviate from the simple tree-type approximations.

This strong influence of the one-loop terms leads to some problems, both on the theoretical and the experimental side. We first consider the density correlation length. It turns out that in the semidilute region ξ_d depends quite strongly on polydispersity, and for a monodisperse system the one-loop correction almost compensates the zero-loop term. Our expansion therefore only allows for a rough estimate of ξ_d. Curiously enough, this calculational problem meets with a serious problem on the experimental side. According to Eq. (19.1) ξ_d has to be measured in the region $qR_g \lesssim 1$. Since $\xi_d/R_g \to 0$ in the semidilute limit, the measuring signal $I_{d,c}^{-1}(\mathbf{q})/I_{d,c}^{-1}(\mathbf{0}) - 1$ for large overlap becomes very small, of order $q\xi_d \sim \xi_d/R_g \ll 1$. This prevents a precise experimental determination of ξ_d. Indeed, taking the semidilute limit with qR_g fixed, we find

$$I_{d,c}^{-1}(\mathbf{q}) \equiv I_{d,c}^{-1}(\mathbf{0}) \ ,$$

independent of $\mathbf{q}$. This result corresponds to the finding that the autocorrelations in that limit simply reduce to the Debye function. In summary, all spatial correlations on scale $r \sim R_g$ are trivial in the semidilute limit.

Also the determination of ξ_E poses some problems. Here the loop expansion is reasonably well behaved, allowing for a reliable estimate. However, due to the $O(q^3)$-contributions the expansion (19.2), which defines ξ_E, does not behave too well. The approximation

$$I_{d,c}^{-1}(\mathbf{q}) \approx I_{d,c}^{-1}(\mathbf{0}) \left(1 + \frac{1}{3}\mathbf{q}^2\xi_E^2\right)$$

is good only for small values $q^2\xi_E^2 \lesssim 0.1$. The singular curvature implies that a precise experimental determination of ξ_E meets with great difficulties. Typically for large overlap the experimental data in the intermediate momentum region (ii) are analyzed with the ansatz

$$I_{d,c}^{-1}(\mathbf{q}) \approx a(c) \left(1 + \frac{1}{3}\mathbf{q}^2\xi^2(c)\right) \ . \tag{19.4}$$

Our quantitative analysis shows that such an ansatz works well for reasonably large momentum intervals, but the parameters $a(c), \xi^2(c)$ are effective quantities, depending on the momentum range. $\xi^2(c)$, in particular can exceed ξ_E^2 by up to 40 %. In summary, for strongly overlapping systems all measurements of ξ_d or ξ_E published so far are likely to give some effective length scale of the correct order of magnitude, which however can be analyzed quantitatively only if the momentum range of the measurement and the method of data analysis are taken into account.

The remainder of this chapter, giving our detailed results, is organized as follows. In Sect. 19.1 we recall some general features of $I_{d,c}(\mathbf{q})$ and we present our one-loop results. The expression for ξ_d is evaluated in Sect. 19.2. The screening length ξ_E is discussed in Sect. 19.3. Section 19.4 is devoted to the momentum dependence. The experimental situation and empirical methods of data analysis are discussed in Sect. 19.5. The full complicated expressions for the one-loop results are given in appendices.

19.1 Formal expressions and general one-loop results

19.1.1 Definitions and scaling laws

The density cumulant has been defined in Sect. 5.1.4 in terms of the positions $\mathbf{r}, \mathbf{r}'$ of the scattering centers. A scattering experiment measures the Fourier transform

$$I_{d,c}(\mathbf{q}) = \frac{1}{\Omega} \int d^d r \, d^d r' \, e^{i\mathbf{q}(\mathbf{r}-\mathbf{r}')} \Big\langle (\hat{\rho}(\mathbf{r}) - c)(\hat{\rho}(\mathbf{r}') - c) \Big\rangle \ . \tag{19.5}$$

In the dilute limit $c \to 0$, N fixed, $I_{d,c}$ coincides with the autocorrelation function

$$I_{d,c}(\mathbf{q}) \to I_{d,a}(\mathbf{q}) \ ,$$

as can be checked by comparing Eqs. (19.5), (18.1). It thus obeys the sum rule

$$\frac{1}{c} I_{d,c}(\mathbf{q} = 0) \underset{c \to 0}{\longrightarrow} \tilde{p}_2 N \ . \tag{19.6}$$

As has been discussed in Sect. 17.1, this is a limiting form of the compressibility sum rule (17.13)

$$I_{d,c}(\mathbf{0}) = \sum_{nn'} nn' \, \frac{\partial \, c_p(n')}{\partial \, \mu_p(n)} \ ,$$

which is valid for all overlap. Recall that the latter for a monodisperse system reduces to

$$I_{d,c}(\mathbf{0}) = c \left(\frac{\partial}{\partial \, c} \frac{\hat{\Pi}}{k_B T} \right)^{-1} \ ,$$

a relation often invoked in experimental or theoretical work.

Equation (19.6) suggests to introduce the renormalization invariant scaling function

$$J_c(\mathbf{q}) = \frac{1}{cN} I_{d,c}(\mathbf{q}) \quad . \tag{19.7}$$

J_c obeys the general scaling law

$$J_c(\mathbf{q}) = \tilde{J}_c(\mathbf{q}^2 \tilde{R}_0^2, \tilde{s}, \tilde{z}, [p]) \quad , \tag{19.8}$$

reducing to

$$J_c(\mathbf{q}) = J_c^*(\mathbf{q}^2 R_g^2(N,0), \hat{s}, [p]) \tag{19.9}$$

in the excluded volume limit.

Taking the semidilute limit with q fixed on the chain length independent scale ξ_E we expect $I_{d,c}$ to become independent of chain length and polydispersity. To find the corresponding scaling predictions we replace $q^2 \tilde{R}_0^2$ (Eq. (19.8)) or $q^2 R_g^2(N,0)$ (Eq. (19.9)) by $q^2 \xi_E^2$, and we replace $\tilde{z}$ by $\tilde{s}/\tilde{z}$ (cf. Eq. (13.31)). We furthermore extract appropriate powers of $\tilde{s}$ or $\hat{s}$ to cancel the explicit factor of $1/N$ in Eq. (19.7). We thus rewrite the scaling functions as

$$\begin{aligned} \tilde{J}_c(q^2 \tilde{R}_0^2, \tilde{s}, \tilde{z}, [p]) &= \tilde{s}^{-2} \tilde{J}_c'(q^2 \xi_E^2, \tilde{s}, \frac{\tilde{s}}{\tilde{z}}, [p]) \\ J_c^*(q^2 \tilde{R}_g^2(N,0), \hat{s}, [p]) &= \hat{s}^{-\frac{1}{3\nu-1}} J_c^{*'}(q^2 \xi_E^2, \hat{s}, [p]) \quad . \end{aligned}$$

We now expect $\tilde{J}_c'$, $J_c^{*'}$ to reduce to finite polydispersity independent functions in the semidilute limit. We thus predict the scaling law

$$I_{d,c}(\mathbf{q}) = \frac{\tilde{\ell}^{-3}}{\tilde{\ell}^3 c} \tilde{J}_{c,sd}\left(q^2 \xi_E^2, \frac{\tilde{\ell}^3 c}{\tilde{v}}\right) \quad , \tag{19.10 i}$$

where we used Eq. (13.31), valid in the semidilute limit: $\tilde{s}/\tilde{z} = s_\ell^3 s_n c = \tilde{\ell}^3 c/\tilde{v}$. In the semidilute excluded volume limit we similarily find

$$I_{d,c}(\mathbf{q}) = B^{-3} \left(B^3 c\right)^{\frac{3\nu-2}{3\nu-1}} J_{c,sd}^*(q^2 \xi_E^2) \quad . \tag{19.10 ii}$$

Clearly these predictions have to be checked by perturbation theory. To one-loop order they are verified by our results.

19.1.2 Loop expansion

The diagrammatic analysis of Sect. 5.2.2 results in the unrenormalized expression (cf. Eq. (5.30))

$$I_{d,c}^{-1}(\mathbf{q}) = \left[K^{(2,0)}(\mathbf{q})\right]^{-1} + u_0 \quad , \tag{19.11}$$

where $K^{(2,0)}$ is defined as the sum of all vertex irreducible diagrams with two density insertions of momenta $\mathbf{q}$ or $-\mathbf{q}$, respectively. We further decompose $K^{(2,0)}$ as

$$K^{(2,0)}(\mathbf{q}) = K_a^{(2,0)}(\mathbf{q}) + K_f^{(2,0)}(\mathbf{q}) \quad . \tag{19.12}$$

The autocorrelation part $K_a^{(2,0)}$, being identical to the autocorrelation function $I_{d,a}$ (cf. Eq. (5.26)), collects the diagrams containing both density insertions in the same polymer line. The remainder $K_f^{(2,0)}$, where the insertions occur in different lines, determines the correlations among different chains. In view of the definition of $J_{a(c)}$ we may thus write

$$J_c^{-1}(\mathbf{q}) = J_a^{-1}(\mathbf{q}) + M_c(\mathbf{q}) \tag{19.13}$$

$$M_c(\mathbf{q}) = cNu_0 - \frac{cNK_f^{(2,0)}(\mathbf{q})}{K_a^{(2,0)}(\mathbf{q})\left(K_a^{(2,0)}(\mathbf{q}) + K_f^{(2,0)}(\mathbf{q})\right)} \quad . \tag{19.14}$$

Note that $K_f^{(2,0)}$ contains no zero-loop term. Its expansion starts with the diagrams of Fig. 19.13.

The representation (19.13,19.14) is rigorous and generalizes RPA (Eq. (5.56)) to arbitrary order of the loop expansion. It conveniently separates the autocorrelation part $J_a^{-1}(\mathbf{q})$ from effects of the intrachain interaction. It may be tempting to approximate $J_c^{-1}(\mathbf{q})$ by keeping the full autocorrelation part, but neglecting the momentum dependence of $M_c(\mathbf{q})$. Such 'improvements' of RPA have been presented in particular by Benoit and coworkers [BB84, BBW90]. However, this is an inconsistent approach, since the corrections $M_c(\mathbf{q}) - M_c(\mathbf{0})$ turn out to be of the same order of magnitude as the first order corrections contained in $J_a^{-1}(\mathbf{q})$. A valid approximation results only from evaluating both $J_a^{-1}(\mathbf{q}), M_c(\mathbf{q})$ to the same order of the loop expansion. This also automatically preserves the compressibility sum rule.

To construct the renormalized one-loop approximation for $J_c^{-1}(\mathbf{q})$ we use our previous one-loop result (18.16), (18.17) for $J_a^{-1}(\mathbf{q})$:

$$J_a^{-1}(\mathbf{q}) = D_p^{-1}(\hat{Q})$$

$$\hat{Q} = Q_R\left[1 - u^* f\left(1 + \frac{1}{2}N_R^{1/2}\mathcal{M}_a(Q_R, W_R)\right)\right] \quad , \tag{19.15}$$

where $Q_R = \bar{\mathbf{q}}^2 N_R, W_R = fc_R N_R$, and $\mathcal{M}_a$ is given in Appendix A 18.1. The additional term $M_c(\mathbf{q})$ to one-loop order can be written as

$$M_c(\mathbf{q}) = W_R\left[1 + u^* f\left(2 - N_R^{1/2}\mathcal{M}_c(Q_R, W_R)\right)\right] \quad . \tag{19.16}$$

The analytical expression for $\mathcal{M}_c(Q_R, W_R)$ is given in the appendix, Eq. (A 19.2). It is definitely found to vary with Q_R, but to attain some finite limit for $Q_R \to \infty$. As a result the large momentum behavior of $J_c^{-1}(\mathbf{q})$ is governed by the divergent contribution $J_a^{-1}(\mathbf{q}) \sim \hat{Q}$ of the autocorrelations, as expected.

19.2 The density correlation length

In the range of small momenta $q^2 R_g^2 \lesssim 1$, $J_c(\mathbf{q})$ can be expanded as

$$J_c^{-1}(\mathbf{q}) = J_c^{-1}(\mathbf{0}) \left[1 + \frac{1}{3}\mathbf{q}^2 \xi_d^2 + O(q^4) \right] \quad . \tag{19.17}$$

(Note that no q^3-term survives on scale R_g!) $J_c(\mathbf{0})$ has been discussed in Chap. 17, where it was found that in the excluded volume limit it varies from $J_c^{*-1}(\mathbf{0}, \hat{s} = 0) = \tilde{p}_2^{-1}$ to $J_c^{*-1}(\mathbf{0}, \hat{s} \to \infty) \to const\ \hat{s}^{1/(3\nu-1)}$. The density correlation length ξ_d obeys the general scaling law

$$\frac{\xi_d^2}{\tilde{R}_0^2} = \tilde{\xi}_d(\tilde{s}, \tilde{z}, [\tilde{p}]) \quad , \tag{19.18}$$

which in the excluded volume limit reduces to

$$\frac{\xi_d^2}{R_g^2(N, 0, [\tilde{p}])} = \xi_d^*(\hat{s}, [\tilde{p}]) \quad . \tag{19.19}$$

Here $R_g^2(N, 0, [\tilde{p}])$ denotes the average radius of an isolated chain for the given chain length distribution, so that $\xi_d^*(0, [p]) \equiv 1$.

To one-loop order our general result for ξ_d^2 follows from Eq. (A 19.6), $d = 3$:

$$\xi_d^2 = \frac{\tilde{p}_3}{(\tilde{p}_2)^2}\, \ell_R^2 N_R \frac{1 - u^* f \left(1 + N_R^{1/2} \xi^{(1)}(W_R, [p])\right)}{J_c^{-1}(0)} \quad . \tag{19.20}$$

The one loop correction $\xi^{(1)}(W_R, [p])$ is given in Appendix A 19.2.1. Note the polydispersity dependent prefactor, which also survives in the semidilute limit.

We first consider the excluded volume limit, following the lines of our discussion of $R_g^2(N, \hat{s})$ (Sect. 18.2).

19.2.1 Excluded volume limit

Combining Eqs. (19.20), (A 19.21) with the result for $J_{c,d}^{-1}(\mathbf{0})$ (Chap. 17) and using the RG mapping we can construct the virial expansion of $\xi_d^*(\hat{s}, [\tilde{p}])$:

$$\xi_d^{*2} = 1 - \begin{Bmatrix} 3.734\hat{s} \\ 5.937\hat{s} \end{Bmatrix} + O(\hat{s}^2) \begin{matrix} [m] \\ [e] \end{matrix} \quad . \tag{19.21}$$

We observe a clear polydispersity effect. Furthermore comparing to the corresponding result for R_g^2, Eq. (18.11), we find that ξ_d decreases with increasing overlap much more rapidly than R_g^2. This clearly was to be expected. We finally note that the one-loop corrections are not particularly large. We may use the zero-loop expressions

$$\begin{aligned} \xi_d^2 &= \frac{\tilde{p}_3}{\tilde{p}_2} \frac{\ell_R^2 N_R}{1 + \tilde{p}_2 f c_R N_R} \\ R_g^2(N, 0, [\tilde{p}]) &= \frac{\tilde{p}_3}{\tilde{p}_2} n_0 B^2 N^{2\nu} \end{aligned}$$

together with the full RG mapping to find

$$\xi_d^* = 1 - \begin{cases} 2.908\hat{s} \\ 5.185\hat{s} \end{cases} + O(\hat{s}^2) \begin{matrix} [m] \\ [e] \end{matrix} .$$

These results are not far from the one-loop expressions.

The situation grossly changes if we consider the semidilute limit. The one-loop expression yields

$$\xi_d^* = \hat{s}^{-\frac{2\nu}{3\nu-1}} \cdot \begin{cases} 0.0422 & [m] \\ 0.0585 & [e] \end{cases} , \tag{19.22}$$

to be compared to the zero-loop result

$$\xi_d^* = \hat{s}^{-\frac{2\nu}{3\nu-1}} \cdot \begin{cases} 0.224 & [m] \\ 0.112 & [e] \end{cases} .$$

For the monodisperse system the one-loop term nearly compensates the zero-loop contribution. For the exponential ensemble the effect is not as drastic. Clearly the one-loop correction sensitively depends on polydispersity. These features can also be seen in the strict ε-expansion. Using Eq. (A 19.19) and the corresponding result for $R_g^2(N, 0, [\tilde{p}])$ we find

$$\begin{aligned} \xi_d^* &= 0.224\hat{s}^{-\frac{2\nu}{3\nu-1}} \left(1 - 0.685\varepsilon + O(\varepsilon^2)\right) \\ &\underset{\varepsilon=1}{=} 0.071\ \hat{s}^{-1.539} \qquad [m] \end{aligned}$$

$$\begin{aligned} \xi_d^* &= 0.112\hat{s}^{-\frac{2\nu}{3\nu-1}} \left(1 - 0.581\varepsilon + O(\varepsilon^2)\right) \\ &\underset{\varepsilon=1}{=} 0.047\ \hat{s}^{-1.539} \qquad [e] . \end{aligned}$$

This clearly exhibits large corrections of order ε.

To examine the problem further, we consider the combination $\xi_d^d \Pi$, which is easily shown to reduce to a universal number in the semidilute excluded volume limit. From Eq. (A 19.19) combined with the results of Appendix A 17.2.1 we find the ε-expansion

$$\xi_d^d \, \Pi = \frac{1}{u^*} \left(\frac{\tilde{p}_3}{(\tilde{p}_2)^2} \frac{d}{12\pi} \right)^{d/2} \left[1 - \frac{\varepsilon}{4} \Big(2 \ln 2 + 2\gamma_{Eu} - 3 + \frac{25}{4} \frac{(\tilde{p}_2)^2}{\tilde{p}_3} \Big) + O(\varepsilon^2) \right] .$$

This yields

$$\xi_d^d \, \Pi = \begin{cases} \frac{1}{u^*} \left(\frac{d}{12\pi} \right)^{d/2} \left[1 - 1.448\, \varepsilon + O(\varepsilon^2) \right] & [m] \\ \frac{1}{u^*} \left(\frac{d}{8\pi} \right)^{d/2} \left[1 - 0.927\, \varepsilon + O(\varepsilon^2) \right] & [e] \end{cases} .$$

The strong sensitivity to polydispersity is obvious. Indeed, as it stands the monodisperse result makes no sense for $\varepsilon = 1$. Related to this problem is the fact that $\xi^d \Pi$, if evaluated directly in $d = 3$, sensitively depends on the choice of the theoretical parameter c_0.

To summarize, we have found that ξ_d in the semidilute excluded volume limit behaves as

$$\xi_d^2 = a_\xi^*[\tilde{p}] \, R_g^2(N, 0, [\tilde{p}]) \, \hat{s}^{-\frac{2\nu}{3\nu-1}} \sim B^2 \left(B^3 \, c \right)^{-\frac{2\nu}{3\nu-1}} , \tag{19.23}$$

independent of chain length. The amplitude $a_\xi^*[\tilde{p}]$ shows a large one-loop correction, which makes an estimate ambiguous, in particular for a monodisperse system. All results, however, agree that $a_\xi^*[\tilde{p}]$ is strongly suppressed compared to a zero-loop estimate.

We now consider the concentration crossover, using our standard approach of evaluating the crossover function directly in three dimensions. Using Eq. (19.20) together with our results for $R_g(N, 0, [\tilde{p}])$ (Sect. 15.1.2) and $J_c^{-1}(\mathbf{0})$ (Appendix A 17.2.2), we find the expressions

$$\xi_d^*(\hat{s}, [m]) = \frac{w^{4\nu-2}}{0.805} \frac{0.636 - 0.265 w^{-1} \xi^{(1)}(W_R, [m])}{1 + W(1.728 - 0.299 w^{-1} J_1)(W_R, [m])}$$

$$\xi_d^*(\hat{s}, [e]) = \frac{w^{4\nu-2}}{1.928} \frac{0.636 - 0.265 w^{-1} \xi^{(1)}(W_R, [e])}{0.5 + W(1.728 - 0.299 w^{-1} J_1)(W_R, [e])} , \tag{19.24}$$

holding for the monodisperse or exponential ensemble, respectively. In evaluating these results we use the relations (see Sect. 13.3)

$$W_R = c_0 n_0 \left(\frac{1}{w^2} - 1 \right) ,$$

$$\frac{n_0}{c_0} \, \tilde{u} \, \hat{s} = (1 - w^2) w^{2-6\nu} .$$

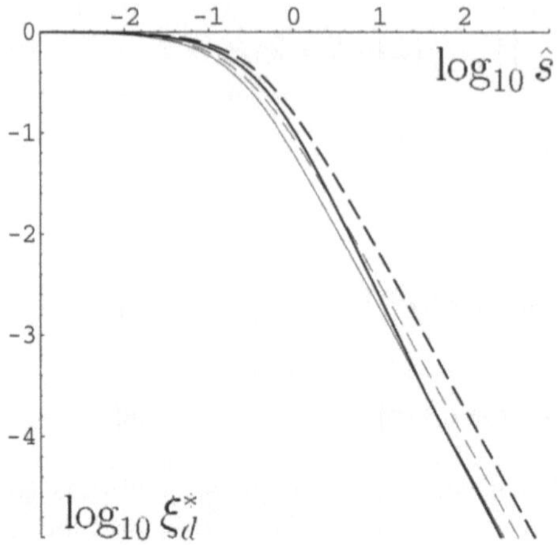

Fig. 19.1. $\log_{10} \xi_d^*$ as function of $\log_{10} \hat{s}$. *Thick lines*: monodisperse ensemble, *thin lines*: exponential ensemble. *Broken lines*: zero-loop approximation; *full lines*: one-loop approximation

The numerical results are shown in Fig. 19.1, together with the results of the zero-loop approximation. For the exponential ensemble the results look quite reasonable, supporting the view that for this ensemble the one-loop approximation can be trusted. For the monodisperse ensemble the one-loop effects stay reasonably small up to $\hat{s} \approx 1$, only. For larger overlap we run into the problems discussed above. Note in particular, that in some intermediate range of $\log \hat{s}$ the slope of $\log \xi_d^*(\hat{s}, [m])$ exceeds the asymptotic slope giving the power law $\xi_d^* \sim \hat{s}^{-\frac{\nu}{3\nu-1}}$. In view of the large first-order correction it is not clear whether this feature is valid.

19.2.2 Crossover outside the excluded volume limit

For fixed $\tilde{z}$ it is easily checked that the ratio $\tilde{\xi}_d(\tilde{s}, \tilde{z}, [p])$ (Eq. (19.18)) in the semidilute limit reduces to a (polydispersity dependent) function of the concentration c and temperature, only. Normalizing ξ_d^2 to its value at zero overlap, i.e. to the radius of gyration of an isolated chain, we show in Fig. 19.2

$$\alpha_\xi^2(\tilde{z}, \tilde{s}) = \frac{\xi_d^2(\tilde{s}, \tilde{z})}{R_g^2(\tilde{s}=0, \tilde{z})} = \bar{\alpha}_g^{-2}(\tilde{z}) \tilde{\xi}_d(\tilde{s}, \tilde{z}) \tag{19.25}$$

as function of $\hat{s} = 2.776\, \tilde{s} \tilde{z}^{0.528}$ (cf. Eq. (17.51)) for a monodisperse ensemble and several values of $\tilde{z}$, both on the weak or the strong coupling branch. For weak coupling and $\tilde{z}$ fixed, increasing overlap drives the system towards the Θ-region, where asymptotically the simple law $\alpha_\xi^2 \sim \tilde{s}^{-1} \sim c^{-1}$ holds. The crossover towards that law is clearly seen. On the strong coupling side the one-loop approximation rapidly breaks down, since the problematic behavior discussed in the previous section becomes more severe with increasing running coupling constant f.

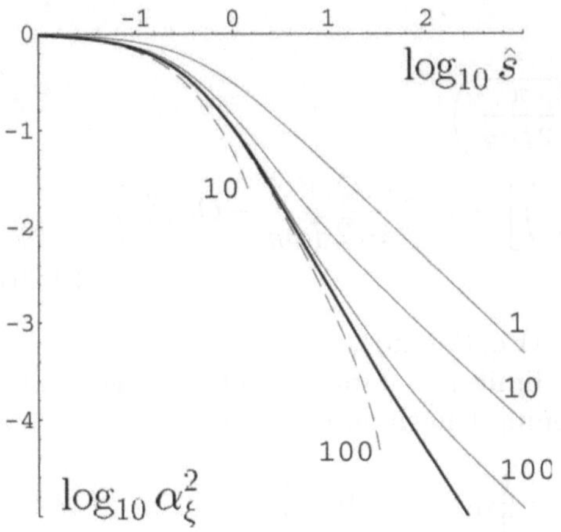

Fig. 19.2. $\log_{10}\alpha_\xi^2(\tilde{z},\tilde{s})$ as function of $\log_{10}\hat{s}$ for the monodisperse ensemble. *Thick line*: excluded volume limit. *Thin lines*: weak coupling. *Broken lines*: strong coupling. $\tilde{z}$-values are given with the curves

19.3 Screening length

Taking the semidilute limit with $\mathbf{q}R_g$ kept fixed, we find a result independent of momentum. Indeed it coincides with the semidilute limit of $J_c(\mathbf{0})$:

$$J_c^{-1}(\mathbf{q}) \underset{\substack{N_R\to\infty \\ \bar{\mathbf{q}}^2 N_R \text{ fixed}}}{\longrightarrow} J_c^{-1}(\mathbf{0}) = W_R\left[1+u^* f\left(2-\sqrt{\frac{\pi}{2fc_R}}\right)\right] \quad . \qquad (19.26)$$

This is in keeping with the fact that ξ_d^2/R_g^2 vanishes in the semidilute limit. Any momentum dependence is found only on scale ξ_d, which means that scattering in the semidilute limit is affected by the blobs, but does not see the individual chains.

To exhibit this momentum dependence we take the semidilute limit, keeping $\mathbf{q}$ fixed on scale ξ_d. This amounts to taking the limit $N_R\to\infty, \bar{\mathbf{q}}^2$ fixed. The general result is given in the appendix, Eq. (A 19.10). In three dimensions it reads

$$\begin{aligned}J_{d,c}^{-1}(\mathbf{q}) = W_R\Big\{1+Q_1+f\,u^*\Big[2-Q_1 \\ +\frac{1}{2}\sqrt{\frac{\pi}{2fc_R}}\left(\frac{\pi}{2}Q_1^{3/2}-2Q_1-2\left(\frac{Q_1}{2}-1\right)^2 Q_1^{-1/2}\arccos\frac{4-Q_1}{4+Q_1}\right)\Big]\Big\} \ ,\end{aligned} \qquad (19.27)$$

where

$$Q_1 = \frac{Q_R}{2W_R} = \frac{\bar{\mathbf{q}}^2}{2\,f\,c_R} \quad . \qquad (19.28)$$

This result is consistent with the scaling laws (19.10).

Expanding for $Q_1 \ll 1$ we find

$$J_{d,c}^{-1}(\mathbf{q}) = W_R\Big\{1 + f\,u^*\left(2 - \sqrt{\frac{\pi}{2fc_R}}\right)$$
$$+Q_1\left[1 - f\,u^*\left(1 - \frac{1}{12}\sqrt{\frac{\pi}{2fc_R}}\right)\right] + f\,u^*\frac{(\pi Q_1)^{3/2}}{4\sqrt{2}\,f\,c_R} + O(Q_1^2)\Big\} \ . \tag{19.29}$$

The leading term reproduces $J_{d,c}^{-1}(\mathbf{0})$ (Eq. (19.26)), so that with respect to this term the semidilute limit and the limit $q \to 0$ may be interchanged. The next to leading term defines the screening length according to

$$J_{d,c}^{-1}(\mathbf{q}) = J_{d,c}^{-1}(0)\left(1 + \frac{1}{3}\mathbf{q}^2\xi_E^2 + O(q^3)\right) \ . \tag{19.30}$$

From Eq. (19.29) (or Eq. (A 19.14), equivalently) we find

$$\xi_E^2 = \frac{3\,\ell_R^2}{2\,f\,c_R}\,\frac{1 + u^*\,f\left(\frac{1}{12}\sqrt{\frac{\pi}{2fc_R}} - 1\right)}{1 + u^*\,f\left(2 - \sqrt{\frac{\pi}{2fc_R}}\right)} \ . \tag{19.31}$$

It is easily checked that this result coincides with the density correlation length for the exponential ensemble in the semidilute limit. It must be stressed, however, that ξ_E is independent of polydispersity, as is the full function (19.27).

In the excluded volume limit Eq. (19.31) together with the RG mapping yields

$$\xi_E^2 = \frac{3}{2c_0}\,B^2\left(\frac{n_0}{c_0}\,\tilde{u}\,B^3\,c\right)^{-\frac{2\nu}{3\nu-1}}\frac{1 + u^*\left(\frac{1}{12}\sqrt{\frac{\pi}{2c_0}} - 1\right)}{1 + u^*\left(2 - \sqrt{\frac{\pi}{2c_0}}\right)}$$
$$= 0.0897B^2(B^3c)^{-1.539} \ . \tag{19.32}$$

Here the prefactor should be fairly reliable since ξ_E equals ξ_d in the exponential ensemble, where the problems pointed out in Sect. 19.2 are not so severe. Outside the excluded volume limit ξ_E^2 is a function of concentration and temperature, which is to be determined by evaluating Eq. (19.31), using the RG mapping in the semidilute limit, as given in Sect. 13.3.2, Eqs. (13.30):

$$\ell_R^2 = \frac{\tilde{R}_0^2}{\tilde{z}^2}\,f^2|1 - f|^{-2/\omega}H_u^2(f)$$
$$\frac{\tilde{z}}{\tilde{s}} = \frac{\tilde{u}}{c_0}\,f^2\Big|1 - f\Big|^{-\frac{1}{\omega}(3-\frac{1}{\nu})}\,H_u(f)H(f) \ .$$

Being interested in the temperature dependence, which essentially is embodied in $\tilde{z} = \tilde{v}(T)N^{1/2}$, we plot in Fig. 19.3

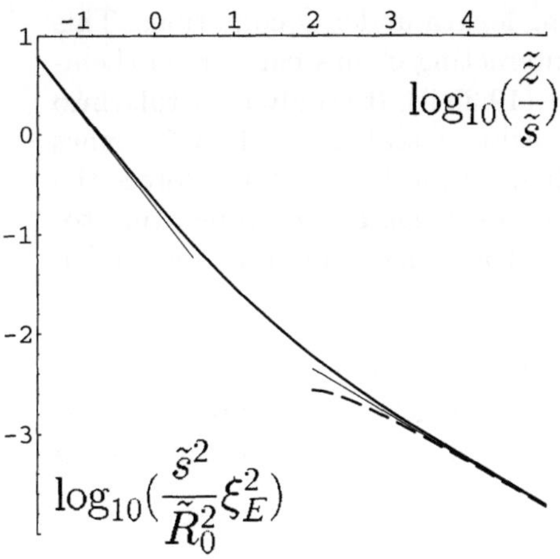

Fig. 19.3. $\log_{10}(\tilde{s}^2\xi_E^2/\tilde{R}_0^2)$ as function of $\log_{10}(\tilde{z}/\tilde{s})$. *Full line*: weak coupling branch; *broken line*: strong coupling branch. *Thin lines*: asymptotes in the excluded volume or Θ-limits

$$\left(\frac{\tilde{s}}{\tilde{R}_0}\right)^2 \xi_E^2 = \tilde{\ell}^4 c^2 \xi_E^2 \tag{19.33}$$

against

$$\frac{\tilde{z}}{\tilde{s}} = \frac{\tilde{v}(T)}{\tilde{\ell}^3 c} \quad .$$

In these variables the excluded volume behavior takes the form

$$\left(\frac{\tilde{s}}{\tilde{R}_0}\right)^2 \xi_E^2 \underset{f\to 1}{\longrightarrow} 0.0368 \left(\frac{\tilde{z}}{\tilde{s}}\right)^{-0.461} \quad . \tag{19.34}$$

As shown in Fig. 19.3 it is closely approached for $\tilde{z}/\tilde{s} \gtrsim 10^3$ both by the weak or the strong coupling branch. In the Θ-limit $f \to 0$ the weak coupling branch approaches the asymptote

$$\left(\frac{\tilde{s}}{\tilde{R}_0}\right)^2 \xi_E^2 \underset{f\to 0}{\longrightarrow} 0.185 \frac{\tilde{s}}{\tilde{z}} \quad , \tag{19.35 i}$$

or

$$\xi_E^2 \to 0.185 \frac{1}{\tilde{v}(T)\tilde{\ell}\, c} \quad , \tag{19.35 ii}$$

equivalently. Since $\tilde{v}(T) \to 0$ for $T \to \Theta$, this apparently implies singular behavior. It must be noted, however, that here we are concerned with the semidilute limit, thus assuming $N_R \gg 1$. For $f \ll 1$ this implies

$$\tilde{v}(T)\tilde{\ell}\, c\, \tilde{R}_0^2 \gg 1 \quad ,$$

so that ξ_E always stays small compared to the radius of gyration $R_g \sim \tilde{R}_0$. It however becomes arbitrarily large compared to the scale $\xi_c \sim const/(\tilde{\ell}c)^{1/2}$, which in some work is mistaken as screening length under Θ-conditions. This identification ignores the fact that only interacting chains can screen themselves, which leads to the factor $\tilde{v}(T)$ in Eq. (19.35 ii). It is only if we take into account three-body forces that in the Θ-region a scale related to ξ_c comes in, which cuts off the singularity of ξ_E. If we neglect three-body forces the singularity is cut off by another mechanism: decreasing the temperature towards $T = \Theta$, but keeping c, N fixed, we leave the semidilute region (cf. Sect. 13.3.3), and Eq. (19.35) does no longer apply.

The role of the excluded volume strength for screening is clearly brought out by the weak coupling branch of Fig. 19.3, which can be interpreted as showing a rapid decrease of ξ_E with increasing $\tilde{z}, \tilde{s}$ fixed, i.e. with increasing excluded volume strength.

19.4 Momentum dependence

We consider here the momentum dependence for general overlap, discussing in turn the different momentum regions. We base our analysis on Eq. (19.13): $J_c^{-1} = J_a^{-1} + M_c$, where $J_a^{-1}(\mathbf{q})$ is taken from Chap. 18, whereas the result for $M_c(\mathbf{q})$ is given in Appendix A 19.2.2. The expressions are quite complicated and in general do not allow for useful simplifications. Only in the semidilute limit can the simpler expression (19.27) be used. To illustrate the effects we restrict ourselves to the excluded volume limit.

19.4.1 Range $qR_g = O(1)$

Figure 19.4 shows $J_c^*(\mathbf{0})/J_c^*(\mathbf{q})$ (monodisperse) in the range $(qR_g(N, \hat{s}))^2 \lesssim 10$ for several values of the overlap. It illustrates how increasing overlap rapidly suppresses the momentum dependence on scale R_g. The straight lines represent the small momentum behavior $J_c^{-1}(\mathbf{q}) \approx J_c^{-1}(\mathbf{0})(1 + \frac{1}{3}q^2\xi_d^2)$, and it becomes quite obvious that for overlap $\hat{s} \gtrsim 1$ a determination of ξ_d^2 requires most precise measurements, asking for a determination of $J_c(\mathbf{q})$ with a precision of 1 % or better. Indeed, the increase of the region, where the plot can be considered as linear, by no means compensates the decrease of the initial slope.

Polydispersity effects survive only on scale $q \sim R_g^{-1}$ and decrease with increasing overlap.

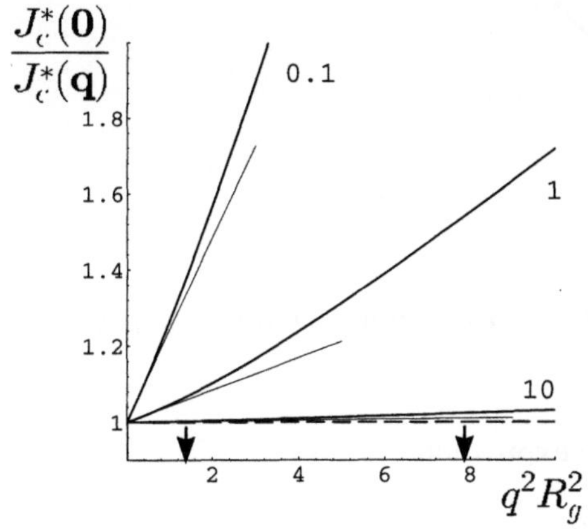

Fig. 19.4. $J_c^*(\mathbf{0})/J_c^*(\mathbf{q})$ as function of $q^2 R_g^2(N, \hat{s})$, for $\hat{s} = 0.1, 1, 10$. Initial slopes defining ξ_d according to Eq. (19.17) are indicated. The *arrows* indicate the values $q^2 = \xi_d^{-2}$ for $\hat{s} = .1, 1$ (from left). For $\hat{s} = 10$ the corresponding value is found at $q^2 R_g^2 = 258$. The *broken line* $J_c^*(0)/J_c^*(q) \equiv 1$ represents the semidilute limit

19.4.2 Range $q\xi_d = O(1)$

In the semidilute limit $J_c^{*-1}(\mathbf{q})$ on scale ξ_d^{-1} rapidly bends away from the linear approximation (19.30), due to a contribution $\sim q^3$. This is illustrated in Fig. 19.5, where it is found that in the semidilute limit deviations from the linear approximations at $q^2\xi_d^2 = 1$ are of the order of 6 % in $J_c^{*-1}(\mathbf{q})$, or of about 25 % in $J_c^{*-1}(\mathbf{q}) - J_c^{*-1}(0)$. Furthermore, over a sizeable range of $q^2\xi_d^2$ the plot indicates approximate linear behavior with some larger slope. The curves for finite $\hat{s}$, included in Fig. 19.5, show that such features are found already for intermediate overlap. This is clearly in keeping with the observations of the previous section.

To bring out these features more clearly, we define an effective q-dependent screening length $\xi_E(\mathbf{q})$ in terms of the slope of $J_c^{-1}(\mathbf{q})$ as indicated in

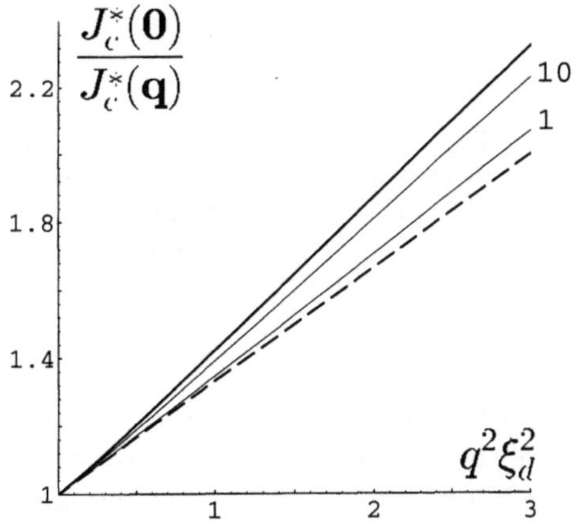

Fig. 19.5. $J_c^*(\mathbf{0})/J_c^*(\mathbf{q})$ (exponential ensemble) as function of $q^2\xi_d^2$ in the semidilute limit (*thick line*) or for the overlap values given (*thin lines*). The *broken line* represents the linear approximation

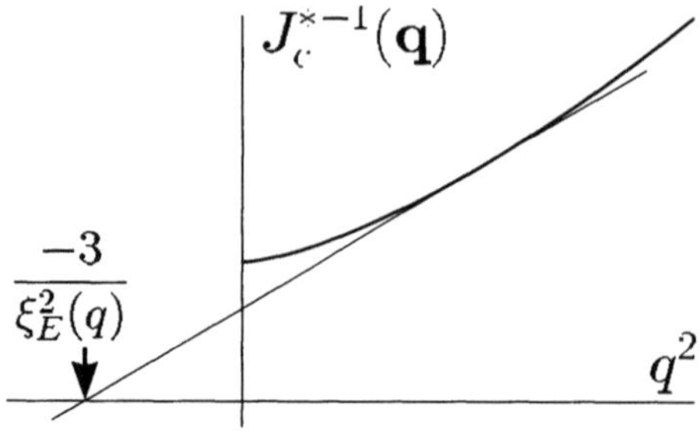

Fig. 19.6. Definition of $\xi_E(q)$

Fig. 19.6. This construction yields the expression

$$\frac{3}{\xi_E^2(q)} = J_c^{-1}(\mathbf{q}) \left(\frac{\partial}{\partial q^2} J_c^{-1}(\mathbf{q}) \right)^{-1} - q^2 \quad . \tag{19.36}$$

It generalizes both the definitions of ξ_d or ξ_E in the sense that for finite overlap $\xi_E^2(0, \hat{s}) = \xi_d^2(\hat{s})$, whereas taking first the semidilute limit we find $\lim_{q\to 0} \xi_E^2(q, \mathbf{s} = \infty) = \xi_E^2$. Figure 19.7 shows $\xi_E^2(q)/\xi_d^2$ as function of $q^2\xi_d^2$ for the exponential ensemble. Recall that for that chain length distribution $\lim_{\hat{s}\to\infty} \xi_d^2 = \xi_E^2$, so that our result in the semidilute limit in fact gives $\xi_E^2(q)/\xi_E^2$. For larger overlap the initial steep rise shown in Fig. 19.7 is due to the q^3-behavior of $J_c^{-1}(\mathbf{q})$. Clearly any experiment, which essentially measures the slope of $J_c^{-1}(\mathbf{q})$ at finite $(q\xi_d)^2$, should overestimate the true screening length.

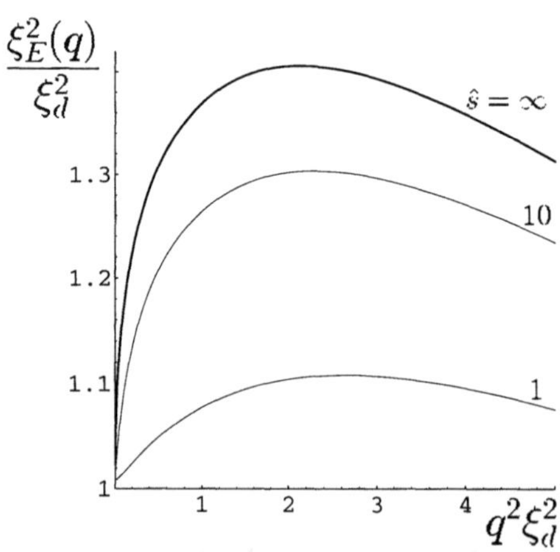

Fig. 19.7. $\xi_E(q)/\xi_d^2$ as function of $q^2\xi_d^2$ for the exponential ensemble. The square root singularity in the semidilute limit (*thick line*) is clearly exhibited

Leaving the excluded volume limit we note that $J_c(\mathbf{q})$, plotted as function of $q^2\xi_d^2$, is not very sensitive to the value of the coupling f and thus to solvent quality or temperature effects. Such effects are important only in the large momentum region $q^2\xi_d^2 \gtrsim 50$ or very close to the Θ-temperature.

19.4.3 Range $q\xi_d \gg 1$, and overall behavior of J_c

As pointed out before, $J_c(\mathbf{q})$ for $q\xi_d \gg 1$ approaches the autocorrelation function of an isolated chain, showing the typical behavior $J_c(\mathbf{q}) \sim q^{-1/\nu}$ in the excluded volume limit. This leading behavior has been discussed in Sect. 15.2, where deviations from the excluded volume limit have also been briefly analyzed. In the semidilute limit we may combine this information with the scaling law (19.10) to find

$$\begin{aligned} J_c(\mathbf{q}) &\sim \hat{s}^{-1/(3\nu-1)} \left(q^2\xi_E^2\right)^{-\frac{1}{2\nu}} \\ &\sim N^{-1} q^{-1/\nu} \quad , \end{aligned}$$

or

$$I_{d,c} \sim c\, q^{-1/\nu} \quad .$$

Using our quantitative one-loop results we find

$$J_c^{-1} \underset{q^2\xi_E^2 \gg 1}{\longrightarrow} 0.439 \left(q^2 B^2 N^{2\nu}\right)^{\frac{1}{2\nu}} \quad ,$$

i.e. the same expression as found for an isolated chain (Eq. (15.34)). Indeed, as we have pointed out before, for $q \to \infty$ the relation $J_c^{-1} \to J_a^{-1} \to J_d^{-1}$ holds independent of the overlap, so that the above result amounts to a consistency check within our analysis.

We are now in the position to discuss the overall behavior of $J_c(\mathbf{q})$ as function of momentum measured on the natural scale ξ_d. Figure 19.8 shows $J_c(\mathbf{0})/J_c(\mathbf{q})$, $\hat{s} = 1, 10, \infty$, (excluded volume limit) for the exponential ensemble. Polydispersity effects are restricted to $q^2 R_g^2 \lesssim 1$ and would not be visible

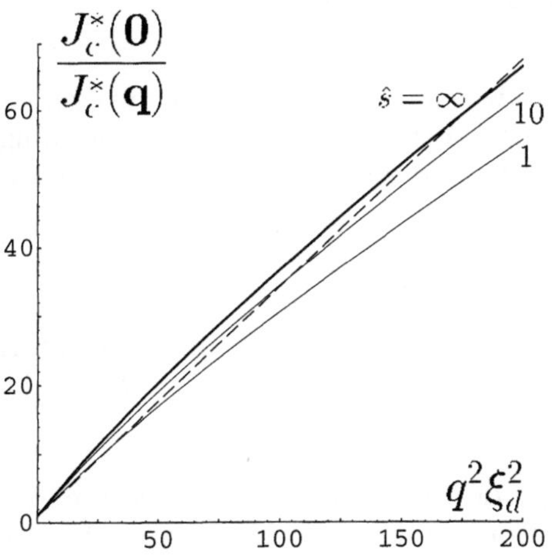

Fig. 19.8. $J_c^*(\mathbf{0})/J_c^*(\mathbf{q})$ (exponential ensemble) as function of $q^2\xi_d^2$ for $\hat{s} = 1, 10, \infty$, over a large range of momenta. The *broken line* represents the RPA-type approximation $J_c^*(\mathbf{0})/J_c^*(\mathbf{q}) \sim 1 + \frac{1}{3}q^2\xi_d^2$

in that plot. Since we normalized by $J_c(\mathbf{0})$, all curves start out at 1 with the same slope, but the amplitude of the asymptotic power law differs. Clearly even the curves for $\hat{s} = 1, \hat{s} = \infty$ look remarkably similar and for large ranges of $q^2 \xi_d^2$ they could be brought to close agreement by rescaling the axes, which amounts to using effective values of $J_c(0), \xi_d^2$. This also does not change if we go to the dilute limit $\hat{s} = 0$ (cf. Fig. 15.6). The qualitative difference among the dilute and semidilute limits can be found only by a precise analyse of the region $q^2 \xi_d^2 \sim 1$, as illustrated in Fig. 19.5.

The finding that the overlap does not dramatically change the form of $J_c(\mathbf{0})/J_c(\mathbf{q})$ is easily understood, if we consider a naive RPA-type approximation:

$$J_c^{-1}(\mathbf{q}) = \frac{1}{D_p\left(\frac{\tilde{p}_2}{\tilde{p}_3} R_g^2 q^2\right)} + v(\hat{s}) \quad . \tag{19.37}$$

This is equivalent to renormalized tree approximation (Sect. 14.3). For the exponential ensemble, where $D_p(x) = 2/(1+x), \tilde{p}_2/\tilde{p}_3 = 1/3$, it yields

$$\frac{J_c(\mathbf{0})}{J_c(\mathbf{q})} = 1 + \frac{q^2}{3} \frac{R_g^2}{1 + 2v(\hat{s})} \quad .$$

Thus this approximation predicts that we can find an overlap dependent length scale $\xi = R_g(1+2v(\hat{s}))^{-1/2}$, so that $J_c(\mathbf{0})/J_c(\mathbf{q})$ is a universal function of $q^2\xi^2$, independent of the overlap. Our full one-loop calculation renders this statement invalid, but the effect is not too severe.

In summary, this discussion shows that a precise and unambiguous identification of nontrivial overlap effects needs an analysis of a large range of q-values.

19.4.4 Corrections to the leading $q^{-1/\nu}$-behavior

The correlations among different chains, which in our formulation are embodied in $M_c(\mathbf{q}) = J_c^{-1}(\mathbf{q}) - J_a^{-1}(\mathbf{q})$, (Eq. (19.13)), contribute only subleading terms to the large q behavior of $J_c(\mathbf{q})$. We, however, may consider the intrachain correlation function

$$I_{d,f}(\mathbf{q}) = I_{d,c}(\mathbf{q}) - I_{d,a}(\mathbf{q}) \quad ,$$

where the leading asymptotic q-dependence $\sim q^{-1/\nu}$ cancels. In principle this quantity is experimentally accessible, and its large q behavior is governed by $M_c(\mathbf{q}) \cdot J_a^2(\mathbf{q})$.

In trying to evaluate the asymptotic behavior of $M_c(\mathbf{q})$ within our approach we run into problems. Considering the loop expansion of

$\mathcal{M}_c(\mathbf{q})$, Eq. (19.16), for $\mathbf{q} \to \infty, \hat{s}$ fixed, we find that the one-loop term $\sim N_R^{1/2} \mathcal{M}_c(Q_R, W)$ dominates over the zero-loop contribution. In strict ε-expansion it develops a logarithmic singularity which is not eliminated by our choice of ℓ_R. Rather in the semidilute limit it takes the form $\varepsilon \ln(q^2 \xi_E^2)$.

This problem is not unexpected: as mentioned in Sect. 15.2.3, field theoretic arguments, based on the short distance expansion predict that correlation functions in the critical limit show an asymptotic expansion in anomalous powers:

$$J_c(\mathbf{q}) \sim a_0\,(\mathbf{q}\xi)^{-x_0} + a_1(\mathbf{q}\xi)^{-x_1} + a_2(\mathbf{q}\xi)^{-x_2} + \dots\ ,$$
$$x_0 < x_1 < x_2 < \dots\ .$$

These powers can be expressed in terms of known exponents (for instance $x_0 = 1/\nu$, in our example), but in general are not simply related to each other. Our method cannot deal with several genuinely different powers simultaneously.

For a discussion of $I_{d,c}(\mathbf{q})$ or $I_{d,a}(\mathbf{q})$, this again is not a serious problem, since, as pointed out in Sect. 15.2.3, all subleading terms have to conspire to produce the smooth crossover towards regular $q \to 0$ behavior. Our results should adequately represent this crossover. However, if we consider $I_{d,f}(\mathbf{q})$, the problem is serious, since there our method cannot reproduce the correct leading behavior.

To solve this problem we have to go back to strict ε-expansion, combined with knowledge of the exact exponents from the short distance expansion. Given that information, we can exponentiate the ε-expansion to produce the desired form. Results of a field theoretic analysis of $I_{d,f}(\mathbf{q})$ have been published by Jannink and coworkers [JPLC94], who in the semidilute limit suggest the large momentum behavior

$$\begin{aligned} I_{d,f}(\mathbf{q}) &= a_1 \frac{c^2}{q^d} + a_2 \frac{c^{\frac{\nu(d+\omega)}{\nu d - 1}}}{q^{2/\nu + \omega}} \\ &= a_1 \frac{c^2}{q^3} + a_2 \frac{c^{2.92}}{q^{4.2}}, \quad (d = 3)\ . \end{aligned} \tag{19.38}$$

Here ω is the correction to scaling exponent ($\omega = 0.8$ in $d = 3$) and a_1, a_2 are constants. Note, that for $d = 4 (\omega = 0, \nu = 1/2)$ both terms reduce to c^2/q^4. The next terms in that expansion are expected to be of canonical order q^{-6}. Our one-loop calculation allows only for a determination of the first two terms in the ε-expansion of $I_{d,f}$. Up to that order the ε-expansion is consistent with any result containing two powers of type $q^{-4+O(\varepsilon)}$ and therefore does not allow for a meaningful check of the proposed form (19.38). We however may raise some doubts concerning this form, since in problems involving the close encounter of segments from *different* chains the exponent ω typically is replaced by some other exponent $\omega_{12} \approx \omega/2$ [MS98]. Since the second term in Eq. (19.38) hardly is of any experimental relevance (see [JPLC94]), this is not a very important issue on the quantitive level. Still it keeps us from extracting approximate values of a_1, a_2.

19.5 The experimental situation

19.5.1 Heuristic methods of data analysis

a) Zimm plot for dilute solutions

Scattering experiments on dilute solutions mostly aim at a determination of the weight averaged chain length N_w, the radius of gyration $R_g^2(N, 0)$ of an isolated chain, or of the second virial coefficient A_2^S. A standard tool to extract this information is the 'Zimm plot', where $c/I_{d,c}(\mathbf{q})$ is plotted against a linear combination of $\mathbf{q}^2$ and c. This allows for a consistent extrapolation to $\mathbf{q} = 0, c = 0$ within a single plot.

To analyze this approach we may take our one-loop results as theoretical input, plotting $J_c^{-1}(\mathbf{q})$ as function of $q^2 R_g^2(N, 0) + \alpha\,\hat{s}$, where the parameter α has to be chosen such that the plot becomes clear and simple. Figure 19.9a shows such a plot for a monodisperse system in the excluded volume limit. We give lines of fixed overlap ($\hat{s} \leq .225$) or of fixed value of momentum ($q^2 R_g^2(N, 0) \leq 1.3$). The 'data points' should be considered to be given by the crossings of lines $\hat{s} > 0, q^2 R_g^2 > 0$. The range of $\hat{s}$ or $q^2 R_g^2$ has been chosen to be adequate for typical light scattering experiments on long ($N \sim 50{,}000$) flexible chains. The plot clearly looks quite flat but still shows some nonnegligible curvature. (We included the lines giving the asymptotic slopes in momentum or overlap.) Thus an unbiased extrapolation towards $q = 0, \hat{s} = 0$ is not quite trivial. Figure 19.9b shows the result of a linear fit according to

$$J_c^{-1}(\mathbf{q}) \approx a_0 + a_1\,\hat{s} + a_2 q^2 R_g^2 \quad , \tag{19.39}$$

where we used the data points of the net of Fig. 19.9a. Clearly this fit looks quite reasonable, but a comparison of the extracted values $a_0 = 0.99; a_1 = 3.43; a_2 = 0.345$ to the exact values ($a_0 = 1, a_1 = 3.07, a_2 = 1/3$) shows that R_g^2 or A_2^S are overestimated by about 3 % or 10 %, respectively.

A systematic analysis shows that the error is quite sensitive to the precise range of $\hat{s}$ or $q^2 R_g^2$, covered by the experiment. In particular the sizeable error of A_2^S can be suppressed only if data for much smaller overlap and

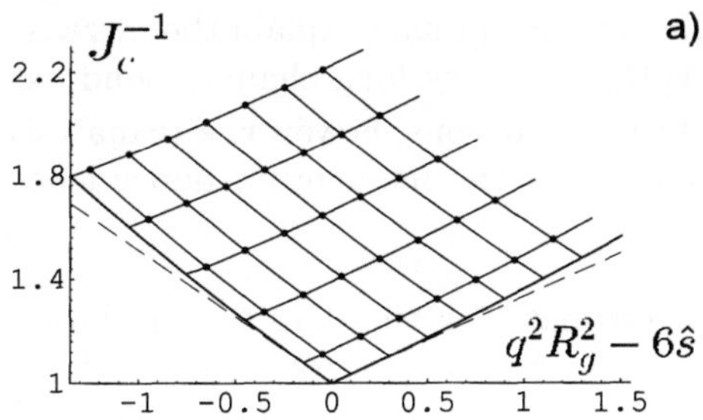

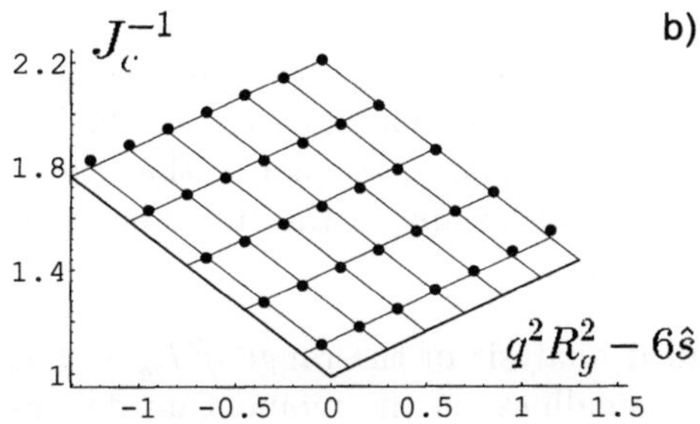

Fig. 19.9. (**a**) $J_c^{*-1}(\mathbf{q}, \hat{s})$ as function of $(q^2 R_g^2(N, \hat{s} = 0) - 6\hat{s})$ for $\hat{s} =$ 0; 0.025; 0.075; 0.125; 0.175; 0.225, or $q^2 R_g^2 =$ 0; 0.1; 0.3; 0.5; 0.7; 0.9; 1.1; 1.3. The *broken lines* give the asymptotic slopes of $J_c^{*-1}(0, \hat{s})$ or $J_c^{-1}(q^2 R_g^2, 0)$, respectively. (**b**) Fit of the 'data points' (*fat dots*), given by the knots of the net to the linear form (Eq. (19.39))

momentum are available. This may suggest to include higher order terms $(\hat{s}^2, (q^2R_g^2)^2, \hat{s}q^2R_g^2)$ into the ansatz for $J_{d,c}^{-1}$. However, as with the standard virial expansion the resulting fit is quite unstable. The sensitivity of the results to the precise range of $\hat{s}, q^2$ increases dramatically.

b) Square root plot for finite overlap

In Sect. 15.5.1 for an isolated chain we have discussed the square root plot, which is often used to extrapolate data to $q = 0$. For sharp chain length distributions we found it to yield good results for chain length N_w and radius of gyration $R_g^2(N, \hat{s} = 0)$, as long as sufficient data in the range $q^2 R_g^2 \lesssim 1.5$ are available. (See Fig. 15.12.) This suggests to test this method also for finite, but small overlap.

An analysis in the range $\hat{s} \leq 0.225$, based on our theoretical expression, shows that the results found for $\hat{s} = 0$ remain valid. The square root plot yields reliable values of $J_c^{-1}(0)$, as long as data in the range $0.1 \lesssim q^2 R_g^2 \lesssim 1.5$ are evaluated. If the chains are so long that this range is not reached, then the results show a peculiar behavior. As found before (Sect. 15.5.1), for $\hat{s} = 0$ $NJ_c(0)$ is underestimated, i.e. $J_c^{-1}(0)$ is overestimated. Beyond some value of $\hat{s}$, which depends on the precise range of $q^2 R_g^2$ available, $J_c^{-1}(0)$ is underestimated. The effect again is not very strong. In the range $\hat{s} \leq 0.225, 1.5 \leq q^2 R_g^2 \leq 5$, it does not exceed a few percent. But still it implies that the *slope* of $J_c^{-1}(0)$ as function of $\hat{s}$, i.e. the second virial coefficient A_2, is seriously underestimated. In our example the effect is of the order of 15 %. Note that this mechanism may explain the increase of the effective ratio $\mathcal{A}^*_{3/2} = A_3/(A_2^2 M_w)$ for very long chains, found experimentally. (Cf. Sect. 17.5.1.) For firm conclusions, however, a reanalysis of the experimental data that fully respects the momentum dependence would be necessary.

The square root plot also yields an estimate for the density correlation length. This is more sensitive to the q-range, a feature also found for $\hat{s} = 0$. For data not covering the range $q^2 R_g^2 \lesssim 0.5$, again the slope of $\xi_d^2(\hat{s})/R_g^2(N,0)$ as function of $\hat{s}$ is seriously underestimated.

In summary, it is most likely that published data in the dilute good solvent range for longer chains ($N \gtrsim 2 \cdot 10^4$) are seriously biased by the method of data analysis. Both the Zimm plot and the square root plot in that range yield effective parameters, which need information on the full functional dependence $J_c^{-1}(\mathbf{q}, \hat{s})$ for their interpretation.

For larger overlap , ($\hat{s} \gtrsim 1$), a detailed analysis of the range $q^2 R_g^2 \sim 1$ is clearly not very relevant, due to the smallness of the momentum dependence (cf. Fig. 19.4).

c) RPA-inspired analysis for strongly overlapping systems

For larger overlap and momenta on scale ξ_d the scattering intensity is often analyzed in terms of an RPA-type ansatz (19.37), where in the range $R_g^{-2} \ll q^2 \lesssim \xi_d^{-2}$ the Debye-function is replaced by $D_p(x) \approx 2/x$ (cf. Eq. (5.51 ii)). This approach, which in particular has been advocated by Benoit and co-workers [BBW90], amounts to the ansatz

$$\frac{J_c(\mathbf{0})}{J_c(\mathbf{q})} = A_0 \left(1 + \frac{1}{3} q^2 \xi^2\right) \quad . \tag{19.40}$$

Of particular interest is ξ^2, which is often identified with the screening length. (We should note that in the definition of the screening length the factor 1/3 in Eq. (19.40) often is omitted.) With such an ansatz both A_0 and ξ^2 are effective parameters, which depend on the momentum and overlap range of the experiment. Indeed, inserting Eq. (19.40) into Eq. (19.36) we find the result

$$\xi^2 = \xi_E^2(\mathbf{q}) \quad .$$

For larger overlap ξ^2 is likely to exceed considerably the true screening length (cf. Fig. 19.7). The effect is less drastic for smaller overlap, but it must be kept in mind that with decreasing overlap and decreasing $q^2\xi_d^2$ polydispersity effects show up. As a result it is unlikely that ξ_d^2 can be measured precisely in the range $\hat{s} \gtrsim 1$. Rather the published data give some effective length of the right order of magnitude, which however cannot be rigorously identified to ξ_d^2 or ξ_E^2.

19.5.2 Comparison between theory and experiment

For the system polystyrene-toluene results of an extensive light scattering experiment [WHCS83] [1] are available, giving both $J_{d,c}^{-1}(\mathbf{0})$ and ξ_d^2. The data cover a large range of chain lengths ($3 \times 10^3 \lesssim N \lesssim 2 \times 10^5$) and overlap ($.01 \lesssim \hat{s} \lesssim 10$). The statistical scatter is not very small, so that these data cannot be used to see any significant deviation from the excluded volume limit. For that reason we did not include the data for $J_c^{-1}(\mathbf{0})$ in our analysis given in Sect. 17.5, where we evaluated a set of more precise data for the same system (cf. Fig. 17.9). It must be noted, however, that the data considered here are perfectly consistent with the results given there, and fitting $J_c^{-1}(\mathbf{0})$ to the excluded volume curve we find a parameter value $B \approx 3.02\,\text{Å}$, consistent with the evaluation of the other data, provided we ignore corrections due to strong coupling behavior.

With the value of B thus determined the results for ξ_d^2 allow for a parameter-free test of the theory. The outcome, as shown in Fig. 19.10, (points), exhibits an excellent fit within the scatter of the data. Despite the problems with the calculation of ξ_d^2, discussed in Sect. 19.2, we may conclude that we have found a reasonable form of the scaling function $\xi^*(\hat{s}, [m])$ at least in the range $\hat{s} \lesssim 10$.

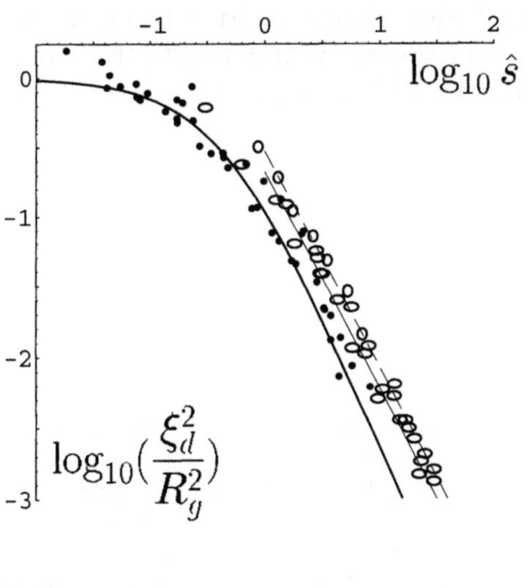

Fig. 19.10. $\log_{10}(\xi_d^2(\hat{s})/R_g^2(N, \hat{s} = 0))$ as function of $\log_{10} \hat{s}$ for a monodisperse system in the excluded volume limit (*fat line*). The *thin line* gives $\log_{10}(\xi_E^2/R_g^2)$. The *thin broken line* gives $\log_{10}(1.4\xi_E^2/R_g^2)$. Data for polystyrene-toluene. *Points*: light scattering [WHCS83], *ellipsoids*: X-ray scattering [HKHN85], *circles*: neutron scattering [KBWU85]

[1] I am indebted to Dr. Cannell for sending me a compilation of the data.

For the same chemical system the screening length has been measured by X-ray or neutron scattering. Since these methods test the scattering function at some higher momentum range, they extract an effective length, as discussed above. In a plot of $\xi^2/R_g^2(N, \hat{s}=0)$ against $\hat{s}$ the data should lie above the scaling function $\xi^*(\hat{s}, [m])$. (Note that in contrast to Fig. 19.1, where $\xi_d^*[e]$ has been defined using $R_g(N, \hat{s}=0, [e])$, we now reduce ξ_E by the *monodisperse* radius of gyration.) As is shown in Fig. 19.10 (open symbols), this is borne out nicely. Indeed, the data scatter about the straight line representing the asymptotic law $\xi_E^2/R_g^2 = 0.210\hat{s}^{-\frac{2\nu}{2\nu-1}}$ (cf. Eqs. (19.32), (15.14), (15.15)) in that plot. The slight tendency to lie above that line is consistent with our discussion of the previous section.

Published results on the momentum dependence of $J_c^{-1}(\mathbf{q})$ for larger overlap are rare. The only more extensive experimental study available seems to be the neutron scattering experiment of [FBC+78]. The results within their large scatter are consistent with our theory, but cannot be used to verify the more detailed quantitative results. We, however, may note a recent simulation of Binder et al. [YMB97], which for intermediate q-values can be nicely fitted to our theory. Deviations for small values of q may be due to the finite size of the simulation volume.

So far we have been concerned with results in good solvents. For Θ-solvents we are aware of only a single experiment pertinent to the present discussion. Chu and Nose [CN80] have measured $J_c^{-1}(\mathbf{0})$, ξ_d^2 for polystyrene in transdecalin ($\Theta \approx 20.5\,°\mathrm{C}$) in a range $20\,°\mathrm{C} \le T \le 40\,°\mathrm{C}$, $N_w \approx$ 1800; 125,000, and for a large range of concentrations. Evaluating independent measurements in the dilute limit for that system we have extracted nonuniversal parameters $\tilde{v} = 0.43\frac{T-\Theta}{\Theta}$; $\tilde{\ell} = 2.9\,\text{Å}$ (cf. Sect. 15.5.2 and Figs. 15.14, 15.15). With these parameters the prediction of our theory is compared to the experimental results in Figs. 19.11 ($\tilde{p}_2 J_c^{-1}(\mathbf{0})$) or 19.12 ($\xi_d^2/R_\Theta^2$). The agreement is

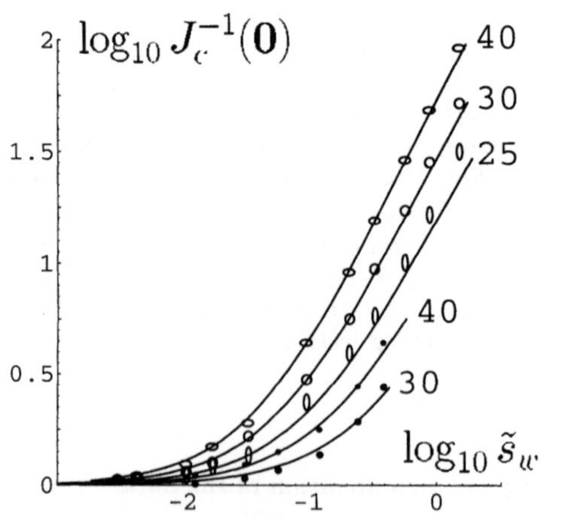

Fig. 19.11. $\log_{10} J_c^{-1}(\mathbf{0})$ as function of $\log_{10} \tilde{s}_w$, on the weak coupling branch. Data for polystyrene-transdecaline [CN80]. *Upper three curves*: sample "TSK", $N_w = 1.16\ 10^5$, *lower two curves*: sample "NBS", $N_w = 1.85\ 10^3$. Temperatures (°C) given with the curves

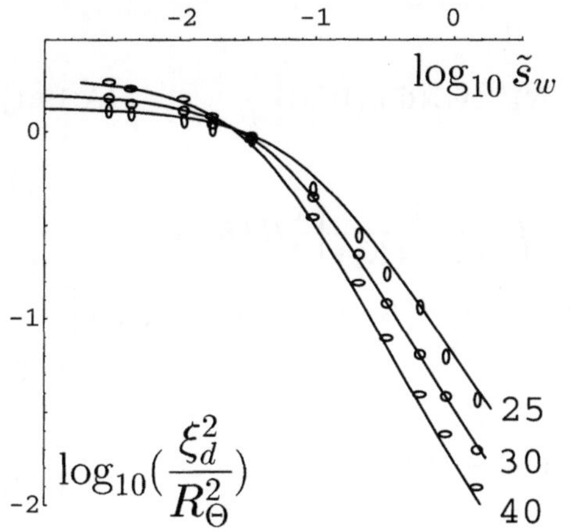

Fig. 19.12. $\log_{10}(\xi_d^2/R_\Theta^2)$ as function of $\log_{10} \tilde{s}_w$. Data for sample 'TSK' (see caption Fig. 19.11)

quite remarkable. Note that the only freedom allowed for these fits is a slight readjustment of the chain lengths which, in particular for the longer chain, are not known to high precision, experimentally.

A 19. Appendix

A 19.1 Full one-loop results

The one-loop contribution to $K_f^{(2,0)}(\mathbf{q})$ is given by the diagrams of Fig. 19.13. The denominator $K_a^{(2,0)}(K_a^{(2,0)}+K_f^{(2,0)})$ in Eq. (19.14) to the order considered is to be replaced by the square of the Debye function. Evaluating the diagrams and renormalizing the result we find the following expression for $M_c(\mathbf{q})$, valid

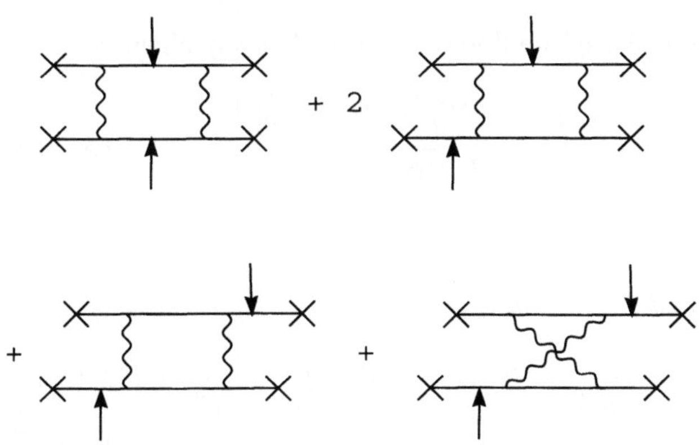

Fig. 19.13. One-loop diagrams, contributing to $K_f^{(2,0)}$

for $d \leq 4$:

$$M_c(\mathbf{q}) = W_R\Big\{1 + u^*\, f\left[\frac{2}{\varepsilon} - N_R^{\varepsilon/2}\mathcal{M}_c(Q_R, W_R)\right]\Big\} \tag{A 19.1}$$

$$\begin{aligned}\mathcal{M}_c(Q_R, W_R) = {} & \frac{1}{2}(4\pi)^{d/2}D_p^{-2}(Q_R)\int_{\hat{k}}\frac{1}{\rho(K)\rho(K_1)}\Big\{\frac{1}{2}H\,(K,K_1)^2 \\ & + H(K,Q_R)\,(2H(K,K_1) + H(K_1,Q_R) + H(K,Q_R))\Big\}\,.\end{aligned} \tag{A 19.2}$$

Here

$$H(z_1, z_2) = \frac{D_p(z_1) - D_p(z_2)}{z_1 - z_2} \quad . \tag{A 19.3}$$

We otherwise use the same notation as in the previous chapter (see Appendix A 18.1):
$\hat{\mathbf{k}} = \mathbf{k}N_R^{1/2}$; $K = \hat{k}^2$; $K_1 = (\mathbf{k} + \bar{\mathbf{q}})^2 N_R$; $\rho(z) = 1 + W_R D_p(z)$.

From these results our expression for the forward scattering intensity takes the form

$$J_{c,d}^{-1}(\mathbf{0}) = \frac{1}{\tilde{p}_2} + W_R\Big\{1 + u^*\, f\left[\frac{2}{\varepsilon} - N_R^{\varepsilon/2}\mathcal{M}_c(0, W_R)\right]\Big\} \tag{A 19.4}$$

$$\mathcal{M}_c(0, W_R) = \frac{1}{\tilde{p}_2^2\,\Gamma\left(\frac{d}{2}\right)}\int_0^\infty dz\,\frac{z^{-3-\varepsilon/2}}{\rho^2(z)}\left(T_1[\tilde{p}'(z)]\right)^2 \quad , \tag{A 19.5}$$

where we used the Taylor operator defined in Eq. (A 18.10). The result (A 19.4) has been analyzed in Chap. 17, where we used the notation

$$J_1(W_R) = \Gamma(\frac{d}{2})\mathcal{M}_c(0, W_R) \quad .$$

Expanding $J_d^{-1}(\mathbf{q})$ to order q^2 we find after some lengthy algebra

$$\xi_d^2 = \frac{d}{3}\ell_R^2\frac{\tilde{p}_3}{(\tilde{p}_2)^2}N_R\frac{1 - u^* f\left[\frac{1}{\varepsilon} + N_R^{\varepsilon/2}\xi^{(1)}(W_R, [p])\right]}{J_d^{-1}(\mathbf{q})} \tag{A 19.6}$$

$$\begin{aligned}\xi^{(1)}(W, [p]) = {} & \frac{-1}{\tilde{p}_3\,\Gamma\left(\frac{d}{2}\right)}\Big[\frac{3}{d}\int_0^\infty dz\,\frac{z^{-\varepsilon/2}}{\rho(z)}\,C_p^{(0)}(z) \\ & + \frac{3}{2}W_R\int_0^\infty dz\,\frac{z^{-\varepsilon/2}}{\rho^2(z)}\,C_p^{(1)}(z) + \frac{4}{d}W_R^2\int_0^\infty dz\,\frac{z^{-\varepsilon/2}}{\rho^3(z)}\,C_p^{(2)}(z)\Big]\end{aligned} \tag{A 19.7}$$

$$C_p^{(1)}(z) = \left[\frac{-4\tilde{p}_3}{3\tilde{p}_2}\frac{1}{z}T_0[\tilde{p}^{'}(z)] - \left(8+\frac{32}{d}\right)\frac{1}{z^3}T_2[\tilde{p}(z)] + \left(\frac{56}{3d}-\frac{1}{3}+\frac{d}{6}\right)\frac{1}{z^2}T_1[\tilde{p}^{'}(z)] - \frac{2}{3z}T_0[\tilde{p}^{''}(z)] - \frac{4}{3d}\tilde{p}^{'''}(z)\right] \cdot \frac{1}{z^2}T_1[\tilde{p}^{'}(z)] - \frac{2}{3d}\left(\frac{1}{z}T_0[\tilde{p}^{''}(z)]\right)^2 \tag{A 19.8}$$

$$C_p^{(2)}(z) = \left[\tilde{p}^{''}(z) - \frac{2}{z^2}T_1[\tilde{p}(z)]\right]\left(\frac{1}{z^2}T_1[\tilde{p}^{'}(z)]\right)^2 \quad . \tag{A 19.9}$$

$C_p^0(z)$ has been defined in Eq. (A 18.9). Equation (A 19.6) results from the loop expansion of $J_c^{-1}(\mathbf{q})$, without further manipulation.

In the semidilute limit, taken as $N_R \to \infty, \bar{\mathbf{q}}^2$ fixed, the appropriate momentum variable has been defined in Sect. A 18.2.2: $Q_1 = Q_R/2W_R$. The general result, valid for all polydispersity, reads

$$J_c^{-1}(\mathbf{q}) = W_R\Big\{1 + 2\frac{u^*}{\varepsilon}f + Q_1\left(1-\frac{u^*}{\varepsilon}f\right) + \frac{u^*}{2}\Gamma\left(\frac{\varepsilon}{2}\right)f\,(2fc_R)^{-\varepsilon/2}\left[\frac{\Gamma^2\left(1-\frac{\varepsilon}{2}\right)}{\Gamma(2-\varepsilon)}\frac{Q_1^{2-\varepsilon/2}}{2} - \frac{Q_1}{1-\varepsilon/2} - 2\left(\frac{Q_1}{2}-1\right)^2 F_1(Q_1)\right]\Big\} \tag{A 19.10}$$

$$F_1(Q_1) = \int_0^1 dx\,[Q_1 x(1-x)+1]^{-\varepsilon/2} \quad . \tag{A 19.11}$$

Expanding for $Q_1 \to 0$ we find

$$J_c^{-1}(\mathbf{q}) = J_c^{-1}(\mathbf{0})\left[1 + \frac{1}{d}\mathbf{q}^2\xi_E^2 + O(q^{4-\varepsilon})\right] \tag{A 19.12}$$

$$J_c^{-1}(\mathbf{0}) = W_R\left[1 + 2\frac{u^*}{\varepsilon}f\left(1 - \frac{\Gamma\left(1+\frac{\varepsilon}{2}\right)}{(2fc_R)^{\varepsilon/2}}\right)\right] \tag{A 19.13}$$

$$\xi_E^2 = d\,\frac{\ell_R^2}{2fc_R}\,\frac{1+\frac{u^*}{\varepsilon}f\left[(2fc_R)^{-\varepsilon/2}\Gamma\left(1+\frac{\varepsilon}{2}\right)\left(2+\frac{\varepsilon}{6}-\frac{1}{1-\varepsilon/2}\right)-1\right]}{1+2\frac{u^*}{\varepsilon}f\left[1-(2fc_R)^{-\varepsilon/2}\Gamma\left(1+\frac{\varepsilon}{2}\right)\right]} \quad . \tag{A 19.14}$$

A 19.2 Evaluations and parameterizations in three dimensions

A 19.2.1 Density correlation length. For the exponential chain length distribution we can evaluate $\xi^{(1)}(W,[p])$ analytically:

$$\xi^{(1)}(W,[e]) = \frac{-\Gamma(\varepsilon/2)}{2-\varepsilon}\,\varepsilon\Big\{(1+2W)^{-1-\varepsilon/2}\left(\frac{9}{8}+\frac{3\varepsilon}{16}-\frac{6}{d}+\frac{3\varepsilon}{2d}\right)$$
$$+(1+2W)^{-\varepsilon/2}\left(\frac{4}{\varepsilon d}+\frac{d}{12}+\frac{1}{3d}-\frac{3}{2}-\frac{\varepsilon}{3d}\right)$$
$$+\left((1+2W)^{-\varepsilon/2}-1\right)\Big[\frac{1}{4W}\left(-\frac{2}{3\varepsilon}-\frac{40}{3\varepsilon d}-\frac{1}{12}+\frac{d}{24}+\frac{22}{3d}-\frac{\varepsilon}{d}\right)$$
$$+\frac{1}{24W^2}\left(\frac{1}{\varepsilon}+\frac{20}{\varepsilon d}+\frac{1}{2}-\frac{7}{d}+\frac{\varepsilon}{2d}\right)+\frac{1}{8\varepsilon W^3}\Big]$$
$$-\frac{1}{4W}\left(\frac{1}{3}+\frac{\varepsilon}{6}-\frac{10}{3d}+\frac{7\varepsilon}{6d}-\frac{\varepsilon^2}{12d}\right)+\frac{1}{8W^2}\Big\} \qquad \text{(A 19.15)}$$

$$\underset{d=3}{=} -\sqrt{\pi}\Big\{-\frac{3}{16}(1+2W)^{-3/2}+(1+2W)^{-1/2}\left(\frac{1}{12}-\frac{71}{96W}+\frac{1}{4W^2}+\frac{1}{8W^3}\right)$$
$$+\frac{77}{96W}-\frac{1}{8W^2}-\frac{1}{8W^3}\Big\}\ . \qquad \text{(A 19.16)}$$

For general polydispersity only the semidilute limit can be evaluated analytically:

$$\xi^{(1)}(W,[p]) \underset{W\to\infty}{=} (2W)^{-\varepsilon/2}\,\frac{\Gamma(\varepsilon/2)}{2-\varepsilon}\,\varepsilon\Big[-\frac{4}{\varepsilon d}-1$$
$$-\frac{(\tilde{p}_2)^2}{\tilde{p}_3}\left(\frac{d}{8}-\frac{15}{4}+\frac{1-\varepsilon}{2d}\right)\Big] \qquad \text{(A 19.17)}$$

$$\underset{d=3}{=} \sqrt{\frac{\pi}{2W}}\left(-\frac{7}{3}+\frac{27}{8}\frac{(\tilde{p}_2)^2}{\tilde{p}_3}\right)\ . \qquad \text{(A 19.18)}$$

We can combine Eqs. (A 19.6), (A 19.17) with the results for $J_c^{-1}(\mathbf{0})$ given in Appendix A 17.2.2 to find the ε-expansion of ξ_d^2 in the semidilute limit:

$$\xi_d^2 = +\frac{d}{3}\,\ell_R^2\,\frac{\tilde{p}_3}{(\tilde{p}_2)^2}\,\frac{1}{f c_R}\Big[1-\frac{\varepsilon}{4}\,f\Big(\frac{3}{2}\ln(2\,f\,c_R)+\frac{3}{2}\gamma_{Eu}-\frac{7}{4}$$
$$+\frac{25}{8}\frac{(\tilde{p}_2)^2}{\tilde{p}_3}\Big)+O(\varepsilon^2)\Big]\ . \qquad \text{(A 19.19)}$$

For the monodisperse ensemble we have evaluated the function $\xi^{(1)}(W,[m])$ numerically for $d=3$. Within 1 % deviation the result can be parameterized as

$$\xi^{(1)}(W,[m]) = \frac{-0.6382 + 0.1059W + 0.00115W^2}{1 + 0.6900W + 0.01647W^2} \cdot (1 + 0.00288W)^{-1/2} \quad . \tag{A 19.20}$$

We finally note the expansions for $W \ll 1$, as derived from Eq. (A 19.7) (monodisperse) or Eq. (A 19.16) (exponential):

$$\begin{aligned} \xi^{(1)}(W,[m]) &= -0.638 + 0.547W + O(W^2) \\ \xi^{(1)}(W,[e]) &= -1.237 + 1.255W + O(W^2) \quad . \end{aligned} \tag{A 19.21}$$

A 19.2.2 $\mathcal{M}_c(Q,W)$ for $d = 3$. For the exponential ensemble in three dimensions the expression (A 19.2) for $\mathcal{M}_c(Q,W)$ can be evaluated analytically:

$$\begin{aligned} \mathcal{M}_c(Q,W,[e]) &= \frac{\sqrt{\pi}}{16W^2}\Big\{8W\left((1+2W)^{1/2}-1\right) \\ &+(1+Q)^2Q^{-1/2}\arccos\frac{4-Q}{4+Q} + (1+Q-4W)^2Q^{-1/2}\arccos\frac{4+8W-Q}{4+8W+Q} \\ &-2(1+Q-2W)^2Q^{-1/2}\arccos\frac{4W^2+4Q\sqrt{1+2W}-Q^2}{4W^2+4Q(1+W)+Q^2}\Big\} \quad . \end{aligned} \tag{A 19.22}$$

In the semidilute limit, taken as $W \to \infty, Q_1 = Q/2W$ fixed, $\mathcal{M}_c(Q,W)$ for all polydispersity tends to the limiting result following from Eq. (A 19.22).

For the monodisperse ensemble we have evaluated $\mathcal{M}_c(Q,W,[m])$ numerically. We parameterize our result using the analytical form of $\mathcal{M}_c(Q,W,[e])$ by making the ansatz

$$\begin{aligned} \mathcal{M}_c(Q,W,[m]) &= \mathcal{M}_c(0,W,[m]) \\ &\quad + a(W)\left[\mathcal{M}_c(b(W)Q,W,[e]) - \mathcal{M}_c(0,W,[e])\right] \quad , \end{aligned} \tag{A 19.23}$$

where

$$\mathcal{M}_c(0,W,[m,e]) = \frac{2}{\sqrt{\pi}}J_1(W,[m,e]) \quad . \tag{A 19.24}$$

The expressions for $J_1(W,[m,e])$ are given in Appendix A 17.2.2. For $a(W)$, $b(W)$ we find the results

$$\begin{aligned} a(W) &= 1 + \frac{0.114 - 0.398W}{(1 + 6.79W + 29.9W^2 + 11.52W^3)^{1/2}} \\ b(W) &= 1 - \frac{0.503 + 0.05W}{(1 + 3.30W + 0.38W^2 + 0.51W^3)^{1/2}} \quad . \end{aligned} \tag{A 19.25}$$

The parameterization (A 19.23)–(A 19.25) yields the correct limits for $Q \to 0$ or $W \to \infty$, and the resulting values of $J_d^{-1}(\mathbf{q})$ differ from the full numerical results by less than 0.4 % in all the parameter region. Otherwise the analytical form of the parameterization has no deeper meaning.

Notes

Calculations of $I_{d,c}(\mathbf{q})$ on the one-loop level may be found in [ON83], [Fre83], [Sch84], [NO85]. The density correlation length has been discussed in detail, but the momentum dependence has not found much interest. Only in [NO85] is a series of plots quite similar to Fig. 19.8 shown. The distinction among ξ_d and ξ_E has been pointed out in [Sch84]. All this work is consistent with the results presented here. It must be noted, however, that the quantitative estimates for ξ_d in the semidilute limit depend somewhat on the precise method used to extrapolate the theory to $d = 3$. Such uncertainties have to be expected in view of the large one-loop correction.

A detailed discussion of the Zimm plot from an experimental point of view may be found in [Kra87].

20. Summarizing Overview

In this part we have discussed an internally consistent evaluation of the theory to one-loop order. It verifies the asymptotic power laws predicted by scaling theory by showing that the assumed limiting behavior of the scaling functions indeed holds to the order considered. More importantly, it provides us with quantitative predictions for the scaling functions in all the crossover regime. As a general rule the crossover functions vary extremely smoothly and show no pronounced structure. This is by no means surprising for a scale invariant system. It, however, implies that asymptotic power laws are attained only slowly. It also implies that parameter values like c^*, where the dilute behavior shall cross over to semidilute behavior, are not well defined. Also the temperature blob model is not meaningful as a model of scaling functions. Data analysis based on such oversimplified pictures is likely to result in effective parameter range dependent exponents. Quite generally it must be noted that power law fits, which have become so popular with the advent of scaling theory, often miss relevant information contained in more precise data and sometimes even have led to a misinterpretation of the results.

The only structural peculiarity of the scaling functions, which does not result from scale invariance, is the occurrence in the semidilute limit of subleading square root singularities. These are a consequence of the residual interaction among the concentration blobs, and for quantities like $I_{d,c(a)}(\mathbf{q})$ in the range $q^2\zeta_E^2 \lesssim 1$, they qualitatively change the results as compared to the renormalized tree approximation. In field theory of critical phenomena such effects are well known as 'Goldstone mode' singularities. To my knowledge they have never been observed, due to experimental difficulties. Thus an experimental verification of predictions like the behavior of $\zeta_E^2(q)$, shown in Fig. 19.7, would be most interesting.

Another feature, which has not received sufficient attention up to now, is the existence of two branches of universal behavior. We have found that typical good solvent systems are described by the strong coupling branch close to the fixed point. It would be quite interesting to find systems in 'very good solvents', where we might be able to follow the strong coupling branch somewhat further, without running into the region of short chains or high concentration, where the theory breaks down. From Fig. 15.19 the systems PMMA-nitroethane or PMMA-chloroform might be good candidates. Also good data for 'moderately good solvent' systems close to the fixed point on the weak coupling side seem to be missing.

We have seen numerous examples which show that theory and experiment agree within the scatter of the data. We can thus expect the theory to be accurate up to deviations of a few percent, holding for flexible chains of lengths $N \gtrsim 500$, and solutions of concentrations less than about 10 wt%. We, however, also have seen that for long chains data analysis is tricky and extrapolation for instance to $c = 0$ or $q = 0$ always implicitly relies on a theoretical model. It would be preferable to compare directly the measured data to scaling functions as evaluated here.

We should finally note another point of concern. Analyzing data for the same chemical system and the same observable, but from different experimental groups, we typically can bring the data on the common scaling curve, but may have to use slightly different nonuniversal parameters. There are quite a few cases where the difference among the parameters exceeds the error bars assigned to each experiment individually. In as far as the same quantity is measured in the same range of concentrations and chain lengths, this is not a problem of the theory. It could be due to uncontrolled polydispersity effects, but also to systematic errors in the data analysis. In any case it clearly hinders a more stringent and comprehensive test of the theory.

We have discussed here only binary solutions in the excluded volume regime, where the theory has been worked out in most detail. In the last part we will consider the tricritical theory of the Θ-point and the extension of the theory to solutions of mixtures of chemically different polymer species. We briefly note here some other extensions of the theory.

In recent years 'living' polymer solutions have found some interest. In such solutions the polymerization reaction is not stopped, but the chains can prolong and shorten according to their surrounding. (For a recent review see [Gre96].) In the simplest case such solutions are described by a chemical potential $\mu_p(n) = \mu_0 + \mu_1 n$, which also underlies the field theoretic polymer ensemble (see Appendix A 7.1). All our results can easily be transcribed to such systems. The chain length distribution in such a system to lowest (tree) approximation is a simple exponential and within the restricted accuracy of the experiments an analysis based on the exponential ensemble might be sufficient. In higher orders the chain length distribution is modified somewhat, depending on the concentration. The necessary changes could easily be implemented in our evaluation of the theory.

The theory can also be extended to solutions of block-copolymers or of macromolecules of more complicated architecture like stars. The only necessary condition is that all strands of the macromolecule are sufficiently long. The general scaling properties are quite similar to those of a binary solution of normal chains and can be worked out easily. (See, for instance, [SvFLD92].) A reliable calculation of the scaling functions is more complicated, due to the richer structure even on the tree level. Only partial results are available. (See, for instance, [CBMF86, AC92, GA96].)

Part V

Special Topics

We consider here two extensions of the theory. As has been mentioned several times, in physical dimension $d = 3$ three-body forces should be taken into account in the Θ-region. This leads to logarithmically decreasing corrections to the Gaussian Θ-behavior, considered so far. As the second topic we discuss solutions containing different polymer species. Even within an approach including only two-body forces, such systems show a very rich crossover behavior, as well as the new phenomenon of a possible phase separation in the excluded volume regime. Both topics have been analyzed extensively in the literature, but due to technical complexity the theory has not been worked out to the same level as binary solution theory.

21. The Θ-Region

Throughout this book we have dealt with the two-parameter model, adequate in a region where the repulsive interaction among pairs of monomers dominates. In Chap. 2 we sketched the steps which in principle must be followed to derive this model. Starting from a realistic microscopic description of polymer and solvent we first have to eliminate all degrees of freedom of the solvent molecules as well as the internal degrees of freedom of the segments. This process results in a tremendously complicated effective interaction, of which we retain only the Gaussian chain part and a local two-segment repulsion. A Flory-type estimate (Sect. 2.2, Eq. (2.10)), equivalent to simple power counting (Sect. 7.2), suggests that all further interactions are irrelevant for $d > 3$. Their contribution to the internal energy of a chain decreases like a negative power of the chain length. (The more precise discussion is given in Chap. 10.) Ignoring these interactions, we treat the Θ-point as a simple Gaussian fixed point, where even the two-body interaction vanishes and where the chain configuration is random-walk like.

Strictly speaking this treatment of Θ-chains is valid only above three dimensions. In $d = 3$ the naive estimates show that the three-body interaction is marginal, its effect apparently being independent of chain length. The renormalization group reveals that (repulsive) three-body interactions in fact are marginally irrelevant: with increasing renormalized segment size their effect decreases, but only logarithmically. (We note in passing that attractive three-body interactions are marginally relevant. For reasons explained below in the sequel we only consider repulsive interactions.) As a consequence of the marginal irrelevance our previous treatment of the Θ-point in $d = 3$ is justified in so far as it is still the Gaussian fixed point that governs the limit of infinitely long chains and vanishing segment concentration. Corrections to the fixed point behavior, due to finite chain length or finite concentration, are strongly modified however. We consider two examples.

For the radius of gyration our previous analysis yields

$$R_g^2 = \tilde{\ell}^2 N \left(1 + O\left(\frac{1}{N}\right)\right), \quad T = \Theta \quad , \tag{21.1}$$

where the $1/N$-corrections result from microscopic features like stiffness, not included in our model. Three-body forces change this result to

$$R_g^2 = \tilde{\ell}^2 N \left(1 - \frac{493}{5808} \ln^{-1}(N) + O\left(\frac{\ln(\ln N)}{(\ln N)^2}\right)\right), T = \Theta \quad . \tag{21.2}$$

Thus, in the presence of three-body forces the corrections to random walk behavior $R_g^2 \sim N$ die out extremely slowly.

A priori, the structure of the result (21.2) is quite surprising. The coefficient of the correction $\sim \ln^{-1} N$ is *universal*, the result being valid for $N \to \infty$ independent of the strength β_3 of the three-body interaction, as long as $\beta_3 > 0$. This immediately raises the question of how Eq. (21.2) crosses over to Eq. (21.1), which must hold for $\beta_3 = 0$. Now it turns out that β_3 determines the region of validity of the asymptotic law (21.2), according to the criterion $\beta_3 \ln N \gg 1$. For N fixed, $\beta_3 \to 0$, Eq. (21.2) breaks down and the trivial behavior (21.1) is recovered. The corresponding statement holds for all the asymptotic results we quote below.

As a second example we consider the virial expansion of the osmotic pressure, written in terms of the overlap $s = c_p R_g^3$:

$$\frac{\hat{\Pi}}{k_B T c_p} = 1 + \frac{a_2}{2} s + \frac{a_3}{6} s^2 + \ldots \quad . \tag{21.3}$$

According to our previous analysis

$$\begin{aligned} a_2(N, T = \Theta) &\equiv 0 \\ a_3(N, T = \Theta) &\equiv 0 \quad , \end{aligned} \tag{21.4}$$

simply because the chains do not interact. For short chains again $1/N$-corrections due to end-effects are expected, but are ignored in our model. Including three-body forces we however find

$$a_2(N, \Theta) = -\frac{8}{11} \frac{(4\pi)^{1/2}}{\ln N} \tag{21.5}$$

$$N \to \infty$$

$$a_3(N, \Theta) = \frac{2}{11} \frac{(4\pi)^2}{\ln N} \quad . \tag{21.6}$$

Again the corrections to naive Θ-point behavior are universal and vanish most slowly.

This result shows that the very definition of the Θ-temperature is more delicate than we thought before. Instead of the simple criterion (21.4) holding for all N beyond the oligomer region, we have to consider very long chains for a precise definition:

$$\lim_{N \to \infty} a_2(N, T = \Theta) = 0 \quad . \tag{21.7}$$

To be sure, for fixed N, there always is a compensation temperature $\Theta(N)$, where

$$a_2(N, \Theta(N)) = 0 \quad ,$$

but this temperature plays no special role for other quantities. For $T = \Theta(N)$ neither the third virial coefficient vanishes, nor is the chain configuration random walk like. Furthermore $\Theta(N)$ will depend on polydispersity. The proper Θ-temperature in the sense defined here is the limit of $\Theta(N)$ for $N \to \infty$.

Such results identify the polymer solution at $T = \Theta$ as a 'tricritical' system at its upper critical dimension $d = 3$. For $d < 3$ the three-body interaction is relevant, and the Θ-point is described by a new nontrivial fixed point of the renormalized three-body interaction. Thus quasi-two-dimensional systems like a layer of polymer solution thin on scale R_g should show Θ-point properties that are *not* random walk like. New nontrivial power- and scaling-laws emerge. We do not review this topic here, but restrict ourselves to $d = 3$ where the logarithmic corrections are the remains of tricriticality.

Results like Eqs. (21.2) or (21.5) hold right at the Θ-point. The logarithmic corrections, however, show up in all the 'tricritical Θ-region', which roughly corresponds to the Θ-region $z \sim \beta_e n^{1/2} \lesssim 1$ of naive two-parameter theory. This region can be analyzed by expanding about $T = \Theta$. The expansion takes the form of the naive z-expansion, but renormalized by three-body effects. This seems to be the only systematic approach, but the method does not reach very far. Even for temperatures $T > \Theta$ very close to the Θ-point with increasing chainlength we rapidly reach the region $z \gg 1$, where the tricritical z-expansion breaks down and should be renormalized by two-body effects. This renormalization, which has been dealt with at length in the previous chapters, is *not included* in the tricritical theory. Indeed, renormalization by two-body interactions basically relies on the expansion about $d = 4$, whereas tricritical effects show up only for $d \leq 3$. No systematic method is known which can deal simultaneously with two critical phenomena of differing upper critical dimensions. Thus either we can analyze the excluded volume problem, ignoring three-body forces and treating the Θ-point as trivial, or set up the tricritical theory, treating two-body effects as small corrections. A systematic matching of the two approaches is not available. It must be stressed that this is purely a technical problem. The general renormalization scenario expounded in Sect. 10.2 well covers the crossover from tricriticality to criticality. The problem is that no systematic perturbative implementation of the RG in all the space spanned by the two-body and the three-body coupling is known.

Turning now to temperatures slightly below Θ we encounter an even more complex situation. We first note that the two-parameter model, including only a local two-body interaction, breaks down. For sufficiently long chains any attractive coupling $\beta_e < 0$ leads to an unphysical collapse of the chain to a microscopically small lump. Additional forces like a many-body repulsion or a hard core of finite range have to be added to transform this unphysical collapse into the physically expected collapse leading to a dense globule of radius $R \sim N^{1/d}$. The simplest model includes a repulsive three-body interaction, leading to the tricritical approach.

As compared to $T > \Theta$ we now have to face the additional complication of the existence of a phase separation in the solution. As mentioned in Chap. 1 for temperatures $T < T_{cr}(N) < \Theta$, there is a two-phase region, where a dense solution is in equilibrium with a dilute solution of collapsed chains. For $N \to \infty$ the critical point $(T_{cr}(N), c_{cr}(N))$ of that phase transition is believed to merge with the Θ-point $(T = \Theta, c = 0)$, and the asymptotic shape of the line $(T_{cr}(N), c_{cr}(N))$ can be calculated within the tricritical theory. For $N < \infty$, however, the physics close to $(T_{cr}(N), c_{cr}(N))$ is governed by a fixed point of its own and cannot be analyzed within the framework of the tricritical approach.

We finally comment on the traditional parameterization of the temperature dependence within the tricritical theory. Clearly both the excluded volume coupling β_e and the three-body coupling β_3 should be analytic functions of the temperature, so that

$$\begin{aligned} \beta_e &= \beta_{e,1}\frac{T-\Theta}{\Theta} + \ldots \\ \beta_3 &= \beta_{3,0} + \beta_{3,1}\frac{T-\Theta}{\Theta} + \ldots \quad . \end{aligned}$$

(Note that β_e is the two-body coupling, subtracted such that it vanishes at Θ; see Sect. 21.1.) Applications of the tricritical theory keep only the leading terms in these expansions. Furthermore it is assumed that $\beta_3 = \beta_{3,0} > 0$, so as to prevent the unphysical collapse. Concerning the latter assumption it must be stated that it is basic to the tricritical approach, but has no deeper justification. For $\beta_3 < 0$ in the full effective interaction there will be higher terms or a hard core, ultimately, to cure the collapsed state. We then can sketch a scenario without a proper Θ-point. Its role is taken over by a critical value β_2^{cr} of the two-body coupling. For $\beta_2 > \beta_2^{cr}$ the long chains are fully swollen, and are collapsed for $\beta_2 < \beta_2^{cr}$. At β_2^{cr} swollen and collapsed configurations are in equilibrium. In fact, recently there are emerging hints that for some models 'Θ-behavior' might be much more complicated than predicted by the tricritical theory [DOST95].

Now, working within the tricritical framework, i.e. taking $\beta_{3,0} > 0$, we still should discuss the neglect of the temperature variation. It is based on the assumption that $\beta_{3,0}$ is large enough, so that in the restricted temperature range covered by the theory the terms $\beta_{3,1,\ldots}$ can be ignored. This clearly is justified in some neighborhood of $T = \Theta$, which for long enough chains covers all the Θ-region. Whether this assumption is valid in the experimental range is an open question, in particular if the experimentally extracted value of β_3 is small.

In the following sections we will work out the details of the tricritical theory. In Sect. 21.1 we consider the renormalization properties. These are found to be similar to but somewhat more complicated than for the excluded volume problem. We have to deal with both the two-body and the three-body interactions, and we have to perform new additive renormalizations, which account for the definition (21.7) of the Θ-point. However, no new techniques must be introduced, the diagrammatic expansion being an obvious generalization of that derived in Chaps. 4 and 5. Section 21.2 is devoted to the RG mapping and the general scaling laws. Section 21.3 reports results derived in the dilute limit. Details of the calculations are collected in appendices. Experimental evidence pertinent to the dilute limit is discussed in Sect. 21.4. Based on the loop expansion briefly discussed in Appendix A 21.3, zero-loop results for the concentration crossover are derived in Sect. 21.5. A brief discussion (Sect. 21.6) of the experimental situation for finite overlap closes this chapter.

It will become clear that on the quantitative level the theory has not been worked out as completely as the theory of the excluded volume region. Some work still needs to be done. Considering also the problems outlined above we conclude that our understanding of the Θ-region is much less advanced than our understanding of the excluded volume limit, a priori a surprising situation.

21.1 Renormalization properties

21.1.1 Model and additive renormalization

Introducing three-body forces we write the (not normalized) configurational weight for a system of M discrete chains as (cf. Sect. 2.1)

$$\exp[-V^{(M)}\{r_j^{(m)}\}] = \exp\left\{-\frac{1}{4\ell^2}\sum_{m=1}^{M}\sum_{j=1}^{n_m}\left(\mathbf{r}_j^{(m)} - \mathbf{r}_{j-1}^{(m)}\right)^2\right\}$$
$$\cdot \prod'_{(m_1 j_1, m_2 j_2)}\left[1 - (4\pi\ell^2)^{d/2}\beta_2\delta^d\left(\mathbf{r}_{j_1}^{(m_1)} - \mathbf{r}_{j_2}^{(m_2)}\right)\right]$$
$$\cdot \prod'_{(m_1 j_1, m_2 j_2, m_3 j_3)}\left[1 - (4\pi\ell^2)^d\beta_3\delta^d\left(\mathbf{r}_{j_1}^{(m_1)} - \mathbf{r}_{j_2}^{(m_2)}\right)\delta^d\left(\mathbf{r}_{j_2}^{(m_2)} - \mathbf{r}_{j_3}^{(m_3)}\right)\right] . \tag{21.8}$$

Again we exclude from the cluster expansion all terms where any segment interacts more than once. In this section we will mainly be concerned with the M-chain partition functions

$$\mathcal{Z}^{(M)}(n_1,\ldots,n_M) = \int\limits_{\Omega} \prod_{m=1}^{M} \prod_{j_m=1}^{n_m} \frac{d^d r_{j_m}^{(m)}}{(4\pi\ell^2)^{d/2}} e^{-V^{(M)}} \tag{21.9}$$

for values $M \leq 3$. To simplify the notation we furthermore consider monodisperse systems: $n_m \equiv n$. The virial coefficients of the osmotic pressure as defined by the expansion

$$\frac{\hat{\Pi}}{k_B T} = c_p + \frac{1}{2} A_2^{\Pi} c_p^2 + \frac{1}{6} A_3^{\Pi} c_p^3 + \ldots$$

can be expressed in terms of the cumulants constructed from the $\mathcal{Z}^{(M)}$. (See Sect. 4.3 and Eq. (5.8) for the definition of the cumulants.)

$$A_2^{\Pi} = -\Omega \frac{\mathcal{Z}_c^{(2)}(n,n)}{[\mathcal{Z}^{(1)}(n)]^2} \tag{21.10}$$

$$A_3^{\Pi} = -2\Omega^2 \frac{\mathcal{Z}_c^{(3)}(n,n,n)}{[\mathcal{Z}^{(1)}(n)]^3} + 6\left(A_2^{\Pi}\right)^2 \quad . \tag{21.11}$$

The cluster expansion proceeds as before, except that we have to include the three-segment vertex. (See Fig. 21.1, for instance.)

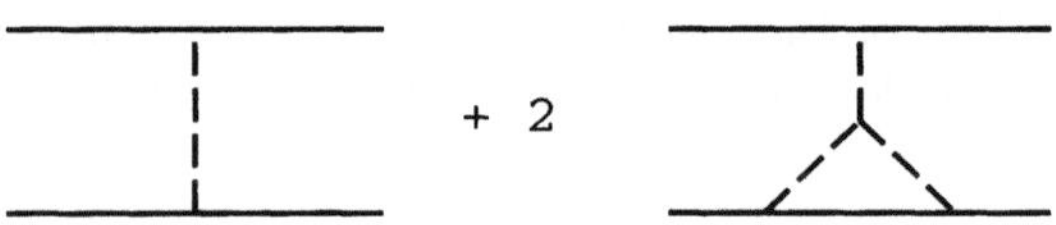

Fig. 21.1. First order diagrams contributing to the two-chain cumulant

In order to determine the Θ-point we have to analyze A_2^{Π}. To first order in β_2, β_3 the diagrammatic expansion of $\mathcal{Z}_c^{(2)}$ is shown in Fig. 21.1. Evaluating these diagrams and taking into account the trivial leading term in $\mathcal{Z}^{(1)}$ we find (see Appendix A 21.1.2)

$$(4\pi\ell^2)^{-d/2} A_2^{\Pi} = n^2 \left(\beta_2 + \frac{2}{n}\beta_3 \sum_{0<j_1<j_2<n} (j_2 - j_1)^{-d/2} + O(\beta_2^2, \beta_3^2, \beta_2\beta_3) \right) .$$

The sum has been evaluated in Appendix A 4.2. Up to terms of order $n^{-d/2}$ the first order expression

$$(4\pi\ell^2)^{-d/2} \frac{A_2^{\Pi}}{n^2} = \beta_2 + \beta_3 \left(2\zeta(\frac{d}{2}) - \frac{8n^{1-d/2}}{(d-2)(4-d)} \right) \tag{21.12}$$

results. For $d > 2$ the leading contribution to the right hand side vanishes for

$$\beta_2^{\Theta}(\beta_3) = -2\beta_3\zeta(\frac{d}{2}) + O(\beta_2^2) \ . \tag{21.13}$$

This identifies the Θ-point, as will be discussed below. We now define the excluded volume coupling β_e as

$$\beta_e = \beta_2 - \beta_2^{\Theta}(\beta_3) \ . \tag{21.14}$$

Note that this 'additive renormalization' just takes into account the two-body part extracted from the three-body interaction, as explained in Sect. 10.1. We finally introduce the radius of a noninteracting chain, $R_0 = \ell\, n^{1/2}$, to write Eq. (21.12) as

$$(4\pi)^{-d/2}\frac{A_2^{\Pi}}{R_0^d} = \beta_e n^{2-d/2} - \frac{8}{(d-2)(4-d)}\beta_3\, n^{3-d} + O(n^{1-d/2}) \ . \tag{21.15}$$

Clearly the three-body term $\sim \beta_3 n^{3-d}$ is consistent with our general expectations: for $d > 3$ it vanishes in the limit $n \to \infty$, and β_3 is irrelevant. For $d < 3$ it blows up, signaling the relevance of three-body forces at the Gaussian fixed point. $d = 3$ is the marginal situation.

We now turn to the precise definition of the Θ-point, which in fact is a delicate matter. On a superficial level Θ often is identified by the vanishing of the second virial coefficient in an expansion in powers of the *segment density* $c = n\, c_p$. Thus

$$\frac{A_2^{\Pi}(n,\Theta)}{n^2} \approx 0 \text{ for } n \gg 1 \ . \tag{21.16}$$

From previous chapters however we know the excluded volume behavior $A_2^{\Pi}(n) \sim n^{\nu d} \sim R_g^d$, so that

$$\frac{A_2^{\Pi}}{n^2} \sim n^{\nu d - 2} \underset{n\to\infty}{\longrightarrow} 0, \quad d < 4 \ ,$$

holds *even in the excluded volume limit.* As mentioned in the introductory section the proper definition of the Θ-point therefore must be based not on Eq. (21.16) but on the vanishing of a contribution $A_2^{\Pi} \sim R_g^d$:

$$a_2(n,\Theta) = \frac{A_2^{\Pi}(n,\Theta)}{R_g^d(\Theta)} \underset{n\to\infty}{\longrightarrow} 0 \ . \tag{21.17}$$

This criterion is fulfilled in a simple way only for $d > 3$: assuming $R_g(\Theta) \sim R_0$ from Eq. (21.15), $d > 3$, we find

$$\beta_e(\Theta) = 0 \ , \tag{21.18}$$

and calculating R_g we find that Eq. (21.18) indeed guarantees that $R_g(\Theta) \to R_0, n \to \infty$. Thus the argument is internally consistent, and the Θ-point is described by the Gaussian fixed point.

Repeating the argument for $d = 3$ we apparently run into problems. Equations (21.15), (21.18) yield

$$\frac{A_2^{\Pi}}{R_0^d} \sim const\, \beta_3 \neq 0 \quad .$$

We must in fact invoke renormalization to show that the definition (21.17) is equivalent to the vanishing of β_e (Eq. (21.18)). As mentioned above, it is found that under renormalization $\beta_3 > 0$ is marginally irrelevant. The three-body contribution in Eq. (21.15) is replaced by *const* / ln n, vanishing for $n \to \infty$. Also $R_g \sim R_0(1 + const\, / \ln\, n)$, so that the equivalence of Eqs. (21.17), (21.18) still holds in $d = 3$. The Θ-point is still identified as the Gaussian fixed point, but corrections due to finite chain length vanish only logarithmically. For $d < 3$, finally, the argument breaks down, $R_g(\Theta) \not\sim R_0$, and the Θ-point converts into a new nontrivial fixed point.

We now turn to the additive renormalization introducing the critical chemical potential μ_s^*. This is a simple generalization of the additive renormalization necessary to define the continuous chain limit in a theory involving only two-body forces (see Sect. 7.2). μ_s^* is specified completely by considering the single chain partition function. To first order the expansion of $\mathcal{Z}^{(1)}$, as given in Appendix A 21.1.1, yields

$$\begin{aligned} \frac{(4\pi\ell^2)^{-d/2}}{\Omega} \mathcal{Z}^{(1)}(n) = 1 &- \beta_2 \sum_{0<j_1<j_2<n} (j_2 - j_1)^{-d/2} \\ &- \beta_3 \sum_{0<j_1<j_2<j_3<n} (j_2 - j_1)^{-d/2} (j_3 - j_2)^{-d/2} \quad . \end{aligned}$$

Using the expressions for the sums given in Eq. (A 21.2) and introducing β_e (Eq. (21.14)) we can write this result explicitly as

$$\begin{aligned} \frac{(4\pi\ell^2)^{-d/2}}{\Omega} \mathcal{Z}^{(1)}(n) = 1 &- \beta_e \left[n\, \zeta(\frac{d}{2}) - \frac{4n^{2-d/2}}{(d-2)(4-d)} \right] \\ &+ \beta_3 \left[n\, \zeta^2(\frac{d}{2}) - \frac{4\Gamma^2(2-\frac{d}{2}) n^{3-d}}{(d-2)^2 \Gamma(4-d)} \right] , \end{aligned} \tag{21.19}$$

where we neglected terms of order $n^{1-d/2}$. Recalling the definition of the segment chemical potential for an infinitely long chain

$$\mu_s^* = - \lim_{n\to\infty} \frac{1}{n} \ln \left(\frac{(4\pi\ell^2)^{-d/2}}{\Omega} \mathcal{Z}^{(1)}(n) \right) \tag{21.20}$$

we thus find

$$\mu_s^* = \beta_e \zeta(\frac{d}{2}) - \beta_3 \zeta^2(\frac{d}{2}) + O(\beta_e^2, \beta_3^2, \beta_e\beta_3) \,. \tag{21.21}$$

The partition function can be written as

$$\frac{(4\pi\ell^2)^{-d/2}}{\Omega}\,\mathcal{Z}^{(1)} = e^{-\mu_s^* n}\Big\{1 + \frac{4}{(d-2)(4-d)}\,\beta_e n^{2-d/2} - \frac{4}{(d-2)^2}\frac{\Gamma^2(2-\frac{d}{2})}{\Gamma(4-d)}\,\beta_3 n^{3-d} + O(\beta_e^2, \beta_3^2, \beta_e\beta_3)\Big\}\,. \tag{21.22}$$

In the curly brackets we can now introduce variables

$$\begin{aligned} R_0^2 &= \ell^2 n \\ \tilde{u}_e &= \beta_e\,\ell^{d-4} \\ \tilde{u}_3 &= \beta_3\,\ell^{2d-6} \quad , \end{aligned}$$

and we can take the continuous chain limit

$$\ell \to 0; \quad R_0, \tilde{u}_e, \tilde{u}_3 \quad \text{fixed.}$$

Indeed, it can be shown that with the additive renormalizations introducing β_e, μ_s^* the continuous chain limit can be taken to all orders perturbation theory in all Greensfunctions for dimensions $2 < d < 3$. This again is the basis of dimensional regularization. It may be noted that in Eq. (21.22) the continuous chain limit can even be taken for $d = 3$, which, however, is a special feature of this low order result.

As has been discussed in Sect. 7.2, separating off the factor $e^{-\mu_s^* n}$ in Eq. (20.22) is equivalent to the 'mass subtraction' of field theory, necessary to make the theory ultraviolet finite for $d < 4$. In a theory involving only two-body forces we furthermore found that for $d = 4 - \varepsilon, \varepsilon > 0$, only a finite number of basic diagrams need the subtractions implied by additive renormalization. Specifically for $d < 3$ only the single contribution of order β_e is to be considered. If we include three-body forces this implies that μ_s^* for $d < 3$ rigorously is of the form

$$\mu_s^* = \mu_s^{(0)}(\beta_3) + \beta_e \mu_s^{(1)}(\beta_3)\,. \tag{21.23}$$

In three dimensions diagrams of order β_e^2 develop new u.v. singularities which in $d > 3$ are to be absorbed into μ_s^*. In $d = 3$ these contributions give rise to an additional multiplicative renormalization, which is considered below (Sect. 21.1.3).

21.1.2 Multiplicative renormalization at $T = \Theta$

At the Θ-temperature, now defined by $\beta_e = 0$, renormalization proceeds in complete analogy to the renormalization in the excluded volume region. The three-body coupling just takes over the role of the two-body coupling. We introduce the renormalized segment size ℓ_R, and we define renormalized variables via

$$\begin{aligned} \beta_3 \ell^{-2\varepsilon_3} &= u_3 \ell_R^{-2\varepsilon_3} Z_3(u_3) \\ n\ell^2 &= n_R \ell_R^2 Z_n^{\Theta}(u_3) \quad , \end{aligned} \tag{21.24}$$

where

$$\varepsilon_3 = 3 - d \tag{21.25}$$

and $d \leq 3$. Renormalized cumulants G_{cR} are defined via the relation (cf. Sect. 11.2)

$$\exp\left\{\mu_s^{(0)}(\beta_3)\sum_{m=1}^{M} n_m\right\}\frac{\ell^d}{\Omega}\,G^{(L,M)}_{c,m_1\ldots m_L}(q_1\ldots p_{2M};n_1,\ldots,n_M;\beta_3,\ell)$$
$$= (4\pi)^{-M\frac{d}{2}}\left(\frac{\ell}{\ell_R}\right)^{d-2L-dM}\left(Z^\Theta(u_3)\right)^M\left(Z_n^\Theta(u_3)\right)^{L-M}$$
$$\cdot G^{(L,M)}_{cR,m_1\ldots m_L}(\bar{\mathbf{q}}_1,\ldots,\bar{\mathbf{p}}_{2M};n_{R1},\ldots,n_{RM};u_3)\left[1+O\left(\frac{\ell}{n_R^{1/2}\ell_R}\right)\right] \tag{21.26}$$

where $\bar{\mathbf{q}} = q\ell_R, \bar{\mathbf{p}} = p\ell_R$, and $Z^\Theta(u_3)$ denotes another renormalization factor. The theorem of renormalizability guarantees the existence of $G^{(L,M)}_{c,R}$ for $d=3$ and identifies the order of the neglected terms as $O\left(\ell/n_R^{1/2}\ell_R\right)$.

Using dimensional regularization and minimal subtraction we take the continuous chain limit for $d < 3$ and fix the renormalization factors to cancel the pole terms showing up for $\varepsilon_3 \to 0$. Specifically Z_3 is determined by renormalizing $\mathcal{Z}_c^{(3)} = G_c^{(0,3)}(0,\ldots,0;n,n,n;\beta_3,\ell)$, whereas Z^Θ and Z_n^Θ follow from the renormalization of $\mathcal{Z}^{(1)}$ or $R_e^2 = -\Delta_q|_0 \ln G_c^{(0,1)}$, respectively. The calculation is sketched in Appendix A 21.2.1. To leading nontrivial order the results

$$Z_3 = \frac{1}{6}\left[1+\frac{22}{3}\,\pi\,\frac{u_3}{\varepsilon_3}+O(u_3^2)\right]$$
$$Z_n^\Theta = 1-\frac{8}{27}\,\pi^2\,\frac{u_3^2}{\varepsilon_3}+O(u_3^3)$$
$$Z^\Theta = 1-\frac{1}{54}\,\pi^2\,\frac{u_3^2}{\varepsilon_3}+O(u_3^3) \tag{21.27}$$

hold, where the prefactor $\frac{1}{6}$ in Z_3 is put in by convention.

21.1.3 Renormalization in the Θ-region

The tricritical theory is concerned with the Θ-region, roughly defined by $z \sim \beta_e n^{1/2} < 1$. (The precise renormalized form of this criterion is given in Sect. 21.2, Eq. (21.59).) It treats the renormalization of the naive two-parameter z-expansion by three-body effects, and it is not concerned with the renormalization due to two-body interactions necessary for $z \gg 1$. We thus renormalize β_e according to

$$\beta_e \ell^{-1-\varepsilon_3} = u_e \ell_R^{-1-\varepsilon_3} Z_e^{\Theta}(u_3) \;, \tag{21.28}$$

a relation quite different from relation (11.1 i) used in the excluded volume region. In particular u_e for $\lambda = \ell/\ell_R \to 0$ will not approach a fixed point but rather runs away to infinity, as does the naive z-parameter.

It finally turns out that for the partition functions besides Z^{Θ} we need an additional renormalization factor that takes into account the logarithmic divergence in $d = 3$ due to single chain subdiagrams of order β_e^2. The theorem of renormalizability for the full theory involving both two- and three-body forces therefore reads

$$\begin{aligned}
&\exp\left\{ \left(\mu_s^{(0)}(\beta_3) + \beta_e \mu_s^{(1)}(\beta_3)\right) \sum_{m=1}^{M} n_m \right\} \frac{\ell^d}{\Omega} \\
&\cdot G_{c,m_1\dots m_L}^{(L,M)}(q_1,\dots,p_{2M}; n_1,\dots,n_m; \beta_3,\beta_2,\ell) \\
&= (4\pi)^{-M\frac{d}{2}} \left(\frac{\ell}{\ell_R}\right)^{d-2L-dM} \left(Z^{\Theta}(u_3)\right)^M \left(Z_n^{\Theta}(u_3)\right)^{L-M} \exp\left[\sigma(u_3) u_e^2 \sum_{m=1}^{M} n_{Rm}\right. \\
&\cdot G_{cR,m_1\dots m_L}^{(L,M)}(\bar{\mathbf{q}}_1,\dots,\bar{\mathbf{p}}_{2M}; n_{R1},\dots,n_{RM}; u_3, u_e) \left[1 + O\left(\frac{\ell}{n_R^{1/2}\ell_R}\right)\right] .
\end{aligned} \tag{21.29}$$

The perturbative calculation of $Z_e^{\Theta}(u_3)$ or $\sigma(u_3)$ to lowest order is again sketched in the appendix. We find

$$\begin{aligned}
Z_e^{\Theta}(u_3) &= \frac{1}{2}\left[1 + \frac{8}{3}\pi \frac{u_3}{\varepsilon_3} + O(u_3^2)\right] \\
\sigma(u_3) &= \frac{\pi}{2\varepsilon_3} + O(u_3) \;.
\end{aligned} \tag{21.30}$$

21.2 Renormalization group mapping and scaling laws

21.2.1 General form of the RG mapping

The flow equations give the change of the renormalized parameters under an infinitesimal change of ℓ_R, for a fixed bare theory. According to the general discussion of Sects. 10.2 or 11.3 they take the form of simple first-order differential equations in the variable $\lambda = \ell/\ell_R$. Furthermore the theorem of renormalizability guarantees that the flow equations are also well defined for $\varepsilon_3 = 0$. Indeed, in the minimal subtraction scheme the dependence on ε_3 reduces to some trivial terms.

We first consider the flow of the three-body coupling. Taking the logarithmic derivative $-\ell_R d/d\ell_R = \lambda\, d/d\lambda$, we find from Eq. (21.24)

$$\lambda \frac{du_3}{d\lambda} = -2\varepsilon_3 u_3 + u_3^2 \hat{W}_3(u_3) \ , \tag{21.31}$$

where

$$\hat{W}_3(u_3) = \frac{2\varepsilon_3 \frac{\partial \ \ln \ Z_3}{d\ u_3}}{1 + u_3 \frac{\partial \ \ln \ Z_3}{d\ u_3}}$$

is known to be independent of ε_3 (see Sect. 12.2) and can be expanded in powers of u_3:

$$\hat{W}_3(u_3) = a_0 + a_1 u_3 + \ldots \tag{21.32}$$

Inserting Eq. (21.27) we find $a_0 = \frac{44}{3}\,\pi$. Now specializing to $d = 3$ we write Eq. (21.31) as

$$du_3 \left[\frac{1}{a_0 u_3^2} - \frac{a_1}{a_0^2 u_3} + p_3(u_3) \right] = d \ \ln \ \lambda \ ,$$

where $p_3(u_3)$ is regular for $u_3 \to 0$. Integrating we find the general result

$$\frac{1}{u_3(\lambda)} + \frac{a_1}{a_0} \ln u_3(\lambda) - P_3(u_3(\lambda)) = s_3(u_3^{(0)}) - a_0 \ \ln \ \lambda \ , \tag{21.33}$$

where $P_3(u_3) = O(u_3)$ is regular for $u_3 \to 0$. The nonuniversal parameter $s_3(u_3^{(0)})$ collects all the dependence on the starting value $u_3^{(0)} = u_3(\lambda = 1)$. In the limit $\lambda \to 0$ Eq. (21.33) reduces to

$$u_3(\lambda) = \frac{1}{a_0 |\ln \ \lambda|} \to 0 \ , \tag{21.34}$$

which shows that three-body forces are indeed marginally irrelevant and are the source of the logarithmic corrections. Note that the nonuniversal parameter $s_3(u_3^{(0)})$ drops out. The next to leading term is due to the contribution $\sim \ln \ u_3(\lambda)$ and yields corrections of order $(\ln \ \lambda)^{-2} \ \ln(\ln \ \lambda)$.

We now turn to the flow equations for the renormalization factors. Denoting by $Z_x^\Theta(u_3)$ any one of $Z^\Theta(u_3), Z_n^\Theta(u_3),, Z_e^\Theta(u_3)$ we write

$$\begin{aligned} \lambda \frac{d}{d\,\lambda} \ \ln \ Z_x^\Theta(u_3) &= \lambda \frac{du_3}{d\lambda} \frac{\partial}{\partial u_3} \ \ln Z_x^\Theta(u_3) \\ &= \left(-2\varepsilon_3 u_3 + u_3^2 \hat{W}_3(u_3)\right) \frac{\partial}{\partial u_3} \ \ln \ Z_x^\Theta(u_3) \\ &= \hat{p}_x(u_3) \ , \end{aligned} \tag{21.35}$$

and we note that $\hat{p}_x(u_3)$ can be expanded in powers of u_3 and is independent of ε_3 by virtue of the theorem of renormalizability. The expansion of $\hat{p}_x(u_3)$ can be derived from the formal expansions (21.27,21.30) of the Z-factors (see also Appendix A 21.2). Specifically we find

$$\begin{aligned}
\hat{p}_n(u_3) &= \frac{32}{27}\,\pi^2 u_3^2 + O(u_3^3) \;\; ; \quad [Z_n^\Theta] \\
\hat{p}(u_3) &= \frac{2}{27}\,\pi^2 u_3^2 + O(u_3^3) \;\; ; \quad [Z^\Theta] \\
\hat{p}_e(u_3) &= -\frac{16}{3}\,\pi\, u_3 + O(u_3^2) \;\; ; \quad [Z_e^\Theta] \; .
\end{aligned} \tag{21.36}$$

The general structure is written as

$$\hat{p}_x(u_3) = b_{x,1}u_3 + b_{x,2}u_3^2 + \ldots$$

Taking now $\varepsilon_3 = 0$ we find

$$\begin{aligned}
\frac{\partial}{\partial\, u_3}\,\ln\, Z_x^\Theta(u_3) &= \frac{\hat{p}_x(u_3)}{u_3^2\hat{W}_3(u_3)} \\
&= \frac{b_{x,1}}{a_0 u_3} + p_x(u_3) \;\; ,
\end{aligned} \tag{21.37}$$

where $p_x(u_3)$ is regular at $u_3 = 0$. Integrating we find the general result

$$\ln\, Z_x^\Theta(u_3) - \ln\, Z_x^\Theta(u_3^{(0)}) = \frac{b_{x,1}}{a_0}\,\ln\,\frac{u_3}{u_3^{(0)}} + P_x(u_3) - P_x(u_3^{(0)})$$

or

$$\left[Z_x^\Theta(u_3)\right]^{-1} = s_x^\Theta(u_3^{(0)})u_3^{-b_{x,1}/a_0}\,\exp(-P_x(u_3)) \tag{21.38}$$

where $s_x^\Theta(u_3^{(0)})$ is a nonuniversal amplitude. We note that a linear term in $\hat{p}_x(u_3)$ gives rise to a prefactor which is some general noninteger power of u_3. In view of Eq. (21.34) this introduces some leading power of $|\ln\,\lambda|$. Equations (21.36) show that this occurs for Z_e^Θ, which asymptotically behaves as

$$[Z_e^\Theta]^{-1} = s_e^\Theta(u_3^{(0)})u_3^{4/11} = const\,|\ln\,\lambda|^{-4/11}, \;\; \lambda \to 0 \; .$$

In contrast, Z_n^Θ and Z^Θ show only subleading corrections regular in u_3.

We finally turn to the calculation of $\sigma(u_3)$, which is a bit more tricky. Taking the derivative $\lambda\, d/d\lambda$ of Eq. (21.29), $L = 0, M = 1$, we note that

$$\lambda\,\frac{d}{d\lambda}\left(\sigma(u_3)u_e^2 n_R\right) = \tilde{\sigma}(u_3, u_e, n_R) \tag{21.39}$$

can be expressed in terms of finite renormalized quantities and thus is a finite function of u_3, u_e, n_R. We can rewrite Eq. (21.39) as

$$\lambda\,\frac{d}{d\lambda}\sigma(u_3) + \sigma(u_3)\lambda\,\frac{d}{d\lambda}\,\ln\,(\tilde{u}_e^2 n_R) = \frac{\tilde{\sigma}(u_3, u_e, n_R)}{u_e^2 n_R} \quad . \tag{21.40}$$

From Eqs. (21.28), (21.24), (21.35) we conclude ($\varepsilon_3 = 0$)

$$\lambda \frac{d}{d\lambda} \ln u_e^2 n_R = -2\hat{p}_e(u_3) - \hat{p}_n(u_3) \quad ,$$

so that the l.h.s. of Eq. (21.40) depends on u_3 only, and $\tilde{\sigma}(u_3, u_e, n_R)$ necessarily is of the structure

$$\tilde{\sigma}(u_3, u_e, n_R) = u_e^2 n_R \hat{p}_\sigma(u_3) \quad .$$

Transforming as usual the derivatives $\lambda\, d/d\lambda$ into derivatives with respect to u_3 we thus find the equation

$$\frac{d}{du_3}\sigma(u_3) + p_{\sigma,1}(u_3)\sigma(u_3) = p_{\sigma,2}(u_3) \tag{21.41}$$

$$\begin{aligned} p_{\sigma,1}(u_3) &= -\frac{2\hat{p}_e(u_3) - \hat{p}_n(u_3)}{u_3^2 \hat{W}_3(u_3)} \\ p_{\sigma,2}(u_3) &= \frac{\hat{p}_\sigma(u_3)}{u_3^2 \hat{W}_3(u_3)} \quad . \end{aligned} \tag{21.42}$$

The general solution reads

$$\begin{aligned} \sigma(u_3) = &\int_{u_3^{(0)}}^{u_3} dy\, p_{\sigma,2}(y) \exp\left\{-\int_y^{u_3} dy' p_{\sigma,1}(y')\right\} \\ &+ \sigma(u_3^{(0)}) \exp\left\{-\int_{u_3^{(0)}}^{u_3} dy\, p_{\sigma,1}(y)\right\} \quad . \end{aligned} \tag{21.43}$$

The function $p_{\sigma,1}(u_3)$ to leading order is found from Eqs. (21.32), (21.36):

$$p_{\sigma,1}(u_3) = \frac{8}{11u_3} + const \quad .$$

Taking $\varepsilon_3 > 0$ we can calculate $\tilde{\sigma}(u_3, u_e, n_R)$ (Eq. (21.39)) from Eqs. (21.24), (21.28), (21.30):

$$\begin{aligned} \tilde{\sigma}(u_3, u_e, n_R) &= \lambda \frac{d}{d\lambda}\left(\frac{\pi}{2\varepsilon_3}\, \beta_e^2 \lambda^{-2-2\varepsilon_3} n\, \lambda^2 + O(u_3)\right) \\ &= -\pi + O(u_3) \quad . \end{aligned}$$

Thus

$$p_{\sigma,2}(u_3) = -\frac{3}{44u_3^2} + O\left(\frac{1}{u_3}\right) \quad .$$

Inserting these results into Eq. (21.43) we find to leading order

$$\sigma(u_3) = \frac{1}{4u_3} + \left(\frac{u_3^{(0)}}{u_3}\right)^{8/11} \left(\sigma(u_3^{(0)}) - \frac{1}{4u_3^{(0)}}\right) \quad . \tag{21.44}$$

21.2.2 Low order approximation of the RG-mapping

For the tricritical theory only the lowest nontrivial order of the RG-mapping has been worked out.[1] We first consider $u_3(\lambda)$. Only the leading term $a_0 = \frac{44}{3}\pi$ of $\hat{W}_3(u_3)$ is known, and ignoring all higher terms we find from Eq. (21.33)

$$u_3(\lambda) = \frac{u_3^{(0)}}{1 + \frac{44}{3}\pi u_3^{(0)} |\ln \lambda|} \quad . \tag{21.45}$$

Note that $s_3(u_3^{(0)}) = 1/u_3^{(0)}$ in that approximation. This equation smoothly interpolates among $u_3(1) = u_3^{(0)}$ and

$$u_3(\lambda) \to \frac{3}{44\pi |\ln \lambda|}, \quad \lambda \to 0 \ . \tag{21.46}$$

We must however note that Eq. (21.45) may be strongly modified by the (unknown) higher order terms of $\hat{W}_3(u_3)$, the leading correction being proportional to $\frac{\ln|\ln \lambda|}{|\ln \lambda|^2}$. Thus only the asymptotic behavior (21.46) is firmly established.

For the renormalization factors Eqs. (21.36), (21.38) yield

$$\begin{aligned}
\left[Z^{\Theta}(u_3)\right]^{-1} &= s^{\Theta}(u_3^{(0)}) \exp\left[-\frac{\pi}{198} u_3 + O(u_3^2)\right] \\
\left[Z_n^{\Theta}(u_3)\right]^{-1} &= s_n^{\Theta}(u_3^{(0)}) \exp\left[-\frac{8}{99} \pi u_3 + O(u_3^2)\right] \\
\left[Z_e^{\Theta}(u_3)\right]^{-1} &= s_e^{\Theta}(u_3^{(0)}) u_3^{4/11} \left[1 + O(u_3)\right] \ .
\end{aligned} \tag{21.47}$$

Introducing parameters

$$\begin{aligned}
\tilde{R}_0^2 &= n\, \ell^2\, s_n^{\Theta}(u_3^{(0)}) \\
\tilde{z}_{\Theta} &= \left(s_n^{\Theta}(u_3^{(0)})\right)^{1/2} s_e^{\Theta}(u_3^{(0)}) \beta_e\, n^{1/2}
\end{aligned} \tag{21.48}$$

we can thus write the mapping for n_R or u_e as

$$\begin{aligned}
n_R \ell_R^2 &= \exp\left(-\frac{8}{99}\pi\, u_3\right) \tilde{R}_0^2 \\
u_e n_R^{1/2} &= u_3^{4/11} \tilde{z}_{\Theta} \ .
\end{aligned} \tag{21.49}$$

These equations correspond to the RG mapping of the excluded volume theory in the form (11.39), (11.40). They clearly exhibit the renormalization of the naive two-parameter theory by three-body effects.

---

[1] Recently J. Hager (1999, in preparation) calculated the next order of $\hat{W}_3$.

Equations (21.48) introduce two nonuniversal parameters

$$\tilde{\ell}_\Theta = s_n^\Theta (u_3^{(0)})^{1/2}\, \ell$$
$$\tilde{v}_\Theta = \beta_e s_e^\Theta (u_3^{(0)}) s_n^\Theta (u_3^{(0)})^{1/2} \quad . \tag{21.50}$$

Clearly $u_3^{(0)}$ or $s_3(u_3^{(0)})$, equivalently, defines a third independent parameter, directly influencing u_3 (Eq. (21.45)). To leading order (21.46) this parameter drops out, so that the three-parameter structure of the tricritical theory reduces to a two-parameter structure. This is analogous to the reduction of two-parameter theory to an effective one-parameter theory at the excluded volume fixed point.

Equations (21.45), (21.49) together with the result (21.44) for $\sigma(u_3)$ specify the RG-mapping to lowest order approximation.

21.2.3 General scaling laws

We consider some dimensionless quantity Q, depending on a set of momenta $\{\mathbf{q}_j \ell\}$, chain concentration $c_p \ell^3$, average chain length N, and interactions β_2, β_3. Explicit dependence on the reduced chain length distribution $p(y)$ will be suppressed, since it is invariant under the RG. The quantity Q shall be properly normalized, so that no explicit renormalization prefactors occur. The theorem of renormalizability (Eq. (21.29)) yields

$$Q(\{\mathbf{q}_j \ell\}; N, c_p \ell^3, \beta_3, \beta_2) = Q_R(\{\bar{\mathbf{q}}_j\}; N_R, c_{pR}, u_3, u_e) \ , \tag{21.51}$$

where

$$c_{pR} = c_p \ell_R^3 \ , \tag{21.52}$$

as before. Setting $N_R = 1$ and exploiting the general structure of the RG mapping we find the general scaling law

$$Q(\{\mathbf{q}_j \ell\}; N, c_p \ell^3, \beta_3, \beta_2) = Q^\Theta(\{\mathbf{q}_j \tilde{R}_0\}; \tilde{s}_\Theta, u_3, \tilde{z}_\Theta) \ . \tag{21.53}$$

Here

$$\tilde{s}_\Theta = c_p \tilde{R}_0^3 \tag{21.54}$$

is the Θ-overlap, and

$$u_3 = u_3 \left(\frac{\tilde{\ell}_\Theta}{\tilde{R}_0}, u_3^{(0)} \right) \tag{21.55}$$

is given by Eq. (21.33).

Clearly Eq. (21.53) differs from the scaling laws derived in Sects. 11.5, 11.6 by the explicit consideration of the three-body coupling. On a more subtle level the variables $\tilde{R}_0, \tilde{s}_\Theta, \tilde{z}_\Theta$ differ from their counterparts introduced in

Eqs. (11.38), (11.83), since they involve nonuniversal factors depending on $u_3^{(0)} \hat{=} \beta_3$ instead of $f_0 \hat{=} \beta_e$. Of course the two theories must approach each other for $\beta_3 \to 0, \beta_e \to 0, n, \ell$ fixed. This is easily checked. For $u_3 = u_3^{(0)}$ Eq. (21.38) yields

$$s_x^{\Theta}(u_3^{(0)}) = \left(Z_x^{\Theta}(u_3^{(0)})\right)^{-1} u_3^{(0)b_{x,1/a_0}} \exp\left(P_x(u_3^{(0)})\right) ,$$

where $P_x(u_3^{(0)} \to 0) \to 0$. We also expect $Z_x^{\Theta}(u_3^{(0)} \to 0) \to 1$, so that

$$s_n^{\Theta}(u_3^{(0)} \to 0) \to 1$$
$$s_e^{\Theta}(u_3^{(0)} \to 0) \to u_3^{(0)-4/11} .$$

Thus $\tilde{R}_0 \to \ell\, n^{1/2}, \tilde{z}_\Theta \to \beta_e n^{1/2} u_3^{(0)-4/11}$. In the limit $u_3^{(0)} \to 0$ we find from Eq. (21.45) $u_3 \to u_3^{(0)}$, so that the combination $\tilde{z}_\Theta u_3^{4/11}$ showing up in u_e (Eq. (21.49)) reduces to the standard z-parameter of two-parameter theory close to the Θ-point. In that limit, then, Eq. (21.53) reduces to the trivial scaling at the Gaussian fixed point.

Even for $u_3^{(0)} > 0$ we find that trivial scaling, provided we can neglect the logarithmically slow variation of u_3. The scaling functions, however, will depend on the value put in for u_3, which is nonuniversal due to its dependence on $u_3^{(0)}$. It is this trivial Gaussian-type scaling that is exhibited in most experiments on Θ-point behavior.

21.3 The dilute limit

As in the excluded volume region, the scaling functions in the tricritical Θ-region have been worked out only to low order approximation. Indeed, technically the calculations in the tricritical theory become even more demanding, since the insertion of an additional three-body vertex introduces two additional loops. Therefore the theory has not been evaluated as comprehensively as excluded volume theory.

21.3.1 The RG-mapping

We review here results obtained for properties involving only a few chains. For such very dilute systems we have to fix the renormalized length scale ℓ_R by the choice $n_R = n_0 = O(1)$. In the excluded volume limit we exploited very precise data for the universal limit of the interpenetration ratio to adjust

n_0. No such data are available in the Θ-region, and we therefore somewhat arbitrarily choose

$$n_R = 1 \ . \tag{21.56}$$

The RG-mapping (21.45), (21.49) takes the form

$$\begin{aligned} u_3 &= \frac{u_3^{(0)}}{1 + \frac{22}{3}\,\pi u_3^{(0)}\,\ln\left(s_n^{\Theta} n\right)} \\ \ell_R &= \exp\left(-\frac{4}{99}\,\pi u_3\right)\tilde{R}_0 \\ u_e &= u_3^{4/11}\,\tilde{z}_{\Theta} \ , \end{aligned} \tag{21.57}$$

where in the first equation we omitted some correction of order u_3, which exceeds the accuracy of that equation. Recall also that only the asymptotic limit of that mapping is rigorously correct:

$$\begin{aligned} u_3 &= \frac{3}{22\pi}\,\ln^{-1}(s_n^{\Theta} n) + O\left(\frac{\ln|\ln s_n^{\Theta} n|}{(\ln s_n^{\Theta} n)^2}\right) \\ \ell_R &= \tilde{R}_0\left(1 - \frac{4}{99}\,\pi u_3 + O(u_3^2)\right) \\ u_e &= \tilde{z}_{\Theta}\,u_3^{4/11}(1 + O(u_3)) \ . \end{aligned} \tag{21.58}$$

We are now in the position to discuss the boundary of the tricritical region in the dilute limit. The tricritical theory assumes $u_e \lesssim 1$. In view of Eqs. (21.58) this condition transforms into

$$\tilde{z}_{\Theta}\left(\frac{22\pi}{3}\,\ln(s_n^{\Theta} n)\right)^{-4/11} \lesssim 1 \ , \tag{21.59}$$

a criterion that replaces the simple estimate $z \lesssim 1$ for the size of the Θ-region. Thus the size of the Θ-region weakly increases with n. To identify properly tricritical effects, the two-body coupling u_e, however, has to be small compared to the three-body coupling u_3. This yields the more restrictive condition

$$\tilde{z}_{\Theta}\left(\frac{22\pi}{3}\,\ln(s_n^{\Theta}\,n)\right)^{-4/11} \lesssim \frac{3}{22\pi}\,\ln^{-1}(s_n^{\Theta} n) \ , \tag{21.60}$$

defining the region, where tricritical effects can be expected to dominate. In the intermediate region

$$\left(\frac{22\pi}{3}\,\ln(s_n^{\Theta}\,n)\right)^{-7/11} \lesssim \tilde{z}_{\Theta} \lesssim \left(\frac{22\pi}{3}\,\ln(s_n^{\Theta}\,n)\right)^{4/11}$$

our previous treatment of the Θ-point, neglecting three-body forces, should be adequate.

21.3.2 Single chain radii

Renormalized first order expressions for R_e^2 or R_g^2 are given in Appendix A 21.2.2, Eqs. (A 21.27), (A 21.28). Taking $n_R = 1$ and using the RG-mapping (21.57) we find for the swelling factors

$$\alpha_e^2 = \frac{R_e^2}{6\tilde{R}_0^2} = e^{-\frac{8}{99}\pi u_3}\left[1 + \frac{2}{3}\tilde{z}_\Theta u_3^{4/11} - \frac{2}{3}\pi u_3\right] \tag{21.61}$$

$$\alpha_g^2 = \frac{R_g^2}{\tilde{R}_0^2} = e^{-\frac{8}{99}\pi u_3}\left[1 + \frac{67}{105}\tilde{z}_\Theta u_3^{4/11} - \frac{13}{24}\pi u_3\right] . \tag{21.62}$$

With Eq. (21.58) the asymptotic behavior takes the form

$$\alpha_e^2 = 1 + \frac{2}{3}\tilde{z}_\Theta\left(\frac{22\pi}{3}\ln(s_n^\Theta\, n)\right)^{-4/11} - \frac{37}{363}\ln^{-1}(s_n^\Theta\, n) \tag{21.63}$$

$$\alpha_g^2 = 1 + \frac{67}{105}\tilde{z}_\Theta\left(\frac{22\pi}{3}\ln(s_n^\Theta\, n)\right)^{-4/11} - \frac{493}{5808}\ln^{-1}(s_n^\Theta\, n) . \tag{21.64}$$

Note that the last equation identifies $\tilde{R}_0$ as the radius of gyration at the Θ-point $\tilde{z}_\Theta = 0, n \to \infty$.

Figure 21.2 shows typical results for α_e^2 as function of $\log_{10}^{-1}(s_n n)$. A plot for α_g^2 looks most similar. At the Θ-point $\tilde{z}_\Theta = 0$ we find for the larger value $u_3^{(0)} = 1$ the asymptotic logarithmic behavior over a large range of $\log_{10}^{-1}(s_n n)$. With decreasing $u_3^{(0)}$ this range rapidly decreases and we see the crossover behavior incorporated in the RG-map (21.57). Taking $\tilde{z}_\Theta = \tilde{v}_\Theta n^{1/2} \neq 0$ even for $u_3^{(0)} = 1$ and very small values of $\tilde{v}_\Theta$ we rapidly find pronounced deviations from the asymptotic law. Note that $|\tilde{z}_\Theta| < 0.05$ in all the curves shown, so that we are deep in the Θ-region.

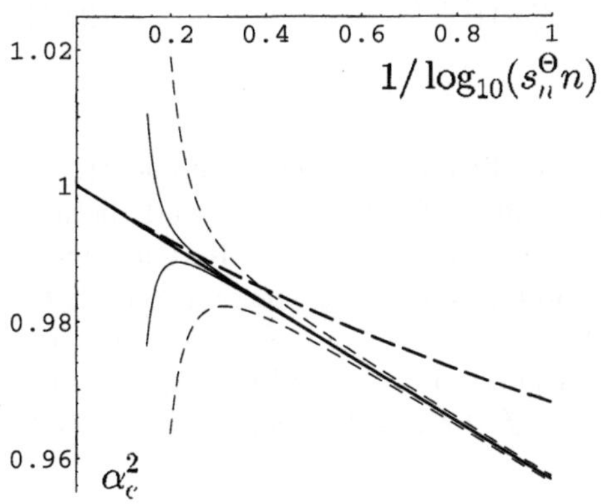

Fig. 21.2. The swelling of the end-to-end distance (Eq. (21.63)) as function of $\log_{10}^{-1}(s_n^\Theta n)$. Parameter values: $\tilde{v}_\Theta = 0, u_3^{(0)} = 1$ (*full thick line*); $\tilde{v}_\Theta = 0, u_3^{(0)} = 0.05$ (*broken thick line*); $\tilde{v}_\Theta = \pm 10^{-4}, u_3^{(0)} = 1$ (*full thin lines*); $\tilde{v}_\Theta = \pm 10^{-3}, u_3^{(0)} = 1$ (*broken thin lines*). Lines for negative $\tilde{v}_\Theta$ bend downwards

21.3.3 The virial coefficients

We write the virial expansion of the osmotic pressure as

$$\frac{\hat{\Pi}}{k_B T} = c_p + \frac{1}{2} A_2^{\Pi} c_p^2 + \frac{1}{6} A_3^{\Pi} c_p^3 + \dots \quad . \tag{21.65}$$

Renormalized low order results for $\mathcal{Z}_{cR}^{(M)}, M \leq 3$, are given in Appendix A 21.2. Calculating the virial coefficients according to Eq. (21.10), (21.11) and taking $n_R = 1$ we find

$$(4\pi)^{-3/2} \frac{A_2^{\Pi}}{\tilde{R}_0^3} = e^{-\frac{4}{33}\pi u_3} \Big[\frac{1}{2} \tilde{z}_\Theta u_3^{4/11} - \frac{4}{3} u_3 - 0.716 \tilde{z}_\Theta^2 u_3^{8/11} + 13.811 \tilde{z}_\Theta u_3^{15/11} \Big] \tag{21.66}$$

$$(4\pi)^{-3} \frac{A_3^{\Pi}}{\tilde{R}_0^6} = e^{-\frac{8}{33}\pi u_3} \left[\frac{u_3}{3} + 25.357 u_3^2 - 2.865 \tilde{z}_\Theta u_3^{15/11} \right] \quad . \tag{21.67}$$

The asymptotic behavior at the Θ-point reads

$$(4\pi)^{-3/2} \frac{A_2^{\Pi}}{\tilde{R}_0^3} = -\frac{2}{11\pi} \ln^{-1}(s_n^{\Theta} n) \tag{21.68}$$

$$(4\pi)^{-3} \frac{A_3^{\Pi}}{\tilde{R}_0^6} = \frac{1}{22\pi} \ln^{-1}(s_n^{\Theta} n) \quad . \tag{21.69}$$

Note that for finite chain length the second virial coefficient at the (infinite chain) Θ-point is negative. Thus $\Theta(n)$, defined by the vanishing of $A_2^{\Pi}(n)$, is larger than Θ. Asymptotically Eq. (21.66) yields

$$\begin{aligned} \Theta(n) - \Theta &\sim \tilde{z}_\Theta(\Theta(n)) n^{-1/2} \sim n^{-1/2} w^{7/11} \\ &\sim n^{-1/2} \left(\ln(s_n^{\Theta} n) \right)^{-7/11} \quad . \end{aligned} \tag{21.70}$$

Also A_3^{Π} does not vanish at $T = \Theta(n)$.

In Fig. 21.3 we plot the second virial coefficient as function of $\log_{10}^{-1}(s_n n)$. Clearly the results look most similar to those for the end-to-end distance (Fig. 21.2). For the third virial coefficient, a somewhat different picture emerges (Fig. 21.4). In view of the large prefactor of the u_3^2-term in Eq. (21.67) the results for $\tilde{z}_\Theta = 0$ reach the asymptotic power law only for extremely long chains. On the other hand, the effect of the two-body coupling is quite small. Note that in this plot the curves for $\tilde{z}_\Theta \neq 0$ extend to values $|u_e| \lesssim 0.25$, much larger than in Figs. 21.2 or 21.3.

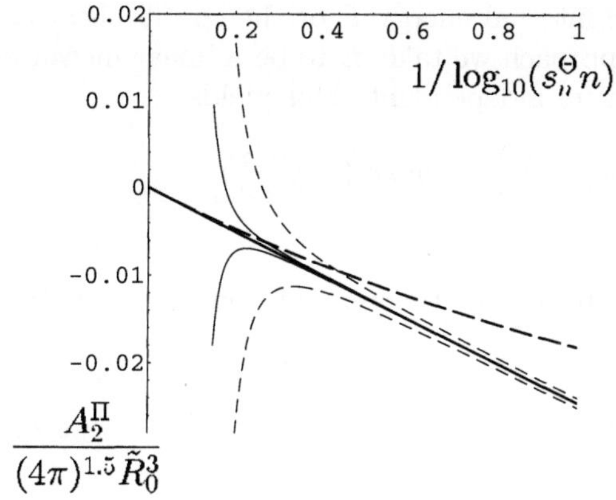

Fig. 21.3. Normalized second virial coefficient $(4\pi)^{-3/2}A_2^{\Pi}/\tilde{R}_0^3$ as function of $\log_{10}^{-1}(s_n^{\Theta}n)$. The symbols are the same as in Fig. 21.2

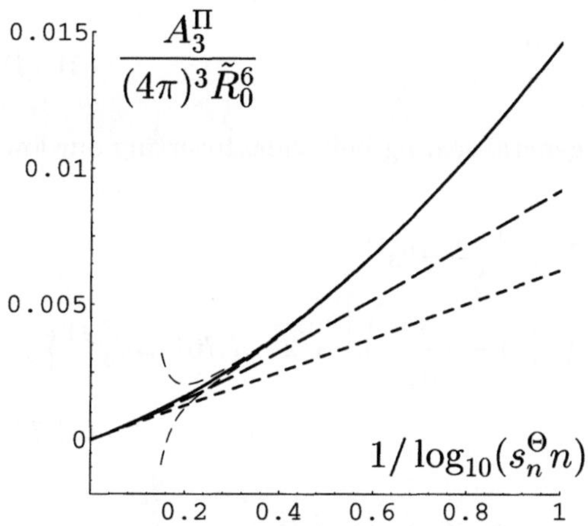

Fig. 21.4. Normalized third virial coefficient $(4\pi)^{-3}A_3^{\Pi}/\tilde{R}_0^6$ as function of $\log_{10}^{-1}(s_n^{\Theta}n)$. Parameter values: $\tilde{v}_{\Theta} = 0, u_3^{(0)} = 1$ (*full thick line*); $\tilde{v}_{\Theta} = 0, u_3^{(0)} = 0.05$ (*long dashes, thick*). Asymptotic law at $T = \Theta$ (*short dashes, thick*). $\tilde{v}_{\Theta} = \pm 10^{-3}$ (*long dashes, thin*)

21.3.4 Binary contacts within a single chain

The average number of spatially close pairs within a single chain can be defined as

$$\begin{aligned}\left\langle n_c^{(2)}\right\rangle &= (4\pi\ell^2)^{d/2}\sum_{j_1<j_2}\left\langle\delta^d(\mathbf{r}_{j_1}-\mathbf{r}_{j_2})\right\rangle \\ &= -\frac{\partial}{\partial\beta_e}\ln\mathcal{Z}^{(1)}\ .\end{aligned} \tag{21.71}$$

The fluctuations of this quantity are given by the cumulant

$$\left\langle (n_c^{(2)})^2\right\rangle_c = \frac{\partial^2}{\partial\beta_e^2}\ln\mathcal{Z}^{(1)}\ . \tag{21.72}$$

These quantities are related to the internal energy E or the specific heat c_v of a single chain. In the simplest approach we take β_e to be a linear measure of temperature and β_3 as temperature independent. This yields

$$E \sim \left\langle n_c^{(2)} \right\rangle ; \quad c_v \sim const \left\langle n_c^{(2)} \right\rangle + const \left\langle (n_c^{(2)})^2 \right\rangle_c \quad .$$

To calculate $\left\langle n_c^{(2)} \right\rangle$ we start from the theorem of renormalizability, Eq. (21.29), to find

$$\left\langle n_c^{(2)} \right\rangle = \mu_s^{(1)}(\beta_3) n - \frac{u_e}{\beta_e} \frac{\partial}{\partial u_e} \left\{ u_e^2 \sigma(u_3) + \ln \mathcal{Z}_R^{(1)}(1, u_3, u_e) \right\} , \tag{21.73}$$

where we put $n_R = 1$ and used the relation (cf. Eq. (21.28))

$$\frac{\partial u_e}{\partial \beta_e} = u_e \frac{\partial \ln u_e}{\partial \beta_e} = \frac{u_e}{\beta_e} \quad . \tag{21.74}$$

Equation (21.73) incorporates the general scaling behavior. Inserting our low order results it reduces to

$$\begin{aligned} \left\langle n_c^{(2)} \right\rangle = {} & \mu_s^{(1)}(\beta_3) n - C_1(u_3^{(0)}) n^{1/2} u_3^{4/11} \Big\{ 2\tilde{z}_\Theta u_3^{4/11} \\ & \cdot \left[\frac{1}{4u_3} + \left(\frac{u_3^{(0)}}{u_3} \right)^{8/11} \left(\sigma(u_3^{(0)}) - \frac{1}{4u_3^{(0)}} \right) \right] + 2 - 7.761 \tilde{z}_\Theta u_3^{4/11} \Big\} , \end{aligned} \tag{21.75}$$

where

$$C_1(u_3^{(0)}) = \frac{\tilde{v}_\Theta}{\beta_e} = s_e^\Theta(u_3^{(0)}) \left(s_n^\Theta(u_3^{(0)}) \right)^{1/2} \quad . \tag{21.76}$$

The dominant term $\sim \mu_s^{(1)} n$ is due to intersections of segments closely spaced along the chain. The leading correction at the Θ-temperature takes the form

$$-2C_1(u_3^{(0)}) n^{1/2} u_3^{4/11} \sim n^{1/2} \left(\ln(s_n^\Theta n) \right)^{-4/11} \quad , \tag{21.77}$$

this expression being valid for $u_3^{(0)} > 0, n \to \infty$. As discussed at the end of Sect. 21.2.3 we can also take the noninteracting chain limit $u_3^{(0)} \to 0, n$ fixed, where

$$C_1(u_3^{(0)}) u_3^{4/11} \to const \quad .$$

Then the well known result for a random walk chain is recovered:

$$\left\langle n_c^{(2)} \right\rangle^{(0)} = const\, n + const\, n^{1/2} + O(n^0) \quad .$$

Taking a further derivative with respect to β_e we find $\left\langle (n_c^{(2)})^2 \right\rangle_c$. Equation (21.73) yields the general scaling behavior

$$\left\langle \left(n_c^{(2)}\right)^2 \right\rangle_c = \left(\frac{u_e}{\beta_e}\right)^2 \left[2\sigma(u_3) + \frac{\partial}{\partial u_e^2} \ln \mathcal{Z}_R^{(1)}(1, u_3, u_e)\right] . \tag{21.78}$$

The low order expression reads

$$\begin{aligned} &\frac{1}{n}\left\langle \left(n_c^{(2)}\right)^2 \right\rangle_c \\ &= C_1^2(u_3^{(0)})u_3^{8/11} \left[\frac{1}{2u_3} + \left(\frac{u_3^{(0)}}{u_3}\right)^{8/11} \left(2\sigma(u_3^{(0)}) - \frac{1}{2u_3^{(0)}}\right) - 7.761\right] . \end{aligned} \tag{21.79}$$

For $u_3^{(0)} > 0, n \to \infty$ this results in the limiting behavior

$$\frac{1}{n}\left\langle \left(n_c^{(2)}\right)^2 \right\rangle_c \sim u_3^{-3/11} \sim \left(\ln\left(s_n^{\Theta} n\right)\right)^{3/11} . \tag{21.80}$$

This is to be compared to the result for a random walk chain in three dimensions:

$$\frac{1}{n}\left\langle \left(n_c^{(2)}\right)^2 \right\rangle_c^{(0)} \sim \ln n . \tag{21.81}$$

Thus three-body interactions weaken the divergence of the specific heat, but do not suppress it completely.

Our result is valid asymptotically right at the Θ-temperature. A determination of corrections due to $\tilde{z}_\Theta \neq 0$ needs a higher order calculation. We note, however, that within the realm of the tricritical theory the corrections take the form of power series in $u_e \approx \tilde{z}_\Theta\, u_3^{4/11}$, which provides a smoothly varying background and cannot compete with the logarithmic divergence.

21.3.5 Critical discussion

Our results clearly show the logarithmic corrections and should be rigorously valid for extremely long chains very close to the Θ-point, provided the Θ-point can be described as tricritical point. Plots like Figs. 21.2 or 21.3 indeed suggest that the logarithmic terms should be observable for chain lengths exceeding $n \approx 100$, a rough estimate. However, these plots may be somewhat misleading, since only the lowest order terms in the expansions have been calculated. For the third virial coefficient, which is the only quantity where terms of order u_3^2 are known, this correction makes a huge effect and indeed for all accessible chain lengths ruins the asymptotic power law (see Fig. 21.4). If this behavior of A_3^{Π} should also be typical for the other quantities, there would be no chance to observe the asymptotic tricritical corrections. We should note here that this result for A_3^{Π}, though being correct to $O(u_3^2)$, suffers from the fact that the RG mapping is known only to lowest order. For large n the term $O(u_3^2)$ yields a correction of order $\ln^{-2}(s_n^{\Theta} n)$, which in fact should be dominated by the unknown term of order $\ln(\ln(s_n^{\Theta} n))/\ln^2(s_n n)$ arising from the RG-mapping. (See Sect. 21.2.2.) It, however, would be a strange coincidence if this term would effectively compensate the large term of order u_3^2 in the range of chain lengths of interest.

Another issue should also be adressed here. As mentioned in the introduction, at present we do not know how to match consistently the tricritical theory with the critical theory holding in the excluded volume regime. So it is not known how, for instance, the tricritical singularity (21.80) of the specific heat crosses over to the behavior (21.81) of the two-parameter model in the Θ-region. Except for the limit $\beta_3 \to 0, \beta_2 \to 0$ also the relation among the z-parameters of the two approaches is not known. This also makes the interpretation of experimental results somewhat ambiguous, as we will discuss in the next section.

21.4 The experimental situation: Dilute limit

On inspecting the figures of the last section one feature is obvious: beyond the region, where nonuniversal $O(1/n^{1/2})$ corrections are to be expected, the logarithmic corrections are very small indeed. For the radii they are less than 2 %, and for A_2^{Π}, for instance, we should compare Fig. 21.3 to Fig. 15.9 and note the large discrepancy of the scales. Furthermore the effects are most sensitive to deviations from the Θ-point.

The situation might be somewhat better for the specific heat of a single chain, where tricritical effects modify the leading term. Here, however, we do not even approximately know the variation in the Θ-region away from the Θ-point, and furthermore this quantity involves a nonuniversal prefactor. Let us analyze this aspect somewhat closer. The nonuniversal amplitude $C_1(u_3^{(0)})$ is defined in Eq. (21.76). Inserting the results

$$s_n^{\Theta} \approx 1, \quad s_e^{\Theta} \approx u_3^{(0)-4/11} \quad \text{(Sect. 21.2.3)},$$

which are expected to hold for small $u_3^{(0)}$, we find from Eq. (21.79) the leading behavior

$$\frac{1}{n}\left\langle \left(n_c^{(2)}\right)^2 \right\rangle_c = \frac{1}{2} u_3^{(0)-8/11} \left(\frac{22\pi}{3} \ln n \right)^{3/11} . \tag{21.82}$$

In an experiment varying the temperature close to Θ, any variations of $u_3^{(0)}$ can strongly modify the effects.

The only unambiguous check of the tricritical approach would be the observation of either the logarithmic corrections with their predicted universal prefactors or of the anomalous power of ln n for the specific heat. All experiments should be carried through at the Θ-point, which must be located with high precision. A rough estimate will illustrate this point. For the system polystyrene-cyclohexane within the framework of the renormalized two-parameter theory we found the parameterization $\tilde{z} = 0.35n^{1/2}(T - \Theta)/\Theta$. Simply equating $\tilde{z}$ and $\tilde{z}_\Theta$ and noting that the logarithmic corrections typically are overrun by the $\tilde{z}_\Theta$-dependence for $\tilde{z}_\Theta \gtrsim 0.01$, for the region dominated by tricritical effects we find the rough estimate

$$\begin{aligned} |T - \Theta| &\lesssim 3 \cdot 10^{-2} n^{-1/2} \Theta \\ &\approx 0.1\,^\circ\text{K} \quad (n = 10^4, \Theta \sim 300\,^\circ\text{K}) \quad . \end{aligned}$$

Neither the Θ-point of the polystyrene-cyclohexane system is known to that precision, nor is the experimental precision sufficient to extract effects in the 1 % range. Furthermore the specific heat of a dilute system, which is the only quantity where larger effects are to be expected, is not accessible experimentally.

Thus only computer experiments have a chance to detect the logarithmic corrections. Simulations in the Θ-region are numereous in the literature, but in almost all studies the range of chain lengths is much too small to address the questions of interest here. A maximum of the specific heat as function of temperature has often been observed (see, for instance [MPB93]), however due to the unknown variation of $u_3^{(0)}$ and the limited range of chain lengths we are not able to interpret properly these data. It should furthermore be noted that most work claiming to see evidence of tricritical behavior in fact only tests the simple Gaussian-type two-parameter scaling near the Θ-point.

The only extended study we are aware of, which properly tests the predictions of the tricritical theory, is due to Grassberger and Hegger [GH95], [Gra97]. On a cubic lattice chains up to $n = 10^4$ have been simulated close to the Θ-point. For the radii there are found corrections which can be interpreted as showing a $1/\ln(n)$ dependence, but the coefficients differ widely from the universal theoretical values. The specific heat shows a proportionality $c_v \sim \ln n$ rather than $c_v \sim (\ln n)^{3/11}$, with weak indications of bending over towards $(\ln n)^x, x < 1$ for extremely long chains. The only feature consistent with the theory is the large sensitivity of the results to small deviations from the Θ-point. At present we do not know whether this failure of the theory to account for the data signals the existence of large u_3^2-terms, as found for A_3^Π, or is due to a violation of basic assumptions of the tricritical theory.

Real experiments which can be used to test the predictions of the tricritical theory in the dilute regime are rare, and for reasons explained above they do not directly test the chain length-dependent logarithmic corrections at $T = \Theta$. Duplantier et al. [DJdC86], reevaluating light scattering data for polystyrene-cyclohexane [PAD82] argued that the temperature dependence of R_g^2 near the Θ-temperature is less by a factor of ~ 0.6 than an estimate based on a determination of the nonuniversal parameters at higher temperatures. They then identified this reduction factor with the tricritical renormalization factor $\left(u_3/u_3^{(0)}\right)^{4/11}$ of the two-body coupling. However, as we stressed above, the proper relation between the nonuniversal parameters of the critical theory ignoring three-body forces and the tricritical theory is not known. So we do not think that this evidence is very strong. Another route has been followed in a neutron scattering experiment on polyethylene dissolved in biphenyl [BRBF92]. Using seven samples spanning about a factor of 20 in chain lengths the authors extracted $A_2(n,T)$ and determined $\Theta(n)$ via $A_2(n,\Theta(n)) = 0$. They then analyzed the chain length dependence of $(\Theta(n) - \Theta)\partial A_2/\partial T$. This approach is interesting, since in that combination the nonuniversal scale of β_e drops out. Indeed, from Eq. (21.66) we find

$$(\Theta(n) - \Theta)\frac{\partial A_2}{\partial T} = \frac{4}{3}\, u_3 (4\pi \tilde{R}_0^2)^{3/2} \quad . \tag{21.83}$$

The data were not precise enough to yield properly the n-dependence of u_3. The experiment essentially only showed the leading behavior $\Theta(n) - \Theta \sim n^{-1/2}$. However, as has been argued in [DJ93] the range of u_3-values found in the experiment allows for a consistent interpretation in terms of the RG-flow of u_3 starting with some fixed $u_3^{(0)}$.

In summary, both computer experiments and physical experiments clearly show that the Θ-point is not just the Gaussian fixed point of two-parameter theory. A dependence $\Theta(n) - \Theta \sim n^{-1/2}$ has been found, as well as slowly varying, presumably logarithmic, behavior for various quantities. In so far as the quantitative prediction of the tricritical theory have been tested, there however exist severe discrepancies among theory and experiments. More work is necessary to resolve that issue.

21.5 Finite overlap

In Appendix A 21.3 we derive the results

$$\frac{\hat{\Pi}}{k_B T c_p} = 1 + \frac{1}{2}(4\pi)^{d/2}\, \beta_2 c_p \ell^d N^2 + \frac{1}{3}(4\pi)^d\, \beta_3 (c_p \ell^d)^2 N^3 \tag{21.84}$$

$$\frac{1}{c_p N^2} I_{d,a}(\mathbf{q}) = D_p(q^2\ell^2 N) \tag{21.85}$$

$$\frac{1}{c_p N^2} I_{d,c}(\mathbf{q}) = D_p(q^2\ell^2 N)\Big[1 + \Big((4\pi)^{d/2}\beta_2 c_p \ell^d N^2 + (4\pi)^d \beta_3 (c_p\ell^d)^2 N^3\Big) D_p(q^2\ell^2 N)\Big]^{-1} , \tag{21.86}$$

valid in unrenormalized tree approximation. Renormalization most simply proceeds by replacing the bare variables by their renormalized counterparts (Eqs. (21.24), (21.28)). The Z-factors are needed only to trivial lowest order, and we note that in tree approximation the identity $\beta_2 = \beta_e$ holds. However, before discussing our final results we need to determine the appropriate choice of the renormalized length scale.

21.5.1 RG-mapping

Specializing to a monodisperse system in $d = 3$ we find from Eq. (21.86) the (unrenormalized) density correlation length at the Θ-point $\beta_2 = \beta_e = 0$ as

$$\begin{aligned}\xi_d^2 &= -\frac{1}{2}\Delta_q|_0 \ \ln I_{d,c}(\mathbf{q}) \\ &= \ell^2\left[\frac{1}{N} + (4\pi)^3\beta_3(c_p\ell^3 N)^2\right]^{-1} .\end{aligned} \tag{21.87}$$

Renormalization yields

$$\frac{\ell_R^2}{\xi_d^2} = \frac{1}{N_R} + \frac{(4\pi)^3}{6}\, u_3 (c_{pR} N_R)^2 \ .$$

In the spirit of our previous approach we now take ℓ_R to be of order ξ_d^2, fixing our renormalization scheme by the condition

$$1 = \frac{1}{N_R} + \frac{(4\pi)^3}{6}\, u_3 (c_{pR} N_R)^2 \ . \tag{21.88}$$

We do not include a scale set by the two-body interaction, since this is treated as a small perturbation in the tricritical approach. We also do not include a scale set by momentum. For large momenta the correlation functions are expected to take their trivial noninteracting form, tricritical corrections showing up only as subleading logarithms which presumably are not observable. We therefore do not consider the large-momentum behavior of the theory any further.

To leading approximation we now can write down the RG-mapping in all the tricritical region. Equations (21.49) yield

$$\ell_R^2 = \frac{\tilde{R}_0^2}{N_R}$$
$$u_e = \tilde{z}_\Theta \, u_3^{4/11} \, N_R^{-1/2} \quad . \tag{21.89}$$

We introduce the Θ-overlap

$$\tilde{s}_\Theta = c_p \tilde{R}_0^3 = c_{pR} N_R^{3/2} \tag{21.90}$$

and use Eq. (21.88) to write

$$\tilde{s}_\Theta^2 = \frac{6}{(4\pi)^3} \frac{N_R - 1}{u_3} \quad . \tag{21.91}$$

Finally u_3 (Eq. (21.45)) takes the form

$$u_3 = \frac{u_3^{(0)}}{1 + \frac{22}{3}\,\pi\, u_3^{(0)} \, \ln\left(\frac{s_n^\Theta N}{N_R}\right)} \quad . \tag{21.92}$$

The equations constitute a parameter representation (parameter N_R) of the RG-mapping, extending Eqs. (21.58) to finite overlap.

We should note that we expect $\tilde{s}_\Theta$ to be monotonous function of N_R, taking its maximal value for the maximal value $N_R = s_n^\Theta N$ allowed for N_R. It turns out that this holds only for $u_3^{(0)}$ small enough: $\frac{22}{3}\,\pi u_3^{(0)} < 1$. For large $u_3^{(0)}$, N_R has to be restricted to $N_R < s_n^\Theta N/e$. It is clear however that for $N_R \approx s_n^\Theta N$, i.e. $\ell_R \approx \ell$, we have to expect strong corrections due to the microstructure of the chains, so that this problem occurs only outside the region, where the theory is expected to be valid.

We now consider the semidilute limit. Taking $N_R \gg 1$ from Eq. (21.91) we find

$$N_R = \frac{(4\pi)^3}{6} \, u_3 \tilde{s}_\Theta^2 \quad , \tag{21.93}$$

so that Eq. (21.92) takes the form

$$u_3 = u_3^{(0)} \left[1 + \frac{22}{3}\,\pi u_3^{(0)} \, \ln\left(\frac{6}{(4\pi)^3} \, u_3^{-1} \tilde{c}_\Theta^{-2}\right)\right]^{-1} \quad ,$$

where we introduced the concentration variable

$$\tilde{c}_\Theta = c_p \ell^3 s_n^\Theta N \quad . \tag{21.94}$$

Omitting terms which exceed the accuracy of our RG flow we reduce this equation to

$$u_3 = \frac{u_3^{(0)}}{1 + \frac{22}{3}\,\pi u_3^{(0)} \, \ln\left(\tilde{c}_\Theta^{-2}\right)} \quad , \tag{21.95}$$

which is the semidilute equivalent of Eq. (21.57). u_e takes the form

$$u_e = s_e^\Theta \left(\frac{6}{(4\pi)^3}\right)^{1/2} \beta_e \tilde{c}_\Theta^{-1} \, u_3^{-3/22} \quad . \tag{21.96}$$

As expected, in the semidilute limit the variable N drops out and is replaced by $\tilde{c}_\Theta$.

Using the criterion $u_e \lesssim 1$ we can now estimate the region of validity of the tricritical theory as

$$s_e^\Theta \beta_e \tilde{c}_\Theta^{-1} \lesssim \left(\frac{(4\pi)^3}{6}\right)^{1/2} u_3^{3/22} \approx const \, |\ln \tilde{c}_\Theta|^{-3/22} \quad . \tag{21.97}$$

Tricritical effects dominate for $u_e \ll u_3$ or

$$s_e^\Theta \beta_e \tilde{c}_\Theta^{-1} \ll const \, |\ln \tilde{c}_\Theta|^{-25/22} \quad . \tag{21.98}$$

21.5.2 The osmotic pressure

Defining the thermodynamic scaling function $\mathcal{P}$ via the relation

$$\frac{\hat{\Pi}}{k_B T c_p} = 1 + \mathcal{P} \quad , \tag{21.99}$$

we find by renormalizing Eq. (21.84)

$$\begin{aligned} \mathcal{P} &= \frac{(4\pi)^{3/2}}{4} \, u_e c_{pR} N_R^2 + \frac{(4\pi)^3}{18} \, u_3 c_{pR}^2 N_R^3 \\ &= \frac{(4\pi)^{3/2}}{4} \, \tilde{z}_\Theta \tilde{s}_\Theta u_3^{4/11} + \frac{(4\pi)^3}{18} \, \tilde{s}_\Theta^2 u_3 \quad . \end{aligned} \tag{21.100}$$

In the semidilute limit the result takes the form

$$\ell^3 \frac{\hat{\Pi}}{k_B T} = \frac{(4\pi)^{3/2}}{4} \, \beta_e s_e^\Theta \tilde{c}_\Theta^2 u_3^{4/11} + \frac{(4\pi)^3}{18} \, u_3 \tilde{c}_\Theta^3 \quad . \tag{21.101}$$

Thus $\hat{\Pi}$ becomes independent of N, as expected. At the Θ-point $\beta_e = 0$ we thus expect to see essentially a cubic concentration dependence, modified by a factor $u_3 \sim |\ln \tilde{c}_\Theta|^{-1}$. Clearly this differs strongly from the predictions of two-parameter theory, where $\frac{\hat{\Pi}}{k_B \Theta} \underset{T \to \Theta}{\longrightarrow} c_p$, rigorously.

21.5.3 Density correlations

In tree-approximation the density autocorrelations are given by the Debye function:

$$\frac{I_{d,a}(\mathbf{q})}{c_p N^2} = D_p\left(\frac{\tilde{p}_2}{\tilde{p}_3}\, q^2 R_g^2\right) \quad . \tag{21.102}$$

We therefore immediately turn to the density cumulant

$$\begin{aligned} \frac{c_p N^2}{I_{d,c}(\mathbf{q})} &= J_c^{-1}(\mathbf{q}) \\ &= D_p^{-1}(\bar{\mathbf{q}}^2 N_R) + \frac{(4\pi)^{3/2}}{2}\, u_e c_{pR} N_R^2 + \frac{(4\pi)^3}{6}\, u_3 c_{pR}^2 N_R^3 \\ &= D_p^{-1}(\bar{\mathbf{q}}^2 N_R) + \frac{(4\pi)^{3/2}}{2} \tilde{z}_\Theta \tilde{s}_\Theta u_3^{4/11} + \frac{(4\pi)^3}{6}\, \tilde{s}_\Theta^2 u_3 \quad . \end{aligned} \tag{21.103}$$

For $\mathbf{q} \to 0$ we find

$$J_c^{-1}(0) = \frac{1}{\tilde{p}_2} + \frac{(4\pi)^{3/2}}{2} \tilde{z}_\Theta \tilde{s}_\Theta u_3^{4/11} + \frac{(4\pi)^3}{6}\, \tilde{s}_\Theta^2 u_3 \tag{21.104}$$

$$\xi_d^2 = \frac{\tilde{p}_3}{(\tilde{p}_2)^2}\, \tilde{R}_0^2 J_c(0) \quad . \tag{21.105}$$

In the semidilute limit these results read

$$J_c^{-1}(0) = s_n n \left[\frac{(4\pi)^{3/2}}{2}\, s_e^\Theta \beta_e \tilde{c}_\Theta u_3^{4/11} + \frac{(4\pi)^3}{6}\, \tilde{c}_\Theta^2 u_3\right] \tag{21.106}$$

$$\xi_d^2 = \frac{\tilde{p}_3}{(\tilde{p}_2)^2}\, \ell^2 \left[\frac{(4\pi)^{3/2}}{2}\, s_e^\Theta \beta_e \tilde{c}_\Theta u_3^{4/11} + \frac{(4\pi)^3}{6}\, \tilde{c}_\Theta^2 u_3\right]^{-1} \quad . \tag{21.107}$$

The factors of u_3 again introduce some logarithmic concentration dependence.

Again these results strongly differ from those of two-parameter theory. Let us consider ξ_d^2, in particular. In Sect. 19.2.2 we found that ξ_d^2 should diverge, as $T \to \Theta$ in the semidilute region of two-parameter theory. This divergence is cut off near $\xi_d^2 \sim \tilde{R}_0^2 \approx R_g^2$, since for $T \to \Theta$, but fixed Θ-overlap $\tilde{s}_\Theta$, we eventually leave the semidilute Θ-region of the two-parameter theory. The present result, Eq. (21.107), shows that this divergence is cut off already on a chain-length independent level $\xi_d^2 \sim \ell^2(\tilde{c}_\Theta^2 u_3)^{-1}$ by the effect of three-body forces. For long chains in the experimentally accessible range of concentrations this cut-off occurs for $\xi_d^2 \ll \tilde{R}_0^2$. Of course, for $\tilde{c}_\Theta \to 0$ we again leave the semidilute region.

In Fig. 21.5 we have plotted $\xi_d^2/\tilde{R}_0^2$ as function of the overlap. For $\tilde{z}_\Theta > 0$ this function starts with some finite initial slope. At the Θ-point this slope vanishes, and for $T < \Theta$, ($\tilde{z}_\Theta < 0$) the maximum is shifted towards ($\tilde{s}_\Theta > 0$), signaling the onset of long-range correlations due to the attractive two-body force.

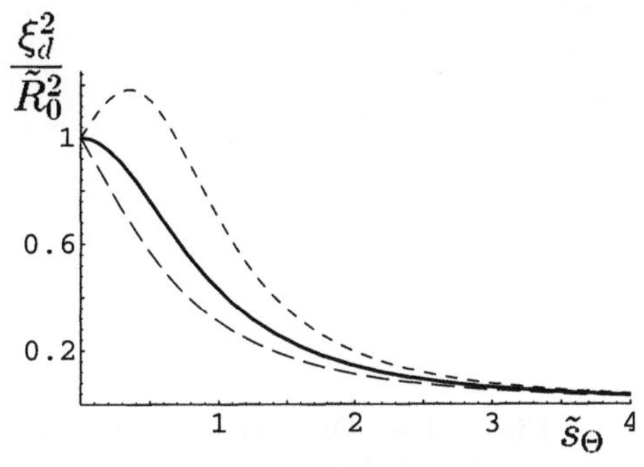

Fig. 21.5. $\xi_d^2/\tilde{R}_0^2$ as function of $\tilde{s}_\Theta$. Parameter values: $s_n N = 10^5, u_3^{(0)} = 1$. *Full line*: $\tilde{z}_\Theta = 0$; *broken lines*: $\tilde{z}_\Theta = 0.3$ (*long dashes*); $\tilde{z}_\Theta = -0.3$ (*short dashes*)

21.6 The experimental situation: concentration crossover

For concentrations in the range of a few weight-percent, data in the Θ-region clearly show effects of the three-body force. Pronounced cubic terms have been observed in the osmotic pressure, and data for the density correlation length cannot be interpreted within the frame of two-parameter theory, without invoking three-body contributions.

Illustrating the presence of repulsive three-body forces this on the one hand is quite encouraging. On the other hand it also is fairly trivial. At higher concentrations we also within the realm of excluded volume theory expect to find 'irrelevant' corrections due to many-body forces. Without a quantitative analysis therefore this finding is no convincing evidence for the validity of the tricritical theory. Indeed, to distinguish the tricritical approach from naive unrenormalized perturbation theory in the Θ-region we again have to identify the logarithmic corrections which modify the simple regular behavior. No experiment reaching the necessary accuracy seems to have been published. Again the data suffice only to establish the trivial Gaussian-type scaling behavior of the theory.

A 21. Appendix

A 21.1 Results of unrenormalized perturbation theory

A 21.1.1 Single chain partition function. The diagrammatic expression for $\mathcal{Z}^{(1)}$ up to order β_2^2, β_3 is shown in Fig. 21.6. The analytic expression reads

$$\frac{(4\pi\ell^2)^{d/2}}{\Omega}\,\mathcal{Z}^{(1)} = 1 - \beta_2 U_1 - \beta_3 U_2 + \beta_2^2 (U_3 + U_4 + U_5)\ , \qquad \text{(A 21.1)}$$

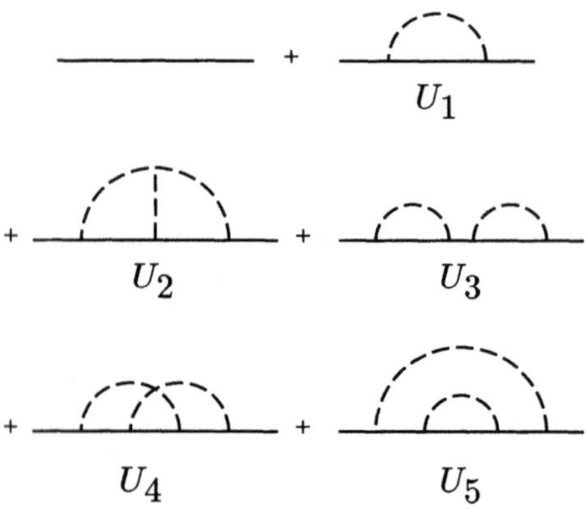

Fig. 21.6. Diagrams for $\mathcal{Z}^{(1)}$ to two-loop order

where

$$
\begin{aligned}
U_1 &= \sum_{0<j_1<j_2<n} (j_2 - j_1)^{-d/2} \\
&= n\zeta(\frac{3-\varepsilon_3}{2}) - \frac{4}{1-\varepsilon_3^2}\, n^{\frac{1+\varepsilon_3}{2}} \\
U_2 &= \sum_{0<j_1<j_2<j_3<n} (j_2 - j_1)^{-d/2}(j_3 - j_2)^{-d/2} \\
&= n\zeta^2(\frac{3-\varepsilon_3}{2}) - \frac{8}{1-\varepsilon_3^2}\,\zeta(\frac{3-\varepsilon_3}{2})\, n^{\frac{1+\varepsilon_3}{2}} + \frac{4}{(1-\varepsilon_3^2)}\frac{\Gamma^2\left(\frac{1+\varepsilon_3}{2}\right)}{\Gamma(1+\varepsilon_3)}\, n^{\varepsilon_3} \\
U_3 &= \sum_{0<j_1<j_2<j_3<j_4<n} (j_2 - j_1)^{-d/2}(j_4 - j_3)^{-d/2} \\
&= \frac{n^2}{2}\zeta^2(\frac{3-\varepsilon_3}{2}) - \frac{16}{(1-\varepsilon_3^2)(3+\varepsilon_3)}\,\zeta(\frac{3-\varepsilon_3}{2})\, n^{\frac{3+\varepsilon_3}{2}} \\
&\quad + \frac{4}{(1-\varepsilon_3^2)(1-\varepsilon_3)}\frac{\Gamma^2\left(\frac{1+\varepsilon_3}{2}\right)}{\Gamma(1+\varepsilon_3)}\, n^{1+\varepsilon_3} \\
U_4 &= \sum_{0<j_1<j_2<j_3<j_4<n} \left[(j_3 - j_1)(j_4 - j_2) - (j_3 - j_2)^2\right]^{-d/2} \\
&= \frac{2}{\varepsilon_3}\frac{\Gamma^2\left(\frac{1+\varepsilon_3}{2}\right)}{\Gamma(1+\varepsilon_3)}\, n^{1+\varepsilon_3} - 2\int_0^1 dxdy\, x^{-\frac{1-\varepsilon_3}{2}}\left[1 - x(1-y+y^2)\right]^{-\frac{1-\varepsilon_3}{2}} n^{1+\varepsilon_3} \\
U_5 &= \sum_{0<j_1<j_2<j_3<j_4<n} (j_3 - j_2)^{-d/2}(j_4 - j_3 + j_2 - j_1)^{-d/2} \\
&= \frac{4}{(1+\varepsilon_3)(3+\varepsilon_3)}\,\zeta\left(\frac{3-\varepsilon_3}{2}\right) n^{\frac{3+\varepsilon_3}{2}} - \frac{2}{1-\varepsilon_3^2}\frac{\Gamma^2\left(\frac{1+\varepsilon_3}{2}\right)}{\Gamma(1+\varepsilon_3)}\, n^{1+\varepsilon_3} \quad .
\end{aligned}
\tag{A 21.2}
$$

Recall $\varepsilon_3 = 3 - d$. We have given only the leading orders in n, needed in the further analysis. Note that the evaluation of U_1 has been discussed in Appendix A 4.2. The other expressions are derived in a similar way, the calculations being a bit lengthy, however.

Carrying through the additive renormalizations introducing β_e and μ_s^* and numerically evaluating the integral in U_4 for $\varepsilon_3 = 0$ we find

$$\frac{(4\pi\ell^2)^{d/2}}{\Omega} e^{\mu_s^* n} \mathcal{Z}^{(1)} = 1 + \frac{4}{1-\varepsilon_3^2} \beta_e n^{\frac{1+\varepsilon_3}{2}} - \frac{4}{(1-\varepsilon_3)^2} \frac{\Gamma^2\left(\frac{1+\varepsilon_3}{2}\right)}{\Gamma(1+\varepsilon_3)} \beta_3 n^{\varepsilon_3} + \left(\frac{2\pi}{\varepsilon_3} + b_1(\varepsilon_3)\right) \beta_e^2 \, n^{1+\varepsilon_3} \qquad \text{(A 21.3)}$$

$$b_1(0) = -7.5224 \ . \qquad \text{(A 21.4)}$$

This result should be viewed as resulting from dimensional regularization and continuous chain limit. Indeed the combinations

$$\beta_e \, n^{\frac{1+\varepsilon_3}{2}} = \tilde{u}_e R_0^{1+\varepsilon_3}$$
$$\beta_3 \, n^{\varepsilon_3} = \tilde{u}_3 R_0^{2\varepsilon_3}$$

are kept fixed in taking the continuous chain limit $\ell \to 0$ for $\varepsilon_3 > 0$. The same remark applies to all further results of this section.

A 21.1.2 Two-chain cumulant. Diagrams for $\mathcal{Z}_c^{(2)}$, including orders $\beta_2^2, \beta_2\beta_3$, are shown in Fig. 21.7, where we enclosed in curly brackets terms which combine to give simpler expressions. The analytic expression reads

$$\frac{(4\pi\ell^2)^{d/2}}{\Omega} \mathcal{Z}_c^{(2)} = -\beta_2 n^2 - 2\beta_3 n U_1 + 2\beta_2^2 (V_1 + n^2 U_1) + \beta_2\beta_3 \Big[2n^2 U_2 + 8V_2 + 4V_3 + 2nU_1^2 + 2nU_5 + 4nU_4 + 4nU_3 + 2nU_5\Big] \ , \qquad \text{(A 21.5)}$$

with new segment sums

$$V_1 = \sum_{\substack{0<j_1<j_2<n \\ 0<j_3<j_4<n}} (j_2 - j_1 + j_4 - j_3)^{-d/2} = 16 \frac{7 + \varepsilon_3 - 8 \cdot 2^{-\frac{1-\varepsilon_3}{2}}}{(1-\varepsilon_3^2)(3+\varepsilon_3)(5+\varepsilon_3)} n^{\frac{5+\varepsilon_3}{2}}$$

$$V_2 = \sum_{\substack{0<j_1<j_2<n \\ 0<j_3<j_4<j_5<n}} (j_2 - j_1 + j_4 - j_3)^{-d/2} (j_5 - j_4)^{-d/2} = \zeta\left(\frac{3-\varepsilon_3}{2}\right) V_1 - \frac{4}{(1-\varepsilon_3^2)(1-\varepsilon_3)} \left(1 - \frac{2^{1+\varepsilon_3} - 1}{2+\varepsilon_3}\right) \frac{\Gamma^2\left(\frac{1+\varepsilon_3}{2}\right)}{\Gamma(1+\varepsilon_3)} n^{2+\varepsilon_3}$$

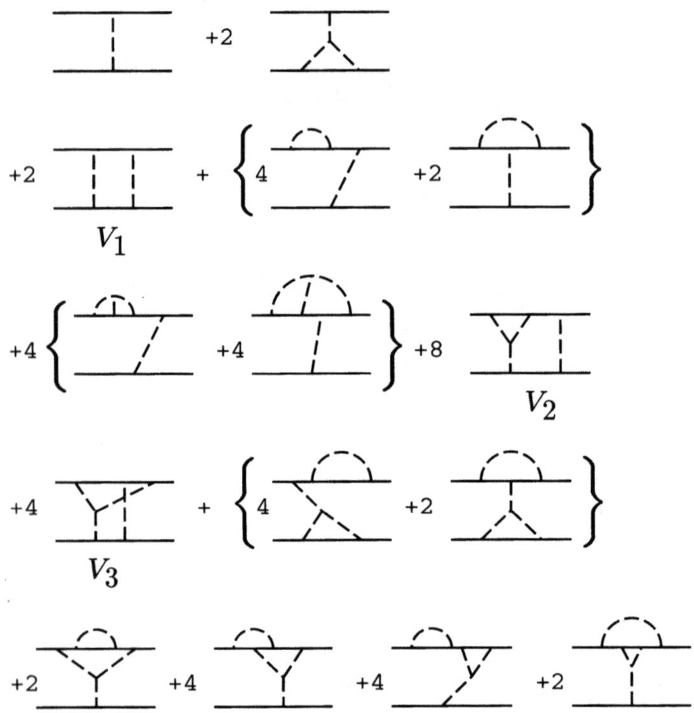

Fig. 21.7. Diagrams contributing to $\mathcal{Z}_c^{(2)}$, up to order $\beta_2^2, \beta_2\beta_3$

$$
\begin{aligned}
V_3 &= \sum_{\substack{0<j_1<j_2<n \\ 0<j_3<j_4<j_5<n}} \left[(j_2-j_1)(j_5-j_4)+(j_5-j_4)(j_4-j_3)+(j_4-j_3)(j_2-j_1)\right]^{-d/2} \\
&= \Bigg[\left(\frac{2}{\varepsilon_3}+\frac{1}{(2+\varepsilon_3)^2}\right)\frac{\Gamma^2\left(\frac{1+\varepsilon_3}{2}\right)}{\Gamma(1+\varepsilon_3)} \\
&\quad - 4\int_0^1 dx\, dy\, x^{-\frac{1-\varepsilon_3}{2}}(1-x)\cdot(1+y(1-y)x)^{\frac{1+\varepsilon_3}{2}}\Bigg]\frac{n^{2+\varepsilon_3}}{1-\varepsilon_3^2} \quad . \qquad \text{(A 21.6)}
\end{aligned}
$$

With the additive renormalizations we find

$$
\begin{aligned}
\frac{(4\pi\ell^2)^{d/2}}{\Omega} e^{2\mu_s^* n}\mathcal{Z}_c^{(2)} = n^{\frac{3-\varepsilon_3}{2}}\Big\{ &-\beta_e n^{\frac{1+\varepsilon_3}{2}} + \frac{8}{1-\varepsilon^2}\beta_3 n^{\varepsilon_3} \\
&+ b_2(\varepsilon_3)\beta_e^2 n^{1+\varepsilon_3} + \left(b_3(\varepsilon_3)+\frac{16\pi}{\varepsilon}\right)\beta_e\beta_3 n^{\frac{1+3\varepsilon_3}{2}}\Big\}
\end{aligned}
$$

$$b_2(0) = -5.1347 \qquad \text{(A 21.7)}$$

$$b_3(0) = -76.6037 \quad . \qquad \text{(A 21.8)}$$

A 21.1.3 Three-chain cumulant. Up to second order in the couplings the diagrammatic expansion of $\mathcal{Z}_c^{(3)}$ is shown in Fig. 21.8. The analytical

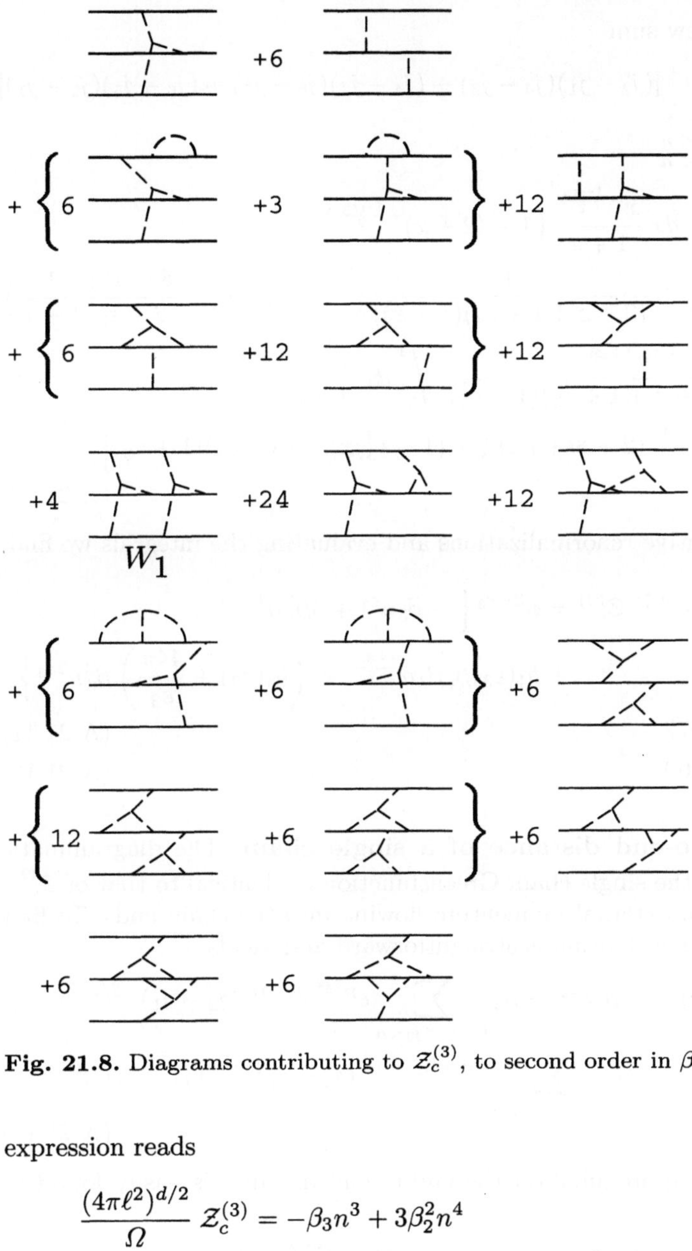

Fig. 21.8. Diagrams contributing to $\mathcal{Z}_c^{(3)}$, to second order in β_2, β_3

expression reads

$$\begin{aligned}\frac{(4\pi\ell^2)^{d/2}}{\Omega}\,\mathcal{Z}_c^{(3)} &= -\beta_3 n^3 + 3\beta_2^2 n^4 \\ &+ \beta_2\beta_3\left[3n^3U_1 + 12nV_1 + 6n^3U_1 + 6n^3U_1\right] \\ &+ \beta_3^2\Big[4W_1 + 24nV_2 + 12nV_3 + 3n^3U_2 \\ &+ 6n^2U_1^2 + 6n^2U_3 + 6n^2U_4 + 6n^2U_5 + 3n^2U_1^2\Big] \ ,\end{aligned} \qquad \text{(A 21.9)}$$

with the single new sum

$$W_1 = \sum_{\substack{0<j_1<j_2<n \\ 0<j_3<j_4<n \\ 0<j_5<j_6<n}} [(j_2-j_1)(j_4-j_3)+(j_4-j_3)(j_6-j_5)+(j_6-j_5)(j_2-j_1)]^{-d/2}$$

$$= 12n^{3+\varepsilon_3}\int_0^1 dx\,\frac{x^{-\frac{1-\varepsilon_3}{2}}}{1+x}\left(1-(2+x)^{-\frac{1-\varepsilon_3}{2}}\right)$$

$$\cdot\left[\frac{1}{\varepsilon_3(1-\varepsilon_3)}+\frac{1}{(3+\varepsilon_3)(1+\varepsilon_3)(1-\varepsilon_3^2)}\left(1-4\varepsilon_3-\varepsilon_3^2-\frac{8-4\varepsilon_3}{2+\varepsilon_3}\frac{1}{1+x}\right)\right]$$

$$+\frac{12n^{3+\varepsilon_3}}{(3+\varepsilon_3)(2+\varepsilon_3)(1+\varepsilon_3)(1-\varepsilon_3^2)}\int_0^1 dx\,\frac{x^{\frac{1+\varepsilon_3}{2}}}{1+x}$$

$$\cdot\left[(2+x)^{-\frac{1-\varepsilon_3}{2}}\left(2+8\varepsilon_3+2\varepsilon_3^2+(1-\varepsilon_3^2)x\right)-(5+\varepsilon_3)(1+\varepsilon_3)\right]\ . \tag{A 21.10}$$

Inserting the additive renormalizations and evaluating the integrals we find

$$\frac{(4\pi\ell^2)^{d/2}}{\Omega}\,e^{3\mu_s^* n}\,\mathcal{Z}_c^{(3)} = n^{3-\varepsilon_3}\Big\{-\beta_3 n^{\varepsilon_3}+3\beta_e^2 n^{1+\varepsilon_3}$$

$$+\,b_4(\varepsilon_3)\beta_e\beta_3 n^{\frac{1+3\varepsilon_3}{2}}+\left(b_5(\varepsilon_3)+\frac{44\pi}{\varepsilon_3}\right)\beta_3^2 n^{2\varepsilon_3}\Big\}$$

$$b_4(0) = -42.8077 \tag{A 21.11}$$

$$b_5(0) = -226.7300\ . \tag{A 21.12}$$

A 21.1.4 End-to-end distance of a single chain. The diagrammatic representation of the single chain Greensfunction is identical to that of $\mathcal{Z}_c^{(1)}$, Fig. 21.6, but with external momentum flowing into the chain ends. To first order in β_2, β_3 the evaluation is straightforward and yields

$$\ln\,G^{(0,1)}(\mathbf{p},-\mathbf{p}) = -\mathbf{p}^2\ell^2 n-\beta_2\sum_{0<j_1<j_2<n} e^{\mathbf{p}^2\ell^2(j_2-j_1)}(j_2-j_1)^{-d/2}$$

$$-\,\beta_3\sum_{0<j_1<j_2<j_3} e^{\mathbf{p}^2\ell^2(j_3-j_1)}(j_2-j_1)^{-d/2}(j_3-j_2)^{-d/2}\ . \tag{A 21.13}$$

In this low-order approximation the end-to-end distance is easily found to take the form

$$\frac{R_e^2}{2dn\ell^2} = 1+\frac{4}{(1+\varepsilon_3)(3+\varepsilon_3)}\beta_e n^{\frac{1+\varepsilon_3}{2}}-\frac{4}{1-\varepsilon_3^2}\frac{\Gamma^2\left(\frac{1+\varepsilon_3}{2}\right)}{\Gamma(1+\varepsilon_3)}\beta_3 n^{\varepsilon_3}\ . \tag{A 21.14}$$

Note that the β_3-term is negative due to the additive renormalization transforming β_2 to β_e. This has raised some discussion in the literature. It of course just illustrates that the leading (repulsive) contribution of β_3 is compensated by the attractive contribution $\beta_2^{\Theta}(\beta_3)$.

A 21.2 Multiplicative renormalization

A 21.2.1 Renormalization factors of the minimal subtraction scheme. From Eqs. (21.24), (21.28) we recall the definition of the renormalized parameters

$$\begin{aligned} \beta_3 \ell^{-2\varepsilon_3} &= u_3 \ell_R^{-2\varepsilon_3} Z_3(u_3) \\ \beta_e \ell^{-1-\varepsilon_3} &= u_e \ell_R^{-1-\varepsilon_3} Z_e^{\Theta}(u_3) \\ n\ell^2 &= n_R \ell_R^2 Z_n^{\Theta}(u_3) \quad . \end{aligned} \tag{A 21.15}$$

The renormalization factors have to absorb the pole terms occurring in the dimensionally regularized theory (cf. Eq. (A 21.11) etc.). In the minimal subtraction scheme they are chosen as

$$\begin{aligned} Z_3(u_3) &= \frac{1}{6}\left(1 + \frac{a_3}{\varepsilon_3}\, u_3 + O(u_3^2)\right) \\ Z_e^{\Theta}(u_3) &= \frac{1}{2}\left(1 + \frac{a_e}{\varepsilon_3}\, u_3 + O(u_3^2)\right) \\ Z_n^{\Theta}(u_3) &= 1 + \frac{a_n^{(1)}}{\varepsilon_3}\, u_3 + \left(\frac{a_n^{(2)}}{\varepsilon_3^2} + \frac{a_n^{(3)}}{\varepsilon_3}\right)\frac{u_3^2}{2} + O(u_3^3) \\ Z^{\Theta}(u_3) &= 1 + \frac{a^{(1)}}{\varepsilon_3}\, u_3 + \left(\frac{a^{(2)}}{\varepsilon_3^2} + \frac{a^{(3)}}{\varepsilon_3}\right)\frac{u_3^2}{2} + O(u_3^3) \\ \sigma(u_3) &= \frac{\sigma^{(0)}}{\varepsilon_3} + O(u_3) \quad . \end{aligned} \tag{A 21.16}$$

The last two quantities are needed as prefactors in the theorem of renormalizability (cf. Eq. (21.29)).

We now first consider the single chain partition function at the Θ-point $\beta_e = 0 = u_e$. From the appropriate form of the theorem of renormalizability, Eq. (21.26) we know that

$$\mathcal{Z}_R^{(1)} = \frac{Z_n^{\Theta}}{Z^{\Theta}} \frac{(4\pi\ell^2)^{d/2}}{\Omega}\, e^{\mu_s^* n}\, \mathcal{Z}^{(1)} \quad , \tag{A 21.17}$$

if expressed in renormalized variables, must be finite for $\varepsilon_3 \to 0$. To leading order we in Eq. (A 21.3), $\beta_e = 0$, can just replace $\beta_3 n^{\varepsilon_3}$ by $\frac{u_3}{6}\, n_R^{\varepsilon_3}$. Thus

$$\mathcal{Z}_R^{(1)} = \frac{Z_n^{\Theta}}{Z^{\Theta}}\left(1 - \frac{2}{3(1-\varepsilon_3)^2}\, \frac{\Gamma^2\left(\frac{1+\varepsilon_3}{2}\right)}{\Gamma(1+\varepsilon_3)}\, u_3 n_R^{\varepsilon_3} + O(u_3^2)\right) \quad .$$

No pole shows up in this expression and thus

$$\frac{Z_n^{\Theta}}{Z^{\Theta}} = 1 + O(u_3^2) \quad .$$

The same analysis, applied to the expression (A 21.14) for R_e^2, shows

$$Z_n^\Theta = 1 + O(u_3^2) \quad .$$

Thus

$$a_n^{(1)} = a^{(1)} = 0 \quad , \tag{A 21.18}$$

and consistency of the renormalization scheme, or more precisely the finiteness of the logarithmic derivatives $d \ln Z^\Theta / d \ln \ell_R$, $d \ln Z_n^\Theta / d \ln \ell_R$ then also requires

$$a_n^{(2)} = a^{(2)} = 0 \quad . \tag{A 21.19}$$

Working only to leading order we can ignore the factors Z_n^Θ, Z^Θ.

Z_3 is determined from the three-chain cumulant at the Θ-point. Equations (21.26), (A 21.11), (A 21.15), (A 21.16) yield

$$\begin{aligned} \mathcal{Z}_{cR}^{(3)} &= \left(\frac{Z_n^\Theta}{Z^\Theta}\right)^3 \left(\frac{\ell_R}{\ell}\right)^{-2d} \frac{(4\pi\ell^2)^{d/2}}{\Omega}\, e^{3\mu_s^* n} \mathcal{Z}_c^{(3)} \\ &= n_R^d \Big\{ -\frac{u_3}{6}\, n_R^{\varepsilon_3} \left(1 + \frac{a_3}{\varepsilon_3}\, u_3\right) + \left(b_5(0) + \frac{44\pi}{\varepsilon_3}\right) \frac{u_3^2}{36}\, n_R^{2\varepsilon_3} \\ &\quad + O(\varepsilon_3 u_3, u_3^3) \Big\} \quad . \end{aligned}$$

To cancel the pole we have to choose

$$a_3 = \frac{22}{3}\,\pi \quad . \tag{A 21.20}$$

The same argument applied to $\mathcal{Z}_c^{(2)}$ for $u \neq 0$ yields

$$a_e = \frac{8}{3}\,\pi \quad . \tag{A 21.21}$$

We finally once more consider the single-chain partition function. Equation (A 21.3) shows a pole in order β_e^2, which must be absorbed by introducing the renormalization factor $\exp[\sigma(u_3) u_e^2 n_R]$ (Eq. (21.29)). Following the standard path the calculation yields

$$\sigma^{(0)} = \frac{\pi}{2} \quad . \tag{A 21.22}$$

Integration of the RG flow equations reveals that the renormalization factors Z^Θ, Z_n^Θ are of the form

$$\begin{aligned} Z^\Theta(w) &= const \,\exp\left(-\frac{a^{(3)}}{a_3}\, u_3 + O(u_3^2)\right) \\ Z_n^\Theta(w) &= const \,\exp\left(-\frac{a_n^{(3)}}{a_3}\, u_3 + O(u_3^2)\right) \end{aligned} \tag{A 21.23}$$

(cf. Eq. (21.38)). To calculate $\mathcal{Z}^{(1)}$ or R_e^2 to order u_3 consistently, we thus need to know $a^{(3)}, a_n^{(3)}$. In bare perturbation theory these coefficients arise from order β_3^2. We have not carried through this most tedious calculation. Rather we used Eqs. (A 21.23) together with our renormalized first order results for $\mathcal{Z}^{(1)}$, R_e^2, and we compared the expressions to results of Duplantier [Dup87]. This allowed us to extract the values

$$\begin{aligned} a^{(3)} &= -\frac{1}{27}\,\pi^2 \\ a_n^{(3)} &= -\frac{16}{27}\,\pi^2 \quad . \end{aligned} \qquad \text{(A 21.24)}$$

This finishes our determination of the renormalization factors.

A 21.2.2 Renormalized scaling functions. We define the renormalized partition functions as

$$\begin{aligned} \mathcal{Z}_{cR}^{(M)} = &\left(\frac{Z_n^{\Theta}}{Z^{\Theta}}\right)^M \exp\left\{-\sigma(u_3)u_e^2 M n_R\right\} \left(\frac{\ell}{\ell_R}\right)^{d(M-1)} \\ &\cdot \frac{(4\pi\ell^2)^{d/2}}{\Omega}\, e^{\mu_s^* M n}\, \mathcal{Z}_c^{(M)} \quad . \end{aligned} \qquad \text{(A 21.25)}$$

Our results yield for $\varepsilon_3 = 0$

$$\begin{aligned} \mathcal{Z}_R^{(1)} &= 1 + 2u_e n_R^{1/2} - \frac{2}{3}\,\pi u_3 + \left(\frac{b_1}{4} + \frac{\pi}{2}\,\ln\, n_R\right) u_e^2 n_R \\ \mathcal{Z}_{cR}^{(2)} &= n_R^{3/2}\Big[-\frac{1}{2}u_e n_R^{1/2} + \frac{4}{3}u_3 + \frac{b_2}{4}u_e^2 n_R \\ &\quad + \left(\frac{b_3}{12} + \frac{4}{3}\pi\,\ln\, n_R\right) u_e u_3 n_R^{1/2}\Big] \\ \mathcal{Z}_{cR}^{(3)} &= n_R^3\Big[-\frac{u_3}{6} + \frac{3}{4}u_e^2 n_R + \frac{b_4}{12}u_e u_3 n_R^{1/2} \\ &\quad + \left(\frac{b_5}{36} + \frac{11}{9}\,\pi\,\ln\, n_R\right) u_3^2\Big] \quad . \end{aligned} \qquad \text{(A 21.26)}$$

Combining our results for R_e^2 we find

$$R_e^2 = 6 n_R \ell_R^2 \left(1 + \frac{2}{3}\,u_e n_R^{1/2} - \frac{2\pi}{3}\,u_3\right) \quad . \qquad \text{(A 21.27)}$$

We also note the result [Dup87] for the radius of gyration, rewritten in our notation

$$R_g^2 = n_R \ell_R^2 \left(1 + \frac{67}{105}\,u_e n_R^{1/2} - \frac{13}{24}\,\pi u_3\right) \quad . \qquad \text{(A 21.28)}$$

A 21.3 Construction of the loop expansion

We decompose the diagrams into vertex-irreducible parts, defined such that they do not separate into disconnected pieces by cutting a further vertex. With two- and three-body vertices, there are three types of cuts, as shown in Fig. 21.9. As before (Sect. 5.2) in this first step we do not allow for cuts which separate external vertices.

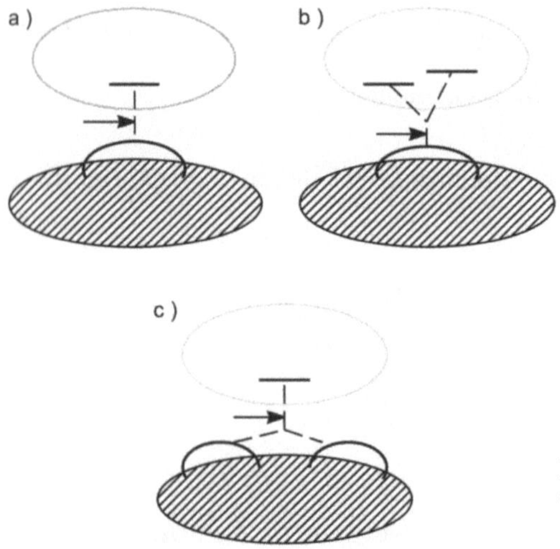

Fig. 21.9a–c. Three different ways to cut vertices, as indicated by the *arrows*. The irreducible part is *shaded*

The two-body cuts of Fig. 21.9a have been considered before. The cut-off part yields a contribution

$$-(4\pi\ell^2)^{d/2}\,\beta_2 c\ ,$$

where $c = c_p N$. The other cuts are specific to the present model. The contribution cut-off in Fig. 21.9b yields

$$\frac{1}{\Omega}(4\pi\ell^2)^{d/2}\,\beta_3 \frac{\partial}{\partial\beta_2}\ \ln\,\mathcal{Z}_G\ ,$$

where $\mathcal{Z}_G$ is the grand partition function.

Contributions of type a), b) occur as insertions into polymer lines of the irreducible part, and summing up any number of these insertions amounts to replacing the chemical potential as

$$\begin{aligned}\exp\left(\frac{\mu_p(n)}{k_BT}\right) &\to \exp(\mu_p(n,c)) \\ &= \exp\left\{\frac{\mu_p(n)}{k_BT} - (4\pi\ell^2)^{d/2}\,\beta_2 cn + (4\pi\ell^2)^{d/2}\,\beta_3 n\,\frac{\partial}{\partial\beta_2}\frac{\ln\,\mathcal{Z}_G}{\Omega}\right\}\ .\end{aligned} \tag{A 21.29}$$

Taking into account the structure of Fig. 21.9c in the irreducible part we replace any two-body vertex by a new vertex

$$(4\pi\ell^2)^{d/2}\,\beta_2(c) = (4\pi\ell^2)^{d/2}\,\beta_2 + (4\pi\ell^2)^{d/2}\,\beta_3 c \ . \qquad \text{(A 21.30)}$$

It is now straightforward to express all observables, except for the osmotic pressure, in terms of irreducible parts and to introduce a screened vertex based on $\beta_2(c)$. With regard to the main text we do not give the general expressions but only note the results in zero-loop approximation. Diagrams with a single density insertion yield

$$c_p(n) = (4\pi\ell^2)^{-d/2}\, e^{\mu_p(n,c)} \ . \qquad \text{(A 21.31)}$$

Density autocorrelations take the same form as before (Sect. 5.4, Eq. (5.54)).

$$I_{d,a}(\mathbf{q}) = cND_p(q^2\ell^2 N) \ . \qquad \text{(A 21.32)}$$

Resumming chains of irreducible parts (cf. Sect. 5.2, Fig. 5.7, where the vertex now stands for $\beta_2(c)$), we find the RPA result for the density cumulant

$$I_{d,c}(\mathbf{q}) = \frac{cND_p(q^2\ell^2 N)}{1 + (4\pi\ell^2)^{d/2}\,\beta_2(c)cND_p(q^2\ell^2 N)} \ . \qquad \text{(A 21.33)}$$

We finally consider the osmotic pressure. Modifying the argument of Appendix A 5.2 to include three-body vertices we easily find the tree approximation

$$\frac{\Pi}{k_B T} = c_p + (4\pi\ell^2)^{d/2}\frac{\beta_2}{2}\,c^2 + (4\pi\ell^2)^{d}\frac{\beta_3}{3}\,c^3 \ . \qquad \text{(A 21.34)}$$

Notes

The tricritical theory of the Θ-region has been suggested by de Gennes [dG75, dG78]. The first quantitative analytic work, using a field theoretic formulation, has been published by Duplantier [Dup82]. Subsequent work on this problem is mainly due to Duplantier [Dup86d, Dup87, Dup88] or to Freed and co-workers (see, for instance, [CDF85]). Contradictory statements on the applicibility of the theory have been published, but the issue has been clarified in [Dup86d, Dup87]. In this chapter we essentially followed the work of Duplantier.

Considerable work has been presented also for $d < 3$ (see, for instance, [Dup86d], [DCF85]). Indeed, for $d = 2$ the tricritical exponents can be calculated exactly [DS87].

22. Multicomponent Solutions

The theory can be extended to cover solutions which contain two ('ternary solutions') or more chemically different polymer species. Here we do not just refer to a difference in molecular weight – polydispersity has been fully taken care off in the binary solution theory of the previous chapters. Rather we deal here with a difference in chemical microstructure, i.e. in the monomer units. Mixtures of different polymers are technically important, and their solution properties deserve, and have found, some interest. Furthermore, solutions containing both protonated and deuterated chains of otherwise identical chemical structure must also be treated as ternary systems, since the polymer-solvent interaction is changed by deuteration. Such isotopic mixtures play a central role in neutron scattering experiments, which aim at a determination of single chain properties in solutions of finite concentration. (See Chap. 18.)

Clearly the phenomenology of multicomponent solutions is much richer than that of binary solutions. It is not only that we can vary more than a single concentration or chain length. Even more important is the fact that the solvent in general is of different quality for the different polymer species. In the critical limit of long chains and small concentrations this leads us to distinguish three classes of ternary solutions.

(i) Both polymer species are above their respective Θ-points, in the excluded volume regime. Then in the critical limit the solution is described by the same excluded volume fixed point u^* as a binary system: the renormalized interactions among all the different species of monomers become identical to u^*. Asymptotically the chemical difference has no effect, a feature, which has been addressed as 'irrelevance of chemical mismatch' [JLB84].

(ii) Only one species is in the excluded volume regime, the other is at its Θ-point. Asymptotically we are confronted with the new situation of excluded volume chains interacting with Θ-chains. The critical limit is governed by a new fixed point with new scaling laws.

(iii) The solvent is a Θ-solvent for both species, but the two species strongly repel each other. The critical limit again is governed by a new fixed point.

Clearly case (i) is the generic situation. To reach case (ii) we have to adjust the temperature to the Θ-point of one species, and case (iii) presupposes that the Θ-temperatures of both species coincide. This latter situation can hardly be reached in physical experiments, but is easily realized in simulations.

There is rich crossover behavior among these different asymptotic cases, and in general a physical system with its finite chain lengths and concentrations will be found somewhere in the crossover regime. This makes the physics of multicomponent polymer solutions a quite involved subject. Indeed, there is even one qualitatively new phenomenon, which in the generic situation not involving Θ-systems is governed by the crossover: different polymer species in general are incompatible, which means that a homogeneous solution of polymer species A and B with increasing concentration separates into an A-rich and a B-rich phase. Such a phase separation cannot occur at the excluded volume fixed point, which from the above is expected to govern the generic critical limit of the system: at this fixed point the chemical difference among species A and B is irrelevant. The phase separation must therefore occur in the crossover regime, maybe close to, but not right at, the excluded volume fixed point.

Since this is a most interesting new effect, specific to ternary or multicomponent solutions, we briefly discuss here the properties of this transition. It shows a standard critical point, where the coexistence line and the spinodal line touch each other. Close to this critical point we find typical critical phenomena like power law and scaling behavior of the scattering intensity as function of temperature and composition. These phenomena, however, are not specific to polymer solutions, but fall into the same class as critical phenomena seen in the demixing of solutions of small molecules. They are due to the divergence of the correlation length ξ_c which describes the range of the composition fluctuations. The size of the solute molecules (R_g) is unimportant as long as $\xi_c \gg R_g$. Our theory, which deals with the effects of increasing polymer size, is not concerned with the critical behavior related to the divergence of ξ_c. With respect to the demixing transition therefore it is on the mean field level.

The standard mean field theory of polymer solutions is the Flory-Huggins theory, which has been extensively applied to the demixing transition in ternary solutions. (See, for instance, [KCG70, Kam90].) It predicts the phase diagram, i.e. the position of the coexistence curve and the spinodal, but as all mean field theories it misses the correct form of the critical singularities. In the simplest zero-loop or tree approximation our theory yields a renormalized form of Flory-Huggins theory. It is found that renormalization strongly influences the results. The most pronounced effects are seen for systems where the solvent is good for both species A,B, and the system is described by the neighborhood of the excluded volume fixed point. As pointed out above, no phase separation can occur right at the fixed point, and the transition is governed by the deviation of the renormalized A-B interaction from the fixed point value. This deviation decreases with increasing chain lengths, which makes the two polymer species more compatible. As a consequence, with increasing chain length the spinodal is shifted deeper and deeper into the semidilute regime, and in the limit of infinitely long chains the phase separation occurs at infinite overlap. For experimentally relevant chain lengths the homogeneous A-B phase regime is much larger than predicted by unrenormalized Flory-Huggins theory. Furthermore, the dependence of the spinodal and the coexistence curve on *chain lengths and total concentration* exhibits anomalous power laws, which however should not be mixed up with the critical behavior of the phase separation itself. As mentioned above, the latter is seen in the dependence of the scattering intensity and susceptibilities on *temperature and composition.*

On a superficial level the theory of multicomponent solutions is a simple extension of binary solution theory. We merely have to replace the single excluded volume coupling β_e by a set of couplings $\beta_e^{(A,A')}$, where $A, A' = 1, 2, \ldots$ distinguishes the different species. The single-species couplings $\beta_e^{(A,A)}$ and the chain lengths $N^{(A)}$ renormalize exactly as for binary solutions. Only $\beta_e^{(A,A')}, A \neq A'$, needs a new renormalization, which follows by straightforward application of the methods worked out in Part III.

Taking a closer look however, we find that our understanding of multicomponent solutions is much less advanced than our understanding of binary solutions. This has both experimental and theoretical reasons, and the theoretical problems are both of a technical and of a more fundamental origin.

On the experimental side a useful study, besides giving data for the ternary A-B-solvent system, should also provide data for the binary subsystems A-solvent, B-solvent, preferably using the same polymer fractions as the ternary experiment. Such data are needed for a determination of the four theoretical parameters $\tilde{\ell}^{(A)}, \tilde{v}^{(A)}, \tilde{\ell}^{(B)}\tilde{v}^{(B)}$. (See Sect. 11.5.) Then only a single parameter, which determines the A-B interaction, has to be extracted from the ternary data. Unfortunately, however, experiments often do not provide enough binary information, leaving up to five more or less free parameters to fit the ternary data.

On the theoretical side we note that the RG mapping for the A-B coupling has not been determined to the same precision as that for the binary couplings. It is known to three-loop order, but due to its more complicated structure, which involves also the A-A and B-B couplings, no resummed crossover form has been given. Only the exponents, which govern the approach to the fixed points, have been determined more precisely [SLK91]. Also, in view of the complicated calculations, only a very restricted set of observables has been calculated to one-loop order of renormalized perturbation theory. These are technical problems, which might be overcome by tedious, but straightforward, work. More fundamental is the observation that here we are dealing with a system possessing at least four independent length scales: the radii of gyration $R_g^{(A)}, R_g^{(B)}$, the screening length ξ_E, which essentially measures the total polymer concentration, and the length ξ_c governing the composition fluctuations. For binary systems we have found a choice of the renormalized length scale ℓ_R that guarantees a smooth crossover among all limits of interest (Sect. 13.2). Now, even ignoring problems of the two-fluid phase transition, which are related to the divergence of ξ_c, we cannot find an equally good and general choice of ℓ_R for a ternary system. In particular the limit of a very unsymmetric system (case (ii) discussed above) is problematic since the radii of a Θ- or an excluded volume chain scale differently with chain lengths. Thus a single prescription for choosing ℓ_R is not appropriate for all observables, and even after renormalization the applicability of perturbation theory outside strict ε-expansion is not guaranteed in all limits. (See the discussion in Sect. 13.2.3.)

Due to all these problems our discussion of multicomponent polymer solutions must be less comprehensive than that of binary systems. Here we mostly concentrate on asymptotic fixed point regions. Though, as stressed above, crossover behavior is most important, it is also so rich that a quantitative analysis is worthwhile only with regard to data for specific chemical systems. I should stress, however, that all our results can be evaluated without problems in the crossover region. Furthermore we evaluate the scaling functions in renormalized tree approximation only. Specifically, in the next section we define the unrenormalized theory, and we discuss the renormalization, the scaling laws, and the choice of ℓ_R in Sect. 22.2. Section 22.3 is devoted to the tree approximation, and an evaluation of renormalized Flory-Huggins theory for the phase separation is given in Sect. 22.4. One-loop results, where available, are mentioned only briefly. The experimental situation is discussed in Sect. 22.5. Full results for the RG flow of the A-B coupling are given in Appendix A 22.1, and Appendix A 22.2 contains some more detailed algebra relevant for strongly unsymmetric systems. The general scaling form of the spinodal in the excluded volume limit is discussed in Appendix A 22.3. We also derive there the full asymptotic form in tree approximation.

22.1 Unrenormalized theory

22.1.1 The model

We describe the multicomponent solution by a straightforward extension of the binary model (Sect. 2.1), where we only add an index $A = 1, \ldots, F$ distinguishing the F different species. The solution contains M_A chains of species A, with lengths $n_m^{(A)}, m = 1, \ldots, M_A$. Segment positions for the m-th chain of species A are denoted as $\mathbf{r}_j^{(A,m)}, j = 0, 1, \ldots, n_m^{(A)}$. With this notation the configurational weight is defined as

$$\mathcal{F} = \left[\mathcal{Z}^{(M_1,\ldots,M_F)}\right]^{-1} \exp\left[-V\left\{\mathbf{r}_j^{(A,m)}\right\}\right] \tag{22.1}$$

$$\exp[-V] = \exp\left[-\frac{1}{4\ell^2}\sum_{A=1}^{F}\sum_{m=1}^{M_A}\sum_{j=1}^{n_m^{(A)}}\left(\mathbf{r}_j^{(A,m)} - \mathbf{r}_{j-1}^{(A,m)}\right)^2\right]$$
$$\cdot\prod\nolimits'_{(Amj,A'm'j')}\left[1 - (4\pi\ell^2)^{d/2}\beta_e^{(A,A')}\delta^d\left(\mathbf{r}_j^{(A,m)} - \mathbf{r}_{j'}^{(A',m')}\right)\right] . \tag{22.2}$$

The partition function $\mathcal{Z}^{(M_1,\ldots,M_F)}$ is defined by integrating Eq. (22.2) over all $r_j^{(A,m)}$, cf. Eq. (2.4). The prime at the product recalls the restriction that each segment is allowed to interact only once. We note that we have taken the same segment size ℓ for all the different species. In view of the irrelevance of the true microscopic structure this simplification does not imply any restriction of our model.

Starting from this canonical description we may transform to the grand canonical ensemble by introducing chemical potentials $\mu_p^{(A)}(n), A = 1, \ldots, F$, and summing over $n_m^{(A)}, M^{(A)}$. The grand canonical partition function, for instance, takes the form

$$\mathcal{Z}\left[\mu_p^{(1)}, \ldots, \mu_p^{(F)}\right] = \sum_{M_1,\ldots,M_F=0}^{\infty}\prod_{A=1}^{F}\frac{1}{M_A!}\sum_{\{n_m^{(A)}\}=0}^{\infty}\prod_{A=1}^{F}\prod_{m_A=0}^{M_A} e^{\mu_p^{(A)}(n_{m_A}^{(A)})}$$
$$\cdot\ \mathcal{Z}^{(M_1,\ldots,M_F)}\{n_m^{(A)}\} , \tag{22.3}$$

where

$$\{n_m^{(A)}\} = \left\{n_1^{(1)}, \ldots, n_{M_1}^{(1)}, n_1^{(2)}, \ldots, n_{M_2}^{(2)}, \ldots, n_1^{(F)}, \ldots, n_{M_F}^{(F)}\right\} . \tag{22.4}$$

The concentration of chains of species A, length n, is found as

$$\begin{aligned} c_p^{(A)}(n) &= \frac{1}{\Omega} \frac{\partial}{\partial \mu_p^{(A)}(n)} \ln \mathcal{Z}\left[\mu_p^{(1)}, \ldots, \mu_p^{(F)}\right] \\ &= c_p^{(A)} P^{(A)}(n) \ . \end{aligned} \tag{22.5}$$

In the last line we introduced the total concentration of species A:

$$c_p^{(A)} = \sum_n c_p^{(A)}(n) \ , \tag{22.6}$$

as well as the corresponding chain length distribution. The number average chain length of species A will be denoted by $N^{(A)} = \Sigma_n n P^{(A)}(n)$, and the segment concentration is $c^{(A)} = c_p^{(A)} N^{(A)}$. We also introduce the total chain concentration

$$c_p = \sum_{A=1}^{F} c_p^{(A)} \ . \tag{22.7}$$

The evaluation of the model now proceeds by perturbation theory in powers of $\beta_e^{(A,A')}$, $A, A' = 1, \ldots F$. As compared to the expansion constructed in Chap. 4 we only have to carry along the index A distinguishing the different species. An interaction line connecting chains of species A and A' yields a factor $-(4\pi\ell^2)^{d/2}\beta_e^{(A,A')}$. This is the only serious modification of the Feynman rules given in Sect. 4.2.

22.1.2 Loop expansion

We now repeat the steps taken in Chap. 5 to derive the loop expansion for the grand canonical ensemble. We consider correlation functions defined as grand canonical expectation values of, for instance, segment density operators of species A:

$$\hat{\rho}^{(A)}(\mathbf{r}) = \sum_m \sum_{j=1}^{n_m^{(A)}-1} \delta^d(\mathbf{r} - \mathbf{r}_j^{(A,m)}) \ . \tag{22.8}$$

The experimentally most relevant example is the density cumulant

$$\hat{I}_{d,c}^{(A,A')}(\mathbf{r} - \mathbf{r}') = \left\langle \left(\hat{\rho}^{(A)}(\mathbf{r}) - c^{(A)}\right) \left(\hat{\rho}^{(A')}(\mathbf{r}') - c^{(A')}\right) \right\rangle \ , \tag{22.9}$$

(cf. Sect. 5.1.4 for the precise definition of grand canonical expectation values). Of course the linked cluster theorem holds, and the expectation values are given by the set of connected grand canonical diagrams (Sect. 5.1.3) with the insertions specified by the expectation value considered.

The insertions identify a distinguished part of the diagram, in which any cut of an interaction line, which separates the diagram into two disconnected pieces, also separates the insertions. This part is decorated with dangling parts, which do not contain insertions and which can be separated from the diagram by cutting one interaction line. Following Sect. 5.2.2 we sum over any number of these dangling parts to find a shift of the chemical potentials, due to the average concentrations:

$$\mu_p^{(A)}(n) \to \mu_p^{(A)}(n) - (4\pi\ell^2)^{d/2}\, n \sum_{A'=1}^{F} \beta_e^{(A,A')} c^{(A')} \quad . \tag{22.10}$$

As a result the expansions for quantities involving a single insertion, like $c^{(A)} = \langle\hat{\rho}^{(A)}(\mathbf{r})\rangle$, or for segment or end-point autocorrelations proceed in terms of vertex irreducible diagrams. We write

$$c_p^{(A)}(n) = 2\exp\left[\mu_p^{(A)}(n) - (4\pi\ell^2)^{d/2}\, n \sum_{A'} \beta_e^{(A,A')}\, c^{(A')}\right] K^{(A)(0,1)}(n) \quad , \tag{22.11}$$

$$I_{d,a}^{(A)} = K_a^{(A)(2,0)}(\mathbf{q}) \quad , \tag{22.12}$$

where in obvious generalization of the notation of Sect. 5.2.2 $K^{(A)(0,1)}(n)$ is given by the set of all vertex irreducible diagrams containing one open line of species A, length n, whereas the diagrams for $K_a^{(A)(2,0)}$ are vertex irreducible and contain two density insertions in the same line of type A.

For analyzing the density cumulant we introduce $K^{(A,A')(2,0)}(\mathbf{q})$, given by the set of all vertex irreducible diagrams containing two density insertions, in lines of species A or A', respectively. $I_{d,c}^{(A,A')}(\mathbf{q})$ is given by a sum of strings of these irreducible parts, as shown in Fig. 5.7. Analytically the generalization of Eqs. (5.28), (5.29) reads

$$\begin{aligned} I_{d,c}^{(A,A')}(\mathbf{q}) = {} & K^{(A,A')(2,0)}(\mathbf{q}) - \sum_{B,B'=1}^{F} K^{(A,B)(2,0)}(\mathbf{q})(4\pi\ell^2)^{d/2}\, \beta_e^{(B,B')} \\ & K^{(B',A')(2,0)}(\mathbf{q}) \\ & + \sum_{B,B',C,C'=1}^{F} K^{(A,B)(2,0)}(\mathbf{q})(4\pi\ell^2)^{d/2}\, \beta_e^{(B,B')} K^{(B',C)(2,0)}(\mathbf{q}) \\ & (4\pi\ell^2)^{d/2}\, \beta_e^{(C,C')} K^{(C',A')(2,0)}(\mathbf{q}) - \ldots \quad . \end{aligned}$$

Introducing a matrix notation

$$(\mathbf{u_0})_{AA'} = (4\pi\ell^2)^{d/2}\, \beta_e^{(A,A')} \tag{22.13}$$

and the corresponding notation for $I_{d,c}^{(A,A')}$, $K^{(A,A')(2,0)}$ we can write this as

$$\begin{aligned}\mathbf{I_{d,c}}(\mathbf{q}) &= \mathbf{K}^{(2,0)}(\mathbf{q}) - \mathbf{K}^{(2,0)}(\mathbf{q})\mathbf{u_0}\mathbf{K}^{(2,0)}(\mathbf{q}) \\ &\quad + \mathbf{K}^{(2,0)}(\mathbf{q})\mathbf{u_0}\mathbf{K}^{(2,0)}(\mathbf{q})\mathbf{u_0}\mathbf{K}^{(2,0)}(\mathbf{q}) - \dots \\ &= \mathbf{K}^{(2,0)}(\mathbf{q}) - \mathbf{K}^{(2,0)}(\mathbf{q})\mathbf{u_0}\mathbf{I_{d,c}}(\mathbf{q}) \ . \end{aligned} \tag{22.14}$$

We thus find

$$\mathbf{I_{d,c}}(\mathbf{q}) = \left[\mathbf{1} + \mathbf{K}^{(2,0)}(\mathbf{q})\mathbf{u_0}\right]^{-1} \mathbf{K}^{(2,0)}(\mathbf{q}) \ , \tag{22.15}$$

which generalizes Eq. (5.30) to an arbitrary number of polymer species.

We finally have to introduce the screened vertex (cf. Sect. 5.2.3). It again is given by the sequence of interacting chains shown in Fig. 5.9. We introduce a diagonal matrix

$$\left(\mathbf{X}(\mathbf{q^2})\right)_{AA'} = \delta_{AA'} \sum_n \frac{2}{(4\pi\ell^2)^{d/2}} \exp\left[\mu_p^{(A)}(n) - \sum_B u_{0,AB}c^{(B)}n\right] \frac{n^2}{2} D(q^2\ell^2 n). \tag{22.16}$$

Figure 5.8 yields the matrix equation

$$\tilde{\mathbf{u}}_\mathbf{0}(\mathbf{q}) = \mathbf{u_0} - \mathbf{u_0}\mathbf{X}(\mathbf{q^2})\tilde{\mathbf{u}}_\mathbf{0}(\mathbf{q}) \ , \tag{22.17}$$

with the solution

$$\tilde{\mathbf{u}}_\mathbf{0}(\mathbf{q}) = \left[\mathbf{1} + \mathbf{u_0}\mathbf{X}(\mathbf{q^2})\right]^{-1} \mathbf{u_0} \ . \tag{22.18}$$

As a result the rules of the loop expansion are not changed, except that a screened vertex connecting lines of species A, A' is to be interpreted as the A, A'-matrix element $\tilde{u}_0(\mathbf{q})_{AA'}$, and each polymer line is to be summed over all species, except if the species is fixed by an insertion.

Again the grand potential has to be considered separately. The analysis of Appendix A 5.2 goes through with only trivial changes of interpretation, and leads to

$$\begin{aligned}\Pi\left[\mu_p^{(1)}, \dots, \mu_p^{(F)}\right] &= \lim_{\Omega\to\infty} \frac{1}{\Omega} \ln \mathcal{Z}\left[\mu_p^{(1)}, \dots, \mu_p^{(F)}\right] \\ &= K^{(0,0)} + \frac{1}{2}\sum_{AA'} c^{(A)} u_{0,AA'} c^{(A')} \ , \end{aligned} \tag{22.19}$$

with $u_{0,AA'}$ given by Eq. (22.13). $K^{(0,0)}$ is the set of all one-vertex irreducible diagrams without any insertion.

22.1.3 Tree approximation

The results of the previous section immediately yield the unrenormalized tree approximation. Inserting the lowest order result $K^{(A)(0,1)}(n) = \frac{1}{2}(4\pi\ell^2)^{-d/2}$ into Eq. (22.11) we find the Flory-Huggins type relation among $c_p^{(A)}$ and $\mu_p^{(A)}$:

$$\mu_p^{(A)}(n) - \frac{d}{2}\ln(4\pi) = \ln(\ell^d c_p^{(A)}(n)) + n\sum_{A'} u_{0,AA'}c^{(A')} \ . \tag{22.20}$$

This generalizes Eq. (5.53).

The density autocorrelation function is identical to that in a binary solution:

$$\begin{aligned} I_{d,a}^{(A)}(\mathbf{q}) &= \sum_n c_p^{(A)}(n)n^2 D(nq^2\ell^2) \\ &= c^{(A)}N^{(A)}D_p^{(A)}(N^{(A)}q^2\ell^2) \ , \end{aligned} \tag{22.21}$$

(cf. Eq. (5.54)). Here $D_p^{(A)}(\ldots)$ is the polydispersity dependent Debye function for species A.

Also the result for the osmotic pressure is most simple. It follows from Eq. (22.19) together with the tree approximation

$$K^{(0,0)} = \sum_A \sum_n c_p^{(A)}(n) = c_p \ .$$

This yields

$$\frac{\hat{\Pi}}{k_B T} = \Pi = c_p + \frac{1}{2}\sum_{AA'} c^{(A)}u_{0,AA'}c^{(A')} \ . \tag{22.22}$$

As for a binary system the unrenormalized expression for Π is quadratic in the concentrations.

Less trivial is the density cumulant. Equation (22.15) yields

$$(\mathbf{I_{d,c}}(\mathbf{q}))^{-1} = \left(\mathbf{K}^{(2,0)}(\mathbf{q})\right)^{-1} + \mathbf{u_0} \ , \tag{22.23}$$

where $\mathbf{K}^{(2,0)}(\mathbf{q})$ in tree approximation reduces to a diagonal matrix:

$$\left(\mathbf{K}^{(2,0)}(\mathbf{q})\right)_{AA'} = \delta_{AA'}c^{(A)}N^{(A)}D_p^{(A)}(N^{(A)}q^2\ell^2) \ . \tag{22.24}$$

For a ternary system ($F = 2$) we will write down the explicit form of $\mathbf{I_{d,c}}$ in renormalized tree approximation in the next section.

Finally we briefly discuss the screened vertex. To leading approximation we can use Eqs. (22.20), (22.21) to replace X_{AA} (Eq. (22.16)) by $c^{(A)} N^{(A)} D_p^{(A)}$. The resulting tree approximation for $\tilde{\mathbf{u}}_0(\mathbf{q})$ will be denoted by $\mathbf{u}_0(\mathbf{q})$. Specializing to a ternary system we find by inverting the matrix in Eq. (22.18)

$$
\begin{aligned}
u_0(\mathbf{q})_{AA} &= \frac{1}{\Delta_0}\left[u_{0,AA} + (u_{0,11}u_{0,22} - u_{0,12}^2) c^{(\bar{A})} N^{(\bar{A})} D_p^{(\bar{A})}\right] \\
u_0(\mathbf{q})_{A\bar{A}} &= \frac{u_{0,A\bar{A}}}{\Delta_0} \quad ,
\end{aligned}
\tag{22.25}
$$

where

$$
\begin{aligned}
\Delta_0 = 1 &+ u_{0,11} c^{(1)} N^{(1)} D_p^{(1)} + u_{0,22} c^{(2)} N^{(2)} D_p^{(2)} \\
&+ (u_{0,11}u_{0,22} - u_{0,12}^2) c^{(1)} c^{(2)} N^{(1)} N^{(2)} D_p^{(1)} D_p^{(2)} \quad .
\end{aligned}
\tag{22.26}
$$

We use the notation

$$
\bar{A} = \begin{cases} 2 \\ 1 \end{cases} \quad \text{if} \quad A = \begin{cases} 1 \\ 2 \end{cases} \quad .
\tag{22.27}
$$

To analyze the screening properties of this vertex we set $\mathbf{q} = 0$, and we take the binary subsystems to be monodisperse: $D_p^{(A)}(q^2 \ell^2 N^{(A)}) \to 1$. We furthermore consider the simplest case of a completely symmetric system: $N^{(1)} = N^{(2)} = N; u_{0,11} = u_{0,22} = u_0; c^{(1)} = c^{(2)} = c/2$. Then Δ_0 factorizes:

$$
\Delta_0 = \left[1 + \frac{1}{2}(u_0 + u_{0,12})cN\right]\left[1 + \frac{1}{2}(u_0 - u_{0,12})cN\right] \quad ,
$$

and $u_{0,11}(\mathbf{0}) = u_{0,22}(\mathbf{0})$ can be written in the form

$$
u_{0,AA}(\mathbf{0}) = \frac{u_0 + u_{0,12}}{1 + \frac{cN}{2}(u_0 + u_{0,12})} - \frac{u_{0,12}}{\Delta_0} \quad .
$$

The first contribution shows screening: for large N it vanishes like $2/(cN)$. The second contribution equals $u_{0,12}(\mathbf{0})$. It behaves like a truly screened vertex only as long as $|u_0 - u_{0,12}|cN \ll 1$. If this condition is violated the behavior crucially depends on the sign of $u_0 - u_{0,12}$. For $u_0 > u_{0,12}$ we can still take the limit of large N to find a strong suppression:

$$
u_{0,12}(\mathbf{0}) \sim \frac{4u_{0,12}}{u_0^2 - u_{0,12}^2} \frac{1}{c^2 N^2} \quad .
$$

However, for $u_{0,12} > u_0$ we run into a zero of Δ_0 leading to a divergence of our vertex. This identifies the spinodal region, where the homogeneous solution is unstable and separates into two phases. In that region our perturbation theory breaks down.

22.2 Renormalization and scaling behavior

22.2.1 General considerations

We recall from the beginning of Chap. 11 that renormalization primarily aims at the construction of a continuous chain limit, where the leading microstructure dependence is absorbed into a redefinition of the model parameters. These leading nonuniversal contributions can be identified as those terms which in the naive continuous chain limit become singular for $d \to 4$. They are due either to self-interactions within a single chain or to interactions among a pair of chains. The latter terms renormalize the coupling β_e.

It is important to note that there are no such leading microstructure corrections which involve more than two chains. (Recall the field theoretic analysis of Appendix A 11.1.) As a consequence, in a multicomponent solution the 'diagonal' couplings $\beta_e^{(A,A)}$ are renormalized by the same contributions as in a binary solution, viz. those containing two chains of species A. Further microstructure terms, which involve only a single chain (type A), are absorbed into the renormalization of $N^{(A)}, \mu_p^{(A)}$, or in the explicit renormalization factor $Z^{(A)}$. Since no chain of type $A' \neq A$ is involved, this renormalization of the binary subsystem (A) is identical to that found in the corresponding binary solution. We can thus take over our previous results for the renormalization factors or the RG flow, adding just an index (A) to indicate the species.

The only new renormalization concerns the coupling $\beta_e^{(A,A')}$ among two chains of different species $A \neq A'$. The analysis is to be carried through in complete analogy to that for the diagonal couplings. Following Eq. (11.1 i) we define the renormalized coupling $u^{(A,A')}$ as

$$\beta_e^{(A,A')}\ell^{-\varepsilon} = u^{(A,A')}\ell_R^{-\varepsilon} Z_u^{(A,A')} \quad . \tag{22.28}$$

The renormalization factor $Z_u^{(AA')}$ has to absorb all terms, which for $d = 4$ would become singular in the naive continuous chain limit, and which involve chains $A \neq A'$. We, for instance, could generalize our calculation of the second virial coefficient within the discrete chain model to chains of different species and extract $Z_u^{(A,A')}$, following the analysis of Sect. 11.1. We omit this lengthy calculation here, and we later give the results for $Z_u^{(A,A')}$ within the standard framework of dimensional regularization and minimal subtraction. We, however, immediately note that $Z_u^{(A,A')}$ must depend on all three couplings $u^{(A,A)}, u^{(A',A')}, u^{(A,A')}$. This is obvious, since the autocorrelations induced by $u^{(A,A)}, u^{(A',A')}$ will influence the effective interaction of the two chains. Furthermore the relation $u^{(A',A)} \equiv u^{(A,A')}$ is evident.

Having also renormalized all off-diagonal couplings $\beta_e^{(A,A')}$, we can take the renormalized continuous chain limit for general multicomponent solutions ($F > 2$) to find results finite for $d \leq 4$. Compared to ternary solutions ($F = 2$), the presence of more than two species does not lead to new dangerous contributions, since – as stressed above – all leading microstructure corrections involve at most two chains and thus are the same as in the corresponding ternary subsystem. The situation changes only if we consider a system where the Θ-temperatures of several species coincide. The *tricritical* theory of such a Θ-point needs renormalization of three-body interactions among chains of three different species. In the following we will not deal with such a situation, and, as for binary systems, we will model Θ-points as Gaussian fixed points.

It is now easy to generalize the theorem of renormalizability (Sect. 11.2) to multicomponent solutions. Clearly it takes the form of Eqs. (11.10), (11.12), with the renormalization prefactors $\exp(\mu_s^* \Sigma n_m)$, Z^M, Z_n^{L-M} replaced by the appropriate product over the F species, and with all couplings $\beta_e^{(A,A')}$, $u^{(A,A')}$ inserted. In the sequel we will not explicitly need these lengthy equations. Rather we concentrate on properly normalized quantities, where all renormalization prefactors drop out. Working with solutions of finite chain concentration $c_p^{(A)}, A = 1, \ldots, F$, we introduce number fractions of chains as

$$x^{(A)} = \frac{c_p^{(A)}}{c_p} \quad ,$$

where c_p (Eq. (22.7)) is the total chain concentration. Thus

$$\sum_{A=1}^{F} x^{(A)} = 1 \quad .$$

Since, except for the dimensional factor ℓ_R^d the $c_p^{(A)}$ need no renormalization factors:

$$c_{pR}^{(A)} = \ell_R^d c_p^{(A)} \quad ,$$

the number fractions $x^{(A)}$ are RG-invariant variables. We now consider some dimensionless observable Q, which depends on L momentum variables $\mathbf{q}_j$ and obeys RG-invariant normalization conditions. Such quantities renormalize according to the relation

$$\begin{aligned} &Q\left(\{\mathbf{q}_j\}; \{c_p^{(A)}\}; \{N^{(A)}\}; \{\beta_e^{(A,A')}\}; \ell\right) \\ &= Q_R\left(\{\mathbf{q}_j\}; \{x^{(A)}\}; c_{pR}; \{N_R^{(A)}\}, \{u^{(A,A')}\}\right)\left(1 + O\left(\frac{\ell^2}{N_R \ell_R^2}\right)\right) \quad . \end{aligned} \tag{22.29}$$

(Recall the notation $\bar{\mathbf{q}}_j = \mathbf{q}_j \ell_R$.) We suppressed the dependence on the reduced chain length distributions $p^{(A)}(n/N^{(A)})$, which again are RG-invariant.

The only quantity of interest here, which does not fall into this category, is the chemical potential $\mu_p^{(A)}(n)$. Its renormalization follows Eq. (11.67):

$$e^{\mu_R^{(A)}(n_R)} = \left(\frac{\ell_R}{\ell}\right)^{d+2} (4\pi)^{-d/2} Z^{(A)} e^{\mu_p^{(A)}(n) - \mu_s^{*(A)} n} \quad . \tag{22.30}$$

22.2.2 RG flow and fixed points

Calculating the new renormalization factor $Z_u^{(A,A')}$ in minimal subtraction to one-loop order we find

$$Z_u^{(A,A')} = \frac{1}{2}\left(1 + \frac{1}{\varepsilon}\left(u^{(A,A)} + u^{(A',A')}\right) + \frac{2}{\varepsilon}\, u^{(A,A')} + O(u^2)\right) \quad , \tag{22.31}$$

where $O(u^2)$ stands for terms quadratic in the couplings. For $A = A'$ this reduces to Eq. (12.19 i), as it should:

$$Z_u^{(A,A)} = \frac{1}{2}\left(1 + \frac{4}{\varepsilon}\, u^{(A,A)} + O\left((u^{(A,A)})^2\right)\right) \quad .$$

Note that we take $b_u(\varepsilon) \equiv 1$ from the outset. For the explicit analysis of this section we mostly will use the result (22.31). A discussion of the RG equation for $u^{(A,A')}$ to three-loop order can be found in Appendix A 22.1.

We now determine the flow equation by taking the derivative $\ell_R d/d\ell_R$ of Eq. (22.28), keeping all bare parameters fixed. We find

$$\begin{aligned} \ell_R \frac{d}{d\ell_R}\, u^{(A,A')} &= -W^{(A,A')}\left(u^{(A,A')}, u^{(A,A)}, u^{(A',A')}\right) \\ &= u^{(A,A')}\left(\varepsilon - 2u^{(A,A')} - u^{(A,A)} - u^{(A',A')} + O(u^2)\right) \quad . \end{aligned} \tag{22.32}$$

In deriving this result we have to use the flow of the diagonal couplings:

$$\begin{aligned} \ell_R \frac{d}{d\ell_R}\, u^{(A,A)} &= -W^{(A,A)}\left(u^{(A,A)}\right) \\ &= u^{(A,A)}\left(\varepsilon - 4u^{(A,A)} + O\left(u^{(AA)}\right)^2\right) \end{aligned} \tag{22.33}$$

(cf. Eqs. (12.14), (12.19 i)). Clearly Eq. (22.32) for $A = A'$ reduces to Eq. (22.33).

In the sequel for simplicity of notation we restrict the analysis to a ternary system: $A = 1, A' = 2$. The results for a multicomponent system $F > 2$ follow from combining the results of all ternary subsystems.

Equation (22.32) shows a rich fixed point structure. We first note that $u^{(1,2)} = 0$ trivially is an invariant plane, for all $u^{(1,1)}, u^{(2,2)}$: If two species do not interact, no interaction is induced by renormalization. Besides this trivial case there are nontrivial fixed points, which according to Eq. (22.32) must obey the relation

$$u^{*(1,2)} = \frac{\varepsilon}{2} - \frac{1}{2}\left(u^{(1,1)} + u^{(2,2)}\right) + \dots \quad , \tag{22.34}$$

where the diagonal couplings must also take fixed point values. Since there are two fixed point values for each diagonal coupling ($u = 0$ or $u = u^*$, respectively), we find four nontrivial fixed points $u^{*(1,2)}$.

S: symmetric fixed point

$$u^{(1,1)} = u^{(2,2)} = u^* = \frac{\varepsilon}{4} + O(\varepsilon^2) \ .$$

Equation (22.34) yields

$$u_S^{*(1,2)} = \frac{\varepsilon}{4} + O(\varepsilon^2) \ .$$

We indeed rigorously have

$$u_S^{*(1,2)} = u^* \ . \tag{22.35}$$

To prove this we note that for identical starting values $u^{(1,1)}(\ell_R^{(0)}) = u^{(2,2)}(\ell_R^{(0)}) = u^{(1,2)}(\ell_R^{(0)})$ the flow functions $W^{(A,A')}$ also take identical values for $\ell_R^{(0)}$. Then the flow equations guarantee $u^{(1,1)}(\ell_R) \equiv u^{(2,2)}(\ell_R) \equiv u^{(1,2)}(\ell_R)$ for all ℓ_R. Indeed, we are just dealing with two identical species, resulting from a fictitious splitting of a binary system into two subsystems. Thus there must exist a fixed point where all couplings equal u^*. This must be fixed point S, which is the only fixed point where both diagonal couplings equal u^*.

U, U′: unsymmetric fixed points

$$u^{(1,1)} = u^*, u^{(2,2)} = 0 \qquad (U)$$

or

$$u^{(1,1)} = 0, u^{(2,2)} = u^* \qquad (U') \ .$$

Here we find

$$u^{*(1,2)}_{U,U'} = \frac{3}{8}\,\varepsilon + O(\varepsilon^2) \ . \tag{22.36}$$

These fixed points describe a Θ-system interacting with an excluded volume system.

G: nontrivial Gaussian fixed point

$$u^{(1,1)} = 0 = u^{(2,2)} \ .$$

Equation (22.34) yields

$$u^{(1,2)}_G = \frac{\varepsilon}{2} + O(\varepsilon^2) \ . \tag{22.37}$$

We are dealing with two mutually repelling Θ-subsystems.

Clearly fixed points $(S), (U, U'), (G)$ correspond to the three different situations described in the introduction.

The next question of relevance is the stability of these fixed points and of their mutually noninteracting ($u^{(1,2)} = 0$) counterparts $(S_0), (U_0, U_0'), (G_0)$. (We note in passing that G_0 is the standard Gaussian fixed point.) Denoting by P any one of the eight fixed points we introduce the deviations

$$\delta u_P^{(A,A')} = u^{(A,A')} - u_P^{*(A,A')}, \quad A = 1,2,\ A' = 1,2 \ , \tag{22.38}$$

which may be combined into a three-component vector

$$(\delta u)_P = \begin{pmatrix} \delta u_P^{(1,1)} \\ \delta u_P^{(2,2)} \\ \delta u_P^{(1,2)} \end{pmatrix} \ . \tag{22.39}$$

Then the flow equations, linearized about fixed point P, can be written in matrix form:

$$\lambda \frac{d}{d\lambda}\,(\delta u)_P = \Omega_P (\delta u)_P \ , \tag{22.40}$$

($\lambda = \ell/\ell_R$). The matrix Ω_P has the general structure

$$\Omega_P = \begin{pmatrix} \omega_P^{(1,1)} & 0 & 0 \\ 0 & \omega_P^{(2,2)} & 0 \\ \alpha_P^{(1,1)} & \alpha_P^{(2,2)} & \omega_P^{(1,2)} \end{pmatrix} \ , \tag{22.41}$$

where

$$\omega_P^{(A,A')} = \frac{\partial}{\partial u^{(A,A')}}\, W^{(A,A')}\Big|_P$$
$$\alpha_P^{(A,A)} = \frac{\partial}{\partial u^{(A,A)}}\, W^{(1,2)}\Big|_P \ . \tag{22.42}$$

This form exploits the information that the flow of the diagonal coupling $u^{(A,A)}$ depends on $u^{(A,A)}$, only, whereas $u^{(1,1)}, u^{(2,2)}$ couple into the flow of $u^{(1,2)}$. The matrix Ω_P has right-hand eigenvectors

$$\vec{Y}_P^{(1,1)} = \begin{pmatrix} 1 \\ 0 \\ \frac{\alpha_P^{(1,1)}}{\omega_P^{(1,1)} - \omega_P^{(1,2)}} \end{pmatrix} \delta u_P^{(1,1)} \tag{22.43 i}$$

$$\vec{Y}_P^{(2,2)} = \begin{pmatrix} 0 \\ 1 \\ \frac{\alpha_P^{(2,2)}}{\omega_P^{(2,2)} - \omega_P^{(1,2)}} \end{pmatrix} \delta u_P^{(2,2)} \tag{22.43 ii}$$

$$\vec{Y}_P^{(1,2)} = \begin{pmatrix} 0 \\ 0 \\ 1 \end{pmatrix} \delta u_P^{(1,2)} \quad , \tag{22.43 iii}$$

with associated eigenvalues $\omega_P^{(A,A')}$. Note that, depending on the fixed point, some eigenvalues may be degenerate.

In the critical region of fixed point P Eq. (22.40) yields the flow of the eigenvectors

$$\vec{Y}_P^{(A,A')}(\lambda) = \lambda^{\omega_P^{(A,A')}} \vec{Y}_P^{(A,A')}(1) \quad . \tag{22.44}$$

Since we are interested in the infrared limit $\lambda = \ell/\ell_R \to 0$ we note that P is stable in directions where $\omega_P^{(A,A')} > 0$, whereas for $\omega_P^{(A,A')} < 0$ the flow is driven away from fixed point P. For $A = A'$ we have the results of the binary subsystems:

$$\begin{aligned} \omega_P^{(A,A)} &= -\varepsilon \quad \text{if} \quad u_P^{(A,A)} = 0 \\ \omega_P^{(A,A)} &= \omega = \varepsilon + O(\varepsilon^2) \quad \text{if} \quad u_P^{(A,A)} = u^* \quad . \end{aligned}$$

The leading order results for $\omega_P^{(1,2)}, \alpha_P^{(A,A)}$, and

$$f_{12}^{*(P)} = \frac{u_P^{*(1,2)}}{u^*} \tag{22.45}$$

are collected in Table 22.1. (Note that the relation $\alpha_P^{(1,1)} = \alpha_P^{(2,2)}$ found to first-order, at the unsymmetric fixed points is invalid in higher orders). At fixed points where the two species do not interact ($u_{P_0}^{*(1,2)} = 0$) there exists an exact exponent relation [SLK91]

$$\omega_{P_0}^{(1,2)} = d - \frac{1}{\nu(u_{P_0}^{*(1,1)})} - \frac{1}{\nu(u_{P_0}^{*(2,2)})} \quad . \tag{22.46}$$

It identifies $\omega_{P_0}^{(1,2)}$ as the fractal dimension of the set of intersection points of two fractal objects of fractal dimensions $1/\nu(u_{P_0}^{*(1,1)}), 1/\nu^*(u_{P_0}^{*(2,2)})$. For $u_{12}^* \neq 0, \omega_P^{(1,2)}, \alpha_P^{(A,A)}$ and $f_P^{*(1,2)}$ have been determined by Borel resummation of three-loop ε-expansions. Table 22.2 collects the results for $d = 3$, where we should also recall the value $\omega = 0.80$ for the binary correction to scaling exponent. It is seen that the higher order terms have no drastic influence. The trend, in particular, always stays the same as in lowest order (cf. Table 22.1).

Table 22.1. Fixed point values to leading order of the ε-expansion. The results for $U_0'(U')$ are identical to those for $U_0(U)$

	G_0	U_0, U_0'	S_0	G	U, U'	S
$\omega^{(1,2)}$	$-\varepsilon$	$-\frac{3}{4}\varepsilon$	$-\frac{\varepsilon}{2}$	ε	$\frac{3}{4}\varepsilon$	$\frac{\varepsilon}{2}$
$\alpha^{(A,A)}$	0	0	0	$\frac{\varepsilon}{2}$	$\frac{3}{8}\varepsilon$	$\frac{\varepsilon}{4}$
f_{12}^*	0	0	0	2	$\frac{3}{2}$	1

Table 22.2. Higher order estimates [SLK91]. The calculation was done within a somewhat different renormalization scheme. This does not affect the universal exponents $\omega^{(1,2)}$, but might weakly influence the other quantities

	G_0	U_0, U_0'	S_0	G	U	U'	S
$\omega^{(1,2)}$	-1	-0.701	-0.401	0.82	0.68	0.68	0.40
$\alpha^{(1,1)}$	0	0	0	0.36	0.25	0.27	0.20
$\alpha^{(2,2)}$	0	0	0	0.36	0.27	0.25	0.20
f_{12}^*	0	0	0	1.83	1.48	1.48	1

Inspecting these results we find that for each fixed point the number of unstable directions equals the number of vanishing couplings. Thus G_0 is totally unstable, G, U_0, U_0' show two unstable directions, S_0, U, U' show a single unstable direction, and only the totally symmetric fixed point S is globally stable. *Generically any ternary system in the critical limit is driven into point* S, where in the renormalized theory any difference among the two species vanishes. This is the 'irrelevance of chemical mismatch'. It is completely consistent with universality of the binary solution behavior: in the limit of long chains renormalization drives all species to the same excluded volume fixed point. Taking a closer look we observe that $\omega_S^{(1,2)} \approx 0.40 < \omega \approx$

0.80, so that generic flow lines asymptotically approach S from directions $\pm(u^{(1,2)} - u^*)$: the coupling among the different species approaches u^* most slowly. This will have important consequences for our discussion of the phase separation in incompatible systems.

In Fig. 22.1 we show the global flow diagram, as evaluated from the one-loop approximation. We have shown those lines which leave fixed point P in one of the j_P unstable directions and connect it to a fixed point with $j_P - 1$ unstable directions. These lines define the edges of a kind of prism, with base equal to the square $(0 \le f_{11} \le 1, 0 \le f_{22} \le 1, f_{12} = 0)$. It is cut off by a curved surface, which is spanned by the lines connecting the nontrivial fixed points S, U, U', G. The interior of the prism is the weak coupling region, where all flow lines are attached to the noninteracting fixed point G_0. Outside the prism for $f_{AA'} > 0; A = 1, 2; A' = 1, 2$, we are in the strong coupling region. Flow lines cannot break through the surface of the prism.

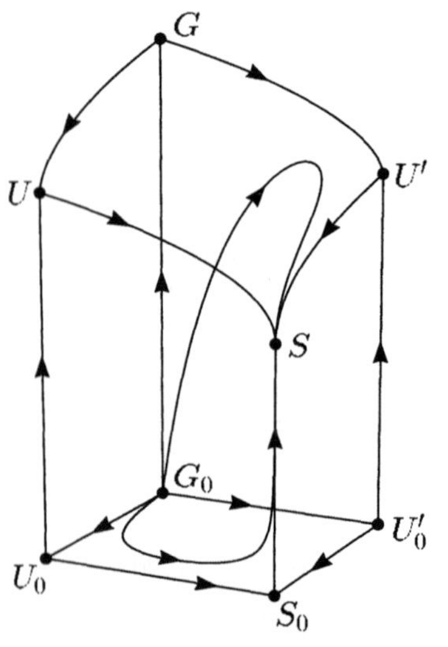

Fig. 22.1. Global flow diagram. The horizontal plane is spanned by f_{11}, f_{22}. f_{12} is plotted in the vertical direction. A subset of the flow lines connecting the fixed points is shown, and two more general flow lines are included

In Fig. 22.1 we included two general flow lines to illustrate the complicated flow that might show up. Qualitatively the figure will stay the same if we use more accurate flow equations.

For special flow lines where two of the couplings take fixed point values, (straight lines in Fig. 22.1), we can analytically integrate the RG flow to find a result of the general structure discussed in Sect. 11.4. Since, however, these are quite special lines, we do not pursue this matter further.

22.2.3 Scaling laws

To write the scaling laws in an appealing form in Eq. (22.29) for the normalized observable Q we replace the chain lengths $N_R^{(A)}$ by $R_g^{(A)2}(0)/\ell_R^2$, where $R_g^{(A)}(0)$ is the radius of gyration of an isolated chain of species A. We furthermore express c_{pR} as

$$c_{pR} = \sum_{A=1}^{F} c_p^{(A)} \ell_R^d = s f_c \left(\{x^{(A)}\}, \left\{ \frac{R_g^{(A)2}(0)}{\ell_R^2} \right\} \right) ,$$

where

$$s = \sum_{A=1}^{F} s^{(A)} = \sum_{A=1}^{F} c_p^{(A)} \left(R_g^{(A)}(0) \right)^d \tag{22.47}$$

is the total geometrical overlap in the solution and $s^{(A)}$ is the partial overlap in the binary subsystem of species A. Choosing $\ell_R = R_g^{(1)}(0)$, we find the general scaling law

$$\begin{aligned} &Q(\{\mathbf{q}_j\}; \{c_p^{(A)}\}; \{N^{(A)}\}; \{\beta_e^{(A,A')}\}; \ell) \\ &= \bar{Q}\left(\{\mathbf{q}_j R_g^{(1)}(0)\}; \{x^{(A)}\}, s; \{R_g^{(A)}(0)/R_g^{(1)}(0)\}; \{f_{AA'}\} \right) , \end{aligned} \tag{22.48}$$

where the normalized couplings

$$f_{AA'} = \frac{u^{(A,A')}}{u^*} \tag{22.49}$$

have to be evaluated at $\ell_R = R_g^{(1)}(0)$.

Clearly Eq. (22.48) gives only one out of many equivalent forms. We, for instance, could replace $\{x^{(A)}\}, s$ by the set of partial overlaps $\{s^{(A)}\}$.

To illustrate the general scaling form we consider the density autocorrelation function $I_{d,a}^{(1)}(\mathbf{q})$ of species (1) in a ternary solution. The normalization is the same as in a binary solution: $I_{d,a}^{(1)}(\mathbf{0}) = c_p^{(1)}(N^{(1)})^2 \tilde{p}_2^{(1)}$ (cf. Eqs. (5.24), (18.2)). We thus again introduce

$$J_a^{(1)}(\mathbf{q}) = \frac{I_{d,a}^{(1)}(\mathbf{q})}{c_p^{(1)}(N^{(1)})^2} , \tag{22.50}$$

which obeys the scaling law

$$J_a^{(1)}(\mathbf{q}) = \bar{J}_a^{(1)}\left((q R_g^{(1)}(0))^2; x^{(1)}, s; \frac{R_g^{(2)}(0)}{R_g^{(1)}(0)}; f_{11}, f_{22}, f_{12} \right) . \tag{22.51}$$

Note that here we deviate from our formulation for binary systems in that we do not introduce the radii $\tilde{R}_0^{(A)}$ of noninteracting chains. Though this would be perfectly possible, it may not be too useful, since the Θ-points of the different species in general will differ.

In the limit of long chains and small concentrations the couplings $f_{AA'}$ will reach some fixed point P, and the 'nonlinear' scaling behavior (22.51) reduces to a 'linear' scaling law:

$$J_a^{(1)}(\mathbf{q}) = J_a^{*(P)}\left((qR_g^{(1)}(0))^2, x^{(1)}, s, \frac{R_g^{(2)}(0)}{R_g^{(1)}(0)}\right) \quad . \tag{22.52}$$

Clearly the scaling function $J_a^{*(P)}$ depends on the fixed point, but otherwise is universal. This is the full limiting result of the RG. Scaling theory makes more specific predictions by postulating simple limiting behavior of the scaling functions. The arguments typically are based on some simple and appealing blob picture. For ternary systems such arguments are only of limited use. We can easily deal with 'quasi-binary' systems: $f_{11} \approx f_{22} \approx f_{12}; R_g^{(1)}(0) \approx R_g^{(2)}(0)$, which on a qualitative level in many respects behave like a binary solution. The only relevant length scales are $R_g(0)$ and a blob size derived from the total overlap, provided we do not come close to a phase separation. In an unsymmetric situation, close to fixed point U, for instance, simple blob arguments may fail, however. The system shows a repulsive interaction within subsystem (1): $f_{11} \approx 1$, but the coupling to a Θ-system: $f_{22} \approx 0$ induces an effective attraction $\sim (f_{12})^2$ of range determined by $R_g^{(2)}$. The competition among these interactions will induce very complicated behavior. Thus simple scaling arguments for general multicomponent systems, which are not supported by a perturbative analysis, must be looked at with reservation.

Equations (22.51), (22.52) illustrate the scaling laws, which hold for all properly normalized observables. Of specific interest in incompatible ternary solutions is the position of the spinodal, of the coexistence-curve, and of the critical point of the phase separation. Even though the critical properties of that transition cannot be calculated perturbatively within the present theory, all the observables of interest must conform to the general scaling laws. The spinodal, for instance, can be taken from the divergence of the forward scattering intensity. It thus must obey an equation of the form

$$0 = F_{sp}\left(x^{(1)}, s, \frac{R_g^{(2)}(0)}{R_g^{(1)}(0)}, f_{11}, f_{22}, f_{12}\right) \quad . \tag{22.53}$$

An equation of the same structure, but with another function F_{co} holds for the coexistence-curve, and the critical point is the simultaneous solution to these two equations. It is now of interest to consider the limiting fixed point behavior of these relations.

At fixed points G, U, U' the system is incompatible. The relation $f_{12}^* > f_{11}^* f_{22}^* = 0$ holds, which in view of Eqs. (22.26) or (22.75) is sufficient to induce a divergence of the screened vertex or the forward scattering intensity. For such systems we can therefore put the coupling constants to their fixed point values and formally solve Eq. (22.53) to find the overlap on the spinodal as

$$s_{sp} = \mathcal{S}_{sp}^{*(P)} \left(x^{(1)}, \frac{R_g^{(2)}(0)}{R_g^{(1)}(0)} \right) \quad . \tag{22.54}$$

The scaling function of course depends on the fixed point, and we always suppress the reduced chain length distributions. A corresponding equation holds for the coexistence line. We thus predict the existence of universal expressions for the spinodal and the coexistence line, which give the overlap as function of the composition, and which for fixed $R_g^{(2)}(0)/R_g^{(1)}(0)$ are independent of the chain lengths. The critical point obeys a universal law

$$\begin{aligned} s_{cr} &= \mathcal{S}_{cr}^{*(P)} \left(\frac{R_g^{(2)}(0)}{R_g^{(1)}(0)} \right) \\ x_{cr}^{(1)} &= \mathcal{X}_{cr}^{*(P)} \left(\frac{R_g^{(2)}(0)}{R_g^{(1)}(0)} \right) \quad . \end{aligned} \tag{22.55}$$

At fixed point S, which for generic systems describes the excluded volume limit, the situation is more complicated. Right at the fixed point the system is compatible, since all couplings are equal. As stressed in the introduction, phase separation is due to corrections to the fixed point behavior. As f_{12} reaches the fixed point most slowly (recall the discussion in the previous section), we can extract the leading behavior in the limit of long chains by taking $f_{11} = f_{22} = 1$, but $f_{12} - 1 > 1$. The analysis of Eq. (22.53), carried through in Appendix A 22.3, yields the scaling behavior

$$s_{sp} = \left[\frac{1}{B_s^2} R_g^{(1)}(0) R_g^{(2)}(0) \right]^{\frac{\omega_s^{(1,2)}}{2} \frac{3\nu - 1}{1 + \nu\omega_s^{(1,2)}}} \mathcal{S}_{sp}^{*(S)} \left(x^{(1)}, \frac{R_g^{(1)}(0)}{R_g^{(2)}(0)} \right) \quad , \tag{22.56}$$

where B_s is a nonuniversal scale, mainly absorbing the starting value of f_{12}. In three dimensions we find

$$\frac{\omega_s^{(1,2)}}{2} \frac{3\nu - 1}{1 + \nu\omega_s^{(1,2)}} \approx .124 \quad ,$$

and, using $R_g(0) \sim N^\nu$ we conclude that the spinodal for $N \to \infty$ is shifted towards infinite overlap: $s_{sp} \sim N^{0.145}$. Results of the very same structure hold for the coexistence line, and as a consequence we find for the critical point

$$
\begin{aligned}
x_{cr}^{(1)} &= \mathcal{X}_{cr}^{*(S)}\left(\frac{R_g^{(2)}(0)}{R_g^{(1)}(0)}\right) \\
s_{cr} &= \left[\frac{1}{B_s^2}\, R_g^{(1)}(0) R_g^{(2)}(0)\right]^{\frac{\omega_s^{(1,2)}}{2}\,\frac{3\nu-1}{1+\nu\omega_s^{(1,2)}}} \mathcal{S}_{cr}^{*(S)}\left(\frac{R_g^{(2)}(0)}{R_g^{(1)}(0)}\right) \quad . \qquad (22.57)
\end{aligned}
$$

Combining Eqs. (22.57), (22.56) we find that also in the present case the spinodal (and the coexistence line), if expressed in terms of s_{sp}/s_{cr}, approaches a universal shape, which only depends on $R_g^{(2)}(0)/R_g^{(1)}(0)$ and on the reduced polydispersities.

Also the line of critical points, with s_{cr} normalized to the critical overlap of a system where the coil sizes are equal and coincide with the geometric mean of $R_g^{(1)}(0), R_g^{(2)}(0)$, is a universal function of $R_g^{(1)}(0)/R_g^{(2)}(0)$.

It is important to note that i) s_{sp}/s_{cr} involves no nonuniversal parameter and ii) universality is valid in all the range of $x^{(1)}$. In other critical systems universality of, for instance, the shape of the coexistence curve is a well known feature, which, however, holds only close to the critical point and up to some nonuniversal coefficient. This is among the well known critical point phenomena governed by the divergence of ξ_c, and thus has a completely different origin than universal laws like (22.57). In the present context it implies a prediction for the $x^{(1)}$-dependence of $\mathcal{S}_{co}^*$ close to $x_{cr}^{(1)}$: $\mathcal{S}_{co}^* \sim |x^{(1)} - x_{cr}^{(1)}|^{\beta_1}, \beta_1 \approx 0.325$.

We should finally stress that these scaling laws are limiting laws, valid in the limit $N \to \infty, c \to 0, s \to \infty$. Corrections for instance due to $f_{AA} \neq 1$ will scale as $R_g^{-\omega} \sim N^{-1/2}$, and may be important in practice.

22.2.4 Choice of the renormalized length scale

For the evaluation of actual perturbative calculations we still have to fix the choice of the renormalized length scale ℓ_R. We recall from the binary solution theory that ℓ_R must be chosen to interpolate smoothly between the radius of gyration (dilute limit) and the screening length (semidilute limit). For small momenta this choice leads to a well-behaved perturbation theory in all the parameter space.

For a ternary system the equivalent to the screening length will be identified in the next section from an analysis of the matrix of density correlations. The resulting form (22.72), (22.73) suggests to fix ℓ_R by the condition

$$2 = \frac{n_0}{N_R^{(1)}} + \frac{n_0}{N_R^{(2)}} + f_{11}\frac{c_R^{(1)}}{c_0} + f_{22}\frac{c_R^{(2)}}{c_0}$$
$$+ \left[\left(\frac{n_0}{N_R^{(1)}} - \frac{n_0}{N_R^{(2)}} + f_{11}\frac{c_R^{(1)}}{c_0} - f_{22}\frac{c_R^{(2)}}{c_0}\right)^2 + 4f_{12}^2\frac{c_R^{(1)}c_R^{(2)}}{c_0^2}\right]^{1/2} . \qquad (22.58)$$

In the dilute limit $\left(c_R^{(1)} \to 0, c_R^{(2)} \to 0\right)$ this yields ℓ_R to be of the order of the radii of gyration, whereas in the semidilute limit $(N_R^{(1)} \to \infty, N_R^{(2)} \to \infty)$ ℓ_R is again found to be of the order of the screening length. We here omitted a term proportional to q^2, since we will not deal with the large momentum behavior. Indeed, in the limit of large q we again see the autocorrelations of small pieces of single chains, which are the same as in a binary solution.

Equation (22.58) has been chosen such that in the binary limit $N_R^{(1)} = N_R^{(2)} = N_R, f_{11} = f_{22} = f_{12} = f$ it reduces to our previous condition (13.21), $q = 0$:

$$1 = \frac{n_0}{N_R} + f\frac{c_R^{(1)} + c_R^{(2)}}{c_0} . \qquad (22.59)$$

We can also reach the binary limit by taking $c_R^{(2)} \to 0$. Then Eq. (22.58) reduces to

$$2 = \frac{n_0}{N_R^{(1)}} + \frac{n_0}{N_R^{(2)}} + f_{11}\frac{c_R^{(1)}}{c_0} + \left|\frac{c_R^{(1)}}{c_0} f_{11} + \frac{n_0}{N_R^{(1)}} - \frac{n_0}{N_R^{(2)}}\right| .$$

This yields the correct binary limit, provided $f_{11}c_R^{(1)}/c_0 + n_0/N_R^{(1)} \geq n_0/N_R^{(2)}$. The latter condition can be seriously violated only if $N_R^{(2)} \ll N_R^{(1)}$, i.e. if we deal with coils of very different size. More precisely, the coil radius of species (2) must be small compared to the density correlation or screening length of subsystem (1). Now, as has been discussed in Sect. 13.2.3, our theory is not constructed to deal with two nontrivial scales of very different size. Thus situations, where the binary limit of condition (22.58) seriously differs from the binary condition (13.21) are outside the domain of the present approach. With this restriction, and provided we stay away from the spinodal, we should expect condition (22.58) to yield well behaved perturbative expressions.

In Eq. (22.58) we included the parameters n_0, c_0, determined within the binary one-loop model. We therefore should use this form here together with a one-loop calculation. Working in tree approximation we might ignore those parameters and put $n_0 = 1 = c_0$. Since the results of the tree approximation in any case should not be expected to be quantitatively correct, in the sequel we will keep these parameters, so that our RG mapping stays consistent with that used for binary solutions. We should note, however, that any comparison to experiment must employ the theory worked out consistently either to one-loop or to tree level, both for the ternary system and the binary subsystems. Recall that the values of the nonuniversal parameters, extracted from experiment, depend on the details of the theoretic approximations.

22.3 Tree approximation

Properly normalized quantities evaluated in tree approximation do depend only on combinations like $q^2\ell^2 N^{(A)}, u_{0,AA'}c^{(A')}N^{(A')}$. Using Eq. (22.28) and the results of Sect. 12.3 we can express these combinations in terms of renormalized variables. To lowest order we find

$$\begin{aligned} q^2\ell^2 N^{(A)} &= \bar{\mathbf{q}}^2 N_R^{(A)} \\ u_{0,AA'}c^{(A')}N^{(A')} &= f_{AA'}c_R^{(A')}N_R^{(A')} \quad , \end{aligned} \tag{22.60}$$

where we recall the notation $\bar{\mathbf{q}}^2 = q^2\ell_R^2, c_R^{(A)} = \tilde{u}c_{pR}^{(A)}N_R^{(A)}, \tilde{u} = (4\pi)^{d/2}u^*/2$ (cf. Eqs. (13.17), (13.18)).

In the sequel we will always consider situations where the binary subsystems are at fixed points ($f_{AA} = 0$ or 1). In more general crossover situations we can evaluate the theory only numerically, and the consideration of fixed point regions is sufficient for working out typical results. At the fixed points we can use the results of Sect. 13.3 to relate the renormalized variables to their observable counterparts. We first consider the Gaussian fixed point.

$\underline{f_{AA} = 0}$

Equations (13.23), (13.24) yield

$$\begin{aligned} \ell_R^2 N_R^{(A)} &= \tilde{R}_0^{(A)2} = R_g^{(A)2}(0) \\ N_R^{(A)1/2}c_R^{(A)} &= \tilde{u}s^{(A)} \end{aligned} \tag{22.61}$$

where $s^{(A)} = c_p^{(A)}(R_g^{(A)}(0))^3$ is the geometrical overlap.

At the nontrivial fixed point the analysis is slightly more complicated.

$\underline{f_{AA} = 1}$

Equations (13.23), (13.24) yield

$$\begin{aligned} \ell_R^2 \left(\frac{N_R^{(A)}}{n_0}\right)^{2\nu} &= B^{(A)2}N^{(A)2\nu} \\ \left(\frac{N_R^{(A)}}{n_0}\right)^{3\nu-1} c_R^{(A)} &= \tilde{u}\, n_0\, c_p^{(A)} B^{(A)3}N^{(A)3\nu} \quad , \end{aligned}$$

where $B^{(A)}$ is the nonuniversal parameter defined in Eq. (13.37). Now in tree approximation we find $R_g^{(A)2}(0) = \ell_R^2 N_R^{(A)}$, to be evaluated with $N_R^{(A)} = n_0$. Thus

$$R_g^{(A)2}(0) = n_0 B^{(A)2} N^{(A)2\nu} \quad ,$$

and the above relations can be written as

$$\ell_R^2 \left(\frac{N_R^{(A)}}{n_0} \right)^{2\nu} = \frac{1}{n_0} R_g^{(A)}(0)^2$$
$$\left(\frac{N_R^{(A)}}{n_0} \right)^{3\nu-1} c_R^{(A)} = \frac{\tilde{u}}{n_0^{1/2}} s^{(A)} \quad . \tag{22.62}$$

Note that in introducing the isolated chain radii we took the systems to be monodisperse. Polydispersity corrections can easily be taken along.

In the sequel we will consider ternary systems. Multicomponent solutions ($F > 2$) give rise to more complicated expressions for the density cumulant, due to the need of inverting an $F \times F$ matrix. Yet, within the range of our theory they involve no basically new physics. For ternary systems we introduce a composition variable

$$x = \frac{c_p^{(1)}}{c_p} \quad , \tag{22.63}$$

and a variable

$$y = \frac{R_g^{(1)}(0)}{R_g^{(2)}(0)} \tag{22.64}$$

which fixes the relation among the molecular weights. We now first consider the density cumulant, which shows the richest structure.

22.3.1 The density cumulant: general considerations

As shown in Sect. 22.2, Eq. (22.15), $I_{d,c}^{(A,A')}$ is best considered as the (A, A') element of a matrix. Of course this matrix is symmetric, a feature masked somewhat in the representation (22.15). At $q = 0$ a sum rule holds in the isolated chain limit:

$$I_{d,c}^{(A,A')}(\mathbf{0}) = \delta_{AA'}\, c_p^{(A)} \left(N^{(A)} \right)^2 \tilde{p}_2^{(A)} \quad . \tag{22.65}$$

(Cf. Sect. 19.1.1.) This suggests to introduce a normalized function $J_c^{(A,A')}(\mathbf{q})$ as

$$\mathbf{I_{d,c}}(\mathbf{q}) = \mathbf{C}\, \mathbf{J_c}(\mathbf{q})\, \mathbf{C} \quad , \tag{22.66}$$

where $\mathbf{C}$ denotes the matrix

$$\mathbf{C}_{AA'} = \delta_{AA'} \sqrt{c_p^{(A)}}\, N^{(A)} \quad . \tag{22.67}$$

Equation (22.66) is written such that the symmetry of $\mathbf{J_c}$ is preserved. From Eqs. (22.23), (22.24), combined with Eq. (22.60) the renormalized expression for $\mathbf{J}_\mathbf{c}^{-1}(\mathbf{q})$ is easily read off:

$$J_c^{-1}(\mathbf{q})_{AA'} = \frac{\delta_{AA'}}{D_p^{(A)}\left(\bar{\mathbf{q}}^2 N_R^{(A)}\right)} + \left(c_R^{(A)} c_R^{(A')} N_R^{(A)} N_R^{(A')}\right)^{1/2} f_{AA'} \ . \qquad (22.68)$$

Clearly the diagonal elements $J_c^{-1}(\mathbf{q})_{AA}$ coincide in structure with the scaling function $J_c^{-1}(\mathbf{q})$ of a binary system. These expressions for the matrix elements are valid also for multicomponent systems ($F > 2$).

For a ternary system the density cumulant involves two length scales, which we determine from the eigenvalues of the matrix

$$H(\mathbf{q})_{AA'} = \left(N_R^{(A)} N_R^{(A')}\right)^{-1/2} J_c^{-1}(\mathbf{q})_{AA'} \ . \qquad (22.69)$$

Here the prefactor $(N_R^{(A)} N_R^{(A')})^{-1/2}$ serves to yield a well-defined semidilute limit. The general expression for the eigenvalues reads

$$\begin{aligned} \Lambda_{1,2}(q) &= \frac{1}{2}\left(H(q)_{11} + H(q)_{22} \pm \tilde{H}(q)\right) \\ \tilde{H}(q) &= \left[(H(q)_{11} - H(q)_{22})^2 + 4H(q)_{12}^2\right]^{1/2} \ . \end{aligned} \qquad (22.70)$$

We now expand in powers of q^2, where for simplicity we take the binary subsystems to be monodisperse:

$$\frac{1}{N_R^{(A)} D_p^{(A)}(\bar{\mathbf{q}}^2 N_R^{(A)})} = \frac{1}{N_R^{(A)}} + \frac{\bar{\mathbf{q}}^2}{3} + O(\bar{\mathbf{q}}^4) \ .$$

Thus

$$\Lambda_{1,2}(q) = \Lambda_{1,2}(0) + \frac{\bar{\mathbf{q}}^2}{3} + O(\bar{\mathbf{q}}^4) \ , \qquad (22.71)$$

where

$$\begin{aligned} \Lambda_{1,2}(0) = \frac{1}{2}\Bigg\{ & \frac{1}{N_R^{(1)}} + \frac{1}{N_R^{(2)}} + c_R^{(1)} f_{11} + c_R^{(2)} f_{22} \\ & \pm \left[\left(\frac{1}{N_R^{(1)}} - \frac{1}{N_R^{(2)}} + c_R^{(1)} f_{11} - c_R^{(2)} f_{22}\right)^2 + 4 c_R^{(1)} c_R^{(2)} f_{12}^2\right]^{1/2} \Bigg\} \ . \end{aligned} \qquad (22.72)$$

Equation (22.71) yields the correlation lengths

$$\xi_{1.2}^2 = \ell_R^2 \Lambda_{1,2}^{-1}(0) \ . \qquad (22.73)$$

In the dilute limit $c_R^{(1)} = 0 = c_R^{(2)}$ the two correlation lengths reduce to the radii of gyration of the two polymer species. With increasing concentration we may reach a point where $\Lambda_2(0)$ vanishes and accordingly ξ_2 diverges. This is one way to identify the spinodal, and ξ_2 is to be interpreted as the length scale ξ_c that governs the critical properties of the phase separation. ξ_1, on the other hand, stays finite and is the generalization of the screening length ξ_E. To show this more clearly, we consider a completely symmetric solution:

$$\begin{aligned} N_R^{(1)} &= N_R^{(2)} = N_R \\ c_R^{(1)} &= c_R^{(2)} = \frac{1}{2}\, c_R \\ f_{11} &= f_{22} = f \quad . \end{aligned}$$

In that special case Eq. (22.72) yields

$$\Lambda_{1,2}(0) = \frac{1}{N_R} + \frac{1}{2}\, c_R(f \pm f_{12}) \quad .$$

As a result $\xi_1^2 = \ell_R^2/\Lambda_1(0)$ for $f_{12} \to f$ reduces to the screening length ξ_E^2 of a binary solution of concentration c_R, chain length N_R, evaluated in tree approximation.

This observation suggests to choose $\ell_R \approx \xi_1$ in the semidilute regime. From Eq. (22.73) this is equivalent to $\Lambda_1(0) \approx 1$, and with the form (22.72) of $\Lambda_1(0)$ this leads to our condition (22.58).

The matrix elements of the density cumulant are easily found by inverting the matrix $\mathbf{J}_{\mathbf{c}}^{-1}(\mathbf{q})$ (Eq. (22.68)), and substituting the result into Eq. (22.66):

$$\begin{aligned} I_{d,c}^{(A,A)}(\mathbf{q}) &= c_p^{(A)}(N^{(A)})^2 \frac{1}{\Delta_R}\left(\frac{1}{D_p^{(\bar{A})}(\bar{\mathbf{q}}^2 N_R^{(\bar{A})})} + c_R^{(\bar{A})} N_R^{(\bar{A})} f_{\bar{A}\bar{A}}\right) \\ I_{d,c}^{(1,2)}(\mathbf{q}) &= -\sqrt{c_p^{(1)} c_p^{(2)}}\, N^{(1)} N^{(2)} \frac{1}{\Delta_R}\left(c_R^{(1)} c_R^{(2)} N_R^{(1)} N_R^{(2)}\right)^{1/2} f_{12} \\ &\equiv I_{d,c}^{(2,1)}(\mathbf{q}) \quad . \end{aligned} \tag{22.74}$$

Here

$$\begin{aligned} \Delta_R &= \operatorname{Det} \mathbf{J}_{\mathbf{c}}^{-1}(\mathbf{q}) \\ &= N_R^{(1)} N_R^{(2)} \Lambda_1(\mathbf{q}) \Lambda_2(\mathbf{q}) \\ &= \frac{1}{D_p^{(1)}(\bar{\mathbf{q}}^2 N_R^{(1)}) D_p^{(2)}(\bar{\mathbf{q}}^2 N_R^{(2)})} + \sum_{A=1}^{2} \frac{c_R^{(A)} N_R^{(A)} f_{AA}}{D_p^{(\bar{A})}(\bar{\mathbf{q}}^2 N_R^{(\bar{A})})} \\ &\quad + c_R^{(1)} c_R^{(2)} N_R^{(1)} N_R^{(2)} (f_{11} f_{22} - f_{12}^2) \quad . \end{aligned} \tag{22.75}$$

The notation (22.27) for $A, \bar{A}$ is understood. On the spinodal $\Lambda_2(0)$ and accordingly $\Delta_R(\mathbf{q} = 0)$ vanish, and all matrix elements $I_{d,c}^{(A,A')}(\mathbf{0})$ diverge.

We now consider the observed scattering intensity. Scattering from a monomer of species A is proportional to the increment $\nu_i^{(A)}$ of the index of refraction. In general the $\nu_i^{(A)}$ are different for the different monomer species, and the scattering intensity takes the form

$$I_{d,c}(\mathbf{q};\{\nu_i^{(A)}\}) = \sum_{A,A'} \nu_i^{(A)} I_{d,c}^{(A,A')} \nu_i^{(A')} \quad . \tag{22.76}$$

For a ternary solution this essentially depends on nine parameters: $\mathbf{q}^2, \nu_i^{(1)}/\nu_i^{(2)}, c_p, x = c_p^{(1)}/c_p, y = R_g^{(1)}(0)/R_g^{(2)}(0), f_{11}, f_{22}, f_{12}$, and some chain length average. Clearly a comprehensive discussion of the full parameter range is not feasable, but the most prominent features, which result from the inherent matrix structure, are easily worked out. We use Eqs. (22.66), (22.67), (22.69) to write $I_{d,c}$ as

$$I_{d,c}(\mathbf{q};\nu_i^{(1)},\nu_i^{(2)}) = \sum_{A,A'=1}^{2} \tilde{\nu}_i^{(A)} H^{-1}_{AA'} \tilde{\nu}_i^{(A')} \quad ,$$

where

$$\tilde{\nu}_i^{(A)} = \left(c^{(A)} \frac{N^{(A)}}{N_R^{(A)}} \right)^{1/2} \nu_i^{(A)} \quad . \tag{22.77}$$

The matrix $\mathbf{H}^{-1}$ has eigenvalues $\Lambda_{1,2}^{-1}(q)$ (Eq. (22.70)), and associated eigenvectors

$$\mathbf{e}^{(1)}(\mathbf{q}) = \begin{pmatrix} a_1 \\ a_2 \end{pmatrix}, \quad \mathbf{e}^{(2)}(\mathbf{q}) = \begin{pmatrix} -a_2 \\ a_1 \end{pmatrix}$$

$$a_{1,2} = \left(\frac{1}{2} \pm \frac{H(q)_{11} - H(q)_{22}}{2\tilde{H}(q)} \right)^{1/2} \quad . \tag{22.78}$$

Equation (22.76) therefore can be written in the form

$$I_{d,c}(\mathbf{q};\nu_i^{(1)},\nu_i^{(2)}) = \sum_{j=1}^{2} \left(\tilde{\nu}_\mathbf{i} \mid \mathbf{e}^{(\mathbf{j})}(\mathbf{q}) \right)^2 \Lambda_j^{-1}(q) \quad , \tag{22.79}$$

where $(\tilde{\nu}_\mathbf{i}|\mathbf{e}^{(\mathbf{j})})$ denotes the scalar product of $\mathbf{e}^{(\mathbf{j})}$ with the vector $\tilde{\nu}_\mathbf{i} = (\tilde{\nu}_i^{(1)}, \tilde{\nu}_i^{(2)})$. This form represents $I_{d,c}$ as a weighted sum of the eigenvalues Λ_j^{-1}.

We now consider incompatible systems, where the spinodal defines a distinguished surface in parameter space, given by $\Lambda_2(0) = 0$. We denote the corresponding eigenvectors $\mathbf{e}^{(\mathbf{j})}(\mathbf{0})$ as $\mathbf{e}_\mathbf{sp}^{(\mathbf{j})}$, with components $a_{sp,A}$ found from Eqs. (22.78), (22.70):

$$a_{sp,A} = \left(\frac{H(0)_{AA}}{H(0)_{11} + H(0)_{22}} \right)^{1/2} \quad . \tag{22.80}$$

These eigenvectors define two distinguished directions for the vector $\tilde{\nu}_\mathbf{i}$.

(i) $\tilde{\nu}_{\mathbf{i}}$ orthogonal to $\mathbf{e}_{\mathbf{sp}}^{(2)}$.

The divergent part in Eq. (22.79) is suppressed and $I_{d,c}(\mathbf{q};\nu_i^{(1)},\nu_i^{(2)})$ stays finite on the spinodal. The observed scattering intensity is similar to that in a binary system.

The explicit form of the criterion reads

$$\frac{\nu_i^{(1)}}{\nu_i^{(2)}} = \left[\frac{H(0)_{11}}{H(0)_{22}}\,\frac{c^{(2)}N^{(2)}N_R^{(1)}}{c^{(1)}N^{(1)}N_R^{(2)}}\right]^{1/2}_{\text{spinodal}} . \tag{22.81}$$

If we consider the special case of a completely symmetric system:

$$c^{(2)} = c^{(1)}, \quad N^{(1)} = N^{(2)}, \quad f_{11} = f_{22}, \quad N_R^{(1)} = N_R^{(2)} ,$$

(this implies $H_{11} \equiv H_{22}, x = 1/2, y = 1$), condition (22.81) reduces to $\nu_i^{(1)} = \nu_i^{(2)} = \nu_i$. We then measure the fluctuations of the total segment density in the solution:

$$\begin{aligned}\frac{1}{\nu_i^2} I_{d,c}(\mathbf{q};\nu_i,\nu_i) &= I_{d,c}^{(1,1)}(\mathbf{q}) + I_{d,c}^{(2,2)}(\mathbf{q}) + 2I_{d,c}^{(1,2)}(\mathbf{q}) \\ &= \int \frac{d^d r d^d r'}{\Omega}\, e^{i\mathbf{q}(\mathbf{r}-\mathbf{r}')} \left\langle \sum_{A=1}^{2} \left(\hat{\rho}^{(A)}(\mathbf{r}) - c^{(A)}\right) \sum_{A'=1}^{2} \left(\hat{\rho}^{(A')}(\mathbf{r}') - c^{(A')}\right)\right\rangle .\end{aligned} \tag{22.82}$$

In the present completely symmetric situation the total segment density is an uncritical mode, decoupled from the phase separation.

(ii) $\tilde{\nu}_{\mathbf{i}}$ orthogonal to $\mathbf{e}_{\mathbf{sp}}^{(1)}$.

In $I_{d,c}(\mathbf{q},\nu_i^{(1)},\nu_i^{(2)})$ we completely suppress the regular contribution and pick out the divergent part. Being dominated by spinodal behavior the observed scattering intensity for small momenta strongly differs from that in a binary system.

The explicit form of the criterion reads

$$\frac{\nu_i^{(1)}}{\nu_i^{(2)}} = -\left[\frac{H(0)_{22}}{H(0)_{11}}\,\frac{c^{(2)}N^{(2)}N_R^{(1)}}{c^{(1)}N^{(1)}N_R^{(2)}}\right]^{1/2}_{\text{spinodal}} . \tag{22.83}$$

In the fully symmetric situation considered above we find $\nu_i^{(1)} = -\nu_i^{(2)} = \nu_i$, and we accordingly measure the fluctuations in the composition $\rho^{(1)}(\mathbf{r}) - \rho^{(2)}(\mathbf{r})$:

$$\frac{1}{\nu_i^2} I_{d,c}(\mathbf{q}; \nu_i, -\nu_i) = I_{d,c}^{(1,1)}(\mathbf{q}) + I_{d,c}^{(2,2)}(\mathbf{q}) - 2I_{d,c}^{(1,2)}(\mathbf{q})$$
$$= \int \frac{d^d r d^d r'}{\Omega} e^{i\mathbf{q}(\mathbf{r}-\mathbf{r}')} \Big\langle \Big(\hat{\rho}^{(1)}(\mathbf{r}) - \hat{\rho}^{(2)}(\mathbf{r}) - c^{(1)} + c^{(2)}\Big)$$
$$\Big(\hat{\rho}^{(1)}(\mathbf{r}') - \hat{\rho}^{(2)}(\mathbf{r}') - c^{(1)} + c^{(2)}\Big) \Big\rangle \ . \tag{22.84}$$

For the symmetric situation the composition fluctuations are the critical mode, driving the phase separation.

For general $\nu_i^{(A)}$ Eq. (22.79) yields $I_{d,c}(\mathbf{q}; \nu_i^{(1)}, \nu_i^{(2)})$ as a superposition of these two distinguished cases. Clearly, the relation among $\nu_i^{(1)}$ and $\nu_i^{(2)}$ is most important for the phenomenology of the observed scattering intensity. In particular, if both scattering increments are of the same sign, then the genuine ternary effects can be expected to be fairly small, as long as we do not come too close to the spinodal. In the opposite case $\nu_i^{(1)}/\nu_i^{(2)} < 0$ we always expect to see strong ternary effects.

We now illustrate these features with an explicit analysis of some simple cases.

22.3.2 The density cumulant: special cases

a) Neighborhood of the symmetric fixed point S:

$$f_{11} = f_{22} = 1; \quad |f_{12} - 1| \ll 1 \ .$$

As stressed above, this is the generic situation in the limit of very long chains. We here simplify the analysis further by taking the binary subsystems to be monodisperse and imposing the symmetry condition $y = R_g^{(1)}(0)/R_g^{(2)}(0) = 1$. From Eq. (22.62) the latter condition implies $N_R^{(1)} = N_R^{(2)} \equiv N_R$ and $x = c_p^{(1)}/c_p \equiv s^{(1)}/s$, where $s = s^{(1)} + s^{(2)}$ is the total overlap. As a further minor simplification we take the nonuniversal parameters $B^{(A)}$ to be equal, so that $y = 1$ implies also $N^{(1)} = N^{(2)} = N$.

We now first consider the case, where the two species are of equal scattering power.

a.1) $\nu_i^{(1)} = \nu_i^{(2)} = \nu_i$

As shown by Eq. (22.82), we measure the fluctuations of the total segment density. Equations (22.74), (22.76) yield for the special case considered

$$\frac{I_{d,c}(\mathbf{q}; \nu_i, \nu_i)}{\nu_1^2 c_p N^2} = \frac{1}{\Delta_R} \left(\frac{1}{D(\bar{\mathbf{q}}^2 N_R)} + 2x(1-x) c_R N_R (1 - f_{12}) \right) \tag{22.85}$$

with $c_R = c_R^{(1)} + c_R^{(2)}$. For $x = 1/2$ this simplifies to

$$\frac{\nu_1^2 c_p N^2}{I_{d,c}(\mathbf{q}; \nu_i, \nu_i)} = \frac{1}{D(\bar{\mathbf{q}}^2 N_R)} + \frac{1}{2} c_R N_R (1 + f_{12}) \quad ,$$

i.e. we find the scattering intensity of a binary system with effective coupling $(1 + f_{12})/2$. As shown above, in this completely symmetric case the scattering intensity does not couple to the critical mode of the phase separation. If $x \neq 1/2$, the condition (22.81) is weakly violated, and as a result $I_{d,c}$ diverges on the spinodal.

To give a quantitative impression of the effects we have evaluated $I_{d,c}(\mathbf{0}; \nu_i, \nu_i)$ (Eq. (22.85)) numerically. Since $f_{12} \approx 1$ with the present symmetry conditions we can replace Eq. (22.58) by Eq. (22.59), $f = 1$ for fixing ℓ_R. Then the RG-mapping is identical to the binary mapping discussed in Sect. 13.3, and can be written in terms of a running parameter w:

$$N_R = \frac{n_0}{w^2}; \quad c_R = c_0(1 - w^2)$$

$$w^{2-6\nu}(1 - w^2) = \frac{\tilde{u}}{c_0 n_0^{1/2}} \, s$$

(cf. Eq. (22.62)). For the flow of f_{12} we can use Eq. (22.44):

$$f_{12} = 1 + \mathit{const}\ \lambda^{\omega_S^{(1,2)}} \quad .$$

With $\lambda = \ell/\ell_R = \mathit{const}\, w^{-2\nu} N^{-\nu} = \mathit{const}\, w^{-2\nu} R_g^{-1}(0)$ (cf. Eq. (13.35 ii)) we find

$$f_{12} = 1 + \left(\frac{B_{12}}{R_g(0)}\right)^{\omega_S^{(1,2)}} w^{-2\nu\omega_S^{(1,2)}}$$

$$\nu\omega_S^{(1,2)} \approx 0.235 \quad (d = 3) \quad . \qquad (22.86)$$

The nonuniversal parameter B_{12} measures the chemical mismatch of the two species.

Figure 22.2a shows the inverse forward scattering intensity for an incompatible system $(B_{12}/R_g(0))^{\omega_S^{(1,2)}} = 0.01 > 0$ as function of the total geometrical overlap $s = c_p R_g^3(0)$. As expected, for $x = 1/2$ essentially binary results are recovered. Indeed, except for ending at the spinodal, the curve on the scale of this plot cannot be distinguished from the normalized inverse scattering intensity of a binary system in the excluded volume limit. For $x = 0.3$ or $x = 0.1$ over a large overlap range we again find essentially binary solution results, but a sharp drop of $I_{d,c}^{-1}(0)$ sets in close to the spinodal. For the same systems we have evaluated the effective length scale

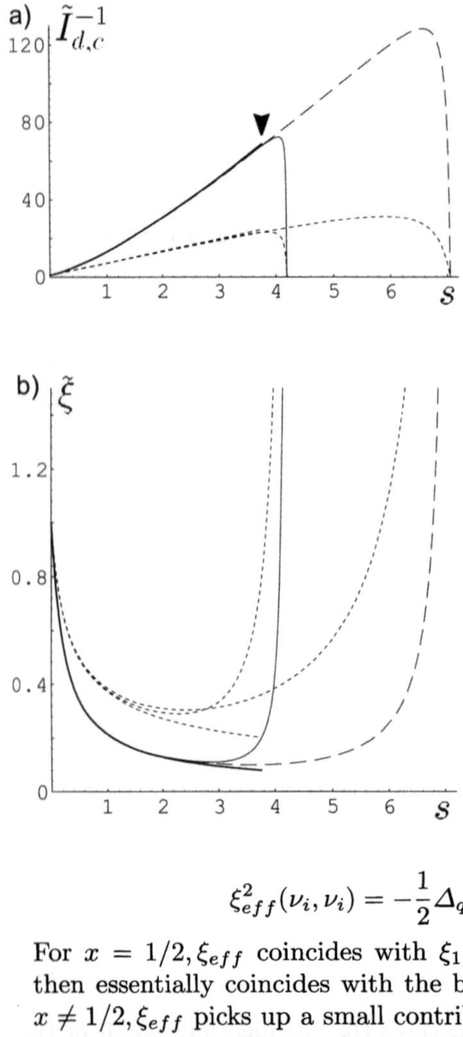

Fig. 22.2. (**a**) $\tilde{I}_{d,c}^{-1}(0,\nu_i,\nu_i) = \nu_i^2 c_p N^2 / I_{d,c}(0,\nu_i,\nu_i)$. (**b**) $\tilde{\xi} = \xi_{eff}(\nu_i,\nu_i)/R_g(0)$. Both as function of the total overlap close to fixed point S. *Fat line*: $x = 1/2$; *thin line*: $x = 0.3$; *long dashes*: $x = 0.1$. For these cases also the results of unrenormalized RPA (*short dashes*), with parameters as explained in the text, are included. In (**a**) the *arrow* points to the end of the line $x = 1/2$, which in that case gives the position of the spinodal

$$\xi_{eff}^2(\nu_i,\nu_i) = -\frac{1}{2}\Delta_q|_0 \ \ln\ I_{d,c}(\mathbf{q};\nu_i,\nu_i) \ . \tag{22.87}$$

For $x = 1/2, \xi_{eff}$ coincides with ξ_1 and stays finite on the spinodal. It then essentially coincides with the binary density correlation length. For $x \neq 1/2, \xi_{eff}$ picks up a small contribution of ξ_2 and accordingly diverges on the spinodal. This is shown in Fig. 22.2b. Since we consider here the fluctuations of the total segment density, this divergence indicates that for $x \neq 1/2$ the two coexisting phases do not only differ in composition, but also in total segment concentration.

In Fig. 22.2 we included results of unrenormalized RPA. It is easily checked that for $N^{(1)} = N^{(2)} = N$, $u_{0,11} = u_{0,22} = u_0$, both $cN/I_{d,c}$ and $\xi_{eff}/R_g(0)$ depend only on the combinations $u_0 cN$ and $u_{0,12} cN$. We thus arbitrarily identified cN with s, and we adjusted $u_{0,12}$ and u_0 to the position of the spinodal and to the initial slope of $cN/I_{d,c}(\mathbf{q};\nu_i,\nu_i)$, as given by the renormalized theory. The latter condition amounts to adjusting the second virial coefficient. Such fits to unrenormalized RPA are often used in experimental work. We note that the resulting parameters here depend on x. As

Fig. 22.2 shows, despite these adjustments the predictions of unrenormalized and renormalized RPA differ widely. This is mainly due here to the failure of RPA to reproduce the scattering intensity of a strictly binary system ($u_{0,12} \equiv u_0$). There RPA predicts $cN/I_{d,c}(\mathbf{0}) = 1 + u_0 cN$, missing of course the correct power law $\sim c^{1/(3\nu-1)} \approx c^{1.309}$ in the semidilute limit. It must be noted, however, that in restricted regions of overlap, which do not cover the dilute limit, the shape of the curves is not too different, and adjusting the unrenormalized parameters in a different way, we could bring the two approaches to closer overall agreement.

We now turn to a situation, where we measure the composition fluctuations.

a.2) $\nu_i^{(1)} = -\nu_i^{(2)} = \nu_i$

The resulting scattering intensity derives from Eq. (22.85) by just changing the sign in front of f_{12}. In particular, we find for $x = 1/2$:

$$\frac{\nu_1^2 c_p N^2}{I_{d,c}(\mathbf{q}; \nu_i, -\nu_i)} = \frac{1}{D(\bar{\mathbf{q}}^2 N_R)} + \frac{1}{2}\, c_R N_R (1 - f_{12}) \quad ,$$

so that for f_{12} very close to 1 and $c_R N_R$ not too large we essentially see the autocorrelations $I_{d,a}^{-1} \sim D^{-1}(\ldots)$ of single chains in the solution. This corresponds to the 'compensation method' discussed in Sect. 18.4. Figure 22.3 shows our results for the inverse forward scattering intensity and the effective correlation length, for the same values of the parameters as used above. Clearly the results now look quite different from those of a binary solution. In particular, the inverse forward scattering intensity rapidly starts decreasing, whereas ξ_{eff}, after an initial small decrease, which is due to standard screening, essentially stays constant until the overlap comes close to the spinodal. Note that for $x = 1/2$, ξ_{eff} coincides with ξ_2. Unrenormalized RPA yields curves which differ not so much in absolute scale, but rather in shape, a feature which could not be corrected for by changing the parameters.

So far we have considered incompatible systems: $f_{12} \gtrsim 1$. For compatible systems, $f_{12} \lesssim 1$, the results for $\nu_i^{(1)} \approx \nu_i^{(2)}$ look very much like in a binary system. Also for $\nu_i^{(1)} \approx -\nu_i^{(2)}$ the scattering intensity looks somewhat like that of a binary system, but the variation of the scaling functions with s is much slower. Thus, both for incompatible and compatible systems the ratio $\nu_i^{(1)}/\nu_i^{(2)}$ is most important in governing the phenomenology of the scattering function, which confirms our general analysis.

b) Strongly repelling Θ-systems: fixed point G:

$$f_{11} = f_{22} = 0; f_{12} = f_{12}^{*(G)} \approx 1.83 \quad .$$

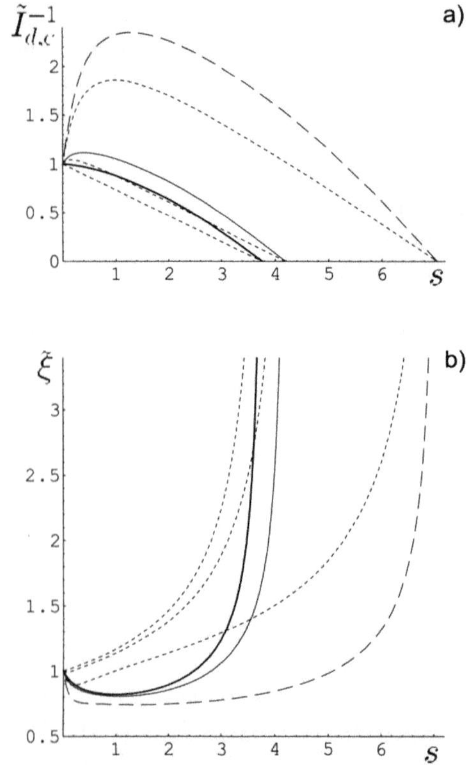

Fig. 22.3a, b. As Fig. 22.2, but for $\nu_i^{(1)} = -\nu_i^{(2)}$

This case is mainly of theoretical interest. In some sense it specifies the most incompatible limit. We again take the binary subsystems to be monodisperse and identical: $R_g^{(1)}(0) = R_g^{(2)}(0), N^{(1)} = N^{(2)} = N$. The matrix elements (22.74) of the density cumulant take the form

$$I_{d,c}^{(1,1)} = \frac{x}{1-x} I_{d,c}^{(2,2)} = \frac{x\, c_p\, N^2}{\Delta_R D(\bar{\mathbf{q}}^2 N_R)}$$
$$I_{d,c}^{(1,2)} = -x(1-x) f_{12,G}^* \frac{c_R N_R}{\Delta_R} c_p N^2$$
$$\Delta_R = D^{-2}(\bar{\mathbf{q}}^2 N_R) - x(1-x) f_{12,G}^{*2} c_R^2 N_R^2 \ .$$

This is to be evaluated with the RG-mapping in the form of Eqs. (22.61), amended by condition (22.58), which reduces to

$$1 = \frac{n_0}{N_R} + f_{12,G}^* \sqrt{x(1-x)}\, \frac{c_R}{c_0} \ . \tag{22.88}$$

The evaluation is straightforward, and we again show the results for $\nu_i^{(1)} = \nu_i^{(2)}$ (Fig. 22.4), or $\nu_i^{(1)} = -\nu_i^{(2)}$ (Fig. 22.5). One important difference as compared to the fixed point region S is in the position of the spinodal.

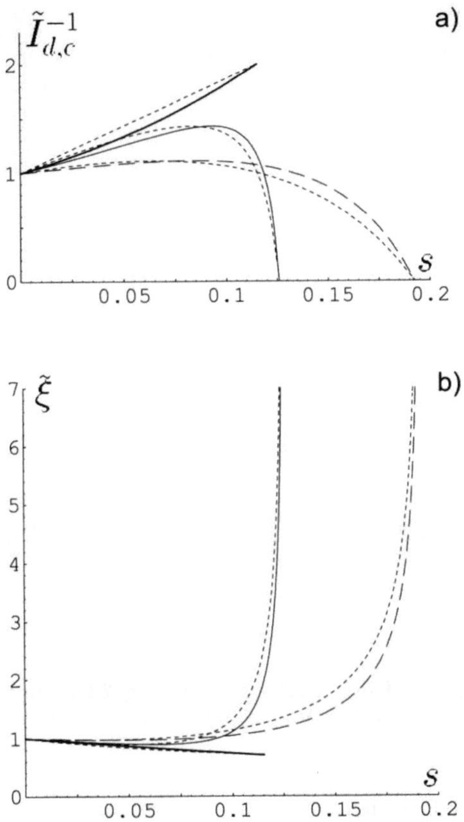

Fig. 22.4a, b. Total scattering intensity and effective correlation length at fixed point G, with $\nu_i^{(1)} = \nu_i^{(2)}$. Symbols as in Fig. 22.2

As discussed in Sect. 22.2.3 the spinodal in the excluded volume region with increasing chain length shifts towards infinite overlap, whereas here it attains a finite limit. For $\nu_i^{(1)} = \nu_i^{(2)}$ the other differences found by comparing Figs. 22.2 and 22.4 are explained by considering the corresponding binary systems: at the excluded volume fixed point with increasing geometrical overlap $I_{d,c}^{-1}$ increases, and ξ_d decreases. In the Θ-limit both quantities stay constant.

Figure 22.5 shows the results for $\nu_i^{(1)} = -\nu_i^{(2)}$. As compared to Fig. 22.4 there now is neither an initial increase of $I_{d,c}^{-1}$, nor an initial decrease of ξ_{eff}.

We again included the results of unrenormalized RPA, where $u_{0,12}$ has been fitted to the position of the spinodal. Clearly the effect of renormalization on the shape of the curves, though visible, is small. It must be kept in mind, however, that the renormalized results are tree approximations to universal scaling functions, which must be reached in the limit of long chains. No such limiting behavior exists in the unrenormalized theory, of course.

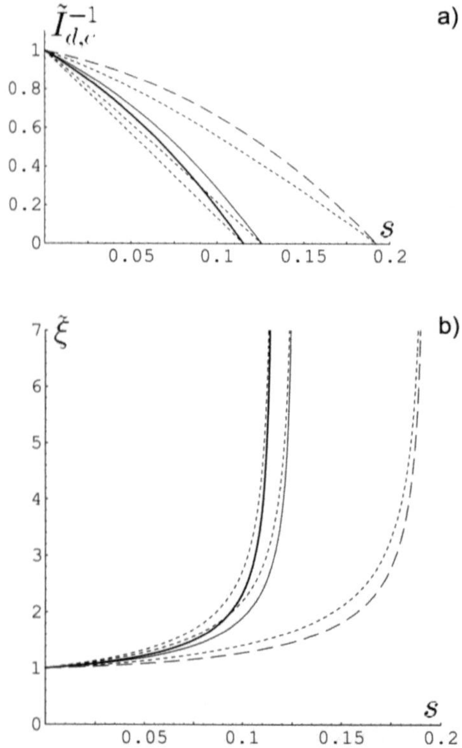

Fig. 22.5a, b. As Fig. 22.4, but with $\nu_i^{(1)} = -\nu_i^{(2)}$

c) Strongly asymmetric systems: fixed point U:

$$f_{11} = 1, f_{22} = 0, f_{12} = f_{12}^{*(U)} \approx 1.48 \quad .$$

Here a Θ-system interacts with an excluded volume system. As with the case of two Θ-systems such mixtures are incompatible for all $f_{12} > 0$, and without losing qualitative features we may set f_{12} to its fixed point value. This, however, is the only essential simplification possible. Since the system intrinsically is unsymmetric, there exists no special symmetry condition which yields particularly simple analytical results. Rather we have to keep the full expressions (22.74), (22.75), combined with the RG-mapping (22.61), (22.62), (22.58). In Appendix A 22.2 we transform the RG-mapping into a more easily evaluated form. Figure 22.6 shows our results for the case $R_g^{(1)}(0) = R_g^{(2)}(0)$. Note that this in general implies that the Θ-chains are longer than the excluded volume chains: $N^{(2)} > N^{(1)}$. We show the inverse forward scattering intensity and the effective correlation length for systems where the solvent is iso-refractive to one species. For $\nu_i^{(2)} = 0$ (full lines) we only see the excluded volume chains ($I_{d,c}^{(1,1)}$). Comparing these results to the previous figures we find that they resemble somewhat the scattering

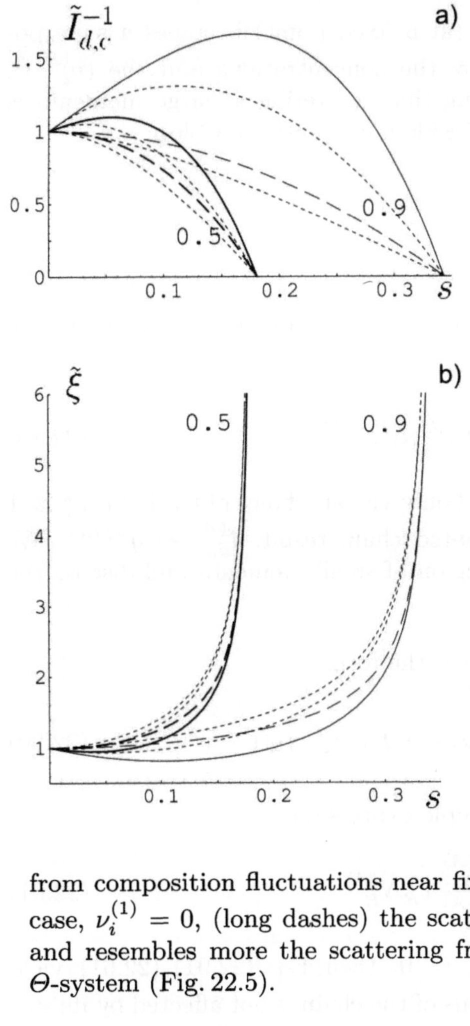

Fig. 22.6. (**a**) $\tilde{I}^{-1}_{d,c}(0,\nu_i^{(1)},\nu_i^{(2)}) = \nu_i^{(A)2} c_p^{(A)} N^{(A)2} / I_{d,c}(0,\nu_i^{(1)},\nu_i^{(2)})$. (**b**) $\tilde{\xi} = (\nu_i,\nu_i)/R_g^{(A)}(0)$. Both as function of the total overlap close to fixed point $U(f_{11}=1, f_{22}=0)$. *Full lines*: $A = 1, \nu_i^{(2)} = 0$; *long dashes*: $A = 2, \nu_i^{(1)} = 0$. Values $x = c_p^{(1)}/c_p = 0.5, 0.9$. *Short dashes* give the results of unrenormalized RPA

from composition fluctuations near fixed point S (Fig. 22.3). In the other case, $\nu_i^{(1)} = 0$, (long dashes) the scattering signal is due to the Θ-chains and resembles more the scattering from composition fluctuations in the Θ-system (Fig. 22.5).

Our discussion has been concerned only with the behavior at small momenta: $q^2R_g^2(0) \lesssim 1$. In that range the observed scattering intensity is most sensitive to $\nu_i^{(1)}/\nu_i^{(2)}$, which determines the weight given to the two eigenvalues $\Lambda^{-1}_{1,2}(q)$ in the representation (22.79). In the limit $q^2\xi_1^2 \gg 1$ the situation is different. It is easily checked that for such large momenta the scattering is dominated by the diagonal terms $I_{d,c}^{(A,A)}$, which in turn approach the asymptotic form factor of a single chain of the corresponding species:

$$I^{-1}_{d,c}(\mathbf{q}) \sim q^{1/\nu_A}, \quad \text{where } \nu_A = \begin{cases} \nu \ ; \ f_{AA} = 1 \\ 1/2 \ ; \ f_{AA} = 0 \ . \end{cases}$$

The observed scattering intensity (at a fixed point) becomes a superposition of such powers, weighted by the concentrations and the $(\nu_i^{(A)})^2$. Clearly this just amounts to stating that scattering at large momenta is due to single chain configurations inside a concentration blob.

22.3.3 Autocorrelations

In tree approximation the density autocorrelation function equals the Debye function (cf. Eq. (22.21)):

$$\frac{I_{d,a}^{(A)}(\mathbf{q})}{c^{(A)}N^{(A)}} = D_p^{(A)}(\bar{\mathbf{q}}^2 N_R^{(A)}) \ . \tag{22.89}$$

Effects of the surrounding are included only via our choice of ℓ_R. For $q^2\xi_1^2 \gg 1$ we of course recover the binary isolated chain result: $I_{d,a}^{(A)} \sim q^{-1/\nu_A}$. We therefore again concentrate on the region of small momenta and discuss the radius of gyration.

The general scaling law for R_g takes the form

$$\frac{R_g^{(A)2}}{R_g^{(A)}(0)^2} = \mathcal{R}_g^{(A)}(x, s, y, f_{11}, f_{22}, f_{12}) \ . \tag{22.90}$$

In tree approximation we find the simple expression

$$R_g^{(A)2} = \frac{\tilde{p}_3^{(A)}}{\tilde{p}_2^{(A)}} \ell_R^2 N_R^{(A)} \ . \tag{22.91}$$

If species A is at its Θ-point: $f_{AA} = 0$, then Eqs. (22.91), (22.61) yield $R_g^{(A)2} \equiv \tilde{R}_0^{(A)2}\tilde{p}_3^{(A)}/\tilde{p}_2^{(A)}$, and the radius of the chain is not affected by its surroundings. For $f_{aa} > 0$, $R_g^{(A)}$ will shrink with increasing overlap, very much like in a binary solution. This is due to screening built into our choice of ℓ_R. Furthermore, if the symmetry conditions $y = 1, f_{11} = 1 = f_{22}$ hold, then Eqs. (22.62), (22.91) guarantee that the radii of the two species stay equal: $R_g^{(1)} = R_g^{(2)}$, independent of the composition of the solution. This result also holds in the more general situation $f_{11} = f_{22} \neq 1$. In summary, in tree approximation the density autocorrelations show no pronounced ternary effects.

These results of the tree approximation give only a rough picture of the behavior of $R_g^{(A)}$. It, for instance, is easily checked that the screened interaction of two mutually repelling Θ-systems for $q \to 0$ induces an attraction within each binary subsystem. Expressed in renormalized variables the corresponding limit of $u_{0,AA}(\mathbf{q})$ (Eq. (22.25)) reads

$$\frac{u^{(AA)}(0)}{u^*} = -\frac{f_{12}^2 c_R^{(\bar{A})} N_R^{(\bar{A})}}{1 - f_{12}^2 c_R^{(1)} c_R^{(2)} N_R^{(1)} N_R^{(2)}} \ .$$

On the spinodal this attraction even diverges, and we have to expect a strong shrinking of $R_g^{(A)}$ with increasing overlap. Clearly this effective self-attraction is quite different from the screening effect. The latter destroys correlations due to self-repulsion and thus leads to shrinking towards Θ-behavior. The self-attraction encountered here decreases the coil radius relative to Θ-behavior. It might even lead to collapse, but it turns out that the spinodal cuts off such dangerous behavior [SK93]. (It is quite obvious that no proper collapse can occur, since collapsed chains would not see each other and thus the origin of the attraction is lost.)

Also the relation $R_g^{(1)} = R_g^{(2)}$ found for $f_{11} = f_{22} = f, y = 1$, in general will break down beyond tree approximation. The effective interaction $u^{(A,A)}(0)$ depends on the relative composition x:

$$\frac{u^{(1,1)}(0)}{u^*} = \frac{1}{\Delta_R(0)} \left[f + (1-x)(f^2 - f_{12}^2) c_R N_R \right] \ .$$

In an incompatible system $(f^2 - f_{12}^2 < 0)$ the effective repulsion for species (1) will therefore decrease with decreasing x, leading to less swelling. Indeed, in simulations it has been found that in such systems the radius of the minority species $(x < \frac{1}{2})$ is smaller than the radius of the majority species $(x > \frac{1}{2})$ [SB88]. Note that this consideration suggests the opposite effect in compatible systems $(f^2 - f_{12}^2 > 0)$.

All such effects ask for a calculation beyond tree approximation. A one-loop analysis can be found in [SK93]. The results are consistent with the above observations and show that the nontrivial structure of the screened vertex induces a rich phenomenology beyond excluded volume screening.

22.3.4 The osmotic pressure

The osmotic pressure of a ternary solution obeys the general scaling law

$$\frac{\hat{\Pi}}{k_B T c_p} = 1 + \mathcal{P}(x, s, y, f_{11}, f_{22}, f_{12}) \ . \tag{22.92}$$

In tree approximation we use Eqs. (22.60) to renormalize Eq. (22.22):

$$\mathcal{P} = \frac{1}{2\tilde{u} c_{pR}} \sum_{A,A'} f_{AA'} c_R^{(A)} c_R^{(A')} \ . \tag{22.93}$$

It is not difficult to evaluate $\mathcal{P}$ in the special cases discussed above for $I_{d,c}$. As an example we consider the neighborhood of the symmetric fixed point S, where $\mathcal{P}$ reduces to

$$\mathcal{P} = \frac{\left(c_R^{(1)} + c_R^{(2)}\right)^2}{2\tilde{u}c_{pR}} + \frac{c_R^{(1)}c_R^{(2)}}{\tilde{u}c_{pR}}\left(f_{12} - 1\right) \ . \tag{22.94}$$

If in addition we impose the symmetry condition $y = 1$, then the first term reduces to the thermodynamic scaling function $\mathcal{P}^*(s)$ of a binary system in the excluded volume limit, where the total overlap $s = s^{(1)} + s^{(2)}$ is calculated from $c_R = c_R^{(1)} + c_R^{(2)}$. Since under these conditions the relations $c_R^{(1)} = xc_R, c_R^{(2)} = (1-x)c_R$ hold, we can also rewrite the second contribution to find

$$\mathcal{P} = \mathcal{P}^*(s)\left[1 + 2x(1-x)(f_{12} - 1)\right] \ . \tag{22.95}$$

With Eq. (22.86) for $f_{12} - 1$ this yields in the semidilute limit

$$\mathcal{P} = const\ s^{1/(3\nu-1)}\left[1 + 2x(1-x)\left(\frac{B_{12}}{R_g(0)}\right)^{\omega_S^{(1,2)}} \cdot \left(\frac{\tilde{u}s}{c_0 n_0^{1/2}}\right)^{\nu\omega_S^{(1,2)}/(3\nu-1)}\right] .$$

The correction term depends on the total segment concentration c, but is independent of N. We thus find the semidilute behavior

$$\frac{\hat{\Pi}}{k_B T} = const\ c^{\frac{3\nu}{3\nu-1}}\left[1 + const\ x(1-x)c^{\nu\omega_S^{(1,2)}/(3\nu-1)}\right]$$
$$\frac{\nu\omega_S^{(1,2)}}{3\nu - 1} \approx 0.308, \quad d = 3 \ . \tag{22.96}$$

(Note that we identified the nonuniversal parameters of the binary subsystems: $B^{(1)} = B^{(2)}$. More generally, c has to be replaced by a weighted sum of $c^{(1)}$ and $c^{(2)}$.)

In the dilute limit Eq. (22.94) yields a correction to the second virial coefficient of the two different species. Again taking $y = 1$ we find in terms of the interpenetration ratio

$$\psi^{(1,2)} = \psi^*\left[1 + const\ N^{-\nu\omega_S^{(1,2)}}\right]$$
$$\nu\omega_S^{(1,2)} = 0.235 \ . \tag{22.97}$$

Note that the form of the corrections in Eqs. (22.96), (22.97) is a consequence of the renormalization group, valid beyond tree approximation.

Clearly all these ternary effects on $\mathcal{P}$ are quite small, and also in other regions of parameter space the thermodynamic scaling functions resembles that of a binary solution. This is illustrated in Fig. 22.7, which shows a doubly logarithmic plot of $\mathcal{P}/s$ as function of s for systems of equal chain concentration: $x = 1/2$, and equal coil radii: $y = 1$. The systems are identical to those used for $I_{d,c}, x = 1/2$.

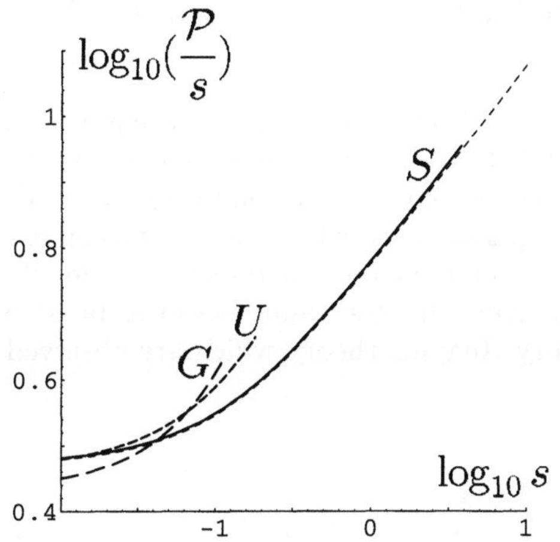

Fig. 22.7. Behavior of the thermodynamic scaling function in fixed point regions, for $x = 1/2, y = 1$. The *thin dashes*, almost coinciding with line 'S', give the binary behavior

The observation that ternary effects do not change $\mathcal{P}$ qualitatively, is quite plausible. The osmotic pressure is governed by the total concentration in the solution, and in this respect is similar to $I_{d,c}$ for $\nu_i^{(1)} = \nu_i^{(2)}$. We may ask whether we can design an experiment where the binary behavior is suppressed, analogous to scattering with $\nu_i^{(1)} = -\nu_i^{(2)}$. Close to fixed point $\mathcal{S}$ this should be possible by measuring $\hat{\Pi}$ not against pure solvent, but against a binary solution adjusted to have the same overlap as the ternary solution: this cancels the leading term in Eq. (22.94). From the theoretical point of view such compensation experiments have been discussed in [KS90]. (See also [Bis91].)

Some one-loop results are available. In particular the second virial coefficient $A_2^{(1,2)}$ or the equivalent interpenetration ratio

$$\psi^{(1,2)} = \left(4\pi R_g^{(1)}(0) R_g^{(2)}(0)\right)^{-3/2} A_2^{(1,2)}$$

have been analyzed [Ste81, JLB84, SK85]. The calculation proceeds by straightforward generalization of the analysis of $A_2(n_1, n_2)$ (Sect. 12.4, and Appendix A 15.3.3), where we only have to insert the different couplings. We, however, have to take care of the fact that for $f_{11} \neq f_{22}$ the ratio $N_R^{(1)}/N_R^{(2)}$ is no RG-invariant and has to be replaced by $y = R_g^{(1)}(0)/R_g^{(2)}(0)$. This is of particular importance close to an unsymmetric fixed point [SLK91]. Qualitatively the one-loop results for $\psi^{(1,2)}/\psi^*$ differ not much from their zero-loop counterpart $\psi^{(1,2)}/\psi^* \sim f_{12}$, and also on the quantitative level the one loop correction to this ratio is not large. Specifically, at fixed points G or U, U', $\psi^{*(1,2)}/\psi^*$ is of the order 2 or 1.5, respectively. Technically this reflects the fixed point value f_{12}^*, and on the physical level it can be interpreted as showing the higher segment density of Θ-coils. However, even at fixed point G, $\psi^{*(1,2)}$ is much smaller than the hard-sphere limit. ($\psi_{hardsphere} \approx 1.6$, see Sect. 15.4.)

A one-loop calculation of the full function $\mathcal{P}$, or rather the equivalent free energy excess, can be found in [CF88]. It is based on ε-expansion, followed by ad hoc exponentiation, with no crossover condition for ℓ_R involved. Results are given in terms of Flory's χ-parameters, which amounts to an experimentally common, but theoretically cumbersome reparameterization. No detailed comparison to experiment is given, but the results seem to be able to explain deviations from simple Flory-Huggins theory which are observed experimentally.

22.4 The spinodal

22.4.1 Basic expressions

On the spinodal the forward scattering intensity diverges. This yields the simple criterion

$$\Delta_R(0) = \text{Det}\, J_c^{-1}(0) = 0 \ . \tag{22.98}$$

In tree approximation, Eq. (22.75), this yields

$$\begin{aligned} & c_R^{(1)} c_R^{(2)} N_{wR}^{(1)} N_{wR}^{(2)} (f_{12}^2 - f_{11} f_{22}) \\ & = 1 + c_R^{(1)} N_{wR}^{(1)} f_{11} + c_R^{(2)} N_{wR}^{(2)} f_{22} \ , \end{aligned} \tag{22.99}$$

where the polydispersity dependence is contained in the weight averages $N_{wR}^{(A)} = \tilde{p}_2^{(A)} N_R^{(A)}$. This expression shows that according to tree approximation a system is incompatible if $f_{12} > \sqrt{f_{11} f_{22}}$.

The more standard derivation of the spinodal equation starts from the chemical potentials. For simplicity we restrict ourselves here to monodisperse systems, where

$$e^{\mu_p^{(A)}(n)} = \delta_{nN^{(A)}}\, e^{\mu_p^{(A)}} \ .$$

Then the spinodal is determined by the equation

$$\text{Det}\left(\partial \mu_p^{(A)} / \partial c_p^{(A')}\right) = 0 \ ,$$

which gives the boundary of the region, where the homogeneous solution is stable against infinitesimally small fluctuations. (In this context we should recall that the spinodal is an ill-defined concept outside mean field theory. In practise the homogeneous state will decay by nucleation, before we can reach the spinodal.) We now insert the appropriate form of the tree approximation (22.20):

$$\mu_p^{(A)} = \frac{d}{2}\ \ln(4\pi) + \ln(\ell^d c_p^{(A)}) + N^{(A)} \sum_{A'} u_{0,AA'} c^{(A')} \quad ,$$

to find after a short calculation

$$(1 + u_{0,11} c^{(1)} N^{(1)})(1 + u_{0,22} c^{(2)} N^{(2)}) - u_{0,12}^2 c^{(1)} c^{(2)} N^{(1)} N^{(2)} = 0 \ . \quad (22.100)$$

This is the Flory-Huggins type result. Renormalization according to Eqs. (22.60) immediately yields the monodisperse version of Eq. (22.99). The generalization of this derivation to arbitrary polydispersity involves the determinant of the derivatives $\partial \mu_p^{(A)}(n)/\partial c_p^{(A')}(n')$ and is somewhat lengthy. It can be found in [Kam90].

For a monodisperse system we can also easily locate the critical point. Besides the spinodal equation, it obeys the condition

$$\mathrm{Det} \begin{pmatrix} \dfrac{\partial D}{\partial c_p^{(1)}} & \dfrac{\partial D}{\partial c_p^{(2)}} \\[2ex] \dfrac{\partial \mu_p^{(2)}}{\partial c_p^{(1)}} & \dfrac{\partial \mu_p^{(2)}}{\partial c_p^{(2)}} \end{pmatrix} = 0 \quad ,$$

where $D = \ \mathrm{Det}\,(\partial \mu_p^{(A)}/\partial c_p^{(A')})$. Evaluating this equation in tree approximation and renormalizing we find

$$\left(1 + c_R^{(2)} N_R^{(2)} f_{22}\right)^2 = \left(1 + c_R^{(1)} N_R^{(1)} f_{11}\right) f_{12} c_R^{(1)} N_R^{(2)} \ . \quad (22.101)$$

For a monodisperse system the critical point is the simultaneous solution to Eqs. (22.99) and (22.101). For general polydispersity we again refer to [Kam90].

These results are to be combined with the RG flow and the condition (22.58). In general the evaluation must proceed numerically, and to illustrate the typical features we again restrict ourselves to fixed point regions.

22.4.2 Fixed point behavior

a) Fixed point G:

$$f_{11} = f_{22} = 0; \quad f_{12} = f_{12,G}^* \ .$$

As has been discussed in Sect. 22.2.3 the spinodal at fixed point G takes a universal form. To evaluate this form in tree approximation for simplicity we again take the system to be monodisperse and the coil radii to be equal: $y = 1$. Then Eqs. (22.99), (22.61) yield

$$\left(\tilde{u} f_{12}^{*(G)}\right)^2 x(1-x)s^2 = \frac{1}{N_R} , \tag{22.102}$$

where $N_R = N_R^{(1)} = N_R^{(2)}, s = s^{(1)} + s^{(2)}$. Condition (22.58) reduces to

$$1 = \frac{n_0}{N_R} + \tilde{u} f_{12}^{*(G)} \sqrt{x(1-x)} \, \frac{s}{c_0 N_R^{1/2}} , \tag{22.103}$$

equivalent to Eq. (22.88). These equations can immediately be solved for N_R to yield

$$N_R = n_0 + \frac{1}{c_0} = const .$$

We thus find a most simple expression

$$\begin{aligned} s_{sp} &= \frac{(x(1-x))^{-1/2}}{\tilde{u} f_{12}^{*(G)} \sqrt{N_R}} \\ &= \frac{0.0577}{(x(1-x))^{1/2}} , \end{aligned} \tag{22.104}$$

where the r.h.s. gives $\mathcal{S}_{sp}^{*(G)}(x, y = 1)$, evaluated in tree approximation. The generalization to $y \neq 1$ can easily be carried through. Note that in the second line of Eq. (22.104) we used our one-loop values $n_0 = 0.53, c_0 = 1.2$. Choosing $n_0 = 1 = c_0$, we would have to replace 0.0577 by 0.0476. Thus the numerical coefficient in this zero-loop approximation should not be taken too seriously. Still, even allowing for some uncertainty in this number we note that the spinodal for $x \approx 1/2$ extends into the crossover region from dilute to semidilute solutions. Equation (22.104) yields $\frac{4\pi}{3} s_{sp}(x = 1/2) = 0.48$, so that the coils roughly fill half the available volume.

The result (22.104) should be compared to the prediction of unrenormalized Flory Huggins theory. In the present case Eq. (22.100) yields

$$s_{sp} = \left(\tilde{\ell}^{-3} u_{0,12} N^{1/2}\right)^{-1} (x(1-x))^{-1/2} . \tag{22.105}$$

Adjusting $u_{0,12}N^{1/2}$, we can bring the two results (22.104), (22.105) into complete agreement. In this sense we are closest here to unrenormalized mean field theory. However, it must be stressed that Eq. (22.104) represents the finite universal limit, attained for $N \to \infty$. In this limit the spinodal of the unrenormalized theory, Eq. (22.105), collapses to $s_{sp} \equiv 0$, a completely unreasonable result. Indeed, unrenormalized Flory Huggins theory is not meant to apply at small concentrations.

Figure 22.8 shows the result (22.104), together with similar results at the other fixed points, to be discussed below. For $y \neq 1$ the form of the curve would only be slightly disturbed, as long as we do not approach $y \to 0$ or $y \to \infty$.

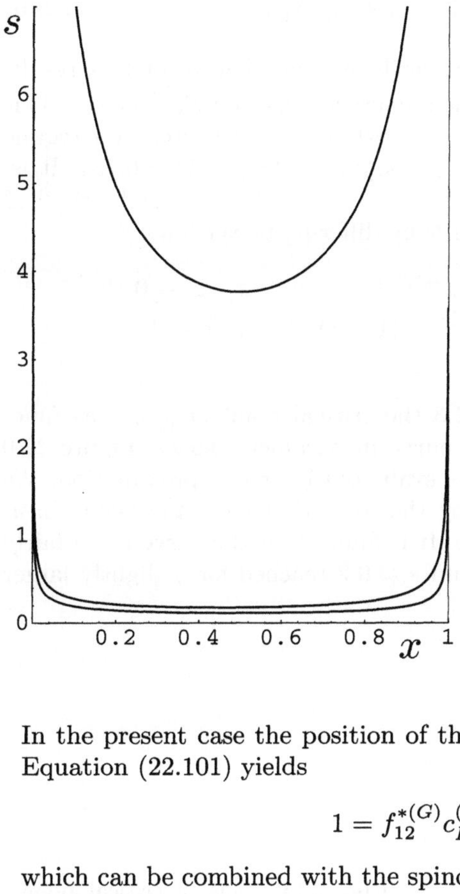

Fig. 22.8. Spinodals $s = s_{sp}(x)$ in fixed point regions $\mathcal{S}, U, G$ (from above); $y = 1$

In the present case the position of the critical point is easily determined. Equation (22.101) yields

$$1 = f_{12}^{*(G)} c_R^{(1)} N_R^{(2)} ,$$

which can be combined with the spinodal equation to yield the symmetric counterpart

$$1 = f_{12}^{*(G)} c_R^{(2)} N_R^{(1)} .$$

Since $c_R^{(A)} = \tilde{u} c_{pR}^{(A)} N_R^{(A)}$, these two equations imply $c_{pR}^{(1)} = c_{pR}^{(2)}$, or $x_{cr} = 1/2$, equivalently. Thus at fixed point G the critical point, evaluated in tree approximation, independent of $y = R_g^{(1)}(0)/R_g^{(2)}(0)$ is found at equal chain concentrations. Beyond tree approximation this result may change, of course.

b) Fixed point U:

$$f_{11} = 1;\ f_{22} = 0;\ f_{12} = f_{12}^{*(U)} \quad .$$

The general results again predict that the spinodal takes a universal form $s_{sp} = \mathcal{S}_{sp}^{*(U)}(x, y)$. In tree approximation it is to be determined from the equation

$$c_R^{(1)} c_R^{(2)} N_{wR}^{(1)} N_{wR}^{(2)} \left(f_{12}^{*(U)}\right)^2 = 1 + c_R^{(1)} N_{wR}^{(1)} \quad . \tag{22.106}$$

In Appendix A 22.2 we present an easily evaluated form of this result, combined with the RG mapping. The numerical result for $\mathcal{S}_{sp}^{*(U)}(x, y = 1)$ is included in Fig. 22.8. Again the spinodal extends into the crossover regime from dilute to semidilute solutions $\left(\frac{4\pi}{3} s_{sp}(x = 1/2, y = 1) \approx 0.75\right)$. It is, however, not symmetric under reflection $x \to 1 - x$. It can easily be checked analytically that the extreme wings obey differing power laws:

$$\mathcal{S}_{sp}^{*(U)}(x, 1) \sim \begin{cases} x^{-1/2} & ; \quad x \to 0 \\ (1-x)^{-\frac{3\nu-1}{4\nu-1}} = (1-x)^{-.565} & ; \quad x \to 1 \ . \end{cases}$$

As has been discussed in Sect. 22.2.3 the critical point (x_{cr}, s_{cr}) as function of y will trace out a universal curve in parameter space. Figure 22.9 shows the locus of critical points as evaluated in tree approximation. We have replaced the variable x here by the somewhat more physically intuitive, but equivalent, variable $s^{(1)}/s$. It is found that the curve has a fairly unsymmetric shape, with a minimum $s \approx 0.2$ reached for y slightly larger than 1.

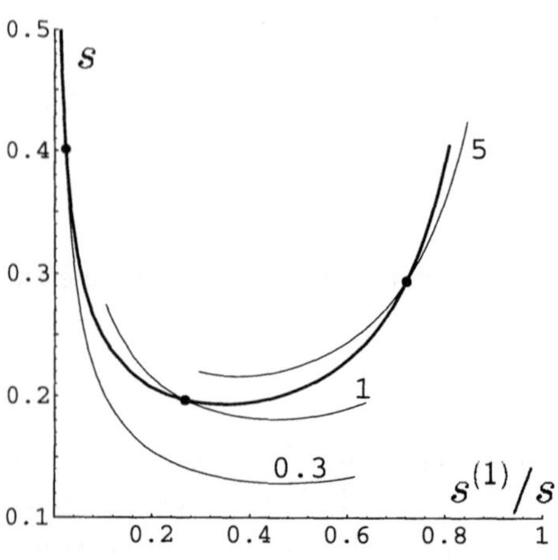

Fig. 22.9. Line of critical points, parameterized by y, in the $s^{(1)}/s - s$ plane (*fat line*). Fixed point U. *Thin lines* give pieces of spinodal curves for the values of y, indicated. Note that the critical points in general are found far from the minimum of the spinodal

c) Critical region of fixed point S:

$$f_{11} = 1 = f_{22}; \quad 0 < f_{12} - 1 \ll 1 \ .$$

As has been stressed repeatedly, close to fixed point S the spinodal does not reach a universal curve but in the limit of infinitely long chains is shifted to infinite overlap:

$$s_{sp} = \left[\frac{1}{B_S^2} R_g^{(1)}(0) R_g^{(2)}(0) \right]^{\frac{\omega_S^{(1,2)}}{2} \frac{3\nu-1}{1+\nu\omega_S^{(1,2)}}} \mathcal{S}_{sp}^{*(S)}(x,y) \ , \qquad (22.107)$$

(Eq. (22.56)). The spinodal concentration, however, still vanishes for $N \to \infty$: $c_{sp} \sim s_{sp} N^{-3\nu+1} \sim N^{-0.619}$, so that the spinodal stays well within the range of our theory. In Fig. 22.8 we included the result of the tree approximation evaluated for the same parameter values $y = 1$, $(B_{12}/R_g(0))^{\omega_S^{(1,2)}} = 0.01$, as used in our evaluation of $I_{d,c}$. Decreasing $B_{12}/R_g(0)$ we would shift it to higher overlap.

This result should be compared to Flory Huggins theory, which for symmetric conditions $N^{(1)} = N^{(2)} = N$, $u_{0,11} = u_{0,22} = u_0$, yields the expression

$$x(1-x)c^2N^2\left(u_{0,12}^2 - u_0^2\right) = 1 + u_0 cN \ .$$

It thus predicts a finite relation among cN and x. Solving for cN and introducing the overlap $s \sim cN^{3\nu-1}$ we find

$$s_{sp} \sim \frac{u_0 N^{3\nu-2}}{2x(1-x)(u_{0,12}^2 - u_0^2)} \left[1 + \left(1 + 4x(1-x) \left(\frac{u_{0,12}^2}{u_0^2} - 1 \right) \right)^{1/2} \right] \ .$$

Again for $N \to \infty$ the spinodal collapses to the line $s \equiv 0$, a behavior opposite to that predicted by the renormalized theory. This is the strongest qualitative consequence of the compatibilizing effect of renormalization.

Since fixed point region S describes the long chain limit of generic ternary systems, it is of interest to determine the asymptotic form of the shape function $\mathcal{S}_{sp}^{*(S)}(x,y)$ for all values of the parameter $y = R_g^{(1)}(0)/R_g^{(2)}(0)$. The somewhat lengthy analysis is carried through in Appendix A 22.3, where we also prove the asymptotic scaling law (22.107). The tree approximation for the universal shape function is found as

$$\mathcal{S}_{sp}^{*(S)}(x,y) = \frac{xy^{3/2} + (1-x)y^{-3/2}}{xy^{1/2\nu} + (1-x)y^{-1/2\nu}} \cdot \left\{ \frac{xy^{1/\nu} + (1-x)y^{-1/\nu}}{x(1-x)} \left(xy^{\frac{1}{2\nu}} + (1-x)y^{-\frac{1}{2\nu}} \right) \right\}^{\frac{3\nu-1}{1+\nu\omega_S^{(1,2)}}} \ . \qquad (22.108)$$

Clearly $\mathcal{S}_{sp}^{*(S)}$ is invariant under an exchange of the two species: $x \to 1 - x, y \to 1/y$. For $y = 1$ the result simplifies to

$$\mathcal{S}_{sp}^{*(S)}(x,1) = [x(1-x)]^{-\frac{3\nu-1}{1+\nu\omega_S^{(1,2)}}} \quad .$$

Compared to Flory Huggins theory the shape of the spinodal is somewhat flatter, due to the anomalous power $(3\nu - 1)/(1 + \nu\omega_S^{(1,2)}) \approx 0.619$, to be compared to $s_{sp} \sim [x(1-x)]^{-1}$ (Flory Huggins), as shown above.

In the limit of long chains we can also evaluate the position of the critical point to find

$$x_{cr} = \left[1 + y^{3/2\nu}\right]^{-1} \quad , \tag{22.109}$$

with s_{cr} following by inserting this result into Eqs. (22.107), (22.108). This yields the tree approximation to the line of critical points, which is universal up to the prefactor in s_{sp} (Eq. (22.107)).

Figure 22.10 shows the shape function $\mathcal{S}_{sp}^{*(S)}(x,y)$ for several values of y, plotted against $s^{(1)}/s$. We note that even for a value $y = 0.1$ the shape is not distorted violently. We indicated the critical points, but omitted the full line of critical points to keep the figure clearer.

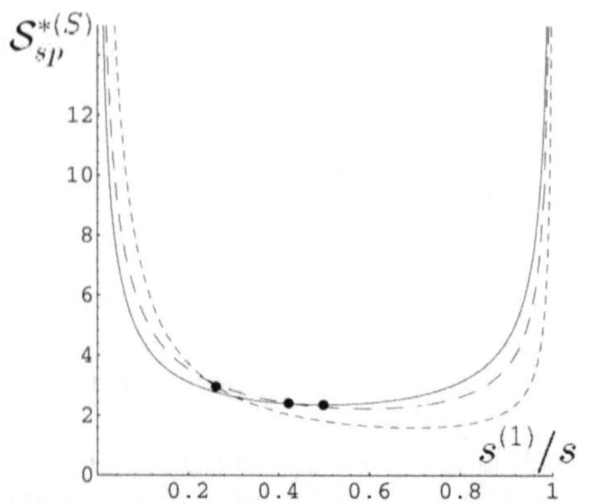

Fig. 22.10. $\mathcal{S}_{sp}^{*(S)}$ as function of $s^{(1)}/s$ for values $y = 1$ (*full line*), .5 (*long dashes*), .1 (*short dashes*). The *dots* indicate the critical points, which shift to the left with decreasing y

22.5 The experimental situation

Experiments in ternary solutions are often analyzed in terms of Flory's χ-parameters. Such an analysis implicitly assumes the validity of Flory Huggins theory or unrenormalized RPA, equivalently. Now it has been amply verified both in physical experiments (see, for instance, [FNI84]) and in computer experiments [SB88] that these theories are inadequate in dilute or even moderately concentrated solutions. For a binary system we have shown in Appendix

A 17.4 that χ depends on chain length, concentration, and renormalized coupling, and it involves all nonuniversal parameters of the theory. Furthermore it depends on the quantity analyzed. In a ternary system the situation only becomes worse. It involves three parameters $\chi_{AA'}$, which each depend on all the variables and all the nonuniversal content of the theory, and χ-parameters extracted for instance from the position of the spinodal will differ from those extracted from second virial coefficients. I therefore omit a discussion. The interested reader easily can construct $\chi_{AA'}$ by applying the standard mean field analysis to the observables expressed in terms of scaling functions, as given in this chapter. An analysis based on renormalized RPA may be found in [SK85, KS90]. In summary, in the range of the present theory the use of χ-parameters amounts to a mean field based reparameterization of the data, which yields quantities of no deeper significance. It should be replaced by an analysis in terms of the nonuniversal parameters of the present theory, where the binary parameters should be taken from binary solution experiments. This leaves a single ternary parameter, which is the starting value of the ternary coupling f_{12}, equivalent to B_{12} (Eq. (22.86)) for long chains in good solvent systems. All nonuniversal parameters should be regular functions of temperature, only.

Extensive experimental studies of ternary systems have been carried through by the group of Strazielle [AKS87, KS87, KAS87]. In particular, Kaddour and Strazielle [KS87] report results on the system polystyrene (PS) – polydimethylsiloxane (PDMS) in tetrahydrofuran, a solvent that is good for both PS and PDMS and is iso-refractive to PDMS. The results were analyzed within the framework of unrenormalized RPA. Good to reasonable agreement among experiment and theory was found by adjusting the parameters. Unfortunately a reanalysis of these data within the present theory meets with problems at the first check: the given values for the isolated chain radii of both PS and PDMS fail to obey the amply verified power law $R_g(0) \sim N^{\nu}, \nu \approx 0.6$. Rather an effective exponent $\nu \approx 0.66$ can be extracted. The source of this discrepancy is unknown, except that it should be noted that at least one sample has a very broad chain length distribution.

Subsequent work [KAS87] analyzes scattering data for several ternary good solvent systems with the aim of testing the renormalized theory. One quantity studied is the chain length dependence of the ternary correction to the second virial coefficient (see Eq. (22.97)). An exponent $\nu\omega_S^{(1,2)} = 0.21 \pm 0.02$ was found, in good agreement with the theoretical prediction ($\nu\omega_S^{(1,2)} = 0.235$). Also the chain length dependence of the critical concentration was studied, which from Eq. (22.107) is predicted to obey the asymptotic power law $c_w \sim N^{-0.619}$. Values of the exponent ranging from 0.585 to 0.615 were found, again in good agreement with the theory. It must be noted, however, that this success should not be overestimated. Predictions like Eq. (22.97) or Eq. (22.107) are valid only asymptotically, and including the next corrections,

Eq. (22.97), for instance, would read

$$\psi^{(1,2)} = \psi^* \Big[1 + const\, N^{-\nu\omega_S^{(1,2)}} + const\, N^{-2\nu\omega_S^{(1,2)}} + const\, N^{-\nu\omega} + \dots \Big] \; .$$

The exponents of both subleading corrections are of order 0.47, and it is not obvious that a neglect of such terms is justified for the experimentally realized chain lengths. Also, of course, the system is never really symmetric, and the coefficients of the power laws depend on $N^{(1)}/N^{(2)}$. Thus changing samples we will also change the coefficients. In summary, the results reported here support the renormalized theory, but give little information on its quantitative performance.

An essential tool for extracting accurate values of the ternary correction to the second virial coefficient in the above experiments was the use of the so-called optical Θ-method. First proposed by Fukuda and Inagaki [FI83], this method uses a solvent, where the scattering increments are of opposite sign: $\nu_i^{(1)}/\nu_i^{(2)} < 0$, and adjusts the concentrations such that in the dilute limit the scattering intensity is sensitive to ternary effects. It is somewhat similar to our condition (22.83) for projecting out the uncritical mode close to the spinodal, and it deserves a more detailed discussion.

The method is based on the virial expansion of the scattering intensity, calculated in unrenormalized RPA. Equations (22.76), (22.23), (22.24) yield for $q = 0$:

$$I_{d,c}(\mathbf{0}; \nu_i^{(1)}, \nu_i^{(2)}) = \frac{1}{\Delta_0} \Big[\nu_i^{(1)2} W^{(1)} \left(1 + u_{0,22} W^{(2)}\right) + \nu_i^{(2)2} W^{(2)} \left(1 + u_{0,11} W^{(1)}\right) - 2\nu_i^{(1)} \nu_i^{(2)} u_{0,12} W^{(1)} W^{(2)} \Big] \; ,$$

where Δ_0 is given in Eq. (22.26) and

$$W^{(A)} = c^{(A)} N_w^{(A)} \; .$$

To second order in the concentrations we thus find

$$\begin{aligned} I_{d,c}(\mathbf{0}; \nu_i^{(1)}, \nu_i^{(2)}) &= \nu_i^{(1)2} W^{(1)} \left(1 - u_{0,11} W^{(1)}\right) \\ &+ \nu_i^{(2)2} W^{(2)} \left(1 - u_{0,22} W^{(2)}\right) - 2\nu_i^{(1)} \nu_i^{(2)} u_{0,12} W^{(1)} W^{(2)} \\ &= \nu_i^{(1)2} W^{(1)} + \nu_i^{(2)2} W^{(2)} \\ &- \left(\nu_i^{(1)} W^{(1)} + \nu_i^{(2)} W^{(2)}\right) \left(\nu_i^{(1)} u_{0,11} W^{(1)} + \nu_i^{(2)} u_{0,22} W^{(2)}\right) \\ &- \nu_i^{(1)} \nu_i^{(2)} \left(2u_{0,12} - u_{0,11} - u_{0,22}\right) W^{(1)} W^{(2)} \; . \end{aligned} \qquad (22.110)$$

We now impose the optical Θ-condition

$$\nu_i^{(1)} W^{(1)} + \nu_i^{(2)} W^{(2)} = 0 \ ,$$

or

$$\nu_i^{(1)} c^{(1)} N_w^{(1)} = -\nu_i^{(2)} c^{(2)} N_w^{(2)} \ , \tag{22.111}$$

equivalently, to find

$$I_{d,c}(\mathbf{0}; \nu_i^{(1)}, \nu_i^{(2)}) = \nu_i^{(1)2} c^{(1)} N_w^{(1)} + \nu_i^{(2)2} c^{(2)} N_w^{(2)} \\ -\nu_i^{(1)} \nu_i^{(2)} \left(2u_{0,12} - u_{0,11} - u_{0,22}\right) c^{(1)} c^{(2)} N_w^{(1)} N_w^{(2)} \ . \tag{22.112}$$

The combination of coupling constants in brackets is proportional to the ternary χ-parameter χ_{12}. Thus, if we assume that unrenormalized RPA adequately describes the scattering intensity, then the optical Θ-condition allows for a direct determination of the deviation of the ternary coupling from the algebraic mean of the binary couplings, with no need of determining the binary couplings separately. Since this deviation in good solvents is expected to be small compared to $(u_{0,11} + u_{0,22})/2$, this is a great achievement.

The method clearly hinges on the validity of unrenormalized RPA. An analysis based on the exact virial expansions of the matrix elements $I_{d,c}^{(A,A')}(\mathbf{0})$ replaces the bracket in Eq. (22.112) by a combination of all five second virial coefficients, weighted with factors like $c^{(1)}/c^{(2)}$ or $N_w^{(1)}/N_w^{(2)}$. This combination has no reason to be simply proportional to a difference of (renormalized or unrenormalized) couplings. However, it must be noted that the optical Θ-condition is typically applied in good solvent systems, where in the dilute range ternary effects typically are small. Indeed, as stressed several times, it is renormalization that suppresses the ternary effects, leading to $f_{12} \approx f_{11} \approx f_{22} \approx 1$. Now, for good solvent systems the optical Θ-condition turns out to be very close to condition (22.83) that focuses on the singular behavior at the spinodal. Close to fixed point S the spinodal is found in the large overlap region, and in Eq. (22.83) we may replace $H(0)_{AA}$ by $c_R^{(A)}$, this approximation being based on renormalized RPA. With $c_R^{(A)} = \tilde{u}\ell_R^d c_p^{(A)} N_R^{(A)}$ Eq. (22.83) then reduces to

$$\nu_i^{(1)} c^{(1)} = -\nu_i^{(2)} c^{(2)} \ .$$

For $N_w^{(1)} \approx N_w^{(2)}$, which is the case in most experiments, this result coincides with the optical Θ-condition. This observation supports the view that the optical Θ-condition is a good criterion for identifying a parameter range where ternary effects are most prominent. Modifications of the condition resulting from the use of renormalized RPA have been proposed in [KS90].

Fukuda and coworkers used the optical Θ-condition in experiments on several good solvent systems [FI83, FNI84, FNI86, FNMI87]. For the system PS-polymethyl-methacrylate [PMMA]-bromobenzene the data have been reanalyzed, using the present renormalized tree approximation [SK90, KSF91].

For that system sufficient independent binary information was available to fix the binary parameters. The single ternary parameter (starting value of $f_{12}(\lambda)$ under the RG flow) then could be determined by fitting to the position of the spinodal. An excellent fit to $I_{d,c}^{-1}(0, \nu_i^{(1)}, \nu_i^{(2)})$ results. With these parameters we can then calculate the effective length scale ξ_{eff}^2 to find very good agreement with experiment. The renormalized theory performs distinctively better than unrenormalized RPA, even though the latter involves a separate set of parameters for each sample used.

Lodge and coworkers [KTL91, KTL92] performed scattering experiments on mixtures of PS and PMMA dissolved in ethyl-benzoate, which is a good solvent iso-refractive to PMMA. They varied the PS-concentration to extract the radius of an isolated PS chain in the presence of PMMA as function of the PMMA concentration and conclude that no specific ternary effects can be seen, but the radius decreases as function of the overlap like in a binary solution. This interpretation has been questioned [SK93], since the overlap reached is far from the semidilute region so that the decrease seen should be due to the reduction of the effective repulsion predicted for a ternary solution, (see Sect. 22.3.3).

So far we have been concerned with good solvent systems, close to the symmetric fixed point. Data on unsymmetric systems, which may be close to fixed point U, have been published in [NFI87]. On the qualitative level the findings are consistent with the present theory. In particular the spinodal extends into the crossover regime $s \lesssim 1$. Due to lack of sufficient information on the binary subsystems a quantitative analysis is not possible, however.

To finish that discussion we should recall from Sect. 18.4 that mixtures of protonated and deuterated chains of the same chemical species in solution must also be treated as ternary systems. This is true in particular for Θ-solvents, and an analysis of the PS-cyclohexane system [Sch93] has shown that the data can be explained only by the ternary theory.

In summary, this discussion shows that more experiments, designed to test the predictions of the renormalized theory, are needed. Such experiments may uncover the deficiences of renormalized RPA and thus may motivate an internally consistent evaluation of the theory to one-loop approximation. To me this is an adequate end to this book.

A 22. Appendix

A 22.1 Flow of the ternary coupling

The renormalization factor $Z_u^{(1,2)}$ and the flow equation for $u^{(1,2)}$ have been calculated to three-loop order in [SLK91], where some standard, but not minimal renormalization scheme has been used. Transformed to minimal subtraction, the results have been given in [Sch93]:

$$\begin{aligned}\lambda \frac{d}{d\lambda} f_{12} = f_{12}\Big\{ &- \varepsilon + u^*(f_{11} + f_{22} + 2f_{12}) \\ &- u^{*2}\left[\frac{5}{4}(f_{11}^2 + f_{22}^2) + 3f_{12}(f_{11} + f_{22}) + 2f_{12}^2\right] \\ &+ u^{*3}\left[\frac{111}{16}(f_{11}^3 + f_{22}^3) + \left(\frac{71}{16} + 9\zeta(3)\right) f_{12}(f_{11}^2 + f_{22}^2)\right. \\ &\left. + \frac{5}{2}f_{12}f_{11}f_{22} + (9 + 6\zeta(3))f_{12}^2(f_{11} + f_{22}) + (3 + 3\zeta(3))f_{12}^3\right] \\ &+ O(u^{*4})\Big\} \ . \end{aligned} \tag{A 22.1}$$

As with the simpler case of the flow of a binary coupling $f_{AA} = f$, this expansion cannot be used as it stands, since subsequent terms rapidly increase in absolute size. The oscillating character furthermore suggests that we are dealing with an asymptotic series, a feature well known from the binary flow equation. For the binary flow this problem found a most satisfactory solution by Borel resummation, employing information on the high orders of the expansion [SD89]. Unfortunately the Borel method is restricted to functions of a single variable and cannot be used here. Also information on the high orders of the expansion in all three variables $f_{AA'}$ is not available. We have only been able to determine the exponents and the fixed points by resummation of the ε-expansion [SLK91]. The results are given in Table 22.2. We thus are confronted with the task of constructing a reasonably simple form of the flow equation, which should obey the following criteria.

i) It should be consistent with the expansion (A 22.1).

ii) It should incorporate the resummed values of $f_{12}^{*(P)}, \omega_P^{(1,2)}$.

iii) For $f_{11} \equiv f_{22} \equiv f_{12}$ it should reproduce the flow of the binary coupling.

Guided by these considerations we write the flow equation in three dimensions as

$$\lambda \frac{df_{12}}{d\lambda} = f_{12} \left(\frac{f_{12}}{f_{12}^*(f_{11}, f_{22})} - 1 \right) \cdot \left[\omega_{12}^{(0)}(f_{11}, f_{22}) \left(\frac{f_{12}}{f_{12}^*(f_{11}, f_{22})} - 1 \right) + \omega_{12}^{(*)}(f_{11}, f_{22}) \frac{f_{12}}{f_{12}^*(f_{11}, f_{22})} \right] , \tag{A 22.2}$$

where

$$\omega_{12}^{(0)}(f_{11}, f_{22}) = \tilde{\nu}(f_{11}) + \tilde{\nu}(f_{22}) - 1 \tag{A 22.3}$$

$$\tilde{\nu}(f) = 0.305 f \left(1 + \frac{0.220(1-f)}{1 + 0.978\, f} \right) \tag{A 22.4}$$

$$\omega_{12}^{(*)}(f_{11}, f_{22}) = 0.820 - 0.140(f_{11} + f_{22} + f_{11} f_{22}) \tag{A 22.5}$$

$$f_{12}^*(f_{11}, f_{22}) = 2 - \frac{1}{2}(f_{11} + f_{22}) - (0.133 + 0.128(f_{11} + f_{22})) \cdot \left(1 - \frac{f_{11} + f_{22}}{2} \right)^2 + \frac{0.016(f_{11} - f_{22})^2}{1 + 0.112(f_{11} + f_{22})} . \tag{A 22.6}$$

The structure of Eq. (A 22.2) guarantees that $f_{12} \equiv 0$ is a fixed plane, and $f_{12} > 0$ approaches fixed point values $f_{12}^* \left(f_{11}^{*(P)}, f_{22}^{*(P)} \right) = f_{12}^{*(P)}$.

The function $\omega_{12}^{(0)}(f_{11}, f_{22})$ interpolates among the ternary correction to scaling exponents for fixed points P_0 with $f_{12}^* = 0$, where we have $\omega_{P_0}^{(1,2)} = 3 - 1/\nu(f_{11}^{*(P_0)}) - 1/\nu(f_{22}^{*(P_0)})$, (cf. Eq. (22.46)). It yields a value $\nu(1) = 0.59$, and the difference to the more precise value $\nu = 0.588$ is negligible in the present context. The function $\omega_{12}^{(*)}(f_{11}, f_{22})$ accordingly interpolates among the correction to scaling exponents at the nontrivial fixed points. The function $f_{12}^*(f_{11}, f_{22})$ parameterizes a nontrivial surface in parameter space, where $\lambda\, d\, f_{12}/d\lambda$ vanishes. The nontrivial fixed points are necessarily located on this surface.

The flow equation (A 22.2) reproduces the expansion (A 22.1) including terms of order u^{*2}. To also reproduce the next term we can multiply the r.h.s. of Eq. (A 22.2) by a correction factor, which is identical to 1 at all fixed points. [Sch93]. It turns out that this factor in all the relevant parameter range is close to 1 and can safely be neglected. Taking the binary limit $f_{AA'} \equiv f$ we furthermore find that Eq. (A 22.2) numerically reproduces the flow of the binary coupling within corrections less than 1 %. In summary, we can be fairly confident to have found a reliable approximation to the flow equation.

The flow equation clearly has to be integrated numerically, and we need the flow of the diagonal couplings $f_{AA}(\lambda)$ as input. Thus, to make the approach work in practice, we need to know the binary parameters to start with. Being confronted with data in the crossover regime, we therefore need good experiments on the binary subsystems. Otherwise we fairly randomly search in a five dimensional parameter space.

A 22.2 Evaluation at an unsymmetric fixed point

For definiteness we consider the RG mapping at fixed point U: $f_{11} = 1; f_{22} = 0; f_{12} = f_{12}^{*(U)}$. We then have the RG mapping in the form of Eqs. (22.61), (22.62):

$$\ell_R^2 \left(\frac{N_R^{(1)}}{n_0} \right)^{2\nu} = \frac{1}{n_0} R_g^{(1)2}(0) \tag{A 22.7}$$

$$\left(\frac{N_R^{(1)}}{n_0} \right)^{3\nu - 1} c_R^{(1)} = \frac{\tilde{u}}{n_0^{1/2}} s^{(1)} \tag{A 22.8}$$

$$\ell_R^2 N_R^{(2)} = R_g^{(2)2}(0) \tag{A 22.9}$$

$$N_R^{(2)1/2} c_R^{(2)} = \tilde{u} s^{(2)} \ , \tag{A 22.10}$$

to be combined with condition (22.58):

$$2 = \frac{n_0}{N_R^{(1)}} + \frac{n_0}{N_R^{(2)}} + \frac{c_R^{(1)}}{c_0} + \left[\left(\frac{n_0}{N_R^{(1)}} - \frac{n_0}{N_R^{(2)}} + \frac{c_R^{(1)}}{c_0} \right)^2 + 4 \left(f_{12}^{*(U)} \right)^2 \frac{c_R^{(1)} c_R^{(2)}}{c_0^2} \right]^{1/2} . \tag{A 22.11}$$

We now consider $N_R^{(1)}/n_0 = \hat{N}$ as running parameter and use Eqs. (A 22.7), (A 22.8) to express $N_R^{(2)}$ in terms of $y = R_g^{(1)}(0)/R_g^{(2)}(0)$:

$$\frac{N_R^{(2)}}{n_0} = \hat{N}^{2\nu} \, y^{-2} \ . \tag{A 22.12}$$

We furthermore solve Eq. (A 22.11) for the square root term and take the square of the result to find

$$\left(f_{12}^{*(U)} \right)^2 \frac{c_R^{(1)} c_R^{(2)}}{c_0^2} + \frac{c_R^{(1)}}{c_0} \left(1 - \frac{n_0}{N_R^{(2)}} \right) = b_1 \ , \tag{A 22.13}$$

where

$$b_1 = 1 - \frac{n_0}{N_R^{(1)}} - \frac{n_0}{N_R^{(2)}} + \frac{n_0^2}{N_R^{(1)} N_R^{(2)}} \quad . \tag{A 22.14}$$

This form is equivalent to Eq. (A 22.11). We now use the relations

$$\begin{aligned} c_R^{(1)} &= \tilde{u}\, x\, c_{pR}\, N_R^{(1)} \\ c_R^{(2)} &= \tilde{u}\, (1-x)\, c_{pR}\, N_R^{(2)} \end{aligned} \tag{A 22.15}$$

to transform Eq. (A 22.13) into a quadratic equation for c_{pR}, which is easily solved:

$$\frac{\tilde{u} n_0}{c_0}\, c_{pR} = \frac{1}{2b_3}\left(\sqrt{b_2^2 + 4b_1 b_3} - b_2\right) \quad . \tag{A 22.16}$$

Here

$$\begin{aligned} b_2 &= x\, \hat{N}\left(1 - y^2 \hat{N}^{-2\nu}\right) \\ b_3 &= \left(f_{12}^{*(U)}\right)^2 x(1-x)\, \hat{N}^{1+2\nu} y^{-2} \\ b_1 &= 1 - \frac{1}{\hat{N}} - y^2 \hat{N}^{-2\nu}\left(1 - \frac{1}{\hat{N}}\right) \quad . \end{aligned} \tag{A 22.17}$$

Equations (A 22.12), (A 22.15)–(A 22.17) yield the renormalized variables as functions of $x = c_p^{(1)}/c_p, y = R_g^{(1)}(0)/R_g^{(2)}(0)$, and of the running parameter $\hat{N}$. The overlap variables are determined from Eqs. (A 22.8), (A 22.10) as

$$\begin{aligned} s^{(1)} &= x\, n_0^{3/2}\, \hat{N}^{3\nu}\, c_{pR} \\ s^{(2)} &= (1-x) n_0^{3/2}\, \hat{N}^{3\nu}\, y^{-3}\, c_{pR} \\ s &= \left(x + \frac{1-x}{y^3}\right) n_0^{3/2}\, \hat{N}^{3\nu}\, c_{pR} \quad . \end{aligned} \tag{A 22.18}$$

In analyzing the spinodal we have to add the spinodal equation (22.99), which can be written here as

$$\left(f_{12}^{*(U)}\right)^2 \frac{c_R^{(1)} c_R^{(2)}}{c_0^2} = \frac{1 + c_R^{(1)} N_R^{(1)}}{c_0^2 N_R^{(1)} N_R^{(2)}} \quad . \tag{A 22.19}$$

(For brevity we take the system to be monodisperse.) Substituting this expression into Eq. (A 22.13) we find a linear equation for $c_R^{(1)}$, which yields

$$c_R^{(1)} = c_0\left(b_1 - \frac{y^2}{n_0^2 c_0^2}\, \hat{N}^{-1-2\nu}\right)\left(1 - \frac{1}{\hat{N}} + \frac{y^2}{n_0 c_0}\, \hat{N}^{-2\nu}\right)^{-1} \quad . \tag{A 22.20}$$

With Eq. (A 22.19) solved for $c_R^{(2)}$:

$$c_R^{(2)} = \frac{1}{\left(f_{12}^{*(U)}\right)^2} \frac{y^2}{n_0} \hat{N}^{-2\nu} \left(1 + \frac{1}{n_0 \hat{N} c_R^{(1)}}\right) \tag{A 22.21}$$

we have a parameter representation of the renormalized spinodal, which is to be transformed to physical variables by Eqs. (A 22.8), (A 22.10).

The critical point obeys the additional equation

$$1 = \left(1 + c_R^{(1)} N_R^{(1)}\right) f_{12}^{*(U)} c_R^{(1)} N_R^{(2)} \quad ,$$

which is solved for $c_R^{(1)}$ to yield

$$c_R^{(1)} = \frac{1}{2n_0 \hat{N}} \left[\left(1 + \frac{4y^2}{f_{12}^{*(U)}} \hat{N}^{1-2\nu}\right)^{1/2} - 1 \right] \quad . \tag{A 22.22}$$

Equating the expressions (A 22.20), (A 22.22) we find an equation, which for given y must be solved numerically to yield the parameter $\hat{N}$ at the critical point.

A 22.3 The spinodal in fixed point region S

A 22.3.1 General scaling behavior. We define the spinodal by the divergence of the forward scattering intensity. So the general equation reads

$$H_{11}(0) H_{22}(0) - H_{12}^2(0) = 0 \quad , \tag{A 22.23}$$

where the matrix $\mathbf{H}$ has been defined in Eq. (22.69). Following Sect. 19.1.2 we may split the diagonal element H_{AA} into an autocorrelation part, where the two density insertions occur in the same chain, and an interchain correlation part, where the insertions are in different chains. By virtue of the sum rule the autocorrelation part evaluated at $q = 0$ yields $1/N_{wR}^{(A)}$. The two-chain part must be proportional to $c_R^{(A)}$, which is the probability to find a segment on a second chain of type A. We thus write

$$H_{AA}(0) = \frac{1}{N_{wR}^{(A)}} + c_R^{(A)} h_{AA}(x, y, s, f_{12}) \quad , \tag{A 22.24}$$

where $x = c_p^{(1)}/c_p, y = R_g^{(1)}(0)/R_g^{(2)}(0)$, and where we put the diagonal couplings equal to the fixed point value: $f_{AA} = 1$. We also suppressed the chain length distributions. Similarly we can write

$$H_{12}(0) = \sqrt{c_R^{(1)} c_R^{(2)}}\, f_{12} h_{12}(x, y, s, f_{12}) \quad . \tag{A 22.25}$$

The occurrence of the square root is due to our definition (22.66), and the factor f_{12} takes into account that we at least once have to connect chains of different species. With these expressions the spinodal equation takes the form

$$c_R^{(1)} c_R^{(2)} N_{wR}^{(1)} N_{wR}^{(2)} \left(f_{12}^2 h_{12}^2 - h_{11} h_{22}\right) = 1 + c_R^{(1)} N_{wR}^{(1)} h_{11} + c_R^{(2)} N_{wR}^{(2)} h_{22} \quad . \tag{A 22.26}$$

Comparing to Eq. (22.99) we note that $h_{AA'} \equiv 1$ holds in tree approximation. We now assume that these functions also beyond tree approximation are regular functions in the neighborhood of $f_{12} = 1$. We furthermore note that $h_{AA'}(x, y, \hat{s}, 1)$ has a finite semidilute limit $s \to \infty, x, y$ fixed. This follows from the scaling of the density correlations in such a quasi binary system, (see Sect. 19.1.1). Using this information we expand the brackets in Eq. (A 22.26) about $f_{12} = 1, s = \infty$ to find

$$\begin{aligned} & f_{12}^2 h_{12}^2 - h_{11} h_{22} \\ & = h_{12}^2(x, y, \infty, 1) - h_{11}^2(x, y, \infty, 1) h_{22}^2(x, y, \infty, 1) \\ & +(f_{12} - 1) g(x, y) + \ldots \quad , \end{aligned} \tag{A 22.27}$$

$$\begin{aligned} g(x, y) = 2h_{12}^2(x, y, \infty, 1) - \frac{\partial}{\partial z}\Big|_1 \Big(& h_{12}^2(x, y, \infty, z) \\ & -h_{11}(x, y, \infty, z) h_{22}(x, y, \infty, x)\Big) \quad . \end{aligned}$$

The leading correction due to finite overlap should be of order $s^{-1/(6\nu-2)}$ and therefore in the semidilute limit is small compared to $f_{12} - 1 \sim c^{\nu\omega_S^{(1,2)}/(3\nu-1)}$. In estimating these corrections we invoke results like Eq. (A 19.16) to argue that corrections to the semidilute limit are of the form $N_R^{\varepsilon/2}/(c_R N_R) \sim w^{2-\varepsilon} \sim s^{-(1-\varepsilon/2)/(\nu d-1)}$. We now further note that in the semidilute limit and for all couplings equal, the scattering is due to Gaussian chains of blobs. Thus the $h_{AA'}(x, y, \infty, 1)$ could depend on the species only by factors of the blob number per chain, which however are extracted explicitly in our definition. As a result the term independent of $(f_{12} - 1)$ in Eq. (A 22.27) vanishes. (This in fact also follows from the observation that at the fixed point the system is compatible.) We thus conclude that Eq. (A 22.26) reduces to

$$\begin{aligned} & c_R^{(1)} c_R^{(2)} N_{wR}^{(1)} N_{wR}^{(2)} (f_{12} - 1) g(x, y) = \\ & c_R^{(1)} N_{wR}^{(1)} h_{11}(x, y, \infty, 1) + c_R^{(2)} N_{wR}^{(2)} h_{22}(x, y, \infty, 1) \quad , \end{aligned} \tag{A 22.28}$$

which differs from the semidilute tree approximation only by the factors g, h_{AA}, which are universal functions of the RG invariants x, y. Inserting the RG mapping we find the scaling law (22.56). This last step proceeds exactly like in tree approximation, to which we turn now.

A 22.3.2 Tree approximation. The mapping from renormalized to physical variables is given by Eqs. (22.62):

$$\ell_R^2 \left(\frac{N_R^{(A)}}{n_0}\right)^{2\nu} = \frac{1}{n_0} R_g^{(A)2}(0)$$

$$\left(\frac{N_R^{(A)}}{n_0}\right)^{3\nu-1} c_R^{(A)} = \frac{\tilde{u}}{n_0^{1/2}} s^{(A)} \quad .$$

We again introduce $y = R_g^{(1)}(0)/R_g^{(2)}(0)$ to find

$$\frac{N_R^{(1)}}{N_R^{(2)}} = y^{1/\nu} \quad . \tag{A 22.29}$$

We also introduce the geometric mean

$$\bar{N}_R = \left(N_R^{(1)} N_R^{(2)}\right)^{1/2} \quad , \tag{A 22.30}$$

and a corresponding concentration variable

$$\bar{c}_R = \tilde{u}\, c_{pR}\, \bar{N}_R \quad . \tag{A 22.31}$$

The $c_R^{(A)}$ then take the form

$$\begin{aligned} c_R^{(1)} &= x y^{\frac{1}{2\nu}}\, \bar{c}_R \\ c_R^{(2)} &= (1-x) y^{-\frac{1}{2\nu}}\, \bar{c}_R \quad . \end{aligned} \tag{A 22.32}$$

Since we are working in the semidilute limit, condition (22.58) reduces to

$$c_R^{(1)} + c_R^{(2)} = c_0 \quad ,$$

which in view of Eq. (A 22.32) yields

$$\bar{c}_R = c_0 \left[x y^{\frac{1}{2\nu}} + (1-x) y^{-\frac{1}{2\nu}}\right]^{-1} \quad . \tag{A 22.33}$$

$\bar{N}_R$ can be expressed via the total overlap as

$$\left(\frac{\bar{N}_R}{n_0}\right)^{3\nu-1} = \frac{\tilde{u}}{n_0^{1/2}} \frac{s}{\bar{c}_R} \left[x y^{3/2} + (1-x) y^{-3/2}\right]^{-1} \quad . \tag{A 22.34}$$

Finally using

$$R_g^{(1)}(0) R_g^{(2)}(0) = n_0 \ell_R^2 \left(\frac{\bar{N}_R}{n_0}\right)^{2\nu} \tag{A 22.35}$$

we can express all renormalized variables $\ell_R, N_R^{(A)}, c_R^{(A)}$ in terms of s, x, y, and $R_g^{(1)}(0) R_g^{(2)}(0)$.

We now turn to the spinodal equation, which in the semidilute limit (and for monodisperse systems) takes the form

$$2(f_{12}-1)c_R^{(1)}c_R^{(2)}N_R^{(1)}N_R^{(2)} = c_R^{(1)}N_R^{(1)} + c_R^{(2)}N_R^{(2)} \quad . \qquad \text{(A 22.36)}$$

Inserting the above results, we find

$$2(f_{12}-1)x(1-x)\bar{c}_R\bar{N}_R = xy^{1/\nu} + (1-x)y^{-1/\nu} \quad . \qquad \text{(A 22.37)}$$

The coupling f_{12} obeys the relation

$$\begin{aligned} f_{12}-1 &= \left(\frac{B_{12}}{n_0^{1/2}}\right)^{\omega_S^{(1,2)}} \ell_R^{-\omega_S^{(1,2)}} \\ &= \left[\frac{B_{12}^2}{R_g^{(1)}(0)R_g^{(2)}(0)}\right]^{\frac{\omega_S^{(1,2)}}{2}} \left(\frac{\bar{N}_R}{n_0}\right)^{\nu\omega_S^{(1,2)}} \quad . \end{aligned} \qquad \text{(A 22.38)}$$

Putting all results together we find the spinodal equation (22.107), (22.108), with

$$B_S = B_{12}\,(2n_0c_0)^{1/\omega_S^{(1,2)}} \left(\frac{\tilde{u}}{n_0^{1/2}c_0}\right)^{\frac{1}{\omega_S^{(1,2)}}\frac{1+\nu\omega_S^{(1,2)}}{3\nu-1}} \quad .$$

We finally turn to the position of the critical point. Equation (22.101) in the present case can be written as

$$\begin{aligned} \left(1+c_R^{(2)}N_R^{(2)}\right)^2 &= \left(1+c_R^{(1)}N_R^{(1)}\right)c_R^{(1)}N_R^{(2)} \\ &+ \left(1+c_R^{(1)}N_R^{(1)}\right)c_R^{(1)}N_R^{(2)}(f_{12}-1) \quad . \end{aligned} \qquad \text{(A 22.39)}$$

We now note that $N_R^{(A)} \sim \bar{N}_R \to \infty$, but according to the spinodal equation $\bar{N}_R(f_{12}-1) = O(1)$. The leading terms in Eq. (A 22.39) are of order $\bar{N}_R^2$, and equating those terms we find

$$\left(c_R^{(2)}N_R^{(2)}\right)^2 = \left(c_R^{(1)}\right)^2 N_R^{(1)}N_R^{(2)} \quad .$$

With Eqs. (A 22.32), (A 22.29) we find

$$(1-x)y^{-1/2\nu} = xy^{1/2\nu}y^{1/2\nu} \quad ,$$

resulting in Eq. (22.109). The first correction in this calculation is of order $\bar{N}_R^{-1} \sim s^{-1/(3\nu-1)}$, and will be overrun by corrections due to $f_{AA} \neq 1$.

Notes

The present presentation of the theory closely follows [SK90]. Unrenormalized RPA for multicomponent systems has been worked out in much detail by Benoit and coworkers. (See, for instance, [BBW90].) Properties of the screened interaction in unrenormalized theory are discussed in [VB90]. An interpretation of the renormalized theory in terms of blob arguments may be found in [OH89], though it must be noted that the blob size, introduced there, is invalid in Θ-systems. Scaling theory of the critical properties of the phase separation, which has not been covered here, may be found in [BLJ87]. Mixtures containing three polymer species in a solvent have been examined experimentally in [SDB95] and theoretically in [BDS$^+$95]. The theory has also been extended to the Θ-point, treated as tricritical point [RM94].

References

[ABA81] A.Z. Akcasu, M. Benmouna, and S. Alkhafaji. *Macromolecules*, 14:147, 1981.

[AC92] J.L. Alessandrini and M.A. Carignano. *Macromolecules*, 25:1157, 1992.

[AD82] J. Amirzadeh and M. E. Mc Donnell. *Macromolecules*, 15:927, 1982.

[AEY94] F. Abe, Y. Einaga, and H. Yamakawa. *Macromolecules*, 27:3262, 1994.

[AEYY93] F. Abe, Y. Einaga, T. Yoshizaki, and H. Yamakawa. *Macromolecules*, 26:1884, 1993.

[AG78] D.J. Amit and Y.Y. Goldschmidt. *Annals of Phys.*, 114:356, 1978.

[AH79] A.Z. Akcasu and C.C. Han. *Macromolecules*, 12:276, 1979.

[AHEY94] F. Abe, K. Horita, Y. Einaga, and H. Yamakawa. *Macromolecules*, 27:725, 1994.

[AKS87] M. Anasagasti, I. Katime, and C. Strazielle. *Makromol. Chem.*, 188:201, 1987.

[Ami84] D.J. Amit. *Field Theory, the Renormalization Group, and Critical Phenomena.* World Scientific, Singapore, 1984.

[AN86] J.A. Aronovitz and D.R. Nelson. *J. Physique*, 47:1445, 1986.

[Bau81] A. Baumgärtner. *Z. Phys. B*, 42:265, 1981.

[BB84] H. Benoit and M. Benmouna. *Polymer*, 25:1059, 1984.

[BB90] C. Bagnuls and C. Bervillier. *Phys. Rev. B*, 41:402, 1990.

[BB94] C. Bagnuls and C. Bervillier. *Phys. Lett. A*, 195:163, 1994.

[BBW90] H. Benoit, M. Benmouna, and W.-L. Wu. *Macromolecules*, 23:1511, 1990.

[BC91] M. Bishop and J.H.R. Clarke. *J. Chem. Phys.*, 94:3936, 1991.

[BCRF91] M. Bishop, J.H.R. Clarke, A. Rey, and J.J. Freire. *J. Chem. Phys.*, 95:4589, 1991.

[BD79] A.J. Barrett and C. Domb. *Proc. R. Soc. Lond. A*, 367:143, 1979.

[BDS+95] M. Benmouna, M. Duval, C. Strazielle, F.-I. Hakem, and E.W. Fisher. *Macromol. Theory Simul.*, 4:53, 1995.

[Ber66] G.C. Berry. *J. Chem. Phys.*, 44:4550, 1966.

[BFK50] C.E.H. Bawn, R.F.J. Freeman, and A.R. Komaliddin. *Trans. Faraday Soc.*, 46:862, 1950.

[BGZJ76] E. Brezin, J.C. Le Guillou, and J. Zinn-Justin. In C. Domb and M.S. Green, editors, *Phase Transitions and Critical Phenomena*, volume 6. Academic Press, London, 1976.

[Bis91] B. Biswas. *Polymer*, 32:1095, 1991.

[BL86] D. Broseta and L. Leibler. *Europhysics Letters*, 2:733, 1986.

[BLJ87] D. Broseta, L. Leibler, and J.-F. Joanny. *Macromolecules*, 20:1935, 1987.

[BM76a] D. Burch and M.A. Moore. *J. Phys. A*, 9:435, 1976.

[BM76b] D. Burch and M.A. Moore. *J. Phys. A*, 9:451, 1976.

[BM85] M. Benhamou and G. Mahoux. *J. Physique*, 46:L 689, 1985.

[BMB91] A.J. Barrett, M. Mansfield, and B.C. Benesch. *Macromolecules*, 24:1615, 1991.

[BNM78] G.A. Baker, B.G. Nickel, and D.I. Meiron. *Phys. Rev. B*, 17:1365, 1978.

[BRBF92] A.T. Boothroyd, A.R. Rennie, C.B. Boothroyd, and L.J. Fetters. *Phys. Rev. Lett.*, 69:426, 1992.

[Bru97] W. Bruns. *Macromolecules*, 30:4429, 1997.

[BS91] M. Bishop and C.J. Saltiel. *J. Chem. Phys.*, 94:6920, 1991.

[CBMF86] B.J. Cherayil, M.G. Bavendi, A. Miyake, and K.F. Freed. *Macromolecules*, 19:2770, 1986.

[CCJ85] J. Des Cloizeaux, R. Conte, and G. Jannink. *J. Physique*, 46:L 595, 1985.

[CDF+74] J.P. Cotton, D. Decker, B. Farnoux, G. Jannink, R. Ober, and C. Picot. *Phys. Rev. Lett.*, 32:1170, 1974.

[CDF85] B.J. Cherayil, J.F. Douglas, and K.F. Freed. *J. Chem. Phys.*, 83:5293, 1985.

[CF88] B.J. Cherayil and K.F. Freed. *Macromolecules*, 21:3204, 1988.

[CJ90] J. Des Cloizeaux and G. Jannink. *Polymers in Solution.* Clarendon Press, Oxford, 1990.

[Clo74] J. Des Cloizeaux. *Phys. Rev. A*, 10:1665, 1974.

[Clo75] J. Des Cloizeaux. *J. Physique*, 36:281, 1975.

[Clo76] J. Des Cloizeaux. *J. Physique*, 37:431, 1976.

[Clo80a] J. Des Cloizeaux. *J. Physique*, 41:749,761, 1980.

[Clo80b] J. Des Cloizeaux. *J. Physique*, 41:223, 1980.

[Clo81] J. Des Cloizeaux. *J. Physique*, 42:635, 1981.

[Clo82] J. Des Cloizeaux. *J. Physique*, 43:1743, 1982.

[CN80] B. Chu and T. Nose. *Macromolecules*, 13:122, 1980.

[CN82] J. Des Cloizeaux and I. Noda. *Macromolecules*, 15:1505, 1982.

[CNB+76] J.P. Cotton, M. Nierlich, F. Boué, M. Daoud, B. Farnoux, G. Jannink, R. Duplessix, and C. Picot. *J. Chem. Phys.*, 65:1101, 1976.

[Cot80] J.P. Cotton. *J. Physique*, 41:L 231, 1980.

[CS80] J.G. Curro and D.W. Schaefer. *Macromolecules*, 13:1199, 1980.

[DB76] C. Domb and A.J. Barrett. *Polymer*, 17:179, 1976.

[DCF+75] M. Daoud, J.P. Cotton, B. Farnoux, G. Jannink, G. Sarma, H. Benoit, R. Duplessix, C. Picot, and P.G. De Gennes. *Macromolecules*, 8:804, 1975.

[DCF85] J.F. Douglas, B.J. Cherayil, and K.F. Freed. *Macromolecules*, 18:2455, 1985.

[DE89] H.W. Diehl and E. Eisenriegler. *J. Phys. A*, 22:L 87, 1989.

[DF85] J.F. Douglas and K.F. Freed. *Macromolecules*, 18:201, 1985.

[dG72] P.G. de Gennes. *Phys. Lett.*, 38 A:339, 1972.

[dG75] P.G. de Gennes. *J. Physique*, 36:L 55, 1975.

[dG78] P.G. de Gennes. *J. Physique*, 39:L 299, 1978.

[dG79] P.G. de Gennes. *Scaling Concepts in Polymer Physics.* Cornell University Press, Ithaca, New York, 1979.

[DJ76] M. Daoud and G. Jannink. *J. Physique*, 37:973, 1976.

[DJ78] M. Daoud and G. Jannink. *J. Physique*, 39:331, 1978.

[DJ93] B. Duplantier and G. Jannink. *Phys. Rev. Lett.*, 70:3174, 1993.

[DJdC86] B. Duplantier, G. Jannink, and J. des Cloizeaux. *Phys. Rev. Lett.*, 56:2080, 1986.

[DLR96] M. Destrée, A. Lyulin, and J.-P. Ryckaert. *Macromolecules*, 29:1721, 1996.

[DOST95] R. Dekeyser, E. Orlandini, A.L. Stella, and M.C. Tesi. *Phys. Rev. E*, 52:5214, 1995.

[DS87] B. Duplantier and H. Saleur. *Phys. Rev. Lett.*, 59:539, 1987.

[Dup82] B. Duplantier. *J. Physique*, 43:991, 1982.

[Dup85] B. Duplantier. *J. Physique*, 46:L 751, 1985.

[Dup86a] B. Duplantier. *J. Physique*, 47:569, 1986.

[Dup86b] B. Duplantier. *J. Physique*, 47:1633, 1986.

[Dup86c] B. Duplantier. *J. Physique*, 47:1865, 1986.

[Dup86d] B. Duplantier. *J. Physique*, 47:745, 1986.

[Dup87] B. Duplantier. *J. Chem. Phys.*, 86:4233, 1987.

[Dup88] B. Duplantier. *Phys. Rev. A*, 38:3647, 1988.

[EAY93] Y. Einaga, F. Abe, and H. Yamakawa. *Macromolecules*, 26:6243, 1993.

[Edw65] S. F. Edwards. *Proc. Phys. Soc.*, 85:613, 1965.

[Edw66] S. F. Edwards. *Proc. Phys. Soc.*, 88:265, 1966.

[Edw75] S.F. Edwards. *J. Phys. A*, 10:1670, 1975.

[Eis93] E. Eisenriegler. *Polymers Near Surfaces.* World Scientific, Singapore, 1993.

[Eld80] D.J. Elderfield. *J. Phys. C*, 13:5883, 1980.

[Eld81] D.J. Elderfield. *J. Phys. A*, 14:1797,3367, 1981.

[Eme75] V.J. Emery. *Phys. Rev. B*, 11:239, 1975.

[Far76] B. Farnoux. *Ann. Physique*, 1:73, 1976.

[FBC+78] B. Farnoux, F. Boué, J.P. Cotton, M. Daoud, G. Jannink, M. Nierlich, and P.G. De Gennes. *J. Physique*, 39:77, 1978.

[FD88] K.F. Freed and J.F. Douglas. *J. Chem. Phys.*, 88:2764, 1988.

[FF51] P.J. Flory and T.G. Fox. *J. Am. Chem. Soc.*, 73:1904, 1951.

[FFKH74] M. Fukuda, M. Fukutomi, Y. Kato, and T. Hashimoto. *J. Pol. Sci., Pol. Phys. Ed.*, 12:871, 1974.

[FI83] T. Fukuda and H. Inagaki. *Pure Appl. Chem.*, 55:1541, 1983.

[Fis66] M.E. Fisher. *J. Chem. Phys.*, 44:616, 1966.

[FKMS62] T.G. Fox, J.B. Kinsinger, H.F. Manson, and E.M. Schuele. *Polymer*, 3:71, 1962.

[Flo49] P.J. Flory. *J. Chem. Phys.*, 17:303, 1949.

[Flo53] P.J. Flory. *Principles of Polymer Chemistry.* Cornell University Press, Ithaca, New York, 1953.

[Flo69] P.J. Flory. *Statistical Mechanics of Chain Molecules.* Wiley Interscience, New York, 1969.

[FN85] H. Fujita and T. Norisuye. *Macromolecules*, 18:1637, 1985.

[FNI84] T. Fukuda, M. Nagata, and H. Inagaki. *Macromolecules*, 17:548, 1984.

[FNI86] T. Fukuda, M. Nagata, and H. Inagaki. *Macromolecules*, 19:1411, 1986.

[FNMI87] T. Fukuda, M. Nagata, H. Miyashita, and H. Inagaki. *Bull. Inst. Chem. Res. Kyoto Univ.*, 65:67, 1987.

[Fre83] K.F. Freed. *J. Chem. Phys.*, 79:6357, 1983.

[Fre87] K.F. Freed. *Renormalization Group Theory of Macromolecules.* Wiley-Interscience, New York, 1987.

[FSW80] J. Francois, T. Schwartz, and G. Weill. *Macromolecules*, 13:564, 1980.

[Fuj70] H. Fujita. *Polymer J.*, 1:537, 1970.

[Fuj90] H. Fujita. *Polymer Solutions.* Elsevier, Amsterdam, 1990.

[GA96] T.S. Grigera and J.L. Alessandrini. *J. Chem. Phys.*, 104:6027, 1996.

[GH95] P. Grassberger and R. Hegger. *J. Chem. Phys.*, 102:6881, 1995.

[GKK82] A.Y. Grosberg, P.G. Khalatur, and A.R. Khokhlov. *Macromol. Chem., Rapid Commun.*, 3:709, 1982.

[Gra97] P. Grassberger. *Phys. Rev. E*, 56:3682, 1997.

[Gre96] S. Greer. *Advances in Chemical Physics*, volume XCIV, page 261. Prigogine and Rice eds, Wiley, New York, 1996.

[Gri83] J. Grigat. Master's thesis, Fakultät für Physik, Universität Hannover, Deutschland, 1983.

[GSS97] P. Grassberger, P. Sutter, and L. Schäfer. *J. Phys. A*, 30:7039, 1997.

[GZJ85] J.C. Le Guillou and J. Zinn-Justin. *J. Physique*, 46:L 137, 1985.

[HKHN85] F. Hamada, S. Kinugasa, H. Hayashi, and A. Nakajima. *Macromolecules*, 18:2290, 1985.

[Jag92] O. Jagodzinski. PhD thesis, University of Essen, Germany, 1992.

[JBJ95] A. Johner, H. Benoit, and J.-F. Joanny. *Macromol. Theory Simul.*, 4:45, 1995.

[JEK92] O. Jagodzinski, E. Eisenriegler, and K. Kremer. *J. Phys. (France)*, 2:2243, 1992.

[JLB84] J.-F. Joanny, L. Leibler, and R. Ball. *J. Chem. Phys.*, 81:4640, 1984.

[JPLC94] G. Jannink, P. Pfeuty, A. Lapp, and J.P. Cotton. *Europhys. Lett.*, 27:47, 1994.

[Jug82] G. Jug. *Annals of Physics*, 142:140, 1982.

[Kad66] L.P. Kadanoff. *Physics (N.Y.)*, 2:263, 1966.

[KAEY95] M. Kamijo, F. Abe, Y. Einaga, and H. Yamakawa. *Macromolecules*, 28:4159, 1995.

[Kam90] K. Kamide. *Thermodynamics of Polymer Solutions.* Polymer Science Library 9, A.D. Jenkins ed., Elsevier Amsterdam, 1990.

[KAS87] L. Ould Kaddour, M. Anasagasti, and C. Strazielle. *Makromol. Chem.*, 188:2223, 1987.

[KBB81] K. Kremer, A. Baumgärtner, and K. Binder. *Z. Phys. B*, 40:331, 1981.

[KBWU85] J.S. King, W. Boyer, G.D. Wignall, and R. Ullman. *Macromolecules*, 18:709, 1985.

[KCG70] R. Koningsveld, H.A.G. Chermin, and M. Gordon. *Proc. Roy. Soc. Lond. A*, 319:331, 1970.

[Kho81] A.R. Khokhlov. *Macromol. Chem., Rapid Commun.*, 2:633, 1981.

[KM71] D.S. Mc Kenzie and M.A. Moore. *J. Phys. A*, 4:L 82, 1971.

[KOK68] N. Kuwahara, T. Okazawa, and M. Kaneko. *J. Pol. Sci. C*, 23:543, 1968.

[Kra87] P. Kratochvil. *Classical Light Scattering from Polymer Solutions, Polymer Science Library 5.* Elsevier, Amsterdam, 1987.

[KS87] L. Ould Kaddour and C. Strazielle. *Polymer*, 28:459, 1987.

[KS90] C. Kappeler and L. Schäfer. *Macromolecules*, 23:2766, 1990.

[KS94] B. Krüger and L. Schäfer. *J. Phys. I (France)*, 4:757, 1994.

[KS96] B. Krüger and L. Schäfer. *Macromolecules*, 29:4737, 1996.

[KSB89] B. Krüger, L. Schäfer, and A. Baumgärtner. *J. Physique*, 50:3191, 1989.

[KSF91] C. Kappeler, L. Schäfer, and T. Fukuda. *Macromolecules*, 24:2715, 1991.

[KSW81] A. Knoll, L. Schäfer, and T.A. Witten. *J. Physique*, 42:767, 1981.

[KTL91] M.S. Kent, M. Tirrell, and T.P. Lodge. *Polymer*, 32:314, 1991.

[KTL92] M.S. Kent, M. Tirrell, and T.P. Lodge. *Macromolecules*, 25:5383, 1992.

[KWP93] S.K. Kehrein, F.J. Wegner, and Y.M. Pismak. *Nucl. Phys. B*, 402:669, 1993.

[LBD78] M. Lax, A.J. Barrett, and C. Domb. *J. Phys. A*, 11:361, 1978.

[LMS+92] A. Lapp, M. Mottin, C. Strazielle, D. Broseta, and L. Leibler. *J. Phys. II (France)*, 2:1247, 1992.

[LMS95] B. Li, N. Madras, and A.D. Sokal. *J. Stat. Phys.*, 80:661, 1995.

[LWXM95] J. Li, Y. Wan, Z. Xu, and J.W. Mays. *Macromolecules*, 28:5347, 1995.

[Ma76] S-K Ma. *Modern Theory of Critical Phenomena*. Benjamin, Inc., London, 1976.

[MEF78] Y. Miyaki, Y. Einaga, and H. Fujita. *Macromolecules*, 11:1180, 1978.

[MF81] Y. Miyaki and H. Fujita. *Macromolecules*, 14:742, 1981.

[MH81] T. Minato and A. Hatano. *Macromolecules*, 14:1035, 1981.

[MN84] M. Muthukumar and B.G. Nickel. *J. Chem. Phys.*, 80:5839, 1984.

[MN87] M. Muthukumar and B.G. Nickel. *J. Chem. Phys.*, 86:460, 1987.

[MNF72] T. Matsumoto, N. Nishioka, and H. Fujita. *J. Pol. Sci A.2*, 10:23, 1972.

[MPB93] A. Milchev, W. Paul, and K. Binder. *J. Chem. Phys.*, 99:4786, 1993.

[MS98] S. Müller and L. Schäfer. *The European Physical Journal B*, 1998.

[Nak71] N. Nakanishi. *Graph Theory and Feynman Integrals*. Gordon and Breach, New York, 1971.

[NAK+92] Y. Nakamura, K. Akasaka, K. Katayama, T. Norisuye, and A. Teramoto. *Macromolecules*, 25:1134, 1992.

[Nel76] D.R. Nelson. *Phys. Rev. B*, 14:1123, 1976.

[NFI87] M. Nagata, T. Fukuda, and H. Inagaki. *Macromolecules*, 20:2173, 1987.

[NHKF84] I. Noda, X. Higo, N. Keno, and T. Fujimoto. *Macromolecules*, 17:1055, 1984.

[Nic91] B.G. Nickel. *Macromolecules*, 24:1358, 1991.

[NIKN83] I. Noda, M. Imai, T. Kitano, and M. Nagasawa. *Macromolecules*, 16:425, 1983.

[NKKN81] I. Noda, N. Kato, T. Kitano, and M. Nagasawa. *Macromolecules*, 14:668, 1981.

[NKTF69] T. Norisuye, K. Kawahara, A. Teramoto, and H. Fujita. *J. Chem. Phys.*, 49:4330, 1969.

[NNA93] T. Norisuye, Y. Nakamura, and K. Akasaka. *Macromolecules*, 26:3791, 1993.

[NNT91] Y. Nakamura, T. Norisuye, and A. Teramoto. *J. Pol. Sci. B*, 29:153, 1991.

[NO85] A. Nakanishi and T. Ohta. *J. Phys. A*, 18:127, 1985.

[OF82] Y. Oono and K.F. Freed. *J. Phys. A*, 15:1931, 1982.

[OH89] A. Onuki and T. Hashimoto. *Macromolecules*, 22:879, 1989.

[Olv74] F.W.J. Olver. *Asymptotics and Special Functions*. Academic Press, London, 1974.

[ON83] T. Ohta and A. Nakanishi. *J. Phys. A*, 16:4155, 1983.

[OO81] Y. Oono and T. Ohta. *Phys. Lett.*, 85 A:480, 1981.

[OO82] T. Ohta and Y. Oono. *Phys. Lett.*, 89 A:460, 1982.

[OOF81a] T. Ohta, Y. Oono, and K.F. Freed. *Macromolecules*, 14:1588, 1981.

[OOF81b] Y. Oono, T. Ohta, and K.F. Freed. *J. Chem. Phys.*, 74:6458, 1981.

[PAD82] R. Perzynski, M. Adam, and M. Delsanti. *J. Physique*, 43:129, 1982.

[RDP87] M. Rawiso, R. Duplessix, and C. Picot. *Macromolecules*, 20:630, 1987.

[RFBC93] A.M. Rubio, J.J. Freire, M. Bishop, and J.H.R. Clarke. *Macromolecules*, 26:4018, 1993.

[RG86] J. Rudnick and G. Gaspari. *J. Phys. A*, 19:L 191, 1986.
[RM94] J.J. Rajasekaran and M. Muthukumar. *Macromolecules*, 27:6418, 1994.
[SB86] L. Schäfer and A. Baumgärtner. *J. Physique*, 47:1431, 1986.
[SB88] A. Sariban and K. Binder. *Macromolecules*, 21:711, 1988.
[Sch76] L. Schäfer. *J. Phys. A*, 9:L 71, 1976.
[Sch82] L. Schäfer. *Macromolecules*, 15:652, 1982.
[Sch84] L. Schäfer. *Macromolecules*, 17:1357, 1984.
[Sch87] L. Schäfer. *Phys. Rev. B*, 35:5184, 1987.
[Sch90] L. Schäfer. *Nuclear Phys. B*, 344:596, 1990.
[Sch92] L. Schäfer. *Phys. Rev. B*, 46:6061, 1992.
[Sch93] L. Schäfer. *Macromolecules*, 26:6425, 1993.
[Sch94] L. Schäfer. *Phys. Rev. E*, 50:3517, 1994.
[Sci96] S.J. Sciutto. *J. Phys. A*, 29:5455, 1996.
[SD89] R. Schloms and V. Dohm. *Nucl. Phys. B*, 328:639, 1989.
[SD90] R. Schloms and V. Dohm. *Phys. Rev. B*, 42:6142, 1990.
[SDB95] C. Strazielle, M. Duwal, and M. Benmouna. *J. Pol. Sci. B: Pol. Phys.*, 33:823, 1995.
[SH78] L. Schäfer and H. Horner. *Z. Phys. B*, 29:251, 1978.
[SK85] L. Schäfer and C. Kappeler. *J. Physique*, 46:1853, 1985.
[SK88] L. Schäfer and B. Krüger. *J. Physique*, 49:749, 1988.
[SK90] L. Schäfer and C. Kappeler. *Colloid Polymer Science*, 268:995, 1990.
[SK93] L. Schäfer and C. Kappeler. *J. Chem. Phys.*, 99:6135, 1993.
[SLK91] L. Schäfer, U. Lehr, and C. Kappeler. *J. Phys. I (France)*, 1:211, 1991.
[Smi91] V.A. Smirnov. *Renormalization and Asymptotic Expansions.* Birkhäuser Verlag, Basel, 1991.
[SN90] F.C. Shanes and B.G. Nickel. *unpublished report*, 1990.
[SNF87] T. Sato, T. Norisuye, and H. Fujita. *J. Pol. Sci. B*, 25:1, 1987.
[Sol71] L. Solc. *J. Chem. Phys.*, 55:335, 1971.
[SS71] L. Solc and W.H. Stockmayer. *J. Chem. Phys.*, 54:2756, 1971.
[Ste81] S. Stepanow. *Acta Polymerica*, 32:98, 1981.
[SvFLD92] L. Schäfer, C. v. Ferber, U. Lehr, and B. Duplantier. *Nucl. Phys. B*, 374:473, 1992.
[SW77] L. Schäfer and T.A. Witten. *J. Chem. Phys.*, 66:2121, 1977.
[SW80] L. Schäfer and T.A. Witten. *J. Physique*, 41:459, 1980.
[tHV72] G. 't Hooft and M. Veltman. *Nucl. Phys. B*, 44:189, 1972.
[TK86] Y. Tsunashima and M. Kurata. *J. Chem. Phys.*, 84:6432, 1986.
[UBK88] R. Ullmann, H. Benoit, and J.S. King. *Macromolecules*, 19:183, 1988.
[UUTK78] H. Utiyama, S. Utsumi, Y. Tsunashima, and M. Kurata. *Macromolecules*, 11:506, 1978.
[Val96] J.P. Valleau. *J. Chem. Phys.*, 104:3071, 1996.
[VB90] T.A. Vilgis and R. Borsali. *Macromolecules*, 23:3173, 1990.
[WC79] G. Weill and J. Des Cloizeaux. *J. Physique*, 40:99, 1979.
[Weg76] F.J. Wegner. In C. Domb and M.S. Green, editors, *Phase Transitions and Critical Phenomena*, volume 6. Academic Press, London, 1976.
[WF72] K.G. Wilson and M.E. Fisher. *Phys. Rev. Lett.*, 28:248, 1972.
[WHCS83] P. Wiltzius, H.R. Haller, D.S. Cannell, and D.W. Schaefer. *Phys. Rev. Lett.*, 51:1183, 1983.
[Wit82] T.A. Witten. *J. Chem. Phys.*, 76:3300, 1982.
[WK74] K.G. Wilson and J.B. Kogut. *Phys. Reports*, 12 C:75, 1974.

[WKG96] M. Wittkop, S. Kreitmeier, and D. Göritz. *J. Chem. Phys.*, 104:351, 1996.

[WP82] T.A. Witten and J.J. Prentis. *J. Chem. Phys.*, 77:4247, 1982.

[WS78] T.A. Witten and L. Schäfer. *J. Phys. A*, 11:1843, 1978.

[WS81] T.A. Witten and L. Schäfer. *J. Chem. Phys.*, 74:2582, 1981.

[WXW+97] X. Wang, Z. Xu, Y. Wan, T. Huang, S. Pispas, J.W. Mays, and C. Wu. *Macromolecules*, 30:7202, 1997.

[YAE93] H. Yamakawa, F. Abe, and Y. Einaga. *Macromolecules*, 26:1898, 1993.

[Yam71] H. Yamakawa. *Modern Theory of Polymer Solutions.* Harper & Row, New York, 1971.

[Yam92] H. Yamakawa. *Macromolecules*, 25:1912, 1992.

[YFTY71] A. Yamamoto, H. Fujii, G. Tanaka, and H. Yamakawa. *Polymer J.*, 2:799, 1971.

[YMB97] V. Yamakov, A. Milchev, and K. Binder. *J. Phys. II (France)*, 7:1123, 1997.

[Zif82] G. Zifferer. PhD thesis, University of Vienna, Austria, 1982.

[ZJ89] J. Zinn-Justin. *Quantum Field Theory and Critical Phenomena.* Clarendon Press, Oxford, 1989.

Frequently Occurring Symbols

References to defining equations are given in parentheses.

d	spatial dimension
$\varepsilon = 4 - d$	
$\Gamma(\alpha)$	Gamma function
$\zeta(\alpha)$	Riemann's zeta function

Unrenormalized theory

β_e (2.3)	excluded volume constant
ℓ (2.3)	segment size
N (5.37)	number averaged chain length (polymerization index)
n	length of a specific chain
c_p	chain concentration
$c = c_p N$	segment concentration
$\hat{\rho}(\mathbf{r})$ (3.14)	local segment density
Ω	system volume
$u_0 = (4\pi\ell^2)^{d/2}\beta_e$ (5.25)	
$z = \beta_e n^{\varepsilon/2}$ (6.1)	
$R_0^2 = \ell^2 N$	
$u_0(\mathbf{q})$ (5.59)	screened interaction in tree approximation
$p(y)$ (5.36)	reduced chain length distribution
$\tilde{p}(x)$ (5.40)	Laplace transform of $p(y)$
$D(x)$ (3.23)	Debye function
$D_p(x)$ (5.49)	polydisperse Debye function
$G_0(\mathbf{p}, n)$ (3.23)	unperturbed Greensfunction
$G(\mathbf{p}, n)$ (4.3)	single chain Greensfunction
$G^{(M)}$ (4.18)	M-chain Greensfunction
$G^{(M,L)}_{m_1,\ldots,m_L}$ (4.24)	Greensfunction with density insertions
$G_c^{(M,L)}$	cumulant, Sect. 4.3

Renormalized theory

u (11.1), (12.16)	renormalized coupling constant
u^* (12.20), (13.4)	nontrivial fixed point
$f = u/u^*$ (11.21)	
ℓ_R (11.1)	renormalized segment size
$\lambda = \ell/\ell_R$ (11.13)	scaling parameter
$\bar{\mathbf{q}} = \ell_R \mathbf{q}$ (11.11)	dimensionless momentum
$c_{pR} = \ell_R^d c_p$ (11.73)	renormalized chain concentration
N_R (11.1), (12.16)	renormalized number averaged chain length
$c_R = \tilde{u} c_p \ell_R^d N_R$	renormalized segment concentration
$\tilde{u} = (4\pi)^{d/2}\, u^*/2$ (13.18)	(with $b_u = 1$)
Z_u (11.1), (12.19)	coupling constant renormalization factor
Z_n (11.1), (12.19)	chain length renormalization factor
Z (11.10), (12.19)	renormalization factor for the partition function
$\mu^*(\beta_e)$ (11.10), (12.18)	chemical potential per segment of an infinite chain
ν (1.3), (8.34), (13.10)	correlation length exponent
ω (11.24), (13.10)	correction to scaling exponent
n_0, c_0, q_0^2 (13.21), (13.22)	parameters fixing the uncritical manifold

Nonuniversal parameters

s_ℓ	(11.28), (13.15)	general crossover
s_n	(11.34), (13.15)	general crossover
$\tilde{v} = s_n^{\varepsilon/2}$	(11.38)	general crossover
$\tilde{\ell}^2 = s_\ell^2 s_n$	(11.38)	general crossover
B	(11.44), (13.37)	excluded volume limit

Physical variables

$\tilde{z} = \tilde{v} n^{\varepsilon/2}$	(11.38)
$\tilde{R}_0^2 = \tilde{\ell}^2 N$	(11.38)
$s = R_g^d c_p$	geometrical overlap
$\tilde{s} = \tilde{R}_0^d c_p$	(11.83)
$\hat{s} = B^d c_p N^{\nu d}$	(13.30) overlap variable in the excluded volume limit

Observables

Scaling functions in the general crossover regime are identified by a tilde. The excluded volume limit is indicated by a star as superscript. Equation numbers not necessarily refer to the first occurence of the quantity, but rather to the chapter where it is discussed in detail.

R_g	(3.26), (15.6)	radius of gyration
R_e	(3.7), (16.8)	end-to-end distance
α_g	(1.16), (15.11)	swelling factor of the radius of gyration
α_e	(16.14)	swelling factor of the end-to-end distance
A_2	(15.36), (15.37)	2nd virial coefficient
ψ	(15.52)	interpenetration ratio
ξ_d	(19.17)	density correlation length
ξ_E	(19.30)	screening length
I_d	(15.1)	isolated chain density correlation function
J_d	(15.4)	scaling function for I_d
$I_{d,a}$	(18.1)	density autocorrelation function
J_a	(18.3)	scaling function for $I_{d,a}$
$I_{d,c}$	(19.5)	density cumulant
J_c	(19.7)	Scaling function for $I_{d,c}$
$\hat{\Pi}$	(5.3), (17.1)	osmotic pressure
$\mathcal{P}$	(17.1)	thermodynamic scaling function

Subject Index

MIX
Papier aus verantwortungsvollen Quellen
Paper from responsible sources
FSC® C105338

If you have any concerns about our products,
you can contact us on
ProductSafety@springernature.com

In case Publisher is established outside the EU,
the EU authorized representative is:
Springer Nature Customer Service Center GmbH
Europaplatz 3, 69115 Heidelberg, Germany

Printed by Libri Plureos GmbH
in Hamburg, Germany